Optical Tweezers

Combining state-of-the-art research with a strong pedagogical approach, this text provides a detailed and complete guide to the theory, practice and applications of optical tweezers. In-depth derivation of the theory of optical trapping and numerical modelling of optical forces are supported by a complete step-by-step design and construction guide for building optical tweezers, with detailed tutorials on collecting and analysing data. Also included are comprehensive reviews of optical tweezers research in fields ranging from cell biology to quantum physics.

Featuring numerous exercises and problems throughout, this is an ideal self-contained learning package for advanced lecture and laboratory courses and an invaluable guide to practitioners wanting to enter the field of optical manipulation.

The text is supplemented by the website www.opticaltweezers.org, a forum for discussion and a source of additional material including free-to-download, customisable, research-grade software (OTS) for calculation of optical forces, digital video microscopy, optical tweezers calibration and holographic optical tweezers.

Philip H. Jones is a Reader in Physics at University College London, where he leads the Optical Tweezers research group.

Onofrio M. Maragò is a Researcher at the Istituto per i Processi Chimico-Fisici (CNR-IPCF) in Messina, Italy, where he leads the Optical Trapping research group.

Giovanni Volpe is an Assistant Professor at Bilkent University, where he is head of the Soft Matter Lab.

Optical Tweezers

Principles and Applications

PHILIP H. JONES

ONOFRIO M. MARAGÒ

GIOVANNI VOLPE

CAMBRIDGE
UNIVERSITY PRESS

University Printing House, Cambridge CB2 8BS, United Kingdom

Cambridge University Press is part of the University of Cambridge.

It furthers the University's mission by disseminating knowledge in the pursuit of education, learning and research at the highest international levels of excellence.

www.cambridge.org
Information on this title: www.cambridge.org/9781107051164

© Philip H. Jones, Onofrio M. Maragò and Giovanni Volpe 2015

This publication is in copyright. Subject to statutory exception and to the provisions of relevant collective licensing agreements, no reproduction of any part may take place without the written permission of Cambridge University Press.

First published 2015

Printed in the United Kingdom by TJ International Ltd. Padstow Cornwall

A catalogue record for this publication is available from the British Library

ISBN 978-1-107-05116-4 Hardback

Additional resources for this publication at www.cambridge.org/9781107051164

Cambridge University Press has no responsibility for the persistence or accuracy of URLs for external or third-party internet websites referred to in this publication, and does not guarantee that any content on such websites is, or will remain, accurate or appropriate.

To Annie, Becky, Andrew, Antonella, Carmen and Joana
For their love and patience

Contents

Preface		*page* xv
1 Introduction		1
	1.1 A brief history of optical manipulation	2
	1.2 Crash course on optical tweezers	4
	1.3 Optical trapping regimes	6
	1.4 Other micromanipulation techniques	8
	1.5 Scope of this book	10
	1.6 How to read this book	11
	1.7 `OTS - the Optical Tweezers Software`	12
	References	13
	Part I Theory	**17**
2 Ray optics		19
	2.1 Optical rays	20
	2.2 Optical forces	24
	2.3 Scattering and gradient forces	26
	2.4 Counter-propagating beam optical trap	29
	2.5 Optical tweezers	31
	2.6 Filling factor and numerical aperture	34
	2.7 Non-uniform beams	36
	2.8 Non-spherical objects and the windmill effect	37
	Problems	40
	References	41
3 Dipole approximation		42
	3.1 The electric dipole in electrostatics	43
	3.2 Polarisability and the Clausius–Mossotti relation	45
	3.3 The electric dipole in an oscillating electric field	50
	3.4 Radiative reaction correction to the polarisability	52
	3.5 Cross-sections	54
	3.6 The optical theorem	56
	3.7 Optical forces	58
	3.7.1 Gradient force	61

		3.7.2 Scattering force	63
		3.7.3 Spin–curl force	64
	3.8	Atomic polarisability	65
	3.9	Plasmonic particles	67
	3.10	Optical binding	70
	Problems		73
	References		75

4 Optical beams and focusing — 76

    4.1 Propagating electromagnetic waves   77
    4.2 Angular spectrum representation   79
    4.3 From near field to far field   81
    4.4 Paraxial approximation   83
        4.4.1 Gaussian beams   83
        4.4.2 Hermite–Gaussian beams   85
        4.4.3 Laguerre–Gaussian beams   87
        4.4.4 Non-diffracting beams   89
        4.4.5 Cylindrical vector beams   92
    4.5 Focusing   92
    4.6 Optical forces near focus   97
    4.7 Focusing near interfaces   100
        4.7.1 Aberrations   102
        4.7.2 Evanescent focusing   102
    Problems   104
    References   105

5 Electromagnetic theory — 106

    5.1 Conservation laws and the Maxwell stress tensor   107
        5.1.1 Angular momentum of light   111
    5.2 Light scattering   116
        5.2.1 Solution of the Helmholtz equation   116
        5.2.2 The scattering problem   123
        5.2.3 Multipole expansion   126
        5.2.4 Transition matrix   131
        5.2.5 Mie scattering   132
    5.3 Optical force and torque   137
        5.3.1 Optical force   137
        5.3.2 Optical torque   139
    5.4 Optical force from a plane wave   139
    5.5 Transfer of spin angular momentum to a sphere   143
    5.6 Optical force in an optical tweezers   146
        5.6.1 Orbital angular momentum   149
    Problems   151
    References   152

6 Computational methods — 154
- 6.1 T-matrix — 155
  - 6.1.1 Optical force — 156
  - 6.1.2 Optical torque — 158
  - 6.1.3 Amplitudes of a focused beam — 160
  - 6.1.4 Translation theorem — 162
  - 6.1.5 Rotation theorem — 165
  - 6.1.6 Clebsch–Gordan coefficients — 168
- 6.2 Metal spheres sustaining longitudinal fields — 170
- 6.3 Radially symmetric spheres — 172
- 6.4 Clusters of spheres — 174
  - 6.4.1 Aggregates of spheres — 174
  - 6.4.2 Inclusions — 176
  - 6.4.3 Convergence — 178
- 6.5 Discrete dipole approximation — 179
- 6.6 Finite-difference time domain — 180
- 6.7 Hybrid techniques — 183
- Problems — 183
- References — 185

7 Brownian motion — 188
- 7.1 The physical picture — 189
- 7.2 Mathematical models — 191
  - 7.2.1 Random walk — 192
  - 7.2.2 Langevin equation — 194
  - 7.2.3 Free diffusion equation — 195
  - 7.2.4 Fokker–Planck equation — 197
- 7.3 Fluctuation–dissipation theorem, potential and equilibrium distribution — 197
- 7.4 Brownian dynamics simulations — 199
  - 7.4.1 White noise — 200
  - 7.4.2 Optically trapped particle — 202
- 7.5 Inertial regime — 205
- 7.6 Diffusion gradients — 207
- 7.7 Viscoelastic media — 211
- 7.8 Non-spherical particles and diffusion matrices — 212
  - 7.8.1 Free diffusion — 213
  - 7.8.2 External forces — 215
- Problems — 216
- References — 217

Part II Practice — 219

8 Building an optical tweezers — 221
- 8.1 The right location — 222
- 8.2 Inverted microscope construction — 222

		8.2.1 Objectives	226
		8.2.2 Illumination schemes	232
	8.3	Sample preparation	236
	8.4	Optical beam alignment	239
		8.4.1 Lasers	244
		8.4.2 Lenses	246
		8.4.3 Mirrors	248
		8.4.4 Filters	248
		8.4.5 Polarisation control	249
	8.5	Optical trapping and manipulation	250
		8.5.1 Steerable optical tweezers	251
	8.6	Alternative set-ups	253
	Problems		253
	References		253

9 Data acquisition and optical tweezers calibration — 255

	9.1	Digital video microscopy	256
		9.1.1 Digital cameras	261
	9.2	Interferometry	262
		9.2.1 Photodetectors	271
		9.2.2 Acquisition hardware	273
	9.3	Calibration techniques: An overview	273
	9.4	Potential analysis	274
	9.5	Equipartition method	276
	9.6	Mean squared displacement analysis	278
	9.7	Autocorrelation analysis	280
		9.7.1 Crosstalk analysis and reduction	280
	9.8	Power spectrum analysis	283
		9.8.1 Analytical least square fitting	286
		9.8.2 Hydrodynamic corrections	288
		9.8.3 Noise tests	290
	9.9	Drag force method	291
	Problems		293
	References		294

10 Photonic force microscope — 296

10.1	Scanning probe techniques	297
10.2	Photonic torque microscope	300
10.3	Force measurement near surfaces	307
	10.3.1 Equilibrium distribution method	308
	10.3.2 Drift method	309
10.4	Relevance of non-conservative effects	311
10.5	Direct force measurement	312
Problems		316
References		317

11 Wavefront engineering and holographic optical tweezers — 319
- 11.1 Basic working principle — 320
- 11.2 Computer-generated holograms — 324
  - 11.2.1 Single steerable trap — 325
  - 11.2.2 Random mask encoding — 327
  - 11.2.3 Superposition of gratings and lenses — 328
  - 11.2.4 Gerchberg–Saxton algorithm — 329
  - 11.2.5 Adaptive–additive algorithm — 331
  - 11.2.6 Direct search algorithms — 332
- 11.3 Higher-order beams and orbital angular momentum — 332
- 11.4 Continuous optical potentials — 334
- 11.5 Set-up implementation — 335
  - 11.5.1 Spatial light modulators — 338
- 11.6 Alternative approaches — 340
  - 11.6.1 Time-shared optical traps — 340
  - 11.6.2 Generalised phase contrast — 341
- Problems — 342
- References — 343

12 Advanced techniques — 345
- 12.1 Spectroscopic optical tweezers — 346
  - 12.1.1 Fluorescence tweezers — 347
  - 12.1.2 Photoluminescence tweezers — 349
  - 12.1.3 Raman tweezers — 349
- 12.2 Optical potentials — 350
  - 12.2.1 Periodic and quasi-periodic potentials — 350
  - 12.2.2 Random potentials and speckle tweezers — 351
- 12.3 Counter-propagating traps and optical fibre traps — 353
  - 12.3.1 Optical stretcher — 354
  - 12.3.2 Longitudinal optical binding — 355
- 12.4 Evanescent wave traps — 356
  - 12.4.1 Evanescent tweezers — 356
  - 12.4.2 Waveguides — 357
  - 12.4.3 Optical binding — 358
  - 12.4.4 Plasmonic traps — 359
- 12.5 Feedback traps — 361
- 12.6 Haptic optical tweezers — 362
- References — 365

Part III Applications — 369

13 Single-molecule biophysics — 371
- 13.1 DNA mechanics: Stretching — 372
- 13.2 DNA mechanics: Thermal fluctuations — 374
- 13.3 DNA mechanics: Torsional properties — 376

	13.4 Motor proteins	379
	13.5 Further reading	382
	References	383

14 Cell biology — 385
    14.1 Cellular adhesion forces — 386
    14.2 Adhesion and structure of bacterial pili — 388
    14.3 Directed neuronal growth — 389
    14.4 Further reading — 392
    References — 393

15 Spectroscopy — 395
    15.1 Absorption and photoluminescence spectroscopy — 396
    15.2 Raman spectroscopy — 398
    15.3 Coherent anti-Stokes Raman spectroscopy — 402
    15.4 Rayleigh spectroscopy and surface-enhanced Raman spectroscopy — 402
    15.5 Further reading — 404
    References — 405

16 Optofluidics and lab-on-a-chip — 409
    16.1 Optical sorting — 410
    16.2 Monolithic integration — 412
    16.3 Photonic crystal cavities — 414
    16.4 Micromachines — 415
    16.5 Further reading — 417
    References — 418

17 Colloid science — 422
    17.1 Hydrodynamic interactions — 423
    17.2 Electrostatic interactions — 425
    17.3 Depletion interactions — 426
    17.4 Further reading — 430
    References — 430

18 Microchemistry — 433
    18.1 Liquid droplets — 434
    18.2 Vesicle and membrane manipulation — 435
    18.3 Vesicle fusion — 438
    18.4 Further reading — 439
    References — 439

19 Aerosol science — 441
    19.1 Optical tweezers in the gas phase — 442
    19.2 Trapping and guiding — 443

19.3	Photophoretic trapping and guiding	444
19.4	Further reading	445
	References	446

20 Statistical physics 448

20.1	Colloids as a model system for statistical physics	449
20.2	Kramers rates	449
20.3	Stochastic resonance	451
20.4	Spurious drift in diffusion gradients	452
20.5	Colloidal crystals and quasicrystals	455
20.6	Random potentials and anomalous diffusion	455
20.7	Further reading	459
	References	459

21 Nanothermodynamics 462

21.1	Violation of the second law	463
21.2	The Jarzynski equality	464
21.3	Information-to-energy conversion	465
21.4	Micrometre-sized heat engine	465
21.5	Further reading	467
	References	468

22 Plasmonics 470

22.1	Plasmonic nanoparticles	471
22.2	Plasmonic substrates	474
22.3	Plasmonic apertures	477
22.4	Further reading	479
	References	480

23 Nanostructures 484

23.1	Metal nanoparticles	485
23.2	Semiconductor nanostructures	486
23.3	Optical force lithography and placement	490
23.4	Prospects for nanotweezers	492
23.5	Further reading	492
	References	493

24 Laser cooling and trapping of atoms 498

24.1	Laser cooling and optical molasses	499
24.2	Atom trapping	504
24.3	Optical dipole traps for cold atoms	506
24.4	The path to quantum degeneracy	507
24.5	Bose–Einstein condensation	508
24.6	Evaporative cooling and Bose–Einstein condensation in dipole traps	513

24.7 Holographic optical traps for cold atoms 514
24.8 Optical lattices 514
24.9 Further reading 518
References 518

25 Towards the quantum regime at the mesoscale — 524
25.1 Cavity optomechanics: The classical picture 525
25.2 Cavity optomechanics: The quantum picture 527
25.3 Laser cooling of levitated particles 528
25.4 Feedback cooling schemes 529
25.5 Below the Doppler limit 531
References 534

Index — 537

Preface

Since the first demonstration of optical tweezers approximately 30 years ago, they have become widespread both as a subject of research in their own right and as an enabling tool in fields as diverse as molecular biology, statistical physics, materials science and quantum physics. Currently the number of active research groups worldwide is in the hundreds – and counting. Furthermore, with the advent of commercially available optical tweezers and low-cost lab kits, optical tweezers experiments can now be found as a common instructional tool in advanced undergraduate and graduate laboratories. This broad interest gives rise to a pressing need for a reference textbook covering the principles and applications of optical tweezers. We began our journey of writing this book with the aim of filling this gap. Therefore we sought to write a textbook with a strong pedagogic approach to both the theory and practice of optical manipulation, supplemented by an overview of the current state of the art in optical manipulation research, and supported by exercises and problems. Eventually, this book saw the light of day.

This book comprises three parts. Part I covers the theory of optical tweezing, providing intuitive and rigorous explanations of the physics behind optical trapping and manipulation, an introduction to the numerical methods most commonly employed in the study of optical forces and torques, and a detailed explanation of the dynamics of optically trapped particles. Part II focuses on the experimental practice of optical manipulation, including both the implementation of a working optical tweezers set-up – complete with detailed step-by-step advice on its construction, on troubleshooting and on the acquisition and analysis of data – and instructions on how to develop more advanced optical manipulation techniques. Parts I and II both include numerous exercises to illustrate the concepts, ideas and techniques discussed, and each chapter ends with problems to solve as a starting point for further investigations. Finally, Part III provides an overview of some of the most exciting applications that optical tweezers have found in various fields, from the study of biological systems to the investigation of the quantum limit for trapped mesoscale objects. Furthermore, we have enhanced this book with an extensive supplementary material, available online for download from the book website at www.opticaltweezers.org. This includes, in particular, the comprehensive `OTS - the Optical Tweezers Software` toolbox, which we encourage readers to download, use and develop further.

Finally, we wish to thank all the colleagues and friends who have contributed to the writing of this book with their advice, input and encouragement. In particular, our special thanks go to Giuseppe Pesce for his help in writing Chapters 8, 9 and 11, Rosalba Saija for her constant advice on scattering theory and computational issues covered in Chapters 5 and 6, Agnese Callegari and S. Masoumeh Mousavi for their help in developing the `OTS` toolbox, Giorgio Volpe for his help in writing Chapter 7, and Juan José (Juanjo) Sáenz for his critical reading of several chapters. We have also received a lot of help and assistance from Ferdinando Borghese, Maria Grazia Donato, Barbara Fazio, Marco Grasso, Pietro Gucciardi, Antonella Iatí, Alessia Irrera, Fatemeh Kalantarifard, Alessandro Magazzù and Mite Mijalkov. We have greatly profited from discussions with Ennio Arimondo, Paolo Denti, Roberto Di Leonardo, Andrea C. Ferrari, Chris Foot, Simon Hanna, Alper Kiraz, Isabel Llorente Garcia, Oliver Morsch, Antonio Alvaro Ranha Neves, Ferruccio Renzoni, Maurizio Righini, Antonio Sasso, Salvatore Savasta, Stephen Simpson and Cirino Vasi. We would also like to acknowledge the students and postdoctoral researchers in our laboratories, whose hard work has permitted us to spare the time needed to write this book. It goes without saying that we claim full ownership of any remaining errors.

Phil Jones
Onofrio Maragò
Giovanni Volpe

1 Introduction

The ability of light to exert forces has been known for quite some time. In fact, Johannes Kepler (1619) recognised that the tails of a comet – an example of which is shown in Fig. 1.1 – are due to the force exerted on the particles surrounding the comet's body by the Sun's rays. However, optical forces are extremely small; so small, in fact, that only in recent years, and only thanks to the advent of the laser, has it been possible to concentrate enough optical power into a small area to significantly affect the motion of microscopic particles, thereby leading to the invention of *optical tweezers*. Optical tweezers are generated by a tightly focused laser beam that can hold and manipulate a particle in the high-intensity region that is the focal spot. Optical tweezers and other optical manipulation techniques have heralded a revolution in the study of microscopic systems, spearheading new and more powerful techniques, e.g., to study biomolecules, to measure forces that act on a nanometre scale and to explore the limits of quantum mechanics. This book provides a comprehensive guide to the theory [Chapters 2–7], practice [Chapters 8–12] and applications [Chapters 13–25] of optical trapping and optical manipulation.

Figure 1.1 Optical forces in the sky. Besides the brightness of its coma, a comet is typically remembered by the length of its tails. Comet Hale–Bopp, which passed perihelion on 1 April 1997, lived up to expectations for it, boasting a blue ion tail (bottom) and a white dust tail (top) of exceptional extent, due respectively to the solar wind and to the Sun's light force. Credits: A. Dimai & D. Ghirardo, Col Druscie Observatory, Associazione Astronomica Cortina.

1.1 A brief history of optical manipulation

Light's ability to exert forces has been recognised at least since 1619, when Kepler's *De Cometis* described the deflection of comet tails by the Sun's rays (Kepler, 1619). In the late nineteenth century, Maxwell's theory of electromagnetism predicted that the momentum flux of a light beam would be proportional to its intensity and could be transferred to illuminated objects, resulting in radiation pressure pushing objects along the beam's direction of propagation (Poynting, 1884). At the dawn of the twentieth century, several exciting early experiments were performed to detect these effects. Lebedev (1901) and Nichols and Hull (1901)[1] first succeeded in detecting radiation pressure on macroscopic objects and absorbing gases: when light from a lamp was focused onto a mirror attached to a torsion balance, the radiation pressure moved the balance from its equilibrium position. A few decades later, Beth (1936) reported the first experimental observation of the torque acting on a macroscopic object as a result of its interaction with light: he observed the deflection of a birefringent quartz wave plate suspended from a thin quartz fibre when circularly polarised light passed through it. These effects were so small, however, that they were not easily detected and any practical application seemed unfeasible. Quoting J. H. Poynting's presidential address to the British Physical Society in 1905:

> a very short experience in attempting to measure these forces is sufficient to make one realize their extreme minuteness – a minuteness which appears to put them beyond consideration in terrestrial affairs.[2]

This 'extreme minuteness' is the reason that optical trapping and manipulation did not exist before the invention of the laser in the 1960s (Townes, 1999).

In the early 1970s, Arthur Ashkin showed that laser-induced optical forces could be used to alter the motion of microscopic particles (Ashkin, 1970a) and neutral atoms (Ashkin, 1970b). In particular, Ashkin (1970a) showed that it was possible to use the radiation forces from a laser beam to significantly affect the dynamics of transparent micrometre-sized neutral particles, thereby identifying two basic light forces, a *scattering force* in the propagation direction of the incident beam and a *gradient force* along the intensity gradient perpendicular to the beam. He showed experimentally that using just these forces one could accelerate, decelerate and even stably trap small micrometre-sized neutral particles using focused laser beams.

---

[1] Nichols and Hull (1901) had found agreement with predictions from Maxwell's theory to within 1%. However, Bell and Green later reanalysed the data from these experiments and found several mathematical errors: Nichols and Hull had used an incorrect value for the mechanical equivalent of heat, taken some logarithms to base 10 instead of base e, and made several mistakes involving units and conversion factors. Once these mistakes were corrected, the results deviated from Maxwell's theory by 10%, which was still a success, but not nearly as good as the original 1% agreement. It is reasonable to believe that Nichols and Hull, finding such good agreement with Maxwell's theory, were happy to publish their results; if the mathematical errors had been in the opposite direction, they presumably would have checked their calculations more carefully. Worrall (1982) presents a fascinating history of this and other light pressure experiments. (Adapted from Jeng (2006).)

[2] Quoted by Ashkin (2000).

Ashkin started his analysis of optical forces by considering a beam of power P and angular frequency ω reflected by a plane mirror. It is well known from quantum mechanics that each photon of the beam carries a momentum, $p = \hbar\omega/c$, where $\hbar = h/(2\pi)$ is the reduced Planck constant, h is the Planck constant and c is the speed of light. In this case, there are $P/(\hbar\omega)$ photons per second striking the mirror, where $\hbar\omega$ is the energy carried by a single photon. If all of them are reflected straight back, the total change in light momentum per second, i.e., the force, is $2 \cdot (P/\hbar\omega) \cdot (\hbar\omega/c) = 2P/c$. By conservation of the total momentum of light and mirror, this implies that the mirror experiences an equal and opposite recoil force in the propagation direction of the light. This is the maximum force that one can extract form the light beam. Quoting Ashkin (2000):

> Suppose we have a laser and we focus our one watt to a small spot size of about a wavelength $\cong 1$ μm, and let it hit a particle of diameter also of 1 μm. Treating the particle as a 100% reflecting mirror of density $\cong 1$ gm/cm^3, we get an acceleration of the small particle $= A = F/m = 10^{-3}$ dynes/10^{-12} gm $= 10^9$ cm/sec^2. Thus, $A \cong 10^6$ g, where g $\cong 10^3$ cm/sec^2, the acceleration of gravity. This is quite large and should give readily observable effects, so I tried a simple experiment... It is surprising that this simple first experiment..., intended only to show forward motion due to laser radiation pressure, ended up demonstrating not only this force but the existence of the transverse force component, particle guiding, particle separation, and stable 3D particle trapping.[3]

Ashkin et al. (1986) reported the first observation of what is now commonly referred to as an *optical tweezers*: a tightly focused beam of light capable of holding microscopic particles in three dimensions. One of Ashkin's co-authors, Steven Chu, would go on to use optical tweezing in his work on cooling and trapping neutral atoms – research that earned Chu, together with Claude Cohen-Tannoudji and William Daniel Phillips, the 1997 Nobel Prize in Physics (Chu, 1998; Cohen-Tannoudji, 1998; Phillips, 1998).

Ghislain and Webb (1993) extended the capabilities of optical tweezers by devising a new kind of scanning probe microscopy that used an optically trapped microsphere as a probe. This technique, which was later called *photonic force microscopy*, provides one with the capability of measuring forces in the range from femtonewtons (10^{-15} N) to piconewtons (10^{-12} N), values well below those reachable with techniques based on microfabricated mechanical cantilevers, such as atomic force microscopy (Weisenhorn et al., 1989).

Since the early 1990s, optical trapping and optical manipulation have been applied to the biological sciences, starting by trapping an individual tobacco mosaic virus and an *Escherichia coli* bacterium (Ashkin et al., 1990). Block et al. (1990), Bustamante et al. (1994) and Finer et al. (1994) pioneered the use of optical force spectroscopy to characterise the mechanical properties of biomolecules and biological motors. These molecular motors are ubiquitous in biology, as they are responsible for locomotion, transport and mechanical action within the cell. Optical traps allowed these biophysicists to observe the forces and dynamics of nanoscale motors at the single-molecule level, leading to a greater

---

[3] In this quotation, we have kept the original cgs units used by Ashkin (2000). Except in direct quotations such as this, in this book we will consistently adopt SI units.

understanding of the workings of these force-generating molecules (Bustamante et al., 2003; Neuman and Nagy, 2008). Since then, optical tweezers have also proven useful in many other areas of physics (Molloy and Padgett, 2002; Grier, 2003; Dholakia et al., 2008; Jonáš and Zemánek, 2008; Dholakia and Čižmár, 2011; Juan et al., 2011; Padgett and Bowman, 2011; Padgett and Di Leonardo, 2011), nanotechnology (Maragò et al., 2013), chemistry (Yi et al., 2006), soft matter (Löwen, 2001; Petrov, 2007) and biology (Müller et al., 2009; Capitanio and Pavone, 2013), as we will see in more detail in Chapters 13–25.

1.2 Crash course on optical tweezers

In the simplest configuration, depicted schematically in Fig. 1.2, an optical tweezers is generated by focusing a laser beam at a diffraction-limited spot using a high-numerical-aperture (NA) objective lens. This objective serves the dual purpose of focusing the trapping light and imaging the trapped object. Samples are often placed in a small (a few microlitres in volume) microfluidic chamber held on a microscope stage, which can be translated either by some manual actuators with sub-micrometric resolution or by piezo-driven actuators with nanometre position resolution. Generally, optical tweezers require little optical power, down to a few milliwatts, so that the risk of photodamage to, e.g., biological specimens is relatively small. The position of the optical trap can be controlled using, e.g., steerable mirrors. It is also possible to generate multiple optical traps by rapidly deflecting a single beam through several positions using, e.g., an acousto–optic deflector.

The focused laser beam acts as an attractive potential well for a particle whose refractive index is higher than that of its surrounding medium. The equilibrium position of the trapped particle lies near the focus.[4] When the object is displaced from this equilibrium position, it experiences an attractive force back towards it. This restoring force is, to a first approximation, proportional to the displacement. This means that along each direction, the optical forces associated with an optical tweezers can be described by Hooke's law, i.e.,

$$\begin{cases} F_x \approx -\kappa_x (x - x_{eq}) \\ F_y \approx -\kappa_y (y - y_{eq}) \\ F_z \approx -\kappa_z (z - z_{eq}), \end{cases} \quad (1.1)$$

where $[x, y, z]$ is the particle's position, $[x_{eq}, y_{eq}, z_{eq}]$ is the equilibrium position, and κ_x, κ_y and κ_z are the optical trap spring constants along the x-, y- and z-directions, usually referred to as *trap stiffnesses*. Therefore, an optical tweezers creates a three-dimensional potential well that can be approximated with three independent harmonic oscillators, one for each direction. If the optical system is well aligned, then for spherical microparticles, which are commonly used in optical tweezers, κ_x and κ_y are roughly equal, whereas κ_z is typically smaller by a factor between 2 and 10. In order to have an idea of the range of forces we are

---

[4] In fact, as we will see in Chapters 2–6, the equilibrium position is typically slightly displaced in the propagation direction of the optical beam because of the presence of scattering optical forces.

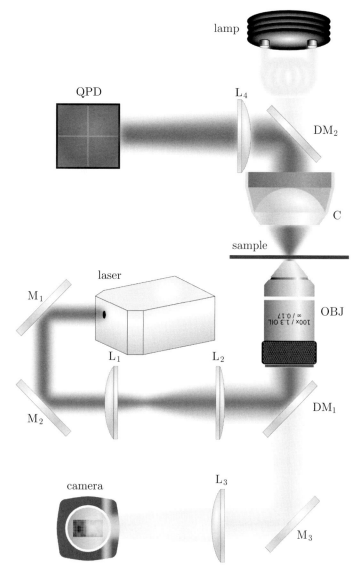

Figure 1.2 Basic experimental design. Optical tweezers are generated by focusing a laser beam on a diffraction-limited spot using a high-numerical-aperture objective lens (OBJ). Additional optics is needed to steer the optical tweezers position (beam-steering mirror M_2 and telescope formed by lenses L_1 and L_2), to image the sample (lamp, dichroic mirrors DM_1 and DM_2, camera lens L_3 and camera) and to track its position (condenser C, lens L_4 and quadrant photodiode QPD). The Brownian motion of the trapped particle can be tracked either by digital video microscopy or by interferometry and the measured particle trajectory can be used to calibrate the optical tweezers stiffness.

talking about, for a focused laser beam with power $P = 10\,\mathrm{mW}$, $\kappa_x \approx \kappa_y$ is on the order of $1\,\mathrm{pN/\mu m} = 1\,\mathrm{fN/nm}$; however, this value can vary by as much as one order of magnitude depending on the experimental conditions, such as particle size and material, particle and medium refractive index, and lens numerical aperture. Finally, optical tweezers can generate not only forces, but also torques, e.g., by the transfer of the *spin angular momentum* associated with circularly polarised light, or of the *orbital angular momentum* associated with structured beams, or as a result of a particle's asymmetric shape (*windmill effect*).

Microscopic and nanoscopic particles undergo permanent *Brownian diffusion* because of collisions with the surrounding fluid molecules. Therefore, an optically trapped particle is, in fact, in a dynamic equilibrium between the thermal noise pushing it out of the trap and the optical forces driving it towards the equilibrium position. For the particle to remain within the optical trap, the optical potential well must be sufficiently deep. The depth of the optical potential is typically characterised in units of the thermal energy $k_B T$, where k_B is the Boltzmann constant and T is the absolute temperature, because this gives a characteristic energy scale for mesoscopic phenomena. The potential well of an optical tweezers should be at least a few $k_B T$ deep to be able to confine a particle. It is useful to keep in mind that at a typical and comfortable room temperature ($T = 300 \text{ K}$) $k_B T = 4.14 \times 10^{-21}$ J.

The presence of Brownian noise does not exhaust the oddities of microscopic environments. In fact, when liquid environments are considered, inertial effects are absent at the microscopic scale because of the overwhelming role of viscosity; i.e., the particle lives at *low Reynolds numbers*. In order to get a feeling for this strange world, we will cite Purcell (1977):

> I want to take you into the world of very low Reynolds number – a world which is inhabited by the overwhelming majority of the organisms in this room. This world is quite different form the one that we have developed our intuitions in ... It helps to imagine under what conditions a man would be swimming at, say, the same Reynolds number as his own sperm. Well you put him in a swimming pool that is full of molasses, and then you forbid him to move any part of his body faster than 1 cm/min. Now imagine yourself in that condition; you're under the swimming pool in molasses, and now you can only move like the hands of a clock. If under those ground rules you are able to move a few metres in a couple of weeks, you may qualify as a low Reynolds number swimmer.

The range of optical tweezer applications has been greatly expanded by the use of advanced beam-shaping techniques, where the structure of a light beam is altered by diffractive optical elements (DOEs) to produce multiple optical traps at definite positions and optical traps capable of imparting torques by the transfer of *orbital angular momentum*, e.g., Laguerre–Gaussian beams (Dholakia and Čižmár, 2011; Padgett and Bowman, 2011). Typically, a DOE is placed in a plane conjugate to the objective back aperture so that the complex field distribution in the trapping plane is the Fourier transform of that in the DOE plane. Often the DOE is a liquid-crystal spatial light modulator and is used to modulate just the phase of the incoming beam, as modulation of the amplitude of the beam would entail a loss of optical power. An optical tweezers incorporating such a device is often referred to as a *holographic optical tweezers* (HOT).

1.3 Optical trapping regimes

Optical trapping of particles is a consequence of the radiation force that stems from the conservation of (electromagnetic and mechanical) momentum. Historically, optical forces

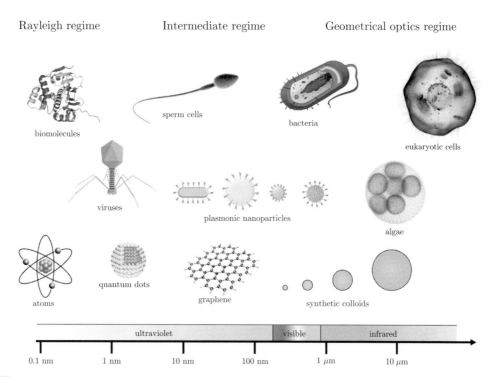

Figure 1.3 Optical trapping regimes. Trapping regimes and typical objects that are trapped in optical manipulation experiments. In defining the regimes of optical trapping, we have assumed trapping wavelengths that fall into in the visible or near-infrared spectral region.

have generally been understood through the use of approximations that depend on the size of the particle. Remarkable simplifications can be made in the calculation of the force exerted by optical tweezers when the particle size is much bigger – the *geometrical optics approximation* or *ray optics regime*, which is the subject of Chapter 2 – or much smaller – the *Rayleigh regime* or *dipole approximation*, which is the subject of Chapter 3 – than the trapping light wavelength. The importance of these approximate approaches is not only historical; they are often a valuable source of physical insight, as they are often amenable to analytical calculation and they can quickly give answers to complex problems. Nevertheless, when the particle's dimensions are comparable to the light wavelength, i.e., in the *intermediate regime*, a complete wave-optical modelling of the particle–light interaction is necessary for calculating the optical trapping forces, as we discuss in detail in Chapters 5 and 6. For homogeneous spherical particles, Mie theory can be used to generate accurate numerical results for essentially any size and refractive index; however, the situation becomes much more complex for non-spherical or non-homogenous particles. In fact, this is the case for most optical trapping experiments, where complex objects from tens of nanometres to tens of micrometres are manipulated, as shown in Fig. 1.3. Thus, in most cases, full electromagnetic calculations need to be used to have accurate predictions of the optical trapping behaviour.

The parameter most often used to determine the range of validity of these approaches is the *size parameter*,

$$k_{\mathrm{m}}a = \frac{2\pi n_{\mathrm{m}}}{\lambda_0}a, \qquad (1.2)$$

where k_{m} is the light wavenumber in the medium surrounding the particle, a is the characteristic dimension of the particle, which in the case of a sphere corresponds to its radius, λ_0 is the trapping wavelength in vacuum, and n_{m} is the refractive index of the surrounding medium, which is typically water ($n_{\mathrm{m}} = 1.33$) or air ($n_{\mathrm{m}} = 1.00$). On one hand, ray optics is valid when $k_{\mathrm{m}}a \gg 1$ and its accuracy increases as the size parameter grows; because exact theories for nonspherical particles become impractical when $k_{\mathrm{m}}a$ exceeds a certain threshold due to the increase of their computational complexity, ray optics is in fact an extremely useful and effective approach for dealing with large particles. On the other hand, the Rayleigh approximation holds when a particle can be approximated as a dipole and the electromagnetic fields can be considered homogeneous inside the particle; this imposes two conditions on the size parameter,

$$\begin{cases} k_{\mathrm{m}}a \ll 1 \\ \left|\dfrac{n_{\mathrm{p}}}{n_{\mathrm{m}}}\right| k_{\mathrm{m}}a \ll 1, \end{cases} \qquad (1.3)$$

where n_{p} is the refractive index of the particle.[5] The second condition in Eq. (1.3) needs to be considered with care, especially when dealing with nanostructures made of materials with high refractive index, e.g., silicon ($n_{\mathrm{p}} \approx 3.7$ at $\lambda_0 \approx 830\,\mathrm{nm}$), or when dealing with materials with a complex refractive index, e.g., plasmonic nanoparticles.

1.4 Other micromanipulation techniques

Apart from optical tweezers, several other micromanipulation techniques have been developed. In this section, we will briefly present *atomic force microscopy, magnetic tweezers, optoelectronic tweezers* and *acoustic tweezers,* focusing on their standing when compared to the optical tweezers technique. In fact, each of these techniques represents a fully fledged research field in its own right, and we will not make any attempt to discuss them in detail in this book.

Binnig et al. (1982) invented the *scanning tunnelling microscope*, which made it possible to resolve at the atomic level crystallographic structures (Binnig et al., 1983) and organic molecules (Smith et al., 1990). The *atomic force microscope* was invented a few years later by Binnig et al. (1986). These instruments have been successfully employed to study biological and nano-fabricated structures, overcoming the traditional diffraction limit of

---

[5] In fact, the dipole approximation can be derived from Mie theory by expanding the Bessel functions in the solutions under the assumption that Eq. (1.3) holds true (Mishchenko et al., 2002).

optical microscopes. Furthermore, they have developed from pure imaging instruments into more general manipulation and measuring tools capable of operating at the level of single atoms or molecules. However, all these techniques require a macroscopic mechanical device to guide their cantilevers, so the force range they can probe is substantially higher (nanonewtons) than that which is accessible with optical tweezers (from piconewtons down to femtonewtons).

A *magnetic tweezers* uses a set of magnets to produce a strong magnetic field gradient that is used to exert a force on a microscopic superparamagnetic particle (Smith et al., 1992; Neuman and Nagy, 2008). The force is varied by modulating the position of the magnets relative to the bead and, because usually the magnets are millimetres in size whereas the bead is only allowed to move by a few micrometres, the applied magnetic force can be regarded as constant for a given position of the magnets. A very interesting feature of the magnetic tweezers technique is its ability to rotate the magnetic bead, e.g., allowing one to easily twist a molecule tethering the bead to the sample. Although the laser beam of an optical tweezers may also interact with other particles of the sample because of contrasts in the refractive index, the magnetic interaction is highly specific to the superparamagnetic microparticles used and therefore practically does not affect the sample. Also, as a consequence of the low trap stiffness associated with magnetic forces, the range of forces accessible with magnetic tweezers is lower than with optical tweezers.

Optoelectronic tweezers use projected images to trap and manipulate large ensembles of particles (Chiou et al., 2005). They are intrinsically two-dimensional and work by pushing the particles against a planar surface: an electric potential difference is applied between the two walls of a sample cell and then light is used to create some reconfigurable electrodes on one of the surfaces. In this way, a non-uniform electromagnetic field is obtained and particles are moved accordingly because of dielectrophoresis. Optoelectronic tweezers have an advantage over conventional optical tweezers in that they can use non-coherent light sources and require several orders of magnitude less power so that several thousands of particles can be manipulated in parallel. They cannot, however, be used to manipulate particles in three dimensions or to measure forces acting on nanoscopic length scales.

Not only electromagnetic waves, but also acoustic waves can generate a radiation force. In fact, the field of a standing wave produced by two collimated ultrasound transducers can trap polymer spheres, biological cells, grass seeds and microbubbles at either nodes or antinodes along the beam axis. This trapping technique is referred to as *acoustic tweezers* (Wu, 1991) and has versatile applications in the separation, concentration and localisation of particles. Based on the concept of optical tweezers, Lee et al. (2005) suggested theoretically that a trapping effect can be produced by a single focused acoustic beam using a highly focusing ultrasonic transducer with a ratio of focal length to transducer diameter near 1, i.e., the acoustic equivalent of a high-NA objective, driven at high frequency ($\approx 100\,\text{MHz}$) and high pressure ($\approx 1.5\,\text{MPa}$). A single-beam acoustic device, with its relatively simple scheme and low intensity, can trap a single lipid droplet in a manner similar to optical tweezers. Forces on the order of hundreds of nanonewtons direct the droplet towards the beam focus, within a range of hundreds of micrometres. This trapping method, therefore, can be a useful tool for particle manipulation in areas where larger particles or forces are involved and for manipulations over longer distances, as Lee et al. (2009) demonstrated experimentally. The main advantage of acoustic tweezers over optical tweezers lies in their being able to exert much larger forces, i.e., nanonewtons, and in acoustic waves having a much greater penetration depth in fluids and biological materials.

1.5 Scope of this book

This book provides a comprehensive guide to the theory, practice and applications of optical trapping and optical manipulation. The material expounded in this book, together with the material provided on the book website (www.opticaltweezers.org), will provide a working knowledge of optical trapping and manipulation sufficient to build an optical tweezers apparatus and operate it to perform an experiment and to calculate optical forces.

Part I [Chapters 2–7] presents the theoretical foundations needed to understand the physics behind optical trapping, including both the theory of optical forces and the theory of Brownian motion. Chapter 2 is dedicated to the geometrical optics approximation and also provides an intuitive overview of many concepts that give insight into the mechanism by which microscopic particles can be optically trapped. Chapter 3 deals with optical forces in the dipole approximation and introduces the concept of optical binding. Chapter 4 provides a detailed introduction to optical beams, their properties and their focusing. Chapters 5 and 6 are dedicated to the exact treatment of optical forces based on electromagnetic light scattering, focusing on the Mie scattering theory and the transition matrix (T-matrix) approach; the discrete dipole approximation (DDA) and finite-difference time domain (FDTD) methods are also briefly introduced. Together, these two chapters cover the core material needed for any accurate electromagnetic study of optical forces. Finally, Chapter 7 reviews the main properties of the Brownian motion that come into play when dealing with optical trapping and manipulation and that are necessary to understand the dynamics of optically trapped particles.

Part II [Chapters 8–12] deals with the experimental practice of optical trapping and manipulation. Chapter 8 presents a step-by-step guide to the realisation of a simple optical tweezers set-up starting from scratch, assuming only a minimum level of experimental knowledge on the part of the reader. Chapter 9 gives a detailed account of how to acquire and analyse data from optical tweezers. Chapter 10 expounds upon how to employ optical tweezers to measure nanoscopic forces and torques, i.e., the use of an optical tweezers as a photonic force microscope. Chapter 11 explains how to work with holographic optical tweezers, including, in particular, the calculation of diffraction patterns that produce a given intensity distribution in the trapping plane. Finally, Chapter 12 reviews a series of alternative and advanced optical manipulation techniques, including spectroscopic tweezers, fibre optical tweezers, evanescent optical traps, feedback optical tweezers and haptic optical tweezers.

Part III [Chapters 13–25] gives detailed reviews of various fields of application for optical trapping and manipulation. First, Chapters 13 and 14 give an overview of how optical tweezers have found application in molecular biology and cell biology, respectively. Then, Chapters 15 and 16 explore how optical tweezers have been successfully combined with spectroscopic and optofluidic techniques. Afterwards, a series of chapters are dedicated to how optical tweezers have been used to study phenomena in soft matter and colloid science [Chapter 17], microchemistry [Chapter 18], aerosol science [Chapter 19], statistical physics [Chapter 20] and nanoscopic thermodynamics [Chapter 21]. Finally, we consider the development of optical tweezers towards the nanoscopic world and, in particular, plasmonic optical tweezers [Chapter 22], nanotweezers [Chapter 23], atom trapping [Chapter 24] and optical trapping of mesoscopic objects towards the quantum regime [Chapter 25].

1.6 How to read this book

While we have striven to keep the treatment of the material at a level accessible to advanced undergraduate or beginning graduate students, we hope that the depth and breath of the topics covered in this book will provide a useful resource even for experienced practitioners. In fact, this book can be read at different levels, from the interested layman level to the specialist level. In order to afford such flexibility without overloading the main text of the book, we have embedded in the body of the chapters of Parts I and II a series of exercises, whose results are often explicitly stated in the exercises themselves, and are in fact part of the content of this book. We are sure every reader will easily find his or her most convenient way through this book; nevertheless, here we give some suggestions on how to read this book depending on the reader profile:

- **Interested scientists**: We suggest starting from the applications explained in Part III and then moving on to Parts I and II, skipping all exercises and Chapters 5 and 6, as, even with this omission, one can get a very good feeling for what optical tweezers are, how they work and in which fields they are employed.
- **Practitioners**: We suggest starting from any of the chapters in Parts I and II they find most useful and perusing (at least) the exercises in the text of the chapters they are planning to use in their research.
- **Theoreticians**: We suggest studying Part I in depth with all exercises and quickly reading Part II (without exercises). In particular, we highlight that the theoretical treatment is accompanied by a software package for the treatment of optical forces and Brownian motion.
- **Experimentalists**: We suggest quickly reading Part I (without exercises and skipping Chapters 5 and 6 at a first reading) and studying Part II in depth. In particular, we highlight that Chapter 8 provides a step-by-step guide to the construction of an optical tweezers from scratch and without assuming any previous knowledge of the field, that Chapter 9 provides a detailed explanation of the technique for acquiring and analysing

optical tweezers data, and that Chapter 11 provides an in-depth explanation of how to operate holographic optical tweezers.

1.7 OTS - the Optical Tweezers Software

Together with this book, we have developed a comprehensive MatLab software toolbox to work with optical tweezers – OTS - the Optical Tweezers Software. OTS is freely available for download from www.opticaltweezers.org. Like this book, OTS covers both the theory and the practice of optical trapping and manipulation. For the theoretical part, we have developed packages to calculate optical forces within the geometrical optics approximation (go), the dipole approximation (da) and the full electromagnetic theory (mie and emt); furthermore, we provide some code to perform Brownian dynamics simulations of optically trapped particles (bm). For the practical part, we provide codes to perform digital video microscopy (dvm), optical tweezers calibration and force measurement (pfm), and holographic optical tweezers (hot). All these codes are based on our research experience and are developed to the point where they can be readily used in research.

OTS is fully documented, accompanied by code examples and ready to be employed to explore more complex situations in learning, teaching and research. In fact, we have implemented OTS using an object-oriented approach so that it can easily be extended and adapted to the specific needs of users; for example, it is possible to create more complex optically trappable particles in the geometrical optics approximation by extending the objects provided for spherical, cylindrical and ellipsoidal particles. In particular, we have used OTS to prepare most of the original figures in this book.

The OTS package contains the following subpackages:

- OTS.m - (file) : Load the software packages
- utility - (directory) : General utility functions
- tools - (directory) : Common tools
- shapes - (directory) : 3D geometrical shapes
- beams - (directory) : Optical beams
- go - (directory) : Geometrical optics
- da - (directory) : Dipole approximation
- mie - (directory) : Mie particles
- emt - (directory) : Electromagnetic theory
- bm - (directory) : Brownian motion
- dvm - (directory) : Digital video microscopy
- pfm - (directory) : Photonic force microscopy
- hot - (directory) : Holographic optical tweezers

All the documentation for the installation and use of the OTS package is available from the book website (www.opticaltweezers.org).

References

Ashkin, A. 1970a. Acceleration and trapping of particles by radiation pressure. *Phys. Rev. Lett.*, **24**, 156–9.

Ashkin, A. 1970b. Atomic-beam deflection by resonance-radiation pressure. *Phys. Rev. Lett.*, **25**, 1321–4.

Ashkin, A. 2000. History of optical trapping and manipulation of small-neutral particles, atoms, and molecules. *IEEE J. Sel. Top. Quant. El.*, **6**, 841–56.

Ashkin, A., Dziedzic, J. M., Bjorkholm, J. E., and Chu, S. 1986. Observation of a single-beam gradient force optical trap for dielectric particles. *Opt. Lett.*, **11**, 288–90.

Ashkin, A., Schütze, K., Dziedzic, J. M., Euteneuer, U., and Schliwa, M. 1990. Force generation of organelle transport measured *in vivo* by an infrared laser trap. *Nature*, **348**, 346–8.

Beth, R. A. 1936. Mechanical detection and measurement of the angular momentum of light. *Phys. Rev.*, **50**, 115–25.

Binnig, G., Rohrer, H., Gerber, C., and Weibel, E. 1982. Surface studies by scanning tunneling microscopy. *Phys. Rev. Lett.*, **49**, 57–61.

Binnig, G., Rohrer, H., Gerber, C., and Weibel, E. 1983. 7×7 reconstruction on Si(111) resolved in real space. *Phys. Rev. Lett.*, **50**, 120–23.

Binnig, G., Quate, C. F., and Gerber, C. 1986. Atomic force microscope. *Phys. Rev. Lett.*, **56**, 930–33.

Block, S. M., Goldstein, L. S. B., and Schnapp, B. J. 1990. Bead movement by single kinesin molecules studied with optical tweezers. *Nature*, **348**, 348–52.

Bustamante, C., Marko, J. F., Siggia, E. D., and Smith, S. 1994. Entropic elasticity of lambda-phage DNA. *Science*, **265**, 1599–1600.

Bustamante, C., Bryant, Z., and Smith, S. B. 2003. Ten years of tension: Single-molecule DNA mechanics. *Nature*, **421**, 423–7.

Capitanio, M., and Pavone, F. S. 2013. Interrogating biology with force: Single molecule high-resolution measurements with optical tweezers. *Biophys. J.*, **105**, 1293–1303.

Chiou, P. Y., Ohta, A. T., and Wu, M. C. 2005. Massively parallel manipulation of single cells and microparticles using optical images. *Nature*, **436**, 370–72.

Chu, S. 1998. The manipulation of neutral particles. *Rev. Mod. Phys.*, **70**, 685–706.

Cohen-Tannoudji, C. N. 1998. Manipulating atoms with photons. *Rev. Mod. Phys.*, **70**, 707–19.

Dholakia, K., and Čižmár, T. 2011. Shaping the future of manipulation. *Nature Photon.*, **5**, 335–42.

Dholakia, K., Reece, P., and Gu, M. 2008. Optical micromanipulation. *Chem. Soc. Rev.*, **37**, 42–55.

Finer, J. T., Simmons, R. M., and Spudich, J. A. 1994. Single myosin molecule mechanics: Piconewton forces and nanometre steps. *Nature*, **368**, 113–19.

Ghislain, L. P., and Webb, W. W. 1993. Scanning-force microscope based on an optical trap. *Opt. Lett.*, **18**, 1678–80.

Grier, D. G. 2003. A revolution in optical manipulation. *Nature*, **424**, 810–16.

Jeng, M. 2006. A selected history of expectation bias in physics. *Am. J. Phys.*, **74**, 578–83.

Jonáš, A., and Zemánek, P. 2008. Light at work: The use of optical forces for particle manipulation, sorting, and analysis. *Electrophoresis*, **29**, 4813–51.

Juan, M. L., Righini, M., and Quidant, R. 2011. Plasmon nano-optical tweezers. *Nature Photon.*, **5**, 349–56.

Kepler, J. 1619. *De Cometis libelli tres.*

Lebedev, P. 1901. Untersuchungen über die Druckkräfte des Lichtes. *Ann. Physik*, **311**, 433–58.

Lee, J., Ha, K., and Shung, K. K. 2005. A theoretical study of the feasibility of acoustical tweezers: Ray acoustics approach. *J. Acoust. Soc. Am.*, **117**, 3273–80.

Lee, J., Teh, S.-Y., Lee, A., et al. 2009. Single beam acoustic trapping. *Appl. Phys. Lett.*, **95**, 073701.

Löwen, H. 2001. Colloidal soft matter under external control. *J. Phys.: Condens. Matt.*, **13**, R415–R432.

Maragò, O. M, Jones, P. H., Gucciardi, P. G., Volpe, G., and Ferrari, A. C. 2013. Optical trapping and manipulation of nanostructures. *Nature Nanotech.*, **8**, 807–19.

Mishchenko, M. I., Travis, L. D., and Lacis, A. A. 2002. *Scattering, absorption, and emission of light by small particles.* Cambridge, UK: Cambridge University Press.

Molloy, J. E., and Padgett, M. J. 2002. Lights, action: Optical tweezers. *Contemp. Phys.*, **43**, 241–58.

Müller, D. J., Helenius, J., Alsteens, D., and Dufrêne, Y. F. 2009. Force probing surfaces of living cells to molecular resolution. *Nature Chem. Biol.*, **5**, 383–90.

Neuman, K. C., and Nagy, A. 2008. Single-molecule force spectroscopy: Optical tweezers, magnetic tweezers and atomic force microscopy. *Nature Methods*, **5**, 491–505.

Nichols, E. F., and Hull, G. F. 1901. A preliminary communication on the pressure of heat and light radiation. *Phys. Rev.*, **13**, 307–20.

Padgett, M., and Bowman, R. 2011. Tweezers with a twist. *Nature Photon.*, **5**, 343–8.

Padgett, M., and Di Leonardo, R. 2011. Holographic optical tweezers and their relevance to lab on chip devices. *Lab Chip*, **11**, 1196–1205.

Petrov, D. V. 2007. Raman spectroscopy of optically trapped particles. *J. Opt. A: Pure Appl. Opt.*, **9**, S139–S156.

Phillips, W. D. 1998. Laser cooling and trapping of neutral atoms. *Rev. Mod. Phys.*, **70**, 721–42.

Poynting, J. H. 1884. On the transfer of energy in the electromagnetic field. *Phil. Trans. Royal Soc. London*, **175**, 343–61.

Purcell, E. M. 1977. Life at low Reynolds number. *Am. J. Phys.*, **45**, 3–11.

Smith, D. P. E., Hörber, J. K. H., Binnig, G., and Nejoh, H. 1990. Structure, registry and imaging mechanism of alkylcyanobiphenyl molecules by tunnelling microscopy. *Nature*, **344**, 641–4.

Smith, S. B., Finzi, L., and Bustamante, C. 1992. Direct mechanical measurements of the elasticity of single DNA molecules by using magnetic beads. *Science*, **258**, 1122–6.

Townes, C. H. 1999. *How the laser happened: Adventures of a scientist.* Oxford, UK: Oxford University Press.

Weisenhorn, A. L., Hansma, P. K., Albrecht, T. R., and Quate, C. F. 1989. Forces in atomic force microscopy in air and water. *Appl. Phys. Lett.*, **54**, 2651–3.

Worrall, J. 1982. The pressure of light: The strange case of the vacillating 'crucial experiment'. *Stud. Hist. Phil. Science Part A*, **13**, 133–71.

Wu, J. R. 1991. Acoustical tweezers. *J. Acoust. Soc. Am.*, **89**, 2140–43.

Yi, C., Li, C.-W., Ji, S., and Yang, M. 2006. Microfluidics technology for manipulation and analysis of biological cells. *Anal. Chim. Acta*, **560**, 1–23.

PART I

THEORY

However, this bottle was *not* marked 'poison', so Alice ventured to taste it, and finding it very nice (it had, in fact, a sort of mixed flavour of cherry-tart, custard, pine-apple, roast turkey, toffee, and hot buttered toast), she very soon finished it off.

'What a curious feeling!' said Alice. 'I must be shutting up like a telescope.'

Alice's Adventures in Wonderland – Lewis Carroll

2 Ray optics

If one shines a laser beam on a large transparent object, such as the prism shown in Fig. 2.1, a fraction of the power of the beam will be reflected back and the rest will go through and exit from the other side. Because the directions of the reflected and transmitted beams are different from that of the incoming beam, the mechanical momentum associated with the light beam changes and, by Newton's action–reaction law, a force acts on the object. As long as one deals with objects that are much larger than the wavelength of light, which is typically around one micrometre for optical tweezing applications, the behaviour of the laser beam can be accurately described by considering it as a collection of *light rays* and employing the tools of *geometrical optics*. One can also make use of geometrical optics for the calculation of optical forces and obtain accurate results when dealing with relatively large objects, such as cells and large colloidal particles, whose size is typically significantly larger than one micrometre. In this chapter, we will therefore study optical forces using the geometrical optics approach, as this permits us to introduce in a more intuitive way many concepts that give insight into the mechanism by which microscopic particles can be optically trapped and that will be treated more rigorously in Chapters 3, 5 and 6.

Figure 2.1 Reflection and transmission on a prism. As a laser light beam impinges on a prism, it is partly reflected and partly transmitted. The directions of the reflected and transmitted beams are different from that of the incoming beam. This change of direction entails a change of the momentum associated with the light beam and, because of the action–reaction law, a force acting on the prism. Picture credits: Marco Grasso and Alessandro Magazzù.

2.1 Optical rays

The density of the energy flux due to an electromagnetic wave is given by the Poynting vector **S** or, equivalently, by the number of photons passing through a unit area per unit time multiplied by the energy per photon.[1] In order to describe how this energy is transported, a series of rays can be associated with an electromagnetic wave. These rays are lines perpendicular to the electromagnetic wavefronts and pointing in the direction of the electromagnetic energy flow, i.e., the direction of **S**. For example, Fig. 2.2 shows the rays associated with a plane electromagnetic wave constructed by dividing the wavefront into several portions of equal area A and by associating with each of these portions a ray with power

$$P = |\mathbf{S}|A. \tag{2.1}$$

When a light ray $\mathbf{r}_i$ impinges on a flat surface between two media with different optical impedances, or *refractive indices*, part of it is reflected and part transmitted, as shown in Fig. 2.3. The *angle of incidence* θ_i is the angle between $\mathbf{r}_i$ and the line n normal to the surface at the incidence point. For the *angle of reflection* θ_r, the law of reflection

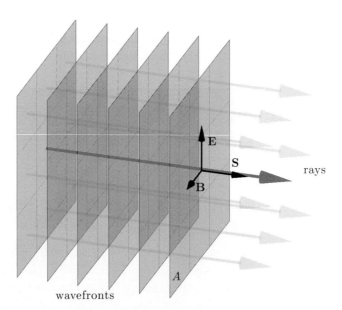

Figure 2.2 From electromagnetic waves to rays. Optical rays (grey arrows) are perpendicular to the wavefronts of an electromagnetic wave (planes) and parallel to the Poynting vector (**S**).

---

[1] Here, and in the rest of this chapter, we are implicitly considering a monochromatic electromagnetic wave. In terms of the physical fields $\mathcal{E}(\mathbf{r}, t)$ and $\mathcal{B}(\mathbf{r}, t)$, $\mathbf{S} = \overline{\mathcal{S}} = \frac{1}{\mu}\overline{\mathcal{E} \times \mathcal{B}}$ [Box 3.6]. In terms of the phasors $\mathbf{E}(\mathbf{r})$ and $\mathbf{B}(\mathbf{r})$ [Box 3.5], $\mathbf{S} = \frac{1}{2\mu}\text{Re}\{\mathbf{E} \times \mathbf{B}^*\}$ [Box 3.6].

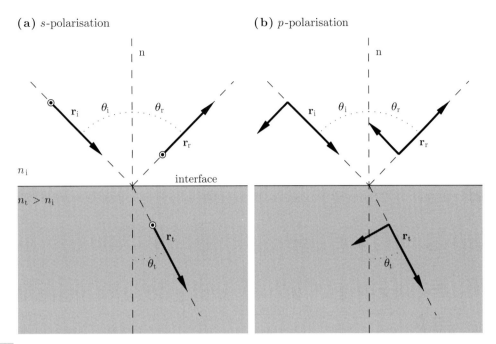

Figure 2.3 Reflection and transmission at a planar interface. As (a) an s-polarised or (b) a p-polarised ray $\mathbf{r}_i$ impinges on a planar interface between two media with refractive indices n_i and n_t, it splits into a reflected ray $\mathbf{r}_r$ and a transmitted ray $\mathbf{r}_t$. The arrows perpendicular to the rays represent the directions of the electric fields, and n is the line perpendicular to the interface at the incidence point.

states that

$$\theta_r = \theta_i, \tag{2.2}$$

and for the *angle of transmission* θ_t, Snell's law states that

$$n_t \sin \theta_t = n_i \sin \theta_i, \tag{2.3}$$

where n_i is the refractive index of the medium in which the incident ray propagates and n_t is that of the medium in which the transmitted ray propagates. Both the reflected and transmitted rays are contained in the *plane of incidence*, i.e., the plane that contains $\mathbf{r}_i$ and n.

As a consequence of energy conservation, the incoming power P_i must be equal to the sum of the reflected power P_r and the transmitted power P_t; i.e.,

$$P_i = P_r + P_t. \tag{2.4}$$

How the power is split between the reflected and transmitted rays can be calculated using Maxwell's laws [Box 3.1] by imposing continuity across the interface of the tangential components of the electric and magnetic fields [Box 3.2]. The result depends on the polarisation of the incoming ray and is expressed by *Fresnel's equations*. For *s*-polarised light, i.e., light whose electric field vector is normal to the plane of incidence [Fig. 2.3a], the intensity reflection coefficient is given by

$$R_s = \left| \frac{n_i \cos\theta_i - n_t \cos\theta_t}{n_i \cos\theta_i + n_t \cos\theta_t} \right|^2 \tag{2.5}$$

and the intensity transmission coefficient by

$$T_s = \frac{4 n_i n_t \cos\theta_i \cos\theta_t}{|n_i \cos\theta_i + n_t \cos\theta_t|^2}. \tag{2.6}$$

For *p*-polarised light, i.e., light whose electric field vector is contained in the plane of incidence [Fig. 2.3b], the intensity reflection coefficient is given by

$$R_p = \left| \frac{n_i \cos\theta_t - n_t \cos\theta_i}{n_i \cos\theta_t + n_t \cos\theta_i} \right|^2 \tag{2.7}$$

and the intensity transmission coefficient by

$$T_p = \frac{4 n_i n_t \cos\theta_i \cos\theta_t}{|n_i \cos\theta_t + n_t \cos\theta_i|^2}. \tag{2.8}$$

It is straightforward to verify that $R_s + T_s = 1$ and $R_p + T_p = 1$, fulfilling the energy conservation statement in Eq. (2.4). For unpolarised or circularly polarised light, one can use the average of the previous coefficients, obtaining

$$R = \frac{R_s + R_p}{2} \tag{2.9}$$

and

$$T = \frac{T_s + T_p}{2}. \tag{2.10}$$

To keep the discussion simple, in the rest of this chapter we will consider all rays to be circularly polarised, unless otherwise stated.

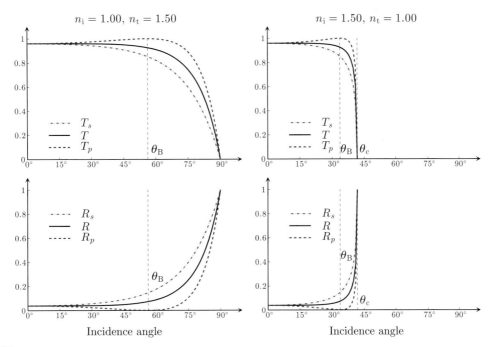

Figure 2.4 Fresnel's coefficients. Fresnel's transmission and reflection coefficients for a ray going from air to glass (left) and from glass to air (right). θ_B and θ_c are Brewster's angle and the critical angle, respectively.

The values of the reflection and transmission coefficients are plotted in Fig. 2.4. For an s-polarised wave the transmission efficiency decreases monotonically as θ_i increases, whereas for a p-polarised wave the transmission efficiency first increases until all incident light is transmitted at *Brewster's angle*,

$$\theta_B = \arctan\left(\frac{n_t}{n_i}\right), \tag{2.11}$$

and then decreases. Going from a medium with a lower refractive index, e.g., air ($n_m = 1.00$), to a medium with higher refractive index, e.g., glass ($n_m = 1.50$), there is always transmission; this is not the case in the opposite direction, e.g., going from glass to water, because above the *critical angle*

$$\theta_c = \arcsin\left(\frac{n_t}{n_i}\right) \tag{2.12}$$

both s-polarised and p-polarised light are completely reflected; this phenomenon is known as *total internal reflection*.

Exercise 2.1.1 Derive Fresnel's equations [Eqs. (2.5), (2.6), (2.7) and (2.8)]. [Hint: Use Maxwell's laws [Box 3.2] and impose continuity across the interface of the tangential components of the electric and magnetic fields [Box 3.4].]

2.2 Optical forces

A photon, i.e., a particle representing a quantum of light, of wavelength λ_0 in vacuum carries energy $u = hc/\lambda_0$, where h is the Planck constant and c is the speed of light in vacuum, and momentum $\mathbf{p} = (h/\lambda_0)\hat{\mathbf{u}}$, where $\hat{\mathbf{u}}$ is a unit vector indicating the photon's direction of motion. When such a photon is elastically scattered by an object, its energy does not change, but its momentum can change direction, resulting in a recoil force acting on the object.

A light ray with power P carries $N = P/u$ photons per second past a fixed point. If, as shown in Fig. 2.5a, such a ray impinges with normal incidence on a mirror, it will be completely reflected back, so that the momentum of each photon changes by $-2\mathbf{p}$ and the total change of momentum per unit time of the ray is $-2N\mathbf{p} = -2(P/c)\hat{\mathbf{u}}$. Thus, by Newton's action–reaction law, the recoil force on the mirror is

$$\mathbf{F}_{\text{reflection}} = \frac{2P}{c}\hat{\mathbf{u}}, \qquad (2.13)$$

which is, in fact, the maximum optical force that can be generated by a ray (or a laser beam) of power P. For example, for a 1 mW laser beam, Eq. (2.13) gives a force of just 7×10^{-12} N, i.e., 7 piconewtons. Although small, this force is comparable to the forces that are relevant in the microscopic and nanoscopic world, e.g., the forces generated by molecular motors, and gives us a first impression of the potential of optical manipulation.

In general, as shown in Fig. 2.5b, a ray will impinge with *non-normal* incidence on a *non-planar* surface and will be partly reflected and partly transmitted. Thus, in order to calculate the forces associated with the reflection and transmission of a ray $\mathbf{r}_i$ of power P_i in direction $\hat{\mathbf{r}}_i$, one typically needs to

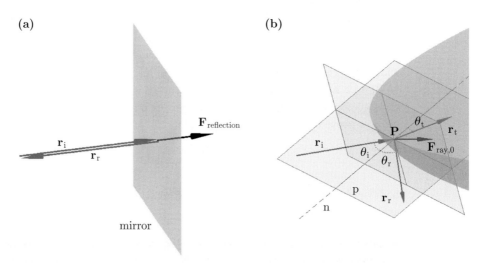

Figure 2.5 Ray optics forces: (a) Force generated by the reflection of a light ray impinging with normal incidence on a mirror. (b) Force generated by the scattering of a light ray by a generic surface with a generic angle of incidence.

> **Box 2.1** **What is the momentum of light?**
>
> What happens to the momentum of a photon when it enters a material medium? Hermann Minkowski (1908) considered that the momentum p carried by a photon is inversely proportional to its wavelength λ_0; therefore, because in a medium of refractive index n the wavelength becomes λ_0/n, the photon momentum must *increase* by a factor n, i.e.,
>
> $$p_M = n\frac{h}{\lambda_0}.$$
>
> One year later, Max Abraham (1909) argued that the momentum of an object is proportional to its velocity; therefore, because the speed of a photon in a medium becomes c/n, the photon momentum must *decrease* by a factor n, i.e.,
>
> $$p_A = \frac{1}{n}\frac{h}{\lambda_0}.$$
>
> Since the beginning of the twentieth century, scientists' opinions on this matter have been swinging between the two positions, leading to the well-known *Abraham–Minkowski dilemma*. Pfeifer et al. (2007) provide an extensive review of this topic.
>
> Experimental evidence in favour of either argument has been inconclusive, because most experiments measure the difference of the momentum carried by a beam before and after the interaction with an object and, because these momenta are calculated in the same medium, the Abraham–Minkowski dilemma does not produce any qualitative difference in the forces. A possibility would be to study what happens at an interface illuminated by a laser beam. Such an experiment was attempted by Ashkin and Dziedzic (1973): they showed that a narrow light beam impinging on a air–water interface produces an outward pull, resulting in the water surface bulging outward; consistent with the sign of the change of the Minkowski momentum, the bulge would be the result of the increased light momentum in water. However, it was later shown that this effect is governed by a radial gradient force – the water tends to collect in the high-intensity region of the laser beam, therefore the bulge – and that it provides no information on the longitudinal force associated with the linear momentum of light. So far, this and similar experiments have proved inconclusive.
>
> A possible solution of this issue has been recently put forward by Stephen Barnett (2010). Surprisingly, both arguments would be reasonable and sound because they could be both right – but they refer to two different momenta, namely the *kinetic momentum*, where Abraham's theory is concerned, and the *canonical momentum*, where Minkowski's theory is concerned. These two momenta are actually the same when light is propagating in vacuum, but they can differ significantly in a medium.

1. find the incidence point **P** where the ray meets the surface, and the normal line n, which is perpendicular to the surface at **P**;
2. find the incidence plane p, i.e., the plane containing $\mathbf{r}_i$ and n;
3. calculate the angle of incidence θ_i, i.e., the angle between $\mathbf{r}_i$ and n, and derive θ_r and θ_t using Eqs. (2.2) and (2.3);
4. determine the directions $\hat{\mathbf{r}}_r$ and $\hat{\mathbf{r}}_t$ of the reflected ray $\mathbf{r}_r$ and the transmitted ray $\mathbf{r}_t$, which are contained in p;

5. calculate the power P_r and P_t of $\mathbf{r}_r$ and $\mathbf{r}_t$, respectively, using Fresnel's equations [Eqs. (2.5), (2.6), (2.7) and (2.8)] and taking into account the polarisation of the incoming ray;
6. calculate the force exerted on the object to which the surface belongs as

$$\mathbf{F}_{\text{ray},0} = \frac{n_i P_i}{c}\hat{\mathbf{r}}_i - \frac{n_i P_r}{c}\hat{\mathbf{r}}_r - \frac{n_t P_t}{c}\hat{\mathbf{r}}_t, \qquad (2.14)$$

where the Minkowski momentum of light in a medium has been used.[2]

In general, if more than one ray interacts with an object, the total force is given by the sum of the forces generated by the reflection and transmission of each ray. In a similar way, it is also possible to consider the mechanical effects produced by the subsequent reflections and transmissions when the rays reach the surface of the object again.

Exercise 2.2.1 Show that the Abraham momentum of a ray of power P is P/c in every medium. How does Eq. (2.14) change when the Abraham momentum is used instead of the Minkowski momentum?

Exercise 2.2.2 Study numerically the force generated by the scattering of a ray at a planar interface as a function of the parameters of the system, e.g., angle of incidence and refractive indices of the media. Assume the ray to be circularly polarised.

Exercise 2.2.3 How do the optical forces due to the scattering of a ray on a planar surface change as a function of the polarisation of the ray?

2.3 Scattering and gradient forces

To understand the forces that act on a trapped microscopic particle, we start with a minimalistic model: the force due to a single ray $\mathbf{r}_i$ of power P_i hitting a dielectric sphere at an angle of incidence θ_i, as shown in Fig. 2.6a. As soon as $\mathbf{r}_i$ hits the sphere, a small amount of power is diverted into the reflected ray $\mathbf{r}_{r,0}$, while most of its power is carried by the transmitted ray $\mathbf{r}_{t,0}$. The ray $\mathbf{r}_{t,0}$ crosses the sphere until it reaches the opposite surface, where again it will be largely transmitted outside the sphere into the ray $\mathbf{r}_{t,1}$, while a small amount will be reflected inside the sphere into the ray $\mathbf{r}_{r,1}$. The ray $\mathbf{r}_{r,1}$ undergoes another scattering event as soon as it reaches the sphere boundary, and the process continues until all light has escaped from the sphere.[3] Repeatedly using Eq. (2.14) at each scattering event,

---

[2] The definition of the momentum of light in a medium is a very complex issue, which goes under the name of the *Abraham–Minkowski dilemma* and is still being debated. The Abraham and Minkowski definitions are equivalent in vacuum. In this book, we will consistently adopt the Minkowski momentum, in agreement with a large part of the optical manipulation literature. For a particle in a homogeneous medium of refractive index n_i, the alternative adoption of the Abraham form would generate forces smaller by a factor of n_i. Some further details are provided in Box 2.1.
[3] Although this is an asymptotic process that requires an infinite number of scattering events, typically virtually all light has escaped from the sphere within less than about 10 scattering events.

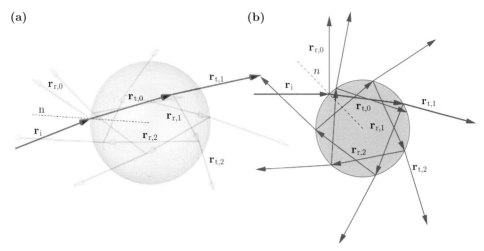

Figure 2.6 Scattering of a ray on a sphere. Multiple scattering of a light ray impinging on a sphere visualised (a) in three dimensions and (b) in the plane of incidence. Note how all the reflected and transmitted rays, as well as the vector of the force acting on the sphere (not shown), are contained in the plane of incidence.

it is possible to calculate the total force acting on the sphere as

$$\mathbf{F}_{\text{ray}} = \frac{n_i P_i}{c}\hat{\mathbf{r}}_i - \frac{n_i P_r}{c}\hat{\mathbf{r}}_{r,0} - \sum_{n=1}^{+\infty} \frac{n_i P_{t,n}}{c}\hat{\mathbf{r}}_{t,n}, \qquad (2.15)$$

where $\hat{\mathbf{r}}_i$, $\hat{\mathbf{r}}_{r,n}$ and $\hat{\mathbf{r}}_{t,n}$ are unit vectors representing the direction of the incident ray and the nth reflected and transmitted rays, respectively. We can also notice that the definition of the momentum of the light beam inside the sphere [Box 2.1] does not matter for the final calculation; this is generally true for virtually all experiments involving optical forces.

Because, as shown in Fig. 2.6b, all the reflected and transmitted rays are contained in the plane of incidence, the force $\mathbf{F}_{\text{ray}}$ also has only components in the plane of incidence. In particular, it is possible to split $\mathbf{F}_{\text{ray}}$ into the optical *scattering force* $\mathbf{F}_{\text{ray,s}}$ that pushes the particle in the direction of the incoming ray ($\hat{\mathbf{r}}_i$) and the optical *gradient force* $F_{\text{ray,g}}$ that pulls the particle in a direction perpendicular to that of the incoming ray ($\hat{\mathbf{r}}_\perp$); i.e.,

$$\mathbf{F}_{\text{ray}} = \mathbf{F}_{\text{ray,s}} + \mathbf{F}_{\text{ray,g}} = F_{\text{ray,s}}\hat{\mathbf{r}}_i + F_{\text{ray,g}}\hat{\mathbf{r}}_\perp. \qquad (2.16)$$

Dividing $F_{\text{ray,s}}$ and $F_{\text{ray,g}}$ by $n_i P_i/c$, it is possible to define the dimensionless quantities known as the *trapping efficiencies* associated with the scattering and gradient forces, i.e.,

$$Q_{\text{ray,s}} = \frac{c}{n_i P_i} F_{\text{ray,s}}, \qquad (2.17)$$

$$Q_{\text{ray,g}} = \frac{c}{n_i P_i} F_{\text{ray,g}}, \qquad (2.18)$$

and the total trapping efficiency as their quadrature sum,

$$Q_{\text{ray}} = \sqrt{Q_{\text{ray,s}}^2 + Q_{\text{ray,g}}^2}. \qquad (2.19)$$

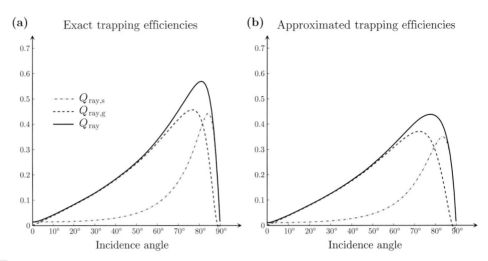

Figure 2.7 Trapping efficiencies for a glass ($n_p = 1.50$) spherical particle in water ($n_m = 1.33$) (a) taking into account all scattering events and (b) considering only the first two scattering events.

The trapping efficiencies permit one to quantify how effectively momentum is transferred from the light ray to the particle. From Eq. (2.13), we can see that the maximum value that they can reach is 2, corresponding to complete reflection of a ray at normal incidence. Fig. 2.7a shows the trapping efficiencies for a glass sphere in water; these reach up to 30% of their theoretical maximum value. $Q_{ray,g}$ grows much faster than $Q_{ray,s}$ as the angle of incidence increases, and the maximum efficiencies are obtained for relatively large angles of incidence ($\approx 80°$). It is interesting to note that these results are independent of the size of the sphere.

Most of the contributions to the optical forces are due to the first two scattering events, i.e., the first reflection of the incoming beam and the first transmission out of the sphere. Fig. 2.7b shows that by calculating the scattering efficiencies accounting only for these two events one obtains a good approximation to the exact results presented in Fig. 2.7a, especially for small angles of incidence. This observation can be very useful in simplifying the numerical calculation of optical forces using ray optics in order to achieve high computational efficiency.

Exercise 2.3.1 Study how the trapping efficiencies $Q_{ray,s}$ [Eq. (2.17)], $Q_{ray,g}$ [Eq. (2.18)] and Q_{ray} [Eq. (2.19)] depend on the refractive index of the particle – e.g., silica ($n_p = 1.42$), polystyrene ($n_p = 1.49$), melamine ($n_p = 1.60$) and air bubble ($n_p = 1.00$) – and the medium – e.g., air ($n_m = 1.00$), water ($n_m = 1.33$) and immersion oil ($n_m = 1.50$). What happens when the refractive index of the particle is lower than that of the medium? How do these results depend on the polarisation of the incoming beam? [Hint: You can use the program `spscattering` from the book website.]

Exercise 2.3.2 Study how the trapping efficiencies depend on the number of scattering events that are considered for various combinations of refractive indices of the particle and of the medium. [Hint: You can adapt the program `spscattering` from the book website.]

Exercise 2.3.3 Demonstrate that, when a ray is scattered by a sphere, all transmitted and reflected rays are contained in the incidence plane. Show that the optical forces are independent of the sphere size. [Hint: Read the original article by Ashkin (1992) studying optical forces on a sphere using geometrical optics.]

Exercise 2.3.4 Ashkin (1992) derived the following theoretical formulas for the scattering efficiencies of a circularly polarised ray on a sphere with angle of incidence θ_i:

$$Q_{\text{ray,s}} = 1 + R\cos 2\theta_i - T^2 \frac{\cos(2\theta_i - 2\theta_t) + R\cos 2\theta_i}{1 + R^2 + 2R\cos 2\theta_t}$$

and

$$Q_{\text{ray,g}} = R\sin 2\theta_i - T^2 \frac{\sin(2\theta_i - 2\theta_t) + R\sin 2\theta_i}{1 + R^2 + 2R\cos 2\theta_t},$$

where R and T are Fresnel's reflection and transmission coefficients and θ_t is the angle of transmission of the incident beam into the sphere. Show that some small corrections to these formulas are needed because of the change of the polarisation of the beam after the first scattering event. [Hint: You can adapt the program `spscattering` from the book website to compare the results of these formulas with the results obtained for the case of a circularly polarised ray.]

Exercise 2.3.5 Show that the gradient force that a ray exerts on a sphere is *conservative*, whereas the scattering force is *non-conservative*. What can you conclude for the force generated by a set of rays? [Hint: Use the definition of conservative force and line integrals.]

2.4 Counter-propagating beam optical trap

It is not possible to achieve stable trapping using a single ray because the particle is permanently pushed by the scattering force in the direction of the incoming ray. This is shown in Fig. 2.8a: a homogenous particle with refractive index higher than the surrounding medium, in this case a glass sphere in water, is attracted towards the propagation axis of the ray by the gradient force and pushed forward by the scattering force. Particles with refractive index lower that the surrounding medium, for example an air bubble in water, are pushed away from the ray by the gradient force, as shown in Fig. 2.8b. The time scales shown in Fig. 2.8 are obtained by considering the overdamped motion of the particle in the viscous medium, i.e., water, as discussed in detail in Chapter 7.

For particles with high refractive index, one possible approach to achieving a stable trap is to use a second counter-propagating light ray, as shown in Fig. 2.8c. In fact, such

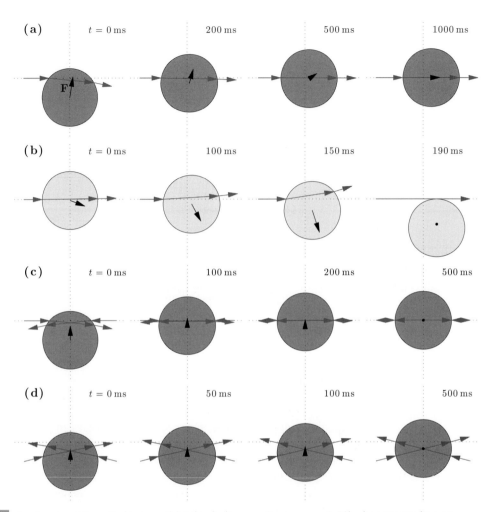

Figure 2.8 Counter-propagating optical tweezers. (a) A glass (radius $a = 3$ μm, $n_p = 1.50$) sphere immersed in water ($n_m = 1.33$) and illuminated by a light ray is attracted towards the ray by the gradient force and pushed forward by the scattering force. (b) An air bubble ($a = 3$ μm, $n_m = 1.00$) in water is pushed away from the ray. (c) The addition of a second counter-propagating ray manages to stabilise the glass sphere position. (d) Two counter-propagating rays at an angle generate a stable trapping position towards which the glass sphere is attracted. The black arrows represent the forces. In all cases, only the deterministic optical forces are taken into account; the Brownian motion of the particles is neglected. The times correspond to the overdamped motion of the particle in the viscous medium (see details in Chapter 7).

a configuration using two laser beams was among the first to be employed to trap and manipulate microscopic particles, and a modern version has been obtained using the light emerging from two optical fibres facing each other, as we will see in Section 12.3. This approach also works if the two beams are not perfectly counter-propagating, but are arranged with a sufficiently large angle, as shown in Figs. 2.8d and 2.9a. If the angle between the two

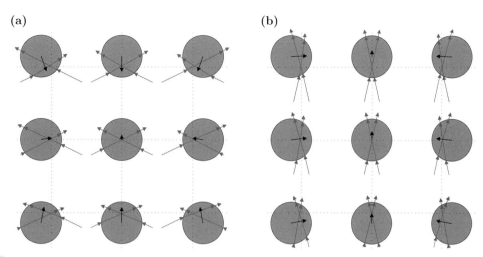

Figure 2.9 Optical trapping by two rays. Optical forces produced by two light rays impinging on a glass ($n_p = 1.50$) sphere immersed in water ($n_m = 1.33$) when they are arranged (a) with a large angle and (b) with a small angle. Only in the first case is there a stable optical trapping position.

rays is not large enough, however, the scattering force prevails and again there is no stable trapping position, as shown in Fig. 2.9b.

Exercise 2.4.1 Study how the trapping of a particle by two counter-propagating rays varies as a function of the particle and medium refractive indices and of the angle between the two rays. What happens when the refractive index of the particle grows? At what angle is trapping lost? [Hint: You can use the program cptrap from the book website.]

Exercise 2.4.2 What configuration of rays can be used to trap a particle with refractive index lower than its surrounding medium? Verify your guess numerically. [Hint: You can adapt the program cptrap from the book website.]

2.5 Optical tweezers

A more convenient alternative to using several counter-propagating light beams is to use a single highly focused light beam. In fact, rays originating from diametrically opposite points of a high-numerical-aperture (high-NA) focusing lens produce in practice a set of rays that converge at very large angles.

As shown in Fig. 2.10 and explained in detail in Chapter 4, a paraxial light beam can be decomposed into a set of rays $\{\mathbf{r}_p^{(m)}\}$ parallel to the optical axis (z), each with appropriate intensity and polarisation. Focusing is then achieved by an objective lens, which has the effect of bending the light rays towards the focal point **O**, obtaining the set of focused rays $\{\mathbf{r}_f^{(m)}\}$. With an ideal aplanatic lens all the rays are made to converge to **O** (this is the

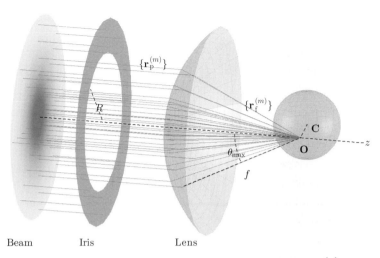

Figure 2.10 Focusing a paraxial light beam. A paraxial light beam is split into a set of parallel light rays $\{\mathbf{r}_p^{(m)}\}$. The focusing objective includes an aperture stop, or iris, with radius R and a focusing lens with focal length f, which bends the rays towards the focus $\mathbf{O}$, obtaining the set of rays $\{\mathbf{r}_f^{(m)}\}$. Here, we have a perfect focus without aberrations; i.e., all rays converge to $\mathbf{O}$. The scattering of $\{\mathbf{r}_f^{(m)}\}$ produces a force $\mathbf{F}_{GO}$ on the sphere whose centre is placed at $\mathbf{C}$. The NA is related to the angle $\theta_{\max}$ over which the rays are focused [Eq. (2.22)].

case shown in Fig. 2.10); however, in practice, often there are aberrations [Box 2.2], which decrease the quality of the focus.

To calculate the optical forces, each ray, $\mathbf{r}_f^{(m)}$, is made to interact with a dielectric sphere placed at point $\mathbf{C} = [x, y, z]$ near $\mathbf{O}$ and the resulting force $\mathbf{F}_{\text{ray}}^{(m)}$ is calculated according to Eq. (2.15). Finally, all the forces $\mathbf{F}_{\text{ray}}^{(m)}$ are summed up to obtain the force acting on the centre of mass of the sphere:

$$\mathbf{F}_{GO} = \sum_m \mathbf{F}_{\text{ray}}^{(m)} = \sum_m \left[\frac{n_i P_i^{(m)}}{c} \hat{\mathbf{r}}_i^{(m)} - \frac{n_i P_r^{(m)}}{c} \hat{\mathbf{r}}_{r,0}^{(m)} - \sum_{n=1}^{+\infty} \frac{n_i P_{t,n}^{(m)}}{c} \hat{\mathbf{r}}_{t,n}^{(m)} \right]. \quad (2.20)$$

As shown in Fig. 2.11, when the object is displaced from its equilibrium position, $\mathbf{C}_{\text{eq}} = [x_{\text{eq}}, y_{\text{eq}}, z_{\text{eq}}]$, which in general does not coincide with $\mathbf{O}$ due to the scattering force, it experiences a *restoring force* proportional to the displacement, at least for relatively small displacements (about up to the particle radius); i.e.,

$$\begin{cases} F_{GO,x} \approx -\kappa_x (x - x_{\text{eq}}), \\ F_{GO,y} \approx -\kappa_y (y - y_{\text{eq}}), \\ F_{GO,z} \approx -\kappa_z (z - z_{\text{eq}}), \end{cases} \quad (2.21)$$

where κ_x, κ_y and κ_z are the spring constants or *trap stiffnesses*. Under the assumption that the beam is circularly polarised, as in Fig. 2.11, the spring constants κ_x and κ_y in the plane perpendicular to the optical axis are exactly equal because the optical forces are rotationally symmetric around the optical axis, whereas the spring constant κ_z in the axial direction is

Box 2.2 Optical aberrations

Chromatic aberrations occur when the refractive index of a material depends on the wavelength (*dispersion*) so that a lens focuses different colours at different focal points. For materials with *normal dispersion*, the refractive index decreases with increasing wavelength, so that longer wavelengths have longer focal lengths. Chromatic aberration can be reduced by increasing the focal length of the lens or by using an achromatic lens (*achromat*), where materials with differing dispersion are assembled together to form a compound lens. Also, many types of glass have been developed to reduce chromatic aberration, such as glasses containing fluorite, which are often employed in the realisation of commercial microscope objectives [Subsection 8.2.1].

Spherical aberrations occur when the light rays that strike a lens near its edge are deflected differently than those that strike the lens nearer the centre. A *positive* (*negative*) *spherical aberration* occurs when peripheral rays are bent too much (are not bent enough).

The *coma*, or *comatic aberration*, is a variation in magnification over the entrance pupil, so that the image of an off-axis object is flared like a comet, hence the name.

The *astigmatism* is due to the fact that rays that propagate in two perpendicular planes have different foci, which are in fact *line foci*, inclined in orthogonal directions and separated by some axial distance.

The *Petzval field curvature*, or *field distortion*, is the optical aberration in which a flat object normal to the optical axis cannot be brought into focus on a flat image plane.

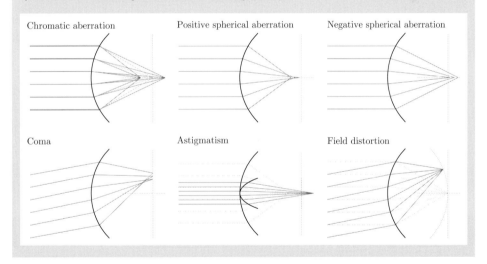

typically smaller. Whereas $\mathbf{F}_{GO}$ is independent of the particle radius, the stiffness, which is given by the force divided by the displacement, is inversely proportional to the particle size. If the beam is not circularly polarised, the optical forces can become significantly asymmetric; this effect is particularly important for particles whose dimensions are smaller than or similar to the light wavelength, which will be discussed in detail in Chapters 3, 5 and 6.

In Fig. 2.11a, the axial trapping efficiencies are plotted as a function of the particle displacement along the optical axis z ($x = y = 0$), where, because of symmetry, only the longitudinal components are different from zero. Whereas the longitudinal gradient efficiency $Q_{GO,g,z}$ (dashed line) reverses its sign around the focal point $z = 0$, the scattering

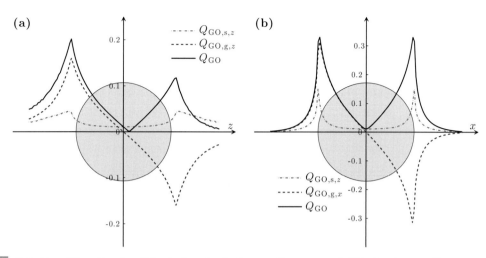

Figure 2.11 Optical trap stiffness. Trapping efficiencies for a glass (radius $a = 3\,\mu\text{m}$, $n_p = 1.50$) sphere immersed in water ($n_m = 1.33$) as a function of the particle displacement (a) along the optical axis z ($x = y = 0$), where both the gradient and the scattering forces are aligned along the z-axis, and (b) along the transverse axis x ($y = z = 0$), where the gradient force is aligned along the x-axis and the scattering force is aligned along the z-axis. The focus is obtained by overfilling a water-immersion objective (NA $= 1.20$).

efficiency $Q_{\text{GO},s,z}$ (dashed-dotted line) is always positive. This scattering force is the reason that the particle equilibrium position in the trap lies just after (not exactly at) the focal point. In Fig. 2.11b, the trapping efficiencies are plotted as a function of the particle displacement along the transverse axis x ($y = z = 0$). In this case, the gradient efficiency has only the x-component, which pulls the particle towards $x = 0$, whereas the scattering efficiency has only the z-component, which pushes the particle along the optical axis. For both longitudinal and transverse displacements, the maximum restoring force is achieved at a displacement of approximately one particle radius.

Exercise 2.5.1 How does the polarisation of the incoming beam affect the trap stiffnesses along the longitudinal and transverse directions? Show that for non-circularly-polarised beams κ_x is generally different from κ_y. [Hint: You can adapt the program `otgo` from the book website.]

Exercise 2.5.2 Show that, in the geometrical optics approximation, the trap stiffness is inversely proportional to the particle radius. [Hint: You can adapt the program `otgo` from the book website.]

Exercise 2.5.3 How do aberrations [Box 2.2] affect the trapping efficiencies? [Hint: You can adapt the program `otgo` from the book website.]

2.6 Filling factor and numerical aperture

The angle $\theta_{\max}$ [Fig. 2.10] over which the rays are focused determines the trapping characteristics of the focus and depends both on the radius R of the aperture stop, or iris, and

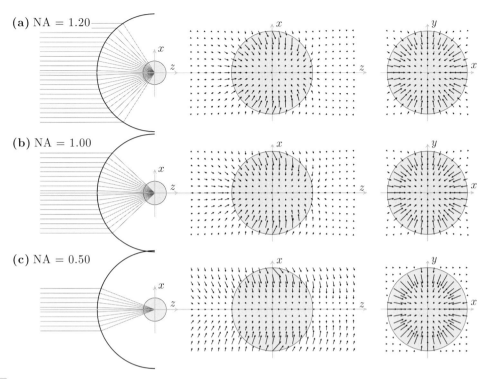

Figure 2.12 Dependence of optical forces on numerical aperture. Optical forces in the longitudinal (zx) and transverse (xy) planes as a function of the NA produced by a focused circularly polarised light beam on a glass (radius $a = 3$ μm, $n_p = 1.50$) sphere immersed in water ($n_m = 1.33$). As the NA decreases, the trapping in the longitudinal plane is lost due to the overwhelming presence of scattering forces. In the transverse plane, trapping is also possible at low NA because it is mainly due to gradient forces.

on the focal length f of the objective lens. A useful parameter that is often employed to characterise θ_{max} is the NA of the objective,

$$\mathrm{NA} = n_m \sin(\theta_{max}) = n_m \frac{R}{f}, \qquad (2.22)$$

where n_m is the refractive index of the medium where the light is focused. In Fig. 2.12, the optical forces on a sphere are presented as a function of the NA. For high NA (Fig. 2.12a, NA = 1.20 in water), there is a stable trapping position both in the longitudinal plane (zx, containing the optical axis) and in the transverse plane (xy, perpendicular to the optical axis). For lower NA (Figs. 2.12b and 2.12c, NA = 1.00 and 0.50, respectively, in water), the particle is still trapped in the transverse plane because of the gradient forces attracting it towards the beam axis, but it is no longer confined along the longitudinal direction. This is in agreement with the fact that, in the presence of only two rays, there is stable trapping only if they are highly convergent, as shown in Fig. 2.9.

Exercise 2.6.1 What is the minimum NA required to trap a glass ($n_p = 1.50$) spherical particle of radius 3 μm in water ($n_m = 1.33$)? How does the minimum NA depend on the material of the particle, e.g., silica ($n_p = 1.42$), polystyrene ($n_p = 1.49$), melamine ($n_p = 1.60$) and air ($n_p = 1.00$)? and on the medium, e.g., air ($n_m = 1.00$) and immersion oil ($n_m = 1.50$)? [Hint: You can adapt the program otgo from the book website.]

2.7 Non-uniform beams

As we have seen in Fig. 2.7, the parts of the beam that contribute most to the trapping efficiencies are the rays propagating with very large angles originating from the edge of the aperture. Therefore, it is reasonable to consider whether beams that do not fill the objective aperture uniformly could be more efficient at achieving stable trapping. We consider some such cases in Fig. 2.13. Fig. 2.13a shows the trapping efficiencies for a Gaussian beam with a waist equal to 30% of the objective pupil (NA = 1.20 in water): although gradient forces ($\propto Q_{GO,g,x}$) allow effective trapping in the transverse plane, the push of the scattering forces

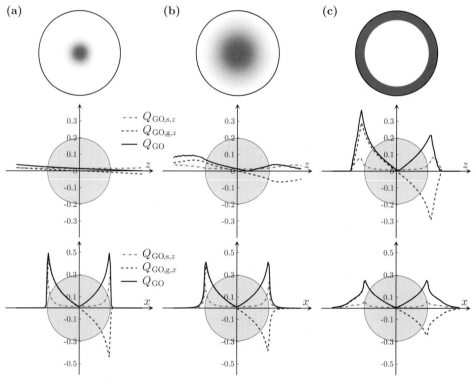

Figure 2.13 Optical traps with non-uniform beams. Trapping efficiencies in the longitudinal (top) and transverse (bottom) planes for a Gaussian beam with waist equal to (a) 30% and (b) 70% of the objective pupil aperture, and (c) for an annular beam. We consider the trapping of a glass (radius $a = 3$ μm, $n_p = 1.50$) sphere immersed in water ($n_m = 1.33$). The beam profile on the water-immersion objective (NA = 1.20) entrance pupil is shown at the top.

($\propto Q_{\text{GO},\text{s},z}$) does not permit effective trapping of the particle along the optical axis. When the waist of the beam is increased to 70% of the objective pupil, as shown in Fig. 2.13b, a longitudinal gradient force ($\propto Q_{\text{GO},\text{g},z}$) large enough to overcome the scattering force along z ($\propto Q_{\text{GO},\text{s},z}$) arises. In fact, as the objective *filling factor* increases, more rays are present at a high angle, where they contribute most to the trapping efficiencies, so that the axial trapping efficiency increases. As we will see in the experimental part of this book, this is the reason that overfilling of the objective aperture is typically a requirement to obtain a stable optical trap, even though it reduces the power available for trapping, as the tails of the Gaussian are cut off by the iris.

Building on the observations made in the previous paragraph, we might consider using an annular beam, i.e., a beam with an annular distribution of light intensity, to enhance axial trapping, as shown in Fig. 2.13c. We note that in this discussion we have not considered the phase of the beam. In fact, as will be discussed in Chapter 4, many beams with a complex intensity profile, e.g., Laguerre–Gaussian beams [Subsection 4.4.3], Bessel beams [Subsection 4.4.4] and cylindrical vector beams [Subsection 4.4.5], also present a complex phase and/or polarisation profile that can have major effects on their trapping properties. In fact, some of the aspects that cannot be easily modelled within the geometrical optics approach are the features related to the presence of spin angular momentum (SAM) and of a nonuniform phase profile, which lead, e.g., to the presence of orbital angular momentum (OAM). The discussion of the mechanical effects of non-Gaussian beams, OAM and SAM will be developed in detail in Chapters 3, 5 and 6 within the electromagnetic theory description of optical forces and torques.

Exercise 2.7.1 What is the optimal beam profile to achieve the highest longitudinal trapping efficiency for a glass ($n_\text{p} = 1.50$) particle of radius 3 μm in water ($n_\text{m} = 1.33$)? What is the optimal beam profile to obtain the highest transverse trapping efficiency? Use a water-immersion objective with NA = 1.20. Note that they are not the same. Why? [Hint: You can use the program otgo from the book website.]

Exercise 2.7.2 Calculate the optical forces on an air bubble immersed in oil. Show that there is no stable equilibrium position when a single Gaussian beam is used, whereas it is possible to trap the bubble stably using an annular beam. What happens when two counter-propagating beams are employed? [Hint: You can use the program otgo from the book website.]

2.8 Non-spherical objects and the windmill effect

A geometrical optics approach similar to the one we followed in the derivation of the forces acting on a microscopic sphere can be employed to study more complex geometries, as long as all the characteristic dimensions of the object under study are significantly larger than the wavelength of the light used for trapping. A simple case is that of a cylindrical object, such as the one shown in Fig. 2.14a. The basic interaction of a ray with the cylinder is the same as for the sphere and the force can be calculated using the formula in Eq. (2.20).

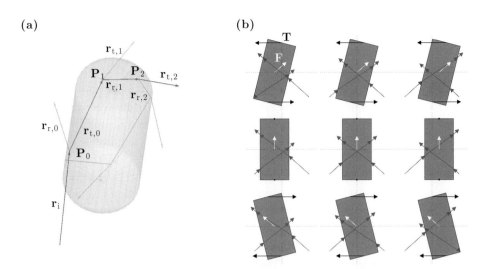

Figure 2.14 Optical force and torque on a cylinder. (a) Multiple scattering of a light ray impinging on a cylinder visualised in three dimensions. Note that, differently from the case of a sphere [Fig. 2.6], the reflected and transmitted rays are not all on the same plane. (b) Optical forces (white arrows) and torques (black arrows) produced by two light rays impinging on a cylinder at a large angle. The torque tends to align the cylinder along the optical axis. In all cases, there is a scattering force pushing the cylinder along the optical axis.

There are two major differences, though. The first is that in the case of nonspherical objects a significant torque can also appear. The torque due to a single ray can be calculated as the difference of the angular momentum associated with the incoming ray and that of the outgoing rays, i.e.,

$$\mathbf{T}_{\text{ray}}^{(m)} = (\mathbf{P}_0 - \mathbf{C}) \times \frac{n_i P_i^{(m)}}{c} \hat{\mathbf{r}}_i^{(m)} - (\mathbf{P}_0 - \mathbf{C}) \times \frac{n_i P_r^{(m)}}{c} \hat{\mathbf{r}}_{r,0}^{(m)} \\ - \sum_{n=1}^{+\infty} (\mathbf{P}_n - \mathbf{C}) \times \frac{n_i P_{t,n}^{(m)}}{c} \hat{\mathbf{r}}_{t,n}^{(m)}, \qquad (2.23)$$

where $\mathbf{C}$ is the centre of mass of the object, $\mathbf{P}_0$ is the incidence point of the incoming ray and $\{\mathbf{P}_n\}$ are the scattering points of the subsequently scattered rays. Then the total torque on the object can be calculated as the sum of the torques due to each ray:

$$\mathbf{T}_{\text{GO}} = \sum_m \mathbf{T}_{\text{ray}}^{(m)}. \qquad (2.24)$$

The black arrows in Fig. 2.14b show the torque arising on a cylinder due to a pair of rays. This torque tends to align the cylinder along the optical axis.

The second difference is that for a spherical particle the radiation pressure of a plane wave, i.e., a set of parallel rays, is always directed in the propagation direction because of symmetry. However, for anisotropic shapes the radiation pressure has a transverse

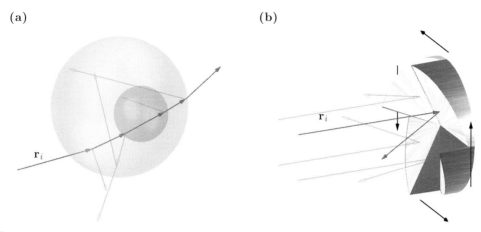

Figure 2.15 Trapping non-convex shapes and the windmill effect. (a) Multiple scattering of a ray on a complex shape constituted by two spheres, one inside the other; this is a simple optical model for a cell containing a nucleus. (b) The asymmetric deflection of the incoming rays on a non-cylindrically-symmetric object produces the torque indicated by the black arrows and the rotation of the object. This phenomenon is known as the *windmill effect*.

component that is responsible for the *optical lift effect*; i.e., non-spherical particles can move transversely to the incident light propagation direction.

More complex objects can be modelled by more complex shapes. For example, as a first approximation, a cell containing a nucleus can be modelled by a sphere (the cytoplasm) containing a smaller sphere of different refractive index (the nucleus), as shown in Fig. 2.15a. It is interesting to notice that in a scattering event a ray can now be split into multiple rays that may not necessarily be able to escape the particle; this is typical of all nonconvex shapes and can lead to an explosion in the number of rays to be taken into account.

In the presence of an optical torque, the object rotates in a certain direction. This effect is known as the *windmill effect* because of its analogy to the motion of a windmill, where the wind in this case is the flow of momentum due to the electromagnetic field, as shown in Fig. 2.15b. This effect should not be confused with the rotation induced by the transfer of SAM and OAM, which will be described in detail in Chapters 5 and 6.

Exercise 2.8.1 A ray scattering on a sphere does not produce any torque on the sphere itself. Verify this fact with some numerical simulations and demonstrate it analytically. Conclude that a beam also cannot produce any torque on a sphere.

Exercise 2.8.2 What are the forces and torques acting on a cylinder illuminated by a plane wave?

Exercise 2.8.3 Calculate the optical forces and torques acting on an object such as the one shown in Fig. 2.15a.

Exercise 2.8.4 Calculate the optical windmill effect on an object such as the one shown in Fig. 2.15b.

Problems

2.1 Given a light beam propagating in the positive z-direction and linearly polarised along the x-direction, calculate numerically the optical trapping efficiencies in the case of a sphere. What do you observe? How do your results compare with the ones presented in the text for the case of a circularly polarised beam?

2.2 Calculate the optical forces and torques acting on an ellipsoid with semi-axes $a_1, a_2, a_3 \gg \lambda_0$ held in an optical tweezers. What is the ellipsoidal equilibrium configuration for the three cases $a_1 > a_2 > a_3$, $a_1 > a_2 = a_3$ (prolate spheroid), and $a_1 = a_2 > a_3$ (oblate spheroid)? How does the force scale with the ratio a_1/a_3 in the case of spheroids?

2.3 Consider a circularly polarised light beam that is reflected by a mirror. Knowing that the spin of a photon is $\hbar$, calculate the optical torque transferred to the mirror upon reflection. What happens if the beam is completely absorbed by the object? What is the maximum torque that can be exerted on the mirror? What are the limits of geometrical optics in the description of this phenomenon?

2.4 Calculate the optical forces arising on a human red blood cell, whose shape is a biconcave disc because of the absence of the nucleus.

2.5 Calculate the optically induced stress profile over the surface of a microscopic bubble. Evaluate also the resulting deformation using elastic membrane theory, e.g., following the approach used by Skelton et al. (2012).

2.6 *Trapping through an interface.* In a standard optical tweezers, the beam is often focused through a planar interface that presents a refractive index mismatch (e.g., between the coverslip glass and the sample water). What kind of aberrations does this introduce? How does the focal point change? How does this affect the optical forces on a spherical particle? How does this depend on the distance between the focal point and the planar interface?

2.7 *Optical binding.* Consider a spherical particle under plane wave illumination. This particle acts as a lens focusing the light beam. Now, consider a second particle placed near the resulting focus. What kind of optical forces does this second particle experience? What happens if the two particles are placed between two counter-propagating plane waves?

2.8 *Layered spheres.* Consider a dielectric layered sphere, constituted by a core covered by a multilayer. Such a model was employed by Chang et al. (2006) to describe the optical properties of a cell. Calculate the resulting optical forces using ray optics under various illumination conditions.

2.9 *Janus particles.* Consider a Janus sphere, i.e., a particle whose hemispheres have different refractive indices. Calculate the optical forces and torques on such objects using ray optics. How does this particle align in an optical tweezers? What happens if the particle has a cylindrical or ellipsoidal shape?

2.10 *Absorbing and reflecting objects.* Consider a sphere made of an absorbing material. What optical forces and torques can a single ray produce on such a sphere? What happens for a focused beam? Under what conditions is it possible to trap it? Repeat the same analysis for the case of a reflecting sphere.

2.11 *Metamaterials.* Metamaterials are artificial materials engineered to have properties that may not be found in nature. For example, it might be possible to have some materials with negative refractive index. Study the optical forces that might arise on a particle made of such a material.

References

Abraham, M. 1909. Zur Elektrodynamik bewegter Körper. *R. C. Circ. Mat. Palermo*, 1–28.

Ashkin, A. 1992. Forces of a single-beam gradient laser trap on a dielectric sphere in the ray optics regime. *Biophys. J.*, **61**, 569–82.

Ashkin, A., and Dziedzic, J. M. 1973. Radiation pressure on a free liquid surface. *Phys. Rev. Lett.*, **30**, 139–42.

Barnett, S. M. 2010. Resolution of the Abraham–Minkowski dilemma. *Phys. Rev. Lett.*, **104**, 070401.

Chang, Y.-R., Hsu, L., and Chi, S. 2006. Optical trapping of a spherically symmetric sphere in the ray-optics regime: A model for optical tweezers upon cells. *Appl. Opt.*, **45**, 3885–92.

Minkowski, H. 1908. Die Grundgleichungen für die elektromagnetischen Vorgänge in bewegten Körpern. *Nachr. Ges. Wiss. Göttingen*, 53–111.

Pfeifer, R. N. C., Nieminen, T. A., Heckenberg, N. R., and Rubinsztein-Dunlop, H. 2007. Momentum of an electromagnetic wave in dielectric media. *Rev. Mod. Phys.*, **79**, 1197–1216.

Skelton, S. E., Sergides, M., Memoli, G., Maragò, O. M., and Jones, P. H. 2012. Trapping and deformation of microbubbles in a dual-beam fibre-optic trap. *J. Opt.*, **14**, 075706.

3 Dipole approximation

If a neutral atom is placed in an electric field, the negative electron cloud surrounding the positive nucleus will be displaced, leading to a separation between the centres of mass of the positive and negative charge distributions, as shown in Fig. 3.1. The resulting *induced dipole* can experience electrostatic forces arising from its interaction with the inducing electric field. An oscillating electromagnetic field, such as that of the laser beam used to generate an optical tweezers, will induce an *oscillating dipole*, which will also experience forces arising from its interaction with the inducing electromagnetic field. Furthermore, an oscillating dipole radiates an electromagnetic field that can produce a mechanical effect on other induced dipoles, leading to, in some cases, the formation of stable ordered structures, a phenomenon known as *optical binding*. In this chapter, we will study how induced dipoles can experience optical forces in an electromagnetic field. This approach is valid for the study of optical forces not only on neutral atoms, but also on other small particles, such as molecules and nanoparticles. Here, 'small' means 'much smaller than the electromagnetic wavelength' and, in this sense, the *dipole approximation* is the antithesis of the geometrical optics thesis described in Chapter 2. Nevertheless, we will come to the synthesis in Chapters 5 and 6, where we describe the general exact approach based on electromagnetic theory.

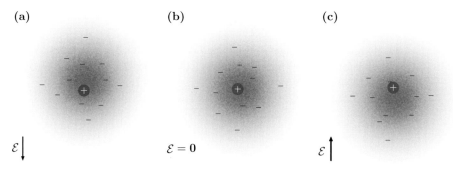

Figure 3.1 Electric dipole induced on an atom. In the presence of an electrostatic field, $\mathcal{E}$, positive and negative charge distributions within an atom (or molecule or nanoparticle) are displaced in opposite directions: the positively charged nucleus parallel to the electric field direction and the negatively charged electronic cloud antiparallel to it. The direction of $\mathcal{E}$ is indicated by the arrow near each atom.

3.1 The electric dipole in electrostatics

A pair of point charges $\pm q$ of equal magnitude but opposite sign separated by a distance l form an *electric dipole*, with *dipole moment*

$$\boldsymbol{p}_\mathrm{d} = q l \hat{\boldsymbol{u}}, \tag{3.1}$$

where the unit vector $\hat{\boldsymbol{u}}$ points from the negative to the positive charge. A schematic of an electric dipole is shown in Fig. 3.2a. If such a dipole is placed in a uniform external electric field $\mathcal{E}_\mathrm{i}$, such that $\boldsymbol{p}_\mathrm{d}$ is at an angle θ with $\mathcal{E}_\mathrm{i}$, then the charges at the ends of the dipole experience oppositely directed forces, whose net effect is the torque

$$\mathbf{T}_\mathrm{d} = \boldsymbol{p}_\mathrm{d} \times \mathcal{E}_\mathrm{i} \tag{3.2}$$

about the dipole centre. The torque $\mathbf{T}_\mathrm{d}$ acts to align the dipole with the electric field. The potential energy of the dipole depends on its orientation with respect to $\mathcal{E}_\mathrm{i}$ and is

$$U_\mathrm{d}(\theta) = -\boldsymbol{p}_\mathrm{d} \cdot \mathcal{E}_\mathrm{i} = -p_\mathrm{d} \mathcal{E}_\mathrm{i} \cos\theta, \tag{3.3}$$

where $p_\mathrm{d} = |\boldsymbol{p}_\mathrm{d}|$, $\mathcal{E}_\mathrm{i} = |\mathcal{E}_\mathrm{i}|$ and we have set $U_\mathrm{d}(\pi/2) = 0$. Mechanical equilibrium is achieved when the dipole has the same orientation as the electric field, i.e., $\theta = 0$ for the *stable equilibrium* at the potential energy minimum [Fig. 3.2b] or $\theta = \pi$ for the *unstable equilibrium* at the potential energy maximum [Fig. 3.2c].

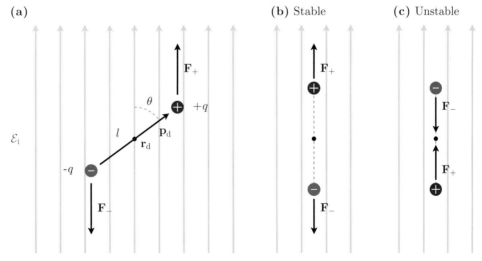

Figure 3.2 Electric dipole in an electrostatic field. (a) An electric dipole $\boldsymbol{p}_\mathrm{d}$, formed by two charges $\pm q$ of equal magnitude and opposite sign with separation l, in a uniform external electromagnetic field $\mathcal{E}_\mathrm{i}$ experiences a torque that tends to align it with the field direction. (b) The *stable equilibrium*, i.e., the minimum of the the potential energy, is reached when $\boldsymbol{p}_\mathrm{d}$ is parallel to $\mathcal{E}_\mathrm{i}$ and (c) the *unstable equilibrium*, i.e., the maximum of the potential energy, when $\boldsymbol{p}_\mathrm{d}$ is antiparallel to $\mathcal{E}_\mathrm{i}$.

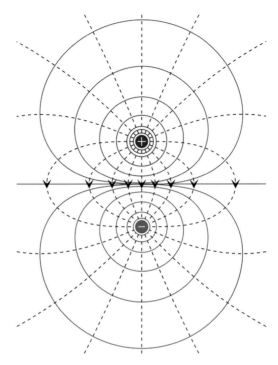

Figure 3.3 Dipole potential and electrostatic field. Equipotential lines (solid lines, Eq. (3.5)) and electric field lines (dashed lines, Eq. (3.6)) generated by an electric dipole.

In a non-uniform electric field, a dipole also experiences a net force due to the spatial gradient in electrical potential energy arising from the gradient of the electric field,

$$\mathbf{F}_d(\mathbf{r}_d) = -\nabla U_d(\mathbf{r}_d), \tag{3.4}$$

where $\mathbf{r}_d$ is the position of the centre of the dipole [Fig. 3.2a] and $U_d(\mathbf{r}_d) = -\mathbf{p}_d \cdot \mathcal{E}_i(\mathbf{r}_d)$.

Both the electric potential and the electric field generated by a dipole are cylindrically symmetric around the axis of the dipole. Consider a dipole placed in vacuum at the origin of the reference system, i.e., $\mathbf{r}_d = \mathbf{0}$. Its electrostatic potential at position $\mathbf{r}$ is the sum of the potentials of the individual charges, i.e.,

$$\phi_d(\mathbf{r}) = \frac{q}{4\pi \varepsilon_0} \left[\frac{1}{|\mathbf{r} + \frac{l}{2}\hat{\mathbf{u}}|} - \frac{1}{|\mathbf{r} - \frac{l}{2}\hat{\mathbf{u}}|} \right], \tag{3.5}$$

where ε_0 is the dielectric permittivity of vacuum. The equipotential lines are shown by the solid lines in Fig. 3.3. The electric field can then be found from the gradient of the electromagnetic potential as

$$\mathcal{E}_d(\mathbf{r}) = -\nabla \phi_d(\mathbf{r}). \tag{3.6}$$

The electric field lines are shown by the dashed lines in Fig. 3.3.

In the far field, i.e., for $r \gg l$, where $r = |\mathbf{r}|$ is the radial distance, Eq. (3.5) becomes

$$\phi_{d,ff}(\mathbf{r}) \approx \frac{\mathbf{p}_d \cdot \hat{\mathbf{r}}}{4\pi\varepsilon_0 r^2}, \tag{3.7}$$

where $\hat{\mathbf{r}}$ is the radial unit vector. The electric field in the far-field approximation can be expressed in spherical coordinates, assuming the dipole to be oriented along the polar axis, as

$$\boldsymbol{\mathcal{E}}_{d,ff}(r,\vartheta,\varphi) = \frac{p}{4\pi\varepsilon_0}\frac{2\cos\vartheta}{r^3}\hat{\mathbf{r}} + \frac{p}{4\pi\varepsilon_0}\frac{\sin\vartheta}{r^3}\hat{\boldsymbol{\vartheta}}, \tag{3.8}$$

where ϑ is the polar angle, $\hat{\boldsymbol{\vartheta}}$ is the polar unit vector and there is no dependence on the azimuthal angle φ because of the rotational symmetry of the dipolar fields around the polar axis.

Exercise 3.1.1 Find an expression for the amount of work that must be done to rotate an electric dipole from angle θ_1 to angle θ_2 in a uniform electric field.

Exercise 3.1.2 Demonstrate that Eq. (3.7) is a good far-field approximation of Eq. (3.5). [Hint: Start by expressing Eq. (3.5) in terms of the ratio l/r.]

Exercise 3.1.3 Calculate the electric field and the electric potential in a dielectric medium, i.e., a medium where the dielectric constant is $\varepsilon_m = \varepsilon_0\varepsilon_{r,m}$.

3.2 Polarisability and the Clausius–Mossotti relation

The separation between the centres of mass of the positive and negative charge distributions in an isolated atom or molecule placed in an external electric field leads to an induced dipole moment, as shown in Fig. 3.1. If the external electric field is not too strong, there is a linear relationship between the *induced polarisation* $\mathbf{p}$ of the atom and the inducing electric field $\boldsymbol{\mathcal{E}}_i$:

$$\mathbf{p} = \alpha_{\text{lin}}\boldsymbol{\mathcal{E}}_i, \tag{3.9}$$

where the proportionality coefficient α_{lin} is the *polarisability* of the atom.[1]

An applied electric field polarises the atoms or molecules throughout a block of dielectric material, resulting in the generation of a polarisation field inside the material,

$$\boldsymbol{\mathcal{P}} = \frac{N}{V}\mathbf{p}, \tag{3.10}$$

where N is the number of atoms or molecules in a volume V of material, each of which has an induced polarisation $\mathbf{p}$ given by Eq. (3.9). As shown in Fig. 3.4a for a rectangular slab of

---

[1] To be precise, α_{lin} is the *linear* polarisability of the atom. Higher-order terms can arise in nonlinear materials in the presence of sufficiently intense fields.

> **Box 3.1** **Vector identities**
>
> If **A**, **B** and **C** are vector functions and f and g are scalar functions,
>
> $$\nabla \cdot (\nabla \times \mathbf{A}) = 0$$
> $$\nabla \times \nabla f = \mathbf{0}$$
> $$\nabla(fg) = f\nabla g + g\nabla f$$
> $$\nabla \cdot (f\mathbf{A}) = f\nabla \cdot \mathbf{A} + \mathbf{A} \cdot \nabla f$$
> $$\nabla \times (f\mathbf{A}) = f\nabla \times \mathbf{A} - \mathbf{A} \times \nabla f$$
> $$\nabla(\mathbf{A} \cdot \mathbf{B}) = \mathbf{A} \times (\nabla \times \mathbf{B}) + (\mathbf{A} \cdot \nabla)\mathbf{B} + \mathbf{B} \times (\nabla \times \mathbf{A}) + (\mathbf{B} \cdot \nabla)\mathbf{A}$$
> $$\nabla \cdot (\mathbf{A} \times \mathbf{B}) = \mathbf{B} \cdot \nabla \times \mathbf{A} - \mathbf{A} \cdot \nabla \times \mathbf{B}$$
> $$\nabla \times (\mathbf{A} \times \mathbf{B}) = (\mathbf{B} \cdot \nabla)\mathbf{A} + \mathbf{A}(\nabla \cdot \mathbf{B}) - (\mathbf{A} \cdot \nabla)\mathbf{B} - \mathbf{B}(\nabla \cdot \mathbf{A})$$
> $$\nabla \times (\nabla \times \mathbf{A}) = \nabla(\nabla \cdot \mathbf{A}) - \nabla^2 \mathbf{A}$$
> $$\mathbf{A} \cdot (\mathbf{B} \times \mathbf{C}) = \mathbf{B} \cdot (\mathbf{C} \times \mathbf{A}) = \mathbf{C} \cdot (\mathbf{A} \times \mathbf{B})$$
> $$\mathbf{A} \times (\mathbf{B} \times \mathbf{C}) = (\mathbf{C} \times \mathbf{B}) \times \mathbf{A} = \mathbf{B}(\mathbf{A} \cdot \mathbf{C}) - \mathbf{C}(\mathbf{A} \cdot \mathbf{B})$$
> $$\mathbf{A} \times (\nabla \times \mathbf{B}) = \sum_{j=x,y,z} A_j(\nabla)B_j - (\mathbf{A} \cdot \nabla)\mathbf{B}.$$
>
> If V is a volume enclosed by a surface S_V with normal unit vector $\hat{\mathbf{n}}$,
>
> $$\oint_{S_V} \mathbf{A} \cdot \hat{\mathbf{n}}\, dS_V = \int_V (\nabla \cdot \mathbf{A})\, dV \qquad \text{(Gauss' theorem).}$$
>
> If S is a surface with normal unit vector $\hat{\mathbf{n}}$ encircled by a line l_S with tangent unit vector $\hat{\mathbf{t}}$,
>
> $$\oint_{l_S} \mathbf{A} \cdot \hat{\mathbf{t}}\, dl_S = \int_S (\nabla \times \mathbf{A}) \cdot \hat{\mathbf{n}}\, dS \qquad \text{(Stokes' theorem).}$$

material and in Fig. 3.4b for a sphere, this process induces net charges on the surface of the dielectric, the so-called *polarisation charges*. For an infinitesimal area dA on the surface of the object, the induced net charge is

$$dQ = \boldsymbol{\mathcal{P}} \cdot \hat{\mathbf{n}}\, dA,$$

where $\hat{\mathbf{n}}$ is the unit vector normal to the surface pointing outwards from the object. The induced charge per unit area is therefore

$$\sigma_{\text{pol}} = \frac{dQ}{dA} = \boldsymbol{\mathcal{P}} \cdot \hat{\mathbf{n}}, \qquad (3.11)$$

which means that the magnitude of the polarisation in the direction perpendicular to the surface is equal to the density of charges appearing at the surface. These charges are *bound charges*: they only appear as a result of the polarisation of the material under the action of the applied electric field. The presence of polarisation charges contributes to the overall

> **Box 3.2** **Maxwell's equations**
>
> In SI units, the *differential form* of Maxwell's equations is:
>
> $$\nabla \times \mathcal{E}(\mathbf{r}, t) = -\frac{\partial \mathcal{B}(\mathbf{r}, t)}{\partial t},$$
>
> $$\nabla \times \mathcal{H}(\mathbf{r}, t) = \frac{\partial \mathcal{D}(\mathbf{r}, t)}{\partial t} + \boldsymbol{j}(\mathbf{r}, t),$$
>
> $$\nabla \cdot \mathcal{D}(\mathbf{r}, t) = \varrho_f(\mathbf{r}, t),$$
>
> $$\nabla \cdot \mathcal{B}(\mathbf{r}, t) = 0,$$
>
> where $\mathcal{E}(\mathbf{r}, t)$ is the *electric field*, $\mathcal{D}(\mathbf{r}, t)$ the *electric displacement*, $\mathcal{H}(\mathbf{r}, t)$ the *magnetic field*, $\mathcal{B}(\mathbf{r}, t)$ the *magnetic induction*, $\boldsymbol{j}(\mathbf{r}, t) = \boldsymbol{j}_s(\mathbf{r}, t) + \boldsymbol{j}_c(\mathbf{r}, t)$ the *total current density* with $\boldsymbol{j}_s(\mathbf{r}, t)$ the *source current density* and $\boldsymbol{j}_c(\mathbf{r}, t)$ the *conduction current density*, and $\varrho_f(\mathbf{r}, t)$ the *free charge density*. The *conservation of charge*, i.e.,
>
> $$\nabla \cdot \boldsymbol{j}(\mathbf{r}, t) = -\frac{\partial \varrho_f(\mathbf{r}, t)}{\partial t},$$
>
> is a straightforward consequence of Maxwell's equations.
>
> The *integral form* of Maxwell's equations can be obtained by applying Gauss' and Stokes' theorems [Box 3.1] to the preceding differential forms,
>
> $$\oint_{l_S} \mathcal{E}(\mathbf{r}, t) \cdot \hat{\mathbf{t}} \, dl_S = -\int_S \frac{\partial \mathcal{B}(\mathbf{r}, t)}{\partial t} \cdot \hat{\mathbf{n}} \, dS,$$
>
> $$\oint_{l_S} \mathcal{H}(\mathbf{r}, t) \cdot \hat{\mathbf{t}} \, dl_S = \int_S \left[\frac{\partial \mathcal{D}(\mathbf{r}, t)}{\partial t} + \boldsymbol{j}(\mathbf{r}, t) \right] \cdot \hat{\mathbf{n}} \, dS,$$
>
> $$\oint_{S_V} \mathcal{D}(\mathbf{r}, t) \cdot \hat{\mathbf{n}} \, dS_V = \int_V \varrho_f(\mathbf{r}, t) dV,$$
>
> $$\oint_{S_V} \mathcal{B}(\mathbf{r}, t) \cdot \hat{\mathbf{n}} \, dS_V = 0,$$
>
> where S denotes an orientable surface with normal unit vector $\hat{\mathbf{n}}$ encircled by the line l_S with tangent unit vector $\hat{\mathbf{t}}$ and V denotes a volume bounded by a closed surface S_V with normal unit vector $\hat{\mathbf{n}}$.

electric field inside the dielectric material, reducing it relative to the total external field by the relative dielectric permittivity ε_r [Box 3.3].

We will now derive the *Clausius–Mossotti relation*, which is an important result that relates the polarisability of a small dielectric sphere to its relative dielectric permittivity ε_r. We will restrict our analysis to a homogeneous, isotropic, non-magnetic spherical particle with radius a. In the presence of an electric field, which can be assumed to be along the polar axis s of a spherical reference system, i.e.,

$$\mathcal{E}_i = \mathcal{E}_i \hat{\mathbf{s}} = \mathcal{E}_i \cos \vartheta \, \hat{\mathbf{r}} - \mathcal{E}_i \sin \vartheta \, \hat{\boldsymbol{\vartheta}},$$

where $\hat{\mathbf{s}}$ is the unit vector along the s-axis, polarisation charges are induced on the surface of the sphere, as illustrated in Fig. 3.4b. If the particle is small, the polarisation may be

> **Box 3.3** **Constitutive relations**
>
> The electromagnetic properties of a material medium are generally cast in terms of *polarisation* $\mathcal{P}(\mathbf{r}, t)$ and *magnetisation* $\mathcal{M}(\mathbf{r}, t)$:
>
> $$\mathcal{D}(\mathbf{r}, t) = \varepsilon_0 \mathcal{E}(\mathbf{r}, t) + \mathcal{P}(\mathbf{r}, t)$$
> $$\mathcal{H}(\mathbf{r}, t) = \mu_0^{-1} \mathcal{B}(\mathbf{r}, t) - \mathcal{M}(\mathbf{r}, t),$$
>
> where ε_0 is the *dielectric permittivity* of vacuum and μ_0 is the *magnetic permeability* of vacuum.
>
> In a homogeneous, isotropic, linear and non-dispersive medium, such as the ones we typically consider in this book,
>
> $$\mathcal{D}(\mathbf{r}, t) = \varepsilon_0 \varepsilon_{r,m} \mathcal{E}(\mathbf{r}, t) = \varepsilon_m \mathcal{E}(\mathbf{r}, t)$$
> $$\mathcal{P}(\mathbf{r}, t) = \varepsilon_0 \chi_e \mathcal{E}(\mathbf{r}, t)$$
> $$\mathcal{B}(\mathbf{r}, t) = \mu_0 \mu_{r,m} \mathcal{H}(\mathbf{r}, t) = \mu_m \mathcal{H}(\mathbf{r}, t)$$
> $$\mathcal{M}(\mathbf{r}, t) = \chi_m \mathcal{H}(\mathbf{r}, t)$$
> $$j_c(\mathbf{r}, t) = \sigma \mathcal{E}(\mathbf{r}, t),$$
>
> where $\varepsilon_{r,m}$ is the *relative dielectric permittivity*, ε_m the *dielectric permittivity*, $\mu_{r,m}$ the *relative magnetic permeability*, μ_m the *magnetic permeability*, χ_e the *dielectric susceptibility*, χ_m the *magnetic susceptibility* and σ the *conductivity* of the medium. Typically, we will only consider non-magnetic media, i.e., media where $\mu_{r,m} = 1$, because this is the case for virtually all materials at optical frequencies.

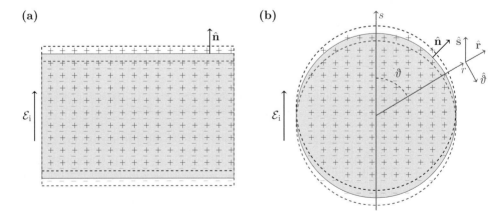

Figure 3.4 Separation of charges due to polarisation. The separation of charges induced by an electric field generates *polarisation charges* on the surface of a dielectric object, e.g., (a) a rectangular slab of dielectric material and (b) a dielectric sphere. In turn, the presence of the polarisation charges contributes to the overall electric field inside the dielectric material, reducing it relative to the total external field by the relative dielectric permittivity ε_r.

> **Box 3.4** **Boundary conditions**
>
> The boundary conditions for the electromagnetic fields at the interface between two homogenous dielectric media, a and b, can be found using the integral form of Maxwell's equations [Box 3.2].
>
> Using Maxwell's equations for the curl of $\mathcal{E}$ and $\mathcal{H}$ with an infinitesimal rectangular surface S perpendicular to the boundary between a and b, as shown below on the left, one obtains
>
> $$\hat{\mathbf{n}} \times (\mathcal{E}_b - \mathcal{E}_a) = \mathbf{0}$$
> $$\hat{\mathbf{n}} \times (\mathcal{H}_b - \mathcal{H}_a) = \boldsymbol{j}_s,$$
>
> where $\boldsymbol{j}_s$ is the *free surface current density* and $\hat{\mathbf{n}}$ is the unit vector normal to S going from a to b. Using Maxwell's equations for the divergence of $\mathcal{D}$ and $\mathcal{B}$ with an infinitesimal rectangular rectangular volume V at the boundary as shown below on the right, one obtains
>
> $$\hat{\mathbf{n}} \cdot (\mathcal{D}_b - \mathcal{D}_a) = \sigma_s$$
> $$\hat{\mathbf{n}} \cdot (\mathcal{B}_b - \mathcal{B}_a) = 0,$$
>
> where σ_s is the *free surface charge density* between the media.
>
> When there are no free charges or free currents, as is often the case, $\boldsymbol{j}_s$ and σ_s vanish.
>
>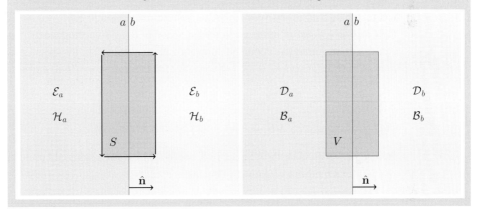

assumed to be uniform throughout the sphere and parallel to the external field, so that the electric field inside the particle is

$$\boldsymbol{\mathcal{E}}_p = \mathcal{E}_p\, \hat{\mathbf{s}} = \mathcal{E}_p \cos\vartheta\, \hat{\mathbf{r}} - \mathcal{E}_p \sin\vartheta\, \hat{\boldsymbol{\vartheta}}.$$

This polarised sphere acts as an electric dipole with dipole moment $\boldsymbol{p}_{sp}$ parallel to the s-axis, whose field is given in spherical coordinates by Eq. (3.8), assuming the sphere to be centred at the origin of the coordinate system. Applying the boundary conditions [Box 3.4] at $r = a$ on the component of the electric displacement perpendicular to the interface, i.e., in the radial direction,

$$\mathcal{D}_{\mathrm{pol},r}(a) = \mathcal{D}_{\mathrm{i},r}(a) + \mathcal{D}_{\mathrm{d,ff},r}(a) \quad \Rightarrow \quad \varepsilon_r \mathcal{E}_{\mathrm{p},r}(a) = \mathcal{E}_{\mathrm{i},r}(a) + \mathcal{E}_{\mathrm{d,ff},r}(a),$$

and the component of electric field parallel to the interface, i.e., the tangential direction,

$$\mathcal{E}_{\mathrm{pol},\vartheta}(a) = \mathcal{E}_{\mathrm{i},\vartheta}(a) + \mathcal{E}_{\mathrm{d,ff},\vartheta}(a),$$

we find that

$$\begin{cases} \varepsilon_r \mathcal{E}_p \cos\vartheta = \mathcal{E}_i \cos\vartheta + \dfrac{2p_{sp} \cos\vartheta}{4\pi\varepsilon_0 a^3}, \\ -\mathcal{E}_p \sin\vartheta = -\mathcal{E}_i \sin\vartheta + \dfrac{p_{sp} \sin\vartheta}{4\pi\varepsilon_0 a^3}, \end{cases} \quad (3.12)$$

and, by eliminating the dipole moment $p_{sp} = |\mathbf{p}_{sp}|$, we find that the magnitude of the field inside the sphere is

$$\mathcal{E}_p = \frac{3}{\varepsilon_r + 2}\mathcal{E}_i. \quad (3.13)$$

Substituting Eq. (3.13) into Eqs. (3.12), we find for the dipole moment

$$p_{sp} = 4\pi\varepsilon_0 a^3 \frac{\varepsilon_r - 1}{\varepsilon_r + 2}\mathcal{E}_i, \quad (3.14)$$

and we can thus define the polarisability $\alpha_{CM} = p_{sp}/\mathcal{E}_i$ of the small sphere as

$$\alpha_{CM} = 3V\varepsilon_0 \frac{\varepsilon_r - 1}{\varepsilon_r + 2}, \quad (3.15)$$

where $V = \frac{4}{3}\pi a^3$ is the volume of the sphere. In the static limit, Eq. (3.15) is known as the *Clausius–Mossotti relation*. In the presence of a time-varying electromagnetic field, an equivalent wavelength-dependent formula can be obtained in the quasi-static limit, i.e., as long as $a \ll \lambda_0$, where λ_0 is the wavelength of the radiation to which the particle is exposed:

$$\alpha_{CM}(\lambda_0) = 3V\varepsilon_0 \frac{\varepsilon_r(\lambda_0) - 1}{\varepsilon_r(\lambda_0) + 2}, \quad (3.16)$$

This formula is often referred to as the *Lorentz–Lorenz relation*.

Exercise 3.2.1 Calculate the Clausius–Mossotti relation for a sphere of dielectric permittivity ε_p in a medium of dielectric permittivity ε_m.

3.3 The electric dipole in an oscillating electric field

We now turn our attention to the oscillating dipoles induced by an oscillating electromagnetic field, e.g., that of the laser beam used to generate an optical trap, and to the corresponding re-radiated electromagnetic fields.

In the presence of a homogeneous time-varying electric field $\mathcal{E}_i(t)$, the polarisation of a small particle will oscillate. In particular, we will consider a harmonic homogeneous electric field at angular frequency ω, which can be expressed in terms of phasors [Box 3.5], i.e., $\mathbf{E}_i(t) = \mathbf{E}_i e^{-i\omega t}$. As long as the amplitude of the electric field is not too large, the induced dipole moment will remain linearly proportional to the electric field, so that it can also be expressed in terms of phasors with the same time dependence, i.e., $\mathbf{p}_d(t) = \mathbf{p}_d e^{-i\omega t}$, and

$$\mathbf{p}_d = \alpha_{lin}(\omega)\mathbf{E}_i, \quad (3.17)$$

> **Box 3.5** **Monochromatic fields and phasors**
>
> For monochromatic fields, the time dependence can easily be separated, so that they can be written as
> $$\mathcal{E}(\mathbf{r}, t) = \mathrm{Re}\{\mathbf{E}(\mathbf{r})e^{-i\omega t}\}.$$
>
> $\mathcal{E}(\mathbf{r}, t)$ is real and represents the physical field, whereas $\mathbf{E}(\mathbf{r})$ is a complex *phasor*, i.e., the complex amplitude of the time-harmonic field. We will generally use $\mathcal{E}(\mathbf{r}, t)$ to indicate a physical (real) field, $\mathbf{E}(\mathbf{r})$ to denote a (complex) phasor and $\mathbf{E}(\mathbf{r}, t) = \mathbf{E}(\mathbf{r})e^{-i\omega t}$ to denote a time-varying (complex) field. Using phasors, Maxwell's equations for a monochromatic field can be rewritten as
> $$\nabla \times \mathbf{E}(\mathbf{r}) = i\omega \mathbf{B}(\mathbf{r})$$
> $$\nabla \times \mathbf{H}(\mathbf{r}) = -i\omega \mathbf{D}(\mathbf{r}) + \mathbf{j}(\mathbf{r})$$
> $$\nabla \cdot \mathbf{D}(\mathbf{r}) = \rho(\mathbf{r})$$
> $$\nabla \cdot \mathbf{B}(\mathbf{r}) = 0.$$
>
> The constitutive relations and the boundary conditions for phasors can be derived straightforwardly from the ones in Boxes 3.3 and 3.4.
>
> In experiments, typically one can only observe time-averaged quantities. Time averages of monochromatic fields vanish; e.g., $\overline{\mathcal{E}(\mathbf{r}, t)} = \overline{\mathrm{Re}\{\mathbf{E}(\mathbf{r})e^{-i\omega t}\}} = 0$. However, their dot and cross products in general do not vanish and can be expressed as, e.g.,
> $$\overline{\mathcal{E}(\mathbf{r}, t) \cdot \mathcal{H}(\mathbf{r}, t)} = \frac{1}{2}\mathrm{Re}\{\mathbf{E}(\mathbf{r}) \cdot \mathbf{H}^*(\mathbf{r})\},$$
> $$\overline{\mathcal{E}(\mathbf{r}, t) \times \mathcal{H}(\mathbf{r}, t)} = \frac{1}{2}\mathrm{Re}\{\mathbf{E}(\mathbf{r}) \times \mathbf{H}^*(\mathbf{r})\}.$$

where the complex polarisability α_{lin} depends on the frequency ω. The real part of the polarisability represents the oscillation of the dipole in phase with the electromagnetic field and the imaginary part represents the oscillation in phase quadrature.

An oscillating dipole emits a radiated field. For simplicity, we will consider a dipole oscillating at an angular frequency ω, oriented along the polar axis s and placed at the origin of a spherical coordinate system, i.e.,

$$\mathbf{p}_\mathrm{d} = p_\mathrm{d} \hat{\mathbf{s}}. \tag{3.18}$$

The radiated electric field in spherical coordinates is

$$\mathbf{E}_\mathrm{d}(r, \vartheta, \varphi) = \frac{p_\mathrm{d} k_0^3}{4\pi \varepsilon_0} \frac{e^{ik_0 r}}{k_0 r} \left\{ 2\cos\vartheta \left[\frac{1}{(k_0 r)^2} - \frac{i}{k_0 r}\right]\hat{\mathbf{r}} \right.$$
$$\left. + \sin\vartheta \left[\frac{1}{(k_0 r)^2} - \frac{i}{k_0 r} - 1\right]\hat{\boldsymbol{\vartheta}} \right\}, \tag{3.19}$$

where k_0 is the vacuum wavenumber, r is the radial distance, ϑ is the polar angle, φ is the azimuthal angle, $\hat{\mathbf{r}}$ is the radial unit vector and $\hat{\boldsymbol{\vartheta}}$ is the polar unit vector.[2] This field is

---

[2] Eq. (3.19) can also be written as $\mathbf{E}_\mathrm{d}(\mathbf{r}) = \mathbb{G}(\mathbf{r}, \mathbf{r}_\mathrm{d})\mathbf{p}_\mathrm{d}$, where $\mathbb{G}(\mathbf{r}, \mathbf{r}_\mathrm{d})$ is the *Green tensor* for a homogenous medium giving the electric field at $\mathbf{r}$ generated by a dipole placed at $\mathbf{r}_\mathrm{d}$.

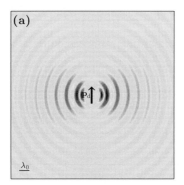 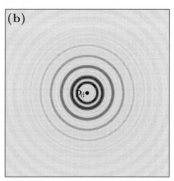

Figure 3.5 Oscillating dipole. Instantaneous fields radiated (a) in the polar and (b) in the transversal plane by an oscillating dipole $\mathbf{p}_\mathrm{d}$, which is represented by the arrow in (a) and by the dot in (b). The real part of the component of the electric field parallel to the dipole is plotted. The bar represents a wavelength.

illustrated in Fig. 3.5. From Eq. (3.19), we notice that the electric field has no azimuthal component. In the far field, i.e., for $k_0 r \gg 1$, Eq. (3.19) becomes

$$\mathbf{E}_{\mathrm{d,ff}}(r, \vartheta, \varphi) = -\frac{p_\mathrm{d} k_0^3}{4\pi\varepsilon_0} \frac{e^{ik_0 r}}{k_0 r} \sin\vartheta \, \hat{\boldsymbol{\vartheta}}, \qquad (3.20)$$

which is purely transverse and shows that the radiated field propagates outwards as a spherical wave.[3]

Exercise 3.3.1 Show that the current associated with an oscillating electric dipole $\mathbf{p}_\mathrm{d}$ placed at $\mathbf{r}_\mathrm{d}$ is

$$\mathbf{j}_\mathrm{d}(\mathbf{r}) = -i\omega \mathbf{p}_\mathrm{d} \delta(\mathbf{r} - \mathbf{r}_\mathrm{d}). \qquad (3.21)$$

Exercise 3.3.2 Calculate the electric field of an oscillating dipole in a dielectric medium with dielectric permittivity ε_m.

3.4 Radiative reaction correction to the polarisability

The polarisation response of a small particle in an external oscillating field depends on its interaction both with the external field and with its own scattered field. The interaction with the self-scattered field in particular introduces some corrections to the Clausius–Mossotti relation [Eq. (3.15)], which was derived for a static incident field where there are no radiated

---

[3] The corresponding magnetic field is

$$\mathbf{B}_\mathrm{d}(r, \vartheta, \varphi) = \frac{1}{c}\frac{p_\mathrm{d} k_0^3}{4\pi\varepsilon_0} \frac{e^{ik_0 r}}{k_0 r} \sin\vartheta \left[-\frac{i}{k_0 r} - 1\right]\hat{\boldsymbol{\varphi}},$$

where $\hat{\boldsymbol{\varphi}}$ is the azimuthal unit vector. The magnetic field has no radial or polar components. Furthermore, because the magnetic field has no terms proportional to $(kr)^{-3}$, the near field is dominated by the electric field. In the far field,

$$\mathbf{B}_{\mathrm{d,ff}}(r, \vartheta, \varphi) = -\frac{1}{c}\frac{p_\mathrm{d} k_0^3}{4\pi\varepsilon_0} \frac{e^{ik_0 r}}{k_0 r} \sin\vartheta \, \hat{\boldsymbol{\varphi}}.$$

fields. We will now proceed to derive such a correction, following the original approach of Draine and Goodman (1993).

We will consider a homogeneous and isotropic small dielectric sphere of radius a and relative dielectric permittivity ε_r. We will assume that the incoming field $\mathbf{E}_i$, the dipole scattered field $\mathbf{E}_d$ and the polarisation $\mathbf{P}$ of the sphere are uniform throughout its volume, which is valid for the case of a small sphere. Using the constitutive relations [Box 3.3], $\mathbf{P}$ is parallel and proportional to the sum of $\mathbf{E}_i$ and $\mathbf{E}_d$, i.e.,

$$\mathbf{P} = \varepsilon_0(\varepsilon_r - 1)(\mathbf{E}_i + \mathbf{E}_d). \tag{3.22}$$

To simplify the following mathematical analysis, we will assume that $\mathbf{E}_i$, $\mathbf{P}$ and $\mathbf{E}_d$ are aligned along the polar axis s and we will place the centre of the sphere at the origin of the system of reference.

We will now proceed to evaluate the scattered field at the centre of the sphere. Each volume element dV of the sphere has a dipole moment $d\mathbf{p} = \mathbf{P} dV$. The electric field scattered by an oscillating dipole with dipole moment $d\mathbf{p}$ placed at the origin of the system of reference and evaluated at position $\mathbf{r}$ in the near field is given by Eq. (3.19). Because we have assumed $d\mathbf{p} = dp\,\hat{\mathbf{s}}$, we need only consider the component parallel to the s-axis,

$$dE_{d,s}(\mathbf{r}) = \frac{k_0^3}{4\pi\varepsilon_0}\frac{e^{ik_0 r}}{k_0 r}\left\{\frac{2\cos^2\vartheta}{(k_0 r)^2} - \frac{2i\cos^2\vartheta}{k_0 r} - \frac{\sin^2\vartheta}{(k_0 r)^2} + \frac{i\sin^2\vartheta}{k_0 r} + \sin^2\vartheta\right\} dp,$$

where we have made use of the fact that $\hat{\mathbf{r}}\cdot\hat{\mathbf{s}} = \cos\vartheta$ and $\hat{\boldsymbol{\vartheta}}\cdot\hat{\mathbf{s}} = -\sin\vartheta$. We can now expand $dE_{d,z}(\mathbf{r})$ as a power series in the small parameter $k_0 r$ as far as the third power:

$$dE_{d,s}(\mathbf{r}) \approx \frac{k_0^3}{4\pi\varepsilon_0}\frac{1}{(k_0 r)^3}\left\{(3\cos^2\vartheta - 1) + \frac{1}{2}(\cos^2\vartheta + 1)(k_0 r)^2 + \frac{2i}{3}(k_0 r)^3\right\} dp, \tag{3.23}$$

where we use $e^{ik_0 r} \approx 1 + ik_0 r - \frac{1}{2}(k_0 r)^2 - \frac{i}{6}(k_0 r)^3$ and $\sin^2\vartheta = 1 - \cos^2\vartheta$.

We can now integrate the contributions of all the volume elements dV to the scattered electric field at the centre of the sphere, which, as a result of spherical symmetry, is equivalent to integrating throughout the sphere the scattering of a dipole element placed at the centre of the sphere; i.e.,

$$\mathbf{E}_d = \int_{\mathbf{r}\in V} dE_{d,s}(\mathbf{r})\,\hat{\mathbf{s}} = \frac{1}{3\varepsilon_0}\left\{-1 + (k_0 a)^2 + \frac{2i}{3}(k_0 a)^3\right\}\mathbf{P}, \tag{3.24}$$

where we have made use of Gauss' divergence theorem [Box 3.1] to help in performing the volume integration.

Substituting Eq. (3.24) into Eq. (3.22) and taking into account that the polarisability of the sphere is $\mathbf{p}_{sp} = \mathbf{P}V$, we obtain the *effective polarisability* of the small sphere, i.e., the proportionality constant between $\mathbf{p}_{sp}$ and $\mathbf{E}_i$,[4]

$$\alpha_{\text{rad}} = \frac{\alpha_{\text{CM}}}{1 - \frac{\varepsilon_r - 1}{\varepsilon_r + 2}\left[(k_0 a)^2 + \frac{2i}{3}(k_0 a)^3\right]}, \tag{3.25}$$

---

[4] Unlike the case in electrostatics, this polarisability is defined with respect to the incoming field and not with respect to the local field.

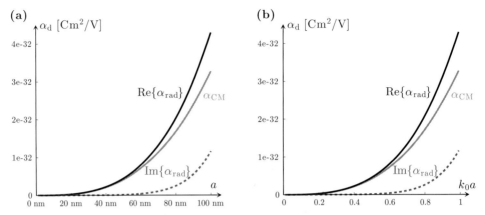

Figure 3.6 Polarisability. Real (black solid line) and imaginary (dashed line) part of the polarisability with radiative correction [Eq. (3.25)] plotted as a function of (a) the particle radius a and (b) the size parameter $k_0 a$ for a particle of relative dielectric permittivity $\varepsilon_p = 2.25$ (typical for glass in the optical spectrum) illuminated by an electromagnetic field with vacuum wavelength $\lambda_0 = 633$ nm. The grey solid line represents the Clausius–Mossotti polarisability [Eq. (3.15)].

where α_{CM} is the polarisability defined in the static limit by the Clausius–Mossotti relation given in Eq. (3.15). Fig. 3.6 plots the real and imaginary parts of Eq. (3.25) and compares them with the polarisability calculated using the Clausius–Mossotti relation, which leads to a purely real polarisability. Eq. (3.25) reveals the presence of an imaginary component of the polarisability, even for a sphere with a purely real dielectric constant, which is lacking in a simple application of the Clausius–Mossotti relation. This imaginary component of the dipole oscillation is in phase quadrature and arises from the interaction of the dipole with its own scattered field at its own location. It is termed the *self-field interaction* or *radiative reaction* and has important consequences for the scattering of light by the sphere and the definition of the extinction cross-section, as we will see in the next section.

In many applications the expression for the effective polarisability given in Eq. (3.25) is further simplified by neglecting the term in $(k_0 a)^2$ in the denominator:

$$\alpha_{\text{rad}} \approx \alpha_{CM} \left\{ 1 - i \frac{k_0^3 \alpha_{CM}}{6\pi \varepsilon_0} \right\}^{-1}. \tag{3.26}$$

Exercise 3.4.1 Calculate the effective polarisability accounting for the radiative reaction of a small dielectric sphere of dielectric permittivity ε_p immersed in a medium of dielectric permittivity ε_m.

3.5 Cross-sections

When an object, e.g., a dipole, is illuminated by an electromagnetic wave, it both scatters and absorbs power from the wave. The *scattering cross-section* σ_{scat} multiplied by the

> **Box 3.6** — **Electromagnetic energy**
>
> The *energy density* stored in an electromagnetic field is given by
>
> $$u(\mathbf{r},t) = \frac{\epsilon}{2}|\mathcal{E}(\mathbf{r},t)|^2 + \frac{1}{2\mu}|\mathcal{B}(\mathbf{r},t)|^2 = \epsilon|\mathcal{E}(\mathbf{r},t)|^2 = \frac{1}{\mu}|\mathcal{B}(\mathbf{r},t)|^2,$$
>
> where the last two equalities show that the energy is evenly distributed between the electric and magnetic fields. The *energy flux density* associated with a propagating wave is given by the *Poynting vector*,
>
> $$\mathcal{S}(\mathbf{r},t) = \frac{1}{\mu}\mathcal{E}(\mathbf{r},t) \times \mathcal{B}(\mathbf{r},t) = \mathcal{E}(\mathbf{r},t) \times \mathcal{H}(\mathbf{r},t).$$
>
> *Poynting's theorem* states that the rate of change of energy density in a volume equals the rate of work done on a charge distribution in that volume plus the energy flux leaving the volume:
>
> $$\frac{\partial u(\mathbf{r},t)}{\partial t} = -\mathbf{j}(\mathbf{r},t) \cdot \mathcal{E}(\mathbf{r},t) - \nabla \cdot \mathcal{S}(\mathbf{r},t).$$
>
> For the case of monochromatic fields expressed by phasors [Box 3.5], the time-averaged quantities are
>
> $$u(\mathbf{r}) = \frac{1}{4}\epsilon|\mathbf{E}(\mathbf{r})|^2 + \frac{1}{4\mu}|\mathbf{B}(\mathbf{r})|^2 = \frac{1}{2}\epsilon|\mathbf{E}(\mathbf{r})|^2 = \frac{1}{2\mu}|\mathbf{B}(\mathbf{r})|^2,$$
>
> $$\mathbf{S}(\mathbf{r}) = \frac{1}{2}\frac{c}{n}\epsilon|\mathbf{E}(\mathbf{r})|^2\hat{\mathbf{k}} = \frac{1}{2}\frac{n}{c\mu}|\mathbf{E}(\mathbf{r})|^2\hat{\mathbf{k}} = \frac{1}{2}\mathrm{Re}\{\mathbf{E}(\mathbf{r}) \times \mathbf{H}^*(\mathbf{r})\},$$
>
> where $\hat{\mathbf{k}}$ is the field direction of propagation.

power density of the incident wave is equivalent to the rate at which energy is removed from the electromagnetic wave by scattering in all directions. The *absorption cross-section* σ_{abs} multiplied by the power density of the incident wave gives the rate at which energy is absorbed, resulting in the heating of the object. The rate at which energy is removed from the electromagnetic wave through scattering and absorption is described by the *extinction cross-section*, $\sigma_{\mathrm{ext}} = \sigma_{\mathrm{scat}} + \sigma_{\mathrm{abs}}$. Therefore, these cross-sections represent the apparent area of the object with respect to the scattering, absorption and extinction processes, respectively. We remark that, generally, the values of these cross-sections do not coincide with the geometric cross-section of the object.

We will now consider the cross-sections of an electric dipole $\mathbf{p}_\mathrm{d}$ of polarisability α_d illuminated by a plane electromagnetic wave $\mathbf{E}_\mathrm{i}$ with intensity, i.e., power per unit area,

$$I_\mathrm{i} = \frac{1}{2}c\varepsilon_0|\mathbf{E}_\mathrm{i}|^2. \qquad (3.27)$$

We will first consider the total power removed from the electromagnetic field by the dipole. This is equal to the work done by $\mathbf{E}_\mathrm{i}$ on the charge distribution of the dipole, in accordance with Poynting's theorem [Box 3.6], i.e., $\frac{1}{2}\int_V \mathrm{Re}\{\mathbf{j}_\mathrm{d} \cdot \mathbf{E}_\mathrm{i}^*\}\,dV$, where $\mathbf{j}_\mathrm{d}$ is the current density of the oscillating dipole, so that, using Eq. (3.21),

$$P_{\mathrm{ext,d}} = \frac{\omega}{2}\mathrm{Im}\left\{\mathbf{p}_\mathrm{d} \cdot \mathbf{E}_\mathrm{i}^*(\mathbf{r}_\mathrm{d})\right\}, \qquad (3.28)$$

where the electric field is evaluated at the dipole origin $\mathbf{r}_d$. The extinction cross-section is defined as the ratio between $P_{ext,d}$ and I_i, i.e.,

$$\sigma_{ext,d} = \frac{k_0}{\varepsilon_0} \operatorname{Im}\{\alpha_d\}. \tag{3.29}$$

Similarly, if we now consider the rate of energy dissipation arising from the work done on the dipole by the scattered field at the location of the dipole, using Eq. (3.24), we find

$$P_{scat,d} = -\frac{\omega}{2} \operatorname{Im}\left\{\mathbf{p}_d \cdot \mathbf{E}_d^*(\mathbf{r}_d)\right\} = \frac{\omega k_0^3 |p_d|^2}{12\pi \varepsilon_0}. \tag{3.30}$$

The scattering cross-section is defined as the ratio of the scattered power $P_{scat,d}$ to the incident intensity I_i,

$$\sigma_{scat,d} = \frac{k_0^4}{6\pi \varepsilon_0^2} |\alpha_d|^2. \tag{3.31}$$

Finally, the absorption cross-section is given by

$$\sigma_{abs,d} = \sigma_{ext,d} - \sigma_{scat,d} = \frac{k_0}{\varepsilon_0} \operatorname{Im}\{\alpha_d\} - \frac{k_0^4}{6\pi \varepsilon_0^2} |\alpha_d|^2. \tag{3.32}$$

For small dielectric particles $\sigma_{ext,d} \approx \sigma_{scat,d}$ and therefore $\sigma_{abs,d} \approx 0$. We note, however, that for small particles with imaginary parts of the refractive index, e.g., metallic particles, the absorption cross-section dominates the scattering cross-section, so that $\sigma_{ext,d} \approx \sigma_{abs,d}$.

Exercise 3.5.1 Show analytically that for small dielectric spheres $\sigma_{ext,d} \approx \sigma_{scat,d}$ when one accounts for the radiative correction [Eq. (3.25)]. What happens if one uses the Clausius–Mossotti relation [Eq. (3.15)] instead? How is it possible to increase the absorption cross-section $\sigma_{abs,d}$? [Hint: Use the approximation given by Eq. (3.26).]

Exercise 3.5.2 Show that the approximate expression for the Clausius–Mossotti relation given by Eq. (3.26) follows directly from Eq. (3.32).

3.6 The optical theorem

The extinction cross-section can be determined independent of the absorption and scattering cross-sections using the *optical theorem*, which relates the extinction cross-section itself to the forward scattering amplitude. This is an interesting derivation in its own right and, as we will see, provides some compelling evidence for the importance of the radiative reaction in the calculation of the polarisability.

For a linearly polarised incoming field $\mathbf{E}_i = E_i \hat{\mathbf{e}}$, where E_i is the field amplitude and $\hat{\mathbf{e}}$ is the polarisation unit vector, the field scattered by a dipole in the far field given by Eq. (3.20) can be written as[5]

$$\mathbf{E}_{\mathrm{d,ff}}(r,\vartheta) = E_\mathrm{i}\frac{e^{ik_0 r}}{r}f_\mathrm{d}(\vartheta)\hat{\mathbf{e}},\qquad(3.33)$$

where we have defined the *scattering amplitude*

$$f_\mathrm{d}(\vartheta) = \frac{k_0^2 \alpha_\mathrm{d}}{4\pi\varepsilon_0}\sin\vartheta.\qquad(3.34)$$

The total field is given by

$$\mathbf{E}_\mathrm{t}(r,\vartheta) = \mathbf{E}_\mathrm{i} + \mathbf{E}_{\mathrm{d,ff}}(r,\vartheta) = E_\mathrm{i}\left[e^{ik_0 z} + \frac{e^{ik_0 r}}{r}f_\mathrm{d}(\vartheta)\right]\hat{\mathbf{e}}.$$

We will now consider the total field intensity close to the forward propagation axis, which we will assume to be along the z-direction, i.e.,

$$I_\mathrm{t} = \frac{1}{2}c\varepsilon_0|\mathbf{E}_\mathrm{t}|^2 \approx I_\mathrm{i}\left|1 + f_\mathrm{d}(\vartheta)\frac{e^{ik_0\frac{x^2+y^2}{2z}}}{z}\right|^2 \approx I_\mathrm{i}\left[1 + \frac{2}{z}\mathrm{Re}\left\{f_\mathrm{d}(\vartheta)e^{ik_0\frac{x^2+y^2}{2z}}\right\}\right],$$

where $I_\mathrm{i} = \frac{1}{2}c\varepsilon_0|E_\mathrm{i}|$ is the intensity of the incoming field, the first approximation is justified because, being close to the propagation axis, i.e., $z^2 \gg x^2+y^2$, we can introduce the approximations $r = \sqrt{z^2+x^2+y^2} \approx z\left(1 + \frac{x^2+y^2}{2z^2}\right)$ in the argument of the exponential and $z \approx r$ in the denominator, and the second approximation is obtained by neglecting the higher-order terms that results from taking the square. We can now integrate I_t over an area A perpendicular to the propagation axis; such an area can correspond, for example, to a photodetector placed along the propagation axis in the far field, as shown in Fig. 3.7. The power impinging on A is

$$P_A = I_\mathrm{i}\left[A + \frac{2}{z}\mathrm{Re}\left\{f_\mathrm{d}\!\left(\frac{\pi}{2}\right)\int_A e^{ik_0\frac{x^2+y^2}{2z}}dA\right\}\right],$$

where we have made the approximation $f_\mathrm{d}(\vartheta) \approx f_\mathrm{d}\left(\frac{\pi}{2}\right)$ justified by the fact that we are close to the propagation axis, i.e., $\vartheta \approx \frac{\pi}{2}$. We can now approximate the integral as

$$\int_A e^{ik_0\frac{x^2+y^2}{2z}}dA \approx \int_{-\infty}^{+\infty} e^{ik_0\frac{x^2}{2z}}dx \int_{-\infty}^{+\infty} e^{ik_0\frac{y^2}{2z}}dy = \frac{2\pi i z}{k_0},$$

where the approximation is valid for $x,y,z \to +\infty$ keeping $\frac{x}{z}, \frac{y}{z} \ll 1$.[6] Therefore, we obtain

$$P_A = I_\mathrm{i}\left[A - \frac{4\pi}{k_0}\mathrm{Im}\left\{f_\mathrm{d}\!\left(\frac{\pi}{2}\right)\right\}\right].\qquad(3.35)$$

---

[5] Note that the unit vector $\hat{\mathbf{e}}$ in Eq. (3.33) is the incoming polarisation vector, which is parallel to the dipole, whereas the unit vector $\hat{\vartheta}$ in Eq. (3.20) is the polar unit vector, which, for $\hat{\mathbf{r}} = \hat{\mathbf{k}}_i$, is antiparallel to the incoming polarisation; thus, $\hat{\vartheta} = -\hat{\mathbf{e}}$.

[6] To get a physical picture of this, think of a detector that moves towards infinity and at the same time becomes larger at a lower rate, as shown in Fig. 3.7. Eventually the detector will occupy all the xy-plane, but it will subtend an infinitesimal solid angle.

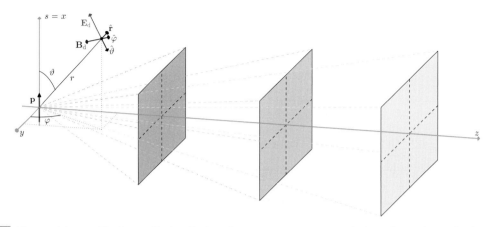

Figure 3.7 The optical theorem. The theorem [Eq. (3.36)] relates the extinction cross-section to the forward scattering amplitude, i.e., the amplitude of the field scattered in the direction of the incoming electromagnetic wave. In this way, the optical theorem permits one to quantify the power removed from the incoming wave. The square surfaces represent the area of a detector that moves towards infinity along the z-direction and, at the same time, becomes larger but at a lower rate: eventually this detector will occupy the whole xy-plane, but it will subtend an infinitesimal solid angle.

We can see from Eq. (3.35) that the power impinging on A in the absence of the scattering dipole, $P_i = I_i A$, is changed by an amount $P_{\text{ext}} = P_i - P_A$ that depends on the scattering amplitude, $f_d\left(\frac{\pi}{2}\right)$. Because the extinction cross-section [Eq. (3.29)] is defined as the ratio of the moduli of P_{ext} to I_i, we obtain the *optical theorem*,

$$\sigma_{\text{ext}} = \frac{4\pi}{k_0} \text{Im}\{f_d(\tfrac{\pi}{2})\}, \tag{3.36}$$

which relates the extinction cross-section to the imaginary part of the forward-scattering amplitude. If one thinks of the extinction as being caused by the interference of the incident wave with the scattered wave, the optical theorem relates the relative phase shift of the two waves to the power removed from the incoming wave. We note that Eq. (3.36) is valid for any particle and not only for a dipolar particle.

We can now fully appreciate the importance of the radiative reaction correction to the Clausius–Mossotti relation [Eq. (3.15)] discussed in Section 3.4. Without radiative reaction, a dipolar particle with a purely real dielectric constant would present zero cross-section, implying that it would not scatter light. Therefore, if the polarisability is found using the Clausius–Mossotti relation only, the extinction cross-section is equal to the absorption cross-section and scattering is unaccounted for.

3.7 Optical forces

We will now calculate the optical force experienced by an electric dipole in the presence of time-varying electric and magnetic fields, $\mathcal{E}_i(\mathbf{r}, t)$ and $\mathcal{B}_i(\mathbf{r}, t)$. As shown in Fig. 3.2a, we

3.7 Optical forces

consider two particles of mass m carrying charges $\pm q$, located at $\mathbf{r}_\pm$, and separated by a distance $\Delta \mathbf{r} = \mathbf{r}_+ - \mathbf{r}_-$, so that they form a dipole of magnitude $\mathbf{p}_d = q\Delta \mathbf{r}$. We implicitly assume that $\Delta r = |\Delta \mathbf{r}| \ll \lambda_0$. Their equations of motion are

$$\begin{cases} m\dfrac{d^2\mathbf{r}_+}{dt^2} = +q\left[\mathcal{E}_i(\mathbf{r}_+, t) + \dfrac{d\mathbf{r}_+}{dt} \times \mathcal{B}_i(\mathbf{r}_+, t)\right] - \nabla U_{\text{int}}(\Delta \mathbf{r}, t) \\ m\dfrac{d^2\mathbf{r}_-}{dt^2} = -q\left[\mathcal{E}_i(\mathbf{r}_-, t) + \dfrac{d\mathbf{r}_-}{dt} \times \mathcal{B}_i(\mathbf{r}_-, t)\right] + \nabla U_{\text{int}}(\Delta \mathbf{r}, t), \end{cases} \quad (3.37)$$

where the terms in square brackets are the Lorentz forces acting on the particles and $U_{\text{int}}(\Delta \mathbf{r}, t)$ is the interaction potential energy that keeps the two particles together.

We can now expand in Taylor series the fields $\mathcal{E}_i(\mathbf{r}, t)$ and $\mathcal{B}_i(\mathbf{r}, t)$ around the centre of mass of the dipole, i.e.,

$$\mathbf{r}_d = \frac{\mathbf{r}_+ + \mathbf{r}_-}{2},$$

obtaining

$$\mathcal{E}_i(\mathbf{r}_\pm, t) \approx \mathcal{E}_i(\mathbf{r}_d, t) + ((\mathbf{r}_\pm - \mathbf{r}_d) \cdot \nabla)\,\mathcal{E}_i(\mathbf{r}_d, t),$$
$$\mathcal{B}_i(\mathbf{r}_\pm, t) \approx \mathcal{B}_i(\mathbf{r}_d, t) + ((\mathbf{r}_\pm - \mathbf{r}_d) \cdot \nabla)\,\mathcal{B}_i(\mathbf{r}_d, t).$$

By substituting these expression into Eqs. (3.37) and summing up the resulting right-hand sides, we obtain the net force acting on the dipole:

$$\mathbf{F}_{\text{DA}}(\mathbf{r}_d, t) = (\mathbf{p}_d \cdot \nabla)\,\mathcal{E}_i(\mathbf{r}_d, t) + \frac{d\mathbf{p}_d}{dt} \times \mathcal{B}_i(\mathbf{r}_d, t) + \frac{d\mathbf{r}_d}{dt} \times (\mathbf{p}_d \cdot \nabla)\,\mathcal{B}_i(\mathbf{r}_d, t). \quad (3.38)$$

The second term in Eq. (3.38) may be re-written as

$$\frac{d\mathbf{p}_d}{dt} \times \mathcal{B}_i = \frac{d}{dt}(\mathbf{p}_d \times \mathcal{B}_i) - \mathbf{p}_d \times \frac{d\mathcal{B}_i}{dt} = \frac{d}{dt}(\mathbf{p}_d \times \mathcal{B}_i) + \mathbf{p}_d \times (\nabla \times \mathcal{E}_i)$$

by making the approximation $\frac{\partial \mathcal{B}_i}{\partial t} \approx \frac{d\mathcal{B}_i}{dt}$, which is justified for nonrelativistic dipoles, i.e., as $|\frac{d\mathbf{r}_d}{dt}| \ll c$, and by invoking Maxwell's equation for the curl of the electric field [Box 3.2]. For non-relativistic dipoles, the third term in Eq. (3.38) is much smaller than the first two and can thus be dropped. Therefore, the total force on the dipole is

$$\mathbf{F}_{\text{DA}}(\mathbf{r}_d, t) = (\mathbf{p}_d \cdot \nabla)\,\mathcal{E}_i(\mathbf{r}_d, t) + \mathbf{p}_d \times (\nabla \times \mathcal{E}_i(\mathbf{r}_d, t)) + \frac{d}{dt}(\mathbf{p}_d \times \mathcal{B}_i(\mathbf{r}_d, t)). \quad (3.39)$$

We are typically interested in the time-averaged value of $\mathbf{F}_{\text{DA}}(\mathbf{r}_d, t)$ because the electromagnetic fields oscillate extremely fast as, e.g., optical frequencies are $\approx 10^{15}$ Hz. When the time-averaged value of Eq. (3.39) is taken the last term is zero, so that we obtain

$$\mathbf{F}_{\text{DA}} = \overline{\mathbf{F}_{\text{DA}}(\mathbf{r}_d, t)} = \overline{(\mathbf{p}_d \cdot \nabla)\,\mathcal{E}_i(\mathbf{r}_d, t)} + \overline{\mathbf{p}_d \times (\nabla \times \mathcal{E}_i(\mathbf{r}_d, t))}, \quad (3.40)$$

which can be re-written as the particularly elegant formula

$$\mathbf{F}_{DA}(\mathbf{r}_d) = \sum_{j=x,y,z} \overline{p_{d,j} \nabla \mathcal{E}_{i,j}(\mathbf{r}_d, t)}. \tag{3.41}$$

For a monochromatic illumination field at angular frequency ω, $\mathcal{E}_i(\mathbf{r}, t)$ can be written in terms of its phasor $\mathbf{E}_i(\mathbf{r})$ [Box 3.5]. In this case the induced dipole moment depends on the electric field and therefore on $\mathbf{r}_d$, i.e.,

$$\mathbf{p}_d(\mathbf{r}_d) = \alpha_d \mathbf{E}_i(\mathbf{r}_d), \tag{3.42}$$

where the complex polarisability α_d is given, e.g., by Eq. (3.25). Using Eq. (3.42), we can now rewrite Eq. (3.40) in terms of phasors, i.e.,

$$\mathbf{F}_{DA} = \frac{1}{2}\alpha'_d \text{Re}\left\{(\mathbf{E}_i \cdot \nabla)\mathbf{E}_i^* + \mathbf{E}_i \times (\nabla \times \mathbf{E}_i^*)\right\} \\ - \frac{1}{2}\alpha''_d \text{Im}\left\{(\mathbf{E}_i \cdot \nabla)\mathbf{E}_i^* + \mathbf{E}_i \times (\nabla \times \mathbf{E}_i^*)\right\}, \tag{3.43}$$

where $\alpha'_d = \text{Re}\{\alpha_d\}$ and $\alpha''_d = \text{Im}\{\alpha_d\}$. Using the vector identity

$$\nabla(\mathbf{E}_i \cdot \mathbf{E}_i^*) = (\mathbf{E}_i \cdot \nabla)\mathbf{E}_i^* + \mathbf{E}_i \times (\nabla \times \mathbf{E}_i^*) + (\mathbf{E}_i^* \cdot \nabla)\mathbf{E}_i + \mathbf{E}_i^* \times (\nabla \times \mathbf{E}_i),$$

we see that $\text{Re}\{(\mathbf{E}_i \cdot \nabla)\mathbf{E}_i^* + \mathbf{E}_i \times (\nabla \times \mathbf{E}_i^*)\} = \frac{1}{2}\nabla(\mathbf{E}_i \cdot \mathbf{E}_i^*) = \frac{1}{2}\nabla(|\mathbf{E}_i|^2)$. Using the vector identity

$$\nabla \times (\mathbf{E}_i \times \mathbf{E}_i^*) = \mathbf{E}_i(\nabla \cdot \mathbf{E}_i^*) - \mathbf{E}_i^*(\nabla \cdot \mathbf{E}_i) + (\mathbf{E}_i^* \cdot \nabla)\mathbf{E}_i - (\mathbf{E}_i \cdot \nabla)\mathbf{E}_i^*$$

and taking into account that $\nabla \cdot \mathbf{E}_i = \nabla \cdot \mathbf{E}_i^* = 0$, we see that $\text{Im}\{(\mathbf{E}_i \cdot \nabla)\mathbf{E}_i^*\} = \frac{i}{2}\nabla \times (\mathbf{E}_i \times \mathbf{E}_i^*)$. Using these vector identities, Eq. (3.43) may be written as

$$\mathbf{F}_{DA} = \frac{1}{4}\alpha'_d \nabla(|\mathbf{E}_i|^2) - \frac{1}{2}\alpha''_d \text{Im}\left\{\mathbf{E}_i \times (\nabla \times \mathbf{E}_i^*)\right\} - \frac{i}{4}\alpha''_d \nabla \times (\mathbf{E}_i \times \mathbf{E}_i^*). \tag{3.44}$$

By using Maxwell's equation for the curl of the electric field [Box 3.5] and introducing the extinction cross-section $\sigma_{\text{ext},d} = k_0 \alpha''_d / \varepsilon_0$ [Eq. (3.29)], we can rewrite Eq. (3.44) as

$$\mathbf{F}_{DA} = \frac{1}{4}\alpha'_d \nabla |\mathbf{E}_i|^2 + \frac{\sigma_{\text{ext},d}}{c}\mathbf{S}_i - \frac{1}{2}\sigma_{\text{ext},d} c \nabla \times \mathbf{s}_d, \tag{3.45}$$

where we have introduced the time-averaged Poynting vector [Box 3.6] of the incoming wave,

$$\mathbf{S}_i = \frac{1}{2}\text{Re}\left\{\mathbf{E}_i \times \mathbf{H}_i^*\right\}, \tag{3.46}$$

and the time-averaged *spin density* of the incoming wave,[7]

$$\mathbf{s}_d = i\frac{\varepsilon_0}{2\omega}\mathbf{E}_i \times \mathbf{E}_i^*. \tag{3.47}$$

---

[7] The derivation and the physical meaning of the spin density will be provided in Subsection 5.1.1 [Eq. (5.24)].

The terms on the right-hand side of Eq. (3.45) can be identified with various kinds of optical forces, as we will see in detail in the next subsections.

Exercise 3.7.1 Find an expression (similar to the one given by Eq. (3.45)) for the optical force on a small dielectric sphere of dielectric permittivity ε_p immersed in a medium of dielectric permittivity ε_m.

Exercise 3.7.2 Show that for a circularly polarised electric field with constant amplitude, i.e., $\mathbf{E}_i(\mathbf{r}) = E_i e^{i\phi(\mathbf{r})}[\hat{\mathbf{x}} + i\hat{\mathbf{y}}]$, the time-averaged spin density $\mathbf{s}_d$ is homogeneous and oriented along $\hat{\mathbf{z}}$. In which cases is $\mathbf{s}_d$ non-homogeneous?

3.7.1 Gradient force

The first term in Eq. (3.45) is the *gradient force*:

$$\mathbf{F}_{\text{DA,grad}}(\mathbf{r}_d) = \frac{1}{4}\alpha'_d \nabla |E_i(\mathbf{r}_d)|^2. \tag{3.48}$$

The gradient force is responsible for confinement in optical tweezers. The gradient force arises from the potential energy of a dipole in the electric field and, thus, is a *conservative* force; i.e., the work done by the force in moving a particle between two points is independent of the path taken. Because the intensity of the electric field is $I_i = \frac{1}{2}c\varepsilon_0 |E_i|^2$, we can rewrite the gradient force in terms of the gradient of intensity as

$$\mathbf{F}_{\text{DA,grad}}(\mathbf{r}_d) = \frac{1}{2}\frac{\alpha'_d}{c\varepsilon_0} \nabla I_i(\mathbf{r}_d). \tag{3.49}$$

Therefore, particles with positive polarisability, i.e., particles whose refractive index is higher than that of the surrounding medium, are attracted towards the high-intensity region of the optical field, and particles with negative polarisability, i.e., particles whose refractive index is lower than that of the surrounding medium, are repelled by the high-intensity regions.

To a first approximation, we can consider the intensity distribution at the focus of a Gaussian beam to be Gaussian, i.e.,[8]

$$I_i(\rho) = I_0 e^{-2\rho^2/w_0^2} \tag{3.50}$$

where ρ is the radial coordinate in the transverse plane, I_0 is the maximum intensity and w_0 is the beam waist. This intensity distribution and the corresponding optical forces on a small particle whose refractive index is higher than the surrounding medium are shown in Fig. 3.8a. For small displacements from the axis,

$$I_i(\rho) \approx I_0 \left(1 - 2\frac{\rho^2}{w_0^2}\right),$$

so that the radial component of the gradient force can be approximated by an elastic restoring force proportional and opposite to the displacement from the origin, i.e.,

$$F_{\text{DA,grad},\rho}(\rho) = -\kappa_\rho \rho, \tag{3.51}$$

---

[8] The precise formulas will be discussed in Chapter 4.

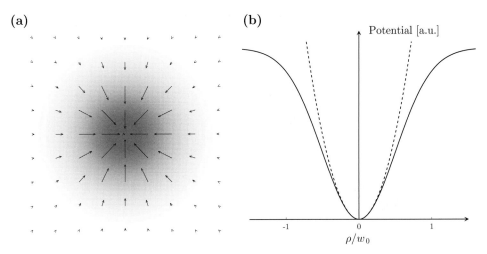

Figure 3.8 Gradient force. (a) Intensity distribution of a Gaussian beam (background) and corresponding optical gradient forces (arrows) on a dielectric particle whose refractive index is higher than that of the surrounding medium. (b) Optical potential along the radial direction (solid line) and its approximation by a harmonic potential (dashed line).

where

$$\kappa_\rho = 2 \frac{\alpha'_d}{c\varepsilon_0} \frac{I_0}{w_0^2}. \tag{3.52}$$

Eq. (3.52) reveals that κ_ρ is proportional to the electric field intensity and to the real part of the polarisability, i.e., to the particle volume, for small dipolar particles. Furthermore, κ_ρ is inversely proportional to the beam area, so, as may be expected, tighter focusing leads to stronger confinement. The corresponding radial potential

$$U_{\text{DA}}(\rho) = \frac{1}{2}\kappa_\rho \rho^2 \tag{3.53}$$

is plotted by the dashed line in Fig. 3.8b, showing that it is a good approximation to the real potential (solid line) for small particle displacements from the potential minimum. A similar analysis may be made for the axial direction, although the spring constant in this direction will be found to be weaker.

Exercise 3.7.3 Calculate the trap stiffness due to the gradient forces generated by a Gaussian beam on a small dielectric sphere of dielectric permittivity ε_p immersed in a medium of dielectric permittivity ε_m. How does this stiffness scale as a function of the parameters of the particle and of the beam? What happens when $\varepsilon_p < \varepsilon_m$?

Exercise 3.7.4 Repeat the analysis of the gradient force in this Subsection along the axial direction of a Gaussian beam. [Hint: You can find the expressions of the fields associated with a Gaussian beam in Chapter 4.]

Exercise 3.7.5 Show that a plane wave, i.e., $\mathbf{E}_i(\mathbf{r}) \equiv E_i \hat{\mathbf{e}}$, exerts no gradient force on a dielectric particle.

Exercise 3.7.6 Given a dielectric particle of real polarisability α_d' and a non-homogeneous linearly polarised electric field $\mathbf{E}_i(\mathbf{r}) = E_i(\mathbf{r})\hat{\mathbf{e}}$, where $\hat{\mathbf{e}}$ is the polarisation unit vector, derive the gradient force [Eq. (3.49)] from the expression for the potential energy of the induced dipole in the electric field. [Hint: Use Eq. (3.4).]

Exercise 3.7.7 Consider a small dielectric particle held by an optical tweezers generated by a strongly focused Gaussian beam, whose intensity distribution is given by Eq. (3.50). What are the stiffnesses of such an optical trap in the radial and axial directions? Investigate the practical cases of a 50 nm particle made of latex, silica, silicon or diamond immersed in water illuminated by a 10 mW beam with $\lambda_0 = 830$ nm and $w_0 = 0.65$ μm.

3.7.2 Scattering force

The second term in Eq. (3.45) is the *scattering force*:

$$\mathbf{F}_{\text{DA,scat}}(\mathbf{r}_d) = \frac{\sigma_{\text{ext},d}}{c}\mathbf{S}_i(\mathbf{r}_d). \tag{3.54}$$

The scattering force is *non-conservative*. It is due to the transfer of momentum from the field to the particle as a result of the scattering and absorption processes, as is revealed by the fact that $\mathbf{F}_{\text{DA,scat}}$ is proportional to the extinction cross-section, $\sigma_{\text{ext},d}$. This force points in the direction of the Poynting vector $\mathbf{S}_i$, i.e., in the direction of the field propagation. Fig. 3.9 shows the scattering force (black arrows) generated by a (collimated) Gaussian beam along the propagation direction. The fact that this force is non-conservative can be seen by taking a path integral along the dashed line and observing that its circulation is not null. The gradient force (grey arrows) is conservative.

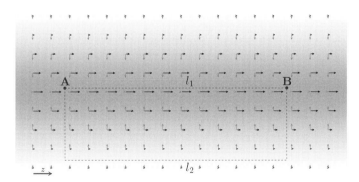

Figure 3.9 Scattering force. Non-conservative scattering force field (black arrows) generated by a collimated Gaussian beam along the propagation z-direction (arrow at the bottom). The grey arrows represent the gradient force field. The total work done by the scattering force depends on the path of integration and is different in the cases of paths l_1 and l_2. Instead, the work done by the gradient force is independent of the path; i.e., the gradient force is conservative.

Exercise 3.7.8 Consider a dielectric dipolar particle with polarisability α_d illuminated by a weakly focused Gaussian beam. Calculate the corresponding scattering and gradient forces along the axial direction. What can you deduce about the conditions required for a stable axial trapping?

3.7.3 Spin–curl force

The third term in Eq. (3.45) is the *spin–curl force*:

$$\mathbf{F}_{\text{DA,sc}}(\mathbf{r}_d) = -\frac{1}{2}\sigma_{\text{ext},d} c \nabla \times \mathbf{s}_d(\mathbf{r}_d). \tag{3.55}$$

The spin–curl force arises from polarisation gradients in the electromagnetic field and is *non-conservative*. In order for this force to arise, the polarisation of the field must be non-homogenous, as in the case shown in Fig. 3.10. This force is relatively small compared with the gradient and scattering forces and therefore does not usually play a major role in optical trapping experiments.

Exercise 3.7.9 Show that the spin–curl force generated by a plane wave, i.e., $\mathbf{E}_i(\mathbf{r}) \equiv E_i \hat{\mathbf{e}}$, is null.

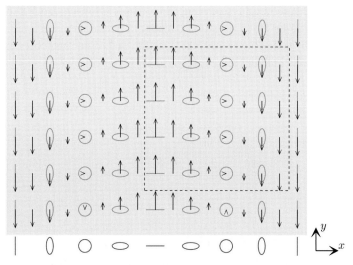

Figure 3.10 Spin–curl force. The spin–curl force (black arrows) emerges in the presence of a gradient of polarisation and is non-conservative, as can be seen by calculating the work done by this force along the dashed line. The polarisation state of the light is indicated by the ellipses on the background: the x- and y-components of the electric field oscillate in phase quadrature and their amplitudes (indicated by the length of the axes of the ellipses) are such that the intensity of the field is constant (so that there is no gradient force).

Exercise 3.7.10 Show that for a linearly polarised electromagnetic field the spin density vanishes, i.e., $\mathbf{s}_d = 0$ [Eq. 3.47], and, therefore, the associated spin–curl force is also null.

Exercise 3.7.11 Given an electromagnetic field

$$\mathbf{E}_i(\mathbf{r}) = E_i \left[A(\mathbf{r})\hat{\mathbf{x}} + i B(\mathbf{r})\hat{\mathbf{y}} \right], \tag{3.56}$$

where $A(\mathbf{r})$ and $B(\mathbf{r})$ are real functions, show that the spin–curl force is

$$\mathbf{F}_{\mathrm{DA,sc}}(\mathbf{r}) = -\frac{\sigma_{\mathrm{ext,d}} c \epsilon_0}{2\omega} |E_i|^2 \left[\frac{\partial}{\partial y} A(\mathbf{r}) B(\mathbf{r}) \hat{\mathbf{x}} - \frac{\partial}{\partial x} A(\mathbf{r}) B(\mathbf{r}) \hat{\mathbf{y}} \right]. \tag{3.57}$$

If $A(\mathbf{r}) = \sin(x)$ and $B(\mathbf{r}) = \cos(x)$, in which direction is the resulting spin–curl force oriented? How large is the corresponding gradient force?

3.8 Atomic polarisability

We start by considering an electron bound to a nucleus in a homogenous time-varying electromagnetic field, as shown in Fig. 3.1. The electromagnetic force on such an electron is given by the Lorentz force

$$\mathbf{F}_e(t) = e_- \left[\mathcal{E}_i(t) + \mathbf{v}_e \times \mathcal{B}_i(t) \right] \approx e_- \mathcal{E}_i(t), \tag{3.58}$$

where e_- is the charge of the electron, $\mathcal{E}_i(t)$ and $\mathcal{B}_i(t)$ are homogeneous time-varying fields, and the approximation is valid for a non-relativistic electron, i.e., an electron whose speed $v_e = |\mathbf{v}_e|$ is much lower than the speed of light c, because in this case $|\mathcal{B}_i(t)| \approx |\mathcal{E}_i(t)|/c \ll |\mathcal{E}_i(t)|$. An approximately harmonic potential binds the electron to the nucleus, so that we can introduce a linear restoring force $-\kappa_e \mathbf{r}_e(t)$ towards the nucleus, where κ_e is the spring constant of the potential and $\mathbf{r}_e(t)$ is the position of the electron with respect to the centre of mass of the atom. The motion of the electron is damped by an additional force $-m_e \gamma_e \frac{d\mathbf{r}_e(t)}{dt}$, where m_e is the mass of the electron and γ_e is the damping coefficient, which can be understood classically in terms of radiative loss, so that $1/\gamma_e$ is the lifetime of the atomic level. The resulting equation of motion is

$$m_e \frac{d^2 \mathbf{r}_e(t)}{dt^2} = -m_e \gamma_e \frac{d\mathbf{r}_e(t)}{dt} - \kappa_e \mathbf{r}_e(t) + e_- \mathcal{E}_i(t). \tag{3.59}$$

Multiplying both sides of Eq. (3.59) by e_-/m_e and considering that the electric dipole moment of the electron is $\mathbf{p}_e(t) = e_- \mathbf{r}_e(t)$, we get a differential equation for the dipole moment,

$$\frac{d^2 \mathbf{p}_e(t)}{dt^2} + \gamma_e \frac{d\mathbf{p}_e(t)}{dt} + \omega_R^2 \mathbf{p}_e(t) = \frac{e_-^2}{m_e} \mathcal{E}_i(t), \tag{3.60}$$

where $\omega_R = \sqrt{\kappa_e/m_e}$ is the natural oscillation frequency of the system, which corresponds to an absorption line of the material.

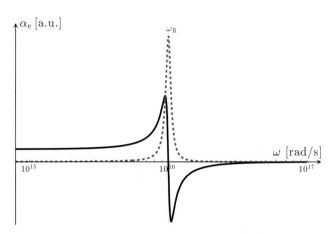

Figure 3.11 Complex polarisability. Real (solid line) and imaginary (dashed line) parts of the complex atomic polarisability $\alpha_e(\omega)$ [Eq. (3.62)] of an atom as a function of the angular frequency ω. $\omega_R = 10^{16}$ rad/s (corresponding to $\lambda_R = 188$ nm) and $\gamma_e = 10^{15}$ rad/s. The order of magnitude of these values is realistic even though they do not correspond to any specific atom. Note that the frequency axis is logarithmic.

A harmonic homogeneous electric field at angular frequency ω can be expressed in terms of phasors as $\mathbf{E}_i(t) = \mathbf{E}_i e^{-i\omega t}$ [Box 3.5]. The induced dipole moment, also expressed in terms of phasors, oscillates at the same frequency, i.e., $\mathbf{p}_e(t) = \mathbf{p}_e e^{-i\omega t}$. Substituting the expressions for $\mathbf{E}_i(t)$ and $\mathbf{p}_e(t)$ into Eq. (3.60), we deduce that

$$\mathbf{p}_e(\omega) = \alpha_e(\omega)\mathbf{E}_i = \frac{e_-^2}{m_e} \frac{1}{\omega_R^2 - \omega^2 - i\gamma_e \omega} \mathbf{E}_i. \tag{3.61}$$

Therefore, a complex polarisability can be assigned to the atom. This complex polarisability depends on the frequency ω as

$$\alpha_e(\omega) = \frac{e_-^2}{m_e} \frac{(\omega_R^2 - \omega^2)}{(\omega_R^2 - \omega^2)^2 + \gamma_e^2 \omega^2} + i \frac{e_-^2}{m_e} \frac{\gamma_e \omega}{(\omega_R^2 - \omega^2)^2 + \gamma_e^2 \omega^2}. \tag{3.62}$$

The real part of the polarisability represents the oscillation of the dipole in phase with the electromagnetic field and the imaginary part represents the oscillation in phase quadrature. As ω approaches ω_R, a resonance in the imaginary part and a zero in the real part occur, as can be seen in Fig. 3.11; ω_R corresponds to a spectroscopic absorption line of the material.

Exercise 3.8.1 Calculate the dielectric constant of a material when illuminated near resonance. [Hint: Start by using the atomic polarisability of the atoms constituting the material to calculate the macroscopic polarisability of the material, assuming that there are N_e atoms in a material volume V_e.]

Exercise 3.8.2 Calculate the gradient and scattering force near a resonance on a small particle in a focused Gaussian beam. Study the frequency dependence of these forces and their scaling behaviour with respect to the width and height of the resonance.

Exercise 3.8.3 Show that the polarisability $\alpha_e(\omega)$ [Eq. (3.61)] can be recast in analogy to Eq. (3.26) as

$$\alpha_e(\omega) = \alpha_{e,0}(\omega) \left\{ 1 - i \frac{k_0^3 \alpha_{e,0}(\omega)}{6\pi \varepsilon_0} \right\}^{-1},$$

where $\alpha_{e,0}(\omega) = \frac{e_-^2}{m}(\omega_R^2 - \omega^2)^{-1}$ and $\gamma_e = \frac{k_0^3 e_-^2}{6\pi \varepsilon_0 m \omega}$. Note that this value of γ_e is in agreement with the radiation damping obtained from quantum mechanics. Show that, as a consequence, an atom illuminated by light blue-shifted (red-shifted) with respect to ω_R has negative (positive) polarisability.

3.9 Plasmonic particles

The distinctive optical properties of metallic particles compared to dielectric particles arise as a result of the free conduction electrons present in metals. This free electron gas can be driven by an applied field and, at optical frequencies, can sustain oscillations in both surface and volume charge density known as *plasmons*. In the case of a small metallic particle the oscillating electrons are also subject to a restoring force due to the positive charge of the nuclei, giving rise to resonances in the optical response. The resulting excitations are known as *localised surface plasmons*.[9]

We first consider the contribution of the free electrons using the *Drude–Sommerfeld model*. We write down the equation of motion for these electrons subject to an electric field oscillating at angular frequency ω, i.e., $\mathbf{E}_i(t) = \mathbf{E}_i e^{-i\omega t}$:

$$m_f \frac{d^2 \mathbf{r}_f(t)}{dt^2} + m_f \gamma_f \frac{d\mathbf{r}_f(t)}{dt} = e_- \mathbf{E}_i(t) e^{-i\omega t}, \quad (3.63)$$

where $\mathbf{r}_f(t)$ is the free electron position, m_f is its effective mass and γ_f is its damping coefficient, i.e., the rate of electron scattering events. Setting $\mathbf{r}_f(t) = \mathbf{r}_f e^{-i\omega t}$, we find that the dipole moment of an individual free electron is

$$\mathbf{p}_f = e_- \mathbf{r}_f = -\frac{e_-^2}{m_f} \frac{1}{\omega^2 + i\omega \gamma_f} \mathbf{E}_i. \quad (3.64)$$

Therefore, the net polarisation of the metal is given by $\mathbf{P}_f = \frac{N_f}{V_f} \mathbf{p}_f$, where N_f is the number of free electrons contained in a volume V_f and the free electron contribution to the dielectric constant is

$$\varepsilon_f(\omega) = 1 - \frac{\omega_f^2}{\omega^2 + i\gamma_f \omega}, \quad (3.65)$$

where $\omega_f = \sqrt{N_f e_-^2 / V_f m_f \varepsilon_0}$ is the *plasma frequency*.

---

[9] A classic text on plasmons was written by Raether (1988). A recent extensive introduction to the field can be found in Maier (2007).

Also, the bound electrons in a metal can contribute to the metal's optical properties, but only at higher (optical) frequencies, where photons have sufficient energy to excite bound electrons into the conduction band. In this case, the equation of motion for the bound electrons subject to the electric field $\mathbf{E}_i e^{-i\omega t}$ is

$$m_b \frac{d^2 \mathbf{r}_b(t)}{dt^2} + m_b \gamma_b \frac{d\mathbf{r}_b(t)}{dt} + \kappa_b \mathbf{r}_b(t) = e_- \mathbf{E}_i e^{-i\omega t}, \qquad (3.66)$$

where $\mathbf{r}_b(t)$ is the bound electron position, m_b is its effective mass, γ_b is its damping coefficient due primarily to radiative damping, and κ_b is the spring constant of the potential holding the electron. Following the same ansatz as in the free electron case, we find that the contribution of such interband transitions to the dielectric constant is

$$\varepsilon_b(\omega) = 1 + \frac{\omega_b^2}{(\omega_{b,R}^2 - \omega^2) - i\gamma\omega}, \qquad (3.67)$$

where $\omega_b = \sqrt{N_b e_-^2 / V_b m_b \varepsilon_0}$ is analogous to the plasma frequency for the free electron case, with N_b the number of bound electrons within a material volume V_b, and $\omega_{b,R} = \sqrt{\kappa_b/m_b}$.

A simple model of the dielectric function for a metal nanoparticle can therefore be constructed from a weighted sum of Lorentzians corresponding to the intraband transitions [Eq. (3.65)] and to the interband transitions [Eq. (3.67)]. For example, in Fig. 3.12, the real and imaginary parts of the dielectric constant of gold are plotted. The experimental data are given according to Palik (1985) (circles) and Johnson and Christy (1972) (crosses).[10] The real part of the dielectric constant is negative over the visible range, so that light can penetrate into the metal, albeit only to a very small extent, typically just a few nanometres, corresponding to the *skin depth* of the metal; this is different from what happens in other frequency ranges, where metals behave as perfect conductors. The imaginary part of the dielectric constant, which describes the dissipation of energy associated with the motion of the electrons in the metal, is always positive. The black continuous lines in Fig. 3.12 fit the Drude–Sommerfeld model [Eq. (3.65)] to the data reported in Johnson and Christy (1972): this model accurately describes the optical properties of gold in the infrared, i.e., for $\lambda_0 > 700$ nm, whereas in the visible it needs to be supplemented by taking into account the response of bound electrons, which can be promoted from lower-lying bands into the conduction band by higher-energy photons [Eq. (3.67)].[11]

Plasmonic nanoparticles can sustain localised plasmonic resonances when the frequency of the incoming electromagnetic wave matches the natural frequency of the nanoparticle electron oscillations. In fact, the optical properties and the resonant wavelength of a plasmonic nanoparticle can be controlled by tailoring its size and shape, as well as the dielectric functions of the metal and the surrounding medium. The physical origin behind localised plasmonic resonances can be better understood by studying the properties of a spherical metallic nanoparticle. The scattering of a sphere can be evaluated analytically by calculating

---

[10] Johnson and Christy (1972) and Palik (1985) are the standard references for the dielectric constants of various materials and, in particular, of metals.

[11] A very good description of the properties of gold with an analytical model including interband transitions can be found in Etchegoin et al. (2006).

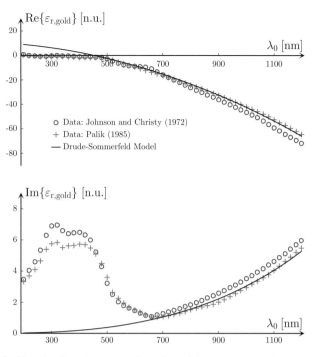

Figure 3.12 Dielectric function of gold. Real and imaginary parts. The circles and the crosses represent experimental values taken respectively from Johnson and Christy (1972) and Palik (1985). The solid line represents the fitting of the Drude–Sommerfeld model for free electrons [Eq. (3.65)] to the data reported in Johnson and Christy (1972). This model accurately describes the experimental optical properties of gold in the infrared regime, whereas in the visible region it needs to be supplemented with additional terms to take into account the response of bound electrons.

its polarisability, which, as long as the particle is much smaller than the wavelength of the incident light, is given by Eq. (3.15), i.e.,

$$\alpha_0 = 3V\varepsilon_m(\omega)\frac{\varepsilon_p(\omega) - \varepsilon_m(\omega)}{\varepsilon_p(\omega) + 2\varepsilon_m(\omega)}, \tag{3.68}$$

where V is the volume of the sphere, and ε_p and ε_m are the dielectric constants of the metal and of the surrounding medium, respectively. When the real part of the dielectric function of the metal is negative and its absolute value matches $2\varepsilon_m$, the polarisability expressed by Eq. (3.68) experiences a resonance, which is responsible for an enhancement in the scattering and absorption cross-sections of the sphere. The increased polarisability of metallic particles gives rise to enhanced optical trapping forces compared to dielectric particles. For example, in the infrared part of the spectrum for a spherical gold nanoparticle the real part of the polarisability is several times greater than for a glass nanoparticle of corresponding size, resulting in an increased optical gradient force and a stronger optical trap. This enhancement can be engineered to more effectively trap nanoparticles, but it can also be a nuisance, as the increased polarisability at resonance may lead to an increased scattering force that may propel nanoparticles away from an optical trap.

Exercise 3.9.1 Study numerically the enhancement in optical forces associated with the plasmonic resonance for a metallic nanoparicle with respect to, e.g., a glass nanoparticle. Compare the effects of the plasmonic resonance on the scattering and gradient forces. Investigate its frequency dependence for the practical cases of gold, silver and aluminium. [Hint: Get the values of the dielectric constants of gold, silver and aluminium from Johnson and Christy (1972) or Palik (1985).]

3.10 Optical binding

Optical binding is the name given to the phenomenon whereby two or more particles spontaneously form ordered structures in an optical field as a result of the forces associated with multiple scattering between the particles themselves. The existence of such an effect was first predicted theoretically by Thirunamachandran (1980) and later demonstrated experimentally by Burns et al. (1989). Such ordering has been observed for particles ranging in size from tens of nanometres to a few micrometres. Here, we will introduce optical binding in a simple manner by considering the forces that occur between just two dipolar particles subject to the same incident field and to the light scattered by each dipole onto the other.

We consider two identical dipoles, $\mathbf{p}_1$ and $\mathbf{p}_2$, with polarisability $\alpha = \alpha' + i\alpha''$ placed at $\mathbf{r}_1$ and $\mathbf{r}_2$, respectively, and subject to the incident electric field $\mathbf{E}_i(\mathbf{r}, t) = \mathbf{E}_i(\mathbf{r})e^{-i\omega t}$, as shown in Fig. 3.13, so that

$$\begin{cases} \mathbf{p}_1 = \alpha \mathbf{E}(\mathbf{r}_1) \\ \mathbf{p}_2 = \alpha \mathbf{E}(\mathbf{r}_2) \end{cases} \tag{3.69}$$

where $\mathbf{E}(\mathbf{r}_1)$ and $\mathbf{E}(\mathbf{r}_2)$ are the electric fields at the locations of the dipoles $\mathbf{p}_1$ and $\mathbf{p}_2$, respectively. If the two particles are not too close to each other, we can neglect the effects of higher-order scattering, e.g., that of the field scattered by particle 1 onto particle 2 and back onto particle 1, so that at each location the electric field contains a contribution from the incident field plus a contribution from the field scattered by the other particle, i.e.,

$$\begin{cases} \mathbf{E}(\mathbf{r}_1) = \mathbf{E}_i(\mathbf{r}_1) + \mathbf{E}_{s,2}(\mathbf{r}_1) \\ \mathbf{E}(\mathbf{r}_2) = \mathbf{E}_i(\mathbf{r}_2) + \mathbf{E}_{s,1}(\mathbf{r}_2) \end{cases} \tag{3.70}$$

where the scattered fields are those of an electric dipole as given by Eq. (3.19).

In order to understand the basis of optical binding, we will consider *longitudinal* optical binding, where two particles are arranged parallel to the direction of propagation of the incoming wave, as shown in Fig. 3.13. Thus, we will assume the particles to be subject to an incident plane wave linearly polarised in the x-direction and propagating in the z-direction with electric field $\mathbf{E}_i(\mathbf{r}) = E_i e^{-ik_0 z}\hat{\mathbf{x}}$, the two particles to be located on the z-axis so that the incident wave encounters first particle 1 and then particle 2, and $\Delta z = |\mathbf{r}_2 - \mathbf{r}_1|$.

We will first calculate the total optical force acting on particle 2. As shown in Fig. 3.13a, the total electric field on particle 2 is the sum of the incident plane wave and the scattered

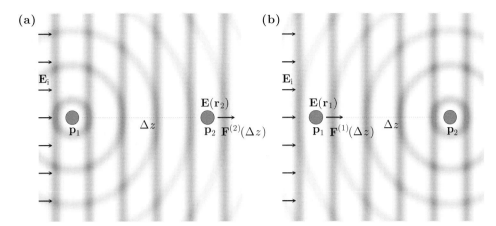

Figure 3.13 Optical binding. Longitudinal binding between two dipoles. (a) The first particle radiates towards the second particle in the same direction as the incident field and with the same wavevector size; therefore, the phase difference between the incident and scattered fields is independent of the inter-particle distance and the optical binding force on the second particle is always directed towards the first particle [Eq. (3.72)]. (b) In contrast, the second particle radiates backwards against the incident wave propagation; thus the phase difference between the incident and scattered waves depends on the inter-particle distance as $2k_0 \Delta z$ and results in an oscillatory behaviour of the binding force [Eq. (3.74)].

dipole field of particle 1. If the distance between the two particles is sufficiently large, i.e., $k_0 \Delta z \gg 1$, we can use the far-field approximation for the scattered field given by Eq. (3.20) with $p = \alpha E_i e^{ik_0 z_1}$ and $\vartheta = \pi/2$, as particle 2 is located on the z-axis, i.e.,

$$\mathbf{E}(\mathbf{r}_2) = E_i e^{ik_0 z_2} \hat{\mathbf{x}} + \alpha E_i e^{ik_0 z_1} \frac{k_0^3}{4\pi \varepsilon_0} \frac{e^{ik_0 \Delta z}}{k_0 \Delta z} \hat{\mathbf{x}} = E_i e^{ik_0 z_2} \left[1 + \frac{k_0^3}{4\pi \varepsilon_0} \frac{\alpha}{k_0 \Delta z} \right] \hat{\mathbf{x}}, \quad (3.71)$$

form which we can see that the incident field and the field scattered by particle 1 are co-propagating with a fixed phase difference independent from Δz. We are now able to calculate the time-averaged optical force acting on particle 2 using Eq. (3.41) written in terms of phasors, i.e.,

$$\overline{\mathbf{F}}^{(2)}(\mathbf{r}_2) = \frac{1}{2} \text{Re} \left\{ \sum_{j=r,\vartheta,z} \alpha E_j(\mathbf{r}_2) \nabla E_j^*(\mathbf{r}_2) \right\},$$

which, in the case of longitudinal binding between dipoles, simplifies because of symmetry to

$$\overline{\mathbf{F}}^{(2)}(\mathbf{r}_2) = \overline{F}_z^{(2)}(\mathbf{r}_2) \hat{\mathbf{z}} = \frac{1}{2} \text{Re} \left\{ \alpha E_x(\mathbf{r}_2) \frac{\partial E_x^*(\mathbf{r}_2)}{\partial z_2} \right\} \hat{\mathbf{z}}.$$

Therefore, the total force exerted on particle 2 is

$$\overline{\mathbf{F}}^{(2)}(\Delta z) = \underbrace{\frac{1}{2}k_0\alpha''|E_i|^2\hat{\mathbf{z}}}_{\text{scattering force}} + \underbrace{\frac{k_0^4}{4\pi\varepsilon_0}\alpha'\alpha''\frac{|E_i|^2}{k_0\Delta z}\hat{\mathbf{z}}}_{\text{optical binding}}, \qquad (3.72)$$

where we have neglected the terms in α^3 because α is small, as it scales with the volume of a dipolar particle, and the terms in $(k_0\Delta z)^{-2}$ because of the far-field approximation. The first term of the force in Eq. (3.72) represents the scattering force exerted by the incident field on particle 2. The second term of the force in Eq. (3.72) is an additional contribution arising from the field scattered by particle 1. Because the incoming field and the field scattered by particle 1 are co-propagating, i.e., they have a fixed phase difference, the optical binding force does not change sign and decreases with the inter-particle distance. If $\alpha'\alpha'' < 0$, the optical binding force acts opposite to the scattering force, so that particle 2 is pushed forward at a lower speed than particle 1 and the two particles tend to approach each other.

We will now consider the force on particle 1, as shown in Fig. 3.13b. Following the same procedure as in the previous case, we obtain the total field at $\mathbf{r}_1$ is

$$\mathbf{E}(\mathbf{r}_1) = E_i e^{ik_0 z_1}\hat{\mathbf{x}} + \alpha E_i e^{ik_0 z_2}\frac{k_0^3}{4\pi\varepsilon_0}\frac{e^{ik_0\Delta z}}{k_0\Delta z}\hat{\mathbf{x}} = E_i e^{ik_0 z_1}\left[1 + \frac{k_0^3}{4\pi\varepsilon_0}\frac{\alpha}{k_0\Delta z}e^{i2k_0\Delta z}\right]\hat{\mathbf{x}}, \qquad (3.73)$$

where, differently from the previous case, the phase difference between the incoming field and the scattered field depends on the particle separation, i.e., on $k_0\Delta z$. We can now calculate the force on particle 1 as

$$\overline{\mathbf{F}}^{(1)}(\mathbf{r}_1) = \frac{1}{2}\text{Re}\left\{\alpha E_x(\mathbf{r}_1)\frac{\partial E_x^*(\mathbf{r}_1)}{\partial z_1}\right\}\hat{\mathbf{z}}.$$

Therefore, the force exerted on particle 1 is

$$\overline{\mathbf{F}}^{(1)}(\Delta z) = \underbrace{\frac{1}{2}k_0\alpha''|E_i|^2\hat{\mathbf{z}}}_{\text{scattering force}} + \underbrace{\frac{k_0^4}{4\pi\varepsilon_0}|E_i|^2\left[\alpha'^2\frac{\sin 2k_0\Delta z}{k_0\Delta z} + \alpha'\alpha''\frac{\cos 2k_0\Delta z}{k_0\Delta z}\right]\hat{\mathbf{z}}}_{\text{optical binding}}, \qquad (3.74)$$

where we have neglected the terms of order α^3 and $(k_0\Delta z)^{-2}$. The first term is the scattering force on particle 1 due to the incoming wave and the second term is the optical binding force due to the field scattered by particle 2, which is counter-propagating. Since the phase difference between the incoming field and the counter-propagating scattered field depends on the particle separation Δz, the instantaneous induced dipole and the instantaneous field with which it is interacting may be parallel (minimum in energy) or anti-parallel (maximum in energy) and the subsequent force is also modulated. It can be seen that these terms are oscillatory, leading to stable binding configurations when the particles are separated by multiples of $\pi/k_0 = \lambda_0/2$. Counter-propagating beams may be used to balance the radiation pressure from the incident beams, which would otherwise propel the optically bound dimer along the direction of beam propagation.

Exercise 3.10.1 Calculate the longitudinal optical binding forces acting on two dipoles illuminated by two counter-propagating plane waves. Under what conditions will the two particles reach an equilibrium configuration?

Exercise 3.10.2 Calculate the optical binding force for two particles in the same plane perpendicular to the direction of propagation of the incident field. This configuration is often referred to as *transverse* optical binding.

Exercise 3.10.3 Calculate the optical binding forces between two dipoles for the cases of longitudinal and transverse optical binding using the exact formula for the dipole radiation [Eq. (3.19)]. What differences do you notice with respect to the calculations where the far-field approximation for the dipole radiation is used?

Exercise 3.10.4 Study numerically the optical forces arising between two dipoles illuminated by a plane wave. Consider also the case in which the particles are illuminated by two counter-propagating waves.

Problems

3.1 *Ellipsoidal particles.* The polarisability of a small ellipsoid with semi-axes $a_1 \geq a_2 \geq a_3$ illuminated along one of the principal axes is given by Bohren and Huffman (2008) as

$$\alpha_{\text{ellipsoid}}^{(i)} = 4\pi a_1 a_2 a_3 \varepsilon_m \frac{\varepsilon_p - \varepsilon_m}{3\varepsilon_m + 3L_i(\varepsilon_p - \varepsilon_m)} = V\varepsilon_m \left[\frac{\varepsilon_m}{\varepsilon_p - \varepsilon_m} + L_i\right]^{-1},$$

where $V = \frac{4}{3}\pi a_1 a_2 a_3$ is the particle volume and

$$L_i = \frac{3V}{8\pi} \int_0^\infty \frac{dq}{(a_i^2 + q)\sqrt{(q + a_1^2)(q + a_2^2)(q + a_3^2)}}$$

is the geometrical factor when the field is polarised along the principal axis i ($L_1 + L_2 + L_3 = 1$, $L_1 \leq L_2 \leq L_3$; what are their values for a sphere?). Calculate the polarisability tensor for a randomly oriented ellipsoid. Calculate the optical torque in the case of a plane wave and a Gaussian beam illumination. Which is the stable orientation?

3.2 *Spheroidal particles.* A special class of ellipsoids are spheroids, which have two semi-axes equal. If the smaller two axes are equal, i.e., $a_1 > a_2 = a_3$, we have a cigar-shaped *prolate spheroid*, in which case

$$L_1 = \frac{1-e^2}{e^2}\left[-1 + \frac{1}{2e}\ln\frac{1+e}{1-e}\right],$$

while when the larger two axes are equal, i.e., $a_1 = a_2 > a_3$, we have a disk-shaped *oblate spheroid*, in which case

$$L_1 = L_2 = \frac{g(e)}{2e^2}\left[\frac{\pi}{2} - \text{atan}(g(e))\right] - \frac{g^2(e)}{2} \quad \text{with} \quad g(e) = \sqrt{\frac{1-e^2}{e^2}},$$

where $e = \sqrt{1 - a_3^2/a_1^2}$ is the particle eccentricity. What are the other geometrical factors? Prolate spheroids range from needles ($e = 1$) to spheres ($e = 0$), while oblate spheroids range from disks ($e = 1$) to spheres ($e = 0$). Calculate the scattering and gradient forces on a prolate or oblate spheroid in the focal spot of a Gaussian beam in its stable orientation. Investigate how the optical forces and torques depend on the particle eccentricity.

3.3 *Effect of particle size on plasmon optical forces.* As metal nanoparticles get larger, at a certain point the quasi-static approximation fails as retardation effects and higher-order corrections must be considered. An expansion of Mie theory for a sphere yields (Meier and Wokaun, 1983)

$$\alpha_{\text{metal}} = V \frac{1 - \frac{1}{10}(\varepsilon_p - \varepsilon_m)x^2 + O(x^4)}{\left[\frac{\varepsilon_m}{\varepsilon_p - \varepsilon_m} + \frac{1}{3}\right] - \frac{1}{30}(\varepsilon_p + 10\varepsilon_m)x^2 - i\frac{4\pi^2\varepsilon_m^{3/2}}{3}\frac{V}{\lambda} + O(x^4)},$$

where $x = 2\pi a/\lambda$ is the size parameter. The x^2 term in the numerator incorporates the effect of retardation, while the x^2 term in the denominator incorporates the effect of retardation in the depolarisation field. Study the shifting and broadening of the plasmon resonance with the increase of particle size. Calculate the optical trapping force constants for a gold particle immersed in water illuminated by a Gaussian beam with $\lambda_0 = 830$ nm and $w_0 = 0.65\,\mu\text{m}$ using the quasi-static polarisability and the size-corrected polarisability given above. From which size do the two approximations deviate?

3.4 *Optical binding.* Study numerically the optical binding between more than two dipolar particles. You can draw inspiration for possible configurations to explore from the review article by Dholakia and Zemánek (2010). In particular, consider the case of longitudinal optical binding in a counter-propagating beam geometry.

3.5 *Binding of anisotropic particles.* Calculate the optical binding force between two spheroids. Distinguish between prolate and oblate spheroids and consider the cases for which the symmetry axis of the spheroids is parallel or orthogonal to the light propagation axis. Generalise the binding forces for a generic orientation. What is the equilibrium orientation? Show explicitly that this orientation minimises the energy of the system.

References

Bohren, C. F., and Huffman, D. R. 2008. *Absorption and scattering of light by small particles*. New York: Wiley.

Burns, M. M., Fournier, J.-M., and Golovchenko, J. A. 1989. Optical binding. *Phys. Rev. Lett.*, **63**, 1233–6.

Dholakia, K., and Zemánek, P. 2010. Gripped by light: Optical binding. *Rev. Mod. Phys.*, **82**, 1767–91.

Draine, B. T., and Goodman, J. 1993. Beyond Clausius-Mosotti: Wave propagation on a polarizable point lattice and the discrete dipole approximation. *Astrophys. J.*, **405**, 685–97.

Etchegoin, P. G., Le Ru, E. C., and Meyer, M. 2006. An analytic model for the optical properties of gold. *J. Chem. Phys.*, **125**, 164705.

Johnson, P. B., and Christy, R. W. 1972. Optical constants of the noble metals. *Phys. Rev. B*, **6**, 4370–79.

Maier, S. A. 2007. *Plasmonics: Fundamentals and applications*. New York: Springer Verlag.

Meier, M., and Wokaun, A. 1983. Enhanced fields on large metal particles: Dynamic depolarization. *Opt. Lett.*, **8**, 581–3.

Palik, E. D. 1985. *Handbook of optical constants of solids*. Orlando, FL: Academic Press.

Raether, H. 1988. *Surface plasmons on smooth surfaces*. Berlin: Springer Verlag.

Thirunamachandran, T. 1980. Intermolecular interactions in the presence of an intense radiation field. *Mol. Phys.*, **40**, 393–9.

4 Optical beams and focusing

The key enabling factor for the rise of optical manipulation was the invention of the laser. In fact, lasers provide easy access to intense electromagnetic fields in the form of *optical beams*. Coherent optical beams propagate approximately along straight lines over large distances. Furthermore, their propagation can be controlled by standard optical elements, e.g., lenses and mirrors, and by more exotic devices, e.g., axicons and gratings. Perhaps more important for optical manipulation, optical beams can be focused, producing the high-intensity optical fields and strong optical field gradients required to exert appreciable forces on microscopic and nanoscopic particles. The most common example, featured in Fig. 4.1, is a *Gaussian beam*, i.e., a laser beam characterised by a Gaussian intensity profile. This beam is the most commonly employed to generate an optical tweezers by focusing it on a tight spot using a high-numerical-aperture objective lens. However, other kinds of beams can also be used, e.g., *Hermite–Gaussian beams*, *Laguerre–Gaussian beams*, *Bessel beams*, *cylindrical vector beams*, to obtain different kinds of focal fields, e.g., focal fields with a dark central spot. In this chapter, we will review the basic theoretical notions needed to understand optical beams, including, in particular, their properties, propagation and focusing.

Figure 4.1 Optical beams and optical components. A collimated Gaussian laser beam is reflected by a mirror and focused by a lens. Picture credits: Marco Grasso and Alessandro Magazzù.

4.1 Propagating electromagnetic waves

We will consider an electromagnetic field in a homogeneous medium with *dielectric permittivity* $\varepsilon_m = \varepsilon_{r,m}\varepsilon_0$ and *magnetic permeability* $\mu_m = \mu_{r,m}\mu_0$ where there are no charges or currents, i.e., $\rho = 0$ and $\mathbf{j} = 0$. By decoupling Maxwell's equations [Box 3.2], we obtain the three-dimensional *homogeneous wave equation* for the electric field $\boldsymbol{\mathcal{E}}(\mathbf{r}, t)$:

$$\nabla^2 \boldsymbol{\mathcal{E}}(\mathbf{r}, t) = \frac{n_m^2}{c^2} \frac{\partial^2}{\partial t^2} \boldsymbol{\mathcal{E}}(\mathbf{r}, t), \qquad (4.1)$$

where ∇^2 is the Laplacian operator, n_m is the refractive index of the medium and $c/n_m = 1/\sqrt{\mu_m \epsilon_m}$ is the speed of light in the medium.

Typically, we will be dealing with monochromatic electromagnetic fields with *angular frequency* ω. It is therefore useful to recast Eq. (4.1) in terms of phasors [Box 3.5], which leads to the three-dimensional *homogeneous Helmholtz equation* for $\mathbf{E}(\mathbf{r})$:

$$\left(\nabla^2 + k_m^2\right) \mathbf{E}(\mathbf{r}) = \mathbf{0}, \qquad (4.2)$$

where $k_m = \omega n_m/c$ is the *wavevector* magnitude in the medium.

The simplest solution of Eq. (4.2) is a *monochromatic plane wave*,

$$\mathbf{E}(\mathbf{r}) = E\, e^{i\mathbf{k}_m \cdot \mathbf{r}}\, \hat{\mathbf{e}}, \qquad (4.3)$$

where $\mathbf{k}_m = k_m \hat{\mathbf{k}}$ is the *wavevector* pointing in the direction of propagation, $\hat{\mathbf{k}}$ is the unit vector in the direction of propagation, E is the *electric field amplitude* and $\hat{\mathbf{e}}$ is the *polarisation vector*, which corresponds to the orientation of the electric field and is always transverse to the propagation direction, i.e., $\hat{\mathbf{k}} \cdot \hat{\mathbf{e}} = 0$. The corresponding magnetic induction is given by

$$\mathbf{B}(\mathbf{r}) = \frac{n_m}{c} \hat{\mathbf{k}} \times \mathbf{E}(\mathbf{r}). \qquad (4.4)$$

The complex vectors $\mathbf{E}(\mathbf{r})$ and $\mathbf{B}(\mathbf{r})$ are in phase and perpendicular to each other and to $\hat{\mathbf{k}}$. The wavevector k_m and the wavelength of light in the medium λ_m are connected by the *dispersion relation*

$$k_m = |\mathbf{k}_m| = \frac{n_m \omega}{c} = \frac{2\pi}{\lambda_m}. \qquad (4.5)$$

The physical fields are obtained by taking the real parts of the phasors expressed by Eqs. (4.3) and (4.4), as explained in Box 3.5. The average power per unit area transported by an electromagnetic wave is its *intensity*:

$$I = |\mathbf{S}(\mathbf{r})| = \frac{1}{2} \frac{c}{n_m} \varepsilon_m |E|^2. \qquad (4.6)$$

The plane wave described by Eqs. (4.3) and (4.4) is *linearly polarised*; i.e., its electric field is always in the direction $\hat{\mathbf{e}} = \hat{\mathbf{e}}_1$. The corresponding physical fields are depicted in Fig. 4.2a. It is possible to have other linearly polarised waves with polarisation vector

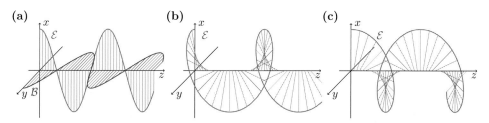

Figure 4.2 Electromagnetic waves. (a) Monochromatic electromagnetic wave linearly polarised along the x-direction and propagating towards the positive z-direction. (b) Left and (c) right circularly polarised waves featuring positive and negative helicity, respectively. For clarity, in (b) and (c), only the electric field is shown.

$\hat{\mathbf{e}}_2 \neq \hat{\mathbf{e}}_1$; usually, $\hat{\mathbf{e}}_1$ and $\hat{\mathbf{e}}_2$ are chosen to be perpendicular, i.e., $\hat{\mathbf{e}}_1 \cdot \hat{\mathbf{e}}_2 = 0$. The most general wave propagating in the direction $\hat{\mathbf{k}}$ is given by the combination of two such waves:

$$\mathbf{E}(\mathbf{r}) = (E_1\,\hat{\mathbf{e}}_1 + E_2\,\hat{\mathbf{e}}_2)\, e^{i\mathbf{k}_m \cdot \mathbf{r}}, \tag{4.7}$$

where the amplitudes E_1 and E_2 are complex numbers to allow a phase difference between the two plane waves. If E_1 and E_2 have the same phase, Eq. (4.7) represents a *linearly polarised* wave. If the phases of E_1 and E_2 are different, in general, Eq. (4.7) represents a left (right) *elliptically polarised* wave, characterised by the fact that, as the electric field rotates at an angular frequency ω, the tip of the electric field sweeps out an ellipse with *positive helicity* (*negative helicity*), i.e., anticlockwise (clockwise) looking at the incoming wave. Fig. 4.2b (Fig. 4.2c) depicts a *left circularly polarised* (*right circularly polarised*) wave, which is a special case of elliptical polarisation where the electric field is constant in magnitude but sweeps around in a circle with positive helicity (negative helicity).

Exercise 4.1.1 Derive the homogeneous wave equations for $\boldsymbol{\mathcal{E}}(\mathbf{r}, t)$ [Eq. (4.1)] and the homogeneous Helmholtz equations for $\mathbf{E}(\mathbf{r})$ [Eq. (4.2)] from Maxwell's equations [Box 3.2]. Derive also the corresponding equations for $\boldsymbol{\mathcal{B}}(\mathbf{r}, t)$ and $\mathbf{B}(\mathbf{r})$. [Hint: Start by applying curl to the curl Maxwell's equations and then use the divergence equations.]

Exercise 4.1.2 Verify that a plane wave [Eq. (4.3)] is a solution of Eq. (4.2).

Exercise 4.1.3 Under what conditions on E_1, E_2, $\hat{\mathbf{e}}_1$ and $\hat{\mathbf{e}}_2$ is the wave represented by Eq. (4.7) circularly polarised?

Exercise 4.1.4 What is the result of the superposition of a left circularly polarised wave and a right circularly polarised wave of the same intensity? And if their intensities are different?

Exercise 4.1.5 Consider two plane waves with equal intensity and wavevectors $\mathbf{k}_{m,1} = -\mathbf{k}_{m,2}$. What is the resulting electromagnetic field? What are its average energy density, Poynting vector and momentum density? What happens if the intensities of the two waves are different?

4.2 Angular spectrum representation

Electromagnetic fields in a homogenous, isotropic, linear and source-free medium can be described as superpositions of *plane waves* and *evanescent waves*. All these waves share the same wavenumber $k_m = |\mathbf{k}_m|$, which for a given wavelength depends solely on the parameters of the medium. For an arbitrary axis, which we will assume to be the z-axis, $\mathbf{k}_m = [k_{m,x}, k_{m,y}, k_{m,z}]$, where $k_{m,z} = \sqrt{k_m^2 - k_{m,x}^2 - k_{m,y}^2}$ can be either real or imaginary. Plane waves are oscillating functions of z and must satisfy the condition $k_{m,x}^2 + k_{m,y}^2 \leq k_m^2$, as shown in Fig. 4.3a, so that $k_{m,z}$ is real. A plane wave propagating along the z-axis, i.e., $\mathbf{k}_m = k_m \hat{\mathbf{z}}$, where $\hat{\mathbf{z}}$ is the unit vector along the z-axis, has no spatial modulation in the transverse plane; i.e., $k_{m,x} = k_{m,y} = 0$ [Fig. 4.3b]. The spatial frequency of the oscillations in the transverse plane increases as the angle between $\mathbf{k}_m$ and $\hat{\mathbf{z}}$ increases [Fig. 4.3c]. The largest number of oscillations in the xy-plane corresponds to a plane wave propagating perpendicular to z, which is characterised by $k_{m,z} = 0$ and, therefore, $k_{m,x}^2 + k_{m,y}^2 = k_m^2$ [Fig. 4.3d]. Even higher spatial frequencies are possible, but they are associated with an imaginary $k_{m,z}$ because $k_{m,x}^2 + k_{m,y}^2 > k_m^2$ and therefore they decay exponentially along the z-axis; these are the evanescent waves, which are thus nonpropagating.

The *angular spectrum representation* permits one to express a generic electric field $\mathbf{E}(\mathbf{r})$ as a superposition of plane waves. The physical idea behind this concept is illustrated in Fig. 4.4. The field on the plane $z = 0$ can be decomposed into plane waves of differing amplitudes propagating at different angles with respect to the optical axis. Due to the linearity of Maxwell's equations, each plane wave can be propagated along the z-axis independent of the other plane waves. Finally, the field on a plane z can be reconstructed by summing the propagated waves. To be precise, this is true only if near-field effects are neglected; to account also for such effects, one should consider both propagating plane waves and evanescent waves.

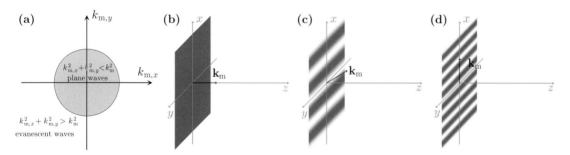

Figure 4.3 Plane waves and evanescent waves. (a) The transverse wavenumbers $k_{m,x}$ and $k_{m,y}$ of plane waves are restricted to the shaded circular area identified by the condition $k_{m,x}^2 + k_{m,y}^2 \leq k_m^2$, whereas evanescent waves fill the space outside. (b) A plane wave propagating along the z-axis ($\mathbf{k}_m = k_m \hat{\mathbf{z}}$) does not present any oscillations in the xy-plane. (c) The spatial frequency in the xy-plane increases as the angle between the propagation direction identified by $\mathbf{k}_m$ and the z-axis increases. (d) The largest number of oscillations in the xy-plane corresponds to a plane wave propagating perpendicular to z (e.g., $\mathbf{k}_m = k_m \hat{\mathbf{x}}$).

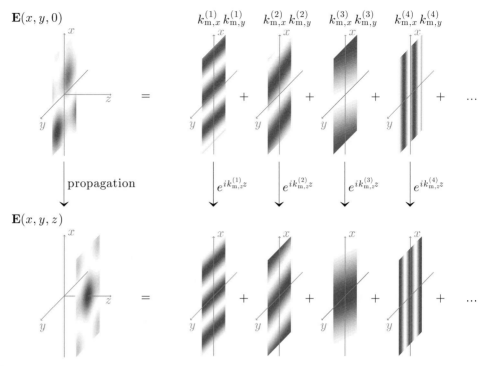

Figure 4.4 Angular spectrum representation. Optical fields are evaluated in planes perpendicular to an arbitrarily chosen axis, in this case the z-axis. The optical fields in the plane $z = 0$ (top left) result from the superposition of various plane waves (top right). The optical fields in another plane with z constant (bottom left) can be evaluated by summing up the propagated plane waves (bottom right).

Considering the field in the plane $z = 0$, i.e., $\mathbf{E}(x, y, 0)$, and calculating the corresponding two-dimensional Fourier transform with respect to x and y, one obtains

$$\check{\mathbf{E}}(k_{\mathrm{m},x}, k_{\mathrm{m},y}; 0) = \frac{1}{4\pi^2} \iint\limits_{-\infty}^{+\infty} \mathbf{E}(x, y, 0) \, e^{-i[k_{\mathrm{m},x} x + k_{\mathrm{m},y} y]} \, dx \, dy, \tag{4.8}$$

which permits one to identify the amplitudes of the various spectral components of $\mathbf{E}(x, y, 0)$. Each Fourier component propagates along the z-axis according to

$$\check{\mathbf{E}}(k_{\mathrm{m},x}, k_{\mathrm{m},y}; z) = \check{\mathbf{E}}(k_{\mathrm{m},x}, k_{\mathrm{m},y}; 0) e^{\pm i k_{\mathrm{m},z} z}, \tag{4.9}$$

where the sign '$\pm$' specifies the direction of the propagation: '+' represents a wave propagating into the $z > 0$ half-space and '−' represents a wave propagating into the $z < 0$ half-space. One can then find $\mathbf{E}(x, y, z)$ on an arbitrary transversal plane $z = $ constant by inverse Fourier transforming the summation of the propagated Fourier components, i.e.,

$$\mathbf{E}(x, y, z) = \iint\limits_{-\infty}^{+\infty} \check{\mathbf{E}}(k_{\mathrm{m},x}, k_{\mathrm{m},y}; 0)\, e^{i[k_{\mathrm{m},x}x + k_{\mathrm{m},y}y \pm k_{\mathrm{m},z}z]}\, dk_{\mathrm{m},x}\, dk_{\mathrm{m},y}, \qquad (4.10)$$

which is the *angular spectrum representation* of the electric field.

Exercise 4.2.1 Explore how an electromagnetic field propagates along the z-axis using the angular spectrum representation [Eq. (4.10)]. How does the spatial evolution of the various components of the field differ depending on their $k_{\mathrm{m},z}$? How is it possible to focus an electromagnetic field on a given plane? [Hint: You can adapt the program `fieldpropagation` from the book website.]

Exercise 4.2.2 Show that the angular spectrum representation of the magnetic induction $\mathbf{B}(x, y, z)$ is

$$\mathbf{B}(x, y, z) = \iint\limits_{-\infty}^{+\infty} \check{\mathbf{B}}(k_{\mathrm{m},x}, k_{\mathrm{m},y}; 0)\, e^{i[k_{\mathrm{m},x}x + k_{\mathrm{m},y}y \pm k_{\mathrm{m},z}z]}\, dk_{\mathrm{m},x}\, dk_{\mathrm{m},y}. \qquad (4.11)$$

[Hint: You can proceed along the same lines as for the derivation of the angular spectrum representation of the electric field.]

4.3 From near field to far field

Building on the results established in the previous section, we will now explore how an electromagnetic field defined in the near field propagates to the far field. As we will see, these results lay the theoretical foundations for the concept of optical rays, which we used in the geometrical optics approach in Chapter 2, and will be pivotal in the study of the focalisation of a paraxial beam, which we will undertake in Section 4.5.

We consider an electromagnetic field defined on a plane $z = 0$, i.e., $\mathbf{E}(x, y, 0)$; this is the near field. The corresponding electric field on a plane $z = s_z$ in the far field, i.e., for $k_{\mathrm{m}} s_z \gg 1$, is given by Eq. (4.10):

$$\mathbf{E}_{\mathrm{ff}}(\mathbf{s}) = \mathbf{E}_{\mathrm{ff}}(s, \hat{\mathbf{s}}) = \iint\limits_{k_{\mathrm{m},x}^2 + k_{\mathrm{m},y}^2 \leq k_{\mathrm{m}}^2} \check{\mathbf{E}}(k_{\mathrm{m},x}, k_{\mathrm{m},y}; 0)\, e^{i k_{\mathrm{m}} s\, \hat{\mathbf{k}} \cdot \hat{\mathbf{s}}}\, dk_{\mathrm{m},x}\, dk_{\mathrm{m},y}, \qquad (4.12)$$

where $\mathbf{s} = s\hat{\mathbf{s}} = [s_x, s_y, s_z]$ is a point in the far field, $\check{\mathbf{E}}(k_{\mathrm{m},x}, k_{\mathrm{m},y}; 0)$ is the angular spectrum representation of the field in the plane $z = 0$ [Eq. (4.8)] and the integration is over the propagating plane waves because the evanescent waves, as a result of their fast decay, do not give any contribution in the far field.

We can now take the far-field plane to infinity, i.e., $s_z \to \infty$, and consider the electric field along the direction $\hat{\mathbf{s}}$, as shown in Fig. 4.5. Asymptotically, we obtain that the field

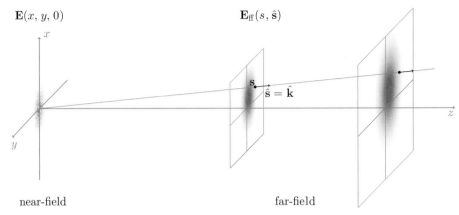

Figure 4.5 From near field to far field. The electromagnetic near field, $\mathbf{E}(x, y, 0)$ defined on the plane $z = 0$, and the far field, $\mathbf{E}_{\mathrm{ff}}(s, \hat{\mathbf{s}})$, are connected by Eq. (4.13), so that the angular component of $\mathbf{E}(x, y, 0)$ in the xy-plane with unit wavenumber $\hat{\mathbf{k}}$ corresponds to a propagating plane wave in the far field along the direction $\hat{\mathbf{s}} = \hat{\mathbf{k}}$.

represented by Eq. (4.12) is

$$\mathbf{E}_{\mathrm{ff}}(s, \hat{\mathbf{s}}) = -2\pi i k_{\mathrm{m}} \hat{s}_z \check{\mathbf{E}}(k_{\mathrm{m}}\hat{s}_x, k_{\mathrm{m}}\hat{s}_y; 0)\frac{e^{ik_{\mathrm{m}}s}}{s}, \qquad (4.13)$$

where $\hat{\mathbf{s}} = [\hat{s}_x, \hat{s}_y, \hat{s}_z] = \hat{\mathbf{k}}$, as only the plane waves in the direction $\hat{\mathbf{s}}$ can propagate along $\hat{\mathbf{s}}$, whereas all other plane waves are cancelled because of interference.[1] Eq. (4.13) has the typical form of the scattered field in the far zone with $\check{\mathbf{E}}$ playing the role of the scattering amplitude and, thus, links the angular spectrum representation of the near field with the corresponding scattering in the far field. In particular, the description of the far field can be associated with a single plane wave, the one that propagates along $\hat{\mathbf{k}}$, which can be identified with an optical ray, whose behaviour is well approximated by the laws of geometrical optics. This fact is what justifies the use of the geometrical optics approach. Eq. (4.13) can be recast as

$$\check{\mathbf{E}}(k_{\mathrm{m},x}, k_{\mathrm{m},y}; 0) = \frac{ise^{-ik_{\mathrm{m}}s}}{2\pi k_{\mathrm{m},z}} \mathbf{E}_{\mathrm{ff}}(k_{\mathrm{m},x}, k_{\mathrm{m},y}), \qquad (4.14)$$

so that the angular spectrum representation [Eq. (4.10)] becomes

$$\mathbf{E}(x, y, z) = \frac{ise^{-ik_{\mathrm{m}}s}}{2\pi} \iint_{k_{\mathrm{m},x}^2 + k_{\mathrm{m},y}^2 \leq k_{\mathrm{m}}^2} \mathbf{E}_{\mathrm{ff}}(k_{\mathrm{m},x}, k_{\mathrm{m},y}) e^{i(k_{\mathrm{m},x}x + k_{\mathrm{m},y}y \pm k_{\mathrm{m},z}z)} \frac{dk_{\mathrm{m},x}\, dk_{\mathrm{m},y}}{k_{\mathrm{m},z}}. \qquad (4.15)$$

---

[1] Mathematically, the asymptotic limit of the integral in Eq. (4.12) can be found through the *stationary phase method*, which also demonstrates the cancellation by interference of the waves not propagating along $\hat{\mathbf{s}}$. The details about this method and the derivation of Eq. (4.13) are provided in Section 3.3 of Mandel and Wolf (1995).

Eq. (4.15) connects the near field to the far field. In particular, it can be interpreted as a description of the near field in terms of optical rays arriving from the far field, or as a description of the optical rays scattered in the far field by a certain electromagnetic field in the near field.

Exercise 4.3.1 Calculate the diffraction of a plane wave illuminating an ideal circular aperture with radius R at normal incidence. [Hint: Use Eq. (4.10) to calculate the Fourier spectrum of the field at the aperture and then determine its far field using Eq. (4.13).]

4.4 Paraxial approximation

The electromagnetic fields of a laser beam propagate mostly in a certain direction, which we will typically assume to be along the z-axis, spreading out only slowly in the transverse direction. In such cases, it is possible to use the *paraxial approximation*, i.e.,

$$k_{m,z} = k_m \sqrt{1 - \frac{k_{m,x}^2 + k_{m,y}^2}{k_m^2}} \approx k_m - \frac{k_{m,x}^2 + k_{m,y}^2}{2k_m}, \qquad (4.16)$$

which significantly simplifies the Fourier integrals in Eqs. (4.10) and (4.11), even though it is not valid for strong focusing, as is required for optical tweezers.

The rest of this section will describe laser beams within the paraxial approximation. We will generally assume the beam waist to be in the plane $z = 0$ and the focal point at $(0, 0, 0)$ and we will often use a cylindrical reference system,

$$\begin{cases} \rho = \sqrt{x^2 + y^2} \\ \varphi = \operatorname{atan}\left(\frac{y}{x}\right) \end{cases}. \qquad (4.17)$$

4.4.1 Gaussian beams

A *Gaussian beam* is a beam of electromagnetic radiation whose transverse electric field and intensity distributions are well approximated by Gaussian functions. This is the most typical beam employed in optical manipulation, because most lasers emit beams that approximate a Gaussian profile.

At the beam waist, the electric field of a Gaussian beam is

$$\mathbf{E}^G(x, y, 0) = \mathbf{E}_0 e^{-\frac{x^2+y^2}{w_0^2}}, \qquad (4.18)$$

where $\mathbf{E}_0$ is a vector in the xy-plane specifying the amplitude, phase and polarisation of the beam, and w_0 is the waist radius. In the plane perpendicular to the propagation axis at the waist, the phase of the beam is constant and the intensity distribution is shown in Fig. 4.6a.

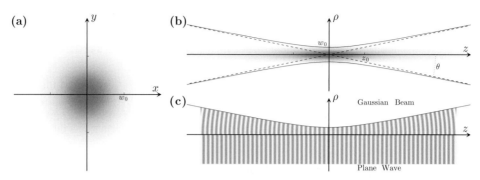

Figure 4.6 Gaussian beam. The intensity distribution of a Gaussian beam is (a) Gaussian in the transverse xy-plane and (b) cylindrically symmetric around the propagation z-axis. (c) As the beam propagates along the z-axis, its phase deviates (top) with respect to that of a reference plane wave (bottom), leading to a phase shift of exactly π as z goes from $-\infty$ to $+\infty$ (Gouy phase shift); the shades of grey represent the phase of the beam from 0 to 2π.

More generally, the paraxial representation of a Gaussian beam is

$$\mathbf{E}^G(\rho, z) = \mathbf{E}_0 \frac{w_0}{w(z)} e^{-\frac{\rho^2}{w(z)^2}} e^{+ik_m z - i\zeta(z) + ik_m \frac{\rho^2}{2R(z)}}, \qquad (4.19)$$

where the coordinate φ does not appear because of the cylindrical symmetry of the Gaussian beam, and we have introduced the *beam width*

$$w(z) = w_0 \sqrt{1 + \frac{z^2}{z_0^2}}, \qquad (4.20)$$

the *wavefront radius*

$$R(z) = z\left(1 + \frac{z_0^2}{z^2}\right), \qquad (4.21)$$

the *phase correction*

$$\zeta(z) = \operatorname{atan}\left(\frac{z}{z_0}\right) \qquad (4.22)$$

and the *Rayleigh range*, which denotes the distance from the beam waist at $z = 0$ to where the beam width has increased by a factor $\sqrt{2}$,

$$z_0 = \frac{k_m w_0^2}{2}. \qquad (4.23)$$

Some of the main characteristics of a Gaussian beam are shown in Fig. 4.6b. The transverse size of the beam is often assumed to be equal to the radial distance from the axis at which the amplitude of the electric field has decayed to $1/e$ the value at the beam centre; this corresponds to a decrease of the intensity by a factor of $1/e^2$. This surface is a hyperboloid with asymptotes $\theta = \frac{2}{k_m w_0}$. Close to the focus the beam is approximately collimated for a distance of about $2z_0$. Finally, along the z-axis the phase of the beam deviates from that of a plane wave so that, as z goes from $-\infty$ to $+\infty$, the beam gets exactly out of phase with a reference plane wave, as described by $\zeta(z)$ and illustrated in Fig. 4.6c. This phase shift is called the *Gouy phase shift* and happens gradually as the beam propagates through its focus; the tighter the focus the faster the phase variation is.

A paraxial Gaussian beam, and actually all the paraxial beams that we will consider, is assumed to have electric and magnetic fields transverse to the propagation direction, i.e., to be a transverse electromagnetic (TEM) beam. However, the only *bona fide* TEM solutions of Maxwell's equations in free space are plane waves.[2] Therefore, even paraxial beams must possess field components polarised in the direction of propagation. These longitudinal fields will typically be negligible unless a beam is tightly focused.

Exercise 4.4.1 Demonstrate that the Fourier spectrum of a Gaussian beam is $\check{\mathbf{E}}(k_{m,x}, k_{m,y}; 0) = \mathbf{E}_0 \frac{w_0}{4\pi} e^{-w_0^2(k_{m,x}^2 + k_{m,y}^2)/4}$, which is also a Gaussian.

Exercise 4.4.2 Derive the paraxial expression of a Gaussian beam in Eq. (4.19) by propagating the angular spectrum representation of the Gaussian beam.

Exercise 4.4.3 Show that for an x-polarised Gaussian beam the field component polarised along the direction of propagation on the waist plane ($z = 0$) is $E_z(x, y, 0) = -i \frac{2x}{k_m w_0^2} E_x(x, y, 0)$, and notice that it increases as the beam waist decreases. [Hint: Use the divergence condition $\nabla \cdot \mathbf{E} = 0$ on the fields of a Gaussian beam given in Eq. (4.18).]

4.4.2 Hermite–Gaussian beams

The *Hermite–Gaussian beams* are light beams whose field distribution can be described by some complex functions based on the Hermite polynomials $H_m(x)$:

$$H_m(x) = (-1)^m e^{x^2} \frac{d^m}{dx^m} e^{-x^2} = e^{x^2/2} \left(x - \frac{d}{dx}\right)^m e^{-x^2/2}, \qquad (4.24)$$

where m is a non-negative integer. In Cartesian coordinates, the analytical three-dimensional expression for these beams reads

$$\mathbf{E}^{\text{HG}}_{m_x m_y}(x, y, z) = \mathbf{E}^{\text{G}}(x, y, z) H_{m_x}\left(\sqrt{2}\frac{x}{w(z)}\right) H_{m_y}\left(\sqrt{2}\frac{y}{w(z)}\right) e^{-i(m_x + m_y)\zeta(z)}. \qquad (4.25)$$

The Hermite–Gaussian beams have a Cartesian symmetry. The indices m_x and m_y determine the shape of the beam profile in the x- and y-direction, respectively: the intensity distribution for these modes, in fact, has m_x nodes (intensity nulls) in the x-direction and m_y nodes in the y-direction, at each of which the phase undergoes a jump of π radians in the direction perpendicular to the node. A Gaussian beam is obtained when $m_x = m_y = 0$. Fig. 4.7 plots intensity and phase profiles of the first few Hermite–Gaussian modes.

---

[2] However, purely transverse electric (TE) and transverse magnetic (TM) modes are possible. They are the azimuthally and radially polarised cylindrical vector beams, respectively, discussed in Subsection 4.4.5.

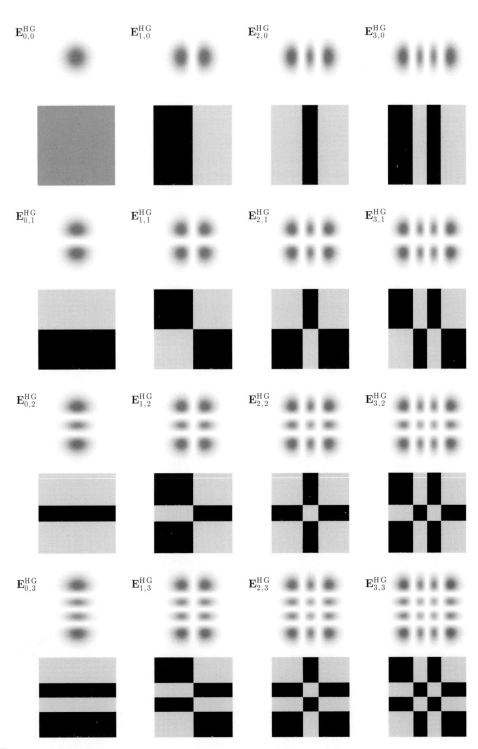

Figure 4.7 Hermite–Gaussian beams. Intensity (top) and phase (bottom) profiles of the first few Hermite–Gaussian beams.

All Hermite–Gaussian beams and their linear combinations satisfy the homogeneous Helmholtz equation [Eq. (4.2)] in the paraxial approximation. Furthermore, the set of Hermite–Gaussian beams forms a complete orthogonal basis of solutions; i.e., any two-dimensional complex light field can be obtained as a superposition of Hermite–Gaussian beams.

Exercise 4.4.4 Calculate the result of the superposition of a $\mathbf{E}_{01}^{HG}$ and $\mathbf{E}_{10}^{HG}$ beam when they are linearly polarised along x. What happens if the $\mathbf{E}_{01}^{HG}$ beam is polarised along x and the $\mathbf{E}_{10}^{HG}$ beam along y? What happens if they are in phase? And if they are in phase quadrature?

Exercise 4.4.5 Show analytically and numerically that, as demonstrated by Zauderer (1986), the complete set of Hermite–Gaussian beams can be generated from the fundamental Gaussian beam as

$$\mathbf{E}_{m_x m_y}^{HG}(x, y, z) = w_0^{m_x + m_y} \frac{\partial^{m_x}}{\partial x^{m_x}} \frac{\partial^{m_y}}{\partial y^{m_y}} \mathbf{E}_G(x, y, z). \tag{4.26}$$

4.4.3 Laguerre–Gaussian beams

The *Laguerre–Gaussian beams* are light beams whose field distribution can be described by some complex functions based on the associated Laguerre polynomials $L_p^l(x)$,

$$L_p^l(x) = \frac{x^{-l} e^x}{p!} \frac{d^p}{dx^p} \left(e^{-x} x^{p+l} \right) = \frac{x^{-l}}{p!} \left(\frac{d}{dx} - 1 \right)^p x^{p+l}, \tag{4.27}$$

where p is a non-negative integer and l is any integer. In cylindrical coordinates the analytical three-dimensional expression for these beams then reads

$$\mathbf{E}_{lp}^{LG}(\rho, \varphi, z) = \mathbf{E}^G(\rho, z) \left(\sqrt{2} \frac{\rho}{w(z)} \right)^l L_p^l \left(2 \frac{\rho^2}{w(z)^2} \right) e^{-i(2p+l)\zeta(z)} e^{il\varphi}. \tag{4.28}$$

The shape of a Laguerre–Gaussian beam is determined by the two indices p and l: p determines the number of nodes along the radial direction, whereas l determines the phase distribution around the azimuthal direction. A phase variation of $2\pi l$ on a closed trajectory winding (once) around the beam axis is due to the term $e^{il\varphi}$ and results in a helix structure of the wavefront; consequently, for $l \neq 0$ there is a phase singularity at the centre of the beam where the intensity must be null. Because of this phase singularity, Laguerre–Gaussian beams are also called *doughnut beams* or *optical vortices*. Fig. 4.8 plots intensity and phase profiles of the first 12 Laguerre–Gaussian modes; $\mathbf{E}_{-lp}^{LG}$ beams have the same intensity profile as $\mathbf{E}_{lp}^{LG}$ beams but a mirror-symmetric phase profile.

Like the Hermite–Gaussian beams, all Laguerre–Gaussian beams and their linear combinations satisfy the homogeneous Helmholtz equation [Eq. (4.2)] in the paraxial approximation and the set of Laguerre–Gaussian beams forms a complete orthogonal basis of solutions of the wave equation in the paraxial approximation.

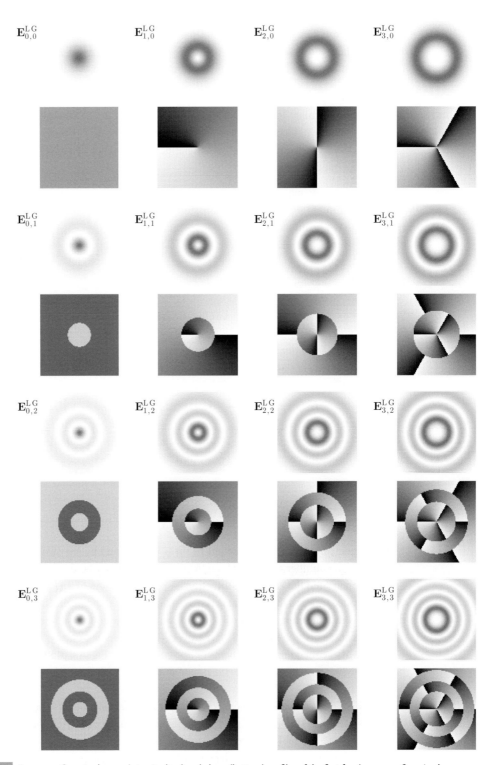

Figure 4.8 Laguerre–Gaussian beams. Intensity (top) and phase (bottom) profiles of the first few Laguerre–Gaussian beams.

Exercise 4.4.6 How is it possible to obtain an $\mathbf{E}_{23}^{\mathrm{LG}}$ using a superposition of Hermite–Gaussian beams?

Exercise 4.4.7 Show analytically and numerically that, as demonstrated by Zauderer (1986), the complete set of Laguerre–Gaussian beams can be generated from the fundamental Gaussian beam as

$$\mathbf{E}_{lp}^{\mathrm{LG}}(x, y, z) = k_{\mathrm{m}}^{p} w_{0}^{2p+l} e^{ikz} \frac{\partial^{p}}{\partial z^{p}} \left(\frac{\partial}{\partial x} + i \frac{\partial}{\partial y} \right)^{l} \left\{ \mathbf{E}_{\mathrm{G}}(x, y, z) e^{-ik_{\mathrm{m}} z} \right\}. \quad (4.29)$$

4.4.4 Non-diffracting beams

Non-diffracting beams are characterised by the fact that their transverse profile does not change as they propagate along the optical axis. In general, they are obtained by the interference of a set of plane waves whose wavevectors lie on a conical surface. All these plane waves share the same axial wavevector $k_{\mathrm{m},z}$ and all their transverse wavevectors $(k_{\mathrm{m},x}, k_{\mathrm{m},y})$ lie on the same circle in the transverse wavevector plane. The angle between these wavevectors and the propagation direction of the beam, i.e., $\xi = \mathrm{asin}(k_{\mathrm{m},\perp}/k_{\mathrm{m}})$, where $k_{\mathrm{m},\perp} = \sqrt{k_{\mathrm{m},x}^{2} + k_{\mathrm{m},y}^{2}}$, is therefore fixed. Their Fourier spectrum on this circle as a function of the azimuthal angle φ can be written as

$$\check{\mathbf{E}}^{\mathrm{ND}}(k_{\mathrm{m},\perp}, \varphi) = \mathbf{E}_{0}\, \delta(k_{\mathrm{m},\perp} - k_{\mathrm{m}} \sin \xi)\, A(\varphi). \quad (4.30)$$

This general spectrum leads to ideal non-diffracting beams whose transverse intensity profiles remain invariant for arbitrary propagation distances. However, these ideal non-diffracting beams carry infinite energy and, therefore, they cannot be realised experimentally. Nevertheless, it is possible to realise reasonably good approximations that are almost non-diffracting over a finite propagation distance. A rigorous theoretical description of these experimentally realisable beams requires convolution of the spectrum given by Eq. (4.30) with the spectrum of their transverse envelope, leading to the class of beams known as *Helmholtz–Gauss beams*.

We will now proceed to briefly give an overview of the four main categories of non-diffracting beams that have been identified, i.e., discrete, Bessel, Mathieu and Weber beams, as described in detail by Bouchal (2003) and by Rose et al. (2012).

Discrete beams are the simplest example of non-diffracting beams. These are the interference patterns resulting from the interference of N discrete beams. Their Fourier spectrum is a discrete function with non-zero values at the corners of a regular N-fold polygon, i.e.,

$$\check{\mathbf{E}}_{m,N}^{\mathrm{D}}(k_{\mathrm{m},\perp}, \varphi) = \mathbf{E}_{0}\, \delta(k_{\mathrm{m},\perp} - k_{\mathrm{m}} \sin \xi) \sum_{n=0}^{N-1} e^{im\varphi} \delta\left(\varphi - \frac{2\pi n}{N}\right), \quad (4.31)$$

where m is the topological charge determining the phase difference between adjacent points. The resulting pattern can be stripes ($N = 2$), periodic plane tilings, i.e., hexagonal

($N = 3$) [Fig. 4.9a], square ($N = 4$) and triangular patterns ($N = 6$), or quasicrystalline plane tilings, e.g., fivefold patterns ($N = 5$) [Fig. 4.9b]. Discrete beams have been employed to produce optical potentials that can lead to the arrangement of microscopic particles in crystalline and quasicrystalline patterns.

Bessel beams are characterised by the Fourier spectrum

$$\check{\mathbf{E}}_m^B(k_{m,\perp}, \varphi) = \mathbf{E}_0\, \delta(k_{m,\perp} - k_m \sin\xi)\, e^{im\varphi}, \tag{4.32}$$

where m represents the order of the beam. Fig. 4.9c shows a zero-order Bessel beam, which was the first non-diffracting beam to be theoretically predicted and experimentally demonstrated (Durnin et al., 1987). Bessel beams are characterised by cylindrical symmetry. These beams present a very narrow core, which is particularly useful in optical guiding applications, and they can reconstruct themselves after being disrupted by an obstacle, which is especially useful for the optical manipulation of multiple microscopic objects. However, because their energy is split almost equally into several rings, only a fraction is available in the central core, which sometimes can be a nuisance for practical applications. Good approximations to Bessel beams are often generated by placing a narrow annular aperture in the far field or by focusing a Gaussian beam with an axicon. Because the deviations of the pseudo-non-diffracting beams from the ideal non-diffracting beams are not significant at the range of the microscopic distances usually considered in optical manipulations, they are frequently called just Bessel beams.

Even and odd *Mathieu beams* have Fourier spectra given by

$$\check{\mathbf{E}}_{m,q}^{eM}(k_{m,\perp}, \varphi) = \mathbf{E}_0\, \delta(k_{m,\perp} - k_m \sin\xi)\, \mathrm{ce}_m(\varphi, q) \tag{4.33}$$

and

$$\check{\mathbf{E}}_{m,q}^{oM}(k_{m,\perp}, \varphi) = \mathbf{E}_0\, \delta(k_{m,\perp} - k_m \sin\xi)\, \mathrm{se}_m(\varphi, q), \tag{4.34}$$

respectively, where $\mathrm{ce}_m(\varphi, q)$ and $\mathrm{se}_m(\varphi, q)$ are the *angular Mathieu functions* with order m, and q is the ellipticity parameter.[3] As can be seen in Fig. 4.9d, Mathieu beams are characterised by an elliptical geometry. In particular, their phase reflects their underlying elliptical character, as can be seen from the fact that the lines separating regions of different phase form ellipses and hyperbolas with joint foci.

Even and odd *Weber beams* have Fourier spectra given by

$$\check{\mathbf{E}}_a^{eW}(k_{m,\perp}, \varphi) = \mathbf{E}_0\, \delta(k_{m,\perp} - k_m \sin\xi)\, \frac{1}{2\sqrt{\pi|\sin\varphi|}}\, e^{ia\ln|\tan(\frac{\varphi}{2})|} \tag{4.35}$$

and

$$\check{\mathbf{E}}_a^{oW}(k_{m,\perp}, \varphi) = \begin{cases} -i\check{\mathbf{E}}_a^{eW}(k_{m,\perp}, \varphi) & \varphi \in [0, \pi] \\ i\check{\mathbf{E}}_a^{eW}(k_{m,\perp}, \varphi) & \varphi \in]\pi, 2\pi] \end{cases}, \tag{4.36}$$

respectively, where a is a continuous parameter. Weber beams present a parabolic symmetry, as can be seen in Fig. 4.9e.

---

[3] A computational toolbox for the calculation of Mathieu functions was developed by Cojocaru (2008).

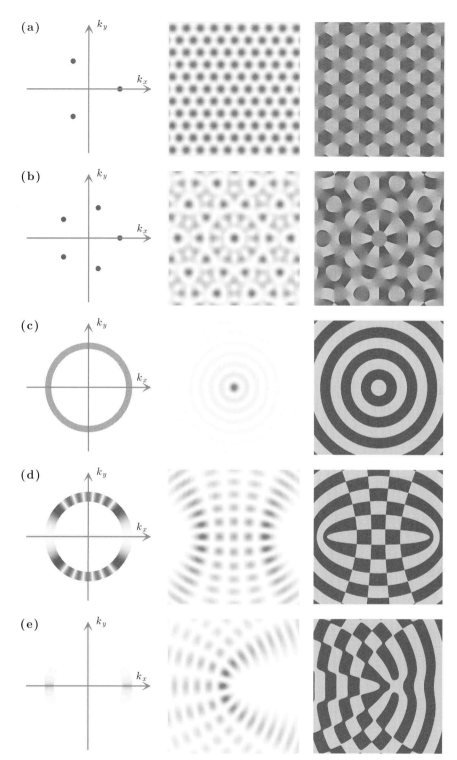

Figure 4.9 Examples of non-diffracting beams. (a) Periodic hexagonal and (b) quasiperiodic fivefold tilings produced by discrete beams with $m = 0$ [Eq. (4.31)]. (c) Bessel beam with $m = 0$ [Eq. (4.32)]. (d) Even Mathieu beam with $m = 4$ and $q = 6.25\pi^2$ [Eq. (4.33)]. (e) Even Weber beam with $a = -2$ [Eq. (4.35)]. From left to right, the columns represent the distribution of the wavevectors, the intensity and the phase.

Exercise 4.4.8 Study what happens if the phase or amplitude of the plane waves interfering to create a Bessel beam changes. How is it possible to obtain a minimum of the intensity at the centre of the beam? [Hint: You can adapt the program `bessel` from the book website.]

4.4.5 Cylindrical vector beams

The beams presented until now have been solutions of the scalar Helmholtz equation in the paraxial limit and, in particular, present a spatially homogeneous polarisation; e.g., they have linear, elliptical or circular polarisation. However, the vector Helmholtz equation in the paraxial limit also admits solutions whose polarisation changes over the beam area, as shown by Hall (1996). One of the simplest realisations of these vector beams is *cylindrical vector beams*, which obey cylindrical symmetry both in amplitude and polarisation. Cylindrical vector beams can be expressed as superpositions of orthogonally polarised Hermite–Gaussian beams, as shown in Fig. 4.10. A *radially polarised beam* can be obtained as

$$\mathbf{E}_r^{\text{CVB}}(x, y, z) = E_{10}^{\text{HG}}(x, y, z)\hat{\mathbf{x}} + E_{01}^{\text{HG}}(x, y, z)\hat{\mathbf{y}}, \tag{4.37}$$

and an *azimuthally polarised beam* as

$$\mathbf{E}_\varphi^{\text{CVB}}(x, y, z) = E_{01}^{\text{HG}}(x, y, z)\hat{\mathbf{x}} - E_{10}^{\text{HG}}(x, y, z)\hat{\mathbf{y}}. \tag{4.38}$$

A generalised form of cylindrical vector beam can be obtained by a linear superposition of radially and azimuthally polarised beams. An exhaustive review of this kind of beams can be found in Zhan (2009).

Exercise 4.4.9 How can radially and azimuthally polarised beams be obtained from the superposition of Laguerre–Gaussian beams?

Exercise 4.4.10 Generalise the concept of cylindrical vector beams to beams of higher orders.

4.5 Focusing

We will now turn our attention to the focusing of a paraxial optical beam by an aplanatic optical lens. We will follow the classical approach proposed by Richards and Wolf (1959) and by Wolf (1959).

We consider the basic focusing configuration shown in Fig. 4.11. An optical beam crosses an aperture stop, or iris, with radius R and then propagates towards the principal plane p_1 of the lens and is transferred to the principal plane p_2, which is a spherical surface with centre at the focal point **O** and with radius equal to the focal length f. The diffraction that occurs inside the objective is modelled by propagating the electromagnetic wave from the aperture

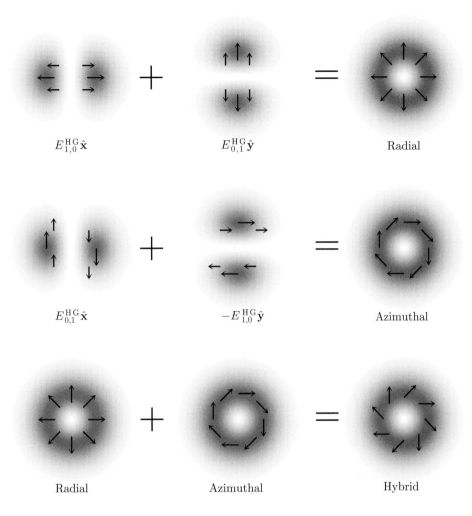

Figure 4.10 Cylindrical vector beams. A radial or azimuthal cylindrical vector beam can be generated as a linear superposition of orthogonally polarised Hermite–Gaussiam beams. A generalised cylindrical vector beam can be obtained as a superposition of a radial and an azimuthal vector beam. The arrows represent the local direction of the polarisation vector.

stop to the principal plane p_1, e.g., by using the angular spectrum representation given in Eq. (4.10). The aperture stop is often placed in the back focal plane, i.e., at a distance f from p_1, which results in a telecentric imaging system.

At p_2, the beam is refracted and focused towards **O**. Because we are considering an aplanatic lens, *Abbe's sine condition* applies, so that the deflection angle at position **R** is

$$\theta = \operatorname{asin}\left(\frac{\rho}{f}\right) = \operatorname{asin}\left(\frac{\rho}{R}\frac{\mathrm{NA}}{n_\mathrm{t}}\right), \tag{4.39}$$

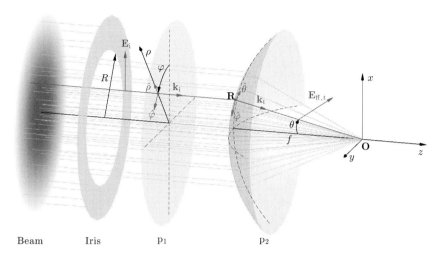

Figure 4.11 Focusing of an optical beam. The beam crosses an aperture stop, or iris, with radius R and then propagates towards the principal plane p_1 and is transferred to the principal plane p_2. At p_2, the beam is refracted and focused towards the focal point **O**. The point **R** is the intersection point of a ray with p_2 and illustrates the relation between the position ρ of the incident ray at p_1 and the propagation angle θ of the transmitted ray at p_2, i.e., Abbe's sine condition [Eq. (4.39)].

where ρ is the radial coordinate of the incident wave, NA is the numerical aperture of the objective [Eq. (2.22)] and n_t is the index of refraction in the region beyond p_2. Furthermore, because the energy of the electromagnetic wave must be conserved, we must also rescale the electric fields according to the *intensity law* of geometrical optics, which states that the energy flux along each ray must remain constant, as illustrated in Fig. 4.12. The power of a ray is given by the intensity [Eq. (4.6)] multiplied by the area, so that if the power transported by an incoming ray is $P_i = \frac{1}{2}\frac{c}{n_i}\varepsilon_i|\mathbf{E}_i|^2 dA_i$, the power of the corresponding transmitted ray is $P_t = \frac{1}{2}\frac{c}{n_t}\varepsilon_t|\mathbf{E}_{\mathrm{ff},t}|^2 dA_t$, where n_i is the refractive index before p_1, ε_i (ε_t) is the dielectric permittivity before p_1 (after p_2), $\mathbf{E}_i$ ($\mathbf{E}_{\mathrm{ff},t}$) is the amplitude of the incoming (transmitted) electric field and dA_i (dA_t) is the cross-section perpendicular to the incoming (transmitted) ray propagation direction. Equating P_i and P_t and taking into account that $dA_i = dA_t \cos\theta$,

$$|\mathbf{E}_{\mathrm{ff},t}| = |\mathbf{E}_i|\sqrt{\frac{n_i}{n_t}}\sqrt{\frac{\mu_t}{\mu_i}}\sqrt{\cos\theta} \approx |\mathbf{E}_i|\sqrt{\frac{n_i}{n_t}}\sqrt{\cos\theta}, \qquad (4.40)$$

where the approximation is usually valid because in the overwhelming majority of cases the magnetic permeability at optical frequencies is equal to that of vacuum, i.e., $\mu_i \approx \mu_t \approx \mu_0$.

The incident field $\mathbf{E}_i(\rho,\varphi)$ can be decomposed into a radial (p-polarised) and an azimuthal (s-polarised) component so that

$$\mathbf{E}_i(\rho,\varphi) = [\mathbf{E}_i(\rho,\varphi)\cdot\hat{\boldsymbol{\rho}}]\,\hat{\boldsymbol{\rho}} + [\mathbf{E}_i(\rho,\varphi)\cdot\hat{\boldsymbol{\varphi}}]\,\hat{\boldsymbol{\varphi}}, \qquad (4.41)$$

where φ is the azimuthal angle around the z-axis,

$$\hat{\boldsymbol{\rho}} = \cos\varphi\,\hat{\mathbf{x}} + \sin\varphi\,\hat{\mathbf{y}} \qquad (4.42)$$

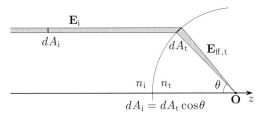

Figure 4.12 Intensity law of geometrical optics. The law states that the energy carried along a ray must remain constant. Therefore, when an electromagnetic field $\mathbf{E}_i$ is transmitted through a lens, its amplitude should be rescaled according to Eq. (4.40).

is the radial unit vector and

$$\hat{\boldsymbol{\varphi}} = -\sin\varphi\,\hat{\mathbf{x}} + \cos\varphi\,\hat{\mathbf{y}} \tag{4.43}$$

is the azimuthal unit vector. Upon refraction at p_2, the polarisation direction associated with the unit vector $\hat{\boldsymbol{\rho}}$ is deflected by θ and becomes aligned to the polar unit vector,[4]

$$\hat{\boldsymbol{\theta}} = \cos\theta\cos\varphi\,\hat{\mathbf{x}} + \cos\theta\sin\varphi\,\hat{\mathbf{y}} + \sin\theta\,\hat{\mathbf{z}}, \tag{4.44}$$

whereas the unit vector $\hat{\boldsymbol{\varphi}}$ is unaffected. Therefore, the transmitted field is

$$\mathbf{E}_{\mathrm{ff,t}}(\theta,\varphi) = t_{\mathrm{p}}(\theta)\left[\mathbf{E}_i(\rho,\varphi)\cdot\hat{\boldsymbol{\rho}}\right]\hat{\boldsymbol{\theta}} + t_{\mathrm{s}}(\theta)\left[\mathbf{E}_i(\rho,\varphi)\cdot\hat{\boldsymbol{\varphi}}\right]\hat{\boldsymbol{\varphi}}, \tag{4.45}$$

where $t_{\mathrm{p}}(\theta)$ and $t_{\mathrm{s}}(\theta)$ are the *transmission coefficients*, also known as *pupil functions* or *apodisations*, for p- and s-polarisation, respectively, which integrate accumulated phase distortions, e.g., aberrations at p_2, as well as attenuations, e.g., the amplitude factor given by Eq. (4.40).

Because p_2 is in the far field of the focal region, the transmitted field $\mathbf{E}_t(\theta,\varphi)$ is the plane wave angular spectrum representation of the focused field $\mathbf{E}_f(x,y,z)$ near $\mathbf{O}$, as explained in Sections 4.2 and 4.3.[5] Therefore, $\mathbf{E}_f(x,y,z)$ can be obtained by integrating the propagated plane waves [Eq. (4.15)], i.e.,

$$\mathbf{E}_f(x,y,z) = \frac{ik_t f e^{-ik_t f}}{2\pi}\int_0^{\theta_{\max}}\sin\theta\int_0^{2\pi}\mathbf{E}_{\mathrm{ff,t}}(\theta,\varphi)e^{i[k_{t,x}x+k_{t,y}y]}e^{ik_{t,z}z}\,d\varphi\,d\theta, \tag{4.46}$$

which is the integral representation of the focused field introduced by Debye (1909). The phase factor $e^{ik_{t,z}z}$ accounts for the phase accumulated by the propagation along the z-axis

---

[4] Note that θ is the angle defining the incoming ray direction and is different from the polar angle in spherical coordinates, which is often employed in the derivation of the focused fields.

[5] We note that, strictly speaking, the angular spectrum representation should be given as a function of $k_{t,x}$ and $k_{t,y}$, which are related to ϑ and φ per Eq. (4.48).

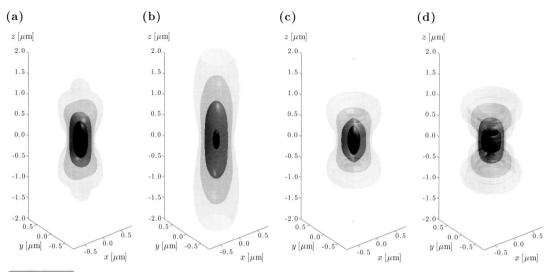

Figure 4.13 Focal fields. Intensity distribution at the focus of a 1.20 NA water-immersion objective with aperture radius $R = 4$ mm for an x-polarised Gaussian beam with beam waist (a) $w_0 = 4$ mm and (b) $w_0 = 2$ mm, and for (c) a radially polarised and (d) an azimuthally polarised cylindrical vector beam. All beams have wavelength $\lambda_0 = 632$ nm and the full laser power after the aperture is 10 mW. The iso-intensity surfaces correspond to $I(x, y, z) = 50, 20, 10, 5 \, \text{GW/m}^2$.

and $e^{i[k_{t,x}x + k_{t,y}y]}$ represents the phase difference of an off-axis point (x, y, z) with respect to the corresponding on-axis point $(0, 0, z)$. The maximum value of θ depends on NA as

$$\theta_{\max} = \operatorname{asin}\left(\frac{\text{NA}}{n_t}\right). \tag{4.47}$$

The wavevector is

$$\mathbf{k}_t(\theta, \varphi) = -k_t \sin\theta \cos\varphi \, \hat{\mathbf{x}} - k_t \sin\theta \sin\varphi \, \hat{\mathbf{y}} + k_t \cos\theta \, \hat{\mathbf{z}}, \tag{4.48}$$

where $k_t = n_t 2\pi/\lambda_0$ is the wavenumber in the medium and λ_0 is the vacuum wavelength of the light.

As we will see in the following sections, the shape of the focal field has a major influence on the optical forces exerted on a particle. Fig. 4.13 illustrates the focal fields of various beams focused by a 1.20 NA water-immersion objective with aperture radius $R = 4$ mm. By using a beam that overfills the objective aperture, one obtains a tighter focus, as can be seen from comparing Figs. 4.13a and 4.13b, which show the focal fields corresponding to a linearly x-polarised Gaussian beam with beam waist $w_0 = 4$ mm and $w_0 = 2$ mm, respectively. It can also be noticed that, when the aperture is overfilled, the resulting focal field is elongated in the polarisation direction. The shape of the focal fields can be further engineered by using non-Gaussian beams. For example, it is possible to obtain a tighter focus by employing radially polarised cylindrical vector beams, as shown in Fig. 4.13c, and it is possible to obtain a null of the intensity along the optical axis by employing azimuthally polarised cylindrical vector beams, as shown in Fig. 4.13d.

Exercise 4.5.1 Numerically calculate the focal fields given by Eq. (4.46). Show that the focal field gets tighter as the size of the optical beam increases and the objective aperture becomes overfilled. [Hint: You can adapt the code `focus` from the book website.]

Exercise 4.5.2 Calculate the focal fields corresponding to some Hermite–Gaussian, Laguerre–Gaussian and cylindrical vector beams. Investigate how the shape of the focal field can be changed by using various linear combinations of these beams. How can you get a dark spot at the focal point? [Hint: You can adapt the code `focus` from the book website.]

Exercise 4.5.3 Derive the analytical expression for a focused magnetic induction field $\mathbf{B}_f(x, y, z)$ and calculate it numerically. Verify that it is related to the focal electric field $\mathbf{E}_f(x, y, z)$ by Maxwell's equation for the curl of the electric field [Box 3.2].

Exercise 4.5.4 Show that the angular spectrum representation at a position $\mathbf{P}$ is

$$\mathbf{E}_{\mathrm{ff},\mathrm{t}}^{(\mathbf{P})}(\theta, \varphi) = \mathbf{E}_{\mathrm{ff},\mathrm{t}}(\theta, \varphi) \, e^{i \mathbf{k}_{\mathrm{t}} \cdot (\mathbf{P} - \mathbf{O})}, \tag{4.49}$$

where $\mathbf{E}_{\mathrm{ff},\mathrm{t}}(\theta, \varphi)$ is the angular spectrum representation of the incoming field at $\mathbf{O}$ given by Eq. (4.45).

Exercise 4.5.5 Show that, if the focal point is at position $\mathbf{Q}$ instead of $\mathbf{O}$, the expression of the focal fields can still be calculated using Eq. (4.46) as long as the angular spectrum representation of the incoming field is adjusted by a phase factor such that

$$\mathbf{E}_{\mathrm{ff},\mathrm{t}}(\theta, \varphi) \to \mathbf{E}_{\mathrm{ff},\mathrm{t}}(\theta, \varphi) \, e^{-i \mathbf{k}_{\mathrm{t}} \cdot (\mathbf{Q} - \mathbf{O})}. \tag{4.50}$$

[Hint: Build on the result obtained in the previous exercise.]

4.6 Optical forces near focus

For particles whose features are much larger than the wavelength of the trapping light, the calculation of the optical forces generated by a focused optical beam can be performed using a ray optics model. This approach is thoroughly explained in Chapter 2 and, in particular, in Section 2.5. As we have seen, the limitations of the geometrical optics approach which neglects the effects of phase and polarisation emerge even for large particles when these effects are considered. In the opposite limit, i.e., for dipolar particles, whose dimensions are much smaller than the wavelength of light, so that the electromagnetic field can be considered approximately uniform over the extent of the particle, one can follow the approach presented in Chapter 3 and, in particular, in Section 3.7, where it was shown that the dominant optical force is the gradient force. This approach works reasonably well for particles up to a size of about a hundred nanometres.[6] In this section, we will focus on this limit. Before proceeding further, we notice that both these approaches are subsumed by the exact treatment of electromagnetic forces based on Maxwell's equations, which is the subject of Chapters 5 and 6.

Figure 4.14 illustrates the optical forces arising on a dipolar particle ($n_p = 1.50$) in water ($n_m = 1.33$) near the focus of a beam. We consider paraxial beams with wavelength $\lambda_0 = 632$ nm focused by a 1.20 NA water-immersion objective with aperture radius 4 mm, whose focal fields have already been shown in Fig. 4.13. In Fig. 4.14a, we consider a Gaussian beam with beam waist $w_0 = 4$ mm so that the objective aperture is overfilled: the tight focus [Fig. 4.13a] generates strong gradient optical forces, which are dominant over scattering and spin–curl forces by several orders of magnitude. In Fig. 4.14b, we relax the overfilling condition by reducing the Gaussian beam waist to $w_0 = 2$ mm; this is sufficient to increase the focal spot size greatly [Fig. 4.13b] and, therefore, to reduce significantly the gradient optical forces acting on the dipole particle. A stronger confinement of the dipole particle can be achieved by employing cylindrical vector beams. In particular, for a radially polarised cylindrical vector beam, tighter confinement of the focus [Fig. 4.13c] translates to stronger confinement of the particle [Fig. 4.14c]. Using an azimuthally polarised cylindrical vector beam permits one to have a minimum of intensity along the optical axis [Fig. 4.13d], so that a dipole particle with positive polarisability can be trapped within a doughnut-shaped volume where the intensity of the focused beam is maximum [Fig. 4.14d]; furthermore, it is possible to use this kind of beams to trap dipole particles with negative polarisability, e.g., particles whose refractive index is lower than that of the surrounding medium, at the minimum of intensity of the focus.

Exercise 4.6.1 Study the optical forces on a dipolar particle arising near the focus of a paraxial beam as a function of its polarisability, of the trapping beam and of the numerical aperture of the objective. Explore, in particular, what happens with Hermite–Gaussian, Laguerre–Gaussian and cylindrical vector beams. [Hint: You can adapt the code `dipoleforces` from the book website.]

Exercise 4.6.2 Calculate the trap stiffness along the x-, y- and z-directions for the cases considered in the previous exercise.

Exercise 4.6.3 Study the optical forces arising in the presence of a non-diffracting beam.

Exercise 4.6.4 A focused azimuthally polarised cylindrical vector beam permits one to trap a low-refractive-index particle at the minimum of the focal intensity in the transversal plane. However, there is no such confinement along the longitudinal axis. What kind of beam should one employ to trap a low refractive index particle in three dimensions? [Hint: You can adapt the code `dipoleforces` from the book website to verify your answers.]

---

[6] To be more precise, as we have seen in Section 1.3, if one defines the size parameter $x = 2\pi a/\lambda_0$, where a is the particle radius, the limits of applicability of the dipole approximation require both $x \ll 1$ and $|m|x \ll 1$, where m is the ratio between the refractive index of the particle and that of the medium.

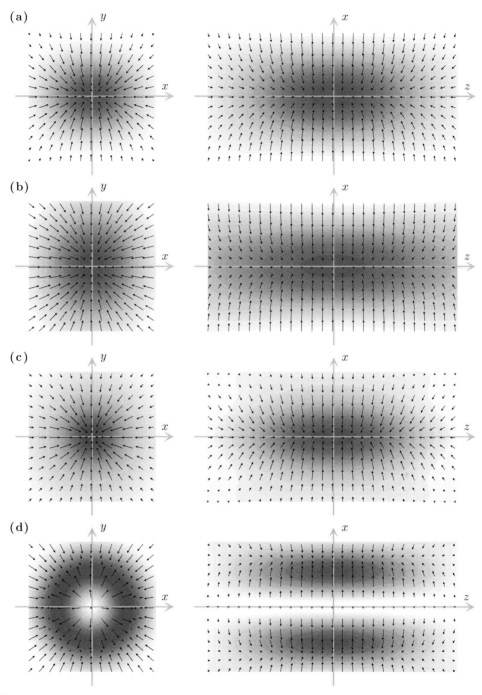

Figure 4.14 Optical forces on a dipole. Forces produced on a dipole particle ($n_p = 1.50$) in water ($n_m = 1.33$) near the focus of an x-polarised Gaussian beam with beam waist (a) $w_0 = 4$ mm and (b) $w_0 = 2$ mm, and for (c) a radially polarised and (d) an azimuthally polarised cylindrical vector beam. All beams have wavelength $\lambda_0 = 632$ nm, have 10 mW power after the aperture and are focused by a 1.20 NA water-immersion objective with aperture radius of 4 mm. The corresponding focal fields are illustrated in Fig. 4.13.

4.7 Focusing near interfaces

In optical trapping applications, optical beams are often focused near planar interfaces. For example, this happens when an oil-immersion objective is used to generate an optical tweezers, as typically the trapping happens in water and, therefore, there is a glass–water interface, as shown in Fig. 4.15a, or when a total internal reflection objective is used in order to excite evanescent waves, as shown in Fig. 4.15b. To analyse these configurations, we will assume that the focal point is placed at the origin **O** of the coordinate system and the interface is placed in a plane $z = z_i$ between two media with refractive indices n_t and n_s, as shown in Fig. 4.15. To calculate the focal field, we need the angular spectrum representation of the field around the focus. This can be obtained by propagating the angular spectrum representation of the optical field $\mathbf{E}_{\text{ff},t}(\theta, \varphi)$ [Eq. (4.45)] up to the interface and then by transmitting it at the interface. The transmission of a ray at the interface entails a change of the amplitude of its electric fields according to Fresnel's coefficients [Box 4.1] and a change of its wavevector according to Snell's law [Eq. (2.3)]. A plane wave approaching the focus at an angle $\theta_t = \theta$ is deflected at the interface to an angle θ_s, which is given by Snell's law, i.e.,

$$n_s \sin \theta_s = n_t \sin \theta_t,$$

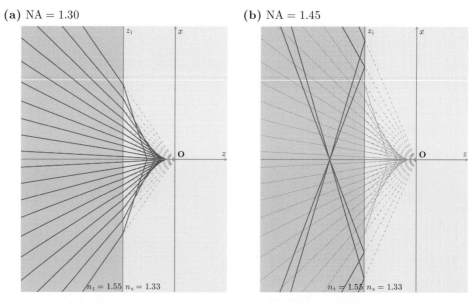

Figure 4.15 Focusing near an interface. Schematic representation of the focusing near a planar interface at $z = z_i$ between glass ($n_t = 1.55$) and water ($n_t = 1.33$) using (a) a 1.30 NA oil-immersion objective and (b) a 1.45 NA total internal reflection objective. The dashed lines represent the light ray directions without interface. The presence of the interface produces spherical aberrations in (a) and evanescent waves in (b).

and acquires an additional phase

$$e^{i(k_{t,z} - k_{s,z})z_i},$$

where $k_{t,z} = k_t \cos\theta_t$, $k_{s,z} = k_s \cos\theta_s$ and k_s is the wavenumber in the sample medium. Finally, the wavevector in the sample medium becomes

$$\mathbf{k}_s(\theta_s, \varphi) = -k_s \sin\theta_s \cos\varphi\,\hat{\mathbf{x}} - k_s \sin\theta_s \sin\varphi\,\hat{\mathbf{y}} + k_s \cos\theta_s\,\hat{\mathbf{z}}, \quad (4.51)$$

which can be expressed more conveniently as a function of θ_t,

$$\mathbf{k}_s(\theta_t, \varphi) = -k_t \sin\theta_t \cos\varphi\,\hat{\mathbf{x}} - k_t \sin\theta_t \cos\varphi\,\hat{\mathbf{y}} + k_t\sqrt{n_s^2/n_t^2 - \sin^2\theta_t}\,\hat{\mathbf{z}}. \quad (4.52)$$

The unit vector for the s-polarisation $\hat{\boldsymbol{\varphi}}$ remains unchanged, whereas that for the p-polarisation $\hat{\boldsymbol{\theta}}$ becomes

$$\hat{\mathbf{s}} = \text{Re}\{\cos\theta_s\}\cos\varphi\,\hat{\mathbf{x}} + \text{Re}\{\cos\theta_s\}\sin\varphi\,\hat{\mathbf{y}} + \min(1, \text{Re}\{\sin\theta_s\})\,\hat{\mathbf{z}}. \quad (4.53)$$

Therefore, the angular spectrum representation in the sample becomes

$$\mathbf{E}_{\text{ff},s}(\theta_t, \varphi) = t_p(\theta_t)[\mathbf{E}_i(r, \varphi) \cdot \hat{\mathbf{r}}]\hat{\mathbf{s}} + t_s(\theta_t)[\mathbf{E}_i(r, \varphi) \cdot \hat{\boldsymbol{\varphi}}]\hat{\boldsymbol{\varphi}}, \quad (4.54)$$

where the apodisation factors now also account for the transmission coefficients and the phase delay. Finally, we get a new formula for the focal fields in the sample,

$$\mathbf{E}_{f,s}(x,y,z) = \frac{ik_t f e^{ik_t f}}{2\pi} \int_0^{\theta_{\max}} \sin\theta_t \int_0^{2\pi} \mathbf{E}_{\text{ff},s}(\theta_t, \varphi) e^{-i[k_{s,x}x + k_{s,y}y]} e^{ik_{s,z}z}\, d\varphi\, d\theta_t. \quad (4.55)$$

Box 4.1 — **Fresnel's coefficients (electric fields)**

Fresnel's coefficients for electric fields give the ratios between the amplitudes of the electric fields reflected and transmitted at an interface and of the incident electric field. The reflection and transmission coefficients for s-polarised and p-polarised light are, respectively,

$$r_s(\theta_t, n_t, n_s) = \frac{n_t \cos\theta_t - n_s \cos\theta_s}{n_t \cos\theta_t + n_s \cos\theta_s},$$

$$t_s(\theta_t, n_t, n_s) = \frac{2n_t \cos\theta_t}{n_t \cos\theta_t + n_s \cos\theta_s},$$

$$r_p(\theta_t, n_t, n_s) = \frac{n_s \cos\theta_t - n_t \cos\theta_s}{n_s \cos\theta_t + n_t \cos\theta_s},$$

$$r_p(\theta_t, n_t, n_s) = \frac{2n_t \cos\theta_t}{n_s \cos\theta_t + n_t \cos\theta_s},$$

where n_t is the refractive index of the medium in which the incident electric field propagates, n_s is that of the medium in which the transmitted ray propagates, θ_t is the incidence angle and θ_s can be calculated using Snell's law [Eq. (2.3)]. Note that whereas $t_s = 1 + r_s$, $t_p \neq 1 + r_p$.

Fresnel's coefficients for the electric field are related, but not equal, to Fresnel's coefficients for the reflected and transmitted power, i.e., R_s, T_s, R_p and T_p, given in Eqs. (2.5), (2.6), (2.7) and (2.8), respectively.

Exercise 4.7.1 Calculate the focal fields in the medium from which the beam propagates. [Hint: Start by calculating the angular spectrum representation of the reflected field.]

Exercise 4.7.2 Generalise Eq. (4.55) to the case of multiple interfaces.

Exercise 4.7.3 Generalise Eq. (4.55) to the case of non-planar interfaces.

4.7.1 Aberrations

When the objective NA is lower than the sample refractive index, i.e., NA $< n_s$, all incoming rays are transmitted at the interface, as shown in Fig. 4.15a. We will consider the case $n_s < n_t$, which occurs, e.g., at a glass–water interface; nonetheless, similar considerations can be extended to the opposite case, i.e., $n_s > n_t$, which occurs, e.g., at a air–water interface. The resulting focal field is similar to the one obtained in the absence of interfaces [Fig. 4.13] but for the presence of spherical aberrations, which reduce the quality of the focus, as shown in Fig. 4.16, and, therefore, of the trapping. When the focus is near the surface, e.g., $z_i = -5\,\mu\text{m}$ [Fig. 4.16a], the aberrations are small and the main effect of the interface is to move the position of the maximum of intensity of the focal field towards lower z with respect to the focus without an interface. As the distance from the focus to the interface increases, e.g., $z_i = -10\,\mu\text{m}$ [Fig. 4.16b] and $z_i = -50\,\mu\text{m}$ [Fig. 4.16c], the quality of the focus deteriorates due to the presence of increasingly strong spherical aberrations and the position of the focus becomes further removed from the position it would have along z in the absence of the interface.

Exercise 4.7.4 Calculate numerically the focal fields corresponding to an optical trap and the relative trap stiffnesses in the presence of an interface as a function of the interface position z_i. Observe the deterioration of the trap stiffnesses due to spherical aberrations as a function of z_i. [Hint: You can adapt the codes `focusinterface` and `dipolestiffness` from the book website.]

Exercise 4.7.5 How is it possible to compensate for the decrease in quality of the focus, and therefore the corresponding decrease in trap qualify, due to the presence of spherical aberrations introduced by the presence of a surface? Verify your conclusions numerically. [Hint: You can adapt the code `focusinterface` and `dipolestiffness` from the book website.]

4.7.2 Evanescent focusing

When a plane wave reaches the interface with an angle larger than the critical angle θ_c [Eq. (2.12)], it is totally internally reflected and gives rise to an evanescent wave on the sample side of the interface, as shown in Fig. 4.15b. In Fig. 4.17a, we calculate the focus fields generated by an objective designed for total internal reflection fluorescence overfilled by an x-polarised Gaussian beam. The objective uses immersion oil with an index of refraction matching that of the coverslip. Its NA of 1.45 is higher than the index of

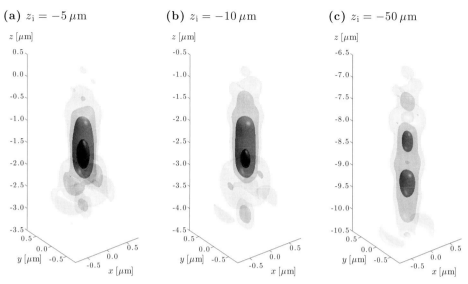

Figure 4.16 Focal fields in the presence of spherical aberrations at an interface. Intensity distribution at the focus of a 1.30 NA oil-immersion objective with aperture radius of 4 mm for a x-polarised Gaussian beam with beam waist $w_0 = 4$ mm in the presence of a glass ($n_t = 1.55$)–water ($n_s = 1.33$) interface placed at $z_i = -5$ μm, -10 μm and -50 μm for (a), (b) and (c), respectively. As (the absolute value of) z_i increases, the quality of the focus deteriorates due to the presence of increasingly strong spherical aberrations. The beam has wavelength $\lambda_0 = 632$ nm and the full laser power after the aperture is 10 mW. The iso-intensity surfaces correspond to $I(x, y, z) = 50, 20, 10, 5$ GW/m².

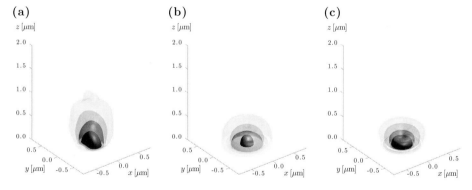

Figure 4.17 Evanescent focus. Intensity distribution at the focus of a 1.45 NA total internal reflection objective with aperture radius of 4 mm for (a) an x-polarised Gaussian beam with beam waist $w_0 = 4$ mm and for (b) a radially polarised and (c) an azimuthally polarised cylindrical vector beam in the presence of a glass ($n_t = 1.55$)–water ($n_s = 1.33$) interface placed at $z_i = 0$ μm. All beams have wavelength $\lambda_0 = 632$ nm and the full laser power after the aperture is 10 mW. The iso-intensity surfaces correspond to $I(x, y, z) = 50, 20, 10, 5$ GW/m².

refraction of the sample ($n_s = 1.33$, aqueous solution). This generates a partly evanescent focus field at the coverslip–sample interface. Depending upon the illumination of the aperture, the focus field can be fully propagating or fully evanescent. In Figs. 4.17b and 4.17c, we calculate the focal field from cylindrical vector beams with radial and azimuthal polarisation, respectively. These beams are very interesting for trapping at surfaces and, in particular, for plasmon-enhanced trapping. They result in rotationally symmetric focal fields and, e.g., the radially polarised beam permits one to obtain a better confinement of the focal spot at the interface. The fine structure of the electric field at the interface is due to the evanescent wave contribution with incidence angles above the critical angle.

Exercise 4.7.6 Calculate the optical forces exerted by an evanescent focal field on a dipole particle for various optical beams.

Problems

4.1 *Polarisation rotator.* Show that the combination of two half-wave plates in series can be used to rotate the polarisation of a beam by any angle. Apply this concept to generate cylindrical vector beams with any assigned polarisation.

4.2 *Airy beams.* Airy beams are a special class of non-diffracting beams that freely accelerate by bending in a parabolic arc as they propagate (Berry and Balazs, 1979). Study the optical forces arising from such beams (Baumgartl et al., 2008).

4.3 *Ince–Gaussian beams.* A third complete family of exact and orthogonal solutions of the paraxial wave equation is constituted by the Ince–Gaussian beams, which were introduced by Bandres and Gutiérrez-Vega (2004). Their transverse structure is described by the Ince polynomials and has an inherent elliptical symmetry. Show that Ince–Gaussian beams constitute the exact and continuous transition modes between Laguerre–Gaussian and Hermite–Gaussian beams and study the optical forces they produce on a Rayleigh particle when focused.

4.4 *Talbot effect.* When a plane wave is incident upon a periodic diffraction grating, the image of the grating is repeated at regular intervals away from the grating plane. This length of these intervals is called the *Talbot length* and the repeated images are called self-images or *Talbot images*. Study the optical forces arising because of the Talbot effect.

4.5 *Focus optimisation.* A superposition of elements belonging to a complete family of orthogonal solutions of the paraxial wave equation can be used to describe any kind of paraxial beam. Using this fact, develop an algorithm to design a laser beam that optimises some desired properties of the focus. What are the limitations?

4.6 *Holographic phase masks.* Study how it is possible to obtain a desired intensity distribution in the focal plane by altering the phase of an input beam in the back focal plane of an objective.

4.7 *Fast focal field calculation.* The numerical calculation of the focal field using Eq. (4.46) by direct integration is straightforward, but not extremely efficient. A computationally improved algorithm hinges on the similarity of Eq. (4.46) to a Fourier transform. In this way, it is possible to make use of standard fast Fourier transform (FFT) routines, leading to an improvement of several orders of magnitude in the computational efficiency. This approach was proposed by Leutenegger et al. (2006). Implement such an approach numerically.

4.8 *Colloidal crystals.* Study the optical forces arising on a Rayleigh particle in the presence of an interference pattern formed by the interference of two, three, four and five beams. What is the density of particles required to form a colloidal crystal?

References

Bandres, M. A., and Gutiérrez-Vega, J. C. 2004. Ince Gaussian beams. *Opt. Lett.*, **29**, 144–6.

Baumgartl, J., Mazilu, M., and Dholakia, K. 2008. Optically mediated particle clearing using Airy wavepackets. *Nature Photon.*, **2**, 675–8.

Berry, M. V., and Balazs, N. L. 1979. Nonspreading wave packets. *Am. J. Phys.*, **47**, 264–7.

Bouchal, Z. 2003. Nondiffracting optical beams: Physical properties, experiments and applications. *Czech. J. Phys.*, **53**, 537–624.

Cojocaru, E. 2008. Mathieu functions computational toolbox implemented in Matlab. *arXiv*, 0811.1970.

Debye, P. 1909. Das Verhalten von Lichtwellen in der Nähe eines Brennpunktes oder einer Brennlinie. *Ann. Physik*, **30**, 755–76.

Durnin, J., Miceli, Jr., J. J., and Eberly, J. H. 1987. Diffraction-free beams. *Phys. Rev. Lett.*, **58**, 1499–1501.

Hall, D. G. 1996. Vector-beam solutions of Maxwell's wave equation. *Opt. Lett.*, **21**, 9–11.

Leutenegger, M., Rao, R., Leitgeb, R. A., and Lasser, T. 2006. Fast focus field calculations. *Opt. Express*, **14**, 11 277–91.

Mandel, L., and Wolf, E. 1995. *Optical coherence and quantum optics*. Cambridge, UK: Cambridge University Press.

Richards, B., and Wolf, E. 1959. Electromagnetic diffraction in optical systems. II. Structure of the image field in an aplanatic system. *Proc. R. Soc. London A*, **253**, 358–79.

Rose, P., Boguslawski, M., and Denz, C. 2012. Nonlinear lattice structures based on families of complex nondiffracting beams. *New J. Phys.*, **14**, 033018.

Wolf, E. 1959. Electromagnetic diffraction in optical systems. I. An integral representation of the image field. *Proc. R. Soc. London A*, **253**, 349–57.

Zauderer, E. 1986. Complex argument Hermite–Gaussian and Laguerre–Gaussian beams. *J. Opt. Soc. Am. A*, **3**, 465–9.

Zhan, Q. 2009. Cylindrical vector beams: From mathematical concepts to applications. *Adv. Opt. Photon.*, **1**, 1–57.

5 Electromagnetic theory

So far we have described optical forces and torques in two extreme regimes, where their expression is greatly simplified: particles either much larger (*ray optics regime*, Chapter 2) or much smaller (*Rayleigh regime*, Chapter 3) than the wavelength of the trapping light. It is now time to turn our attention to the *intermediate regime*, where, as is often the case in experiments, the particle dimensions are comparable to the wavelength. Here, the previously used approximations break down and a complete wave-optical modelling of the particle–light interaction based on electromagnetic scattering theory is necessary to obtain accurate results, as shown in Fig. 5.1. In this chapter, we set out the general equations describing optical forces and torques within the framework of electromagnetic scattering theory. We start by discussing the general *scattering problem* and the *Maxwell stress tensor*, which describes the mechanical interaction of light with matter. Then we derive the optical force and torque exerted by a plane wave in terms of the Maxwell stress tensor. Finally, we discuss optical trapping, where the incident light is focused through an objective lens, by expanding the optical fields using the angular spectrum representation described in Chapter 4. We will leave the detailed discussion of the numerical methods practically used to compute optical forces for complex particles to Chapter 6.

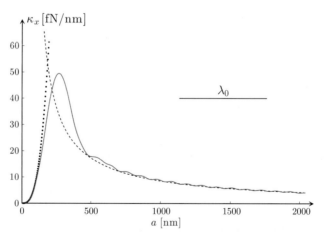

Figure 5.1 Comparison of optical forces calculated in various trapping regimes. Transverse trap stiffness produced by a 10 mW laser beam of wavelength $\lambda_0 = 632$ nm focused by a 1.20 NA objective on a dielectric sphere of radius a ($n_p = 1.50$) in water ($n_m = 1.33$). The solid line represents the exact electromagnetic calculation. The dipole approximation (dotted line) works for small spheres ($a \ll \lambda_0$). The geometrical optics approximation (dashed line) works for large spheres ($a \gg \lambda_0$).

5.1 Conservation laws and the Maxwell stress tensor

Particles illuminated by a radiation field experience a radiation force $\mathbf{F}_{rad}$ and torque $\mathbf{T}_{rad}$, which contribute to determining their dynamical behaviour. Because the interaction between radiation and matter is regulated by conservation laws, it is possible to derive $\mathbf{F}_{rad}$ and $\mathbf{T}_{rad}$ using the conservation of linear and angular momentum.

In relation to an electromagnetic field, a particle is a distribution of charges with density $\varrho = \varrho(\mathbf{r}, t)$ and currents $\boldsymbol{j} = \boldsymbol{j}(\mathbf{r}, t)$ contained in a volume V bounded by a surface S. Therefore, the force produced by the electromagnetic field on the particle is equal to the force it produces on ϱ and $\boldsymbol{j}$, which is a straightforward generalisation of the Lorentz force:

$$\frac{d}{dt}\mathbf{P}_{mech} = \int_V [\varrho \boldsymbol{\mathcal{E}} + \boldsymbol{j} \times \boldsymbol{\mathcal{B}}] \, dV, \tag{5.1}$$

where $\mathbf{P}_{mech} = \mathbf{P}_{mech}(t)$ is the mechanical momentum of the particle, $\boldsymbol{\mathcal{E}} = \boldsymbol{\mathcal{E}}(\mathbf{r}, t)$ is the electric field and $\boldsymbol{\mathcal{B}} = \boldsymbol{\mathcal{B}}(\mathbf{r}, t)$ is the magnetic induction field. Assuming the particle to be in vacuum, we can now eliminate ϱ and $\boldsymbol{j}$ in Eq. (5.1) using Maxwell's equations [Box 3.2] and the constitutive relations in vacuum [Box 3.3],

$$\varrho = \varepsilon_0 \left(\nabla \cdot \boldsymbol{\mathcal{E}} \right),$$

$$\boldsymbol{j} = \frac{1}{\mu_0} \nabla \times \boldsymbol{\mathcal{B}} - \varepsilon_0 \frac{\partial \boldsymbol{\mathcal{E}}}{\partial t},$$

where ε_0 and μ_0 are the dielectric permittivity and the magnetic permeability of vacuum, so that we obtain

$$\frac{d}{dt}\mathbf{P}_{mech} = \int_V \varepsilon_0 \left[(\nabla \cdot \boldsymbol{\mathcal{E}}) \boldsymbol{\mathcal{E}} + \boldsymbol{\mathcal{B}} \times \frac{\partial \boldsymbol{\mathcal{E}}}{\partial t} - c^2 \boldsymbol{\mathcal{B}} \times (\nabla \times \boldsymbol{\mathcal{B}}) \right] dV, \tag{5.2}$$

where $c = 1/\sqrt{\varepsilon_0 \mu_0}$ is the speed of light in vacuum. By using the equality

$$\boldsymbol{\mathcal{B}} \times \frac{\partial \boldsymbol{\mathcal{E}}}{\partial t} = \boldsymbol{\mathcal{E}} \times \frac{\partial \boldsymbol{\mathcal{B}}}{\partial t} - \frac{\partial}{\partial t} (\boldsymbol{\mathcal{E}} \times \boldsymbol{\mathcal{B}})$$

together with Maxwell's equation for the curl of the electric field [Box 3.2],

$$\frac{\partial \boldsymbol{\mathcal{B}}(\mathbf{r}, t)}{\partial t} = -\nabla \times \boldsymbol{\mathcal{E}}(\mathbf{r}, t),$$

and by adding the null term $c^2 (\nabla \cdot \boldsymbol{\mathcal{B}}) \boldsymbol{\mathcal{B}}$, we finally obtain the symmetric form of Eq. (5.1):

$$\frac{d}{dt}\mathbf{P}_{mech} = \int_V \varepsilon_0 \left[(\nabla \cdot \boldsymbol{\mathcal{E}}) \boldsymbol{\mathcal{E}} + c^2 (\nabla \cdot \boldsymbol{\mathcal{B}}) \boldsymbol{\mathcal{B}} - \boldsymbol{\mathcal{E}} \times (\nabla \times \boldsymbol{\mathcal{E}}) - c^2 \boldsymbol{\mathcal{B}} \times (\nabla \times \boldsymbol{\mathcal{B}}) \right] dV$$

$$- \int_V \varepsilon_0 \frac{\partial}{\partial t} (\boldsymbol{\mathcal{E}} \times \boldsymbol{\mathcal{B}}) \, dV. \tag{5.3}$$

The last term on the right-hand side of Eq. (5.3) is the time derivative of the linear momentum density $c^{-2}\mathcal{S} = \varepsilon_0(\mathcal{E} \times \mathcal{B})$ [Box 3.6] integrated over the volume V, i.e., the time derivative of the linear momentum of the field $\mathbf{P}_{\text{field}} = \mathbf{P}_{\text{field}}(t)$, so that we can write

$$\int_V \varepsilon_0 \frac{\partial}{\partial t}(\mathcal{E} \times \mathcal{B}) \, dV = \frac{d}{dt} \int_V \frac{\mathcal{S}}{c^2} \, dV = \frac{d}{dt} \mathbf{P}_{\text{field}}. \tag{5.4}$$

In the integrand of the first term on the right-hand side of Eq. (5.3), by means of the vector identity [Box 3.1]

$$\mathcal{E} \times (\nabla \times \mathcal{E}) = -(\mathcal{E} \cdot \nabla)\mathcal{E} + \frac{1}{2}\nabla(\mathcal{E} \cdot \mathcal{E}),$$

the electric field terms become

$$(\nabla \cdot \mathcal{E})\mathcal{E} - \mathcal{E} \times (\nabla \times \mathcal{E}) = (\nabla \cdot \mathcal{E})\mathcal{E} + (\mathcal{E} \cdot \nabla)\mathcal{E} - \frac{1}{2}\nabla(\mathcal{E} \cdot \mathcal{E})$$

$$= \nabla \cdot \left[\mathcal{E} \otimes \mathcal{E} - \frac{1}{2}(\mathcal{E} \cdot \mathcal{E})\mathsf{I} \right],$$

where we have introduced the dyadic product and the unit dyadic [Box 5.1] to express the tensor form in the square brackets. An analogous expression can be obtained for the magnetic induction field terms. Therefore, Eq. (5.3) can be re-written to explicitly state the conservation of total linear momentum as

$$\frac{d}{dt}(\mathbf{P}_{\text{mech}} + \mathbf{P}_{\text{field}}) = \int_V \nabla \cdot \mathsf{T}_{\text{M}} \, dV, \tag{5.5}$$

where we have introduced the *Maxwell stress tensor*

$$\mathsf{T}_{\text{M}} = \varepsilon_0 \left[\mathcal{E} \otimes \mathcal{E} + c^2 \mathcal{B} \otimes \mathcal{B} - \frac{1}{2}(\mathcal{E} \cdot \mathcal{E} + c^2 \mathcal{B} \cdot \mathcal{B})\mathsf{I} \right]. \tag{5.6}$$

Box 5.1 — **Dyadics**

Given two vectors, $\mathbf{A} = A_x \hat{\mathbf{x}} + A_y \hat{\mathbf{y}} + A_z \hat{\mathbf{z}}$ and $\mathbf{B} = B_x \hat{\mathbf{x}} + B_y \hat{\mathbf{y}} + B_z \hat{\mathbf{z}}$, their *dyadic product* is

$$\mathbf{A} \otimes \mathbf{B} = \begin{bmatrix} A_x \\ A_y \\ A_z \end{bmatrix} \begin{bmatrix} B_x & B_y & B_z \end{bmatrix} = \begin{bmatrix} A_x B_x & A_x B_y & A_x B_z \\ A_y B_x & A_y B_y & A_y B_z \\ A_z B_x & A_z B_y & A_z B_z \end{bmatrix}.$$

The *unit dyadic* is

$$\mathsf{I} = \begin{bmatrix} 1 & 0 & 0 \\ 0 & 1 & 0 \\ 0 & 0 & 1 \end{bmatrix}.$$

Using Gauss' theorem [Box 3.1], we can substitute the integral of $\nabla \cdot \mathsf{T}_M$ over the volume V in Eq. (5.5) with an integral of its flux, i.e., $\mathsf{T}_M \cdot \hat{\mathbf{n}}$, over the bounding surface S, which we assume to be regular and orientable with outward normal unit vector $\hat{\mathbf{n}}$:

$$\frac{d}{dt}(\mathbf{P}_{\text{mech}} + \mathbf{P}_{\text{field}}) = \oint_S \mathsf{T}_M \cdot \hat{\mathbf{n}} \, dS, \tag{5.7}$$

where the term on the right-hand side represents the flux of linear momentum that leaves the surface S.[1] Finally, by time-averaging the force acting on the particle, we obtain

$$\mathbf{F}_{\text{rad}} = \oint_S \overline{\mathsf{T}}_M \cdot \hat{\mathbf{n}} \, dS. \tag{5.8}$$

The same type of analysis, but based on the conservation of angular momentum, yields the general expression for the radiation torque $\mathbf{T}_{\text{rad}}$. In fact, from the definition of angular momentum, $\mathbf{J} = \mathbf{r} \times \mathbf{P}$, and the conservation law for linear momentum expressed in Eq. (5.7), we directly obtain

$$\frac{d}{dt}(\mathbf{J}_{\text{mech}} + \mathbf{J}_{\text{field}}) = -\oint_S (\mathsf{T}_M \times \mathbf{r}) \cdot \hat{\mathbf{n}} \, dS, \tag{5.9}$$

where we have used the fact that $\mathbf{r} \times \mathsf{T}_M = -\mathsf{T}_M \times \mathbf{r}$, $\mathbf{J}_{\text{mech}}$ is the mechanical angular momentum of the particle,

$$\mathbf{J}_{\text{field}} = \int_V \mathbf{r} \times \frac{\mathbf{S}}{c^2} \, dV \tag{5.10}$$

is the angular momentum of the field and the term on the right-hand side represents the flux of angular momentum that enters the surface S. The time-averaged radiation torque is, therefore,

$$\mathbf{T}_{\text{rad}} = -\oint_S (\overline{\mathsf{T}}_M \times \mathbf{r}) \cdot \hat{\mathbf{n}} \, dS. \tag{5.11}$$

Eqs. (5.8) and (5.11) are quite general and apply to particles of any shape and composition in the presence of any electromagnetic field, at least as long as Maxwell's equations remain valid. However, the actual calculation of $\mathbf{F}_{\text{rad}}$ and $\mathbf{T}_{\text{rad}}$ requires calculation of how the electromagnetic field is affected by the presence of the particle itself. This is known as

---

[1] Eq. (5.7) can also be written in a form more similar to the equations for the conservation of angular momentum [Eq. (5.9)] and energy [Eq. (5.13)] as

$$\frac{d}{dt}(\mathbf{P}_{\text{mech}} + \mathbf{P}_{\text{field}}) = -\oint_S (-\mathsf{T}_M) \cdot \hat{\mathbf{n}} \, dS,$$

where $\oint_S (-\mathsf{T}_M) \cdot \hat{\mathbf{n}} \, dS$ is the flux of linear momentum entering the surface S.

the *scattering problem* and can be rather complex for anything other than homogeneous spheres. We will deal with this topic in the next sections and more deeply in Chapter 6.

Exercise 5.1.1 Show that the expression for (the Minkowski form of) the Maxwell stress tensor in a homogeneous medium is[2]

$$\mathsf{T}_M = \mathcal{E} \otimes \mathcal{D} + \mathcal{H} \otimes \mathcal{B} - \frac{1}{2}(\mathcal{E} \cdot \mathcal{D} + \mathcal{H} \cdot \mathcal{B}) \mathsf{I}. \qquad (5.12)$$

What is the expression for the Maxwell stress tensor in a homogeneous, linear and non-dispersive medium with dielectric permittiviy ε_m and magnetic permeability μ_m? [Hint: Use the constitutive relations given in Box 3.3.]

Exercise 5.1.2 Derive the law of energy conservation in electrodynamics (*Poynting's theorem*),

$$\frac{d}{dt}(U_{\text{mech}} + U_{\text{field}}) = -\oint_S \mathcal{S} \cdot \hat{\mathbf{n}} \, dS, \qquad (5.13)$$

where U_{mech} is the mechanical energy, U_{field} is the field energy and $-\oint_S \mathcal{S} \cdot \hat{\mathbf{n}} \, dS$ is the energy entering the volume enclosed by S. [Hint: Start from Poynting's theorem [Box 3.6] expressed in integral form.]

Exercise 5.1.3 Show that for harmonic fields the expression of the time-averaged Maxwell stress tensor in a homogeneous, linear and non-dispersive medium is[3]

$$\overline{\mathsf{T}}_M = \frac{1}{2}\varepsilon_m \text{Re}\left[\mathbf{E} \otimes \mathbf{E}^* + \frac{c^2}{n_m^2}\mathbf{B} \otimes \mathbf{B}^* - \frac{1}{2}\left(|\mathbf{E}|^2 + \frac{c^2}{n_m^2}|\mathbf{B}|^2\right)\mathsf{I}\right], \qquad (5.14)$$

where $\mathbf{E} = \mathbf{E}(\mathbf{r})$ and $\mathbf{B} = \mathbf{B}(\mathbf{r})$ are phasors, ε_m is the dielectric permittivity of the medium and n_m is the refractive index of the medium [Box 3.5].

Exercise 5.1.4 Calculate the optical force of a plane wave on a partly reflecting mirror with reflectance R using the Maxwell stress tensor and show that $F_{\text{rad}} = 2RP/c$, where P is the light power and c the speed of light. [Hint: Apply the definition of radiation force to a finite volume containing the mirror surface.]

Exercise 5.1.5 The integration of Eqs. (5.8) and (5.11) is often performed using a spherical surface of radius r enclosing the particle and centred at the origin of the reference system. Show that in this case Eqs. (5.8) and (5.11) become

$$\mathbf{F}_{\text{rad}} = r^2 \oint_\Omega \overline{\mathsf{T}}_M \cdot \hat{\mathbf{r}} \, d\Omega \qquad (5.15)$$

---

[2] The Minkowski form of the Maxwell stress tensor is not universally accepted, especially because it is non-symmetric in an anisotropic medium and a consequence of this asymmetry is the lack of conservation of the angular momentum. There have been several attempts to put the Maxwell stress tensor in a symmetrised form, the most well-known being that proposed by Abraham. No experiment has given final evidence in favour of the form of Minkowski or of that of Abraham. Because most results in optical trapping and manipulation do not depend qualitatively on the momentum definition, in this book we employ the Minkowski momentum definition, which in fact is the most often employed in optical tweezers studies. For a more detailed discussion of this issue, we refer the reader to Box 2.1 and references therein.

[3] Note that in a non-magnetic medium $\varepsilon_m = \epsilon_0 n_m^2$.

and

$$\mathbf{T}_{\text{rad}} = -r^3 \oint_\Omega \left(\overline{\mathsf{T}}_M \times \hat{\mathbf{r}} \right) \cdot \hat{\mathbf{r}} \, d\Omega, \tag{5.16}$$

where $\hat{\mathbf{r}}$ is the radial unit vector and the integration is over the full solid angle Ω.

5.1.1 Angular momentum of light

The expression for the angular momentum of an electromagnetic field obtained in Eq. (5.10) can be further developed by explicitly considering the irrotational and rotational components of the field.[4] On one hand, any electric field can always be decomposed into its irrotational, $\mathcal{E}_\parallel$, and rotational, $\mathcal{E}_\perp$, components using the Helmholtz decomposition [Box 5.2].[5] On the other hand, the magnetic field is always purely rotational, i.e., $\mathcal{B} = \mathcal{B}_\perp$, because $\nabla \cdot \mathcal{B} = 0$.

To proceed with the analysis of the angular momentum of the optical field, it is convenient to introduce the electromagnetic potentials ϕ and $\mathcal{A}$ in the Coulomb gauge [Box 5.3], so that the magnetic induction field is $\mathcal{B} = \nabla \times \mathcal{A}$ and the electric field components are $\mathcal{E}_\parallel = -\nabla \phi$ and $\mathcal{E}_\perp = -\partial \mathcal{A}/\partial t$. Thus, we can rewrite Eq. (5.10) as

$$\mathbf{J}_{\text{field}} = \int_V \varepsilon_0 \mathbf{r} \times \left[\mathcal{E}_\parallel \times (\nabla \times \mathcal{A}) + \mathcal{E}_\perp \times (\nabla \times \mathcal{A}) \right] dV, \tag{5.17}$$

where we have made use of the fact that $c^{-2}\mathcal{S} = \varepsilon_0 \mathcal{E} \times \mathcal{B}$ [Box 3.6].

Box 5.2   **Helmholtz decomposition**

The *Helmholtz decomposition*, or *fundamental theorem of vector calculus*, states that any sufficiently smooth, rapidly decaying vector field in three dimensions can be resolved into the sum of an irrotational (also known as conservative, curl-free, longitudinal) vector field and a rotational (also known as solenoidal, divergence-free, transverse) vector field. This implies that any three-dimensional vector field $\mathbf{F}$ can be generated by a pair of potentials, i.e., a scalar potential ϕ and a vector potential $\mathbf{A}$, such that

$$\mathbf{F} = -\nabla \phi + \nabla \times \mathbf{A}.$$

In particular, any irrotational vector field can always be written as the gradient of some scalar field (which enables us to replace a vector field by a much simpler scalar field) and any rotational vector field can always be written as the curl of some other vector field.

---

[4] A good review of this topic is Piccirillo et al. (2013).
[5] The use of $\parallel$ and $\perp$ to represent the irrotational and rotational components is standard in the literature and refers to the properties of their Fourier components as longitudinal, $\mathbf{k} \times \mathcal{E}_\parallel(\mathbf{k}) = 0$, and transverse, $\mathbf{k} \cdot \mathcal{E}_\perp(\mathbf{k}) = 0$, to $\mathbf{k}$. For further details see, e.g., Cohen–Tannoudji et al. (1995).

Box 5.3 — Electromagnetic potentials and gauge invariance

Maxwell's equations [Box 3.1] describe the dynamical behaviour of the electromagnetic fields $\mathcal{E}$ and $\mathcal{B}$. It is often convenient to rewrite these fields in term of a *scalar potential*, $\phi(\mathbf{r}, t)$, and a *vector potential*, $\mathcal{A}(\mathbf{r}, t)$ [Box 5.2], as

$$\mathcal{B} = \nabla \times \mathcal{A},$$
$$\mathcal{E} = -\nabla \phi - \frac{\partial \mathcal{A}}{\partial t}.$$

An advantage of using the electromagnetic potentials is that the two homogeneous Maxwell's equations are automatically satisfied.

The original set of four first-order partial differential equations can then be rewritten as two second-order partial differential equations describing the dynamics of the potentials:

$$\nabla^2 \phi + \frac{\partial}{\partial t}(\nabla \cdot \mathcal{A}) = -\varrho/\varepsilon_0,$$

$$\nabla^2 \mathcal{A} - \frac{1}{c^2}\frac{\partial^2}{\partial t^2}\mathcal{A} - \nabla\left(\nabla \cdot \mathcal{A} + \frac{1}{c^2}\frac{\partial^2}{\partial t^2}\phi\right) = -\mu_0 \mathbf{j}.$$

The potentials, however, are not uniquely defined. In fact, the physical fields, and hence Maxwell's equations, are left unchanged by a so-called *gauge transformation*:

$$\mathcal{A} \rightarrow \mathcal{A}' = \mathcal{A} + \nabla \psi,$$
$$\phi \rightarrow \phi' = \phi - \frac{\partial}{\partial t}\psi,$$

where $\psi(\mathbf{r}, t)$ is an arbitrary function. Thus, there is freedom to choose a set of potentials that satisfy a certain *gauge* condition. The two most commonly used gauges are the *Lorenz gauge*, for which $\nabla \mathcal{A} + c^{-2}\partial \phi/\partial t = 0$, and the *Coulomb gauge* (also called *transverse gauge*), for which $\nabla \cdot \mathcal{A} = 0$. In the Lorenz gauge the equations for the potentials are more symmetric and relativistically invariant. In the Coulomb gauge, the transverse radiation fields are given by the vector potential alone, whereas the scalar potential contributes only to the near fields. More details can be found, e.g., in Cohen–Tannoudji et al. (1995).

The first term on the right-hand side of Eq. (5.17) is related to the canonical angular momentum and is associated with the charge density[6] ϱ, i.e.,

$$\mathbf{J}_{\text{field, canonical}} = \int_V \varepsilon_0 \mathbf{r} \times \left[\mathcal{E}_\parallel \times (\nabla \times \mathcal{A})\right] dV = \int_V \varrho \mathbf{r} \times \mathcal{A}\, dV. \tag{5.18}$$

To derive Eq. (5.18), we start by using the vectorial identity [Box 3.1]

$$\mathbf{r} \times [\mathcal{E}_\parallel \times (\nabla \times \mathcal{A})] = \sum_{j=x,y,z} \mathcal{E}_{\parallel,j}(\mathbf{r} \times \nabla)\mathcal{A}_j - \mathbf{r} \times (\mathcal{E}_\parallel \cdot \nabla)\mathcal{A}$$

---

[6] Note that in Eq. (5.18), we consider only the canonical components of the moment relative to the field. To obtain the total canonical angular momentum, we need also consider the mechanical momenta of the charges. To do this, we can rewrite Eq. (5.18) in terms of charges as $\sum_i \mathbf{r} \times q_i \mathcal{A}(\mathbf{r}_i)$ and then find the total canonical angular momentum of the system, i.e., $\mathbf{J}_{\text{canonical}} = \sum_i \mathbf{r} \times (\mathbf{p}_i + q_i \mathcal{A}(\mathbf{r}_i))$.

to rewrite Eq. (5.18) as

$$\mathbf{J}_{\text{field,canonical}} = \int_V \varepsilon_0 \sum_{j=x,y,z} \mathcal{E}_{\|,j}(\mathbf{r}\times\nabla)\mathcal{A}_j \, dV - \int_V \varepsilon_0 \mathbf{r}\times(\mathcal{E}_\| \cdot \nabla)\mathcal{A} \, dV, \quad (5.19)$$

and then we proceed to integrate by parts both terms on the right-hand side of Eq. (5.19). For the first term, we have that

$$\sum_{j=x,y,z} \mathcal{E}_{\|,j}(\mathbf{r}\times\nabla)\mathcal{A}_j = -\sum_{j=x,y,z} \nabla_j \phi (\mathbf{r}\times\nabla)\mathcal{A}_j$$

$$= -\nabla \cdot [\phi(\mathbf{r}\times\nabla)\mathcal{A}] - \phi(\nabla\times\mathcal{A}) + \underbrace{\phi(\mathbf{r}\times\nabla)(\nabla\cdot\mathcal{A})}_{=0},$$

where $\nabla\cdot\mathcal{A} = 0$, because, as we are in Coulomb gauge, when we perform the integral for a volume going to infinity, the (first) divergence term gives a null contribution because the longitudinal fields vanish in the far field faster than r^{-2}, and we are left with the asymptotic result

$$\int_V \varepsilon_0 \sum_{j=x,y,z} \mathcal{E}_{\|,j}(\mathbf{r}\times\nabla)\mathcal{A}_j \, dV \longrightarrow -\int_V \varepsilon_0 \phi(\nabla\times\mathcal{A}) \, dV.$$

For the second term, we can use the fact that, for any arbitrary constant vector $\mathbf{a}$,

$$-[\mathbf{r}\times(\mathcal{E}_\| \cdot \nabla)\mathcal{A}]\cdot\mathbf{a} = -\nabla\cdot[\mathcal{E}_\|(\mathbf{r}\times\mathcal{A}\cdot\mathbf{a})] + (\nabla\cdot\mathcal{E}_\|)(\mathbf{r}\times\mathcal{A})\cdot\mathbf{a} + (\mathcal{E}_\|\times\mathcal{A})\cdot\mathbf{a}$$

$$= -\nabla\cdot[\mathcal{E}_\|(\mathbf{r}\times\mathcal{A}\cdot\mathbf{a})] + \frac{\varrho}{\varepsilon_0}(\mathbf{r}\times\mathcal{A})\cdot\mathbf{a} - (\nabla\phi\times\mathcal{A})\cdot\mathbf{a},$$

where we have used the facts that $\mathcal{E}_\| = -\nabla\phi$ and $\nabla\cdot\mathcal{E}_\| = \varrho/\varepsilon_0$, so that, because again the divergence term does not make any asymptotic contribution to the volume integral, we are left with

$$-\int_V \varepsilon_0 \mathbf{r}\times(\mathcal{E}_\| \cdot \nabla)\mathcal{A} \, dV \longrightarrow \int_V \varrho(\mathbf{r}\times\mathcal{A}) \, dV - \int_V \varepsilon_0(\nabla\phi\times\mathcal{A}) \, dV.$$

Combining these results, Eq. (5.19) becomes

$$\mathbf{J}_{\text{field,canonical}} = \int_V \varrho(\mathbf{r}\times\mathcal{A}) \, dV - \int_V \varepsilon_0 \nabla\times(\phi\mathcal{A}) \, dV,$$

where the second term is asymptotically null, and thus we are left with Eq. (5.18). Therefore, in a source-free space or in the case of radiating fields rapidly vanishing at infinity, the canonical part of the electromagnetic angular momentum is negligible.

We now consider the second term on the right-hand side of Eq. (5.17), which is related to the rotational fields and can be further split into orbital and spin components:

$$\int_V \varepsilon_0 \mathbf{r}\times[\mathcal{E}_\perp \times(\nabla\times\mathcal{A})] \, dV = \int_V \varepsilon_0 \Big[\underbrace{\sum_{j=x,y,z}\mathcal{E}_{\perp,j}(\mathbf{r}\times\nabla)\mathcal{A}_j}_{\text{orbital}} \underbrace{-\mathbf{r}\times(\mathcal{E}_\perp\cdot\nabla)\mathcal{A}}_{\text{spin}}\Big] dV,$$

(5.20)

where we have used the same vector identity as for Eq. (5.19) but applied to $\mathcal{E}_\perp$. The spin term can be further integrated by parts following a procedure similar to the one employed in the derivation of Eq. (5.18) and using the fact that the divergence term is null because $\mathcal{E}_\perp$ is transverse to the surface normal and $\nabla \cdot \mathcal{E}_\perp = 0$, so that we get

$$-\mathbf{r} \times (\mathcal{E}_\perp \cdot \nabla)\mathcal{A} = \mathcal{E}_\perp \times \mathcal{A}.$$

Finally, we obtain the following expressions for the orbital, **L**, and spin, **s**, components of the light angular momentum:

$$\mathbf{L} = \int_V \varepsilon_0 \sum_{j=x,y,z} \mathcal{E}_{\perp,j}(\mathbf{r} \times \nabla)\mathcal{A}_j \, dV, \tag{5.21}$$

$$\mathbf{s} = \int_V \varepsilon_0 \mathcal{E}_\perp \times \mathcal{A} \, dV. \tag{5.22}$$

For a monochromatic transverse electromagnetic field, we can use phasors [Box 3.5] to express the electric field in terms of the vector potential, i.e., $\mathbf{E} = \mathbf{E}_\perp = i\omega\mathbf{A}$. Thus, we can write the orbital, $\mathbf{L}_d$, and spin, $\mathbf{s}_d$, averaged angular momentum densities in a form that is useful in many practical cases:

$$\mathbf{L}_d = i\frac{\varepsilon_0}{2\omega} \sum_{j=x,y,z} E_j(\mathbf{r} \times \nabla)E_j^*, \tag{5.23}$$

$$\mathbf{s}_d = i\frac{\varepsilon_0}{2\omega} \mathbf{E} \times \mathbf{E}^*. \tag{5.24}$$

Exercise 5.1.6 Consider a generic vector field $\mathbf{F}(\mathbf{r})$ whose curl $\nabla \times \mathbf{F}$ and divergence $\nabla \cdot \mathbf{F}$ are known. Demonstrate that $\mathbf{F}(\mathbf{r})$ can be written as $\mathbf{F} = \mathbf{F}_\parallel + \mathbf{F}_\perp$, where $\mathbf{F}_\parallel$ is the longitudinal (irrotational) component and $\mathbf{F}_\perp$ is the transverse (rotational) component [Box 5.2]. Explicitly write the components in terms of the known curl and divergence. [Hint: Use the vector identity $\nabla \times (\nabla \times \mathbf{F}) = \nabla(\nabla \cdot \mathbf{F}) - \nabla^2 \mathbf{F}$ and the relation $\nabla^2(1/|\mathbf{r} - \mathbf{r}'|) = -4\pi\delta(\mathbf{r} - \mathbf{r}')$.]

Exercise 5.1.7 Show that it is possible to decompose the linear momentum of light into canonical, spin and orbital parts. [Hint: Use the Helmholtz decomposition in the expression for the linear momentum of an optical field given in Eq. (5.4).]

Exercise 5.1.8 Demonstrate that the canonical part of the electromagnetic angular momentum can also be written as a surface integral:

$$\mathbf{J}_{\text{field,canonical}} = \varepsilon_0 \oint_S (\mathbf{r} \times \mathcal{A})\mathcal{E} \cdot \hat{\mathbf{n}} \, dS, \tag{5.25}$$

where S is a surface containing the sources of the field. Show that for radiative fields this expression is negligible in the far field. [Hint: Use Gauss' theorem on Eq. (5.18).]

Exercise 5.1.9 Calculate the averaged angular momentum density of a circularly polarised plane wave. What are the values of the orbital and spin components? Show that the angular momentum-to-energy ratio is given by

$$\frac{J_{\text{field}}}{U_{\text{field}}} = \frac{1}{\omega},$$

where ω is the angular frequency of the plane wave. Note that the same result is obtained by considering the light as composed by a stream of photons each carrying quantised spin, $\hbar$, and energy, $\hbar\omega$. [Hint: Use the definition of orbital and spin angular momentum densities and apply them to a circular polarised plane wave.]

Exercise 5.1.10 Consider a generic elliptically polarised plane wave. This can be expressed in terms of the *circular unit vectors*,

$$\hat{\mathbf{e}}_\pm = \frac{1}{\sqrt{2}}(\hat{\mathbf{x}} \pm i\hat{\mathbf{y}}), \qquad (5.26)$$

where $\hat{\mathbf{e}}_+$ corresponds to left-handed polarisation and $\hat{\mathbf{e}}_-$ to right-handed polarisation, so that

$$\mathbf{E} = E_0 e^{ik_0 z}(\alpha_+ \hat{\mathbf{e}}_+ + \alpha_- \hat{\mathbf{e}}_-),$$

where k_0 is the light wavenumber in vacuum. Write the spin angular momentum density and show that it is characterised by a polarisation content $s = |\alpha_+|^2 - |\alpha_-|^2$, so that $-1 \leq s \leq +1$, with $s = \pm 1$ for left- and right-handed circular polarisation.

Exercise 5.1.11 Show that in the paraxial approximation the angular momentum of a light beam along the (propagation) z-direction becomes

$$J_z = \mathbf{J} \cdot \hat{\mathbf{z}} = \frac{\varepsilon_0}{2\omega} \int_V (E_+^2 - E_-^2) dV + \frac{\varepsilon_0}{2i\omega} \int_V \sum_{j=x,y,z} \left(E_j^* \frac{\partial}{\partial \varphi} E_j \right) dV, \qquad (5.27)$$

where E_+ and E_- are the left and right circular polarisation components of the field. [Hint: Consider the z-component of the generic expression of the angular momentum in terms of its spin and orbital parts and apply the paraxial approximation.]

Exercise 5.1.12 Consider a circularly polarised Laguerre–Gaussian beam [Subsection 4.4.3] carrying both orbital and spin angular momentum. Calculate, within the paraxial approximation, the spin and orbital angular momentum density of the beam. Show that the angular-momentum-to-energy ratio is given by

$$\frac{L_{\text{field}}}{U_{\text{field}}} = \frac{l \pm s}{\omega},$$

where l is the azimuthal index of the Laguerre–Gaussian beam and s denotes the polarisation state of the beam. Generalise the problem to a non-paraxial laser beam. [Hint: Study the papers by Allen et al. (1992) and Barnett and Allen (1994).]

5.2 Light scattering

To calculate radiation forces and torques acting on a particle using Eqs. (5.8) and (5.11), it is necessary first to find the electromagnetic fields scattered by the particle itself. This is known as the *scattering problem*.

In this section, we recall the essential results of light-scattering theory. We start with a brief digression on the solution of the Helmholtz equation [Subsection 5.2.1], which permits us to introduce the mathematical background of light scattering. Then we proceed to introduce the scattering problem and the *scattering amplitude* [Subsection 5.2.2]. Finally, after having introduced the multipole expansion of the electric field [Subsection 5.2.3], we present the fundamentals of one of the methods most commonly used to compute the scattering amplitude in the calculation of optical forces, i.e., the *transition matrix* (T-matrix) method [Subsection 5.2.4], and we apply it to the classical problem of scattering by a homogenous sphere, i.e., *Mie scattering* [Subsection 5.2.5]. More details on how to actually perform the computations will be provided in Chapter 6. A full account of light-scattering theory goes well beyond the scope of this book and can be found in, e.g., Mishchenko et al. (2002) and Borghese et al. (2007).

Before proceeding further, we note that in the rest of this chapter we will always consider harmonic electromagnetic fields with angular frequency ω expressed in phasor form [Box 3.5] and, to simplify the notation, we will not explicitly indicate their time dependence.

5.2.1 Solution of the Helmholtz equation

We consider the solution of the *scalar Helmholtz equation* [Eq. (4.2)], i.e.,

$$(\nabla^2 + k^2)F(\mathbf{r}) = 0, \tag{5.28}$$

which in spherical coordinates reads

$$\frac{1}{r^2}\frac{\partial}{\partial r}\left[r^2\frac{\partial F(\mathbf{r})}{\partial r}\right] + \frac{1}{r^2 \sin\vartheta}\frac{\partial}{\partial \vartheta}\left[\sin\vartheta \frac{\partial F(\mathbf{r})}{\partial \vartheta}\right] + \frac{1}{r^2 \sin^2\vartheta}\frac{\partial^2 F(\mathbf{r})}{\partial \varphi^2} + k^2 F(\mathbf{r}) = 0, \tag{5.29}$$

where $F(\mathbf{r}) = F(r, \vartheta, \varphi)$, r is the radial distance, ϑ is the polar angle and φ is the azimuthal angle. The standard approach to the solution of Eq. (5.29) is to separate the variables, i.e., to write

$$F(r, \vartheta, \varphi) = R(r)\Theta(\vartheta)\Phi(\varphi)$$

and to divide Eq. (5.28) by $R(r)\Theta(\vartheta)\Phi(\varphi)/r^2$, obtaining

$$\frac{1}{R}\frac{d}{dr}\left[r^2\frac{dR}{dr}\right] + k^2 r^2 + \frac{1}{\Theta \sin\vartheta}\frac{d}{d\vartheta}\left[\sin\vartheta \frac{d\Theta}{d\vartheta}\right] + \frac{1}{\Phi \sin^2\vartheta}\frac{d^2\Phi}{d\varphi^2} = 0. \tag{5.30}$$

Angular dependence and spherical harmonics

When Eq. (5.30) is multiplied by $\sin^2 \vartheta$, the last term depends only on φ and can therefore be equated to a constant, i.e.,

$$\frac{1}{\Phi(\varphi)} \frac{d^2 \Phi(\varphi)}{d\varphi^2} = -m^2,$$

whose solution is

$$\Phi(\varphi) = e^{im\varphi}, \tag{5.31}$$

where m must be an integer in order to satisfy the periodic boundary condition $\Phi(\varphi + 2\pi) = \Phi(\varphi)$.

Substituting Eq. (5.31) into Eq. (5.30) gives

$$\frac{1}{R} \frac{d}{dr}\left[r^2 \frac{dR}{dr} \right] + k^2 r^2 + \frac{1}{\Theta \sin \vartheta} \frac{d}{d\vartheta}\left[\sin \vartheta \frac{d\Theta}{d\vartheta} \right] - \frac{m^2}{\sin^2 \vartheta} = 0.$$

The last two terms in the previous equation are dependent only on ϑ and must therefore be a constant, which we will assume to be $-l(l+1)$ with l a non-negative integer, i.e.,

$$\frac{1}{\Theta \sin \vartheta} \frac{d}{d\vartheta}\left[\sin \vartheta \frac{d\Theta}{d\vartheta} \right] - \frac{m^2}{\sin^2 \vartheta} = -l(l+1).$$

By substituting $z = \cos \vartheta$ and rearranging, we obtain

$$\frac{d}{dz}\left[(1 - z^2) \frac{d\Theta(z)}{dz} \right] + \left[l(l+1) - \frac{m^2}{1 - z^2} \right] \Theta(z) = 0, \tag{5.32}$$

which is the *associated Legendre equation* and whose solutions are the *associated Legendre functions* $P_{lm}(z)$ [Box 5.4]. Therefore,

$$\Theta(\vartheta) = P_{lm}(\cos \vartheta), \tag{5.33}$$

where $l = 0, 1, 2, 3, \ldots$ and $m = -l, \ldots, 0, \ldots, l$.

Box 5.4   **Associated Legendre functions**

The *associated Legendre functions* $P_{lm}(z)$ are the solutions of the associated Legendre equation [Eq. (5.32)]. They are defined for $l = 0, 1, 2, 3, \ldots$ and $m = -l, \ldots, 0, \ldots, l$ as

$$P_{lm}(z) = \begin{cases} \dfrac{(-1)^m}{2^l l!} (1 - z^2)^{m/2} \dfrac{d^{l+m}}{dz^{l+m}} (z^2 - 1)^l & \text{for } m \geq 0, \\[2ex] (-1)^{|m|} \dfrac{(l - |m|)!}{(l + |m|)!} P_{l|m|}(z) & \text{for } m < 0. \end{cases}$$

The first few associated Legendre functions are given in Table 5.1. For $m = 0$, the associated Legendre functions reduce to the *Legendre functions* $P_l(z)$.

Table 5.1 First few associated Legendre functions and spherical harmonics

$P_{0,0} = 1$	$Y_{0,0}(\hat{\mathbf{r}}) = \sqrt{\dfrac{1}{4\pi}}$
$P_{1,1}(z) = -\sqrt{1-z^2}$	$Y_{1,1}(\hat{\mathbf{r}}) = -\sqrt{\dfrac{3}{8\pi}}\sin\vartheta\, e^{i\varphi}$
$P_{1,0}(z) = z$	$Y_{1,0}(\hat{\mathbf{r}}) = \sqrt{\dfrac{3}{4\pi}}\cos\vartheta$
$P_{1,-1}(z) = -\dfrac{1}{2}P_{1,1}(z)$	$Y_{1,-1}(\hat{\mathbf{r}}) = \sqrt{\dfrac{3}{8\pi}}\sin\vartheta\, e^{-i\varphi}$
$P_{2,2}(z) = 3(1-z^2)$	$Y_{2,2}(\hat{\mathbf{r}}) = \sqrt{\dfrac{15}{32\pi}}\sin^2\vartheta\, e^{2i\varphi}$
$P_{2,1}(z) = -3z\sqrt{1-z^2}$	$Y_{2,1}(\hat{\mathbf{r}}) = -\sqrt{\dfrac{15}{8\pi}}\sin\vartheta\cos\vartheta\, e^{i\varphi}$
$P_{2,0}(z) = \dfrac{1}{2}(3z^2-1)$	$Y_{2,0}(\hat{\mathbf{r}}) = \sqrt{\dfrac{5}{16\pi}}(3\cos^2\vartheta - 1)$
$P_{2,-1}(z) = -\dfrac{1}{6}P_{2,1}(z)$	$Y_{2,-1}(\hat{\mathbf{r}}) = \sqrt{\dfrac{15}{8\pi}}\sin\vartheta\cos\vartheta\, e^{-i\varphi}$
$P_{2,-2}(z) = \dfrac{1}{24}P_{2,2}(z)$	$Y_{2,-2}(\hat{\mathbf{r}}) = \sqrt{\dfrac{15}{32\pi}}\sin^2\vartheta\, e^{-2i\varphi}$

Often the angular portion of the solution of the Helmholtz equation [Eq. (5.29)] is given in terms of *spherical harmonics*, i.e.,

$$Y_{lm}(\hat{\mathbf{r}}) = Y_{lm}(\vartheta, \varphi) = \Theta(\vartheta)\Phi(\varphi) = \sqrt{\frac{2l+1}{4\pi}\frac{(l-m)!}{(l+m)!}}\, P_{lm}(\cos\vartheta)\, e^{im\varphi}, \quad (5.34)$$

where $\hat{\mathbf{r}}$ is the radial unit vector and the prefactor is chosen so that $Y_{lm}(\hat{\mathbf{r}})$ form a complete orthonormal set of scalar functions on the unit sphere, i.e.,

$$\oint_\Omega Y_{lm}(\hat{\mathbf{r}})\, Y^*_{l'm'}(\hat{\mathbf{r}})\, d\Omega = \delta_{ll'}\delta_{mm'}, \quad (5.35)$$

where the integration is over the full solid angle Ω. The first few spherical harmonics are shown in Fig. 5.2 and their formulas are given in Table 5.1.

Exercise 5.2.1 Write a program to visualise the real part, imaginary part and modulus of spherical harmonics and explore their properties. [Hint: You can use the program `sphericalharmonics` from the book website.]

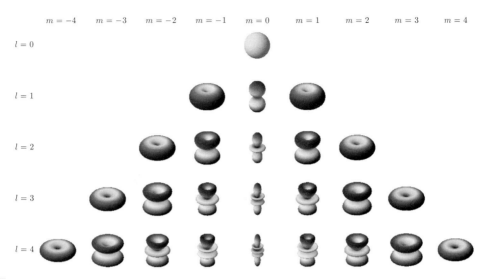

Figure 5.2 Spherical harmonics. First few spherical harmonics $Y_{lm}(\hat{\mathbf{r}})$ [Eq. (5.34)]. The radial distance from the origin is proportional to $|Y_{lm}(\hat{\mathbf{r}})|$ and the shades of grey represent the value of $\text{Re}\{Y_{lm}(\hat{\mathbf{r}})\}$.

Exercise 5.2.2 Demonstrate that

$$Y_{l,-m}(\hat{\mathbf{r}}) = (-1)^m\, Y^*_{lm}(\hat{\mathbf{r}}). \tag{5.36}$$

Exercise 5.2.3 Demonstrate that $Y_{lm}(\hat{\mathbf{r}})$ for l odd (even) are odd (even) functions of $\hat{\mathbf{r}}$, i.e.,

$$Y_{lm}(-\hat{\mathbf{r}}) = (-1)^l\, Y_{lm}(\hat{\mathbf{r}}). \tag{5.37}$$

Exercise 5.2.4 Prove analytically and verify numerically the *addition theorem of spherical harmonics*, i.e.,

$$P_l(\hat{\mathbf{r}} \cdot \hat{\mathbf{k}}) = \frac{4\pi}{2l+1} \sum_{m=-l}^{l} Y_{lm}(\hat{\mathbf{r}})\, Y^*_{lm}(\hat{\mathbf{k}}). \tag{5.38}$$

Exercise 5.2.5 Show that any scalar function $G(\hat{\mathbf{r}})$ on the unit sphere can be expanded in spherical harmonics as

$$G(\hat{\mathbf{r}}) = \sum_{l=0}^{+\infty} \sum_{m=-l}^{+l} G_{lm}\, Y_{lm}(\hat{\mathbf{r}}), \tag{5.39}$$

where

$$G_{lm} = \oint_\Omega G(\hat{\mathbf{r}})\, Y^*_{lm}(\hat{\mathbf{r}})\, d\Omega. \tag{5.40}$$

Exercise 5.2.6 Show that the expansion of $\delta(\hat{\mathbf{r}} - \hat{\mathbf{k}})$ in spherical harmonics is

$$\delta(\hat{\mathbf{r}} - \hat{\mathbf{k}}) = \sum_{l=0}^{+\infty} \sum_{m=-l}^{+l} Y^*_{lm}(\hat{\mathbf{k}})\, Y_{lm}(\hat{\mathbf{r}}). \tag{5.41}$$

Exercise 5.2.7 Show that the angular part of the Helmholtz equation [Eq. (5.29)] can be also written as

$$L^2 Y_{lm}(\hat{\mathbf{r}}) = l(l+1) Y_{lm}(\hat{\mathbf{r}}), \qquad (5.42)$$

where the differential operator $L^2 = \mathbf{L} \cdot \mathbf{L} = L_x^2 + L_y^2 + L_z^2$ is defined in terms of

$$\mathbf{L} = -i\mathbf{r} \times \nabla, \qquad (5.43)$$

i.e., $\hbar^{-1}$ times the angular momentum operator of quantum mechanics. Note that $\mathbf{L}$ is independent of $\mathbf{r}$, operates only on angular variables and, from its definition, $\mathbf{r} \cdot \mathbf{L} = 0$.

Exercise 5.2.8 Show that the spherical radial unit vector $\hat{\mathbf{r}}$ can be written in terms of spherical harmonics with $l = 1$ as

$$\hat{\mathbf{r}} = -\sqrt{\frac{4\pi}{3}} Y_{1,-1}(\hat{\mathbf{r}}) \, \hat{\boldsymbol{\xi}}_+ + \sqrt{\frac{4\pi}{3}} Y_{1,0}(\hat{\mathbf{r}}) \, \hat{\boldsymbol{\xi}}_0 - \sqrt{\frac{4\pi}{3}} Y_{1,1}(\hat{\mathbf{r}}) \, \hat{\boldsymbol{\xi}}_-, \qquad (5.44)$$

where

$$\hat{\boldsymbol{\xi}}_\pm = \mp \frac{1}{\sqrt{2}} (\hat{\mathbf{x}} \pm i\hat{\mathbf{y}}) \quad \text{and} \quad \hat{\boldsymbol{\xi}}_0 = \hat{\mathbf{z}} \qquad (5.45)$$

constitute the *spin-1 eigenvectors*.[7] [Hint: Write the radial unit vector in terms of polar and azimuthal angles in a Cartesian basis. Then find the expansion coefficients in terms of the spin-1 eigenvectors. Finally, use the expression of the $l = 1$ spherical harmonics in Table 5.1 to get Eq. (5.44).]

Exercise 5.2.9 Show that the components of the angular momentum operator, $\mathbf{L}$, can be written as (*ladder operators*)

$$\begin{cases} L_+ = L_x + iL_y = e^{+i\varphi} \left(\frac{\partial}{\partial \theta} + i \cot \vartheta \frac{\partial}{\partial \varphi} \right) \\ L_- = L_x - iL_y = e^{-i\varphi} \left(-\frac{\partial}{\partial \theta} + i \cot \vartheta \frac{\partial}{\partial \varphi} \right) \\ L_z = -i \frac{\partial}{\partial \varphi} \end{cases} \qquad (5.46)$$

and that the following relations can be established:

$$\begin{cases} L_+ Y_{l,m} = \sqrt{(l-m)(l+m+1)} Y_{l,m+1} \\ L_- Y_{l,m} = \sqrt{(l+m)(l-m+1)} Y_{l,m-1} \\ L_z Y_{l,m} = m Y_{l,m} \end{cases} \qquad (5.47)$$

[Hint: Use the definition of $\mathbf{L}$ in Eq. (5.43).]

---

[7] Note that $\hat{\boldsymbol{\xi}}_+ = -\hat{\mathbf{e}}_+$ and $\hat{\boldsymbol{\xi}}_- = +\hat{\mathbf{e}}_-$, where $\hat{\mathbf{e}}_\pm$ are defined in Eq. (5.26).

Radial dependence and spherical Bessel functions

We can now turn our attention to the radial dependence $R(r)$. Substituting Eqs. (5.31) and (5.33) into Eq. (5.30), we obtain an equation for $R(r)$,

$$r^2 \frac{d^2 R(r)}{dr^2} + 2r \frac{dR(r)}{dr} + \left[k^2 r^2 - l(l+1)\right] R(r) = 0. \tag{5.48}$$

By substituting $z = kr$ in Eq. (5.48), we obtain the spherical Bessel equation

$$z^2 \frac{d^2 R(z)}{dz^2} + 2z \frac{dR(z)}{dz} + \left[z^2 - l(l+1)\right] R(z) = 0. \tag{5.49}$$

Two linearly independent solutions of Eq. (5.49) are given by the *spherical Bessel function* (or spherical Bessel function of the first kind)

$$j_l(z) = \sqrt{\frac{\pi}{2z}} J_{l+\frac{1}{2}}(z) = (-z)^l \left(\frac{1}{z}\frac{d}{dz}\right)^l \frac{\sin z}{z}, \tag{5.50}$$

where $J_{l+\frac{1}{2}}(z)$ is a Bessel function of the first kind, and by the *spherical Neumann function* (or spherical Bessel function of the second kind)

$$y_l(z) = \sqrt{\frac{\pi}{2z}} Y_{l+\frac{1}{2}}(z) = -(-z)^l \left(\frac{1}{z}\frac{d}{dz}\right)^l \frac{\cos z}{z}, \tag{5.51}$$

where $Y_{l+\frac{1}{2}}(z)$ is a Neumann function (or Bessel function of the second kind). Alternatively, the solution of Eq. (5.48) can also be expressed as a linear combination of a spherical Bessel function and a spherical Neumann function, i.e., as *spherical Hankel functions* of the first and second kind:

$$h_l(z) = h_l^{(1)}(z) = j_l(z) + i y_l(z) \tag{5.52}$$

and

$$h_l^{(2)}(z) = j_l(z) - i y_l(z). \tag{5.53}$$

The first few of these functions are plotted in Fig. 5.3 and their formulas are given in Table 5.2. It can be seen that the functions $j_l(z)$ are regular at the origin, whereas the functions $y_l(z)$ and $h_l(z)$ diverge at the origin. Therefore, we will typically express $R(r)$ as a function of $j_l(kr)$ to describe an electromagnetic field that is finite at the origin, and of $h_l(kr)$ to describe a field that satisfies the radiation condition at infinity, i.e., that is constituted by outgoing spherical waves towards infinity. Some important asymptotic properties of these functions are given in Table 5.3.

Table 5.2 First few spherical Bessel functions and spherical Hankel functions

$$j_0(z) = \frac{\sin(z)}{z} = \text{sinc}(z) \qquad h_0(z) = -ie^{iz}\frac{1}{z}$$

$$j_1(z) = \frac{\sin(z)}{z^2} - \frac{\cos(z)}{z} \qquad h_1(z) = -ie^{iz}\frac{z+i}{z^2}$$

$$j_2(z) = \left[\frac{3}{z^2} - 1\right]\frac{\sin(z)}{z} - \frac{3\cos(z)}{z^2} \qquad h_2(z) = -ie^{iz}\frac{z^2 + 3iz - 3}{z^3}$$

$$j_3(z) = \left[\frac{15}{z^3} - \frac{6}{z}\right]\frac{\sin(z)}{z} - \left[\frac{15}{z^2} - 1\right]\frac{\cos(z)}{z} \qquad h_3(z) = -ie^{iz}\frac{z^3 + 6iz^2 - 15z - 15i}{z^4}$$

Table 5.3 Asymptotic properties of spherical Bessel functions and spherical Hankel functions

$$j_l(kr) \approx \frac{(kr)^l}{(2l+1)!!} \qquad h_l(kr) \approx \frac{(2l-1)!!}{(kr)^{l+1}} \qquad \text{for } kr \to 0$$

$$j_l(kr) \approx \frac{1}{kr}\sin(kr - l\pi/2) \qquad h_l(kr) \approx (-i)^{l+1}\frac{e^{ikr}}{kr} \qquad \text{for } kr \to \infty$$

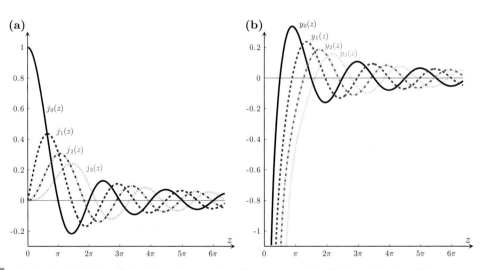

Figure 5.3 Spherical Bessel functions. The first few (a) spherical Bessel functions $j_l(z)$ [Eq. (5.50)] and (b) spherical Neumann functions $y_l(z)$ [Eq. (5.51)]. When z is real, $j_l(z)$ and $y_l(z)$ are respectively equal to the real and imaginary part of the Hankel function $h_l(z)$ [Eq. (5.52)].

Exercise 5.2.10 Show that

$$j_l(kr) \approx \frac{i^{l+1}}{2} \left[\frac{e^{-ikr}}{kr} - (-1)^l \frac{e^{ikr}}{kr} \right] \quad \text{for} \quad kr \to \infty. \tag{5.54}$$

[Hint: Note that $j_l(kr) = \frac{1}{2}[h_l^*(kr) + h_l(kr)]$ and use the asymptotic expression for $h_l(kr)$ given in Table 5.3.]

General solution of the Helmholtz equation

Finally, the general solution of the Helmholtz equation that is regular at the origin is

$$F_j(\mathbf{r}) = \sum_{l=0}^{+\infty} \sum_{m=-l}^{+l} B_{lm} \, j_l(kr) \, Y_{lm}(\hat{\mathbf{r}}) \tag{5.55}$$

and the general solution that satisfies the radiation condition at infinity is

$$F_h(\mathbf{r}) = \sum_{l=0}^{+\infty} \sum_{m=-l}^{+l} C_{lm} \, h_l(kr) \, Y_{lm}(\hat{\mathbf{r}}), \tag{5.56}$$

where B_{lm} and C_{lm} are the amplitudes corresponding to each mode.

Exercise 5.2.11 Both plane waves and spherical waves represent complete sets of solutions of the vector Helmholtz equation. Thus, one can be expanded in terms of the other. Show that a plane wave can be expanded as (*Bauer's expansion*)

$$e^{i\mathbf{k}\cdot\mathbf{r}} = 4\pi \sum_{l=0}^{+\infty} i^l j_l(kr) \sum_{m=-l}^{+l} Y_{lm}^*(\hat{\mathbf{k}}) \, Y_{lm}(\hat{\mathbf{r}}). \tag{5.57}$$

Verify this result numerically. [Hint: Start by assuming that the plane wave is propagating along the z-axis and prove that $e^{ikz} = e^{ikr\cos\vartheta} = \sum_{l=0}^{+\infty}(2l+1)i^l j_l(kr) P_l(\cos\vartheta)$ by comparing the coefficients of the powers of $r\cos\vartheta$ on both sides of the equation.]

Exercise 5.2.12 Show that a plane wave can be expressed in the far field as

$$e^{i\mathbf{k}\cdot\mathbf{r}} = \frac{2\pi i}{kr} \left[e^{-ikr} \delta(\hat{\mathbf{k}} + \hat{\mathbf{r}}) - e^{ikr} \delta(\hat{\mathbf{k}} - \hat{\mathbf{r}}) \right]. \tag{5.58}$$

[Hint: Substitute the far-field expression of $j_l(kr)$ [Eq. (5.54)] into the expansion of a plane wave [Eq. (5.57)] and then use Eqs. (5.37) and (5.41). See also Jackson (1999).]

5.2.2 The scattering problem

The general scattering problem aims at describing the electromagnetic fields scattered by a particle when it is illuminated by an incoming electromagnetic wave. The geometry of the problem is illustrated in Fig. 5.4. We consider a homogeneous particle of refractive index n_p in a medium of refractive index n_m. If the incident electric field is $\mathbf{E}_i(\mathbf{r})$ and we denote the scattered electric field as $\mathbf{E}_s(\mathbf{r})$, the total electric field outside the particle is

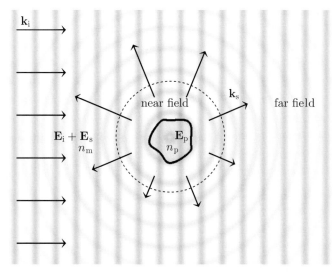

Figure 5.4 The scattering problems. Pictorial view of a scattering process. Scattering theory studies how an incoming electromagnetic wave is scattered by a particle. In general, when light impinges on an object, the object emits a scattered electromagnetic field, which in the far field is a spherical wave. In particular, given an incoming linearly polarised plane electromagnetic wave ($\mathbf{E}_i$) in a medium of refractive index n_m impinging on a particle of homogeneous refractive index n_p, one wants to determine the electromagnetic field inside the particle ($\mathbf{E}_p$) and the scattered electromagnetic field ($\mathbf{E}_s$), both in the near field and in the far field.

$\mathbf{E}_t(\mathbf{r}) = \mathbf{E}_i(\mathbf{r}) + \mathbf{E}_s(\mathbf{r})$. We will indicate the total electric field inside the particle with $\mathbf{E}_p(\mathbf{r})$. All these electric fields satisfy the vector Helmholtz equation in their respective media, i.e.,

$$\begin{cases} \left(\nabla^2 + k_m^2\right) \mathbf{E}_i(\mathbf{r}) = \mathbf{0}, \\ \left(\nabla^2 + k_m^2\right) \mathbf{E}_s(\mathbf{r}) = \mathbf{0}, \\ \left(\nabla^2 + k_p^2\right) \mathbf{E}_p(\mathbf{r}) = \mathbf{0}, \end{cases} \qquad (5.59)$$

where $k_m = n_m k_0$, $k_p = n_p k_0$ and k_0 is the vacuum wavenumber. The corresponding magnetic induction fields are given by

$$\begin{cases} \mathbf{B}_i(\mathbf{r}) = -\frac{i}{\omega} \nabla \times \mathbf{E}_i(\mathbf{r}), \\ \mathbf{B}_s(\mathbf{r}) = -\frac{i}{\omega} \nabla \times \mathbf{E}_s(\mathbf{r}), \\ \mathbf{B}_p(\mathbf{r}) = -\frac{i}{\omega} \nabla \times \mathbf{E}_p(\mathbf{r}), \end{cases} \qquad (5.60)$$

where ω is the angular frequency of the electromagnetic wave. The electromagnetic fields outside and inside the particle are related by the boundary conditions across the surface between the particle and the medium [Box 3.4].

Because any electromagnetic field can be described as a superposition of plane waves, e.g., by its angular spectrum representation described in Section 4.2, and Maxwell's equations are linear, it suffices to consider the scattering produced by a single linearly polarised incoming plane wave:

$$\mathbf{E}_i(\mathbf{r}) = E_i e^{i \mathbf{k}_i \cdot \mathbf{r}} \hat{\mathbf{e}}_i, \qquad (5.61)$$

where $\hat{\mathbf{e}}_i$ is the unit vector indicating the polarisation direction and $\mathbf{k}_i = k_m \hat{\mathbf{k}}_i$ is the wavevector along the incidence propagation direction.

Because $\mathbf{E}_s(\mathbf{r}) = [E_{s,x}(\mathbf{r}), E_{s,y}(\mathbf{r}), E_{s,z}(\mathbf{r})]$ satisfies the vector Helmholtz equation, its Cartesian components must satisfy the scalar Helmholtz equation, which in the case of $E_{s,x}(\mathbf{r})$ is

$$(\nabla^2 + k_m^2) E_{s,x}(\mathbf{r}) = 0. \tag{5.62}$$

As shown in Subsection 5.2.1, the solution of Eq. (5.62) that satisfies the radiation condition at infinity is given by Eq. (5.56), i.e.,

$$E_{s,x}(\mathbf{r}) = E_{s,x}(r, \hat{\mathbf{k}}_s) = \sum_{l=0}^{+\infty} h_l(k_m r) \sum_{m=-l}^{+l} C_{lm,x}(\hat{\mathbf{k}}_i) \, Y_{lm}(\hat{\mathbf{k}}_s), \tag{5.63}$$

where $\hat{\mathbf{k}}_s$ is the radial unit vector indicating the direction of the scattered wave, $\mathbf{r} = r \hat{\mathbf{k}}_s$ and the amplitudes $C_{lm,x}(\hat{\mathbf{k}}_i)$, which depend on the direction of the incident wave, are determined by the boundary conditions at the surface of the particle.

Using the asymptotic properties of $h_l(k_m r)$ for $k_m r \to \infty$ given in Table 5.3, the asymptotic form of $E_{s,x}(\mathbf{r})$ can be written as

$$E_{s,x}(\mathbf{r}) = E_i \frac{e^{ik_m r}}{r} f_x(\hat{\mathbf{k}}_s, \hat{\mathbf{k}}_i), \tag{5.64}$$

where we define the x-component of the *normalised scattering amplitude* as

$$f_x(\hat{\mathbf{k}}_s, \hat{\mathbf{k}}_i) = k_m^{-1} \sum_{l=0}^{+\infty} \sum_{m=-l}^{+l} (-i)^{l+1} C_{lm,x}(\hat{\mathbf{k}}_i) \, Y_{lm}(\hat{\mathbf{k}}_s). \tag{5.65}$$

Repeating the same procedure on $E_{s,y}(\mathbf{r})$ and $E_{s,z}(\mathbf{r})$, we obtain the *normalised scattering amplitude*, $\mathbf{f}(\hat{\mathbf{k}}_s, \hat{\mathbf{k}}_i) = [f_x(\hat{\mathbf{k}}_s, \hat{\mathbf{k}}_i), f_y(\hat{\mathbf{k}}_s, \hat{\mathbf{k}}_i), f_z(\hat{\mathbf{k}}_s, \hat{\mathbf{k}}_i)]$, and the *asymptotic form of the scattered field*,

$$\mathbf{E}_s(\mathbf{r}) = \mathbf{E}_s(r, \hat{\mathbf{k}}_s) = E_i \, \mathbf{f}(\hat{\mathbf{k}}_s, \hat{\mathbf{k}}_i) \, \frac{e^{ik_m r}}{r}. \tag{5.66}$$

In dealing with scattering problems, the scattering amplitude $\mathbf{f}(\hat{\mathbf{k}}_s, \hat{\mathbf{k}}_i)$ is the most important single quantity, to which we can relate all other optical properties such as the cross-sections presented in Section 3.5. For example, the *differential scattering cross-section*, which represents the ratio between the energy scattered by the particle into the unit solid angle around $\hat{\mathbf{k}}_s$ and the incident energy, can be expressed as

$$\frac{d\sigma_{\text{scat}}}{d\Omega} = |\mathbf{f}(\hat{\mathbf{k}}_s, \hat{\mathbf{k}}_i)|^2, \tag{5.67}$$

and the *scattering cross-section*, which is the ratio between the total scattered energy and the incident energy, is given by

$$\sigma_{\text{scat}} = \oint_\Omega \frac{d\sigma_{\text{scat}}}{d\Omega} d\Omega = \oint_\Omega |\mathbf{f}(\hat{\mathbf{k}}_s, \hat{\mathbf{k}}_i)|^2 d\Omega. \tag{5.68}$$

The asymmetry of the scattering with respect to the incoming wave direction and polarisation can be quantified by the asymmetry parameters and, in particular, by the *asymmetry parameter* in the direction of the incoming wave, defined as

$$g_\text{i} = \frac{1}{\sigma_\text{scat}} \oint_\Omega \frac{d\sigma_\text{scat}}{d\Omega} \hat{\mathbf{r}} \cdot \hat{\mathbf{k}}_\text{i} d\Omega, \tag{5.69}$$

and the *transverse asymmetry parameters*, defined as

$$g_1 = \frac{1}{\sigma_\text{scat}} \oint_\Omega \frac{d\sigma_\text{scat}}{d\Omega} \hat{\mathbf{r}} \cdot \hat{\mathbf{u}}_1 \, d\Omega \tag{5.70}$$

and

$$g_2 = \frac{1}{\sigma_\text{scat}} \oint_\Omega \frac{d\sigma_\text{scat}}{d\Omega} \hat{\mathbf{r}} \cdot \hat{\mathbf{u}}_2 \, d\Omega, \tag{5.71}$$

where $\hat{\mathbf{u}}_1 = \hat{\mathbf{e}}_\text{i}$ and $\hat{\mathbf{u}}_2 = \hat{\mathbf{k}}_\text{i} \times \hat{\mathbf{e}}_\text{i}$. The *extinction cross-section* can be obtained using the *optical theorem* [Eq. (3.36)] as

$$\sigma_\text{ext} = \frac{4\pi}{k_\text{m}} \text{Im}[\mathbf{f}(\hat{\mathbf{k}}_\text{s} = \hat{\mathbf{k}}_\text{i}, \hat{\mathbf{k}}_\text{i}) \cdot \hat{\mathbf{e}}_\text{i}]. \tag{5.72}$$

Recalling that extinction is the sum of scattering and absorption, we obtain the *absorption cross-section* as

$$\sigma_\text{abs} = \sigma_\text{ext} - \sigma_\text{scat}. \tag{5.73}$$

Exercise 5.2.13 Show that the normalised scattering amplitude $\mathbf{f}(\hat{\mathbf{k}}_\text{s}, \hat{\mathbf{k}}_\text{i})$ is transverse to the scattering propagation direction, i.e., $\hat{\mathbf{k}}_\text{s}$. What does this mean for the scattered electromagnetic field in the far field?

Exercise 5.2.14 Show that in vacuum the scattered magnetic induction field in the far field is

$$\mathbf{B}_\text{s}(\mathbf{r}) = \mathbf{B}_\text{s}(r, \hat{\mathbf{k}}_\text{s}) = \frac{E_\text{i}}{c} \hat{\mathbf{k}}_\text{s} \times \mathbf{f}(\hat{\mathbf{k}}_\text{s}, \hat{\mathbf{k}}_\text{i}) \frac{e^{ik_\text{m} r}}{r}, \tag{5.74}$$

and hence

$$\mathbf{E}_\text{s} \times \hat{\mathbf{k}}_\text{s} = -c\mathbf{B}_\text{s}, \quad c\mathbf{B}_\text{s} \times \hat{\mathbf{k}}_\text{s} = \mathbf{E}_\text{s}. \tag{5.75}$$

[Hint: Start from the fact that $\mathbf{B}_\text{s}(\mathbf{r}) = -\frac{i}{\omega} \nabla \times \mathbf{E}_\text{s}(\mathbf{r})$ and use Eq. (5.66).]

5.2.3 Multipole expansion

The *vector spherical harmonics* are an extension of the scalar spherical harmonics $Y_{lm}(\hat{\mathbf{r}})$ [Eq. (5.34)] to deal with vector fields, e.g., the solutions of the vector Helmholtz equation.

The vector spherical harmonics are defined as

$$\begin{cases} \mathbf{Y}_{lm}(\hat{\mathbf{r}}) = Y_{lm}(\hat{\mathbf{r}})\hat{\mathbf{r}}, \\ \mathbf{Z}^{(1)}_{lm}(\hat{\mathbf{r}}) = -\dfrac{i}{\sqrt{l(l+1)}}\, \hat{\mathbf{r}} \times \nabla Y_{lm}(\hat{\mathbf{r}}), \\ \mathbf{Z}^{(2)}_{lm}(\hat{\mathbf{r}}) = \mathbf{Z}^{(1)}_{lm}(\hat{\mathbf{r}}) \times \hat{\mathbf{r}}, \end{cases} \quad (5.76)$$

where r is the radial distance and $\hat{\mathbf{r}}$ is the radial unit vector, whereas the polar angle ϑ and the azimuthal angle φ can be expressed as functions of $\hat{\mathbf{r}}$. The first few vector spherical harmonics are shown in Fig. 5.5.

The set of *radial*, i.e., $\mathbf{Y}_{lm}(\hat{\mathbf{r}})$, and *transversal*, i.e., $\mathbf{Z}^{(1)}_{lm}(\hat{\mathbf{r}})$ and $\mathbf{Z}^{(2)}_{lm}(\hat{\mathbf{r}})$, vector spherical harmonics constitute an orthonormal basis for the vectors on the unit sphere, i.e.,

$$\oint_\Omega \mathbf{Y}_{lm} \cdot \mathbf{Y}^*_{l'm'}\, d\Omega = \oint_\Omega \mathbf{Z}^{(1)}_{lm} \cdot \mathbf{Z}^{(1)*}_{l'm'}\, d\Omega = \oint_\Omega \mathbf{Z}^{(2)}_{lm} \cdot \mathbf{Z}^{(2)*}_{l'm'}\, d\Omega = \delta_{ll'}\delta_{mm'},$$

$$\oint_\Omega \mathbf{Y}_{lm} \cdot \mathbf{Z}^{(1)*}_{l'm'}\, d\Omega = \oint_\Omega \mathbf{Z}^{(1)}_{lm} \cdot \mathbf{Z}^{(2)*}_{l'm'}\, d\Omega = \oint_\Omega \mathbf{Z}^{(2)}_{lm} \cdot \mathbf{Y}^*_{l'm'}\, d\Omega = 0,$$

where the integration is over the full solid angle Ω. Therefore, any vector field $\mathbf{F}(r, \hat{\mathbf{r}})$ can be expanded in spherical coordinates as a series of vector spherical harmonics, i.e.,

$$\mathbf{F}(r, \hat{\mathbf{r}}) = \sum_{l=0}^\infty \sum_{m=-l}^l F^{(r)}_{lm}(r)\mathbf{Y}_{lm}(\hat{\mathbf{r}}) + F^{(1)}_{lm}(r)\mathbf{Z}^{(1)}_{lm}(\hat{\mathbf{r}}) + F^{(2)}_{lm}(r)\mathbf{Z}^{(2)}_{lm}(\hat{\mathbf{r}}), \quad (5.77)$$

where r is the radial distance and

$$F^{(r)}_{lm}(r) = \oint_\Omega \mathbf{F}(r, \hat{\mathbf{r}}) \cdot \mathbf{Y}^*_{lm}(\hat{\mathbf{r}})\, d\Omega$$

is the radial dependence of the radial component of the vector field, whereas

$$F^{(1)}_{lm}(r) = \oint_\Omega \mathbf{F}(r, \hat{\mathbf{r}}) \cdot \mathbf{Z}^{(1)*}_{lm}(\hat{\mathbf{r}})\, d\Omega$$

and

$$F^{(2)}_{lm}(r) = \oint_\Omega \mathbf{F}(r, \hat{\mathbf{r}}) \cdot \mathbf{Z}^{(2)*}_{lm}(\hat{\mathbf{r}})\, d\Omega$$

are the radial dependence of the transverse components of the vector field. The expansion in Eq. (5.77) separates the radial dependence of $\mathbf{F}(r, \hat{\mathbf{r}})$ from its angular dependence.

In the remainder of this subsection, we will consider the expansion in Eq. (5.77) for the case of electric fields $\mathbf{E}(r, \hat{\mathbf{r}})$ in a homogenous source-free medium. Differently from the generic vector field $\mathbf{F}(r, \hat{\mathbf{r}})$, $\mathbf{E}(r, \hat{\mathbf{r}})$ must satisfy Maxwell's equations; in particular, it must be divergence-free and satisfy the vector Helmholtz equation, $(\nabla^2 - k^2)\mathbf{E}(r, \hat{\mathbf{r}}) = \mathbf{0}$. We will only present the key results without going into the details of the derivations, which can be found in Borghese et al. (2007).[8]

---

[8] A different notation for the multipole fields is also in use (see, e.g., Mishchenko et al. (2002)), which builds on the notation introduced by Hansen (1935) for the outgoing multipole fields, $\mathbf{M}_{lm}$ and $\mathbf{N}_{lm}$, and their regularised

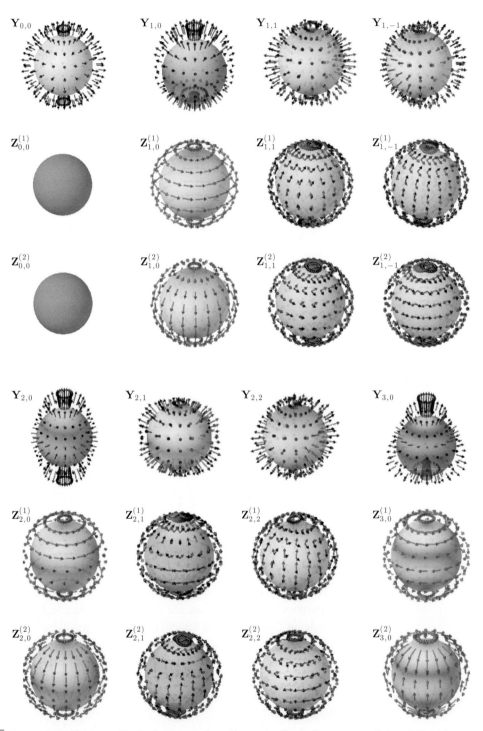

Figure 5.5 Vector spherical harmonics. The first few vector spherical harmonics. The shading represents their moduli and the arrows of different shades of grey represent the components in phase quadrature.

Multipoles regular at the origin

For electric fields regular at the origin, e.g., incident electric fields and electric fields inside particles, the radial dependence can be given in terms of the spherical Bessel functions $j_l(kr)$ as

$$\begin{cases} E^{(r)}_{lm}(r) = E \, \dfrac{i}{kr} \sqrt{l(l+1)} \, j_l(kr), \\ E^{(1)}_{lm}(r) = E \, j_l(kr), \\ E^{(2)}_{lm}(r) = -E \, \dfrac{1}{kr} \left[j_l(kr) + r \dfrac{dj_l(kr)}{dr} \right], \end{cases}$$

where k is the wavenumber and E is a constant that determines the amplitude of the vector field. We can, therefore, introduce the **J**-multipoles

$$\begin{cases} \mathbf{J}^{(1)}_{lm}(kr, \hat{\mathbf{r}}) = j_l(kr) \, \mathbf{Z}^{(1)}_{lm}(\hat{\mathbf{r}}), \\ \mathbf{J}^{(2)}_{lm}(kr, \hat{\mathbf{r}}) = \dfrac{i}{kr} \sqrt{l(l+1)} \, j_l(kr) \, \mathbf{Y}_{lm}(\hat{\mathbf{r}}) - \dfrac{1}{kr} \left[j_l(kr) + r \dfrac{dj_l(kr)}{dr} \right] \mathbf{Z}^{(2)}_{lm}(\hat{\mathbf{r}}) \end{cases} \quad (5.78)$$

and obtain the decomposition of $\mathbf{E}(r, \hat{\mathbf{r}})$ in **J**-multipoles,

$$\mathbf{E}(r, \hat{\mathbf{r}}) = E \sum_{l=0}^{\infty} \sum_{m=-l}^{l} W^{(1)}_{lm} \mathbf{J}^{(1)}_{lm}(kr, \hat{\mathbf{r}}) + W^{(2)}_{lm} \mathbf{J}^{(2)}_{lm}(kr, \hat{\mathbf{r}}), \quad (5.79)$$

where $W^{(1)}_{lm}$ and $W^{(2)}_{lm}$ are numerical coefficients. The superscript '1' ('2') refers to multipolar components of the magnetic (electric) kind, i.e., to magnetic (electric) transverse radiant modes with the components aligned along the magnetic (electric) field.[9] The simplest ones, i.e., the ones corresponding to a dipole, are discussed in Section 3.3, where it can be noted that the electric field of an electric dipole has a radial component in the near field, whereas the magnetic field is always transverse.

In the far field, i.e., for $kr \gg 1$, the **J**-multipoles in Eqs. (5.78) become purely transverse, as the radial near field term of $\mathbf{J}^{(2)}_{lm}(kr, \hat{\mathbf{r}})$ vanishes:

$$\begin{cases} \mathbf{J}^{(1)}_{lm}(kr, \hat{\mathbf{r}}) \longrightarrow j_l(kr) \, \mathbf{Z}^{(1)}_{lm}(\hat{\mathbf{r}}), \\ \mathbf{J}^{(2)}_{lm}(kr, \hat{\mathbf{r}}) \longrightarrow -\dfrac{1}{k} \dfrac{dj_l(kr)}{dr} \mathbf{Z}^{(2)}_{lm}(\hat{\mathbf{r}}). \end{cases} \quad (5.80)$$

Multipoles satisfying the radiation condition at infinity

For an electric field that satisfies the radiation condition at infinity, e.g., a scattered electromagnetic field, the radial dependence can be given in terms of spherical Hankel functions

---

forms at the origin, $\mathrm{Rg}\mathbf{M}_{lm}$ and $\mathrm{Rg}\mathbf{N}_{lm}$. This alternative notation is completely equivalent to the one used in this book, as the following identities hold: $\mathbf{H}^{(1)}_{lm} \equiv \mathbf{M}_{lm}$, $\mathbf{H}^{(2)}_{lm} \equiv \mathbf{N}_{lm}$, $\mathbf{J}^{(1)}_{lm} \equiv \mathrm{Rg}\mathbf{M}_{lm}$ and $\mathbf{J}^{(2)}_{lm} \equiv \mathrm{Rg}\mathbf{N}_{lm}$.

[9] The numerical values of the superscripts are linked to the parity operator; i.e., by reflecting a magnetic (electric) vector, its sign changes (does not change), so that the eigenvalue is $(-1)^1$ ($(-1)^2$). This can be clearly noticed in the case of the magnetic induction and electric field of a dipole, discussed in Section 3.3.

$h_j(kr)$ as

$$\begin{cases} E^{(r)}_{lm}(r) = E \,\dfrac{i}{kr}\sqrt{l(l+1)}\, h_l(kr) \\ E^{(1)}_{lm}(r) = E\, h_l(kr) \\ E^{(2)}_{lm}(r) = -E\, \dfrac{1}{kr}\left[h_l(kr) + r\,\dfrac{dh_l(kr)}{dr}\right], \end{cases}$$

where k is the wavenumber and E is a constant that determines the amplitude of the vector field. We can therefore introduce the **H**-multipoles

$$\begin{cases} \mathbf{H}^{(1)}_{lm}(kr,\hat{\mathbf{r}}) = h_l(kr)\, \mathbf{Z}^{(1)}_{lm}(\hat{\mathbf{r}}), \\ \mathbf{H}^{(2)}_{lm}(kr,\hat{\mathbf{r}}) = \dfrac{i}{kr}\sqrt{l(l+1)}\, h_l(kr)\, \mathbf{Y}_{lm}(\hat{\mathbf{r}}) - \dfrac{1}{kr}\left[h_l(kr) + r\,\dfrac{dh_l(kr)}{dr}\right] \mathbf{Z}^{(2)}_{lm}(\hat{\mathbf{r}}) \end{cases} \quad (5.81)$$

and obtain the decomposition of $\mathbf{E}(r,\hat{\mathbf{r}})$ into **H**-multipoles,

$$\mathbf{E}(r,\hat{\mathbf{r}}) = \sum_{l=0}^{\infty}\sum_{m=-l}^{l} A^{(1)}_{lm}\mathbf{H}^{(1)}_{lm}(kr,\hat{\mathbf{r}}) + A^{(2)}_{lm}\mathbf{H}^{(2)}_{lm}(kr,\hat{\mathbf{r}}), \quad (5.82)$$

where $A^{(1)}_{lm}$ and $A^{(2)}_{lm}$ are numerical coefficients. The superscripts (1) and (2) refer to magnetic and electric multipoles, respectively, as in the case of the **J**-multipoles.

In the far field, i.e., for $kr \gg 1$, the **H**-multipoles in Eqs. (5.81) become purely transverse, as the radial near-field term of $\mathbf{H}^{(2)}_{lm}(kr,\hat{\mathbf{r}})$ vanishes:

$$\begin{cases} \mathbf{H}^{(1)}_{lm}(kr,\hat{\mathbf{r}}) \longrightarrow h_l(kr)\, \mathbf{Z}^{(1)}_{lm}(\hat{\mathbf{r}}), \\ \mathbf{H}^{(2)}_{lm}(kr,\hat{\mathbf{r}}) \longrightarrow -\dfrac{1}{k}\dfrac{dh_l(kr)}{dr}\, \mathbf{Z}^{(2)}_{lm}(\hat{\mathbf{r}}). \end{cases} \quad (5.83)$$

Multipole expansion of a plane wave

We now consider the multipole expansion of a linearly polarised plane wave of the form given in Eq. (5.61), i.e.,

$$\mathbf{E}_i(\mathbf{r}) = E_i e^{i\mathbf{k}_i \cdot \mathbf{r}} \hat{\mathbf{e}}_i.$$

Because a plane wave is finite at the origin, we will use the **J**-multipole expansion given in Eq. (5.79), i.e.,

$$\mathbf{E}_i(r,\hat{\mathbf{r}}) = E_i \sum_{l=0}^{\infty}\sum_{m=-l}^{l} W^{(1)}_{i,lm} \mathbf{J}^{(1)}_{lm}(kr,\hat{\mathbf{r}}) + W^{(2)}_{i,lm} \mathbf{J}^{(2)}_{lm}(kr,\hat{\mathbf{r}}), \quad (5.84)$$

where the amplitudes are

$$W^{(1)}_{i,lm} = 4\pi i^l\, \hat{\mathbf{e}}_i \cdot \mathbf{Z}^{(1)*}_{lm}(\hat{\mathbf{k}}_i) \quad (5.85)$$

and

$$W^{(2)}_{i,lm} = 4\pi i^{l+1}\, \hat{\mathbf{e}}_i \cdot \mathbf{Z}^{(2)*}_{lm}(\hat{\mathbf{k}}_i). \quad (5.86)$$

Exercise 5.2.15 Write a program to visualise the vector spherical harmonics and explore their properties. [Hint: You can use the program `vectorsphericalharmonics` from the book website.]

Exercise 5.2.16 Write a program to visualise the multipoles and explore their properties. [Hint: You can use the program `multipoles` from the book website.]

Exercise 5.2.17 Verify numerically the validity of the expansion of a plane wave in terms of multipoles given by Eq. (5.84). [Hint: You can adapt the program `planewaveexpansion` from the book website.]

Exercise 5.2.18 Show that for a plane wave propagating along the z-direction, i.e., $\hat{\mathbf{k}}_i = \hat{\mathbf{z}}$, and polarised along the x-direction, i.e., $\hat{\mathbf{e}}_i = \hat{\mathbf{x}}$, the amplitudes are

$$W^{(1)}_{i,lm} = \begin{cases} i^l \sqrt{\pi(2l+1)} & m = \pm 1 \\ 0 & m \neq \pm 1 \end{cases} \tag{5.87}$$

and

$$W^{(2)}_{i,lm} = \begin{cases} m\, i^l \sqrt{\pi(2l+1)} & m = \pm 1 \\ 0 & m \neq \pm 1. \end{cases} \tag{5.88}$$

Verify these results numerically.

Exercise 5.2.19 Show that for a circularly polarised plane wave with helicity ± 1, i.e., such that $\mathbf{E}_{i,\pm} = E_i e^{ikz}(\hat{\mathbf{x}} \pm i\hat{\mathbf{y}})$, the amplitudes are

$$W^{(1)}_{\pm,lm} = W^{(2)}_{\pm,lm} = i^l \sqrt{4\pi(2l+1)}\, \delta_{m,\pm 1}. \tag{5.89}$$

Note that, because such a circularly polarised plane wave carries ± 1 unit of angular momentum per photon, its expansion can only involve vector spherical harmonics with $m = \pm 1$. [Hint: Start by writing a general expansion in terms of transverse harmonics, considering the terms $W^{(1)}_{\pm,lm} j_l(kr)$ and $W^{(2)}_{\pm,lm} j_l(kr)$ as expansion coefficients, and then use Eqs. (5.46) and (5.47) to express the field and Bauer's expansion [Eq. (5.57)] to obtain the final result.]

5.2.4 Transition matrix

Because of the linearity of Maxwell's equations [Box 3.2] and of the boundary conditions [Box 3.4], the scattering process can be considered as a linear operator $\mathbb{T}$ (*transition operator*) such that

$$\mathbf{E}_s = \mathbb{T}\mathbf{E}_i, \tag{5.90}$$

where $\mathbf{E}_i$ is the incoming electric field and $\mathbf{E}_s$ is the scattered electric field. Therefore, if both $\mathbf{E}_i$ and $\mathbf{E}_s$ are expanded in suitable bases (not necessarily the same), it is possible to find a *transition matrix* $\mathbb{T}$ that relates the coefficients of such expansions, encompassing all the information on the morphology and orientation of the particle with respect to the incident field (Waterman, 1971).

Because $\mathbf{E}_i$ is in general finite at the origin, its expansion is conveniently given in terms of **J**-multipoles [Eqs. (5.78)] using Eq. (5.79) with amplitudes $W_{i,lm}^{(1)}$ and $W_{i,lm}^{(2)}$. In the important case of a linearly polarised plane wave the expansion is given by Eq. (5.84) with amplitudes given by Eqs. (5.85) and (5.86).

Because $\mathbf{E}_s$ must satisfy the radiation condition at infinity, it is convenient to expand it in terms of **H**-multipoles [Eqs. (5.81)] using Eq. (5.82) with amplitudes $A_{s,lm}^{(1)}$ and $A_{s,lm}^{(2)}$. These amplitudes are determined by imposing the boundary conditions across the surface of the scattering particle.

The transition matrix $\mathbb{T} = \{T_{l'm'lm}^{(p'p)}\}$ of the scattering particle acts on the known multipole amplitudes of the incident field $W_{i,lm}^{(p)}$, where $p = 1, 2$, to give the unknown amplitudes of the scattered field $A_{s,l'm'}^{(p')}$, where $p' = 1, 2$:

$$A_{s,l'm'}^{(p')} = \sum_{p=1,2} \sum_{l=0}^{+\infty} \sum_{m=-l}^{l} T_{l'm'lm}^{(p'p)} W_{i,lm}^{(p)}. \tag{5.91}$$

Exercise 5.2.20 Show that, for a particle illuminated by a linearly polarised plane wave $\mathbf{E}_i(\mathbf{r}) = E_i e^{i\mathbf{k}_i \cdot \mathbf{r}} \hat{\mathbf{e}}_i$, the scattering amplitudes can be expressed in terms of the transition matrix as

$$\mathbf{f}(\hat{\mathbf{k}}_s) = \frac{1}{k_m} \sum_{plm} \sum_{p'l'm'} (-i)^{l+p} \, \mathbf{Z}_{lm}^{(p)}(\hat{\mathbf{k}}_s) \, T_{lml'm'}^{(pp')} \, W_{i,l'm'}^{(p')}. \tag{5.92}$$

[Hint: Use the asymptotic form of the **H**-multipole fields [Eqs. (5.83)].]

5.2.5 Mie scattering

A classical result in scattering theory is the complete solution to the problem of the scattering of a linearly polarised plane wave by a homogeneous sphere of arbitrary size and refractive index. This result was obtained by Gustav Mie (1908) and is therefore known as *Mie theory*. We will now apply the general formalism of scattering theory described in the previous subsection to the derivation of Mie theory. We will first proceed to derive the *Mie coefficients* a_l and b_l for a generic incoming field, and we will then discuss in detail the case of an incoming linearly polarised plane wave.

We consider a dielectric sphere of radius a and refractive index n_p surrounded by a medium of refractive index n_m. We assume the origin of the coordinate system to be at the centre of the particle. The incoming field is expanded in **J**-multipoles as in Eq. (5.84), i.e.,

$$\mathbf{E}_i(r, \hat{\mathbf{r}}) = E_i \sum_{l=0}^{\infty} \sum_{m=-l}^{l} W_{i,lm}^{(1)} \mathbf{J}_{lm}^{(1)}(k_m r, \hat{\mathbf{r}}) + W_{i,lm}^{(2)} \mathbf{J}_{lm}^{(2)}(k_m r, \hat{\mathbf{r}}).$$

The scattered wave is expanded in **H**-multipoles as in Eq. (5.82), i.e.,

$$\mathbf{E}_s(r, \hat{\mathbf{r}}) = E_i \sum_{l=0}^{\infty} \sum_{m=-l}^{l} A_{s,lm}^{(1)} \mathbf{H}_{lm}^{(1)}(k_m r, \hat{\mathbf{r}}) + A_{s,lm}^{(2)} \mathbf{H}_{lm}^{(2)}(k_m r, \hat{\mathbf{r}}).$$

Because the field within the sphere must be regular at the origin, it can be expanded in **J**-multipoles [Eq. (5.79)] as

$$\mathbf{E}_p(r, \hat{\mathbf{r}}) = E_i \sum_{l=0}^{\infty} \sum_{m=-l}^{l} W_{p,lm}^{(1)} \mathbf{J}_{lm}^{(1)}(k_p r, \hat{\mathbf{r}}) + W_{p,lm}^{(2)} \mathbf{J}_{lm}^{(2)}(k_p r, \hat{\mathbf{r}}). \tag{5.93}$$

The corresponding magnetic fields, $\mathbf{B}_i(r, \hat{\mathbf{r}})$, $\mathbf{B}_s(r, \hat{\mathbf{r}})$ and $\mathbf{B}_p(r, \hat{\mathbf{r}})$, can be obtained using Eqs. (5.60).

If we assume that the material of the sphere and that of the surrounding medium are non-magnetic, the boundary conditions reduce to the requirement of continuity of the tangential components of both the electric and magnetic induction fields [Box 3.4]:

$$\begin{cases} \hat{\mathbf{r}} \times (\mathbf{E}_i + \mathbf{E}_s) = \hat{\mathbf{r}} \times \mathbf{E}_p, \\ \hat{\mathbf{r}} \times (\mathbf{B}_i + \mathbf{B}_s) = \hat{\mathbf{r}} \times \mathbf{B}_p. \end{cases} \tag{5.94}$$

Taking into account that the vector spherical harmonics form a complete orthonormal set of functions tangent to the surface of the unit sphere, we can express these boundary conditions using the projections of the fields on the transversal vector spherical harmonics, i.e., $\mathbf{Z}_{lm}^{(1)}(\hat{\mathbf{r}})$ and $\mathbf{Z}_{lm}^{(2)}(\hat{\mathbf{r}})$ [Eqs. (5.76)], as

$$\begin{cases} \oint_\Omega (\mathbf{E}_i + \mathbf{E}_s) \cdot \mathbf{Z}_{lm}^{(1)*}(\hat{\mathbf{r}}) \, d\Omega = \oint_\Omega \mathbf{E}_p \cdot \mathbf{Z}_{lm}^{(1)*}(\hat{\mathbf{r}}) \, d\Omega, \\ \oint_\Omega (\mathbf{E}_i + \mathbf{E}_s) \cdot \mathbf{Z}_{lm}^{(2)*}(\hat{\mathbf{r}}) \, d\Omega = \oint_\Omega \mathbf{E}_p \cdot \mathbf{Z}_{lm}^{(2)*}(\hat{\mathbf{r}}) \, d\Omega, \\ \oint_\Omega (\mathbf{B}_i + \mathbf{B}_s) \cdot \mathbf{Z}_{lm}^{(1)*}(\hat{\mathbf{r}}) \, d\Omega = \oint_\Omega \mathbf{B}_p \cdot \mathbf{Z}_{lm}^{(1)*}(\hat{\mathbf{r}}) \, d\Omega, \\ \oint_\Omega (\mathbf{B}_i + \mathbf{B}_s) \cdot \mathbf{Z}_{lm}^{(2)*}(\hat{\mathbf{r}}) \, d\Omega = \oint_\Omega \mathbf{B}_p \cdot \mathbf{Z}_{lm}^{(2)*}(\hat{\mathbf{r}}) \, d\Omega, \end{cases} \tag{5.95}$$

where the integration is over the full solid angle Ω. Therefore, for each l and m, we get a set of four independent equations:

$$\begin{cases} j_l(\rho_m) W_{i,lm}^{(1)} + h_l(\rho_m) A_{s,lm}^{(1)} = j_l(\rho_p) W_{p,lm}^{(1)}, \\ \dfrac{1}{\rho_m} \{[\rho_m j_l(\rho_m)]' W_{i,lm}^{(2)} + [\rho_m h_l(\rho_m)]' A_{s,lm}^{(2)}\} = \dfrac{1}{\rho_p} [\rho_p h_l(\rho_p)]' W_{p,lm}^{(2)}, \\ \dfrac{n_m}{\rho_m} \{[\rho_m j_l(\rho_m)]' W_{i,lm}^{(1)} + [\rho_m h_l(\rho_m)]' A_{s,lm}^{(1)}\} = \dfrac{n_p}{\rho_p} [\rho_p h_l(\rho_p)]' W_{p,lm}^{(1)}, \\ n_m \{j_l(\rho_m) W_{i,lm}^{(2)} + h_l(\rho_m) A_{s,lm}^{(2)}\} = n_p j_l(\rho_p) W_{p,lm}^{(2)}, \end{cases} \tag{5.96}$$

where the prime denotes differentiation with respect to the argument,

$$\rho_m = n_m k_0 a$$

and

$$\rho_p = n_p k_0 a.$$

The elimination of the internal field amplitudes $W^{(1)}_{p,lm}$ and $W^{(2)}_{p,lm}$ in Eqs. (5.96) yields the amplitudes of the scattered field, i.e.,

$$\begin{cases} A^{(1)}_{s,lm} = -\dfrac{n_p u'_l(\rho_p) u_l(\rho_m) - n_m u_l(\rho_p) u'_l(\rho_m)}{n_p u'_l(\rho_p) w_l(\rho_m) - n_m u_l(\rho_p) w'_l(\rho_m)} W^{(1)}_{i,lm}, \\ A^{(2)}_{s,lm} = -\dfrac{n_m u'_l(\rho_p) u_l(\rho_m) - n_p u_l(\rho_p) u'_l(\rho_m)}{n_m u'_l(\rho_p) w_l(\rho_m) - n_p u_l(\rho_p) w'_l(\rho_m)} W^{(2)}_{i,lm}, \end{cases} \quad (5.97)$$

where the prime denotes differentiation with respect to the argument and

$$u_l(\rho) = \rho j_l(\rho) \quad (5.98)$$

and

$$w_l(\rho) = \rho h_l(\rho) \quad (5.99)$$

are Riccati–Bessel and Riccati–Hankel functions, respectively. The Mie coefficients can thus be defined as

$$\begin{cases} a_l = -\dfrac{A^{(2)}_{s,lm}}{W^{(2)}_{i,lm}} = \dfrac{n_m u'_l(\rho_p) u_l(\rho_m) - n_p u_l(\rho_p) u'_l(\rho_m)}{n_m u'_l(\rho_p) w_l(\rho_m) - n_p u_l(\rho_p) w'_l(\rho_m)}, \\ b_l = -\dfrac{A^{(1)}_{s,lm}}{W^{(1)}_{i,lm}} = \dfrac{n_p u'_l(\rho_p) u_l(\rho_m) - n_m u_l(\rho_p) u'_l(\rho_m)}{n_p u'_l(\rho_p) w_l(\rho_m) - n_m u_l(\rho_p) w'_l(\rho_m)}. \end{cases} \quad (5.100)$$

We note that the Mie coefficients a_l and b_l depend only on l and not on m and that only the Mie coefficients with $l \geq 1$ contribute to the far field. To have an accurate representation of the scattered field, a sufficient number of terms must be considered. In fact, there is a threshold L above which the values of the Mie coefficients drop almost to zero, as shown in Figs. 5.6a, 5.6b and 5.6c. Typically, this threshold can be estimated as

$$L_{\text{Mie}} = k_m a, \quad (5.101)$$

where $k_m a$ is the size parameter of the particle. A more precise empirical formula found by Wiscombe (1979) is

$$L_W = \begin{cases} \text{floor}\{k_m a + 4(k_m a)^{\frac{1}{3}} + 1\} & k_m a \leq 8, \\ \text{floor}\{k_m a + 4.05(k_m a)^{\frac{1}{3}} + 2\} & 8 < k_m a < 4200, \\ \text{floor}\{k_m a + 4(k_m a)^{\frac{1}{3}} + 2\} & k_m a \geq 4200. \end{cases} \quad (5.102)$$

In practice, for small particles, i.e., $k_m a < 0.1$, one need only consider $l = 1$ (Rayleigh approximation) with either formula.

From Eqs. (5.97) and (5.100), we understand that the transition matrix for a sphere illuminated by a plane wave must be diagonal and that its diagonal elements can be

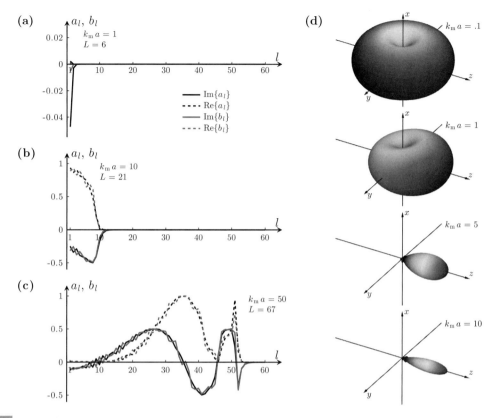

Figure 5.6 Mie coefficients and Mie scattering. (a–c) Imaginary (solid lines) and real (dashed lines) parts of the Mie coefficients a_l and b_l [Eq. (5.100)] as a function of l for dielectric ($n_p = 1.50$) particles in water ($n_m = 1.33$). The values of a_l and b_l drop almost to zero above a certain threshold L_{Mie} [Eq. (5.101)] or L_W [Eq. (5.102)], depending on the particle size parameter $k_m a$. (d) Scattering amplitudes from spheres with various size parameters $k_m a$ illuminated by a plane wave propagating along the z-axis and polarised along the x-axis. For small particles, the scattering is symmetric around the particle, whereas as the size parameter increases, the scattering becomes more directed towards the direction of propagation. All the scattering amplitudes are normalised to the forward scattering amplitude.

expressed in terms of the Mie coefficients as

$$T_{l'm'lm}^{(p'p)} = \begin{cases} -b_l & p = p' = 1 \text{ and } l = l' \text{ and } m = m', \\ -a_l & p = p' = 2 \text{ and } l = l' \text{ and } m = m', \\ 0 & \text{otherwise.} \end{cases} \qquad (5.103)$$

We now consider the incoming wave to be a linearly polarised plane wave; i.e., $W_{i,lm}^{(1)}$ and $W_{i,lm}^{(2)}$ are given by Eqs. (5.85) and (5.86), which is the case originally considered by Mie (1908). The resulting scattering amplitudes are shown in Fig. 5.6d for particles with various size parameters. The observation that the scattered field of small particles is essentially the same as that of a scattering dipole further justifies the Rayleigh approximation discussed in Chapter 3. For larger particles, the scattered fields become more asymmetric with a larger component of the scattering in the forward direction.

Exercise 5.2.21 Show that the magnetic induction field scattered by a Mie particle is

$$\mathbf{B}_s(r, \hat{\mathbf{r}}) = -\frac{i n_m}{c} E_i \sum_{l=0}^{\infty} \sum_{m=-l}^{l} A_{s,lm}^{(2)} \mathbf{H}_{lm}^{(1)}(k_m r, \hat{\mathbf{r}}) + A_{s,lm}^{(1)} \mathbf{H}_{lm}^{(2)}(k_m r, \hat{\mathbf{r}}). \quad (5.104)$$

Exercise 5.2.22 Write a program to calculate the Mie scattering from a sphere. Explore how the scattered fields change as a function of the incoming wavelength, the particle size, the particle refractive index and the medium refractive index. [Hint: You can use the program `miescattering` from the book website.]

Exercise 5.2.23 Show that the scattering amplitude of a spherical particle can be given as a function of the Mie coefficients a_l and b_l as

$$\mathbf{f}(\hat{\mathbf{k}}_s) = -\frac{1}{4\pi k_m} \sum_{l=1}^{+\infty} \left\{ a_l \sum_m |W_{i,lm}^{(2)}|^2 + b_l \sum_m |W_{i,lm}^{(1)}|^2 \right\}. \quad (5.105)$$

What does this expression become in the case of a linearly polarised incoming plane wave (assume $\hat{\mathbf{k}}_i = \hat{\mathbf{z}}$ and $\hat{\mathbf{e}}_i = \hat{\mathbf{x}}$)?

Exercise 5.2.24 Show that for a spherical dielectric particle illuminated by a linearly polarised plane wave it is possible to express the extinction cross-section [Eq. (5.72)] in terms of the Mie coefficients as

$$\sigma_{\text{ext}} = \frac{2\pi}{k_m^2} \sum_{l=1}^{\infty} (2l+1) \operatorname{Re}\{a_l + b_l\}, \quad (5.106)$$

the scattering cross-section [Eq. (5.68)] as

$$\sigma_{\text{scat}} = \frac{2\pi}{k_m^2} \sum_{l=1}^{\infty} (2l+1) \left(|a_l|^2 + |b_l|^2 \right), \quad (5.107)$$

and the asymmetry parameter [Eq. (5.69)] as

$$g_i = \frac{4\pi}{\sigma_{\text{scat}} k_m^2} \operatorname{Re} \left\{ \sum_{l=1}^{+\infty} \left[\frac{l(l+2)}{l+1} \left(a_l a_{l+1}^* + b_l b_{l+1}^* \right) + \frac{2l+1}{l(l+1)} a_l b_l^* \right] \right\}, \quad (5.108)$$

and that g_1 [Eq. (5.70)] and g_2 [Eq. (5.71)] are null.

Exercise 5.2.25 Show that the particle polarisability α_0 given by the Clausius Mosotti relation [Eq. (3.15)] for a small spherical particle of radius a and refractive index n_m in vacuum or air is

$$\alpha_0 = \frac{3}{2} i \lim_{k_m \to 0} \frac{a_1}{k_m^3},$$

where a_1 is the first Mie coefficient [Eq. (5.100)].

Exercise 5.2.26 Show that when the extinction cross-section of spheres is plotted versus the size parameter, some peaks (*Mie resonances*) occur when either denominator of the Mie coefficients [Eq. (5.100)] becomes zero, i.e.,

$$n_m u_l'(\rho_p) w_l(\rho_m) - n_p u_l(\rho_p) w_l'(\rho_m) = 0$$

or

$$n_p u'_l(\rho_p) w_l(\rho_m) - n_m u_l(\rho_p) w'_l(\rho_m) = 0.$$

These equations can be satisfied by complex values of ρ_p and, because the radius a is real, the wavevector must be complex, thus yielding a damping of the field; provided the imaginary part of ρ_p is small, the damping itself will be small, and a resonance peak of finite height and width can occur.

5.3 Optical force and torque

In this section, we will consider the force and torque associated with a scattering process. As usual, we will denote the incoming fields as $\mathbf{E}_i$ and $\mathbf{B}_i$, the scattered fields as $\mathbf{E}_s$ and $\mathbf{B}_s$, and the total fields as $\mathbf{E}_t = \mathbf{E}_i + \mathbf{E}_s$ and $\mathbf{B}_t = \mathbf{B}_i + \mathbf{B}_s$. As we have seen, from the knowledge of these fields it is possible to calculate force and torque using Eqs. (5.8) and (5.11).

5.3.1 Optical force

The force acting on a particle illuminated by an electromagnetic wave can be calculated by integrating the Maxwell stress tensor over a spherical surface of radius r which contains the particle, as indicated by Eq. (5.15). For a harmonic field, the Maxwell stress tensor is given by Eq. (5.14). The resulting expression is

$$\mathbf{F}_{\text{rad}} = \frac{1}{2}\varepsilon_m r^2 \text{Re} \left\{ \oint_\Omega \left[\mathbf{E}_t \otimes \mathbf{E}_t^* + \frac{c^2}{n_m^2} \mathbf{B}_t \otimes \mathbf{B}_t^* - \frac{1}{2}\left(|\mathbf{E}_t|^2 + \frac{c^2}{n_m^2}|\mathbf{B}_t|^2\right) \mathbf{I} \right] \cdot \hat{\mathbf{r}} \, d\Omega \right\}, \quad (5.109)$$

where ε_m is the dielectric permittivity of the medium and n_m is the refractive index of the medium.

The expression for the optical force given by Eq. (5.109) can be significantly simplified by finding it in the far field, i.e., $r \to \infty$. The far-field expressions of the incident and scattered fields are given by

$$\mathbf{E}_i \approx E_i \sum_{p=1,2} \sum_{l=0}^{\infty} \sum_{m=-l}^{l} \mathbf{Z}_{lm}^{(p)}(\hat{\mathbf{r}}) W_{i,lm}^{(p)} \frac{(-1)^{p-1}}{k_m r} \sin[k_m r - (l+1-p)\pi/2], \quad (5.110)$$

$$\mathbf{E}_s \approx E_i \sum_{p=1,2} \sum_{l=0}^{\infty} \sum_{m=-l}^{l} \mathbf{Z}_{lm}^{(p)}(\hat{\mathbf{r}}) A_{s,lm}^{(p)} \frac{\exp(ik_m r)}{k_m r} i^{-l-p}. \quad (5.111)$$

The first integral on the right-hand side of Eq. (5.109) is

$$\oint_\Omega \mathbf{E}_t \otimes \mathbf{E}_t^* \cdot \hat{\mathbf{r}} \, d\Omega = \oint_\Omega (\hat{\mathbf{r}} \cdot \mathbf{E}_t) \mathbf{E}_t \, d\Omega = 0, \quad (5.112)$$

because the far field of both $\mathbf{E}_i$ and $\mathbf{E}_s$ can be expressed in terms of transverse harmonics $\mathbf{Z}_{lm}^{(p)}$ [Eqs. (5.110) and (5.111)], which are by definition orthogonal to $\hat{\mathbf{r}}$. Analogously, the second term on the right-hand side of Eq. (5.109) is

$$\oint_\Omega \frac{c^2}{n_m^2} \mathbf{B}_t \otimes \mathbf{B}_t^* \cdot \hat{\mathbf{r}} \, d\Omega = \oint_\Omega \frac{c^2}{n_m^2} (\hat{\mathbf{r}} \cdot \mathbf{B}_t) \mathbf{B}_t \, d\Omega = 0. \quad (5.113)$$

Therefore, the optical force is given only by the third term in Eq. (5.109). Using the fact that $\mathbf{I} \cdot \hat{\mathbf{r}} = \hat{\mathbf{r}}$, we can rewrite it as

$$\mathbf{F}_{\text{rad}} = -\frac{1}{4}\varepsilon_m r^2 \text{Re}\left\{\oint_\Omega \left[|\mathbf{E}_t|^2 + \frac{c^2}{n_m^2}|\mathbf{B}_t|^2\right] \hat{\mathbf{r}} \, d\Omega\right\}, \quad (5.114)$$

where

$$|\mathbf{E}_t|^2 = |\mathbf{E}_i + \mathbf{E}_s|^2 = |\mathbf{E}_i|^2 + |\mathbf{E}_s|^2 + 2\text{Re}\left\{\mathbf{E}_i \cdot \mathbf{E}_s^*\right\}$$

and

$$|\mathbf{B}_t|^2 = |\mathbf{B}_i + \mathbf{B}_s|^2 = |\mathbf{B}_i|^2 + |\mathbf{B}_s|^2 + 2\text{Re}\left\{\mathbf{B}_i \cdot \mathbf{B}_s^*\right\}.$$

Noticing that

$$\oint_\Omega |\mathbf{E}_i|^2 \hat{\mathbf{r}} \, d\Omega = 0 \quad (5.115)$$

and

$$\oint_\Omega |\mathbf{B}_i|^2 \hat{\mathbf{r}} \, d\Omega = 0, \quad (5.116)$$

as can be seen from the fact that they are the only remaining terms in the absence of the particle, we can further simplify Eq. (5.114) and rewrite it as

$$\mathbf{F}_{\text{rad}} = -\frac{1}{4}\varepsilon_m r^2 \oint_\Omega \left[|\mathbf{E}_s|^2 + \frac{c^2}{n_m^2}|\mathbf{B}_s|^2 + 2\text{Re}\left\{\mathbf{E}_i \cdot \mathbf{E}_s^* + \frac{c^2}{n_m^2}\mathbf{B}_i \cdot \mathbf{B}_s^*\right\}\right] \hat{\mathbf{r}} \, d\Omega. \quad (5.117)$$

Eq. (5.109) can be used for the direct calculation of the optical force acting on a particle by integrating the contribution arising from the optical fields on a surface enclosing the particle in the near field. It is important to note that the mesh of the surface of integration should be sufficiently small to give a good approximation of the optical field. In practice, it is enough that the largest mesh size is significantly smaller than the optical wavelength in the medium. We will employ Eq. (5.109) for the direct calculation of the optical forces arising in an optical tweezers generated by a focused laser beam in Section 5.6. Eq. (5.117) is particularly useful for the theoretical analysis of optical forces. In particular, we will employ it in Section 5.4 for the calculation of the optical forces exerted on a particle by a plane wave and in Subsection 6.1.1 in the development of the T-matrix method for the calculation of optical forces.

Exercise 5.3.1 Show that a generic incident (scattered) field can be expanded in the far field as Eq. (5.110) (Eq. (5.111)). [Hint: Start from the multipole expansion for a generic incident (scattered) field in terms of the **J**-multipoles (**H**-multipoles); then use the

5.3.2 Optical torque

The torque acting on a particle illuminated by an electromagnetic wave can be calculated by integrating the angular momentum of the Maxwell stress tensor over a spherical surface of radius r that contains the particle, as indicated by Eq. (5.16), i.e.,

$$\mathbf{T}_{\text{Rad}} = -r^3 \oint_{\Omega} \left(\overline{\mathsf{T}}_M \times \hat{\mathbf{r}} \right) \cdot \hat{\mathbf{r}} \, d\Omega.$$

This integration requires a well-established origin. If possible, it is convenient to choose a system of reference with origin located at the particle centre of mass and axes coinciding with the particle's principal axes of inertia, because this greatly simplifies the study of the dynamics. For a harmonic field, the Maxwell stress tensor is given by Eq. (5.14), i.e.,

$$\overline{\mathsf{T}}_M = \frac{1}{2}\varepsilon_m \text{Re} \left\{ \mathbf{E}_t \otimes \mathbf{E}_t^* + \frac{c^2}{n_m^2} \mathbf{B}_t \otimes \mathbf{B}_t^* - \frac{1}{2}\left(|\mathbf{E}_t|^2 + \frac{c^2}{n_m^2}|\mathbf{B}_t|^2 \right) \mathsf{I} \right\},$$

where, because $\hat{\mathbf{r}} \times \mathsf{I} \cdot \hat{\mathbf{r}} \equiv \mathbf{0}$, the last two terms do not make any contribution to the torque. Therefore, the resulting expression for the torque is

$$\mathbf{T}_{\text{Rad}} = -\frac{\varepsilon_m r^3}{2} \text{Re} \left\{ \oint \left[(\hat{\mathbf{r}} \cdot \mathbf{E}_t)(\mathbf{E}_t^* \times \hat{\mathbf{r}}) + \frac{c^2}{n_m^2}(\hat{\mathbf{r}} \cdot \mathbf{B}_t)(\mathbf{B}_t^* \times \hat{\mathbf{r}}) \right] d\Omega \right\}. \quad (5.118)$$

It is important to note that the far-field expression for the scattered fields in terms of the scattering amplitude cannot be used in the calculation of the torque. In fact, in the far field, the scattering amplitude is orthogonal to the radial unit vector $\hat{\mathbf{r}}$, so that such a limiting procedure would lead to a vanishing result. For a correct calculation of the radiation torque one has to retain the radial terms of the fields, and therefore the integration must be performed on a sphere of finite radius but still enclosing the whole particle.

No torque can be exerted by a linearly polarised plane wave on a spherical particle, but some torque can arise if a microsphere is made of an absorbing material in the presence of an elliptically polarised plane wave, as will be explained in Section 5.5.

The radiation torque can produce alignment of an asymmetric particle with the polarisation of an incoming plane wave or, if the torque is not aligned with the particle inertial axes, it can induce a constant spinning of the particle, such as the windmill effect described in Section 2.8.

5.4 Optical force from a plane wave

We will now turn our attention to the force and torque associated with the scattering process of a linearly polarised plane wave whose electric field is $\mathbf{E}_i = E_i e^{i \mathbf{k}_i \cdot \mathbf{r}} \hat{\mathbf{e}}_i$ and whose magnetic induction field is given by $\mathbf{B}_i = -\frac{i}{\omega} \nabla \times \mathbf{E}_i$ [Box 3.5].

For the first two terms on the right-hand side of Eq. (5.117), because we are assuming $r \to \infty$, we can use the far-field expressions for the scattered fields given by Eqs. (5.66) and (5.74), obtaining respectively

$$\oint_\Omega |\mathbf{E}_s|^2 \hat{\mathbf{r}} \, d\Omega = \frac{|E_i|^2}{r^2} \oint_\Omega |\mathbf{f}(\hat{\mathbf{r}}, \hat{\mathbf{k}}_i)|^2 \hat{\mathbf{r}} \, d\Omega$$

and

$$\oint_\Omega |\mathbf{B}_s|^2 \hat{\mathbf{r}} \, d\Omega = \frac{n_m^2}{c^2} \frac{|E_i|^2}{r^2} \oint_\Omega |\mathbf{f}(\hat{\mathbf{r}}, \hat{\mathbf{k}}_i)|^2 \hat{\mathbf{r}} \, d\Omega,$$

where $\hat{\mathbf{k}}_s = \hat{\mathbf{r}}$. Thus, we can rewrite this sum as

$$-\frac{1}{4}\varepsilon_m r^2 \oint_\Omega \left[|\mathbf{E}_s|^2 + \frac{c^2}{n_m^2}|\mathbf{B}_s|^2\right] \hat{\mathbf{r}} \, d\Omega = -\frac{1}{2}\varepsilon_m |E_i|^2 \oint_\Omega |\mathbf{f}(\hat{\mathbf{r}}, \hat{\mathbf{k}}_i)|^2 \hat{\mathbf{r}} \, d\Omega. \quad (5.119)$$

We now come to the last term on the right-hand side of Eq. (5.117),

$$-\frac{1}{2}\varepsilon_m r^2 \mathrm{Re}\left\{\oint_\Omega \left(\mathbf{E}_i \cdot \mathbf{E}_s^* + \frac{c^2}{n_m^2}\mathbf{B}_i \cdot \mathbf{B}_s^*\right) \hat{\mathbf{r}} \, d\Omega\right\}.$$

Making use of the far-field expression for the incoming plane wave given by Eq. (5.58),

$$\mathbf{E}_i = E_i \frac{2\pi i}{k_m} \left[\delta(\hat{\mathbf{k}}_i + \hat{\mathbf{r}})\frac{e^{-ik_m r}}{r} - \delta(\hat{\mathbf{k}}_i - \hat{\mathbf{r}})\frac{e^{ik_m r}}{r}\right] \hat{\mathbf{e}}_i,$$

and of the far-field expression for the scattered field in terms of the scattering amplitude given by Eq. (5.66), we can write

$$\mathbf{E}_i \cdot \mathbf{E}_s^* = |E_i|^2 \frac{2\pi i}{k_m r^2} \left[\delta(\hat{\mathbf{k}}_i + \hat{\mathbf{r}})e^{2ik_m r} - \delta(\hat{\mathbf{k}}_i - \hat{\mathbf{r}})\right] \hat{\mathbf{e}}_i \cdot \mathbf{f}^*(\hat{\mathbf{r}}, \hat{\mathbf{k}}_i). \quad (5.120)$$

Analogously, making use of the fact that $\mathbf{B}_i = -\frac{i}{\omega} \nabla \times \mathbf{E}_i$, where

$$\nabla \times \mathbf{E}_i = E_i \frac{2\pi i}{k_m} \nabla \times \left[\left(\delta(\hat{\mathbf{k}}_i + \hat{\mathbf{r}})\frac{e^{-ik_m r}}{r} - \delta(\hat{\mathbf{k}}_i - \hat{\mathbf{r}})\frac{e^{ik_m r}}{r}\right) \hat{\mathbf{e}}_i\right],$$

and of the far-field expression for $\mathbf{B}_s$ given in Eq. (5.74),

$$\mathbf{B}_s(\mathbf{r}) = \mathbf{B}_s(r, \hat{\mathbf{r}}) = \frac{n_m}{c} E_i \, \hat{\mathbf{r}} \times \mathbf{f}(\hat{\mathbf{r}}, \hat{\mathbf{k}}_i) \, \frac{e^{ik_m r}}{r},$$

we can write

$$\frac{c^2}{n_m^2} \mathbf{B}_i \cdot \mathbf{B}_s^*$$

$$= |E_i|^2 \frac{2\pi}{k_m^2} \frac{e^{-ik_m r}}{r} \nabla \times \left\{\left[\delta(\hat{\mathbf{k}}_i + \hat{\mathbf{r}})\frac{e^{-ik_m r}}{r} - \delta(\hat{\mathbf{k}}_i - \hat{\mathbf{r}})\frac{e^{ik_m r}}{r}\right] \hat{\mathbf{e}}_i\right\} \cdot [\hat{\mathbf{r}} \times \mathbf{f}^*(\hat{\mathbf{r}}, \hat{\mathbf{k}}_i)].$$

We can now use the vector identities [Box 3.1] and the fact that $\nabla \times \hat{\mathbf{e}}_i = \mathbf{0}$ because $\hat{\mathbf{e}}_i$ is constant to simplify

$$\nabla \times \left[\delta(\hat{\mathbf{k}}_i \mp \hat{\mathbf{r}})\frac{e^{\pm ik_m r}}{r} \hat{\mathbf{e}}_i\right] = \nabla\left[\delta(\hat{\mathbf{k}}_i \mp \hat{\mathbf{r}})\frac{e^{\pm ik_m r}}{r}\right] \times \hat{\mathbf{e}}_i$$

and
$$\nabla\left[\delta(\hat{\mathbf{k}}_i \mp \hat{\mathbf{r}})\frac{e^{\pm ik_m r}}{r}\right] \approx \pm ik_m \frac{e^{\pm ik_m r}}{r}\delta(\hat{\mathbf{k}}_i \mp \hat{\mathbf{r}})\hat{\mathbf{r}},$$

where the approximation is justified for $r \to \infty$. Thus, we obtain

$$\frac{c^2}{n_m^2}\mathbf{B}_i \cdot \mathbf{B}_s^* = -|E_i|^2 \frac{2\pi i}{k_m r^2}\left[\delta(\hat{\mathbf{k}}_i + \hat{\mathbf{r}})e^{-2ik_m r} + \delta(\hat{\mathbf{k}}_i - \hat{\mathbf{r}})\right]\hat{\mathbf{e}}_i \cdot \mathbf{f}^*(\hat{\mathbf{r}}, \hat{\mathbf{k}}_i), \quad (5.121)$$

where we have made use of the fact that $(\hat{\mathbf{r}} \times \hat{\mathbf{e}}_i) \cdot (\hat{\mathbf{r}} \times \mathbf{f}^*) = \hat{\mathbf{e}}_i \cdot \mathbf{f}^*$. Therefore, from Eqs. (5.120) and (5.121) it follows that

$$\mathbf{E}_i \cdot \mathbf{E}_s^* + \frac{c^2}{n_m^2}\mathbf{B}_i \cdot \mathbf{B}_s^* = -|E_i|^2 \frac{4\pi i}{k_m r^2}\delta(\hat{\mathbf{k}}_i - \hat{\mathbf{r}})\hat{\mathbf{e}}_i \cdot \mathbf{f}^*(\hat{\mathbf{r}}, \hat{\mathbf{k}}_i)$$

and

$$-\frac{1}{2}\varepsilon_m r^2 \mathrm{Re}\left\{\oint_\Omega\left(\mathbf{E}_i \cdot \mathbf{E}_s^* + \frac{c^2}{n_m^2}\mathbf{B}_i \cdot \mathbf{B}_s^*\right)\hat{\mathbf{r}}\,d\Omega\right\} = \varepsilon_m |E_i|^2 \frac{2\pi}{k}\mathrm{Im}\{\hat{\mathbf{e}}_i \cdot \mathbf{f}(\hat{\mathbf{k}}_i, \hat{\mathbf{k}}_i)\}\hat{\mathbf{k}}_i. \quad (5.122)$$

Finally, by combining Eqs. (5.119) and (5.122), we obtain (Debye, 1909; Mishchenko, 2001)

$$\mathbf{F}_{\mathrm{rad}} = \frac{n_m}{c}I_i\left[\sigma_{\mathrm{ext}}\hat{\mathbf{k}}_i - \oint_\Omega \frac{d\sigma_{\mathrm{scat}}}{d\Omega}\hat{\mathbf{r}}\,d\Omega\right], \quad (5.123)$$

where the differential scattering cross-section is [Eq. (5.67)]

$$\frac{d\sigma_{\mathrm{scat}}}{d\Omega} = |\mathbf{f}(\hat{\mathbf{r}}, \hat{\mathbf{k}}_i)|^2,$$

the extinction cross-section is [Eq. (5.72)]

$$\sigma_{\mathrm{ext}} = \frac{4\pi}{k_m}\mathrm{Im}\left\{\hat{\mathbf{e}}_i \cdot \mathbf{f}(\hat{\mathbf{k}}_i, \hat{\mathbf{k}}_i)\right\}$$

and the intensity of the incident plane wave is [Eq. (4.6)]

$$I_i = \frac{\varepsilon_m}{2}\frac{c}{n_m}E_i^2.$$

The scattering force on a sphere is presented in Figs. 5.7a and 5.7b as a function of the sphere radius a and size parameter $k_m a$, respectively.

Eq. (5.123) relates the radiation force to the optical properties of the particle. Whereas the first term on the right-hand-side of Eq. (5.123) is a force in the direction of the incident wave $\hat{\mathbf{k}}_i$, the second term can also present transverse components perpendicular to $\hat{\mathbf{k}}_i$. The projection of the force along $\hat{\mathbf{k}}_i$ is the *radiation pressure*,

$$\mathbf{F}_{\mathrm{rad}}^{\parallel} = \frac{n_m}{c}I_i\left[\sigma_{\mathrm{ext}} - g_i\sigma_{\mathrm{scat}}\right]\hat{\mathbf{k}}_i = \frac{n_m}{c}I_i\sigma_{\mathrm{rad}}\hat{\mathbf{k}}_i, \quad (5.124)$$

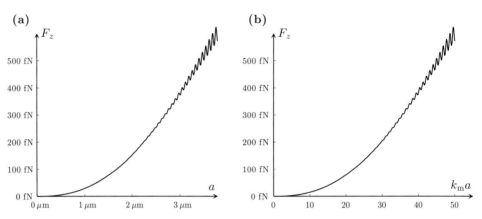

Figure 5.7 Radiation force of a plane wave on a sphere. Radiation force [Eq. (5.123)] produced on a sphere by a linearly polarised plane wave plotted as a function of (a) the sphere radius a and (b) the sphere size parameter $k_m a$. The medium refractive index is $n_m = 1.33$ and the particle refractive index is $n_p = 1.50$. The vacuum wavelength is $\lambda_0 = 633$ nm and the electric field amplitude is $E_i = 10^5$ V/m, corresponding to $1.8 \cdot 10^7$ W/m^2.

where the asymmetry parameter in the direction of the incoming wave is [Eq. (5.69)]

$$g_i = \frac{1}{\sigma_{\text{scat}}} \oint_\Omega \frac{d\sigma_{\text{scat}}}{d\Omega} \hat{\mathbf{r}} \cdot \hat{\mathbf{k}}_i \, d\Omega,$$

the scattering cross-section is [Eq. (5.68)]

$$\sigma_{\text{scat}} = \oint_\Omega \frac{d\sigma_{\text{scat}}}{d\Omega} d\Omega = \oint_\Omega |\mathbf{f}(\hat{\mathbf{k}}_s, \hat{\mathbf{k}}_i)|^2 d\Omega$$

and the radiation cross-section is

$$\sigma_{\text{rad}} = \sigma_{\text{ext}} - g_i \sigma_{\text{scat}}.$$

The transverse component of the force is

$$\mathbf{F}_{\text{rad}}^\perp = -\frac{n_m}{c} I_i \, g_1 \sigma_{\text{scat}} \hat{\mathbf{u}}_1 - \frac{n_m}{c} I_i \, g_2 \, \sigma_{\text{scat}} \hat{\mathbf{u}}_2, \quad (5.125)$$

where $\hat{\mathbf{u}}_1 = \hat{\mathbf{e}}_i$, $\hat{\mathbf{u}}_2 = \hat{\mathbf{k}}_i \times \hat{\mathbf{e}}_i$ and the transverse asymmetry parameters are [Eqs. (5.70) and (5.71)]

$$g_1 = \frac{1}{\sigma_{\text{scat}}} \oint_\Omega \frac{d\sigma_{\text{scat}}}{d\Omega} \hat{\mathbf{r}} \cdot \hat{\mathbf{u}}_1 \, d\Omega$$

and

$$g_2 = \frac{1}{\sigma_{\text{scat}}} \oint_\Omega \frac{d\sigma_{\text{scat}}}{d\Omega} \hat{\mathbf{r}} \cdot \hat{\mathbf{u}}_2 \, d\Omega.$$

For small (dipolar) particles g_i, g_1, $g_2 \approx 0$, as can be seen from the first of the scattering diagrams in Fig. 5.6d. Furthermore, if the particle has a spherical symmetry, g_1 and g_2 are always 0, whereas g_i can be nonzero.

Exercise 5.4.1 Calculate the radiation pressure on a sphere using the cross-sections and the asymmetry factor expressed in terms of the Mie coefficients [Eqs. (5.106), (5.107) and (5.108)]. Explore how the radiation force scales with the sphere size. Compare these results with the direct integration of Eq. (5.109). Which method is numerically more stable? [Hint: You can adapt the program `mieplaneforce` from the book website.]

Exercise 5.4.2 Demonstrate analytically that a linearly polarised plane wave does not produce any radiation torque on a spherical particle. Verify this numerically by integrating Eq. (5.118).

5.5 Transfer of spin angular momentum to a sphere

A linearly polarised plane wave does not produce any radiation torque on a spherically symmetric homogeneous particle made of any material. A torque is possible, however, if the illuminating light is elliptically polarised, if the spherical particle is made of an absorbing material. This result was first rigorously proved by Marston and Crichton (1984) using the multipole expansion. Here, we present a derivation following that original work.

We start by considering a left-handed circularly polarised plane wave incident on a homogeneous sphere. The incident fields are

$$\begin{cases} \mathbf{E}_i = E_i e^{ik_m z}(\hat{\mathbf{x}} + i\hat{\mathbf{y}}) \\ \dfrac{c}{n_m}\mathbf{B}_i = -i\mathbf{E}_i. \end{cases}$$

Because of the cylindrical symmetry of the problem, the only component of the torque that can be different from zero is $T_{\text{Rad},z}$, whereas $T_{\text{Rad},x} = T_{\text{Rad},y} = 0$. By expressing the total field as $\mathbf{E}_t = \mathbf{E}_i + \mathbf{E}_s$, we can write $T_{\text{Rad},z}$ as the sum of four terms related to the combination of incident and scattered field contributions,

$$T_{\text{Rad},z} = T^{\text{ii}} + T^{\text{is}} + T^{\text{si}} + T^{\text{ss}}, \tag{5.126}$$

where, from Eq. (5.118),

$$T^{\text{ii}} = -\frac{\varepsilon_m r^3}{2}\text{Re}\left\{\oint_\Omega \left[(\hat{\mathbf{r}}\cdot\mathbf{E}_i)(\mathbf{E}_i^*\times\hat{\mathbf{r}}) + \frac{c^2}{n_m^2}(\hat{\mathbf{r}}\cdot\mathbf{B}_i)(\mathbf{B}_i^*\times\hat{\mathbf{r}})\right]\cdot\hat{\mathbf{z}}\,d\Omega\right\} = 0,$$

because the averaged angular momentum transported across S by the incident wave in the absence of the scatterer must vanish:

$$T^{\text{is}} = -\frac{\varepsilon_m r^3}{2}\text{Re}\left\{\oint_\Omega \left[(\hat{\mathbf{r}}\cdot\mathbf{E}_i)(\mathbf{E}_s^*\times\hat{\mathbf{r}}) + \frac{c^2}{n_m^2}(\hat{\mathbf{r}}\cdot\mathbf{B}_i)(\mathbf{B}_s^*\times\hat{\mathbf{r}})\right]\cdot\hat{\mathbf{z}}\,d\Omega\right\},$$

$$T^{\text{si}} = -\frac{\varepsilon_m r^3}{2}\text{Re}\left\{\oint_\Omega \left[(\hat{\mathbf{r}}\cdot\mathbf{E}_s)(\mathbf{E}_i^*\times\hat{\mathbf{r}}) + \frac{c^2}{n_m^2}(\hat{\mathbf{r}}\cdot\mathbf{B}_s)(\mathbf{B}_i^*\times\hat{\mathbf{r}})\right]\cdot\hat{\mathbf{z}}\,d\Omega\right\}$$

and

$$T^{ss} = -\frac{\varepsilon_m r^3}{2} \text{Re} \left\{ \oint_\Omega \left[(\hat{\mathbf{r}} \cdot \mathbf{E}_s)(\mathbf{E}_s^* \times \hat{\mathbf{r}}) + \frac{c^2}{n_m^2}(\hat{\mathbf{r}} \cdot \mathbf{B}_s)(\mathbf{B}_s^* \times \hat{\mathbf{r}}) \right] \cdot \hat{\mathbf{z}} \, d\Omega \right\}.$$

We now evaluate these terms in the far field, i.e., neglecting contributions that vanish as $r \to \infty$. For the incident field, we make use of the expansion of the circularly polarised plane wave given in Eq. (5.89), whereas the scattered field is expanded in terms of the vector spherical harmonics using the Mie coefficients as

$$\begin{cases} \mathbf{E}_s = -E_i \sum_{l=1}^{\infty} i^l \sqrt{4\pi(2l+1)} \left\{ b_l h_l \mathbf{Z}_{l1}^{(1)} + \frac{a_l}{k_m} \nabla \times [h_l \mathbf{Z}_{l1}^{(1)}] \right\} \\ \frac{c}{n_m} \mathbf{B}_s = E_i \sum_{l=1}^{\infty} i^{l+1} \sqrt{4\pi(2l+1)} \left\{ \frac{b_l}{k_m} \nabla \times [h_l \mathbf{Z}_{l1}^{(1)}] + a_l h_l \mathbf{Z}_{l1}^{(1)} \right\}. \end{cases} \quad (5.127)$$

The second term in Eq. (5.126), T^{is}, is simplified using the far field relations between $\mathbf{E}_s$ and $\mathbf{B}_s$ [Eqs. (5.75)] as

$$T^{is} = -\frac{\varepsilon_m r^3}{2} \text{Re} \left\{ \oint_\Omega (\hat{\mathbf{r}} \cdot \mathbf{E}_i) \left(\frac{c}{n_m} \mathbf{B}_s^* + i \mathbf{E}_s^* \right) \cdot \hat{\mathbf{z}} \, d\Omega \right\},$$

where we can use the plane wave expansion in Eq. (5.89) to rewrite the incoming field factor as

$$\hat{\mathbf{r}} \cdot \mathbf{E}_i = E_i \sum_{l'=1}^{\infty} \frac{i^{l'+1}}{k_m r} \sqrt{4\pi(2l'+1)l'(l'+1)} \, j_{l'} \, Y_{l'1}$$

and we can write the scattered field factor in terms of the Mie coefficients using Eqs. (5.127) as

$$\left(\frac{c}{n_m} \mathbf{B}_s^* + i \mathbf{E}_s^* \right) \cdot \hat{\mathbf{z}} = E_i \sum_{l=1}^{\infty} (-i)^{l+1} \sqrt{4\pi(2l+1)} (a_l^* + b_l^*) h_l^* \left[\mathbf{Z}_{l1}^{(1)*} + i \mathbf{Z}_{l1}^{(2)*} \right] \cdot \hat{\mathbf{z}}.$$

Thus, to evaluate T^{is}, we need to calculate the integrals

$$\oint_\Omega \hat{\mathbf{z}} \cdot \mathbf{Z}_{l1}^{(1)*} Y_{l'1} \, d\Omega \quad \text{and} \quad \oint_\Omega i\hat{\mathbf{z}} \cdot \mathbf{Z}_{l1}^{(2)*} Y_{l'1} \, d\Omega.$$

The first integral can be solved using the definition of transverse spherical harmonics and the orthogonality property of spherical harmonics, so that we get

$$\oint_\Omega \hat{\mathbf{z}} \cdot \mathbf{Z}_{l1}^{(1)*} Y_{l'1} \, d\Omega = \oint_\Omega \frac{\hat{\mathbf{z}} \cdot (\mathbf{L} Y_{l1})^*}{\sqrt{l(l+1)}} Y_{l'1} \, d\Omega = \frac{\delta_{l'l}}{\sqrt{l(l+1)}}. \quad (5.128)$$

The second integral can be solved using the fact that (Rose, 1955)

$$i\hat{\mathbf{z}} \cdot \mathbf{Z}_{l1}^{(2)*} = \sqrt{\frac{l^2(l+2)}{(2l+1)(l+1)(2l+3)}} Y_{l+1,1}^* - \sqrt{\frac{(l+1)^2(l-1)}{(2l-1)l(2l+1)}} Y_{l-1,1}^*$$

and then using the orthogonality property of spherical harmonics to get

$$\oint_\Omega i\hat{\mathbf{z}} \cdot \mathbf{Z}_{l1}^{(2)*} Y_{l'1} \, d\Omega = \sqrt{\frac{l^2(l+2)}{(2l+1)(l+1)(2l+3)}} \delta_{l+1,l'} - \sqrt{\frac{(l+1)^2(l-1)}{(2l-1)l(2l+1)}} \delta_{l-1,l'}.$$
(5.129)

Finally, by using these integrals and the far-field expansion of j_l and h_l [Table 5.3], we get

$$T^{\text{is}} = \frac{2\pi \varepsilon_{\text{m}} E_{\text{i}}^2}{k_{\text{m}}^3} \sum_{l=1}^{\infty} (2l+1)\text{Re}\,\{a_l + b_l\} = +\frac{\varepsilon_{\text{m}} E_{\text{i}}^2}{k_{\text{m}}} \sigma_{\text{ext}},$$
(5.130)

where we have used the expression for the extinction cross-section σ_{ext} in terms of Mie coefficients [Eq. (5.106)].

The third term in Eq. (5.126), T^{si}, can be simplified using vector algebra to

$$T^{\text{si}} = -\frac{\varepsilon_{\text{m}} r^3}{2} \text{Re}\left\{\oint_\Omega i\frac{c}{n_{\text{m}}} \left[(\hat{\mathbf{r}} \cdot \mathbf{E}_{\text{s}})(\mathbf{B}_{\text{i}}^* \cdot \hat{\mathbf{r}}) - (\hat{\mathbf{r}} \cdot \mathbf{B}_{\text{s}})(\mathbf{E}_{\text{i}}^* \cdot \hat{\mathbf{r}})\right] d\Omega\right\} = 0,$$

as can be proved by explicitly writing the fields into the equation.

Finally, we evaluate the fourth term in Eq. (5.126), T^{ss}, which can be simplified using vector algebra to

$$T^{\text{ss}} = -\frac{\varepsilon_{\text{m}} r^3}{2} \text{Re}\left\{\oint_\Omega \frac{c}{n_{\text{m}}} \left[-(\hat{\mathbf{r}} \cdot \mathbf{E}_{\text{s}})(\hat{\mathbf{z}} \cdot \mathbf{B}_{\text{s}}^*) + (\hat{\mathbf{r}} \cdot \mathbf{B}_{\text{s}})(\hat{\mathbf{z}} \cdot \mathbf{E}_{\text{s}}^*)\right] d\Omega\right\},$$

where, using multipole expansions in the far field, we get

$$\hat{\mathbf{r}} \cdot \mathbf{E}_{\text{s}} = -E_{\text{i}} \sum_{l'=1}^{\infty} \frac{i^{l'+1}}{k_{\text{m}} r} \sqrt{4\pi(2l'+1)l'(l'+1)} a_{l'} h_{l'} Y_{l'1},$$

$$\frac{c}{n_{\text{m}}} \hat{\mathbf{r}} \cdot \mathbf{B}_{\text{s}} = -E_{\text{i}} \sum_{l'=1}^{\infty} \frac{i^{l'}}{k_{\text{m}} r} \sqrt{4\pi(2l'+1)} b_{l'} h_{l'} Y_{l'1},$$

$$\frac{c}{n_{\text{m}}} \hat{\mathbf{z}} \cdot \mathbf{B}_{\text{s}}^* = E_{\text{i}} \sum_{l=1}^{\infty} (-i)^{l+1} \sqrt{4\pi(2l+1)} \left[i b_l^* h_l^* \mathbf{Z}_{l1}^{(2)*} \cdot \hat{\mathbf{z}} + a_l^* h_l^* \mathbf{Z}_{l1}^{(1)*} \cdot \hat{\mathbf{z}}\right],$$

$$\hat{\mathbf{z}} \cdot \mathbf{E}_{\text{s}}^* = -E_{\text{i}} \sum_{l=1}^{\infty} (-i)^l \sqrt{4\pi(2l+1)} \left[b_l^* h_l^* \mathbf{Z}_{l1}^{(1)*} \cdot \hat{\mathbf{z}} + i a_l^* h_l^* \mathbf{Z}_{l1}^{(2)*} \cdot \hat{\mathbf{z}}\right].$$

Finally, using Eqs. (5.128) and (5.129), we obtain

$$T^{\text{ss}} = -\frac{2\pi \varepsilon_{\text{m}} E_{\text{i}}^2}{k_{\text{m}}^3} \sum_{l=1}^{\infty} (2l+1) \left(|a_l|^2 + |b_l|^2\right)$$

$$- \frac{2\pi \varepsilon_{\text{m}} E_{\text{i}}^2}{k_{\text{m}}^3} \sum_{l=1}^{\infty} l(l+2)\text{Re}\,\left\{\left(a_l b_{l+1}^* + b_l a_{l+1}^*\right) - \left(a_l^* b_{l+1} + b_l^* a_{l+1}\right)\right\}$$

and, because $\mathrm{Re}\{a_l b_{l+1}^* + b_l a_{l+1}^*\} = \mathrm{Re}\{a_l^* b_{l+1} + b_l^* a_{l+1}\}$, using the expression for the scattering cross-section σ_{scat} [Eq. (5.107)], we get

$$T^{\mathrm{ss}} = -\frac{\varepsilon_{\mathrm{m}} E_i^2}{k_{\mathrm{m}}} \sigma_{\mathrm{scat}}. \tag{5.131}$$

Thus, summing up Eqs. (5.130) and (5.131), we obtain the expression for the radiation torque on a sphere,

$$T_{\mathrm{Rad},z} = +\frac{\varepsilon_{\mathrm{m}} E_i^2}{k_{\mathrm{m}}} \sigma_{\mathrm{abs}}, \tag{5.132}$$

which shows that the radiation torque on a homogeneous sphere is associated only with the absorption cross-section. Because $k_{\mathrm{m}} = n_{\mathrm{m}} \omega/c$, we can also write the latter equation in terms of the intensity of the incident field, so that

$$T_{\mathrm{Rad},z} = +\frac{I_i}{\omega} \sigma_{\mathrm{abs}}. \tag{5.133}$$

Exercise 5.5.1 Show that for a right-handed circularly polarised plane wave, the radiation torque is

$$T_{\mathrm{Rad},z} = -\frac{\varepsilon_{\mathrm{m}} E_i^2}{k_{\mathrm{m}}} \sigma_{\mathrm{abs}} = -\frac{I_i}{\omega} \sigma_{\mathrm{abs}}.$$

Exercise 5.5.2 Generalise the result obtained in Eq. (5.132) for a generic elliptical polarisation and show that

$$T_{\mathrm{Rad},z} = s \frac{\varepsilon_{\mathrm{m}} E_i^2}{k_{\mathrm{m}}} \sigma_{\mathrm{abs}} = s \frac{I_i}{\omega} \sigma_{\mathrm{abs}},$$

where s is the polarisation content of the plane wave defined in Exercise 5.1.10.

Exercise 5.5.3 Generalising further the results obtained in the previous exercise, show that the optical torque produced on a spherical particle by a plane wave propagating along the $\hat{\mathbf{k}}_i$ direction is

$$\mathbf{T}_{\mathrm{Rad}} = s \frac{\varepsilon_{\mathrm{m}} E_i^2}{k_{\mathrm{m}}} \sigma_{\mathrm{abs}} \hat{\mathbf{k}}_i = s \frac{I_i}{\omega} \sigma_{\mathrm{abs}} \hat{\mathbf{k}}_i.$$

5.6 Optical force in an optical tweezers

We are finally ready to calculate the forces in an optical tweezers using the full power of electromagnetic theory. The geometry of the problem is depicted in Fig. 5.8. The trapping beam is focused by a high-NA lens to the point **O**. For the analysis we choose a reference frame fixed in the laboratory and centred at **O**, with respect to which the centre of mass of the particle is at position **C**. The coincidence of **C** with the centre of mass is not strictly necessary, but it simplifies the analysis and the calculations when one is interested both in the translational and in the rotational dynamics of the particle.

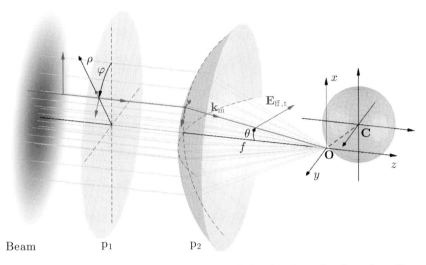

Figure 5.8 Reference frames for a focused beam. Reference frames used to calculate the radiation force from a focused beam. The scattering problem is solved in the reference frame of the particle, centred at **C**, whereas we want to calculate the radiation force and torque with respect to the laboratory frame centred at the laser beam focus **O**.

The procedure for the calculation of the radiation forces and torques can be summarised in the following steps:

1. determine the angular spectrum representation of the incoming field, e.g., $\mathbf{E}_{\text{ff},t}(\theta, \varphi)$ [Eq. (4.45)] for an aplanatic lens in a homogenous medium or $\mathbf{E}_{\text{ff},s}(\theta, \varphi)$ [Eq. (4.54)] when focusing through an interface;
2. correct the angular spectrum representation of the incoming field by a phase factor $e^{i\mathbf{k}_m \cdot (\mathbf{C}-\mathbf{O})}$ [Eq. (4.49)] in order to account for the phase accumulated by the propagation from **O** to **C**;
3. calculate the incoming focus field $\mathbf{E}_i(x, y, z)$ using Eq. (4.46) together with the angular spectrum representation of the incoming field, e.g., $\mathbf{E}_{\text{ff},t}(\theta, \varphi)$ or $\mathbf{E}_{\text{ff},s}(\theta, \varphi)$;
4. calculate the scattered field for each plane wave component of the angular spectrum representation of the incoming field and sum the resulting fields up to obtain the scattered field $\mathbf{E}_s(x, y, z)$;
5. use Eqs. (5.109) and (5.118) to calculate the radiation force and torque acting on the particle from $\mathbf{E}_t(x, y, z) = \mathbf{E}_i(x, y, z) + \mathbf{E}_s(x, y, z)$.

The most challenging part of this procedure is the calculation of the scattered field. This problem will be dealt with in detail in Chapter 6, but, as an important example, we present in Fig. 5.9 the forces arising on a spherical dielectric particle, for which the solution of the scattering problem was already obtained in Subsection 5.2.5 from Mie theory. A crucial point to keep in mind in the numerical implementation of this procedure is that the focal field must be calculated with a sufficiently small mesh; in practice, it is enough that the largest mesh size is significantly smaller than the optical wavelength in the medium. Furthermore, it is important that the incoming focal field is well approximated over a volume larger than

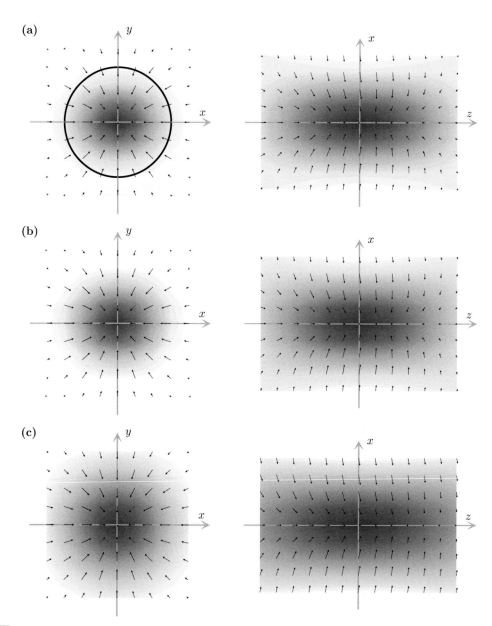

Figure 5.9 Radiation force on a sphere in an optical tweezers. Optical forces produced on a particle ($n_p = 1.50$) with radius $a = 300$ nm (solid line in (a)) in water ($n_m = 1.33$) near the focus of (a) an x-polarised beam uniform over the objective aperture and of a Gaussian beam with beam waist (b) $w_0 = 4$ mm and (c) $w_0 = 2$ mm. Note how the trap position along the z-direction is shifted from the focal point at the origin of the axes. All beams have wavelength $\lambda_0 = 632$ nm, are focused by a 1.20 NA water-immersion objective with aperture radius 4 mm and have 10 mW power after the aperture.

the particle under consideration; this practically means that one has to divide the incoming laser beam into a sufficiently high number of plane waves.

Exercise 5.6.1 Consider the optical trapping of a dielectric ($n_p = 1.50$) spherical particle of radius $a = 1\,\mu\text{m}$ with a microscope objective with NA $= 1.20$ and a wavelength $\lambda_0 = 830\,\text{nm}$. Calculate the force constants as a function of the overfilling of the objective. How does the maximum force constant change when a changes in the range from $0.2\,\mu\text{m}$ to $2\,\mu\text{m}$? [Hint: You can adapt the program ot from the book website.]

Exercise 5.6.2 Study numerically the optical forces arising on a dielectric sphere immersed in a fluid as a function of its radius a, its refractive index n_p, the medium refractive index n_m, the trapping wavelength λ_0 and the beam waist w_0. [Hint: You can adapt the program ot from the book website.]

Exercise 5.6.3 Study numerically the optical forces arising on optically trapped spherical particles of various sizes. In particular, consider the conditions on the mesh size of the surface of integration used for the calculation of the optical forces and the discretisation of the optical beam. How do these parameters affect the quality of the final result? [Hint: You can adapt the program ot from the book website.]

Exercise 5.6.4 Study numerically the optical forces arising from focusing various kinds of optical beams. Consider, in particular, the case of cylindrical vector beams. Under what conditions is it possible to trap particles whose refractive index is lower than their surrounding medium? [Hint: You can adapt the program ot from the book website.]

Exercise 5.6.5 Compare the trap stiffness produced by a focused laser beam on a spherical particle obtained from exact electromagnetic calculations, the dipole approximation and geometrical optics as a function of the particle size parameter. Reproduce, in particular, the results shown in Fig. 5.1. How do these results change as a function of the kind of optical beam, the laser wavelength λ_0, the objective NA, the refractive index of the particle n_p and that of the medium n_m? [Hint: You can adapt the program theorycomparison from the book website.]

5.6.1 Orbital angular momentum

In the presence of optical beams that carry orbital angular momentum, such as higher-order Laguerre–Gaussian beams, an angular momentum transfer to the optically trapped particle is also possible. This is illustrated in Fig. 5.10. In Fig. 5.10a, a highly focused Laguerre–Gaussian beam with azimuthal index $l = +1$ is considered: the transfer of orbital angular momentum entails a bending of the force field lines and therefore a rotation imposed on the optically trapped particle. As the helicity of the beam is reversed, i.e., $l = -1$, the orbital angular momentum transfer is also reversed so that the particle is set in rotation in the opposite direction [Fig. 5.10b].

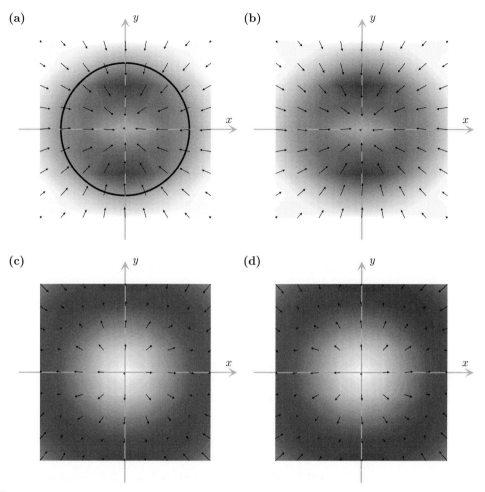

Figure 5.10 Orbital angular momentum on a sphere. Optical forces produced on a particle ($n_p = 1.50$) with radius $a = 300$ nm (solid line in (a)) in water ($n_m = 1.33$) near the focus of a Laguerre–Gaussian beam with (a) $l = +1$ and (b) $l = -1$ ($p = 1$, $w_0 = 4$ mm). One can observe how the force fields bend because of the transfer of orbital angular momentum from the light beam to the particle. When the size of the focal spot is increased by decreasing the beam waist to $w_0 = 1.2$ mm, the particle is attracted towards the annular region where the focal intensity is maximum, as can be seen in (c) for $l = +1$ and (d) for $l = -1$ ($p = 1$). All beams have wavelength $\lambda_0 = 632$ nm, are focused by a 1.20 NA water-immersion objective with aperture radius 4 mm and have power 10 mW after the aperture.

When the waist of the Laguerre–Gaussian beam decreases the focal spot becomes larger and a minimum of intensity forms at the centre. Under such conditions, the particle moves along a circle around the focus, as can be seen from the force fields shown in Figs. 5.10c and 5.10d.

Exercise 5.6.6 Study numerically the optical forces arising on a dielectric sphere immersed in a fluid when illuminated by a focused Laguerre–Gaussian beam of various indices.

How does the azimuthal index *l* influence the trapping properties of the beam? Consider the resulting forces as a function of the particle radius a, the particle refractive index n_p, the medium refractive index n_m and the trapping wavelength λ_0. What happens if the polarisation is circular? [Hint: You can adapt the program ot from the book website.]

Problems

5.1 *Magnetic particles.* Use the general expression of the Maxwell stress tensor to calculate optical forces on a small magneto-dielectric particle, i.e., for $\mu_m \neq 1$. Consider only the electric and magnetic dipole fields and investigate their role in the behaviour of the optical force. Study the paper by Nieto-Vesperinas et al. (2010).

5.2 *Resonant emission.* Calculate the optical force that an incident plane wave with frequency ω_i exerts on a spherical particle that absorbs light at frequency ω_a and re-emits it at a frequency ω_e (assume Lorentzian lineshapes for absorption and emission). Compare the cases when the emission is isotropic and directional. How does the emission anisotropy affect the optical force? Consider a particle trapped in an optical tweezers, calculate the force constants and show how they are affected by the orientation of the emitted radiation.

5.3 *Ripples and resonances.* In Fig. 5.7, it is possible to see ripples building up in the optical force as the particle size increases. Calculate the radiation force for plane wave illumination on a dielectric particle with increasing size. Show that ripples and resonances occur for large size parameter, $k_m a \gg 1$, and compare these features with those appearing in the extinction cross-section with size as calculated from Mie theory. These resonances in the radiation force were first observed by Ashkin and Dziedzic (1977). Now consider a dielectric particle trapped in an optical tweezers with a high-NA objective lens. Calculate the radiation pressure along the propagation axis and the transverse and axial force constants as a function of the particle size and compare the features you observe (ripples, resonances) with the case of plane wave illumination.

5.4 *Birefringence.* Calculate the optical torque transferred to a birefringent microparticle by a circularly polarised field. Fix a coordinate system aligned with the birefringence optical axes, the effect of the birefringence is related to phase factors on the field components propagating through the particle along the ordinary and extraordinary axes with two different refractive index, n_o and n_e, respectively. Study the articles by Beth (1936) and by Friese et al. (1998).

5.5 *Optical activity.* Consider a spherical particle made of an isotropic, nonmagnetic, optically active medium. In such a medium only circularly polarised waves (with either handedness) can propagate without change in their polarisation state but with different complex refractive indices for left, $n_+ = n/(1 - \beta n)$, and right, $n_- = n/(1 + \beta n)$, circular polarisation, with n being the average refractive index and β a complex chiral factor. Following the formulation developed by Bohren and Huffman (1983), calculate the radiation force and torque exerted by a circularly polarised plane wave on such a particle both analytically and computationally. Calculate the force constants when the particle is trapped by an optical tweezers created by a high-NA objective.

5.6 *Optical pulling force.* Consider a dielectric particle illuminated by two plane waves with the same intensity that propagate with wavevectors $\mathbf{k}_1$ and $\mathbf{k}_2$ at an angle 2θ. Show that, if z is the symmetry axis directed as $\mathbf{k}_1 + \mathbf{k}_2$, an optical pulling force directed against z occurs for particular values of wavelength and particle size. Calculate the optical pulling force in the case of a Bessel beam. Study the paper (including supplementary information) by Chen et al. (2011).

5.7 *Chiral particles.* Consider an elliptically polarised plane wave incident on a partly reflecting mirror. Show that upon reflection the optical torque component along the propagation direction transferred to the mirror is zero. Now, consider a partly reflecting left chiral mirror (e.g., a mirror based on cholesteric liquid crystal). This mirror partly reflects left-handed circularly polarised light with reflectance R_+ without changing the light helicity upon reflection, but fully transmits right-handed circular polarised light. Show that the radiation torque transferred to the left-handed chiral mirror by an elliptically polarised plane wave is $2\alpha_+^2 R_+ P/\omega$. Use the definition of spin angular momentum density and calculate the torque transferred to the mirror from the difference of angular momentum before and after reflection. Study the paper by Donato et al. (2014).

References

Allen, L., Beijersbergen, M. W., Spreeuw, R. J. C., and Woerdman, J. P. 1992. Orbital angular momentum of light and the transformation of Laguerre–Gaussian laser modes. *Phys. Rev. A*, **45**, 8185–9.

Ashkin, A., and Dziedzic, J. M. 1977. Observation of resonances in the radiation pressure on dielectric spheres. *Phys. Rev. Lett.*, **38**, 1351–4.

Barnett, S. M., and Allen, L. 1994. Orbital angular momentum and nonparaxial light beams. *Opt. Commun.*, **110**, 670–78.

Beth, R. A. 1936. Mechanical detection and measurement of the angular momentum of light. *Phys. Rev.*, **50**, 115–25.

Bohren, C. F., and Huffman, D. R. 1983. *Absorption and scattering of light by small particles*. New York: Wiley.

Borghese, F., Denti, P., and Saija, R. 2007. *Scattering from model nonspherical particles*. Berlin: Springer Verlag.

Chen, J., Ng, J., Lin, Z., and Chan, C. T. 2011. Optical pulling force. *Nature Photon.*, **5**, 531–4.

Cohen–Tannoudji, C., Dupont-Roc, J., and Gilbert, G. 1995. *Photons and atoms I. Introduction to quantum electrodynamics*. New York: Wiley.

Debye, P. 1909. Der Lichtdruck auf Kugeln von beliebigem Material. *Ann. Phys.*, **30**, 57–136.

Donato, M. G., Hernandez, J., Mazzulla, A., Provenzano, C., Saija, R., Sayed, R., Vasi, S., Magazzù, A., Pagliusi, P., Bartolino, R., Gucciardi, P. G., Maragò, O. M., and Cipparrone, G. 2014. Polarization-dependent optomechanics mediated by chiral microresonators. *Nature Commun.*, **5**, 3656.

Friese, M. E. J., Nieminen, T. A., Heckenberg, N. R., and Rubinsztein-Dunlop, H. 1998. Optical alignment and spinning of laser-trapped microscopic particles. *Nature*, **394**, 348–50.

Hansen, W. W. 1935. A new type of expansion in radiation problems. *Phys. Rev.*, **47**, 139–43.

Jackson, J. D. 1999. *Classical electrodynamics*. New York: Wiley.

Marston, P. L., and Crichton, J. H. 1984. Radiation torque on a sphere caused by a circulaly-polarized electromagnetic wave. *Phys. Rev. A*, **30**, 2508–16.

Mie, G. 1908. Beiträge zur Optik trüber Medien, speziell kolloidaler Metallösungen. *Ann. Phys.*, **25**, 377–445.

Mishchenko, M. I. 2001. Radiation force caused by scattering, absorption, and emission of light by nonspherical particles. *J. Quant. Spectrosc. Radiat. Transfer*, **70**, 811–16.

Mishchenko, M. I., Travis, L. D., and Lacis, A. A. 2002. *Scattering, absorption, and emission of light by small particles*. Cambridge, UK: Cambridge University Press.

Nieto-Vesperinas, M., Sáenz, J. J., Gómez-Medina, R., and Chantada, L. 2010. Optical forces on small magnetodielectric particles. *Opt. Express*, **18**, 11428–43.

Piccirillo, B., Slussarenko, S., Marrucci, L., and Santamato, E. 2013. The orbital angular momentum of light: Genesis and evolution of the concept and of the associated photonic technology. *Rivista Nuovo Cimento*, **36**, 501–55.

Rose, M. E. 1955. *Multipole fields*. New York: Wiley.

Waterman, P. C. 1971. Symmetry, unitarity and geometry in electromagnetic scattering. *Phys. Rev. D*, **3**, 825–39.

Wiscombe, W. J. 1979. Mie scattering calculations: Advances in techniques and fast vector-speed computer codes. NCAR Technical Note NCAR_TN-140_STR, National Center for Atmospheric Research, Boulder, CO.

6 Computational methods

As we have seen in the previous chapters, the computation of optical forces and torques requires the solution of the optical scattering problem in order to evaluate the rate of change of linear and angular optical momenta. There is no one single right approach for all situations; rather, the most appropriate computational technique depends on the system under study. For particles that are large compared with the optical wavelength, the geometrical optics approach described in Chapter 2 may be the most appropriate, whereas for particles much smaller than the wavelength, the dipole approximation dealt with in Chapter 3 may be a computationally efficient option. At intermediate length scales, though, rigorous solutions of Maxwell's equations are required, as we have seen in Chapter 5. The special and extremely important case of a homogeneous spherical particle can be solved analytically thanks to Mie theory, as we have seen again in Chapter 5. Nevertheless, non-homogeneous and non-spherical particles, such as the ones shown in Fig. 6.1, are of crucial importance for optical manipulation applications. This chapter is dedicated to exploring computational techniques adequate to deal with these particles and, in particular, the T-matrix method. Other approaches will also be introduced – in particular, the discrete dipole approximation (DDA) method and the finite-difference time-domain (FDTD) method, which have the advantage of being able to model particles of arbitrary shape and material, but are typically regarded as extremely demanding in terms of computing resources.

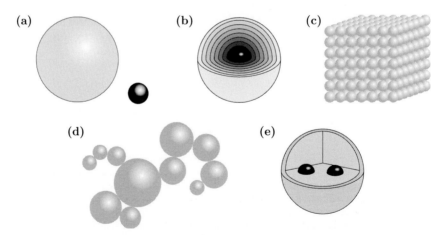

Figure 6.1 Complex non-spherical particles. A variety of computational methods are required to calculate the optical forces on objects other than (a) homogeneous spheres, such as (b) layered spheres, (c) cubes (here made of an array of spheres), (d) aggregates of spheres and (e) particles with inclusions of different materials.

6.1 T-matrix

The *transition matrix*, or *T-matrix*, formalism was derived by Waterman (1971), starting from the integral equation formulation of electromagnetic scattering to solve the scattering problem [Subsection 5.2.2]. The T-matrix was calculated by expanding a field into a series of spherical multipole fields and by imposing boundary conditions across the surfaces of the particles. This formulation of the T-matrix method, which is known as the *extended boundary condition method* (EBCM), can then be regarded as a generalisation of Mie theory [Subsection 5.2.5], to which it reduces for a single homogeneous spherical particle. Although the analytical approach of the multipole expansion is relatively simple and general, the computational methods needed to calculate the expansion coefficients from the imposition of the boundary conditions can be quite complex (Mishchenko et al., 2002). Here, we will follow the formulation provided by Borghese et al. (2007b), which has the advantage that most calculations are carried out analytically and the resulting algorithms are therefore particularly computationally efficient and accurate.[1]

As we have seen in Subsection 5.2.4, the T-matrix $\mathbb{T} = \{T_{l'm'lm}^{(p'p)}\}$ characterises how the scattering particle acts on the known multipole amplitudes of the incident field $W_{i,lm}^{(p)}$ to give the unknown amplitudes of the scattered field $A_{s,l'm'}^{(p')}$, i.e.,[2]

$$A_{s,l'm'}^{(p')} = \sum_{plm} T_{l'm'lm}^{(p'p)} W_{i,lm}^{(p)}, \qquad (6.1)$$

or, in a more compact form,

$$\mathbf{A}_s = \mathbb{T}\mathbf{W}_i, \qquad (6.2)$$

where $\mathbf{A}_s = \{A_{s,l'm'}^{(p')}\}$ and $\mathbf{W}_i = \{W_{i,lm}^{(p)}\}$. For example, the T-matrix for a homogenous spherical particle is diagonal, independent of m and connected to the Mie coefficients a_l and b_l [Eqs. (5.100)], i.e.,

$$\mathbf{A}_s = -\mathbb{R}\mathbf{W}_i, \qquad (6.3)$$

---

[1] An alternative, and equivalent, way to generalise the Mie approach using the multipole expansion is given by the so-called *generalised Lorenz–Mie theories* (GLMTs), where a generic laser beam is expanded on a base of functions, e.g., vector spherical harmonics, and the scattering problem is solved for symmetric scatterers, e.g., spheres, so that separation of variables can be used to find the expansion coefficients of the scattered fields (Gouesbet and Gréhan, 2011). The precise connection between the T-matrix formulation and GLMTs can be found in Gouesbet (2010), whereas a description of the use of GLMTs for calculations in optical tweezers can be found, e.g., in Neves et al. (2006a, 2007).

[2] The indices, as we have seen in Subsection 5.2.4, are $p = 1, 2$, $l = 0, 1, \ldots$ and $m = -l, \ldots, 0, \ldots, l$ (and analogously for p', l' and m'). In this chapter, we will use the simplified expressions $\sum_{plm}$ to signify $\sum_{p=1,2} \sum_{l=0}^{+\infty} \sum_{m=-l}^{l}$ and $\sum_{lm}$ to signify $\sum_{l=0}^{+\infty} \sum_{m=-l}^{l}$.

where $\mathbb{R} = \{R^{(p'p)}_{l'm'lm}\}$ and

$$R^{(p'p)}_{l'm'lm} = \begin{cases} b_l & p = p' = 1 \quad \text{and} \quad l = l' \quad \text{and} \quad m = m' \\ a_l & p = p' = 2 \quad \text{and} \quad l = l' \quad \text{and} \quad m = m' \\ 0 & \text{otherwise.} \end{cases} \quad (6.4)$$

In the following sections, we will derive the T-matrix coefficients for particles or structures with more complex composition, i.e., metallic spheres [Section 6.2], layered spheres [Section 6.3] and clusters of spheres [Section 6.4]. We note at this point that by an appropriate choice of the radii and indices of refraction of a cluster of spheres, it is possible to mimic many kinds of particles so that reasonably accurate numerical results can be obtained within a reasonable computational time for a vast class of particles. Before proceeding further, we provide in this section the formulas for the radiation force and torque (together with a sketch of their derivation) as a function of the T-matrix coefficients and the incident field amplitudes [Subsections 6.1.1 and 6.1.2]; then we derive the incident field amplitudes for the particularly important case of a focused beam [Subsection 6.1.3]; and finally, we consider the properties of the T-matrix under rotation and translation of the reference frame in which the multipole expansion is performed [Subsections 6.1.4, 6.1.5 and 6.1.6].

Exercise 6.1.1 Show that any T-matrix satisfies the symmetry property

$$T^{(p'p)}_{l'm'lm} = (-1)^{m+m'} T^{(pp')}_{l,-m,l',-m'}. \quad (6.5)$$

Exercise 6.1.2 Show that the T-matrix of a scatterer with cylindrical symmetry (assuming the symmetry axis to be the z-axis) has the property

$$T^{(p'p)}_{l'm'lm} = 0 \quad \text{for} \quad m \neq m' \quad (6.6)$$

and that the T-matrix of a scatterer with spherical symmetry is diagonal, i.e.,

$$T^{(p'p)}_{l'm'lm} = 0 \quad \text{for} \quad p \neq p' \quad \text{or} \quad l \neq l' \quad \text{or} \quad m \neq m'. \quad (6.7)$$

6.1.1 Optical force

The starting point to calculate the optical force is Eq. (5.117), i.e.,

$$\mathbf{F}_{\text{rad}} = -\frac{1}{4}\varepsilon_{\text{m}} r^2 \oint_\Omega \left[|\mathbf{E}_{\text{s}}|^2 + \frac{c^2}{n_{\text{m}}^2}|\mathbf{B}_{\text{s}}|^2 + 2\text{Re}\left\{ \mathbf{E}_{\text{i}} \cdot \mathbf{E}_{\text{s}}^* + \frac{c^2}{n_{\text{m}}^2}\mathbf{B}_{\text{i}} \cdot \mathbf{B}_{\text{s}}^* \right\}\right] \hat{\mathbf{r}}\, d\Omega.$$

By substituting the expansions of the incident and scattered waves in terms of multipoles given by Eqs. (5.110) and (5.111) into Eq. (5.117), we obtain the expression for the radiation force along the direction of a unit vector $\hat{\mathbf{u}}$, i.e., $F_{\text{rad}}(\hat{\mathbf{u}}) = \mathbf{F}_{\text{rad}} \cdot \hat{\mathbf{u}}$:

$$F_{\text{rad}}(\hat{\mathbf{u}}) = -\frac{\varepsilon_{\text{m}} E_{\text{i}}^2}{2 k_{\text{m}}^2} \text{Re}\left\{ \sum_{plm} \sum_{p'l'm'} i^{l-l'} I^{(pp')}_{lml'm'}(\hat{\mathbf{u}}) \left[A^{(p)*}_{\text{s},lm} A^{(p')}_{\text{s},l'm'} + W^{(p)*}_{\text{i},lm} A^{(p')}_{\text{s},l'm'} \right] \right\}, \quad (6.8)$$

where the amplitudes $A_{s,l'm'}^{(p)}$ of the scattered field are given in terms of the elements of the T-matrix, the amplitudes $W_{i,lm}^{(p)*}$ of the incident field by Eq. (6.1) and

$$I_{lml'm'}^{(pp')}(\hat{\mathbf{u}}) = \oint_\Omega (\hat{\mathbf{r}} \cdot \hat{\mathbf{u}}) i^{p-p'} \mathbf{Z}_{lm}^{(p)*}(\hat{\mathbf{r}}) \cdot \mathbf{Z}_{l'm'}^{(p')}(\hat{\mathbf{r}}) \, d\Omega. \quad (6.9)$$

The integrals $I_{lml'm'}^{(pp')}(\hat{\mathbf{u}})$ can be expressed in closed form (Borghese et al., 2007a) as

$$I_{lml'm'}^{(pp')}(\hat{\mathbf{u}}) = \frac{4\pi}{3} \sum_{\mu=-1,0,1} Y_{1\mu}^*(\hat{\mathbf{u}}) K_{\mu;lml'm'}^{(pp')}, \quad (6.10)$$

where we used Eq. (5.44) to express the unit vectors in terms of spherical harmonics,

$$K_{\mu;lml'm'}^{(pp')} = \oint_\Omega Y_{1\mu}(\hat{\mathbf{r}}) i^{p-p'} \mathbf{Z}_{lm}^{(p)*}(\hat{\mathbf{r}}) \cdot \mathbf{Z}_{l'm'}^{(p')}(\hat{\mathbf{r}}) \, d\Omega = \sqrt{\frac{3}{4\pi}} C_1(l',l;\mu,m-\mu) O_{ll'}^{(pp')},$$

$$O_{ll'}^{(pp')} = \begin{cases} \sqrt{\dfrac{(l-1)(l+1)}{l(2l+1)}} & l' = l-1 \text{ and } p = p' \\ -\dfrac{1}{\sqrt{l(l+1)}} & l' = l \text{ and } p \neq p' \\ -\sqrt{\dfrac{l(l+2)}{(l+1)(2l+1)}} & l' = l+1 \text{ and } p = p' \\ 0 & \text{otherwise,} \end{cases}$$

and $C_1(l',l;\mu,m-\mu)$ are Clebsch–Gordan coefficients [Table 6.1]. The coefficients $K_{\mu;lml'm'}^{(pp')}$ obey the following symmetry properties:

$$K_{\mu;lml'm'}^{(11)} = K_{\mu;lml'm'}^{(22)}, \quad K_{\mu;lml'm'}^{(12)} = K_{\mu;lml'm'}^{(21)}.$$

The force expressed by Eq. (6.8) can be separated into two parts,

$$F_{\text{rad}}(\hat{\mathbf{u}}) = -F_{\text{scat}}(\hat{\mathbf{u}}) + F_{\text{ext}}(\hat{\mathbf{u}}), \quad (6.11)$$

where

$$F_{\text{scat}}(\hat{\mathbf{u}}) = \frac{\varepsilon_m E_i^2}{2k_m^2} \operatorname{Re}\left\{\sum_{plm}\sum_{p'l'm'} A_{s,lm}^{(p)*} A_{s,l'm'}^{(p')} i^{l-l'} I_{lml'm'}^{(pp')}(\hat{\mathbf{u}})\right\} \quad (6.12)$$

and

$$F_{\text{ext}}(\hat{\mathbf{u}}) = -\frac{\varepsilon_m E_i^2}{2k_m^2} \operatorname{Re}\left\{\sum_{plm}\sum_{p'l'm'} W_{i,lm}^{(p)*} A_{s,l'm'}^{(p')} i^{l-l'} I_{lml'm'}^{(pp')}(\hat{\mathbf{u}})\right\}. \quad (6.13)$$

$F_{\text{scat}}(\hat{\mathbf{u}})$ depends on the amplitudes $A_{s,lm}^{(p)}$ of the scattered field only, whereas $F_{\text{ext}}(\hat{\mathbf{u}})$ depends both on $A_{s,lm}^{(p)}$ and on the amplitudes $W_{i,lm}^{(p)}$ of the incident field. This dependence is analogous to that on the scattering and extinction cross-sections for the force exerted by a plane wave [Eq. (5.123)]; hence the subscripts.

Exercise 6.1.3 Use the general expression for the force obtained in Eq. (6.8) to get, as a special case, the radiation force on a spherical particle by a linearly polarised plane wave, i.e., Eq. (5.123). Verify these results with numerical simulations. [Hint: Use the expression of the T-matrix in terms of Mie coefficients [Subsection 5.2.5] and the expansion of a plane wave [Eqs. (5.87) and (5.88)]. You can use the program `forcecomparison` from the book website.]

6.1.2 Optical torque

For the calculation of the radiation torque, we start from Eq. (5.118), i.e.,

$$\mathbf{T}_{\text{Rad}} = -\frac{\varepsilon_{\text{m}} r^3}{2} \text{Re} \left\{ \oint \left[(\hat{\mathbf{r}} \cdot \mathbf{E}_{\text{t}}) \left(\mathbf{E}_{\text{t}}^* \times \hat{\mathbf{r}} \right) + \frac{c^2}{n_{\text{m}}^2} (\hat{\mathbf{r}} \cdot \mathbf{B}_{\text{t}}) \left(\mathbf{B}_{\text{t}}^* \times \hat{\mathbf{r}} \right) \right] d\Omega \right\}.$$

Similarly to the derivation in Section 5.5, by expressing the total fields as $\mathbf{E}_{\text{t}} = \mathbf{E}_{\text{i}} + \mathbf{E}_{\text{s}}$ and $\mathbf{B}_{\text{t}} = \mathbf{B}_{\text{i}} + \mathbf{B}_{\text{s}}$, we can write the torque as the sum of four terms related to the combination of incident and scattered fields contributions,

$$\mathbf{T}_{\text{rad}} = \mathbf{T}^{\text{ii}} + \mathbf{T}^{\text{is}} + \mathbf{T}^{\text{si}} + \mathbf{T}^{\text{ss}}, \tag{6.14}$$

where $\mathbf{T}^{\text{ii}} = 0$ because the averaged angular momentum transported across a surface S by the incident electromagnetic wave in the absence of the scatterer must vanish and the remaining terms are

$$\mathbf{T}^{\text{is}} = -\frac{\varepsilon_{\text{m}} r^3}{2} \text{Re} \left\{ \oint \left[(\hat{\mathbf{r}} \cdot \mathbf{E}_{\text{i}}) \left(\mathbf{E}_{\text{s}}^* \times \hat{\mathbf{r}} \right) + \frac{c^2}{n_{\text{m}}^2} (\hat{\mathbf{r}} \cdot \mathbf{B}_{\text{i}}) \left(\mathbf{B}_{\text{s}}^* \times \hat{\mathbf{r}} \right) \right] d\Omega \right\},$$

$$\mathbf{T}^{\text{si}} = -\frac{\varepsilon_{\text{m}} r^3}{2} \text{Re} \left\{ \oint \left[(\hat{\mathbf{r}} \cdot \mathbf{E}_{\text{s}}) \left(\mathbf{E}_{\text{i}}^* \times \hat{\mathbf{r}} \right) + \frac{c^2}{n_{\text{m}}^2} (\hat{\mathbf{r}} \cdot \mathbf{B}_{\text{s}}) \left(\mathbf{B}_{\text{i}}^* \times \hat{\mathbf{r}} \right) \right] d\Omega \right\}$$

and

$$\mathbf{T}^{\text{ss}} = -\frac{\varepsilon_{\text{m}} r^3}{2} \text{Re} \left\{ \oint \left[(\hat{\mathbf{r}} \cdot \mathbf{E}_{\text{s}}) \left(\mathbf{E}_{\text{s}}^* \times \hat{\mathbf{r}} \right) + \frac{c^2}{n_{\text{m}}^2} (\hat{\mathbf{r}} \cdot \mathbf{B}_{\text{s}}) \left(\mathbf{B}_{\text{s}}^* \times \hat{\mathbf{r}} \right) \right] d\Omega \right\}.$$

The terms $\mathbf{T}^{\text{is}}$ and $\mathbf{T}^{\text{si}}$ are the extinction terms, as they involve both the incoming and scattered fields, whereas $\mathbf{T}^{\text{ss}}$ is the scattering term, as it only involves the scattered field.

As we already explained in Subsection 5.3.2, the far-field expression for the scattered fields in terms of the scattering amplitude cannot be used in the calculation of the torque, but one needs to retain the radial terms of the fields and thus expand the electromagnetic fields in terms of multipoles [Eqs. (5.79) and (5.82)]. Substituting the asymptotic expressions for $j_l(k_{\text{m}}r)$ and $h_j(k_{\text{m}}r)$ [Table 5.3] in the definitions of **J**-multipoles [Eq. (5.78)] and **H**-multipoles [Eq. (5.81)], we get the dot and cross products that enter the integrals in

Eq. (6.14):

$$\hat{\mathbf{r}} \cdot \mathbf{J}_{lm}^{(1)}(k_\mathrm{m}r, \hat{\mathbf{r}}) = 0, \qquad (6.15)$$

$$\hat{\mathbf{r}} \cdot \mathbf{J}_{lm}^{(2)}(k_\mathrm{m}r, \hat{\mathbf{r}}) = i\sqrt{l(l+1)}\frac{\sin(k_\mathrm{m}r - l\pi/2)}{(k_\mathrm{m}r)^2} Y_{lm}(\hat{\mathbf{r}}), \qquad (6.16)$$

$$\hat{\mathbf{r}} \cdot \mathbf{H}_{lm}^{(1)}(k_\mathrm{m}r, \hat{\mathbf{r}}) = 0, \qquad (6.17)$$

$$\hat{\mathbf{r}} \cdot \mathbf{H}_{lm}^{(2)}(k_\mathrm{m}r, \hat{\mathbf{r}}) = (-i)^l \sqrt{l(l+1)}\frac{e^{ik_\mathrm{m}r}}{(k_\mathrm{m}r)^2} Y_{lm}(\hat{\mathbf{r}}), \qquad (6.18)$$

$$\mathbf{J}_{lm}^{(1)*}(k_\mathrm{m}r, \hat{\mathbf{r}}) \times \hat{\mathbf{r}} = \frac{\sin(k_\mathrm{m}r - l\pi/2)}{k_\mathrm{m}r} \mathbf{Z}_{lm}^{(2)*}(\hat{\mathbf{r}}), \qquad (6.19)$$

$$\mathbf{J}_{lm}^{(2)*}(k_\mathrm{m}r, \hat{\mathbf{r}}) \times \hat{\mathbf{r}} = \frac{\cos(k_\mathrm{m}r - l\pi/2)}{k_\mathrm{m}r} \mathbf{Z}_{lm}^{(1)*}(\hat{\mathbf{r}}), \qquad (6.20)$$

$$\mathbf{H}_{lm}^{(1)*}(k_\mathrm{m}r, \hat{\mathbf{r}}) \times \hat{\mathbf{r}} = i^{l+1} \frac{e^{-ik_\mathrm{m}r}}{k_\mathrm{m}r} \mathbf{Z}_{lm}^{(2)*}(\hat{\mathbf{r}}), \qquad (6.21)$$

$$\mathbf{H}_{lm}^{(2)*}(k_\mathrm{m}r, \hat{\mathbf{r}}) \times \hat{\mathbf{r}} = i^l \frac{e^{-ik_\mathrm{m}r}}{k_\mathrm{m}r} \mathbf{Z}_{lm}^{(1)*}(\hat{\mathbf{r}}), \qquad (6.22)$$

where we have neglected the terms that vanish at infinity to higher order.

We will first calculate the expression for the axial z-component of the radiation torque, i.e., $T_{\mathrm{rad},z} = \mathbf{T}_\mathrm{rad} \cdot \hat{\mathbf{z}} = \mathbf{T}^\mathrm{is} \cdot \hat{\mathbf{z}} + \mathbf{T}^\mathrm{si} \cdot \hat{\mathbf{z}} + \mathbf{T}^\mathrm{ss} \cdot \hat{\mathbf{z}}$. Using Eqs. (6.15)–(6.22) to express each contribution in terms of the transverse harmonics $\mathbf{Z}_{lm}^{(1)*}$ and $\mathbf{Z}_{lm}^{(2)*}$, and using Eqs. (5.128) and (5.129), we obtain

$$T_{\mathrm{rad},z} = -\frac{\varepsilon_\mathrm{m} E_\mathrm{i}^2}{2k_\mathrm{m}^3} \underbrace{\sum_{plm} m \operatorname{Re}\left\{W_{\mathrm{i},lm}^{(p)} A_{\mathrm{s},lm}^{(p)*}\right\}}_{\text{extinction}} - \frac{\varepsilon_\mathrm{m} E_\mathrm{i}^2}{2k_\mathrm{m}^3} \underbrace{\sum_{plm} m |A_{\mathrm{s},lm}^{(p)}|^2}_{\text{scattering}}, \qquad (6.23)$$

where we have distinguished the extinction and scattering contributions so that

$$T_{\mathrm{rad},z} = T_{\mathrm{ext},z} - T_{\mathrm{scat},z}. \qquad (6.24)$$

The transversal components of the radiation torque, i.e., $T_{\mathrm{rad},x} = \mathbf{T}_\mathrm{rad} \cdot \hat{\mathbf{x}}$ and $T_{\mathrm{rad},y} = \mathbf{T}_\mathrm{rad} \cdot \hat{\mathbf{y}}$, can be calculated in a similar way (Borghese et al., 2006), obtaining

$$\begin{aligned}
T_{\mathrm{rad},x} = & -\frac{\varepsilon_\mathrm{m} E_\mathrm{i}^2}{4k_\mathrm{m}^3} \underbrace{\sum_{plm} \operatorname{Re}\left\{s_{lm}^{(-)} W_{\mathrm{i},l,m+1}^{(p)} A_{\mathrm{s},lm}^{(p)*} + s_{lm}^{(+)} W_{\mathrm{i},l,m-1}^{(p)} A_{\mathrm{s},lm}^{(p)*}\right\}}_{\text{extinction}} \\
& -\frac{\varepsilon_\mathrm{m} E_\mathrm{i}^2}{4k_\mathrm{m}^3} \underbrace{\sum_{plm} \operatorname{Re}\left\{s_{lm}^{(-)} A_{\mathrm{s},l,m+1}^{(p)} A_{\mathrm{s},lm}^{(p)*} + s_{lm}^{(+)} A_{\mathrm{s},l,m-1}^{(p)} A_{\mathrm{s},lm}^{(p)*}\right\}}_{\text{scattering}}
\end{aligned} \qquad (6.25)$$

and

$$T_{\text{rad},y} = -\underbrace{\frac{\varepsilon_m E_i^2}{4k_m^3} \sum_{plm} \text{Im}\left\{-s_{lm}^{(-)} W_{i,l,m+1}^{(p)} A_{s,lm}^{(p)*} + s_{lm}^{(+)} W_{i,l,m-1}^{(p)} A_{s,lm}^{(p)*}\right\}}_{\text{extinction}}$$
$$-\underbrace{\frac{\varepsilon_m E_i^2}{4k_m^3} \sum_{plm} \text{Im}\left\{-s_{lm}^{(-)} A_{s,l,m+1}^{(p)} A_{s,lm}^{(p)*} + s_{lm}^{(+)} A_{s,l,m-1}^{(p)} A_{s,lm}^{(p)*}\right\}}_{\text{scattering}}, \quad (6.26)$$

where $s_{lm}^{(-)} = \sqrt{(l-m)(l+1+m)}$ and $s_{lm}^{(+)} = \sqrt{(l+m)(l+1-m)}$. Also, for these components, we have distinguished the extinction and scattering contributions.

Exercise 6.1.4 Use the general expression for the torque obtained in Eqs. (6.23), (6.25) and (6.26) to get, as a special case, the radiation torque on a spherical particle from a plane wave with circular or elliptical polarisation, i.e., the results of Section 5.5. Verify these results with numerical simulations. [Hint: Use the expression of the T-matrix in terms of Mie coefficients given in Subsection 5.2.5. You can use the program torquecomparison from the book website.]

6.1.3 Amplitudes of a focused beam

A particularly important case for the scope of this book is that of a focused beam. To calculate the multipole amplitudes $\mathcal{W}_{i,lm}^{(p)}$ of a focused beam, we can exploit the expansion of the incoming beam into plane waves and its focusing as described in Section 4.5.[3]

The expansion of the focused beam around the focal point is given by Eq. (4.46), i.e.,

$$\mathbf{E}_f(x,y,z) = \frac{ik_t f e^{-ik_t f}}{2\pi} \int_0^{\theta_{\max}} \sin\theta \int_0^{2\pi} \mathbf{E}_{\text{ff},t}(\theta,\varphi) e^{i[k_{t,x}x + k_{t,y}y]} e^{ik_{t,z}z} \, d\varphi \, d\theta,$$

where we have taken into account that each plane wave transmitted through the objective lens $\mathbf{E}_{\text{ff},t}(\theta,\varphi)$ can be expanded into multipoles according to Eq. (5.84), i.e.,

$$\mathbf{E}_{\text{ff},t}(\theta,\varphi) \equiv \mathbf{E}_i(r,\hat{\mathbf{r}}) = E_i \sum_{lm} W_{i,lm}^{(1)}(\hat{\mathbf{k}}_i, \hat{\mathbf{e}}_i) \mathbf{J}_{lm}^{(1)}(k_m r, \hat{\mathbf{r}}) + W_{i,lm}^{(2)}(\hat{\mathbf{k}}_i, \hat{\mathbf{e}}_i) \mathbf{J}_{lm}^{(2)}(k_m r, \hat{\mathbf{r}}),$$

---

[3] Alternative ways to describe the amplitudes of focused laser beams have been developed. For example, a Gaussian laser beam can be approximated by the Davis formulation (Davis, 1979) through a localised approximation (Gouesbet et al., 2011). Another accurate computational method relies on the point-matching of the fields either at the focus or in the far field in order to obtain multipole expansion equivalents of focused paraxial scalar (Gaussian or Laguerre–Gaussian) beams (Nieminen et al., 2003b). However, for an accurate and quantitative description of optical tweezers without approximations, the detailed description of focal fields in the angular spectrum representation is crucial (Fontes et al., 2005; Neves et al., 2006a). The calculations of the amplitudes for a focused beam through the angular spectrum representation within the GLMT formalism can be found in Neves et al. (2006b) and Moreira et al. (2010). These amplitudes are equivalent to the ones described in Subsection 6.1.3.

with the amplitudes given by Eqs. (5.85) and (5.86), i.e.,

$$W_{i,lm}^{(1)}(\hat{\mathbf{k}}_i, \hat{\mathbf{e}}_i) = 4\pi i^l \, \hat{\mathbf{e}}_i \cdot \mathbf{Z}_{lm}^{(1)*}(\hat{\mathbf{k}}_i) \quad \text{and} \quad W_{i,lm}^{(2)}(\hat{\mathbf{k}}_i, \hat{\mathbf{e}}_i) = 4\pi i^{l+1} \, \hat{\mathbf{e}}_i \cdot \mathbf{Z}_{lm}^{(2)*}(\hat{\mathbf{k}}_i).$$

Therefore, we obtain that the amplitudes of the focused field are

$$\mathcal{W}_{i,lm}^{(p)} = \frac{ik_t f e^{-ik_t f}}{2\pi} \int_0^{\theta_{\max}} \sin\theta \int_0^{2\pi} E_i(\theta, \varphi) \, W_{i,lm}^{(p)}(\hat{\mathbf{k}}_i, \hat{\mathbf{e}}_i) \, d\varphi \, d\theta. \quad (6.27)$$

If the centre around which the expansion is performed is displaced by **P** with respect to the focal point **O**, the multipole expansion coefficients can be obtained from Eq. (6.27) using Eq. (4.49), so that we have

$$\mathcal{W}_{i,lm}^{(p)}(\mathbf{P}) = \frac{ik_t f e^{-ik_t f}}{2\pi} \int_0^{\theta_{\max}} \sin\theta \int_0^{2\pi} E_i(\theta, \varphi) \, W_{i,lm}^{(p)}(\hat{\mathbf{k}}_i, \hat{\mathbf{e}}_i) \, e^{i\mathbf{k}_t \cdot \mathbf{P}} \, d\varphi \, d\theta. \quad (6.28)$$

The amplitudes $\mathcal{W}_{lm}^{(p)}(\mathbf{P})$ define the focal field and can be numerically calculated once the characteristics of the optical system are known.

The radiation force and torque are calculated through the knowledge of the scattered amplitudes $\mathcal{A}_{s,lm}^{(p)}$, e.g., by using the T-matrix [Eq. (6.1)]. In particular, the radiation force on the particle is given by

$$F_{\text{rad}}(\hat{\mathbf{u}}) = -\frac{\varepsilon_m}{2k_m^2} \text{Re} \left\{ \sum_{plm} \sum_{p'l'm'} i^{l-l'} I_{lml'm'}^{(pp')}(\hat{\mathbf{u}}) \left[\mathcal{A}_{s,lm}^{(p)*} \mathcal{A}_{s,l'm'}^{(p')} + \mathcal{W}_{i,lm}^{(p)*} \mathcal{A}_{s,l'm'}^{(p')} \right] \right\}. \quad (6.29)$$

In practice, the expression of the force in Eq. (6.29) is obtained from Eq. (6.8) by changing $E_i W_{i,lm}^{(p)} \to \mathcal{W}_{i,lm}^{(p)}(\mathbf{P})$ and $E_i A_{s,lm}^{(p)} \to \mathcal{A}_{s,lm}^{(p)}$. Analogous considerations hold true for the calculation of the radiation torque, so that the expressions for the radiation torque by a focused field are obtained by applying the same substitutions to Eqs. (6.23), (6.25) and (6.26).

Exercise 6.1.5 Calculate numerically the forces produced on a spherical particle by a focused beam using the T-matrix approach. Compare these results to the ones obtained in Section 5.6 by using Eq. (5.109) and the explicit expressions of the total fields around the particle. [Hint: You can use the program `focustm` from the book website.]

Exercise 6.1.6 Show analytically and computationally that the amplitudes $\mathcal{W}_{i,lm}^{(p)}$ for a focused linearly polarised Gaussian beam are equivalent to the beam shape coefficients described in Neves et al. (2006b).

Exercise 6.1.7 Write the expression for the components of the optical force on a spherical particle with respect to the laser beam main axes (propagation direction and polarisation) in terms of Mie coefficients and beam amplitudes.

Exercise 6.1.8 Calculate the amplitudes and optical forces on a spherical particle for a Gaussian beam focused through an interface. [Hint: Start from Section 4.7 and compare your results with the ones reported in Borghese et al. (2007a) and Neves et al. (2007).]

6.1.4 Translation theorem

We will now consider how multipoles and T-matrices transform under translation. We consider two reference frames, Σ and Σ', with parallel axes and different origins, $\mathbf{O}$ and $\mathbf{O}'$ with $\mathbf{O}' - \mathbf{O} = \mathbf{R} = R\hat{\mathbf{R}}$, so that a point with vector position $\mathbf{r}$ in Σ has a vector position $\mathbf{r}' = \mathbf{r} - \mathbf{R}$ in Σ'.

In order to derive the formulas that express how spherical harmonics are transformed under translation, often referred to as *addition formulas* (Nozawa, 1966), we start by considering a scalar plane wave $e^{i\mathbf{k}_m \cdot \mathbf{r}}$ in Σ and its translated version $e^{i\mathbf{k}_m \cdot \mathbf{r}'}$ in Σ', for which the following identity holds:

$$e^{i\mathbf{k}_m \cdot \mathbf{r}} = e^{i\mathbf{k}_m \cdot \mathbf{R}} e^{i\mathbf{k}_m \cdot \mathbf{r}'}. \tag{6.30}$$

Expanding both sides of Eq. (6.30) in terms of spherical harmonics using Bauer's expansion [Eq. (5.57)], we obtain

$$4\pi \sum_{l''m''} i^{l''} j_{l''}(k_m r) Y^*_{l''m''}(\hat{\mathbf{k}}_m) Y_{l''m''}(\hat{\mathbf{r}})$$
$$= \left[4\pi \sum_{l'm'} i^{l'} j_{l'}(k_m r') Y^*_{l'm'}(\hat{\mathbf{k}}_m) Y_{l'm'}(\hat{\mathbf{r}}') \right] \left[4\pi \sum_{LM} i^L j_L(k_m R) Y_{LM}(\hat{\mathbf{k}}_m) Y^*_{LM}(\hat{\mathbf{R}}) \right],$$

where the third relation follows from the symmetry between $\hat{\mathbf{k}}$ and $\hat{\mathbf{R}}$ in Bauer's expansion. Now, multiplying both sides by $Y_{lm}(\hat{\mathbf{k}}_m)$, integrating over the solid angle and using the orthonormality properties of spherical harmonics [Eq. (5.35)], we obtain

$$j_l(k_m r) Y_{lm}(\hat{\mathbf{r}}) = \sum_{l'm'} j_{l'}(k_m r') Y_{l'm'}(\hat{\mathbf{r}}') G^j_{l'm'lm}(k_m R, \hat{\mathbf{R}}), \tag{6.31}$$

where

$$G^j_{l'm'lm}(k_m R, \hat{\mathbf{R}}) = 4\pi \sum_{L=|l-l'|}^{l+l'} i^{l'+L-l} j_L(k_m R) Y^*_{L,m'-m}(\hat{\mathbf{R}}) I(l, m, L, m'-m, l', m'), \tag{6.32}$$

with $G^j_{l'm'lm}(k_m, \hat{\mathbf{R}}) \neq 0$ only if $l + l' + L$ is even, and

$$I(l, m, L, M, l', m') = \oint_\Omega Y^*_{l'm'}(\hat{\mathbf{k}}_m) Y_{LM}(\hat{\mathbf{k}}_m) Y_{lm}(\hat{\mathbf{k}}_m) d\Omega \tag{6.33}$$

are *Gaunt integrals*,[4] which can be expressed in terms of Clebsch–Gordan coefficients [Subsection 6.1.6] as

$$I(l, m, L, M, l'm') = \sqrt{\frac{(2l+1)(2L+1)}{4\pi(2l'+1)}} C(l, L, l'; m, M, m') C(l, L, l'; 0, 0, 0). \quad (6.34)$$

In an analogous way, a similar expression is found for the case of the Hankel function of the first kind, i.e.,

$$h_l(k_m r) Y_{lm}(\hat{\mathbf{r}}) = \sum_{l'm'} h_{l'}(k_m r') Y_{l'm'}(\hat{\mathbf{r}}') G^h_{l'm'lm}(k_m R, \hat{\mathbf{R}}), \quad (6.35)$$

where

$$G^h_{l'm'lm}(k_m R, \hat{\mathbf{R}}) = 4\pi \sum_{L=|l-l'|}^{l+l'} i^{l'+L-l} h_L(k_m R) Y^*_{L, m'-m}(\hat{\mathbf{R}}) I(l, m, L, m'-m, l', m'), \quad (6.36)$$

with $G^h_{l'm'lm}(k_m R, \hat{\mathbf{R}}) \neq 0$ only if $l + l' + L$ is even.

The addition formulas given by Eqs. (6.31) and (6.35) can be generalised to vector multipole fields (Borghese et al., 1980) by expressing the **J** and **H**-multipoles in Σ in terms of the **J** or **H**-multipoles in Σ' and introducing the translation matrices $\mathbb{J}(k_m R, \hat{\mathbf{R}}) = \{J^{(p'p)}_{l'm'lm}(k_m R, \hat{\mathbf{R}})\}$ and $\mathbb{H}(k_m R, \hat{\mathbf{R}}) = \{H^{(p'p)}_{l'm'lm}(k_m R, \hat{\mathbf{R}})\}$. For the **J**-multipoles [Eq. (5.78)],

$$\mathbf{J}^{(p)}_{lm}(k_m r, \hat{\mathbf{r}}) = \sum_{p'l'm'} J^{(p'p)}_{l'm'lm}(k_m R, \hat{\mathbf{R}}) \mathbf{J}^{(p')}_{l'm'}(k_m r', \hat{\mathbf{r}}'), \quad (6.37)$$

where

$$\begin{cases} J^{(11)}_{l'm'lm} = \sum_{\mu=-1,0,1} C_1(l', l'; -\mu, m'+\mu) C_1(l, l; -\mu, m+\mu) G^j_{l', m'+\mu, l, m+\mu}, \\ J^{(22)}_{l'm'lm} = J^{(11)}_{l'm'lm}, \\ J^{(12)}_{l'm'lm} = i \sqrt{\frac{2l'+1}{l'+1}} \\ \qquad \times \sum_{\mu=-1,0,1} C_1(l'-1, l'; -\mu, m'+\mu) C_1(l, l; -\mu, m+\mu) G^j_{l'-1, m'+\mu, l, m+\mu}, \\ J^{(21)}_{l'm'lm} = J^{(12)}_{l'm'lm} \end{cases} \quad (6.38)$$

and $\mathbb{J}(0, 0)$ is diagonal, i.e., $J^{(p'p)}_{l'm'lm}(0, 0) = \delta_{p'p} \delta_{l'l} \delta_{mm'}$.

For the **H**-multipoles [Eq. (5.81)], we have to distinguish whether the position at which we expand the fields is close to $\mathbf{O}'$, i.e., $r' < R$, or if it is further away, i.e., $r' > R$. In the former case we need to expand the translated fields in terms of **J**-multipoles (regular at the

---

[4] Note that $\int_\Omega Y^*_{l'm'}(\hat{\mathbf{k}}) Y_{LM}(\hat{\mathbf{k}}) Y_{lm}(\hat{\mathbf{k}}) d\Omega = 0$ and $C(l, L, l'; m, M, m') = 0$ for $M \neq m' - m$.

origin), whereas in the latter case we use **H**-multipoles (regular at infinity). Thus, we obtain

$$\begin{cases} \mathbf{H}_{lm}^{(p)}(k_{\mathrm{m}}r, \hat{\mathbf{r}}) = \sum_{p'l'm'} H_{l'm'lm}^{(p'p)}(k_{\mathrm{m}}R, \hat{\mathbf{R}}) \mathbf{J}_{l'm'}^{(p')}(k_{\mathrm{m}}r', \hat{\mathbf{r}}') & \text{for } r' < R, \\ \mathbf{H}_{lm}^{(p)}(k_{\mathrm{m}}r, \hat{\mathbf{r}}) = \sum_{p'l'm'} J_{l'm'lm}^{(p'p)}(k_{\mathrm{m}}R, \hat{\mathbf{R}}) \mathbf{H}_{l'm'}^{(p')}(k_{\mathrm{m}}r', \hat{\mathbf{r}}') & \text{for } r' > R, \end{cases} \quad (6.39)$$

where

$$\begin{cases} H_{l'm'lm}^{(11)} = \sum_{\mu=-1,0,1} C_1(l', l'; -\mu, m' + \mu) C_1(l, l; -\mu, m + \mu) G_{l',m'+\mu,l,m+\mu}^{\mathrm{h}}, \\ H_{l'm'lm}^{(22)} = H_{l'm'lm}^{(11)}, \\ H_{l'm'lm}^{(12)} = i\sqrt{\frac{2l'+1}{l'+1}} \\ \qquad \times \sum_{\mu=-1,0,1} C_1(l'-1, l'; -\mu, m' + \mu) C_1(l, l; -\mu, m + \mu) G_{l'-1,m'+\mu,l,m+\mu}^{\mathrm{h}}, \\ H_{l'm'lm}^{(21)} = H_{l'm'lm}^{(12)}, \end{cases} \quad (6.40)$$

and $H_{l'm'lm}^{(p'p)}(0, \mathbf{0}) \equiv 0$.

The incident field can be expanded independently in the two reference frames, Σ and Σ'. Since its actual value is invariant from the reference frame, by using Eq. (5.84) and Eq. (6.37) we obtain the translated amplitudes for the incident field,

$$\tilde{W}_{l'm'}^{(p')} = \sum_{plm} J_{l'm'lm}^{(p'p)}(k_{\mathrm{m}}R, \hat{\mathbf{R}}) W_{\mathrm{i},lm}^{(p)}, \quad (6.41)$$

or, in more compact form,

$$\tilde{\mathbf{W}}_{\mathrm{i}} = \mathbb{J}(k_{\mathrm{m}}R, \hat{\mathbf{R}}) \mathbf{W}_{\mathrm{i}}. \quad (6.42)$$

Similarly, using Eq. (5.82) and Eq. (6.39), we find the translated amplitudes for the scattered field,

$$\begin{cases} \tilde{A}_{\mathrm{s},l'm'}^{(p')} = \sum_{plm} H_{l'm'lm}^{(p'p)}(k_{\mathrm{m}}R, \hat{\mathbf{R}}) A_{\mathrm{s},lm}^{(p)} & \text{for } r' < R, \\ \tilde{A}_{\mathrm{s},l'm'}^{(p')} = \sum_{plm} J_{l'm'lm}^{(p'p)}(k_{\mathrm{m}}R, \hat{\mathbf{R}}) A_{\mathrm{s},lm}^{(p)} & \text{for } r' > R, \end{cases} \quad (6.43)$$

or, in more compact form,

$$\begin{cases} \tilde{\mathbf{A}}_{\mathrm{s}} = \mathbb{H}(k_{\mathrm{m}}R, \hat{\mathbf{R}}) \mathbf{A}_{\mathrm{s}} & \text{for } r' < R, \\ \tilde{\mathbf{A}}_{\mathrm{s}} = \mathbb{J}(k_{\mathrm{m}}R, \hat{\mathbf{R}}) \mathbf{A}_{\mathrm{s}} & \text{for } r' > R. \end{cases} \quad (6.44)$$

Finally, using the definition of T-matrix [Eq. 6.2], we obtain for the case $r' < R$

$$\tilde{\mathbf{A}}_{\mathrm{s}} = \mathbb{H}\mathbf{A}_{\mathrm{s}} = \mathbb{H}\mathbb{T}\mathbf{W}_{\mathrm{i}} = \mathbb{H}\mathbb{T}\mathbb{J}^{-1}\tilde{\mathbf{W}}_{\mathrm{i}} = \tilde{\mathbb{T}}\tilde{\mathbf{W}}_{\mathrm{i}}.$$

An analogous expression can be obtained for the case $r' > R$. Thus, the translated T-matrix is

$$\begin{cases} \tilde{\mathbb{T}} = \mathbb{H}(k_{\mathrm{m}} R, \hat{\mathbf{R}})\,\mathbb{T}\,\mathbb{J}^{-1}(k_{\mathrm{m}} R, \hat{\mathbf{R}}) & \text{for } r' < R, \\ \tilde{\mathbb{T}} = \mathbb{J}(k_{\mathrm{m}} R, \hat{\mathbf{R}})\,\mathbb{T}\,\mathbb{J}^{-1}(k_{\mathrm{m}} R, \hat{\mathbf{R}}) & \text{for } r' > R. \end{cases} \quad (6.45)$$

Exercise 6.1.9 Calculate numerically the optical forces on a spherical particle in a focused laser beam using its translated T-matrix. Compare the results with the ones obtained by translating the incoming field amplitudes. [Hint: You can use the program `tmatrixtranslation` from the book website.]

Exercise 6.1.10 Show that $\mathbb{J}^{-1}(k_{\mathrm{m}} R, \hat{\mathbf{R}}) = \mathbb{J}(k_{\mathrm{m}} R, -\hat{\mathbf{R}})$ and $\mathbb{H}^{-1}(k_{\mathrm{m}} R, \hat{\mathbf{R}}) = \mathbb{H}(k_{\mathrm{m}} R, -\hat{\mathbf{R}})$.

Exercise 6.1.11 Show that when the translation of the reference system occurs along $\hat{\mathbf{z}}$, i.e., $\mathbf{R} = \pm R\hat{\mathbf{z}}$, $\mathbb{H}(k_{\mathrm{m}} R, \hat{\mathbf{R}})$ and $\mathbb{J}(k_{\mathrm{m}} R, \hat{\mathbf{R}})$ are diagonal in m, i.e., their elements are not vanishing only for $m = m'$. [Hint: Use the definition of $\mathbb{H}$ and $\mathbb{J}$ in Eqs. (6.37) and (6.39) and the fact that the spherical harmonics for $\hat{\mathbf{R}} = \hat{\mathbf{z}}$ ($\theta = 0$) are $Y_{lm}(0, \varphi) = \sqrt{\frac{2l+1}{4\pi}} \delta_{m0}$.]

6.1.5 Rotation theorem

To find how multipole fields and T-matrices transform under rotation, we consider two reference frames, Σ and $\tilde{\Sigma}'$, with a common origin, but with the coordinate systems rotated about a fixed axis. The rotation that transforms Σ into $\tilde{\Sigma}'$ entails first a rotation of γ about the z-axis, then a rotation of β about the y-axis and finally a rotation of α about the z-axis, where $[\alpha, \beta, \gamma]$ are the *Euler angles*.

We start by considering how spherical harmonics $Y_{lm}(\hat{\mathbf{r}})$ [Section 5.2.1] are transformed under rotation. Because $Y_{lm}(\hat{\mathbf{r}})$ are eigenfunctions of L^2 [Eqs. (5.42)] and L^2 is invariant under rotations, the transformed spherical harmonics in the rotated reference frame $\tilde{Y}_{lm'}(\hat{\mathbf{r}}')$ are linear combinations of $Y_{lm}(\hat{\mathbf{r}})$ with the same l, i.e.,

$$\tilde{Y}_{lm'}(\hat{\mathbf{r}}') = \sum_{m=-l}^{l} D^{(l)}_{mm'}(\alpha, \beta, \gamma) Y_{lm}(\hat{\mathbf{r}}), \quad (6.46)$$

where $D^{(l)}_{mm'}(\alpha, \beta, \gamma)$ are the elements of the *Wigner rotation matrices* [Box 6.1]. The inverse of Eq. (6.46) is

$$Y_{lm}(\hat{\mathbf{r}}) = \sum_{m'=-l}^{l} D^{(l)*}_{mm'}(\alpha, \beta, \gamma) \tilde{Y}_{lm'}(\hat{\mathbf{r}}'). \quad (6.47)$$

The same transformation rules hold for the vector spherical harmonics [Section 5.2.3], so that for the rotated multipoles we have

$$\tilde{\mathbf{J}}^{(p)}_{lm'}(k_{\mathrm{m}} r', \hat{\mathbf{r}}') = \sum_{m=-l}^{l} D^{(l)}_{mm'}(\alpha, \beta, \gamma) \mathbf{J}^{(p)}_{lm}(k_{\mathrm{m}} r, \hat{\mathbf{r}}) \quad (6.48)$$

> **Box 6.1**   **Wigner rotation matrices**
>
> The *Wigner rotation matrices* $D^{(l)}_{mm'}(\alpha, \beta, \gamma)$ are obtained as products of the three rotation matrices corresponding to the Euler angles (Rose, 1955), i.e.,
>
> $$D^{(l)}_{mm'}(\alpha, \beta, \gamma) = e^{-im\alpha} d^{(l)}_{mm'}(\beta) e^{-im'\gamma},$$
>
> with
>
> $$d^{(l)}_{mm'}(\beta) = \sqrt{(l+m)!(l-m)!(l+m')!(l-m')!}$$
> $$\times \sum_s \frac{(-1)^{m-m'+s} \left(\cos\frac{\beta}{2}\right)^{2l+m'-m-2s} \left(\sin\frac{\beta}{2}\right)^{m-m'+2s}}{(l+m'-s)! \, s! \, (m-m'+s)! \, (l-m-s)!},$$
>
> where the sum is over the values of s such that the arguments of the factorials are nonnegative. Alternatively, $d^{(l)}_{mm'}(\beta)$ is given explicitly in differential form in Varshalovich et al. (1988). Since the inverse of the rotation with Euler angles $[\alpha, \beta, \gamma]$ is $[-\gamma, -\beta, -\alpha]$, we have that
>
> $$\left[D^{(l)}_{m'm}(\alpha, \beta, \gamma)\right]^{-1} = D^{(l)*}_{m'm}(\alpha, \beta, \gamma) = D^{(l)}_{mm'}(-\gamma, -\beta, -\alpha).$$
>
> The following orthogonality relations hold:
>
> $$\sum_m D^{(l)*}_{mm'}(\alpha, \beta, \gamma) D^{(l)}_{mm''}(\alpha, \beta, \gamma) = \delta_{m'm''},$$
> $$\sum_m D^{(l)*}_{m'm}(\alpha, \beta, \gamma) D^{(l)}_{m''m}(\alpha, \beta, \gamma) = \delta_{m'm''}.$$
>
> Furthermore, the following relations hold:
>
> $$d^{(l)}_{mm'}(\beta) = d^{(l)}_{-m',-m}(\beta) = (-1)^{m-m'} d^{(l)}_{-m,-m'}(\beta) \quad \text{for} \quad m \geq m',$$
> $$d^{(l)}_{mm'}(\beta) = d^{(l)}_{m'm}(-\beta) = (-1)^{m'-m} d^{(l)}_{m'm}(\beta) \quad \text{for} \quad m \leq m'.$$

and

$$\tilde{\mathbf{H}}^{(p)}_{lm'}(k_m r', \hat{\mathbf{r}}') = \sum_{m=-l}^{l} D^{(l)}_{mm'}(\alpha, \beta, \gamma) \mathbf{H}^{(p)}_{lm}(k_m r, \hat{\mathbf{r}}). \tag{6.49}$$

The corresponding inverses are

$$\mathbf{J}^{(p)}_{lm}(k_m r, \hat{\mathbf{r}}) = \sum_{m'=-l}^{l} D^{(l)*}_{mm'}(\alpha, \beta, \gamma) \tilde{\mathbf{J}}^{(p)}_{lm'}(k_m r', \hat{\mathbf{r}}') \tag{6.50}$$

and

$$\mathbf{H}^{(p)}_{lm}(k_m r, \hat{\mathbf{r}}) = \sum_{m'=-l}^{l} D^{(l)*}_{mm'}(\alpha, \beta, \gamma) \tilde{\mathbf{H}}^{(p)}_{lm'}(k_m r', \hat{\mathbf{r}}'). \tag{6.51}$$

We can now express the incident field [Eq. (5.84)] in $\tilde{\Sigma}'$ as

$$\begin{aligned}\mathbf{E}_i &= E_i \sum_{plm} \tilde{W}^{(p)}_{i,lm} \tilde{\mathbf{J}}^{(p)}_{lm}(k_m r', \hat{\mathbf{r}}') \\ &= E_i \sum_{plm} \tilde{W}^{(p)}_{i,lm} \sum_{M=-l}^{l} D^{(l)}_{Mm}(\alpha,\beta,\gamma) \mathbf{J}^{(p)}_{lM}(k_m r, \hat{\mathbf{r}}) \\ &= E_i \sum_{plM} W^{(p)}_{i,lM} \mathbf{J}^{(p)}_{lM}(k_m r, \hat{\mathbf{r}}),\end{aligned}$$

so that the transformation of the incident wave amplitudes is given by

$$W^{(p)}_{i,lM} = \sum_{m=-l}^{l} \tilde{W}^{(p)}_{i,lm} D^{(l)}_{Mm}(\alpha,\beta,\gamma). \tag{6.52}$$

In a similar way, but using the scattered field expansion [Eq. (5.82)], we obtain the transformation of the scattered amplitudes,

$$\tilde{A}^{(p)}_{s,lm} = \sum_{M=-l}^{l} A^{(p)}_{s,lM} D^{(l)*}_{Mm}(\alpha,\beta,\gamma). \tag{6.53}$$

Finally, using the definition of the T-matrix [Eq. (6.1)], we get

$$\begin{aligned}\tilde{A}^{(p)}_{s,lm} &= \sum_{M=-l}^{l} D^{(l)*}_{Mm}(\alpha,\beta,\gamma) A^{(p)}_{s,lM} \\ &= \sum_{M=-l}^{l} D^{(l)*}_{Mm}(\alpha,\beta,\gamma) \sum_{p'l'M'} T^{(pp')}_{lMl'M'} W^{(p')}_{i,l'M'} \\ &= \sum_{p'l'm'} \sum_{M=-l}^{l} \sum_{M'=-l'}^{l'} D^{(l)*}_{Mm}(\alpha,\beta,\gamma) T^{(pp')}_{lMl'M'} D^{(l')}_{M'm'}(\alpha,\beta,\gamma) \tilde{W}^{(p')}_{i,l'm'},\end{aligned}$$

from which the T-matrix in the rotated reference frame follows:

$$\tilde{T}^{(pp')}_{lml'm'} = \sum_{M=-l}^{l} \sum_{M'=-l'}^{l'} D^{(l)*}_{Mm}(\alpha,\beta,\gamma) T^{(pp')}_{lMl'M'} D^{(l')}_{M'm'}(\alpha,\beta,\gamma), \tag{6.54}$$

or, in more compact form,

$$\tilde{\mathbb{T}} = \mathbb{D}^*(\alpha,\beta,\gamma)\,\mathbb{T}\,\mathbb{D}(\alpha,\beta,\gamma), \tag{6.55}$$

where $\mathbb{D}^*(\alpha,\beta,\gamma) = \{D^{(l)*}_{Mm}(\alpha,\beta,\gamma)\}$ and $\mathbb{D}(\alpha,\beta,\gamma) = \{D^{(l')}_{M'm'}(\alpha,\beta,\gamma)\}$.

Exercise 6.1.12 Given a T-matrix corresponding to a non-spherically-symmetric object (e.g., consider the T-matrix of a spherical particle displaced from the origin of the reference frame), calculate numerically the corresponding rotated T-matrix as a function of the rotation angles α, β and γ. Calculate the corresponding optical forces produced by a beam whose focus is at the origin of the coordinate system. [Hint: You can use the program tmatrixrotation from the book website.]

Exercise 6.1.13 Show that the independent elements of $d^{(1)}_{mm'}(\beta)$ are

$$d^{(1)}_{0,0}(\beta) = \cos\beta,$$

$$d^{(1)}_{1,1}(\beta) = \frac{1}{2}(1 + \cos\beta),$$

$$d^{(1)}_{1,0}(\beta) = -\frac{\sqrt{2}}{2}(1 - \cos^2\beta)^{1/2},$$

$$d^{(1)}_{1,-1}(\beta) = \frac{1}{2}(1 - \cos\beta),$$

and that the remaining elements can be found by symmetry. Derive the corresponding rotation matrix $D^{(1)}_{mm'}(\alpha, \beta, \gamma)$.

6.1.6 Clebsch–Gordan coefficients

In many physical problems there is a need to couple two or more angular momenta. Standard examples in quantum mechanics are the coupling of orbital and spin angular momenta and the calculation of the total angular momentum of two particles. If we consider two angular momenta $\mathbf{J}_1$ and $\mathbf{J}_2$ with associated eigenfunctions $\psi_{j_1 m_1}$ and $\psi_{j_2 m_2}$ and eigenvalues j_1 and j_2 (so that $J_1^2 \psi_{j_1 m_1} = j_1(j_1+1)\psi_{j_1 m_1}$, $J_{1,z}\psi_{j_1 m_1} = m_1 \psi_{j_1 m_1}$, $J_2^2 \psi_{j_2 m_2} = j_2(j_2+1)\psi_{j_2 m_2}$ and $J_{2,z}\psi_{j_2 m_2} = m_1 \psi_{j_2 m_2}$), their coupling is the total angular momentum $\mathbf{J} = \mathbf{J}_1 + \mathbf{J}_2$ with associated eigenfunction ψ_{jm}. The eigenfunction ψ_{jm} can be expressed in terms of $\psi_{j_1 m_1}$ and $\psi_{j_2 m_2}$ using the Clebsch–Gordan coefficients,

$$\psi_{jm} = \sum_{m_1=-j_1}^{j_1} \sum_{m_2=-j_2}^{j_2} C(j_1, j_2, j; m_1, m_2, m)\psi_{j_1 m_1}\psi_{j_2 m_2}. \tag{6.56}$$

Therefore, the Clebsch–Gordan coefficients are the elements of the unitary matrix that transforms the set of $(2j_1+1)\cdot(2j_2+1)$ product vectors into the set of the simultaneous eigenfunctions of J^2 and J_z. The Clebsch–Gordan coefficients vanish unless their arguments satisfy the relations

$$\begin{cases} m_1 + m_2 = m \\ |j_1 - j_2| \leq j \leq j_1 + j_2 \\ -j_1 \leq m_1 \leq j_1 \\ -j_2 \leq m_2 \leq j_2 \end{cases} \tag{6.57}$$

and their explicit expression is obtained within group theory (Rose, 1995) as

$$C(j_1, j_2, j; m_1, m_2, m) = \sqrt{\frac{(2l_3+1)\prod_{a=1}^{3}(j_1+j_2+j_3-2j_a)!(j_j+m_j)!(j_j-m_j)!}{(j_1+j_2+j_3+1)!}}$$

$$\times \sum_{\nu} \frac{(-1)^{\nu}}{\nu!(l_1+l_2-l_3-\nu)!(l_1-m_1-\nu)!(l_2+m_2-\nu)!(l_3-l_2+m_1+\nu)!(l_3-l_1-m_2+\nu)!}, \tag{6.58}$$

where ν runs over all values that do not make the arguments of the factorials negative, i.e., from $\max(0, j_2 - j - m_1, j_1 - j - m_2)$ to $\min(j_1 + j_2 - j, j_1 - m_1, j_2 + m_2)$. Some

Table 6.1 Clebsch–Gordan coefficients $\mathcal{C}_1(l, l'; \mu, m' - \mu)$

	$\mu = -1$	$\mu = 0$	$\mu = 1$
$l = l' - 1$	$\sqrt{\frac{(l'-1-m')(l'-m')}{2l'(2l'-1)}}$	$\sqrt{\frac{(l'-m')(l'+m')}{l'(2l'-1)}}$	$\sqrt{\frac{(l'-1+m')(l'+m')}{2l'(2l'-1)}}$
$l = l'$	$-\sqrt{\frac{(l'-m')(l'+1+m')}{2l'(2l'+1)}}$	$-\frac{m'}{\sqrt{l'(l'+1)}}$	$\sqrt{\frac{(l'+m')(l'+1-m')}{2l'(l'+1)}}$
$l = l' + 1$	$\sqrt{\frac{(l'+2+m')(l'+1+m')}{2(l'+1)(2l'+3)}}$	$-\sqrt{\frac{(l'+1-m')(l'+1+m')}{(l'+1)(2l'+3)}}$	$\sqrt{\frac{(l'+1-m')(l'+2-m')}{2(l'+1)(2l'+3)}}$

$\mathcal{C}_1(l, l'; \mu, m' - \mu) = 0$ for $|m' - l| > l$ as a consequence of the relations in Eq. (6.57).

useful symmetry properties of the Clebsh–Gordan coefficients are

$$C(j_1, j_2, j; m_1, m_2, m) = \frac{C(j_1, j_2, j; -m_1, -m_2, -m)}{(-1)^{j_1+j_2-j}}, \tag{6.59}$$

$$C(j_2, j_1, j; m_2, m_1, m) = \frac{C(j_1, j_2, j; -m_1, -m_2, -m)}{(-1)^{j_1+j_2-j}}, \tag{6.60}$$

$$C(j, j_2, j_1; -m, m_2, -m_1) = \sqrt{\frac{2j_1+1}{2j+1}} \frac{C(j_1, j_2, j; -m_1, -m_2, -m)}{(-1)^{j_2+m_2}}, \tag{6.61}$$

$$C(j_1, j, j_2; m_1, -m, m_2) = \sqrt{\frac{2j_2+1}{2j+1}} \frac{C(j_1, j_2, j; -m_1, -m_2, -m)}{(-1)^{j_1-m_1}}, \tag{6.62}$$

$$C(j, j_1, j_2; m_1, -m, m_2) = \sqrt{\frac{2j_2+1}{2j+1}} \frac{C(j_1, j_2, j; -m_1, -m_2, -m)}{(-1)^{j_1-m_1}}, \tag{6.63}$$

$$C(j_2, j, j_1; -m_2, m, m_1) = \sqrt{\frac{2j_1+1}{2j+1}} \frac{C(j_1, j_2, j; -m_1, -m_2, -m)}{(-1)^{j_2+m_2}}. \tag{6.64}$$

A particularly important case is the coupling between orbital and spin angular momenta, i.e., between **L** and **s**. For the case of light, the spin is 1, i.e., $j_1 = 1$, and the corresponding Clebsch–Gordan coefficients C_1 are tabulated in Table 6.1.

Exercise 6.1.14 In principle, the Clebsch–Gordan coefficients can be calculated using their explicit formula given in Eq. (6.58); however, this presents some numerical challenges because of the presence of products of large factorials. Show that the logarithm of the factorial function can be computed easily for very large arguments and can be utilised to avoid overflow when factorials of large integers are needed. Note, however, that on a standard computer this leads to incorrect results for j_1, j_2 and j larger than about 50.[5]

---

[5] This is due to the finite precision of computer arithmetic (typically limited to 64 bits). A possible approach to avoid this problem is to calculate the coefficients to a very high precision (e.g., calculating them with an accuracy of 100 bits permits one to estimate correctly coefficients of the order of at least 300). Furthermore, several recursive algorithms have been proposed to solve the problem (Brock, 2001).

6.2 Metal spheres sustaining longitudinal fields

Mie theory is based on the assumption that the material of the scattering sphere can sustain the propagation of transverse waves only. This came into question when Ferrell (1958) and Ferrell and Stern (1962) were able to show that thin metal foils may sustain the propagation of longitudinal polarisation waves when the frequency of the exciting field exceeds the plasma frequency (*surface plasmon polaritons*). The presence of longitudinal waves implies a revision of the boundary conditions that must be satisfied by the electromagnetic field. Eventually, Ruppin (1975) extended Mie theory to the case of small metal spheres whose material sustains the propagation of longitudinal waves: when the frequency of the incident plane wave exceeds the plasma frequency, the optical properties of the scattering sphere may be noticeably affected, whereas, at lower frequencies, they are practically indistinguishable from those predicted by the classical Mie theory. This can have non-trivial consequences for the optical trapping of metal particles as demonstrated, e.g., by Saija et al. (2009).

Following Ruppin (1975), we consider a homogeneous sphere of radius a centred at the origin of the coordinate frame. The external field is, as usual, the superposition of the incident field and the scattered field. The incoming field is expanded in **J**-multipoles [Eq. (5.79)] with amplitudes $W^{(1)}_{i,lm}$ and $W^{(2)}_{i,lm}$ and the scattered wave is expanded in **H**-multipoles [Eq. (5.82)] with amplitudes $A^{(1)}_{s,lm}$ and $A^{(2)}_{s,lm}$ to be determined by the boundary conditions at the surface of the sphere. The expansion of the field within the sphere requires some further comment. When longitudinal waves can propagate, the equation $\nabla \cdot \mathbf{E} = 0$ is no longer valid so that the waves are not transverse. As a result, the usual divergenceless **J**-multipoles are not sufficient for the expansion, but they must be supplemented by the longitudinal multipole fields

$$\mathbf{J}_{\mathrm{L},lm}(k_\mathrm{L} r, \hat{\mathbf{r}}) = \frac{1}{k_\mathrm{L}\sqrt{l(l+1)}} \nabla j_l(k_\mathrm{L} r) Y_{lm}(\hat{\mathbf{r}}), \tag{6.65}$$

where $k_\mathrm{L} = n_\mathrm{L} k_0$ is the propagation constant of the longitudinal waves and n_L is the corresponding refractive index,[6] whereas the refractive index for the propagation of the transverse waves is denoted by n_p, as usual. Therefore, the internal electric field can be written as a superposition of the transverse field with expansion in **J**-multipoles [Eq. (5.79)] and of a longitudinal field:[7]

$$\mathbf{E}_\mathrm{p}(k_\mathrm{p} r, k_\mathrm{L} r, \hat{\mathbf{r}}) = E_\mathrm{i} \sum_{lm} W^{(1)}_{\mathrm{p},lm}\mathbf{J}^{(1)}_{lm}(k_\mathrm{p} r, \hat{\mathbf{r}}) + W^{(2)}_{\mathrm{p},lm}\mathbf{J}^{(2)}_{lm}(k_\mathrm{p} r, \hat{\mathbf{r}}) + W^{(\mathrm{L})}_{\mathrm{p},lm}\mathbf{J}_{\mathrm{L},lm}(k_\mathrm{L} r, \hat{\mathbf{r}}). \tag{6.66}$$

Besides the continuity of the tangential components of the electric and magnetic fields, the continuity of the normal component of the electric displacement must also now be

---

[6] The dielectric properties of the metal can be assumed to be well described, e.g., by the simplified form reported by Pack et al. (2001), which includes non-local effects crucial when modelling coupled metal particles.

[7] Because the expansion in Eq. (6.65) includes curl-free multipole fields only, there is no magnetic field associated with the longitudinal waves. Therefore, the expansion of the total magnetic field is not affected by the occurrence of longitudinal waves, as can be expected in any nonmagnetic material.

satisfied. Taking into account that the vector harmonics form a complete orthonormal set, we can express these conditions using the projections of the fields on the vector harmonics,

$$\begin{cases} \oint_\Omega (\mathbf{E}_i + \mathbf{E}_s) \cdot \mathbf{Z}_{lm}^{(1)*}(\hat{\mathbf{r}})\, d\Omega = \oint_\Omega \mathbf{E}_p \cdot \mathbf{Z}_{lm}^{(1)*}(\hat{\mathbf{r}})\, d\Omega, \\ \oint_\Omega (\mathbf{E}_i + \mathbf{E}_s) \cdot \mathbf{Z}_{lm}^{(2)*}(\hat{\mathbf{r}})\, d\Omega = \oint_\Omega \mathbf{E}_p \cdot \mathbf{Z}_{lm}^{(2)*}(\hat{\mathbf{r}})\, d\Omega, \\ \oint_\Omega (\mathbf{B}_i + \mathbf{B}_s) \cdot \mathbf{Z}_{lm}^{(1)*}(\hat{\mathbf{r}})\, d\Omega = \oint_\Omega \mathbf{B}_p \cdot \mathbf{Z}_{lm}^{(1)*}(\hat{\mathbf{r}})\, d\Omega, \\ \oint_\Omega (\mathbf{B}_i + \mathbf{B}_s) \cdot \mathbf{Z}_{lm}^{(2)*}(\hat{\mathbf{r}})\, d\Omega = \oint_\Omega \mathbf{B}_p \cdot \mathbf{Z}_{lm}^{(2)*}(\hat{\mathbf{r}})\, d\Omega, \\ \oint_\Omega (\mathbf{E}_i + \mathbf{E}_s) \cdot \mathbf{Z}_{lm}^{(1)*}(\hat{\mathbf{r}})\, d\Omega = \oint_\Omega \mathbf{E}_p \cdot \mathbf{Y}_{lm}^*(\hat{\mathbf{r}})\, d\Omega, \end{cases}$$

where the integration is over the full solid angle Ω. By substituting the appropriate expression, we obtain for each l and m the set of five independent equations

$$\begin{cases} j_l(\rho_m) W_{i,lm}^{(1)} + h_l(\rho_m) A_{s,lm}^{(1)} = j_l(\rho_p) W_{p,lm}^{(1)}, \\ \dfrac{1}{\rho_m} \left\{ [\rho_m j_l(\rho_m)]' W_{i,lm}^{(2)} + [\rho_m h_l(\rho_m)]' A_{s,lm}^{(2)} \right\} = \dfrac{1}{\rho_p} [\rho_p h_l(\rho_p)]' W_{p,lm}^{(2)} \\ \qquad\qquad - \dfrac{i}{\rho_L} j_l(\rho_L) W_{p,lm}^{(L)}, \\ \dfrac{n_m}{\rho_m} \left\{ [\rho_m j_l(\rho_m)]' W_{i,lm}^{(2)} + [\rho_m h_l(\rho_m)]' A_{s,lm}^{(2)} \right\} = \dfrac{n_p}{\rho_p} [\rho_p h_l(\rho_p)]' W_{p,lm}^{(1)}, \\ n_m \left\{ j_l(\rho_m) W_{i,lm}^{(2)} + h_l(\rho_m) A_{s,lm}^{(2)} \right\} = n_p j_l(\rho_p) W_{p,lm}^{(2)}, \\ \dfrac{i\sqrt{l(l+1)}}{\rho_m} j_l(\rho_m) W_{p,lm}^{(2)} + \dfrac{i\sqrt{l(l+1)}}{\rho_m} h_l(\rho_m) A_{p,lm}^{(2)} = \dfrac{i\sqrt{l(l+1)}}{\rho_p} j_l(\rho_p) W_{p,lm}^{(2)} \\ \qquad\qquad + \dfrac{1}{\sqrt{l(l+1)}} j_l'(\rho_L) W_{p,lm}^{(L)}, \end{cases}$$

where $\rho_L = k_L a$ and the definitions of the other symbols are given in Subsection 5.2.5. The elimination of the amplitudes $W_{p,lm}^{(1)}$, $W_{p,lm}^{(2)}$ and $W_{p,lm}^{(L)}$ yields the amplitudes of the scattered fields,

$$\begin{cases} A_{s,lm}^{(1)} = -\dfrac{n_p u_l'(\rho_p) u_l(\rho_m) - n_m u_l(\rho_p) u_l'(\rho_m)}{n_p u_l'(\rho_p) w_l(\rho_m) - n_m u_l(\rho_p) w_l'(\rho_m)} W_{i,lm}^{(1)}, \\ A_{s,lm}^{(2)} = -\dfrac{j_l'(\rho_L) \left[n_m u_l'(\rho_p) u_l(\rho_m) - n_p u_l(\rho_p) u_l'(\rho_m) \right] + k_0 j_l(\rho_m) d_l}{j_l'(\rho_L) \left[n_m u_l'(\rho_p) w_l(\rho_m) - n_p u_l(\rho_p) w_l'(\rho_m) \right] + k_0 h_l(\rho_m) d_l} W_{i,lm}^{(2)}, \end{cases} \qquad (6.67)$$

where

$$d_l = (n_p^2 - n_m^2) \frac{l(l+1)}{k_L} j_l(\rho_p) j_l(\rho_L),$$

$u_l(\rho)$ are Riccati–Bessel functions [Eq. (5.98)] and $w_l(\rho)$ are Riccati–Hankel functions [Eq. (5.99)].[8]

Exercise 6.2.1 Write the T-matrix $\mathbb{T}$ of a sphere sustaining longitudinal fields as a function of the coefficients $A^{(1)}_{\text{p},lm}$ and $A^{(2)}_{\text{p},lm}$ given by Eqs. (6.67). Note that in this case also spherical symmetry implies the independence from m of the elements of $\mathbb{T}$.

6.3 Radially symmetric spheres

Wyatt (1962) extended Mie theory to the case of a radially symmetric sphere with radius a and complex refractive index $n_\text{p} = n_\text{p}(r)$, where r is the radial distance from the centre of the sphere. As usual, in the region external to the sphere the field is the superposition of the incoming field and the scattered field. The incoming field is expanded in **J**-multipoles as in Eq. (5.84) with amplitudes given by $W^{(1)}_{\text{i},lm}$ and $W^{(2)}_{\text{i},lm}$, and the scattered wave is expanded in **H**-multipoles as in Eq. (5.82) with amplitudes given by $A^{(1)}_{\text{s},lm}$ and $A^{(2)}_{\text{s},lm}$. Because the medium inside the sphere is not homogenous, the internal electric and magnetic fields do not satisfy two independent Helmholtz equations, but rather the coupled equations

$$\begin{cases} \nabla \times \nabla \times \mathbf{E}_\text{p} - n_\text{p}(r)^2 k_0^2 \mathbf{E}_\text{p} = 0, \\ \nabla \times \nabla \times \mathbf{B}_\text{p} - n_\text{p}(r)^2 k_0^2 \mathbf{B}_\text{p} = -ik_0 \nabla \left[n_\text{p}(r)^2 \right] \times \mathbf{E}_\text{p}. \end{cases} \quad (6.68)$$

Because of the spherical symmetry of the scatterer, however, the internal field can still be expanded in a series of vector spherical harmonics, whereas the inhomogeneity of the refractive index suggests that a single radial function is not sufficient. Indeed, the internal fields can still be expanded in a series of vector spherical harmonics in the forms

$$\mathbf{E}_\text{p} = E_\text{i} \sum_{lm} W^{(1)}_{\text{p},lm} \Phi_l(r) \mathbf{Z}^{(1)}_{lm}(\hat{\mathbf{r}}) + W^{(2)}_{\text{p},lm} \frac{1}{n_\text{p}(r)^2 k_0} \nabla \times \left[\Psi_l(r) \mathbf{Z}^{(1)}_{lm}(\hat{\mathbf{r}}) \right] \quad (6.69)$$

and

$$i\mathbf{B}_\text{p} = E_\text{i} \sum_{lm} W^{(1)}_{\text{p},lm} \frac{1}{k_0} \nabla \times \left[\Phi_l(r) \mathbf{Z}^{(1)}_{lm}(\hat{\mathbf{r}}) \right] + W^{(2)}_{\text{p},lm} \Psi_l(r) \mathbf{Z}^{(1)}_{lm}(\hat{\mathbf{r}}), \quad (6.70)$$

which, for any choice of the radial functions and for any radial dependence of the refractive index, satisfy Maxwell's equations $\nabla \cdot \mathbf{B} = 0$ and $\nabla \cdot n_\text{p}(r)^2 \mathbf{E} = 0$. By substituting Eqs. (6.69) and (6.70) into Eq. (6.68), we obtain the following equations for the radial

---

[8] In the case of metallic particles, there is not a formula such as the Wiscombe formula [Eq. (5.102)] to estimate the number of significant coefficients, even though, as a rule of thumb, one can expect the number of significant coefficients to be greater than the corresponding number used for a dielectric particle of the same size. In any case, it is necessary to verify in each case the actual convergence of the solution in order to establish the number of coefficients that need to be retained.

functions $\Phi(r)$ and $\Psi(r)$ (regular at the origin):

$$\begin{cases} \left[\dfrac{d^2}{dr^2} - \dfrac{l(l+1)}{r^2} + k_0^2 n_p(r)^2 - \dfrac{2}{n_p(r)}\dfrac{dn_p(r)}{dr}\dfrac{d}{dr}\right][r\Psi_l(r)] = 0, \\ \left[\dfrac{d^2}{dr^2} - \dfrac{l(l+1)}{r^2} + k_0^2 n_p(r)^2\right][r\Phi_l(r)] \qquad\qquad\qquad = 0, \end{cases} \qquad (6.71)$$

which can be integrated numerically. Next, we impose the boundary conditions following a procedure similar to the one outlined for homogeneous spheres in Subsection 5.2.5. Thus, we take advantage of the mutual independence of the vector spherical harmonics to obtain, for each l and m, four equations among which the amplitudes of the internal field $W^{(1)}_{p,lm}$ and $W^{(2)}_{p,lm}$ can be eliminated. Finally, we get the amplitudes of the scattered field in the form

$$\begin{cases} A^{(1)}_{s,lm} = -\dfrac{G^{(1)'}_l(\rho_m) u_l(\rho_m) - G^{(1)}_l(\rho_m) u'_l(\rho_m)}{G^{(1)'}_l(\rho_m) w_l(\rho_m) - G^{(1)}_l(\rho_m) w'_l(\rho_m)} W^{(1)}_{i,lm}, \\[1em] A^{(2)}_{s,lm} = -\dfrac{n_m^2 G^{(2)'}_l(\rho_m) u_l(\rho_m) - n_p(a)^2 G^{(2)}_l(\rho_m) u'_l(\rho_m)}{n_m^2 G^{(2)'}_l(\rho_m) w_l(\rho_m) - n_p(a)^2 G^{(2)}_l(\rho_m) w'_l(\rho_m)} W^{(2)}_{i,lm}, \end{cases} \qquad (6.72)$$

where $\rho_m = n_m k_0 a$, $G^{(1)}_l(k_m r) = k_m r \Phi_l(r)$, $G^{(2)}_l(k_m r) = k_m r \Psi_l(r)$, and $u_l(\rho)$ and $w_l(\rho)$ are Riccati–Bessel functions [Eq. (5.98)] and Riccati–Hankel functions [Eq. (5.99)], respectively.

Exercise 6.3.1 Show that, when $n_p(r) \equiv n_p$ (constant), Eqs. (6.68) are identical to one another and to the Helmholtz equation, that Eqs. (6.71) are identical to one another, and that the normalisation can be chosen so that $\Phi_l(r) = \Psi_l(r) = j_l(n_p k_0 r)$. Show also that Eqs. (6.72) reduce to Eqs. (5.100).

Exercise 6.3.2 Write the T-matrix $\mathbb{T}$ of a radially symmetric sphere as a function of the coefficients $A^{(1)}_{s,lm}$ and $A^{(2)}_{s,lm}$ given by Eqs. (6.72). Note that also in this case the spherical symmetry implies the independence from m of the elements of $\mathbb{T}$.

Exercise 6.3.3 Apply the theory for radially symmetric spheres to spheres composed of concentric homogenous layers of different refractive indices (layered spheres). To preserve the continuity of the refractive index and its radial derivative between two contiguous layers, place a thin transition layer at the interface.[9] For example, in a layer defined by the radial interval $r_- \leq r \leq r_+$, the refractive index may vary from n_- to n_+ according to the rule

$$n_p^2(r) = n_-^2 + (3s^2 - 2s^3)\Delta n^2,$$

where $\Delta n^2 = n_+^2 - n_-^2$ and $s = \frac{r-r_-}{r_+-r_-}$.

---

[9] The introduction of a transition layer of appropriate thickness may also account for the fact that the external layer is actually formed by successive deposition of atoms and molecules. It is reasonable to think that the first deposited atoms are adsorbed at the surface of the internal layer, thus giving rise to a smooth transition from the properties of the internal layer to those of the external layer.

6.4 Clusters of spheres

The T-matrix approach is suitable to calculate the scattering amplitudes of composite non-spherical particles and, in particular, of particles modelled as aggregates of spheres (Borghese et al., 1984). To do this, we will use the rotation and translation theorems described in Subsections 6.1.4 and 6.1.5, respectively, to calculate the T-matrices of clusters of spheres [Subsection 6.4.1] and spheres with spherical inclusions [Subsection 6.4.2]. Once the amplitudes of the scattered fields and, hence, the T-matrices are calculated, the optical force can be found using Eq. (6.8) and the optical torque using Eqs. (6.23), (6.25) and (6.26).

6.4.1 Aggregates of spheres

We consider a cluster composed of N (homogeneous) spheres of radius a_α and refractive index n_α, with $\alpha = 1, \ldots, N$, immersed in a medium with refractive index n_m. The spheres are centred at $\mathbf{R}_\alpha$ with respect to the common origin $\mathbf{O}$. The field scattered by the whole aggregate is the superposition of the fields scattered by each sphere, i.e.,[10]

$$\mathbf{E}_\mathrm{s}(\mathbf{r}) = \sum_\alpha \sum_{plm} \mathcal{A}^{(p)}_{\mathrm{s},\alpha lm} \mathbf{H}^{(p)}_{lm}(k_\mathrm{m} r_\alpha, \hat{\mathbf{r}}_\alpha), \qquad (6.73)$$

where $\mathbf{r}_\alpha = r_\alpha \hat{\mathbf{r}}_\alpha = \mathbf{r} - \mathbf{R}_\alpha$, $\mathcal{A}^{(p)}_{\mathrm{s},\alpha lm}$ are the translated amplitudes and $k_\mathrm{m} = n_\mathrm{m} k_0$ is the wavenumber in the medium. The field inside the αth sphere is expanded as

$$\mathbf{E}_{\mathrm{p},\alpha}(\mathbf{r}) = \sum_{plm} \mathcal{W}^{(p)}_{\mathrm{p},\alpha lm} \mathbf{J}^{(p)}_{lm}(k_\alpha r_\alpha, \hat{\mathbf{r}}_\alpha), \qquad (6.74)$$

with $k_\alpha = n_\alpha k_0$, so that it is regular everywhere inside the sphere. The magnetic field, necessary for imposing the boundary conditions, is given by Eq. (5.60) both inside and outside the sphere.

The scattered field is given by a linear combination of multipole fields that have different origins, whereas the incident field is given by a combination of multipole fields centred at the origin of the coordinates. Since the boundary conditions must be imposed at the surface of each constituent sphere, we exploit the addition theorem for multipole fields [Subsection (6.1.4)] to express the incident field at the surface of the αth sphere as

$$\mathbf{E}_\mathrm{i}(\mathbf{r}) = \sum_{plm} \sum_{p'l'm'} J^{(p,p')}_{lml'm'}(k_\mathrm{m} R_\alpha, \hat{\mathbf{R}}_\alpha)\, W^{(p')}_{\mathrm{i},l'm'}\, \mathbf{J}^{(p)}_{lm}(k_\mathrm{m} r_\alpha, \mathbf{r}_\alpha), \qquad (6.75)$$

where the quantities $J^{(p,p')}_{lml'm'}(k_\mathrm{m} R_\alpha, \hat{\mathbf{R}}_\alpha)$ are again defined by Eqs. (6.38). Analogously, we exploit the addition theorem for multipole fields to express $\mathbf{E}_\mathrm{s}(\mathbf{r})$ at the surface of the αth

---

[10] In the rest of this section we will use the notation $\sum_\alpha$ to signify $\sum_{\alpha=1}^{N}$.

sphere in terms of multipole fields centred at $\mathbf{R}_\alpha$,

$$\mathbf{E}_s(\mathbf{r}) = \sum_{plm} \mathcal{A}^{(p)}_{s,\alpha lm} \mathbf{H}^{(p)}_{lm}(k_m r_\alpha, \hat{\mathbf{r}}_\alpha)$$
$$+ \sum_{plm} \sum_{\alpha' \neq \alpha} \sum_{p'l'm'} H^{(pp')}_{lml'm'}(k_m \Delta R_{\alpha\alpha'}, \Delta\hat{\mathbf{R}}_{\alpha\alpha'}) \mathcal{A}^{(p')}_{\alpha'l'm'} \mathbf{J}^{(p)}_{lm}(k_m r_\alpha, \hat{\mathbf{r}}_\alpha), \qquad (6.76)$$

where the quantities $H^{(pp')}_{lml'm'}(k_m \Delta R_{\alpha\alpha'}, \Delta\hat{\mathbf{R}}_{\alpha\alpha'})$ are defined by Eqs. (6.40) and $\Delta\mathbf{R}_{\alpha\alpha'} = \Delta R_{\alpha\alpha'} \Delta\hat{\mathbf{R}}_{\alpha\alpha'} = \mathbf{R}_\alpha - \mathbf{R}_{\alpha'}$.

Applying the boundary conditions, we get four equations for each α, l and m, among which the elimination of the amplitudes of the internal fields yields as final result a system of linear non-homogeneous equations (Borghese et al., 2007b),

$$\sum_{\alpha'} \sum_{p'l'm'} M^{(pp')}_{\alpha lm\alpha'l'm'} \mathcal{A}^{(p')}_{s,\alpha'l'm'} = -\mathcal{W}^{(p)}_{i,\alpha lm}, \qquad (6.77)$$

where we have defined the shifted amplitudes of the incident field for each sphere with respect to the origin,

$$\mathcal{W}^{(p)}_{i,\alpha lm} = \sum_{p'l'm'} J^{(pp')}_{lml'm'}(k_m R_\alpha, \mathbf{R}_\alpha) W^{(p')}_{i,l'm'},$$

$$M^{(pp')}_{\alpha lm\alpha'l'm'} = \frac{1}{R^{(p)}_{\alpha l}} \delta_{\alpha\alpha'} \delta_{pp'} \delta_{ll'} \delta_{mm'} + H^{(pp')}_{lml'm'}(k_m \Delta R_{\alpha\alpha'}, \Delta\hat{\mathbf{R}}_{\alpha\alpha'}),$$

$R^{(p)}_{\alpha l}$ are the Mie coefficients for the αth sphere given by Eq. (6.3), and the matrix $\mathbb{H} = \left\{ H^{(pp')}_{lml'm'}(k_m \Delta R_{\alpha\alpha'}, \Delta\hat{\mathbf{R}}_{\alpha\alpha'}) \right\}$ describes the multiple scattering processes occurring among the spheres in the aggregate. The formal solution to the system given by Eq. (6.77) is

$$\mathcal{A}^{(p)}_{s,\alpha lm} = -\sum_{p'l'm'} [\mathbb{M}^{-1}]^{(pp')}_{\alpha lm\alpha'l'm'} \mathcal{W}^{(p')}_{i,\alpha'l'm'}, \qquad (6.78)$$

where the matrix $\mathbb{M}^{-1}$ relates the amplitudes of the incident field to those of the fields scattered by each sphere in the aggregate.

The addition theorem allows us to write the scattered field in terms of multipole fields with origin at $\mathbf{O}$ as

$$\mathbf{E}_s(\mathbf{r}) = \sum_{plm} A^{(p)}_{s,lm} \mathbf{H}^{(p)}_{lm}(k_m r, \hat{\mathbf{r}}), \qquad (6.79)$$

where we have defined

$$A^{(p)}_{s,lm} = \sum_{\alpha'} \sum_{p'l'm'} J^{(pp')}_{lml'm'}(k_m R_{\alpha'}, -\hat{\mathbf{R}}_{\alpha'}) \mathcal{A}^{(p')}_{s,\alpha'l'm'} \qquad (6.80)$$

and the quantities $J^{(pp')}_{lml'm'}(k_m R_{\alpha'}, -\hat{\mathbf{R}}_{\alpha'})$ are the elements of the matrix that translates the origin of the $\mathbf{H}$-multipole fields from $\mathbf{R}_{\alpha'}$ to the origin of the coordinates at $\mathbf{O}$. We note that Eq. (6.79) is valid outside the smallest sphere that contains the whole aggregate and is thus appropriate to get the scattered field in the far-field zone. Finally, the T-matrix elements of the whole aggregate are

$$T^{(pp')}_{lml'm'} = -\sum_{\alpha\alpha'}\sum_{qLM}\sum_{q'L'M'} J^{(pq)}_{lmLM}(k_m R_\alpha, -\hat{\mathbf{R}}_\alpha)\left[\mathrm{M}^{-1}\right]^{(qq')}_{\alpha L M \alpha' L' M'} J^{(q'p')}_{L'M'l'm'}(k_m R_{\alpha'}, \hat{\mathbf{R}}_{\alpha'}). \tag{6.81}$$

Exercise 6.4.1 Derive the T-matrix for an aggregate of radially non-homogeneous spheres. [Hint: Note that the expansion of the internal field is given by Eq. (6.69) and that the quantities $R^{(p)}_{\alpha l}$ are obtained from Eq. (6.72).]

6.4.2 Inclusions

An interesting extension of the cluster scattering model is that of a homogeneous sphere containing one or more off-centre spherical inclusions (Borghese et al., 1992). This problem is exhaustively treated in Borghese et al. (2007b) as an extension of the cluster model presented in the previous subsection. Here, we only present the final results, together with an outline of their derivation.

We divide the space into three regions: the external region, which is filled by a homogeneous non-dispersive medium of refractive index n_m ($k_m = n_m k_0$); the interstitial region inside the sphere, of radius a_p and centred at $\mathbf{R}_p$, filled by a homogeneous medium of refractive index n_p ($k_p = n_p k_0$); and the spherical inclusions, each of radius a_α and refractive index $n_\alpha(r_\alpha)$ ($k_\alpha = n_\alpha k_0$), where r_α is the radial coordinate centred on the αth sphere centre at $\mathbf{r}_\alpha$ and $\alpha = 1, \ldots, N$. As usual, the field in the external region is the superposition of the incident field and the field scattered by the whole object, whose respective multipole expansions are given by Eqs. (5.79) and (5.82). The field within the inclusions can be expanded in the form of Eq. (6.74), if they are homogeneous, or Eq. (6.69), if they are radially symmetric. Finally, we assume that the field in the interstitial region is the superposition of the field that would exist without any inclusion and of the field scattered by the inclusions themselves. Hence, within the interstitial region, the multipole expansion of the field is

$$\mathbf{E}_p(\mathbf{r}) = E_i \sum_{plm}\left[\mathcal{N}^{(p)}_{plm}\mathbf{J}^{(p)}_{lm}(k_m r_p, \hat{\mathbf{r}}_p) + \sum_\alpha \mathcal{N}^{(p)}_{\alpha lm}\mathbf{H}^{(p)}_{lm}(k_m r_\alpha, \hat{\mathbf{r}}_\alpha)\right], \tag{6.82}$$

where $\mathbf{r}_p = \mathbf{r} - \mathbf{R}_p$, $\mathbf{r}_\alpha = \mathbf{r} - \mathbf{R}_\alpha$, $\mathcal{N}^{(p)}_{plm}$ are the amplitudes of the interstitial field without inclusions and $\mathcal{N}^{(p)}_{\alpha lm}$ are the amplitudes of the fields scattered by the inclusions.

In calculating the multipole amplitudes of the scattered field in the external region, the first step is to impose the boundary conditions to the internal field and interstitial field for each inclusion. Thus, we can eliminate the amplitudes of the internal fields inside the inclusions and determine the amplitudes of the interstitial field. Then, as a second step, we impose the boundary conditions between the particle and the external region in order to get the amplitudes of the scattered field in the external region. In both cases, we need to use the translation theorem to obtain the multipole expansion with fields centred at the same origin.

By imposing the first set of boundary conditions, a system of linear non-homogeneous equations is found that relates the multipole amplitudes $\mathbf{N} = \{\mathcal{N}_{\alpha lm}^{(p)}\}$ and $\mathbf{N}_p = \{\mathcal{N}_{plm}^{(p)}\}$ to the incident amplitudes $\mathbf{W}_i = \{W_{lm}^{(p)}\}$, i.e.,

$$\mathcal{M}\mathcal{N} = \begin{bmatrix} \mathbb{R}^{-1} + \mathbb{H} & \mathbb{J}_{\leftarrow p} \\ \mathbb{R}_W \mathbb{J}_{p\leftarrow} & \mathbb{R}_p^{-1} \end{bmatrix} \begin{bmatrix} \mathbf{N} \\ \mathbf{N}_p \end{bmatrix} = \begin{bmatrix} \mathbf{0} \\ \mathbf{W}_i \end{bmatrix} = \mathcal{W}_i. \qquad (6.83)$$

The matrix $\mathbb{R}$ is related to the spherical inclusions and its elements are

$$\mathcal{R}_{\alpha lm\alpha'l'm'}^{(pp')} = \delta_{\alpha\alpha'}\delta_{pp'}\delta_{ll'}\delta_{mm'} R_{\alpha l}^{(p)},$$

where in the case of homogeneous inclusions the elements $R_{\alpha l}^{(p)}$ are given by Eq. (6.3), whereas for layered inclusions the elements are modified as in Section 6.3. The matrix $\mathbb{R}_p$ is related to the embedding sphere, and its elements are given by

$$\mathcal{R}_{p,lml'm'}^{(pp')} = \delta_{pp'}\delta_{ll'}\delta_{mm'} R_{p,l}^{(p)},$$

where

$$\begin{cases} R_{p,l}^{(1)} = \dfrac{in_p}{n_m u_l(k_p a_p) w_l'(k_m a_p) - n_p u_l'(k_p a_p) w_l(k_m a_p)}, \\ R_{p,l}^{(2)} = \dfrac{in_p}{n_p u_l(k_p a_p) w_l'(k_m a_p) - n_m u_l'(k_p a_p) w_l(k_m a_p)}. \end{cases}$$

The matrix $\mathbb{H} = \{H_{\alpha lm\alpha'l'm'}^{(pp')}\}$ describes the multiple scattering process among the inclusions, whereas the matrices $\mathbb{J}_{\leftarrow p} = \{J_{\alpha lmpl'm'}^{(pp')}\}$ and $\mathbb{J}_{p\leftarrow} = \{J_{plm\alpha l'm'}^{(pp')}\}$ are related to multiple scattering processes between the inclusions and the embedding sphere; their elements are obtained as for the case of a cluster discussed in the previous subsection. Finally, the elements of the matrix $\mathbb{R}_W$ are

$$\mathcal{R}_{W,lml'm'}^{(pp')} = \delta_{pp'}\delta_{ll'}\delta_{mm'} R_{Wl}^{(p)},$$

where

$$\begin{cases} R_{W,l}^{(1)} = \dfrac{n_m w_l(k_p a_p) w_l'(k_m a_p) - n_p w_l'(k_p a_p) w_l(k_m a_p)}{in_p}, \\ R_{Wl}^{(2)} = \dfrac{n_p w_l(k_p a_p) w_l'(k_m a_p) - n_m w_l'(k_p a_p) w_l(k_m a_p)}{in_p}. \end{cases}$$

By inverting Eq. (6.83), we can find the $\mathcal{N}$ coefficients using the matrix $\mathcal{M}^{-1}$. In fact, because of the zeros in $\mathcal{W}$, we only need to consider certain partitions of $\mathcal{M}^{-1}$,

$$\mathcal{M}^{-1} = \begin{bmatrix} \mathcal{Z}_1 & \mathcal{Z}_{1p} \\ \mathcal{Z}_{p1} & \mathcal{Z}_p \end{bmatrix},$$

and, thus,

$$\mathcal{N} = \begin{bmatrix} \mathcal{Z}_{1p} \\ \mathcal{Z}_p \end{bmatrix} \mathbf{W}_i = \mathbb{Z}\mathbf{W}_i.$$

Once the coefficients $\mathcal{N}$ have been calculated, we consider the boundary conditions between the particle and the external region and obtain the amplitudes of the scattered field in the external region,

$$\mathbf{A}_s = \begin{bmatrix} \mathbb{M}_\mathrm{W} \mathbb{J}_{p\leftarrow} & \mathbb{M}_p \end{bmatrix} \mathcal{N} = \mathbb{T} \mathbf{W}_i,$$

where $\mathbb{T}$ is the T-matrix of the system

$$\mathbb{T} = \begin{bmatrix} \mathbb{M}_\mathrm{W} \mathbb{J}_{p\leftarrow} \mathcal{Z}_{\mathrm{l}p} + \mathbb{M}_p \mathcal{Z}_p \end{bmatrix}. \tag{6.84}$$

The matrices $\mathbb{M}_\mathrm{W} = \{\delta_{pp'}\delta_{ll'}\delta_{mm'} M^{(p)}_{\mathrm{W},l}\}$ and $\mathbb{M}_p = \{\delta_{pp'}\delta_{ll'}\delta_{mm'} M^{(p)}_{\mathrm{p},l}\}$ are diagonal with elements

$$\begin{cases} M^{(1)}_{\mathrm{W},l} = \dfrac{n_\mathrm{m} w'_l(k_\mathrm{p} a_\mathrm{p}) u_l(k_\mathrm{m} a_\mathrm{p}) - n_\mathrm{p} w_l(k_\mathrm{p} a_\mathrm{p}) u'_l(k_\mathrm{m} a_\mathrm{p})}{i n_\mathrm{m}}, \\[6pt] M^{(2)}_{\mathrm{W},l} = \dfrac{n_\mathrm{p} w'_l(k_\mathrm{p} a_\mathrm{p}) u_l(k_\mathrm{m} a_\mathrm{p}) - n_\mathrm{m} w_l(k_\mathrm{p} a_\mathrm{p}) u'_l(k_\mathrm{m} a_\mathrm{p})}{i n_\mathrm{m}} \end{cases}$$

and

$$\begin{cases} M^{(1)}_{\mathrm{p},l} = \dfrac{n_\mathrm{m} u'_l(k_\mathrm{p} a_\mathrm{p}) u_l(k_\mathrm{m} a_\mathrm{p}) - n_\mathrm{p} u_l(k_\mathrm{p} a_\mathrm{p}) u'_l(k_\mathrm{m} a_\mathrm{p})}{i n_\mathrm{m}}, \\[6pt] M^{(2)}_{\mathrm{p},l} = \dfrac{n_\mathrm{p} u'_l(k_\mathrm{p} a_\mathrm{p}) u_l(k_\mathrm{m} a_\mathrm{p}) - n_\mathrm{m} u_l(k_\mathrm{p} a_\mathrm{p}) u'_l(k_\mathrm{m} a_\mathrm{p})}{i n_\mathrm{m}}. \end{cases}$$

Exercise 6.4.2 Derive the T-matrix for a spherical particle with a centred spherical inclusion. Compare your result with the one obtained for a radially symmetric sphere in Section 6.3.

6.4.3 Convergence

Solving the scattering problem for a cluster and calculating its T-matrix requires the inversion of the matrix $\mathbb{M}$. In principle this has infinite elements, being the result of a multipole expansion of the fields. Thus, we need to rely on a truncation of the system in Eq. (6.81) to some finite order by including in the multipole expansion all the terms up to a multipole order l_T, which should be chosen to ensure the numerical stability of the calculated quantities (e.g., optical forces and torques) at a reasonable computational effort. Practically, one should check for the existence of a minimum l-value, l_M, such that when $l_\mathrm{T} > l_\mathrm{M}$ the observable quantities do not change within numerical accuracy. Thus, for a cluster of N spheres, a matrix $\mathbb{M}$ of order $d = 2Nl_\mathrm{T}(l_\mathrm{T} + 2)$ needs to be inverted, which might be quite large. The inversion of the matrix $\mathbb{M}$ is a process that accounts for a large share of the entire light-scattering calculations. Once the matrix $\mathbb{M}^{-1}$ has been obtained, the T-matrix of the aggregate is obtained by inserting its elements into Eq. (6.81). The order of the truncation that must be considered in the multipole expansion of a cluster depends not only on the individual particle size, but also on the geometrical packing and on the smallest surface-to-surface distance occurring in the aggregate. The case of metal particles is of particular interest because the value of l_M and hence the truncation order, l_T, is related to the internal fields, which include the longitudinal fields described in Section 6.2. This generally implies that truncation at much larger orders, which can reach $l_T = 100$ to 1000 (Ruppin, 1975), is needed for clusters of metal particles with respect to clusters of dielectric particles. Similar considerations apply also to the case of spheres with inclusions.

6.5 Discrete dipole approximation

The *discrete dipole approximation* (DDA), which is also known as the *coupled dipole method* (CDM), is a finite element method originally devised by Purcell and Pennypacker (1973) and later improved by Draine and Flatau (1994). In the DDA method, a particle is split into a series of dipoles, as shown in Fig. 6.2, each of which interacts with the incident electromagnetic wave and with the electromagnetic waves re-radiated by all the other dipoles. In this section, we briefly review the basics of the DDA technique. We refer the interested reader to, e.g., Girard (2005) and Yurkin and Hoekstra (2007) for a detailed description of this method and to Simpson and Hanna (2010a, 2011) for studies where the DDA method has been applied to optical trapping problems.

A particle is approximated by an array of N dipoles with polarisability α_n and located at positions $\mathbf{r}_n$, where $n = 1, \ldots, N$, within a homogeneous host medium, as shown in Fig. 6.2. The polarisabilities can be given by the Lorentz–Lorenz formula [Eq. (3.15)] or, better, by the formula including the radiative correction [Eq. (3.25)]. These dipoles acquire a dipole moment

$$\mathbf{p}_n = \alpha_n \mathbf{E}_t(\mathbf{r}_n), \tag{6.85}$$

where $\mathbf{E}_t(\mathbf{r}_n)$ is the total electric field at $\mathbf{r}_n$, including both the incoming field $\mathbf{E}_i(\mathbf{r}_n)$ and the fields re-radiated by all other dipoles. The electric field associated with the nth dipole including the near-field contributions is given by Eq. (3.19), which can be recast as

$$\mathbf{E}_n(\boldsymbol{\rho}_n) = \frac{e^{ik\rho_n}}{4\pi \varepsilon_{r,m} \varepsilon_0} \left\{ \left[\frac{k^2}{\rho_n} - \frac{1}{\rho_n^3} + \frac{ik}{\rho_n^2} \right] \mathbf{p}_n + \left[-\frac{k^2}{\rho_n} + \frac{3}{\rho_n^3} - \frac{3ik}{\rho_n^2} \right] (\hat{\boldsymbol{\rho}}_n \cdot \mathbf{p}_n) \hat{\boldsymbol{\rho}}_n \right\}, \tag{6.86}$$

where $\boldsymbol{\rho}_n = \rho_n \hat{\boldsymbol{\rho}}_n = \mathbf{r} - \mathbf{r}_n$, $\mathbf{r}$ is a point in space and $\varepsilon_{r,m}$ is the relative dielectric permittivity of the medium. Therefore, in order to find the value of $\mathbf{p}_n$ for all dipoles, one needs to solve

Figure 6.2 Discrete dipole approximation: The particles with which the incoming electromagnetic wave interacts are approximated by a series of discrete dipoles, which must be much smaller than the incoming wavelength. In this picture, we show (a) a spherical nanoparticle, (b) a nanorod and (c) a bow tie nanoantenna approximated using spherical dipoles (spheres).

the system of N linear vectorial equations, or $3N$ linear scalar equations,

$$\mathbf{p}_n = \alpha_n \left[\mathbf{E}_i(\mathbf{r}_n) + \sum_{m \neq n} \mathbf{E}_m(\mathbf{r}_n - \mathbf{r}_m) \right]. \tag{6.87}$$

Once $\mathbf{p}_n$ is known for each dipole, the electric field can be found using Eq. (6.86) and the superposition of effects. The magnetic induction $\mathbf{B}(\mathbf{r})$ can be calculated using Eq. (5.60). Finally, one can evaluate $\mathbf{E}(\mathbf{r})$ and $\mathbf{B}(\mathbf{r})$ on a surface enclosing the particle in order to calculate the optical forces and torques acting on the particle using Eqs. (5.15) and (5.16).

The DDA method permits one to straightforwardly approximate particles of any shape and material by an appropriate choice of the dipole locations and polarisabilities. However, a major problem connected with the practical use of the DDA method is the large number of equations that need to be solved in order to approximate a given particle adequately. In fact, to obtain reliable results, one needs to use dipoles significantly smaller than the light wavelength, which effectively limits the range of particles that can be studied, typically to a few hundreds nanometres in size. Moreover, the inter-particle distance, d, must satisfy the condition $|n_p| k_m d \leq 1$, where n_p is the refractive index of the particle and k_m is the light wavenumber in the medium. With non-spherical particles, another issue related to the DDA method is the lack of any simple relation between the particle orientation and the scattered field. Thus, unlike the case with T-matrix methods, when the DDA technique is used a full calculation is required for each position and orientation of the particle within the incoming field to obtain the optical force or torque acting on the particle.

6.6 Finite-difference time domain

The methods discussed so far for the solution of Maxwell's equations, including the T-matrix and DDA methods, operate in the frequency domain, as they assume a harmonic time-dependence of the fields. The *finite-difference time-domain* (FDTD) method, instead, works in the time domain by sampling the electric and magnetic fields at discrete times and positions and, therefore, does not assume a harmonic time-dependence of the fields. In this section, we will give a brief overview of the FDTD technique. We refer the interested readers to more specialised textbooks for an in-depth discussion; in particular, Elsherbeni and Demir (2009) give a good overview of the FDTD technique with an emphasis on its computational aspects. For applications of FDTD to optical trapping problems, we refer readers to Gauthier (2005) and Benito et al. (2008).

From Maxwell's time-dependent curl equations [Box 3.2] and the constitutive relations [Box 3.3], it can be seen that the time derivative of $\mathcal{E}(\mathbf{r}, t)$ depends on the curl of $\mathcal{H}(\mathbf{r}, t)$ and that, vice versa, the time derivative of $\mathcal{H}(\mathbf{r}, t)$ depends on the curl of $\mathcal{E}(\mathbf{r}, t)$.[11] This

---

[11] Typically, in the FDTD method the fundamental fields are taken to be $\mathcal{E}(\mathbf{r}, t)$ and $\mathcal{H}(\mathbf{r}, t)$. On one hand, this is due to the fact that Maxwell's curl equations are computationally more symmetric for these fields than for $\mathcal{E}(\mathbf{r}, t)$ and $\mathcal{B}(\mathbf{r}, t)$. On the other hand, historically the FDTD method was originally developed by electrical engineers, and engineers prefer to work with $\mathcal{E}(\mathbf{r}, t)$ and $\mathcal{H}(\mathbf{r}, t)$, differently from physicists, who would rather

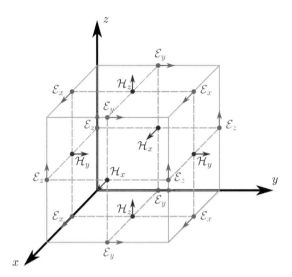

Figure 6.3 Yee cell for finite-difference time domain. The electric field components are aligned along the edges of this cubic voxel and the magnetic field components are normal to the voxel faces. A three-dimensional Yee lattice consists of a tessellation of such Yee cells filling the whole space.

observation is the fundamental stepping stone on which the FDTD is built: iteratively updating $\mathcal{E}(\mathbf{r}, t)$ and $\mathcal{H}(\mathbf{r}, t)$ results in a marching-in-time process wherein discrete field analogues of the continuous electromagnetic waves propagate in a numerical grid stored in the computer memory. This permits one to develop the FDTD method in one, two and three dimensions. Once $\mathcal{E}(\mathbf{r}, t)$ and $\mathcal{H}(\mathbf{r}, t)$ are computed, the instantaneous optical forces and torques acting on a particle can be calculated using Eqs. (5.8) and (5.11).[12] If the sources are periodic, e.g., time-harmonic fields, these forces and torques then need to be time-averaged over a period.

Yee (1966) proposed an algorithm that is at the core of most FDTD implementations. One of the critical choices in any FDTD implementation is that of the grid to be adopted. In Yee's proposal, the vector components of $\mathcal{E}(\mathbf{r}, t)$ and $\mathcal{H}(\mathbf{r}, t)$ are spatially staggered about the cubic unit cells of a Cartesian computational grid so that each Cartesian component of $\mathcal{E}(\mathbf{r}, t)$ is located midway between a pair of Cartesian components of $\mathcal{H}(\mathbf{r}, t)$, and vice versa, as shown in Fig. 6.3. Then the space is filled with these *Yee cells*, constructing what is now referred to as a *Yee lattice*. Furthermore, Yee proposed a leapfrog scheme to calculate the evolution of $\mathcal{E}(\mathbf{r}, t)$ and $\mathcal{H}(\mathbf{r}, t)$, where $\mathcal{E}(\mathbf{r}, t)$ is updated midway during each time-step between successive updates of $\mathcal{H}(\mathbf{r}, t)$; this time-stepping scheme has the advantage of avoiding the need to solve simultaneous equations for $\mathcal{E}(\mathbf{r}, t)$ and $\mathcal{H}(\mathbf{r}, t)$.

---

deal with $\mathcal{E}(\mathbf{r}, t)$ and $\mathcal{B}(\mathbf{r}, t)$. Ultimately, this is just a question of personal taste that has no consequences whatsoever for the purposes of this book.

[12] It is also possible to use directly the volumetric equation Eq. (5.2), which can make direct use of the fields $\mathcal{E}(\mathbf{r}, t)$ and $\mathcal{B}(\mathbf{r}, t)$ calculated on the Yee lattice and is often numerically more stable.

In the implementation of an FDTD solution of Maxwell's equations, one must first establish the computational domain (i.e., the physical region over which the simulation will be performed) and specify the material (i.e., the electric permittivity, magnetic permeability and conductivity) of each cell within the computational domain. Once the computational domain and the grid materials are established, some sources are specified. The sources can be, e.g., an impinging plane wave, an applied electric field or a current in a wire. In particular, using a plane wave, the FDTD method can be used to simulate light scattering from arbitrarily shaped objects or planar periodic structures at various incident angles as well as the photonic band structure of infinite periodic materials. Even though the FDTD technique computes electromagnetic fields within a compact spatial region, scattered and radiated far fields, which are of special interest for optical trapping applications, can be obtained via near-to-far-field transformations, such as the one described in Section 4.3.

The key strength of the FDTD technique is that it is very straightforward and intuitive, as one can easily understand what to expect from a given model; for example, because FDTD calculates $\mathcal{E}(\mathbf{r}, t)$ and $\mathcal{H}(\mathbf{r}, t)$ everywhere in the computational domain as they evolve in time, it lends itself to providing animated displays of the electromagnetic field evolution. Furthermore, just like the DDA technique described in the previous section, the FDTD technique permits one to study the electromagnetic interaction with objects of arbitrary geometrical shape and material composition, at least as long as they can be adequately approximated by the voxels employed in the simulation. FDTD being a time-domain technique, it is possible to use a single FDTD simulation to obtain the response of a system over a wide range of frequencies by using a broadband pulse, e.g., a Gaussian pulse; this is useful, for example, in applications where resonant frequencies are not exactly known. Furthermore, because the FDTD technique allows the user to specify the material at all points within the computational domain, a wide variety of linear and nonlinear dielectric and magnetic materials can be naturally and easily modelled.

The main weaknesses of FDTD simulations emerge from their computational requirements. Because the entire computational domain must be gridded and the grid spatial discretisation must be sufficiently fine to resolve both the smallest electromagnetic wavelength and the smallest geometrical feature in the model, very large computational domains are often required, which results in very long computation times. Moreover, to ensure the stability of the numerical integration of Maxwell's equations, the time step Δt must satisfy the Courant–Friedrichs–Lewy (CFL) condition (Courant et al., 1967),

$$\Delta t \leq \frac{1}{c\sqrt{\Delta x^{-2} + \Delta y^{-2} + \Delta z^{-2}}}, \qquad (6.88)$$

where Δx, Δy and Δz are the sizes of the edges of a Yee cell. The CFL condition means that a wave cannot be allowed to travel more than one cell size in a single time step. Furthermore, because the computational domain must necessarily be finite, whereas the scattering process occurs in infinite space, it is necessary to introduce appropriate boundary conditions across the surface that limits the region of integration to prevent unphysical reflections of the field that would otherwise modify, even severely, the near field. Although there are a number of available highly effective *absorbing boundary conditions* to simulate an infinite unbounded computational domain, most modern FDTD implementations instead use a special absorbing material, called a *perfectly matched layer*, to implement absorbing boundaries. Finally, because FDTD is solved by propagating the fields forwards in the time domain, the electromagnetic time response of the medium must be modelled explicitly; for an arbitrary response, this involves a computationally expensive time convolution, although in most cases the time response of the medium can be adequately and simply modelled using the *recursive convolution* technique, the *auxiliary differential equation* technique or the *Z-transform* technique.

6.7 Hybrid techniques

As shown in the previous sections, each method used to solve the scattering problem for non-spherical particles has its own advantages and disadvantages. For example, calculating the T-matrix in optical trapping problems is useful and computationally effective because it is possible to exploit its rotation and translation properties to obtain at once optical forces and torques for different positions and orientations of the trapped particles (Nieminen et al., 2001, 2011; Simpson and Hanna, 2006, 2007; Borghese et al., 2007a, 2008; Simpson et al., 2007). The DDA and FDTD methods, although more computationally intensive than the T-matrix, can be readily applied to particles of any shape and composition, and to any light field configuration. To take advantage of the complementary properties of different methods, hybrid methods have also been developed that make use, e.g., of the T-matrix obtained by *point-matching* the fields at the particle surface (Mackowski, 2002; Nieminen et al., 2003a) or the near fields calculated with the DDA method (Loke et al., 2009; Cao et al., 2012) to get the radiation force and torque on non-spherical scatterers. These mixed approaches are well suited to the calculation of optical forces and torques on optically trapped non-spherical particles and composites. An accurate computational comparison of optical forces on cylinders calculated using the T-matrix formulation with different methods (extended boundary condition, point-matching, and DDA) can be found in Qi et al. (2014).

Problems

6.1 *Transfer of angular momentum.* The radiation torque transferred to a particle can be calculated from the difference of the angular momentum of incoming and outgoing waves (Nieminen et al., 2009). Calculate the torque about the z-axis and show that this procedure yields the same result as the one given in Eq. (6.23). Calculate the spin and orbital contributions to the radiation torque about the z-axis in terms of incident and scattering amplitudes.

6.2 *Core–shell particles.* Consider a core–shell particle with a polymer ($n_\mathrm{p} = 1.50$) core of radius a_c and a gold shell of thickness d. Study how the extinction and scattering cross-sections change as a function of wavelength, λ_0, and of the particle parameters,

a_c and d. For what parameters do the plasmon resonances fall into the visible range? Calculate the radiation force exerted by a plane wave on these particles as a function λ_0. Finally, consider an optical tweezers made with a Gaussian beam and an objective lens with NA = 1.30; for which wavelength range is the optical trapping of the core–shell particles in water stable in three dimensions?

6.3 *Optical tweezers with Laguerre–Gaussian beams.* Consider an optical tweezers made with a Laguerre–Gaussian beam trapping a dielectric absorbing spherical particle. Calculate the optical force and torque on the particle. Study the behaviour of the force as a function of the particle size. In which range of parameters (refractive index, size) is the particle trapped off-axis? Calculate the optical force and orbital torque for off-axis trapping. Compare your results with the work by Simpson and Hanna (2009, 2010b).

6.4 *Dielectric linear chain.* Consider a linear cluster of N dielectric ($n_p = 1.50$) spherical particles with radius a_0 (in the Rayleigh range) and touching each other. Calculate the radiation force and torque exerted by a Gaussian beam focused in water by an objective lens with NA = 1.30 on the chain as a function of its orientation. How does the optical force scale with N at a fixed wavelength? Study how the results change as the subunit size increases.

6.5 *Plasmonic linear chain.* Consider a linear cluster of N gold spherical particles with radius a_0 (in the Rayleigh range) and evenly spaced by Δr. Calculate the radiation force and torque from a plane wave on the chain as a function of its orientation. How does the maximum force scale with N at a fixed wavelength? Study how the results change as the subunit size increases and compare your results with the work of Albaladejo et al. (2011). Finally, consider an optical tweezers in water made with a Gaussian beam and an objective lens with NA = 1.30. For which range of parameters (trapping wavelength, orientation, number of subunits) is the optical trapping of such a gold chain stable in three dimensions?

6.6 *Hybrid nanostructures.* Consider a dimer particle composed of a silver sphere of radius a_0 and a resonant emitter nanoparticle (*quantum dot*) of fixed radius $a_{qd} = 5$ nm placed at a distance Δr. The dielectric constant of the resonant emitter is modelled as

$$\varepsilon_{qd} = \varepsilon_b + \frac{p^2/(\varepsilon_0 V)}{\omega_0^2 - \omega^2 - 2i\omega\gamma_0},$$

where $\hbar\gamma_0 = 0.1$ meV is the broadening of the excitonic resonance, $\varepsilon_b = 2.8$ is a background dielectric constant, V is the quantum emitter volume, and p is the dipole moment of the emitter with a value such that $p/e_- = 0.3$ nm, with e_- the electron charge. Study how the extinction and scattering cross-sections change with wavelength for an emission frequency ω_0 smaller or larger than the plasmon resonance of the silver nanoparticle. Investigate the behaviour of the cross-sections as a function of the dimer parameters. The shape of the narrow linewidth that appears in the cross-sections on top of the plasmonic resonance is known as the *Fano lineshape* and is the result of the *hybridisation* of the plamonic resonance with the emitter resonance. Investigate the radiation force and torque from a plane wave on the hybrid particle. For which orientation is the force maximum? Which is the stable orientation? How does the radiation force change with wavelength? What happens if the resonant emitter is placed in the middle of two silver nanoparticles?

References

Albaladejo, S., Saenz, J. J., and Marques, M. I. 2011. Plasmonic nanoparticle chain in a light field: A resonant optical sail. *Nano Lett.*, **11**, 4597–4600.

Benito, D. C., Simpson, S. H., and Hanna, S. 2008. FDTD simulations of forces on particles during holographic assembly. *Opt. Express*, **16**, 2942–57.

Borghese, F., Denti, P., Toscano, G., and Sindoni, O. I. 1980. An addition theorem for vector Helmholtz harmonics. *J. Math. Phys.*, **21**, 2754–55.

Borghese, F., Denti, P., Saija, R., Toscano, G., and Sindoni, O. I. 1984. Multiple electromagnetic scattering from a cluster of spheres. I. Theory. *Aerosol Sci. Technol.*, **3**, 227–35.

Borghese, F., Denti, P., Saija, R., and Sindoni, O. I. 1992. Optical properties of spheres containing a spherical eccentric inclusion. *J. Opt. Soc. Am. A*, **9**, 1327–35.

Borghese, F., Denti, P., Saija, R., and Iatì, M. A. 2006. Radiation torque on nonspherical particles in the transition matrix formalism. *Opt. Express*, **14**, 9508–21.

Borghese, F., Denti, P., Saija, R., and Iatì, M. A. 2007a. Optical trapping of nonspherical particles in the T-matrix formalism. *Opt. Express*, **15**, 11,984–98.

Borghese, F., Denti, P., and Saija, R. 2007b. *Scattering from model nonspherical particles*. Berlin: Springer Verlag.

Borghese, F., Denti, P., Saija, R., Iatì, M. A., and Maragò, O. M. 2008. Radiation torque and force on optically trapped linear nanostructures. *Phys. Rev. Lett.*, **100**, 163903.

Brock, B. C. 2001. Using vector spherical harmonics to compute antenna mutual impedance from measured or computed fields. Sandia report, SAND2000–2217–Revised. Sandia National Laboratories, Albuquerque, NM.

Cao, Y., Stilgoe, A. B., Chen, L., Nieminen, T. A., and Rubinsztein-Dunlop, H. 2012. Equilibrium orientations and positions of non-spherical particles in optical traps. *Opt. Express*, **20**, 12,987–96.

Courant, R., Friedrichs, K., and Lewy, H. 1967. On the partial difference equations of mathematical physics. *IBM J. Res. Develop.*, **11**, 215–34.

Davis, L. W. 1979. Theory of electromagnetic beams. *Phys. Rev. A*, **19**, 1177–9.

Draine, B. T., and Flatau, P. J. 1994. Discrete-dipole approximation for scattering calculations. *J. Opt. Soc. Am. A*, **11**, 1491–9.

Elsherbeni, A. Z., and Demir, V. 2009. *The finite-difference time-domain method for electromagnetics with MATLAB simulations*. Raleigh, NC: Scitech.

Ferrell, R. A. 1958. Predicted radiation of plasma oscillations in metal films. *Phys. Rev.*, **111**, 1214–22.

Ferrell, R. A., and Stern, E. 1962. Plasma resonance in the electrodynamics of metal films. *Am. J. Phys.*, **30**, 810–12.

Fontes, A., Neves, A. A. R., Moreira, W. L., et al. 2005. Double optical tweezers for ultrasensitive force spectroscopy in microsphere Mie scattering. *Appl. Phys. Lett.*, **87**, 221109.

Gauthier, R. C. 2005. Computation of the optical trapping force using an FDTD based technique. *Opt. Express*, **13**, 3707–18.

Girard, C. 2005. Near fields in nanostructures. *Rep. Prog. Phys.*, **68**, 1883–1933.

Gouesbet, G. 2010. T-matrix formulation and generalized Lorenz–Mie theories in spherical coordinates. *Opt. Commun.*, **283**, 517–21.

Gouesbet, G., and Gréhan, G. 2011. *Generalized Lorenz–Mie theories*. Berlin: Springer Verlag.

Gouesbet, G., Lock, J. A., and Gréhan, G. 2011. Generalized Lorenz–Mie theories and description of electromagnetic arbitrary shaped beams: Localized approximations and localized beam models, a review. *J. Quant. Spectrosc. Radiat. Transfer*, **112**, 1–27.

Loke, V. L. Y., Nieminen, T. A., Heckenberg, N. R., and Rubinsztein-Dunlop, H. 2009. T-matrix calculation via discrete dipole approximation, point matching and exploiting symmetry. *J. Quant. Spectrosc. Radiat. Transfer*, **110**, 1460–71.

Mackowski, D. W. 2002. Discrete dipole moment method for calculation of the T-matrix for non-spherical particles. *J. Opt. Soc. Am. A*, **19**, 881–93.

Mishchenko, M. I., Travis, L. D., and Lacis, A. A. 2002. *Scattering, absorption, and emmision of light by small particles*. Cambridge, UK: Cambridge University Press.

Moreira, W. L., Neves, A. A. R., Garbos, M. K., et al. 2010. Expansion of arbitrary electromagnetic fields in terms of vector spherical wave functions. *ArXiv*, 1003.2392.

Neves, A. A. R., Fontes, A., Pozzo, L. de Y., et al. 2006a. Electromagnetic forces for an arbitrary optical trapping of a spherical dielectric. *Opt. Express*, **14**, 13,101–6.

Neves, A. A. R., Fontes, A., Padilha, L. A., et al. 2006b. Exact partial wave expansion of optical beams with respect to an arbitrary origin. *Opt. Lett.*, **31**, 2477–9.

Neves, A. A. R., Fontes, A., Cesar, C. L., et al. 2007. Axial optical trapping efficiency through a dielectric interface. *Phys. Rev. E*, **76**, 061917.

Nieminen, T. A., Rubinsztein-Dunlop, H., Heckenberg, N. R., and Bishop, A. I. 2001. Numerical modelling of optical trapping. *Comput. Phys. Commun.*, **142**, 468–71.

Nieminen, T. A., Rubinsztein-Dunlop, H., and Heckenberg, N. R. 2003a. Calculation of the T-matrix: General considerations and application of the point-matching method. *J. Quant. Spectrosc. Radiat. Transfer*, **79-80**, 1019–29.

Nieminen, T. A., Rubinsztein-Dunlop, H., and Heckenberg, N. R. 2003b. Multipole expansion of strongly focussed laser beams. *J. Quant. Spectrosc. Radiat. Transfer*, **79**, 1005–17.

Nieminen, T. A., Asavei, T., Loke, V. L., Heckenberg, N. R., and Rubinsztein-Dunlop, H. 2009. Symmetry and the generation and measurement of optical torque. *J. Quant. Spectrosc. Radiat. Transfer*, **110**, 1472–82.

Nieminen, T. A., Loke, V. L., Stilgoe, A. B., Heckenberg, N. R., and Rubinsztein-Dunlop, H. 2011. T-matrix method for modelling optical tweezers. *J. Mod. Opt.*, **58**, 528–44.

Nozawa, R. 1966. Bipolar expansion of screened Coulomb potentials, Helmholtz' solid harmonics, and their addition theorems. *J. Math. Phys.*, **7**, 1841–60.

Pack, A., Hietschold, M., and Wannemacher, R. 2001. Failure of local Mie theory: Optical spectra of colloidal aggregates. *Opt. Commun.*, **194**, 277–87.

Purcell, E. M., and Pennypacker, C. R. 1973. Scattering and absorption of light by non-spherical dielectric grains. *Astrophys. J.*, **186**, 705–14.

Qi, X., Nieminen, T. A., Stilgoe, A. B., Loke, V. L. Y., and Rubinsztein-Dunlop, H. 2014. Comparison of T-matrix calculation methods for scattering by cylinders in optical tweezers. *Opt. Lett.*, **39**, 4827–30.

Rose, M. E. 1955. *Multipole fields*. New York: Wiley.

Rose, M. E. 1995. *Elementary theory of angular momentum*. New York: Dover.

Ruppin, R. 1975. Optical properties of small metal spheres. *Phys. Rev. B*, **11**, 2871–6.

Saija, R., Denti, P., Borghese, F., Maragò, O. M., and Iatì, M. A. 2009. Optical trapping calculations for metal nanoparticles: Comparison with experimental data for Au and Ag spheres. *Opt. Express*, **17**, 10,231–41.

Simpson, S. H., and Hanna, S. 2006. Numerical calculation of inter-particle forces arising in association with holographic assembly. *J. Opt. Soc. Am. A*, **23**, 1419–31.

Simpson, S. H., and Hanna, S. 2007. Optical trapping of spheroidal particles in Gaussian beams. *J. Opt. Soc. Am. A*, **24**, 430–43.

Simpson, S. H., and Hanna, S. 2009. Optical angular momentum transfer by Laguerre–Gaussian beams. *J. Opt. Soc. Am. A*, **26**, 625–38.

Simpson, S. H., and Hanna, S. 2010a. Holographic optical trapping of microrods and nanowires. *J. Opt. Soc. Am. A*, **27**, 1255–64.

Simpson, S. H., and Hanna, S. 2010b. Orbital motion of optically trapped particles in Laguerre–Gaussian beams. *J. Opt. Soc. Am. A*, **27**, 2061–71.

Simpson, S. H., and Hanna, S. 2011. Application of the discrete dipole approximation to optical trapping calculations of inhomogeneous and anisotropic particles. *Opt. Express*, **19**, 16,526–41.

Simpson, S. H., Benito, D. C., and Hanna, S. 2007. Polarization-induced torque in optical traps. *Phys. Rev. A*, **76**, 043408.

Varshalovich, D. A., Moskalev, A. N., and Khersonskii, V. K. 1988. *Quantum theory of angular momentum*. Singapore: World Scientific Publishing.

Waterman, P. C. 1971. Symmetry, unitarity, and geometry in electromagnetic scattering. *Phys. Rev. D*, **3**, 825–39.

Wyatt, P. J. 1962. Scattering of electromagnetic plane waves from inhomogeneous spherically symmetric objects. *Phys. Rev.*, **127**, 1837–43.

Yee, K. 1966. Numerical solution of initial boundary value problems involving Maxwell's equations in isotropic media. *IEEE Trans. Antennas Propag.*, **14**, 302–7.

Yurkin, M. A., and Hoekstra, A. G. 2007. The discrete dipole approximation: An overview and recent developments. *J. Quant. Spectrosc. Radiat. Transfer*, **106**, 558–89.

7 Brownian motion

An important aspect of optical trapping and manipulation is the ubiquitous presence of *Brownian motion*. In fact, as shown in Fig. 7.1, microscopic particles undergo perpetual random motion because of collisions with the molecules of the fluid in which they are immersed. The motion of an optically trapped particle is, therefore, the result of the interplay between this random motion and the deterministic optical forces we have studied in the previous chapters: eventually the particle settles down near an equilibrium position in the optical force field, but never completely, as instead it keeps on jiggling because of Brownian motion. In this chapter, we will review the main properties of Brownian motion that come into play in dealing with optical trapping and manipulation.

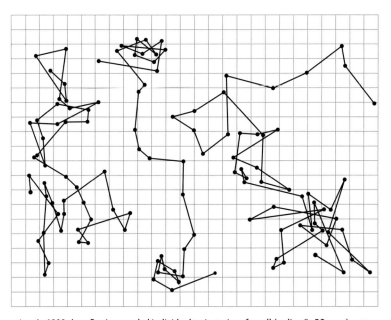

Figure 7.1 Brownian motion. In 1908, Jean Perrin recorded individual trajectories of small (radius $0.53\,\mu$m) putty particles in water at 30 s intervals. In the three trajectories shown, successive positions are joined by straight line segments. The observed random walk is due to collisions with water molecules. The mesh size is $3.2\,\mu$m. Adapted from Perrin, Ann. Chim. Physique **18**, 5–114 (1909).

7.1 The physical picture

In the early nineteenth century, the botanist Robert Brown gave the first detailed account of Brownian motion (Brown, 1828). While he was examining aqueous suspensions of pollen grains, he found that these grains were always in rapid oscillatory motion. This movement had been observed previously, but it had been explained by supposing that the particles were alive. Brown ruled out this possibility with a simple, yet brilliant, experiment: he repeated his observations on some ashes from his chimney, which he could safely assume not to be alive, and the motion was still there.

Some examples of Brownian trajectories are shown in Fig. 7.1. The main features of such trajectories can be explained by assuming that the erratic motion of a microscopic object immersed in a fluid is the result of continuous collisions with the fluid molecules, which are moving because of their thermal energy. Therefore, Brownian motion never ceases even for a system isolated from external perturbations; i.e., Brownian motion is a phenomenon that happens at thermodynamic equilibrium and is not due to external perturbations. Brownian motion increases as the particle becomes smaller, as the viscosity of the fluid decreases and as the temperature increases, whereas, at least in first approximation, it does not depend on the composition and mass of the particle. Brownian particles move independently (to first order), even when they approach one another to within less than their diameter. Finally, Brownian trajectories are very irregular, composed of translations and rotations, to the point that they appear to have no tangent, their velocity is not well defined and the motion of a particle at one particular instant is independent of the motion of that particle at any other instant.

We can make these observations more quantitative. The motion of the object and of the molecules can be described by the set of Newton's equations

$$m_n \frac{d^2}{dt^2}\mathbf{r}_n = F_n(\mathbf{r}_1, \ldots, \mathbf{r}_N) \quad \text{for } n = 1, \ldots, N, \tag{7.1}$$

where N is the total number of particles, including the microscopic object and the fluid molecules, $\mathbf{r}_n = \mathbf{r}_1, \ldots \mathbf{r}_N$ is the position of the nth particle and $\mathbf{F}_n = \mathbf{F}_1, \ldots, \mathbf{F}_N$ is the force acting on the nth particle, depending both on the interactions with the other particles and on external forces. Because Eq. (7.1) is deterministic, the resulting motion of the microscopic object is also deterministic. In fact, knowing the initial positions $\mathbf{r}_n(0)$ and velocities $\mathbf{v}_n(0) = \frac{d}{dt}\mathbf{r}_n(0)$, it is in principle possible to *deterministically* determine the motion of all particles over time, as is done in molecular dynamics simulations [Box 7.1]. Nevertheless, as shown in Fig. 7.2, the resulting motion of the microscopic object appears to be random, especially when one has no access to the exact positions and velocities of the fluid molecules.

Exercise 7.1.1 Write a code to perform a molecular dynamics simulation where one of the particles is significantly more massive then the others, as in the simulation shown in Fig. 7.2. Study how the resulting trajectory of the massive particle changes as a function of the simulation parameters. [Hint: You can adapt the code md from the book website.]

> **Box 7.1** **Molecular dynamics simulations**
>
> In a molecular dynamics simulation, Eq. (7.1) is recursively integrated to determine the evolution of the position of all particles in the system at a discrete sequence of times, typically, $t_i = i\Delta t$, where Δt is the time step.
>
> Given the positions $\mathbf{r}_{n,i}$ and velocities $\mathbf{v}_{n,i}$ of particle n at time t_i, $\mathbf{r}_{n,i+1}$ and $\mathbf{v}_{n,i+1}$ at time t_{i+1} can be found using the *leapfrog algorithm*, which consists of propagating first the positions by half time step to $\mathbf{r}_{n,i+\frac{1}{2}}$, then the velocities by a full time step to $\mathbf{v}_{n,i+1}$ using $\mathbf{r}_{n,i+\frac{1}{2}}$, and finally the positions by another half time step to $\mathbf{r}_{n,i}$ using $\mathbf{v}_{n,i+1}$, i.e.,
>
> 1. $\mathbf{r}_{n,i+\frac{1}{2}} = \mathbf{r}_{n,i} + \mathbf{v}_{n,i}\dfrac{\Delta t}{2}$,
> 2. $\mathbf{v}_{n,i+1} = \mathbf{v}_{n,i} + m_n^{-1}\mathbf{F}_n(\mathbf{r}_{1,i+\frac{1}{2}}, \ldots, \mathbf{r}_{N,i+\frac{1}{2}})\Delta t$,
> 3. $\mathbf{r}_{n,i+1} = \mathbf{r}_{n,i+\frac{1}{2}} + \mathbf{v}_{n,i+1}\dfrac{\Delta t}{2}$.
>
> Molecular dynamics algorithms are deterministic. However, in most cases, they feature chaotic behaviour, where the future state of a system sensitively depends on the initial conditions. A detailed introduction to molecular dynamics simulation can be found in Frenkel and Smit (2002).

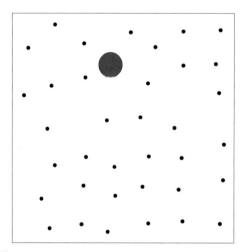
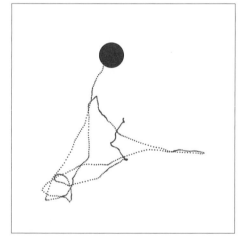

Figure 7.2 Deterministic randomness. A microscopic particle (large circle) immersed in a fluid undergoes continuous collisions with the fluid molecules (dots on the left). The resulting motion (on the right, obtained from a molecular dynamics simulation), despite being deterministic, appears to be random, especially if we do not have access to the exact positions and velocities of the fluid molecules.

Exercise 7.1.2 The motion of the massive particle in Exercise 7.1.1 goes along almost straight lines for a while before taking random turns. What are their characteristic correlation time and length? How do they change with the simulation parameters?

Exercise 7.1.3 What are the problems you encounter if you set out to perform a realistic molecular dynamics simulation of a Brownian particle?

7.2 Mathematical models

Even though in principle it would be possible to construct a model of Brownian motion by writing down Newton's equation of motion for each particle [Eq. (7.1)], this is a practically impossible task because of the huge number of molecules in any real situation — a number on the order of the Avogadro number 6.02×10^{23}. Thus, in order to reduce the number of effective degrees of freedom, many theories of Brownian motion have been developed during the past century. These theories lie along two main lines:

1. the first approach focuses on the *stochastic trajectory* $r(t)$ of a single particle, whose motion is modeled with a differential equation to which a stochastic force term is added to account for the interaction of the particle with its environment [Fig. 7.3a];
2. the second approach focuses on the *probability density distribution* $\rho(r, t)$ of an ensemble of Brownian particles, whose *deterministic* evolution is modeled using partial differential equations [Fig. 7.3b].

Not surprisingly, these approaches are strongly connected and, in fact, they can be seen as two sides of the same coin. On one hand, probability density distributions can be obtained by averaging over many trajectories and, on the other hand, the statistical properties of the random forces used to calculate the trajectories depend on the probability density distributions.

In the rest of this section, we will give an overview of these theories. Even though we will deal with one-dimensional Brownian motion, the generalisation of these results to higher dimensions is straightforward and will be dealt with within the exercises. More details on the subject can be found in Karatzas and Shreve (1998), Nelson (1967) and Øksendal (2003).

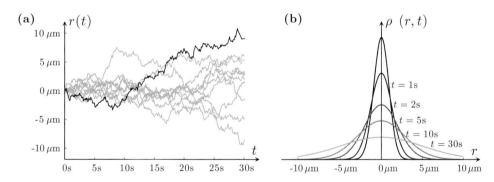

Figure 7.3 Theories of Brownian motion: Trajectories and probability distributions. Brownian motion can be modelled by focusing on (a) the stochastic trajectories $r(t)$ of single particles (random walks and Langevin equations) or on (b) the deterministic evolution of the probability density function $\rho(r, t)$ of an ensemble of particles (diffusion equations). The data shown correspond to 0.53 μm radius particles in water, released at $t = 0$ s from $r = 0$ μm. The trajectories in (a) were obtained by a Brownian dynamics simulation and the probability distributions in (b) are the solution of a free diffusion equation.

7.2.1 Random walk

A *random number* can easily be generated by flipping a coin and taking the number -1 if heads is thrown and the number $+1$ if tails is thrown. By flipping the coin repeatedly, one can generate a *sequence* of *binary random numbers* $w_i \in \{-1, +1\}$, where $i = 1, 2, 3, \ldots$. One such sequence is shown in Fig. 7.4a. A one-dimensional *random walk* is obtained by summing the terms of the sequence w_i, i.e.,

$$r_i = \sum_{j=1}^{i} w_j. \tag{7.2}$$

For example, summing the sequence in Fig. 7.4a, one obtains the random walk shown in Fig. 7.4b. Instead of these binary random numbers, it is possible to construct a random walk employing random numbers with any probability distribution. For a sufficiently large number of steps, the resulting random walk has universal properties that do not depend on the details of this probability distribution, at least as long as the random numbers have the same mean and variance.

Brownian motion can be described as a random walk. When a particle in a fluid receives a blow due to a collision with a molecule, its velocity changes, but, if the fluid is very viscous, this change is quickly dissipated, so that the net result of an impact is a displacement of the particle; this kind of behaviour is typical of systems in the *low Reynolds number regime* [Box 7.2]. The cumulative effect of multiple collisions is to produce a random walk of the particle. If the particle is at position r_i at time $t_i = i \Delta t$, where Δt is the time step, r_i evolves according to

$$r_{i+1} = r_i + \xi_i, \tag{7.3}$$

where ξ_i is a random displacement whose probability distribution $p_\xi(\xi)$ has zero mean and standard deviation σ_ξ depending on Δt.

The precise form of a Brownian motion is obviously not predictable, because it depends on a sequence of random events. However, analysing the Brownian motion of several particles, it is possible to identify some *average properties*, which are deterministic. For

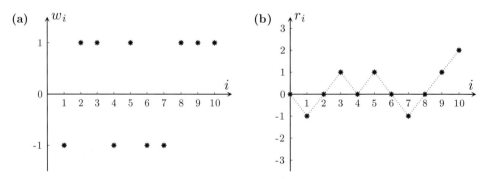

Figure 7.4 A random walk. (a) A sequence of random numbers generated by repeatedly flipping a coin. (b) A random walk resulting from the partial summations of the random numbers in (a).

> **Box 7.2** — **Life at low Reynolds numbers**
>
> The *Reynolds number* of an object moving in a fluid is the ratio between the inertial and viscous forces acting on the object itself. If the object has a characteristic dimension a and is moving at velocity v through a fluid with viscosity η and density ρ, its Reynolds number is $\mathrm{Re} = av\rho/\eta$. In the *low Reynolds number regime*, i.e., for $\mathrm{Re} \lesssim 100$, viscosity dominates over inertia. Considering, for example, an *E. coli* bacterium swimming in water, $a \approx 1\,\mu\mathrm{m}$, $v \approx 30\,\mu\mathrm{m/s}$, $\eta = 0.001\,\mathrm{Pa\,s}$ and $\rho = 1000\,\mathrm{kg/m^3}$, so $\mathrm{Re} = 3 \times 10^{-5} \ll 1$. One of the most striking aspects of low-Reynolds-number phenomena is that the speed of an object is solely determined by the forces acting on it *at the moment*. A good introduction to life at low Reynolds numbers was given by Purcell (1977). In general, most optical manipulation experiments take place at low Reynolds numbers, the exceptions being experiments in fluids with very low viscosity such as air at low pressure or in vacuum.

example, the average particle displacement after h time steps is zero because in each time step the displacement has zero mean, i.e.,

$$\Delta r(h) = \langle x_h - x_0 \rangle = \overline{x_{i+h} - x_i} = 0, \qquad (7.4)$$

where the brackets denote an *ensemble average*, i.e., an average over the different Brownian particles, and the overbar denotes a *time average*, i.e., an average in time of the Brownian motion of one given particle. In dealing with *ergodic* systems, as is most often the case, the two averages coincide.

The mean in Eq. (7.4) does not deliver a lot of information about the random walk, but there are other, more informative, average properties. In particular, the *mean squared-displacement* (MSD) after h time steps quantifies how a particle moves away from its initial position. For example, for ballistic motion, the MSD is proportional to t^2. The MSD of a Brownian particle, instead, is proportional to t, i.e.,

$$\mathrm{MSD}(h) = \langle (x_h - x_0)^2 \rangle = \overline{(x_{i+h} - x_i)^2} = 2Dt, \qquad (7.5)$$

where $t = h\Delta t$ and, again, the ensemble and time averages coincide. D is the diffusion constant and can be related to σ_ξ by $D = \sigma_\xi^2/(2\Delta t)$.

In many cases there is also an average *drift* pushing a Brownian particle during its random walk. The resulting motion is a *biased random walk*. For example, in the low Reynolds number regime [Box 7.2], a uniform external force F pushes a Brownian particle with a constant drift velocity $v_{\mathrm{drift}} = F/\gamma$, where γ is the particle friction coefficient, producing a displacement $v_{\mathrm{drift}}\Delta t$ in a time step, so that Eq. (7.3) becomes

$$r_{i+1} = r_i + v_{\mathrm{drift}}\Delta t + \xi_i, \qquad (7.6)$$

where the second and third terms on the right-hand side of Eq. (7.6) are responsible respectively for the drift and the diffusion of the Brownian particle.

Exercise 7.2.1 Simulate random walks using random numbers with various distributions (e.g., binary, uniform, exponential, Gaussian). Show that, as the number of steps increases, they all share some universal properties (e.g., appearance, MSD). [Hint: You can adapt the code `randomwalk` from the book website.]

Exercise 7.2.2 Demonstrate Eqs. (7.4) and (7.5). [Hint: You can proceed by induction.]

Exercise 7.2.3 Generalise Eqs. (7.3), (7.4) and (7.5) to the multidimensional case. [Hint: Use independent random numbers for each dimension. In l dimensions, $\langle |\Delta \mathbf{r}(t)|^2 \rangle = 2lDt$.]

Exercise 7.2.4 Estimate the Reynolds number of an olympic swimmer, of a whale and of a coastal tanker. Estimate the Reynolds number of the Brownian particles shown in Fig. 7.1. In each case, write down the equation of motion under the action of a constant force and calculate the characteristic time to reach the terminal velocity. What happens in each case if the force is abruptly switched off?

Exercise 7.2.5 What is the value of the diffusion constant D of the particles used to produce the data shown in Fig. 7.1?

Exercise 7.2.6 What is the average displacement and the MSD of a biased random walk [Eq. (7.6)]? Verify your conclusion with numerical simulations. [Hint: You can adapt the code `biasedrw` from the book website.]

7.2.2 Langevin equation

By adding a fluctuating force to Newton's equation of motion for a particle of mass m in a viscous fluid, one obtains the *Langevin equation*,

$$m\frac{d^2}{dt^2}r(t) = -\gamma \frac{d}{dt}r(t) + \chi(t), \qquad (7.7)$$

where γ is the particle friction coefficient, which for a spherical particle of radius a moving in a fluid of viscosity η, is determined by *Stokes' law*,

$$\gamma = 6\pi \eta a, \qquad (7.8)$$

and $\chi(t)$ is a random force with zero mean, i.e., $\langle \chi(t) \rangle = 0$, uncorrelated with the actual particle position, i.e., $\langle \chi(t)x(t) \rangle = 0$, and fluctuating much faster than the particle position, i.e., $\langle \chi(t)\chi(t+\tau) \rangle = 2S\delta(\tau)$, where $2S$ is the intensity of the noise.[1] Because of these three properties, $\chi(t) = \sqrt{2S}\,W(t)$, where $W(t)$ is a *white noise* [Box 7.3].

In the presence of a potential $U(r)$, and therefore of a force $F(r) = -\frac{d}{dr}U(r)$ acting on the particle, Eq. (7.7) becomes

$$m\frac{d^2}{dt^2}r(t) = -\frac{d}{dr}U(r) - \gamma \frac{d}{dt}r(t) + \chi(t). \qquad (7.9)$$

The fluid damps the colloidal particle motion as in the free-diffusive case, but now the confining potential limits the particle displacement, so that the particle explores only a limited region. A particularly important case, which was first studied by Uhlenbeck and Ornstein (1930), is when the potential is harmonic.

---

[1] In Section 7.3, we will see that $S = \gamma k_B T$, with k_B the Boltzmann constant and T the absolute temperature.

> **Box 7.3** **White noise**
>
> Brownian motion inspired the development of the mathematical concept of *white noise* and, in fact, of the whole field of *stochastic differential equations*. A white noise $W(t)$ is characterised by the following properties:
>
> 1. $\langle W(t) \rangle = 0$ for each t;
> 2. $\langle W(t)^2 \rangle = 1$ for each t;
> 3. $W(t_1)$ and $W(t_2)$ are independent of each other for $t_1 \neq t_2$.
>
> Because of these properties, a white noise is not a standard function. In particular, $W(t)$ is *almost everywhere discontinuous* and has *infinite variation*. In an intuitive picture, it can be seen as the continuous-time equivalent of a discrete sequence of independent random numbers.

In the low Reynolds number regime [Box 7.2], it is possible to drop the inertial term in Eq. (7.9), obtaining the *overdamped Langevin equation*

$$\frac{d}{dt}r(t) = -\frac{1}{\gamma}\frac{d}{dr}U(r) + \xi(t), \tag{7.10}$$

where $\xi(t) = \sqrt{2D}\,W(t)$ is a white noise with intensity $2D$, where D is the diffusion coefficient.[2]

Exercise 7.2.7 Write Eqs. (7.7) and (7.9) for a particle moving in two and three dimensions.

Exercise 7.2.8 Write the Langevin equation for an optically trapped particle in one, two and three dimensions. [Hint: Assume an elastic restoring force.]

7.2.3 Free diffusion equation

Given an ensemble of Brownian particles with probability density distribution $\rho(r, t)$, we want to calculate $\rho(r, t + \Delta t)$ as a function of Δt. For example, in Fig. 7.3b, we see the evolution of the probability density starting from $\rho(r, 0) = \delta(r)$.

Each Brownian particle in the ensemble performs a random walk according to Eq. (7.3). Thus, the probability that a particle arrives at r at time $t + \Delta t$ from $r - \xi$ at time t is equal to the probability $p_\xi(\xi)$ that the particle displacement is ξ times the probability $\rho(r - \xi, t)$ that the particle was at $r - \xi$ in the first place. Integrating over all the initial positions, we obtain

$$\rho(r, t + \Delta t) = \int_{-\infty}^{+\infty} \rho(r - \xi, t) p_\xi(\xi) d\xi. \tag{7.11}$$

---

[2] In Section 7.3, we will see that $D = k_\mathrm{B} T/\gamma$.

Expanding in Taylor series, we obtain $\rho(r, t + \Delta t) = \rho(r, t) + \Delta t \frac{\partial}{\partial t}\rho(r, t) + \mathcal{O}(\Delta t^2)$ and $\rho(r - \xi, t) = \rho(r, t) - \xi \frac{\partial}{\partial r}\rho(r, t) + \frac{1}{2}\xi^2 \frac{\partial^2}{\partial r^2}\rho(r, t) + \mathcal{O}(\xi^3)$ and, inserting these expansions into Eq. (7.11), we obtain to lowest order

$$\rho + \Delta t \frac{\partial \rho}{\partial t} = \rho \underbrace{\int p_\xi(\xi)d\xi}_{=1} - \frac{\partial \rho}{\partial r} \underbrace{\int \xi\, p_\xi(\xi)d\xi}_{=0} + \frac{1}{2}\frac{\partial^2 \rho}{\partial r^2} \underbrace{\int \xi^2\, p_\xi(\xi)d\xi}_{=\sigma_\xi^2 = 2D\Delta t}, \quad (7.12)$$

where the explicit values of the moments of ξ are shown below the underbraces and, to simplify the notation, we have written $\rho = \rho(r, t)$ and the extremes of integration have been omitted. Finally, the *free diffusion equation* is obtained by rearranging Eq. (7.12) as

$$\frac{\partial \rho(r, t)}{\partial t} = D \frac{\partial^2 \rho(r, t)}{\partial r^2}. \quad (7.13)$$

If $\rho(r, 0) = \delta(r)$, i.e., all the particles are initially at position $r = 0$ at time $t = 0$, the solution of Eq. (7.13) is a Gaussian with zero mean and variance $2Dt$, i.e.,

$$\rho(r, t) = \frac{1}{\sqrt{4\pi D t}} \exp\left(-\frac{r^2}{4Dt}\right), \quad (7.14)$$

as shown in Fig. 7.3b.

On the average, freely diffusing Brownian particles move from more crowded regions to less crowded ones, producing a *particle current* due to the particle diffusion $J_{\text{diff}}(r, t)$. Because the total number of particles is conserved, we can introduce the *continuity relation*

$$\frac{\partial \rho(r, t)}{\partial t} = -\frac{\partial J_{\text{diff}}(r, t)}{\partial r}. \quad (7.15)$$

The comparison of Eqs. (7.15) and (7.13) leads to

$$J_{\text{diff}}(r, t) = -D \frac{\partial \rho(r, t)}{\partial x}, \quad (7.16)$$

which represents the particle *diffusion current* due to the particle diffusion from crowded to less crowded regions.

Exercise 7.2.9 Show how the solution of the free diffusion equation for $\rho(r, 0) = \delta(r)$ can be used to construct a solution of the free diffusion for all possible initial conditions $\rho(r, 0)$. [Hint: Use the fact that $\rho(r, 0) = \int \rho(r', 0)\delta(r - r')dr'$.]

Exercise 7.2.10 Study how $\rho(r, t)$ evolves for various initial conditions. In particular, notice how density fluctuations with higher spatial frequency decay faster.

Exercise 7.2.11 Generalise Eq. (7.13) to the cases of two and three dimensions, and calculate the corresponding solution. For the two-dimensional case, compare your results with the outcome of numerical simulations. [Hint: You can use the code `freediffusion` from the book website.]

7.2.4 Fokker–Planck equation

The free diffusion equation [Eq. (7.13)] can be generalised to the case where some external forces $F(r,t)$ are acting on the particles. In the low-Reynolds-number regime [Box 7.2], $F(r,t)$ induces on the particles a drift velocity $v(r,t) = F(r,t)/\gamma$, where γ is the particle friction coefficient. The total particle current is then given by

$$J(r,t) = J_{\text{drift}}(r,t) + J_{\text{diff}}(r,t) = v(r,t)\rho(r,t) - D\frac{\partial \rho(r,t)}{\partial x}, \tag{7.17}$$

where $J_{\text{drift}}(r,t) = v(r,t)\rho(r,t)$ is the *drift current* due to the presence of external forces and $J_{\text{diff}}(r,t)$ is the diffusion current given in Eq. (7.16). Using the continuity equation [Eq. (7.15)], one obtains the *Fokker–Planck equation*, which describes the evolution of the probability density in the presence of both drift and diffusion:

$$\frac{\partial \rho(r,t)}{\partial t} = -\frac{\partial [F(r,t)\rho(r,t)]}{\partial r} + D\frac{\partial^2 \rho(r,t)}{\partial r^2}. \tag{7.18}$$

Exercise 7.2.12 Derive the Fokker–Planck equation [Eq. (7.18)] with an alternative approach starting with a biased random walk [Eq. (7.6)] and proceeding along the lines of the derivation of Eq. (7.13) [Subsection 7.2.3].

Exercise 7.2.13 Generalise the Fokker–Planck equation [Eq. (7.18)] to two and three dimensions.

7.3 Fluctuation–dissipation theorem, potential and equilibrium distribution

The diffusion and friction coefficients, D and γ, respectively, are closely related to each other and to the average kinetic energy of a particle in a heat bath, i.e., $\frac{1}{2}k_BT$ [Box 7.4], where k_B is the Boltzmann constant and T is the absolute temperature. This can be seen by considering a particle coupled to a heat bath in a potential, e.g., an optically trapped particle, where the optical trap generates a harmonic potential and the strong coupling to the surrounding fluid molecules keeps the particle at thermodynamic equilibrium.

Box 7.4 — **Equipartition theorem**

For a system at thermodynamic equilibrium at absolute temperature T, the energy associated with each harmonic degree of freedom is equal to $\frac{1}{2}k_BT$, where k_B is the Boltzmann constant. For example, the total average energy of a Brownian particle in a three-dimensional harmonic optical trap, including kinetic and potential degrees of freedom, is $3k_BT$.

> **Box 7.5** **Maxwell–Boltzmann distribution**
>
> The probability that a system at thermodynamic equilibrium at absolute temperature T is in a state s with energy E_s is given by the *Maxwell–Boltzmann distribution*,
>
> $$p(s) = \exp(-E_s/(k_B T))/Z,$$
>
> where k_B is the Boltzmann constant and the *partition function* of the system is
>
> $$Z = \sum_s \exp(-E_s/(k_B T)).$$
>
> For example, for a particle of mass m moving in one dimension, the state s is defined by the particle position r and the particle velocity v so that $E_s = U(r) + \frac{1}{2}mv^2$, $p(s) = \exp(-U(r)/(k_B T))\exp(-\frac{1}{2}mv^2/(k_B T))/Z$ and $p(r) \propto \exp(-U(r)/(k_B T))$.
>
> The fact that at thermal equilibrium the momentum distribution is uncoupled from the position distribution permits one to factor out the kinetic energy distribution, which means that the kinetic energy distribution of a particle does not depend on the position of the particle itself or on the potential energy at that position.

In the presence of a force $F(r)$ and an associated potential $U(r) = -\int F(r)dr$, using the *Maxwell–Boltzmann distribution* [Box 7.5], the equilibrium probability density is

$$\rho(r) = \rho_0 \exp\left\{-\frac{U(r)}{k_B T}\right\}, \tag{7.19}$$

where $\rho_0 = [\int \exp\{-\frac{U(r)}{k_B T}\} dr]^{-1}$ is the probability density normalisation factor.

At thermodynamic equilibrium there is no net particle current; i.e., $J(r, t) = J(r) \equiv 0$.[3] Therefore, using Eq. (7.17), i.e.,

$$v(r)\rho(r) = J_{\text{drift}}(r) = -J_{\text{diff}}(r) = D\frac{\partial \rho(r)}{\partial r},$$

Stokes' law,

$$v(r, t) = \frac{F(r)}{\gamma},$$

and the derivative with respect to r of the logarithm of Eq. (7.19),

$$\frac{\partial \rho(r)}{\partial r} = \frac{F(r)}{k_B T}\rho(r),$$

one obtains that

$$D = \frac{k_B T}{\gamma}. \tag{7.20}$$

---

[3] At equilibrium, more generally, all quantities do not depend on time.

Eq. (7.20) is the simplest statement of the *fluctuation–dissipation theorem*, which relates the intensity of the fluctuations (D) to the rate of energy dissipation (γ) in a system at thermal equilibrium. There are various statements of the fluctuation–dissipation theorem, which apply to different situations. The crucial points to keep in mind are that it applies to systems that are at thermal equilibrium and that it relates the intensity of the thermal noise and the dynamical response of a system.

Exercise 7.3.1 Repeat the demonstration of Eq. (7.20) using a gravitational force field, which was the force field originally used by Einstein (1905) to derive the fluctuation–dissipation theorem, and a harmonic trapping potential.

Exercise 7.3.2 Derive the fluctuation–dissipation theorem for a particle moving in an optical trap in two and three dimensions. [Hint: The demonstration will only work for conservative force fields. Why?]

7.4 Brownian dynamics simulations

Almost all the terms in a Langevin equation, e.g., Eqs. (7.7), (7.9) and (7.10), can be approximated using standard techniques for the approximation of ordinary differential equations with finite-difference equations, such as the first-order approach described in Box 7.6. The only exception is the noise term, i.e., $\chi(t) = \sqrt{2S}\, W(t)$ or $\xi(t) = \sqrt{2D}\, W(t)$.

Box 7.6   **Finite-difference simulations**

Finite-difference simulations of ordinary differential equations are straightforward: the continuous-time solution $x(t)$ of an ordinary differential equation is approximated by a discrete-time sequence x_i, which is the solution of the corresponding finite-difference equation evaluated at regular time steps $t_i = i\,\Delta t$; if Δt is sufficiently small, $x_i \approx x(t_i)$.

Using a first-order integration method (Euler method), a finite-difference equation is obtained from the corresponding ordinary differential equation by replacing

1. $r(t) \to r_i$;
2. $\dfrac{dr(t)}{dt} \to (r_i - r_{i-1})/\Delta t$;
3. $\dfrac{d^2 r(t)}{dt^2} \to [(r_i - r_{i-1})/\Delta t - (r_{i-1} - r_{i-2})/\Delta t]/\Delta t = (r_i - 2r_{i-1} + r_{i-2})/\Delta t^2$;

and so on for higher derivatives. The solution is obtained by solving the resulting finite difference equation recursively for r_i, using the values $r_{i-1}, r_{i-2}, \ldots$ obtained from previous iterations.

Higher-order algorithms can also be employed to obtain faster convergence of the solution. An exhaustive reference book with a focus on stochastic differential equations such as Langevin equations is Kloeden and Platen (1999).

The problem with the noise terms arises because of the properties of the white noise $W(t)$ [Box 7.3]: as $W(t)$ is almost everywhere discontinuous and has infinite variation, it cannot be approximated by its instantaneous values at times t_i, because these values are not well defined (because of the lack of continuity) and their magnitude varies wildly (because of the infinite variation). In this section, we will see how a white noise can be treated within the finite difference approach and then how this method can be employed to simulate the motion of an optically trapped particle. We will follow the approach in Volpe and Volpe (2013); more advanced techniques for the numerical simulation of stochastic differential equations can be found in Kloeden and Platen (1999).

7.4.1 White noise

To understand how to treat a white noise $W(t)$ within a finite-difference approach, we will start by considering the continuous-time random walk described by the equation

$$\frac{dr(t)}{dt} = W(t). \tag{7.21}$$

We need a discrete sequence of random numbers W_i that mimics the properties of $W(t)$ [Box 7.3]:

1. because $W(t)$ is stationary with zero mean, we will take W_i to be random numbers with zero mean;
2. to account for the property that $W(t)$ has unitary power, we will also impose the condition that $\langle (W_i \Delta t)^2 \rangle / \Delta t = 1$ so that the W_i have variance $1/\Delta t$;
3. and because $W(t)$ is uncorrelated, we will assume W_i and W_j to be independent for $i \neq j$.

We will therefore use a sequence of uncorrelated random numbers with zero mean and variance $1/\Delta t$. Practically, we generate a sequence w_i of Gaussian random numbers with zero mean and unitary variance, and we then rescale it to obtain the sequence $W_i = w_i/\sqrt{\Delta t}$ with variance $1/\Delta t$.[4] Figs. 7.5a, 7.5b and 7.5c show that the values of w_i increase and diverge as $\Delta t \to 0$.

The finite-difference equation corresponding to Eq. (7.21) is

$$r_i = r_{i-1} + \sqrt{\Delta t}\, w_i. \tag{7.22}$$

Some examples of the resulting free diffusion trajectories r_i are plotted by the lines in Figs. 7.5d, 7.5e and 7.5f for $\Delta t = 1.0, 0.5$, and 0.1, respectively. The numerical solutions become more jagged as Δt decreases. The solutions shown in Figs. 7.5d, 7.5e and 7.5f differ because they are specific realisations of a random process, but their statistical properties do not change, as can be seen by averaging over many realisations. In fact, the shaded areas in Figs. 7.5d, 7.5e and 7.5f, which represent the variance around the mean position

---

[4] Some languages have built-in functions that directly generate a sequence of Gaussian random numbers with zero mean and unit variance. Alternatively, it is possible to employ various algorithms to generate Gaussian random numbers using uniform random numbers between 0 and 1, such as the Box–Muller algorithm or the Marsaglia polar algorithm.

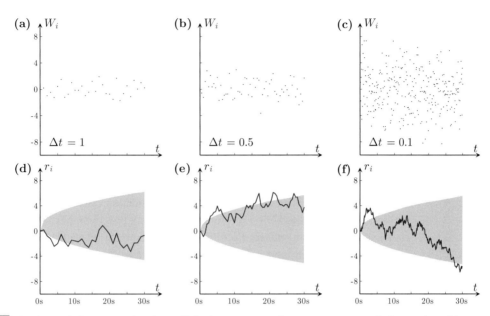

Figure 7.5 Simulation of white noise and random walk. As the time step Δt decreases, one must employ larger values of the noise W_i to approximate the solution of the free diffusion equation [Eq. (7.21)] accurately: (a) $\Delta t = 1$, (b) 0.5, and (c) 0.1. The corresponding solutions of the finite-difference free diffusion equation [Eq. (7.22)] in (d)–(f) for r_i (lines) behave similarly: although these solutions differ because they are specific realisations of a random process, their statistical properties do not change, as can be seen by comparing the shaded areas, which show the regions within one standard deviation of the mean of 10 000 realisations. Adapted with permission from Volpe and Volpe, *Am. J. Phys.* **81**, 224–30. Copyright 2013, American Association of Physics Teachers.

of the freely diffusing random walker obtained by averaging over 10 000 trajectories, are roughly the same, independent of Δt; the small differences are due to the finite number of trajectories used in the averaging.

The time step Δt should be much smaller than the characteristic time scales of the stochastic process to be simulated. If Δt is comparable to or larger than the smallest time scale, the numerical solution typically will not converge to the correct solution and may show unphysical oscillatory or diverging behaviour. The case of free diffusion treated in this subsection is special because there is no characteristic time scale, as can be seen from the fact that Eq. (7.21) is self-similar under a rescaling of time, and therefore there is no optimal choice of Δt.

Exercise 7.4.1 In Eq. (7.21), for a sufficiently large number of time steps, w_i can be any random number with zero mean and unitary variance. This is an example of universality of random walks. Show this numerically using various random variables (e.g., binary, uniform) and analytically using some statistical analysis measures (e.g., MSD). [Hint: You can use the code `brownian` from the book website as a starting point.]

Exercise 7.4.2 Demonstrate that Eq. (7.21) is scale-free, i.e., it is invariant under an appropriate re-scaling of the time and space variables.

Exercise 7.4.3 Generalise Eq. (7.21) to two and three dimensions and write down the corresponding finite-difference equations. Write also a code to numerically solve them. [Hint: You can use the code `brownian` from the book website as a starting point.]

7.4.2 Optically trapped particle

A fundamental example for the scope of this book is the simulation of an optically trapped particle. A Brownian particle in an optical trap is in a dynamic equilibrium where the thermal noise tries to push it out of the trap and optical forces drive it towards the centre of the trap. The time scale on which the restoring force acts is given by the ratio $\tau_{ot} = \gamma/\kappa$. Typically, τ_{ot} is significantly greater than the momentum relaxation time τ_m, which will be discussed in Section 7.5. Therefore, it is possible to ignore inertial effects and use an overdamped equation such as Eq. (7.10), where the only relevant time scale is τ_{ot}. This approach has the advantage that one can employ a relatively large time step, $\Delta t \gg \tau_m$. The time step Δt should still be significantly smaller than τ_{ot}, because, if $\Delta t \gtrsim \tau_{ot}$, the numerical solution does not converge and typically shows an unphysical oscillatory or diverging behaviour.

For a three-dimensional optical trap one can employ a set of three independent overdamped Langevin equations [Eq. (7.10)] with a harmonic restoring force,

$$\begin{cases} \dfrac{dx(t)}{dt} = -\dfrac{\kappa_x}{\gamma}x(t) + \sqrt{2D}\, W_x(t), \\ \dfrac{dy(t)}{dt} = -\dfrac{\kappa_y}{\gamma}y(t) + \sqrt{2D}\, W_y(t), \\ \dfrac{dz(t)}{dt} = -\dfrac{\kappa_z}{\gamma}z(t) + \sqrt{2D}\, W_z(t), \end{cases} \quad (7.23)$$

where x and y represent the position of the particle in the plane perpendicular to the beam propagation direction, z represents the position of the particle along the propagation direction, κ_x, κ_y and κ_z are the stiffnesses of the trap, γ is the particle friction coefficient and $W_x(t)$, $W_y(t)$ and $W_z(t)$ are independent white noises. The corresponding system of finite difference equations is

$$\begin{cases} x_i = x_{i-1} - \dfrac{\kappa_x}{\gamma}x_{i-1}\Delta t + \sqrt{2D\Delta t}\, w_{x,i}, \\ y_i = y_{i-1} - \dfrac{\kappa_y}{\gamma}y_{i-1}\Delta t + \sqrt{2D\Delta t}\, w_{y,i}, \\ z_i = z_{i-1} - \dfrac{\kappa_z}{\gamma}z_{i-1}\Delta t + \sqrt{2D\Delta t}\, w_{z,i}, \end{cases} \quad (7.24)$$

where x_i, y_i and z_i represent the position of the particle at time t_i, and $w_{i,x}$, $w_{i,y}$ and $w_{i,z}$ are independent Gaussian random numbers with zero mean and unitary variance.

The line in Fig. 7.6a shows a simulated trajectory of a Brownian particle in an optical trap with $\kappa_x = \kappa_y = 1.0\,\text{fN/nm}$ and $\kappa_z = 0.2\,\text{fN/nm}$. As we have seen in Chapters 2–6, the fact that the trapping stiffness along the beam propagation axis (z) is lower than in the

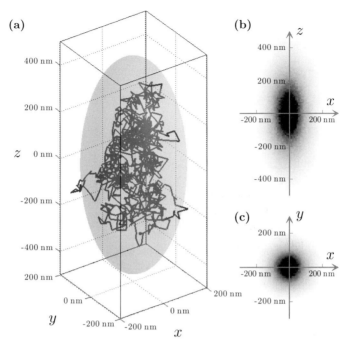

Figure 7.6 Simulation of the motion of an optically trapped particle. (a) Trajectory of a Brownian particle in an optical trap ($\kappa_x = \kappa_y = 1.0$ fN/nm and $\kappa_z = 0.2$ fN/nm). The particle explores an ellipsoidal volume around the centre of the trap, as evidenced by the shaded area, which represents an equiprobability surface. The probability densities of finding the particle in the (b) xz and (c) xy planes follow a two-dimensional Gaussian distribution around the trap centre. The simulations correspond to a silica microparticle in water with radius $a = 1$ μm, mass $m = 11$ pg, viscosity $\eta = 0.001$ Ns/m^2, $\gamma = 6\pi\eta a$, temperature $T = 300$ K and $\tau_m = 0.6$ μs. Adapted with permission from Volpe and Volpe, *Am. J. Phys.* **81**, 224–30. Copyright 2013, American Association of Physics Teachers.

perpendicular plane (xy) is commonly observed in experiments and is mainly due to the different intensity distribution along the different axes and to the presence of scattering forces along the optical axis. Thus, the particle explores an ellipsoidal volume around the centre of the trap, represented in Fig. 7.6a by the shaded grey equiprobability surface. In Figs. 7.6b and 7.6c, we show the projections of the probability density of finding the particle onto the xz- and xy-planes, respectively.

The time scale τ_{ot}, which characterises how a particle falls into the trap, can be seen in the position autocorrelation function (ACF) [Fig. 7.7a]:

$$C_x(\tau) = \overline{x(t+\tau)x(t)}. \tag{7.25}$$

As the stiffness increases, the particle undergoes a stronger restoring force and the correlation time decreases, because the particle explores a smaller phase space. Unlike the free diffusion case, the MSD [Fig. 7.7b], i.e.,

$$\text{MSD}_x(\tau) = \overline{[x(t+\tau) - x(t)]^2}, \tag{7.26}$$

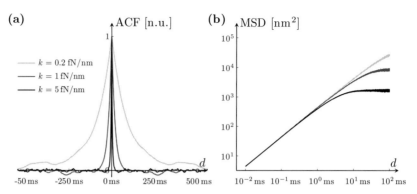

Figure 7.7 Autocorrelation function and mean squared displacement of an optically trapped particle. (a) The position autocorrelation function (ACF) of a trapped particle [Eq. (7.25)] gives information about the effect of the trap restoring force on the particle motion. As the trap stiffness and, therefore, the restoring force are increased, the characteristic decay time of the position ACF decreases. (b) The mean squared displacement (MSD), unlike that in the free diffusion case, does not increase indefinitely but reaches a plateau, which also depends on the trap stiffness – the stronger the trap, the sooner the plateau is reached. The simulation parameters are the same as in Fig. 7.6. Adapted with permission from Volpe and Volpe, *Am. J. Phys.* **81**, 224–30. Copyright 2013, American Association of Physics Teachers.

does not increase indefinitely, but reaches a plateau because of the confinement imposed by the trap. The transition from linear growth (corresponding to the free diffusion behaviour) to the plateau (due to the confinement) occurs at about τ_{ot}.

Exercise 7.4.4 In Eq. (7.24), the time step Δt should be significantly smaller than the trapping time scale τ_{ot}. Explore the numerical solutions for the case when this condition is not fulfilled. What goes wrong? [Hint: You can adapt the code `ot` from the book website.]

Exercise 7.4.5 It is possible to increase the stiffness of the trap, and, therefore, the confinement of the particle by increasing the optical power. This can be quantified by measuring the variance of the particle position around the trap centre, for example, in the xy plane, σ_{xy}^2. Assuming that $\kappa_x = \kappa_y$, show numerically that the variance σ_{xy}^2 is inversely proportional to κ_x. [Hint: You can adapt the code `variance` from the book website.]

Exercise 7.4.6 The position ACF and the MSD of an optically trapped particle can be calculated as time averages [Eqs. (7.25) and (7.26)] and as ensemble averages. Show that both procedures give the same results. [Hint: You can generate trajectories of optically trapped particles using the code `ot` from the book website, where you also find the functions `acf` and `msd` to obtain the ACF and MSD.]

Exercise 7.4.7 Show that the position ACF of an optically trapped particle is

$$C_x(\tau) = \frac{k_B T}{\kappa} e^{-|\tau|/\tau_{to}}, \tag{7.27}$$

where κ is the trap stiffness and $\tau_{to} = \gamma/\kappa$ is the trap characteristic time.

Exercise 7.4.8 Show that the MSD of an optically trapped particle is

$$\text{MSD}_x(\tau) = 2\frac{k_B T}{\kappa}\left[1 - e^{-|\tau|/\tau_{to}}\right], \qquad (7.28)$$

where κ is the trap stiffness and $\tau_{to} = \gamma/\kappa$ is the trap characteristic time. What is its behaviour for small and large τ?

7.5 Inertial regime

The Langevin equation describing the free diffusion of a Brownian particle [Eq. (7.7)] includes an inertial term depending on the particle mass m:

$$m\frac{d^2}{dt^2}r(t) = -\gamma\frac{d}{dt}r(t) + \sqrt{2k_B T \gamma}\, W(t). \qquad (7.29)$$

However, inertial effects decay on a very short time scale, the *momentum relaxation time* $\tau_m = m/\gamma$, producing a transition from smooth ballistic behaviour to diffusive behaviour. The time τ_m is very short, typically on the order of a fraction of a microsecond. For example, $\tau_m = 0.6\,\mu\text{s}$ for a silica microsphere with radius $a = 1\,\mu\text{m}$ ($m = 11$ pg) in water ($\eta = 0.001\,\text{N s m}^{-2}$) at temperature $T = 300$ K. We remark that τ_m is orders of magnitude smaller than the time scales of typical experiments. Thus, it is often possible to drop the inertial term, i.e., to set $m = 0$ in Eq. (7.29), obtaining

$$\frac{dr(t)}{dt} = \sqrt{2D}\, W(t). \qquad (7.30)$$

Eq. (7.30) is a very good approximation to Brownian motion for long time steps, i.e., for $\Delta t \gg \tau_m$, but it shows clear deviations from Eq. (7.29) on short time scales, i.e., for $\Delta t \lesssim \tau_m$. Figs. 7.8a and 7.8b compare two trajectories with and without inertia using the same realisation of white noise. For short times [Fig. 7.8a] the trajectory of a particle with inertia (solid line) appears smooth with a well-defined velocity that also changes smoothly, whereas in the absence of inertia (dashed line) the trajectory is ragged and discontinuous with a velocity that is not well defined. For long times, shown in Fig. 7.8b, both the trajectory with inertia (solid line) and that without inertia (dashed line) show behaviour typical of the diffusion of a Brownian particle, i.e., they appear jagged; this is true also for the trajectory with inertia because the microscopic details are not resolvable.

To better understand the free diffusion of a Brownian particle and the differences between the inertial and non-inertial regimes, we analyse some statistical quantities that are derived from the trajectories, namely, the velocity ACF and the MSD of the particle position. The velocity ACF provides a measure of the time it takes for the particle to 'forget' its initial velocity and is defined as

$$C_v(\tau) = \overline{v(t+\tau)v(t)}, \qquad (7.31)$$

where $v(t) = \frac{d}{dt}r(t)$.

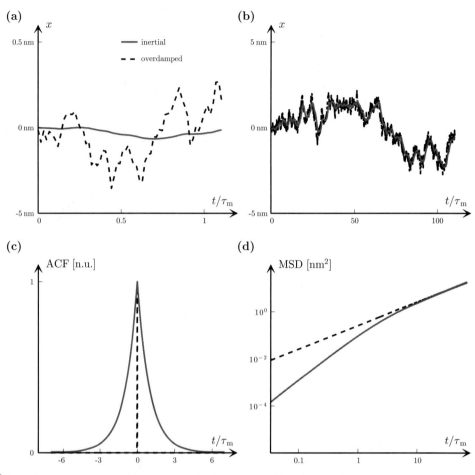

Figure 7.8 Inertial and diffusive regimes. (a) For times smaller than or comparable to the inertial time τ_m the trajectory of a particle with inertia (solid line) appears smooth. In contrast, in the absence of inertia (dashed line) the trajectory is ragged and discontinuous. (b) For times significantly longer than τ_m, both the trajectory with inertia (solid line) and that without inertia (dashed line) are jagged, because the microscopic details are not resolvable. These trajectories are computed using Eqs. (7.29) and (7.30) with $\Delta t = 10$ ns and the same realisation of white noise so that the two trajectories can be compared. (c) The velocity autocorrelation function (ACF) for a particle with inertia (solid line) decays to zero with the time constant τ_m, whereas that for a particle without inertia (dashed line) drops immediately to zero, demonstrating that the particle velocity is not correlated and does not have a characteristic time scale. (d) A log–log plot of the mean squared displacement (MSD) for a particle with inertia (solid line) shows a transition from quadratic behaviour at short times to linear behaviour at long times, whereas for a particle without inertia (dashed line) it is always linear. The simulation parameters are the same as in Fig. 7.6. Adapted with permission from Volpe and Volpe, *Am. J. Phys.* **81**, 224–30. Copyright 2013, American Association of Physics Teachers.

The solid line in Fig. 7.8c depicts the velocity ACF for a Brownian particle with inertia and shows that $C_v(\tau)$ decays to zero with time constant τ_m, demonstrating the time scale over which the velocity of the particle becomes uncorrelated to its initial value. The dashed line in Fig. 7.8c represents $C_v(\tau)$ for a trajectory without inertia, which drops immediately to zero, demonstrating that, in the absence of inertia, the velocity is uncorrelated over time and does not have a characteristic time scale.

For ballistic motion the MSD is proportional to t^2, whereas for diffusive motion it is proportional to t. The solid line in Fig. 7.8d shows the MSD in the presence of inertia. At short times, i.e., $t \lesssim \tau_m$, the MSD is quadratic in t and for longer times, i.e., $t \gg \tau_m$, the MSD becomes linear. This transition from ballistic to diffusive motion occurs on a time scale τ_m. In the absence of inertia (dashed line), the MSD is always linear.

Exercise 7.5.1 Derive the finite difference equation corresponding to the Langevin equation with inertia [Eq. (7.29)] and solve it numerically. Do the same for the overdamped Langevin equation [Eq. (7.30)]. Calculate and compare the velocity ACF and the position MSD corresponding to each case. [Hint: You can adapt the codes ot, acf and msd from the book website.]

7.6 Diffusion gradients

Diffusion gradients emerge naturally when a Brownian particle is in a complex or crowded environment. For example, diffusion gets hindered when a particle is close to a wall because of hydrodynamic interactions, as shown in Fig. 7.9a, where the diffusion coefficient increases with the particle–wall distance approaching the corresponding bulk value at a distance of several particle radii from the wall. This is an example of the fact that

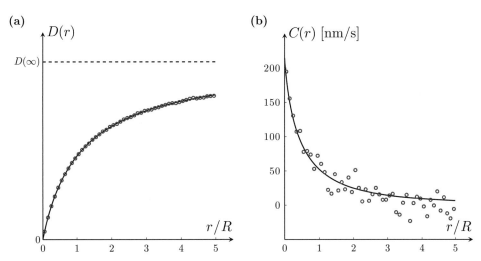

Figure 7.9 Brownian particle in a diffusion gradient. (a) Diffusion coefficient $D(r)$, normalised to the bulk diffusion $D(\infty)$ (dashed line), and (b) drift $C(r)$ for a particle performing Brownian motion next to a planar wall, assuming a hard-wall interaction. The lines represent the theoretical expectation and the symbols [Eqs. (7.36) and (7.37)] are calculated from a simulated series of data. Because no external forces are acting on the particle, $C(r)$ coincides with the spurious drift.

hydrodynamic interactions are extremely long-ranged and therefore must often be taken into account. Diffusion gradients are often encountered in the practice of optical manipulation, e.g., when particles are optically trapped near a coverslip or near other particles.

We will again consider a one-dimensional case, but our conclusions can be straightforwardly generalised to the multidimensional case. We consider the one-dimensional Langevin equation

$$\frac{d}{dr}r(t) = \frac{F(r)}{\gamma(r)} + \sqrt{2D(r)}\,W(t) \tag{7.32}$$

with a position-dependent diffusion coefficient $D(r)$. $D(r)$ and $\gamma(r)$ are related by the fluctuation–dissipation relation, which generalises Eq. (7.20),

$$D(r) = \frac{k_B T}{\gamma(r)}. \tag{7.33}$$

Unlike the Langevin equations we have encountered so far, the integration of Eq. (7.32) presents some difficulties because of the irregularity of the Wiener process [Boxes 7.3 and 7.7]. This, in particular, leads to the need to take into account the presence of a *spurious drift*, which emerges in the presence of diffusion gradients and is necessary to preserve the relation between the external forces $F(r)$ acting on the particle and the Maxwell–Boltzmann probability distribution $\rho(r)$ given by Eq. (7.19).

The diffusion constant of a spherical particle of radius a near a flat wall is of particular importance for optical tweezers experiments. Its derivation can be found in Happel and Brenner (1983). In particular, the diffusion coefficient in the direction parallel to the flat wall can be approximated by the Faxén formula (Faxén, 1922),

$$\frac{D_\parallel(h)}{D(\infty)} = 1 - \frac{9}{16}\left(\frac{a}{h}\right) + \frac{1}{8}\left(\frac{a}{h}\right)^3 - \frac{45}{256}\left(\frac{a}{h}\right)^4 - \frac{1}{16}\left(\frac{a}{h}\right)^5 + \mathcal{O}\left(\left(\frac{a}{h}\right)^6\right), \tag{7.34}$$

where $D(\infty)$ is the bulk diffusion coefficient and h is the distance between the centre of the particle and the flat wall. There is no second-order term in the denominator, so this formula remains good to within 1% for $h > 3a$ if one ignores all but the first-order term, i.e.,

$$\frac{D_\parallel(h)}{D(\infty)} \approx 1 - \frac{9}{16}\frac{a}{h}.$$

The diffusion coefficient in the vertical direction [Fig. 7.9a] is

$$\frac{D_\perp(h)}{D(\infty)}$$
$$= \left\{\frac{4}{3}\text{Sh}(\tilde{h})\sum_{n=1}^{\infty}\frac{n(n+1)}{(2n-1)(2n+3)}\left[\frac{2\text{Sh}((2n+1)\tilde{h}) + (2n+1)\text{Sh}(2\tilde{h})}{4\text{Sh}^2((n+0.5)\tilde{h}) - (2n+1)^2\text{Sh}^2(\tilde{h})} - 1\right]\right\}^{-1}, \tag{7.35}$$

where $\tilde{h} = \text{Ch}^{-1}(1 + \frac{h}{a})$, and Ch and Sh are the hyperbolic consine and sine, respectively. This formula can be approximated to first order as

$$\frac{D_\perp(h)}{D(\infty)} \approx 1 - \frac{9}{8}\frac{a}{h}.$$

> **Box 7.7** — **Stochastic integrals**
>
> The notion of a *stochastic integral* generalises the concept of the Riemann–Stieltjes integral to the case where the integration is done with respect to a Wiener process. In general, a stochastic integral is defined as the limit
>
> $$\int_0^t \sqrt{2D(r(s))} \circ_\alpha dW(s) \equiv \lim_{N\to\infty} \sum_{n=0}^{N-1} \sqrt{2D(r(t_n))} \Delta W_n,$$
>
> where $\Delta W_n = W\left(\frac{n+1}{N}T\right) - W\left(\frac{n}{N}T\right)$, $t_n = \frac{n+\alpha}{N}T$ and $\alpha \in [0,1]$. The symbol '$\circ_\alpha$' is used to indicate the value of α used in the integration. Differently from a standard integral, the choice parameterised by α of the point of integration within each infinitesimal interval may lead to different values of the integral, whenever the integrand depends on the state of the system, i.e., whenever in the presence of *multiplicative noise* such as in a diffusion gradient. Common choices are the *Itô integral* with $\alpha = 0$; the *Stratonovich integral* with $\alpha = 0.5$; and the *anti-Itô* or *isothermal integral* with $\alpha = 1$.
>
> In order to be able to integrate a Langevin equation in the presence of a diffusion gradient, one must specify the value of α. The convention $\alpha = 1$ emerges naturally in physical systems in equilibrium with a heat bath such as a Brownian particle. The reason is that this convention permits one to preserve the relation between the external forces $F(r)$ acting on the particle and the Maxwell–Boltzmann probability distribution $\rho(r)$ given by Eq. (7.19).
>
> We note that the different choices of α are connected to each other by a precise mathematical relationship. Therefore, one can change the convention, but only by adding an appropriate drift term at the same time, following the prescriptions in Table 7.1. In particular, Eq. (7.32) with $\alpha = 1$ is equivalent to the Itô equation ($\alpha = 0$),
>
> $$\frac{d}{dr}r(t) = \frac{F(z)}{\gamma(z)} + \frac{d}{dr}D(r) + \sqrt{2D(r)}\, W(t)|_{\alpha=0},$$
>
> where the term $\frac{d}{dz}D(z)$ is called *spurious drift*. This Itô form ($\alpha = 0$) of the equation is particularly convenient for numerical simulations with the finite difference method.
>
> Further information on this topic can be found in Brettschneider et al. (2011), Ermak and McCammon (1978) and Lau and Lubensky (2007).

Any stochastic system can be characterised by its diffusion $D(r)$ and its drift $C(r)$. If the system is allowed to evolve from an initial state r for an infinitesimal time step, $D(r)$ is proportional to the variance of the system's state change and $C(r)$ to its average. $D(r)$ and $C(r)$ can be obtained from an experimental discrete time series $\{r_1, \ldots, r_N\}$ sampling the output signal at intervals Δt as

$$D(r) = \frac{1}{2\Delta t}\left\langle (r_{n+1} - r_n)^2 \mid r_n \cong r \right\rangle \tag{7.36}$$

and

$$C(r) = \frac{1}{\Delta t}\langle r_{n+1} - r_n \mid r_n \cong r\rangle = \frac{F(r)}{\gamma(r)} + \underbrace{\frac{d}{dr}D(r)}_{\text{spurious drift}}. \tag{7.37}$$

Table 7.1 Correct use of Langevin equations and Fokker–Planck equations for Brownian particles

α	Langevin equation	Fokker–Planck equation
0	$\dfrac{d}{dt}r(t) = -\dfrac{F(r)}{\gamma(r)} + \underbrace{\dfrac{\partial}{\partial r}D(r)}_{\text{spurious drift}} dt + \sqrt{2D(r)}\,W(t)$	$\dfrac{\partial}{\partial t}\rho(r,t) = \left[\dfrac{\partial}{\partial r}\left(\dfrac{F(r)}{\gamma(r)} - \underbrace{\dfrac{\partial}{\partial r}D(r)}_{\text{spurious drift}} \right) + \underbrace{\dfrac{\partial}{\partial r}\left(\dfrac{\partial}{\partial r}D(r) \right)}_{\text{diffusion}} \right]\rho(r,t)$
1/2	$\dfrac{d}{dr}r(t) = -\dfrac{F(r)}{\gamma(r)} + \underbrace{\dfrac{1}{2}\dfrac{\partial}{\partial r}D(r)}_{\text{spurious drift}} + \sqrt{2D(r)}\,W(t)$	$\dfrac{\partial}{\partial t}\rho(r,t) = \left[\dfrac{\partial}{\partial r}\left(\dfrac{F(r)}{\gamma(r)} - \underbrace{\dfrac{1}{2}\dfrac{\partial}{\partial r}D(r)}_{\text{spurious drift}} \right) + \underbrace{\dfrac{\partial}{\partial r}\left(D(r)^{\frac{1}{2}}\dfrac{\partial}{\partial r}D(r)^{\frac{1}{2}} \right)}_{\text{diffusion}} \right]\rho(r,t)$
1	$\dfrac{d}{dt}r(t) = -\dfrac{F(r)}{\gamma(r)} + \sqrt{2D(r)}\,W(t)$	$\dfrac{\partial}{\partial r}\rho(r,t) = \left[\dfrac{\partial}{\partial r}\dfrac{F(r)}{\gamma(r)} + \underbrace{\dfrac{\partial}{\partial r}\left(D(r)\dfrac{\partial}{\partial r} \right)}_{\text{diffusion}} \right]\rho(r,t)$

Eqs. (7.36) and (7.37) are strictly true in the limit $\Delta t \to 0$; therefore, in experiments Δt should be much smaller than the relaxation time of the system.[5] Figs. 7.9a and 7.9b show the values of $D(r)$ and $C(r)$ for the case of a particle diffusing in front of a wall in the absence of external forces, i.e., $F(r) \equiv 0$, so that $C(r)$ coincides with the spurious drift.

Exercise 7.6.1 Write a Langevin equation for a Brownian particle diffusing near a wall and solve it numerically. For an electrically charged dielectric colloidal sphere suspended in a solvent, the interaction forces are $F(z) = Be^{z/\lambda_D} - G_{\text{eft}}$, where the first term is due to double-layer forces with λ_D the Debye length and B a prefactor depending on the surface charge densities of the particle and the wall, and the second term describes the effective gravitational contributions. [Hint: You can adapt the code `diffusiongradient` from the book website and you can find some physical values for the parameters λ_D and B in Brettschneider et al. (2011).]

Exercise 7.6.2 Write the appropriate Langevin equation for a Brownian particle optically trapped next to a wall and simulate its motion. Verify that the probability distribution in the optical trap is unaffected by the presence of the diffusion gradient.

Exercise 7.6.3 Calculate drift and diffusion from the trajectory of a Brownian particle in front of a wall using Eqs. (7.36) and (7.37). Verify the presence of the spurious drift. [Hint: You can use the function `driffusion` from the book website.]

7.7 Viscoelastic media

Until now we have considered only Brownian particles moving in a viscous medium. However, for many optical tweezers applications, *viscoelastic media* are important. Common examples of viscoelastic media are polymer solutions, many biomaterials and cytoplasm. These materials are considered complex fluids because they present mechanical behaviour halfway between that of a solid, which responds elastically to external stresses, and that of a liquid, whose response has a viscous character. Therefore, under shear stress, these materials give rise to a viscoelastic rheological response, where the energy is partly stored (elasticity) and partly dissipated (viscosity).

The viscoelastic response of a complex medium can be expressed in terms of the stress relaxation modulus $G_{\text{ve}}(t)$. The viscoelastic force acting on the particle is given by the convolution of $G_{\text{ve}}(t)$ and the particle velocity,

$$F_{\text{ve}}(t) = \int_{-\infty}^{t} G_{\text{ve}}(t-s) v(s) \, ds = \int_{0}^{\infty} G_{\text{ve}}(s) v(t-s) \, ds, \qquad (7.38)$$

---

[5] Furthermore, in the limit $\Delta t \to 0$ inertial effects come into play and, therefore, in practice Eqs. (7.36) and (7.37) should only be used in the overdamped limit, i.e., for $\Delta t \gg \tau_m$, where τ_m is the inertial characteristic time scale of the particle.

where $v(t) = \frac{d}{dt}r(t)$. Furthermore, in order to satisfy the fluctuation–dissipation theorem, the driving noise term $\chi_{\text{ve}}(t)$ must satisfy the relation

$$\langle \chi_{\text{ve}}(t+\tau)\chi_{\text{ve}}(t)\rangle = k_B T G_{\text{ve}}(|\tau|). \tag{7.39}$$

Eq. (7.39) shows, in particular, that the driving noise in the presence of a viscoelastic medium is a correlated noise. The resulting Langevin equation is

$$m\frac{d^2}{dt^2}r(t) = -\int_0^\infty G_{\text{ve}}(s)\, v(t-s)\, ds + \chi_{\text{ve}}(t). \tag{7.40}$$

One of the simplest models for a viscoelastic fluid is a simple *Maxwell fluid*, characterised by a single relaxation time τ_{ve}. In this case, we have

$$G_{\text{ve}}(t) = \frac{\gamma}{\tau_{\text{ve}}} \exp\left(-\frac{t}{\tau_{\text{ve}}}\right) \Theta(t), \tag{7.41}$$

where $\Theta(t)$ is the Heaviside step function. However, in general, more complex models are used in order to take into account that the relaxation does not occur at a single time, but at a distribution of times. This can be due, e.g., to the presence in solution of molecular segments of different lengths with shorter ones contributing less than longer ones. Therefore, whereas in the Maxwell model the material is considered as a purely viscous damper and a purely elastic spring connected in series, in the generalised Maxwell model there are as many spring–damper Maxwell elements as are necessary to accurately represent the distribution of the relaxation times.

Exercise 7.7.1 How is Eq. (7.40) changed in the presence of an optical trap?

Exercise 7.7.2 Show that for a Brownian particle in a viscous fluid $G_{\text{ve}}(t) = \gamma\delta(t)$. What form do Eqs. (7.38), (7.39) and (7.40) take in a viscous fluid?

Exercise 7.7.3 Generalise Eq. (7.40) to the multidimensional case. [Hint: Note that the driving noise terms along different directions are uncorrelated.]

Exercise 7.7.4 By taking the Fourier transform of $G_{\text{ve}}(t)$, one obtains the frequency-dependent complex shear modulus $\check{G}_{\text{ve}}(f)$. Show that its real part, $\check{G}'_{\text{ve}}(f) = \text{Re}\{\check{G}_{\text{ve}}(f)\}$, is related to the elastic storage modulus and its imaginary part, $\check{G}''_{\text{ve}}(f) = \text{Im}\{\check{G}_{\text{ve}}(f)\}$, is related to the viscous loss modulus. Observe that, because $\check{G}'_{\text{ve}}(f)$ and $\check{G}''_{\text{ve}}(f)$ derive from the same function, they are related by the Kramers–Kronig relations.

7.8 Non-spherical particles and diffusion matrices

Until now we have only considered spherical particles, which are in fact the ones most commonly employed in optical manipulation experiments. However, it is interesting also to consider other shapes, from elongated, e.g., bacteria, to more complex particles, e.g., structures produced by two-photon polymerisation.

Just as in the case of a spherical particle, the motion of a non-spherical particle is influenced by thermal noise, resistance of the fluid and external deterministic forces. However, a scalar diffusion coefficient is not enough to describe the statistics of the random motion of such a particle; one needs, instead, a symmetric 6×6 diffusion tensor, which depends on the particle shape and orientation,

$$\mathbf{D} = \begin{bmatrix} \mathbf{D}_{tt} & \mathbf{D}_{tr} \\ \mathbf{D}_{rt} & \mathbf{D}_{rr} \end{bmatrix}, \tag{7.42}$$

where $\mathbf{D}_{tt}$, $\mathbf{D}_{rr}$ and $\mathbf{D}_{tr} = \mathbf{D}_{rt}^T$ are 3×3 blocks and the subscripts 't' and 'r' refer to the translational and rotational degrees of freedom of the particle, respectively. Analytical expressions for $\mathbf{D}$ exist for simple shapes. For example, in the case of a sphere,

$$\mathbf{D}_{\text{sphere}} = \begin{bmatrix} \frac{k_B T}{6\pi \eta a} \mathbf{I}_3 & 0 \\ 0 & \frac{k_B T}{8\pi \eta a^3} \mathbf{I}_3 \end{bmatrix}, \tag{7.43}$$

where $\mathbf{I}_3$ represents the 3×3 identity matrix and the absence of off-diagonal terms shows that all the degrees of freedom are independent of each other. Analytical formulas exist also for symmetric shapes such as revolution ellipsoids or cylinders. However, for more complex shapes it is necessary to resort to numerical approaches such as the bead modelling technique developed by De La Torre et al. (1994), which is implemented in widely used and publicly available computer programs such as HYDRO++. We note that $\mathbf{D}$ is typically calculated with reference to the particle centre of diffusion, which does not necessarily coincide with the particle centre of mass.

Because of the fluctuation–dissipation theorem, the friction properties of the particle are also described by a 6×6 tensor, $\boldsymbol{\gamma}$, which satisfies the property

$$\mathbf{D}\boldsymbol{\gamma} = k_B T \mathbf{I}_6, \tag{7.44}$$

where $\mathbf{I}_6$ represents the 6×6 identity matrix.

Exercise 7.8.1 Calculate the friction tensor for a spherical particle and verify that it satisfies Eq. (7.44).

Exercise 7.8.2 Use HYDRO++ to calculate the diffusion tensor of an ellipsoidal particle. [Hint: You can find HYDRO++ online and a brief guide on how to use it is provided on the book website.]

7.8.1 Free diffusion

We will now consider how the motion of a non-spherical Brownian particle can be simulated. In the laboratory frame of reference, Σ_l, which is centred at $\mathbf{0} = [0, 0, 0]$ and whose axes are oriented along $\hat{\mathbf{x}}$, $\hat{\mathbf{y}}$ and $\hat{\mathbf{z}}$, the particle centre of diffusion lies at $\mathbf{r}_{CD}(t) = [x_{CD}(t), y_{CD}(t), z_{CD}(t)]$ and the particle orientation is described by the direction cosines

$\alpha_l(t)$, $\beta_l(t)$ and $\gamma_l(t)$ defined with respect to the particle unit vectors $\hat{\mathbf{x}}_p(t) = [\hat{x}_{p,x}(t), \hat{x}_{p,y}(t), \hat{x}_{p,z}(t)]$, $\hat{\mathbf{y}}_p(t) = [\hat{y}_{p,x}(t), \hat{y}_{p,y}(t), \hat{y}_{p,z}(t)]$ and $\hat{\mathbf{z}}_p(t) = [\hat{z}_{p,x}(t), \hat{z}_{p,y}(t), \hat{z}_{p,z}(t)]$. Typically, for computational efficiency, $\mathbf{D}$ is calculated in a frame of reference fixed with respect to the particle, $\Sigma_p(t)$, which is centred at $\mathbf{r}_{CD}(t)$ and whose axes are oriented along $\hat{\mathbf{x}}_p(t)$, $\hat{\mathbf{y}}_p(t)$ and $\hat{\mathbf{z}}_p(t)$. We will now describe the steps necessary in order to simulate the free diffusion of an arbitrarily shaped particle from time t to time $t + \Delta t$, where Δt is the time step.

First, one has to calculate the increments of the particle position and orientation in $\Sigma_p(t)$, i.e.,[6]

$$\begin{bmatrix} \Delta x_p \\ \Delta y_p \\ \Delta z_p \\ \Delta \alpha_p \\ \Delta \beta_p \\ \Delta \gamma_p \end{bmatrix} = \sqrt{2\Delta t} \begin{bmatrix} w_x \\ w_y \\ w_z \\ w_\alpha \\ w_\beta \\ w_\gamma \end{bmatrix}, \qquad (7.45)$$

where $[w_x, w_y, w_z, w_\alpha, w_\beta, w_\gamma]^T$ is a set of random numbers extracted from a multivariate normal distribution with mean zero and covariance $\mathbf{D}$.

As a second step, the increments of the position of the particle centre of diffusion obtained from Eq. (7.45) need to be transformed from $\Sigma_p(t)$ to Σ_l. This transformation is a three-dimensional rotation around the centre of diffusion given by the transformation matrix

$$\mathbf{M}_{\Sigma_p \to \Sigma_l}(t) = \begin{bmatrix} \hat{\mathbf{x}}_p(t), \hat{\mathbf{y}}_p(t), \hat{\mathbf{z}}_p(t) \end{bmatrix} = \begin{bmatrix} \hat{x}_{p,x}(t) & \hat{y}_{p,x}(t) & \hat{z}_{p,x}(t) \\ \hat{x}_{p,y}(t) & \hat{y}_{p,y}(t) & \hat{z}_{p,y}(t) \\ \hat{x}_{p,z}(t) & \hat{y}_{p,z}(t) & \hat{z}_{p,z}(t) \end{bmatrix}. \qquad (7.46)$$

Thus, the finite difference equation to update the particle position in Σ_l is

$$\begin{bmatrix} x_{CM}(t+\Delta t) \\ y_{CM}(t+\Delta t) \\ z_{CM}(t+\Delta t) \end{bmatrix} = \begin{bmatrix} x_{CM}(t) \\ y_{CM}(t) \\ z_{CM}(t) \end{bmatrix} + \mathbf{M}_{\Sigma_p \to \Sigma_l}(t) \begin{bmatrix} \Delta x_p \\ \Delta y_p \\ \Delta z_p \end{bmatrix}. \qquad (7.47)$$

As the final step, one has to transform from $\Sigma_p(t)$ to Σ_l the change of the orientation of the particle, which is effectively a rotation of the particle unit vectors. This rotation is expressed in Σ_p by the rotation matrix

$$\mathbf{R}_p(\Delta \alpha_p, \Delta \beta_p, \Delta \gamma_p) = \mathbf{R}_{p,x}(\Delta \alpha_p) \mathbf{R}_{p,y}(\Delta \beta_p) \mathbf{R}_{p,z}(\Delta \gamma_p), \qquad (7.48)$$

---

[6] At time t, the particle position and orientation in $\Sigma_p(t)$ are clearly $[0, 0, 0, 0, 0, 0]^T$.

where

$$\mathbf{R}_{p,x}(\Delta\alpha_p) = \begin{bmatrix} 1 & 0 & 0 \\ 0 & \cos\Delta\alpha_p & -\sin\Delta\alpha_p \\ 0 & \sin\Delta\alpha_p & \cos\Delta\alpha_p \end{bmatrix},$$

$$\mathbf{R}_{p,y}(\Delta\beta_p) = \begin{bmatrix} \cos\Delta\beta_p & 0 & \sin\Delta\beta_p \\ 0 & 1 & 0 \\ -\sin\Delta\beta_p & 0 & \cos\Delta\beta_p \end{bmatrix},$$

$$\mathbf{R}_{p,z}(\Delta\gamma_p) = \begin{bmatrix} \cos\Delta\beta_p & -\sin\Delta\beta_p & 0 \\ \sin\Delta\beta_p & \cos\Delta\beta_p & 0 \\ 0 & 0 & 1 \end{bmatrix}.$$

Transforming this rotation matrix to Σ_l, we obtain the unit vectors representing the orientation of the particle at the end of the time step,

$$[\hat{\mathbf{x}}_p(t+\Delta t), \hat{\mathbf{y}}_p(t+\Delta t), \hat{\mathbf{z}}_p(t+\Delta t)] = [\hat{\mathbf{x}}_p(t), \hat{\mathbf{y}}_p(t), \hat{\mathbf{z}}_p(t)] \, \mathbf{R}_p(\Delta\alpha_p, \Delta\beta_p, \Delta\gamma_p) \quad (7.49)$$

and

$$\mathbf{M}_{\Sigma_p \to \Sigma_l}(t+\Delta t) = \mathbf{M}_{\Sigma_p \to \Sigma_l}(t) \, \mathbf{R}_p(\Delta\alpha_p, \Delta\beta_p, \Delta\gamma_p). \quad (7.50)$$

Exercise 7.8.3 Show that the matrix that describes the transformation from Σ_l to Σ_p is given by

$$\mathbf{M}_{\Sigma_l \to \Sigma_p} = \left(\mathbf{M}_{\Sigma_p \to \Sigma_l}\right)^{-1} = \left(\mathbf{M}_{\Sigma_p \to \Sigma_l}\right)^{\mathrm{T}}. \quad (7.51)$$

Exercise 7.8.4 Simulate the free diffusion of an ellipsoidal particle.

7.8.2 External forces

In general, it is also necessary to consider forces and torques acting on the particle, e.g., optical forces and torques. Assuming a low-Reynolds-number regime, in Σ_p the increments of the particle position and orientation, taking into account also the effect of a force $\mathbf{F}_p = [F_{p,x}, F_{p,y}, F_{p,z}]^{\mathrm{T}}$ and a torque $\mathbf{T}_p = [T_{p,x}, T_{p,y}, T_{p,z}]^{\mathrm{T}}$, are

$$\begin{bmatrix} \Delta x_p \\ \Delta y_p \\ \Delta z_p \\ \Delta\alpha_p \\ \Delta\beta_p \\ \Delta\gamma_p \end{bmatrix} = \frac{\mathbf{D}}{k_{\mathrm{B}}T} \Delta t \begin{bmatrix} F_{x,p} \\ F_{y,p} \\ F_{z,p} \\ T_{x,p} \\ T_{y,p} \\ T_{z,p} \end{bmatrix} + \sqrt{2\Delta t} \begin{bmatrix} w_x \\ w_y \\ w_z \\ w_\alpha \\ w_\beta \\ w_\gamma \end{bmatrix}, \quad (7.52)$$

which then need to be transformed into the laboratory system of reference. In general, $\mathbf{F}_p$ and $\mathbf{T}_p$ are applied to the particle centre of mass, $\mathbf{r}_{CM}$. However, the particle reference frame is centred on the particle centre of diffusion, $\mathbf{r}_{CD}$. If $\mathbf{r}_{CM} \neq \mathbf{r}_{CD}$, the torque needs to be adjusted to

$$\mathbf{T}_{p,\text{total}} = \mathbf{T}_p + (\mathbf{r}_{CD} - \mathbf{r}_{CM}) \times \mathbf{F}. \tag{7.53}$$

Typically, forces and torques are given in the laboratory system of reference, i.e., $\mathbf{F}_l = [F_{l,x}, F_{l,y}, F_{l,z}]^T$ and $\mathbf{T}_l = [T_{l,x}, T_{l,y}, T_{l,z}]^T$. Therefore, they need to be transformed into the particle system of reference. This can be accomplished by the transformations

$$\begin{bmatrix} F_{p,x} \\ F_{p,y} \\ F_{p,z} \end{bmatrix} = \mathbf{M}_{\Sigma_l \to \Sigma_l} \begin{bmatrix} F_{l,x} \\ F_{l,y} \\ F_{l,z} \end{bmatrix} \tag{7.54}$$

and

$$\begin{bmatrix} T_{p,x} \\ T_{p,y} \\ T_{p,z} \end{bmatrix} = \mathbf{M}_{\Sigma_l \to \Sigma_l} \begin{bmatrix} T_{l,x} \\ T_{l,y} \\ T_{l,z} \end{bmatrix}. \tag{7.55}$$

Exercise 7.8.5 Simulate the free diffusion of an ellipsoidal particle in an optical trap, taking into account also the optical forces and torques acting on the particle.

Problems

7.1 *Hydrodynamic interactions.* Using the Oseen tensor, study the hydrodynamic interactions between multiple particles held close together by optical traps. How do they affect each other's Brownian motion?

7.2 *Ellipsoids.* Reproduce the calculations of the friction tensor for ellipsoids performed by Perrin (1934). Extrapolate the friction results in the limit of low-dimensional particles, i.e., oblate and prolate ellipsoids. Now, consider the optical trapping of dielectric ellipsoids in water. Calculate analytically and numerically the expected translational and orientational relaxation frequencies in the trap and discuss their scaling behaviour with respect to the lengths of the semiaxes.

7.3 *Cylinders.* The friction tensor of cylinders is given analytically with respect to their main symmetry axes and can be expressed in terms of the coefficients $\gamma_{\parallel} = 2\pi \eta L/(\ln p + \delta_{\parallel})$, $\gamma_{\perp} = 4\pi \eta L/(\ln p + \delta_{\perp})$ (for translations parallel and perpendicular to the cylinder main axis, respectively) and $\gamma_{\theta} = \pi \eta L^3/[3(\ln p + \delta_{\theta})]$ (for rotations), where $p = L/d$ is the ratio between the cylinder length L and diameter d, η is the fluid viscosity and $\delta_{\parallel,\perp}$ are end corrections calculated as a polynomial function of $(\ln 2p)^{-1}$ (Tirado et al., 1984). Compare these analytical results with those obtained by a computational approach and explore their range of validity. Consider the optical trapping ($\lambda_0 = 830$ nm) of silicon ($n_p = 3.7$) nanowires in water with fixed diameter ($d = 10\,\text{nm}$) and variable length ($L = 0.1\,\mu\text{m}$ to $5\,\mu\text{m}$), as in Irrera et al. (2011). Investigate the behaviour of the force and torque constants as a function of L. Obtain the scaling for the relaxation frequencies in the trap for translational and orientational fluctuations.

7.4 *Rotational force field.* Study the Brownian motion of a particle held in an optical trap in the presence of a rotational force field, as studied in Volpe and Petrov (2006). How can a recording of the fluctuations of such a particle be used to measure the rotational component of the force field?

7.5 *Matters of shape.* Consider a sphere, a cylinder and a disk held by an optical tweezers. Assuming they have the same volume and material, which one is confined more strongly and stably?

7.6 *Kramers rates.* Following the experiment in McCann et al. (1999), study the Kramers rates of a Brownian particle held between two optical tweezers placed near to each other as a function of the particle, medium and optical tweezers parameters. Compare the results you obtain with particles of various shapes (e.g., spheres, ellipsoids, cylinders).

References

Brettschneider, T., Volpe, G., Helden, L., Wehr, J., and Bechinger, C. 2011. Force measurement in the presence of Brownian noise: Equilibrium-distribution method versus drift method. *Phys. Rev. E*, **83**, 041113.

Brown, R. 1828. A brief account of microscopical observations made in the months of June, July and August, 1827, on the particles contained in the pollen of plants; and on the general existence of active molecules in organic and inorganic bodies. *Phil. Mag.*, **4**, 161–73.

De La Torre, J. G., Navarro, S., Lopez Martinez, M. C., Diaz, F. G., and Lopez Cascales, J. J. 1994. HYDRO: A computer program for the prediction of hydrodynamic properties of macromolecules. *Biophys. J.*, **67**, 530–31.

Einstein, A. 1905. Über die von der molekularkinetischen Theorie der Wärme geforderte Bewegung von in ruhenden Flüssigkeiten suspendierten Teilchen. *Ann. Physik*, **322**, 549–60.

Ermak, D. L., and McCammon, J. A. 1978. Brownian dynamics with hydrodynamic interactions. *J. Chem. Phys.*, **69**, 1352–60.

Faxén, H. 1922. Der Widerstand gegen die Bewegung einer starren Kugel in einer zähen Flüssigkeit, die zwischen zwei parallelen ebenen Wänden eingeschlossen ist. *Ann. Physik*, **373**, 89–119.

Frenkel, D., and Smit, B. T. 2002. *Understanding molecular simulations: From algorithms to applications*. 2nd ed. Waltham, MA: Academic Press.

Happel, J., and Brenner, H. 1983. *Low Reynolds number hydrodynamics*. New York: Springer.

Irrera, A., Artoni, P., Saija, R., et al. 2011. Size-scaling in optical trapping of silicon nanowires. *Nano Lett.*, **11**, 4879–84.

Karatzas, I., and Shreve, S. 1998. *Brownian motion and stochastic calculus*. New York: Springer Verlag.

Kloeden, P. E., and Platen, E. 1999. *Numerical solution of stochastic differential equations*. Heidelberg: Springer Verlag.

Lau, A. W. C., and Lubensky, T. C. 2007. State-dependent diffusion: Thermodynamic consistency and its path integral formulation. *Phys. Rev. E*, **76**, 011123.

McCann, L. I., Dykman, M., and Golding, B. 1999. Thermally activated transitions in a bistable three-dimensional optical trap. *Nature*, **402**, 785–7.

Nelson, E. 1967. *Dynamical theories of Brownian motion*. Princeton, NJ: Princeton University Press.

Øksendal, B. 2003. *Stochastic differential equations*. Berlin: Springer Verlag.

Perrin, F. 1934. Mouvement brownien d'un ellipsoide–I. Dispersion diélectrique pour des molécules ellipsoidales. *J. Phys. Radium*, **5**, 497–511.

Perrin, J. 1909. Mouvement Brownien et réalité moléculaire. *Ann. Chimie Physique*, **18**, 5–114.

Purcell, E. M. 1977. Life at low Reynolds number. *Am. J. Phys*, **45**, 3–11.

Tirado, M. M., Martinez, C. L., and de la Torre, J. G. 1984. Comparison of theories for the translational and rotational diffusion coefficients of rod-like macromolecules. Application to short DNA fragments. *J. Chem. Phys.*, **81**, 2047–52.

Uhlenbeck, G. E., and Ornstein, L. S. 1930. On the theory of the Brownian motion. *Phys. Rev.*, **36**, 823–41.

Volpe, G., and Petrov, D. 2006. Torque detection using Brownian fluctuations. *Phys. Rev. Lett.*, **97**, 210603.

Volpe, G., and Volpe, G. 2013. Simulation of a Brownian particle in an optical trap. *Am. J. Phys.*, **81**, 224–30.

PART II

PRACTICE

'...and what is the use of a book,' thought Alice, 'without pictures or conversations?'

Alice's Adventures in Wonderland – Lewis Carroll

8 Building an optical tweezers*

It is finally time to build our first optical tweezers. In this chapter, we shall do just that: we will build the optical tweezers shown in Fig. 8.1, proceeding step by step and explaining all the details and tricks. In a nutshell, a simple optical tweezers is a laser beam coupled to an optical microscope, which can be either commercial or homemade. For completeness, we will first proceed to realise a homemade microscope. Then we will couple it to a laser beam, appropriately prepared. Along the way, we will explain all the necessary tricks and tips, and we will also present some basic information on how to achieve steerable and multiple optical tweezers. By the end of this chapter, we will have built the optical tweezers shown in Fig. 8.1, apart from the position detection part, which will be explained in Chapter 9, and we will be able to trap and manipulate microscopic particles. In the following chapters, we will explain how to acquire and analyse optical trapping data [Chapter 9], how to measure nanoscopic forces and torques [Chapter 10], and how to realise more complex optical trapping configurations using holographic optical tweezers [Chapter 11] and alternative trapping geometries [Chapter 12].

Figure 8.1 Homemade optical tweezers: Photograph of the optical tweezers that we will build in this and the following chapters.

* This chapter was written together with Giuseppe Pesce.

8.1 The right location

In the realisation of an optical tweezers, the first crucial step is to choose an appropriate location. In fact, this choice will largely determine the stability of the set-up and, ultimately, the quality of the experimental results. For simple experiments that do not require very high stability or reproducibility, e.g., experiments where the optical tweezers are used to trap and displace cells, it can be sufficient to mount the optical tweezers in a standard optical lab on a standard optical table. For more complex experiments, e.g., experiments where the optical tweezers are used to measure or exert forces, it might be necessary to choose a laboratory with stabilised temperature and humidity; in most cases, it is also a good idea to use a laboratory located on the ground floor, as this contributes to limiting the building vibrations transmitted to the optical set-up. The optical table should also use a passive or active vibration-insulation system.[1] For even more demanding experiments, e.g., the accurate measurement of forces generated by biomolecules, the highest level of experimental stability and reproducibility can be achieved by eliminating one of the greatest sources of noise in any optical lab: the human factor. Thus, it might be necessary to perform the experiment on a remote-controlled platform in a closed and acoustically insulated room.

Suggestion: Eliminate noise sources. Common sources of vibrations are table-mounted devices (e.g., the fan of the laser head), as well as vacuum pumps and other acoustic sources present in the laboratory. The best way to reduce these vibrations is to eliminate them at the origin, using baffles on air ducts, compliant pads under instruments, and acoustic enclosures.

Suggestion: Enclose the set-up. The presence of air flow, dust, or thermal gradients can affect the laser pointing stability. Thus, enclosing the set-up within a thermally, acoustically and electrically insulated box greatly increases both the stability of the laser and the quality of the acquired signals. We have done this with our set-up, although for clarity the enclosure is not shown in the pictures.

8.2 Inverted microscope construction

Once an appropriate location has been identified, we can start building an inverted microscope. The microscope will be built on several levels. We start by constructing a two-level structure, as shown in Figs. 8.2a and 8.2b. The lower level is fixed directly to the optical table; here is where the optical components necessary to prepare the optical beam and to

---

[1] An alternative idea is to place the optical table on a bed of sand, because this effectively damps the transmission of ground vibrations to the optical bench.

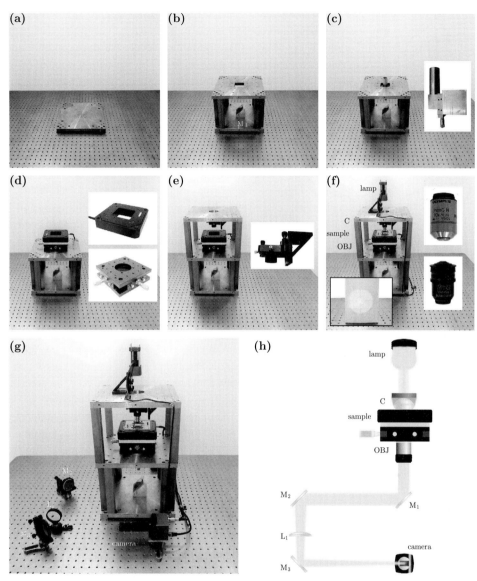

Figure 8.2 Homemade inverted microscope. Construction: (a) the base is firmly fixed to the optical table; (b) a 45° mirror M_1 is fixed on the base and a second stage is built; (c) the vertical translation stage (inset) for the objective is fixed; (d) a manual translation stage (lower inset) for rough alignment and a piezoelectric automatised translation stage (upper inset) for fine adjustment of the sample are fixed on the second stage; (e) the lateral translation stage (inset) for the condenser is fixed; (f) the objective (OBJ; lower right inset) and the condenser (C; upper right inset) are placed on the respective stages, a calibration glass slide is placed on the sample holder and the illumination LED lamp is fixed on the third level (lower left inset: image of the sample on a screen placed on the optical table); (g) the image is focused on a camera using lens L_1 and mirrors M_1, M_2 and M_3. (h) Corresponding schematic of the microscope.

focus the sample image on the camera will be housed. The second level is a breadboard supported on four columns; here is where the stages to hold and manipulate the sample will be housed. It is important that both these levels are stable and horizontal, as can be verified during their construction using a level. The translation stage to control the vertical position of the objective with respect to the sample is also attached to the lower side of the second-level breadboard, as shown in Fig. 8.2c.

> **Suggestion: Objective position control.** For the control of the objective, it can be often sufficient to use a one-axis (vertical) manual translation stage, as the fine adjustment of the relative position between the objective and the sample can be realised using the piezoelectric translation stage on which the sample is to be placed. However, for some set-ups, it might be useful to have more degrees of freedom for lateral alignment and/or for tilting. In other cases, it might be better to fix the position of the objective in all directions, e.g., to maintain the alignment between objective and condenser when interferometric measurements are performed [Section 9.2].

We can now mount the sample translation stage, which permits us to position the sample in three dimensions, on the second-level breadboard, as shown in Fig. 8.2d. This translation stage can be either manual or automatised. Manual translation stages typically offer precision down to about a fraction of a micrometre over a range that can easily reach several millimetres. Automatised translation stages, which are often driven by mechanical or piezoelectric actuators, permit one to achieve subnanometre position precision over a range that typically does not exceed some tens or hundreds of micrometres[2] and have the advantage that they can be controlled remotely. Often, a good approach is to combine a manual translation stage for the rough positioning of the sample and an automatised translation stage for fine adjustment; such an approach is implemented in our set-up [Fig. 8.2d].

At this point, we proceed to build a third level, as shown in Fig. 8.2e, where we will host the illumination and detection components. Furthermore, also shown in Fig. 8.2e, we fix a translation stage on which to mount the condenser. This stage permits us to align the condenser position along the lateral (horizontal) directions and, therefore, to centre the position of the collected beam.

Finally, we can now proceed to mount the objective (OBJ), which will permit us to collect the image of the sample (and later to focus a laser beam into an optical tweezers), and the condenser (C), as shown in Fig. 8.2f. The main parameters that describe an objective are discussed in Subsection 8.2.1. To start, it is convenient to use a low-magnification (e.g., $10\times$ or $20\times$) objective because it permits a simpler alignment.

Because a microscope serves to magnify the image of a small object, it is appropriate to start by choosing a simple object to look at. For the moment, we can use a calibration glass slide.[3] This sample will be mounted on the second-level sample stage, as shown in Fig. 8.2f.

---

[2] In order to achieve subnanometre position *accuracy* and *reproducibility*, these stages need to be actuated with closed-loop position control, as in open-loop they are subject to significant hysteresis.

[3] Alternatively, one can use a sample of microspheres stuck to a coverslip or a glass coverslip to which some colour has been applied on one side, e.g., using an indelible marker.

Table 8.1 Objective specifications

Manufacturer	Parfocal length	Tube lens	Thread
Leica	45 mm	200 mm	M25 × 0.75
Nikon	60 mm	200 mm	M25 × 0.75
Olympus	45 mm	180 mm	RMS[a]
Zeiss	45 mm	165 mm	M27 × 0.7

[a] The RMS (Royal Microscopical Society) standard for the objective thread was first set down in 1866 and revised in 1896. The RMS standard is a 55° Whitworth thread, with a diameter of 0.7652–0.7982 in. (19.44–19.51mm) at the top of the objective lens thread and 36 turns per inch (TPI), i.e., a pitch of 0.706 mm. Often the RMS thread diameter is quoted as 0.8 in., which refers to the diameter at the bottom of the female thread in the microscope turret.

To capture the image of the sample, an illumination source is necessary. This can be realised in a first simple configuration using a LED lamp focused on the sample by the condenser, as shown again in Fig. 8.2f. It is possible to see the resulting image with the naked eye by placing a screen along the path of the light collected by the objective. At this point it is possible to optimise the illumination train in order to maximise the illumination intensity by adjusting the condenser position. For infinity-corrected objectives, it is necessary to add a tube lens (whose focal length is typically about 150 to 200 mm [Table 8.1]) in order to form an image of the sample. More advanced illumination configurations are also possible and will be discussed in Subsection 8.2.2.

Troubleshooting: I cannot mount the objective at the right distance to image the sample. Start by aligning your system using objectives with low magnification (e.g., 10× or 20×). These objectives can be aligned much more easily because they have a longer working distance and focal depth. Only once your system is aligned and working should you switch to higher magnification objectives (e.g., 40×, 60× water-immersion, 100× oil-immersion). Note that, as long as you use objectives with the same parfocal distance, you can switch objectives, keeping the sample in focus so that, when moving to higher-magnification objectives, you do not need to adjust the vertical focal position.

Troubleshooting: I cannot focus on the sample. If you cannot focus on your sample, it might be that the coverslip is too thick for the working distance of the objective. In this case, you will also see that the objective pushes the coverslip or that (for spring-loaded objectives) the objective spring is pushed. Use a thinner coverslip, or reverse the coverslip and look at its lower side. The coverslip thickness is typically indicated by the coverslip number, as shown in Table 8.2.

Table 8.2 Microscope coverslip thickness

Number	Thickness
0	0.085 to 0.13 mm
1	0.13 to 0.16 mm
1.5	0.16 to 0.19 mm
1.5H	0.17 to 0.18 mm
2	0.19 to 0.23 mm
3	0.25 to 0.35 mm
4	0.43 to 0.64 mm

We have finally built our first inverted microscope. We only need to add a camera at the image plane to start observing our sample. The final set-up is shown in Fig. 8.2g and the corresponding schematic in Fig. 8.2h.

8.2.1 Objectives

In an optical tweezers, the objective serves the dual function of imaging the sample and focusing the laser beam in order to generate the optical trap. For optical manipulation applications, a high-numerical-aperture objective is used in order to generate the tight focal spot needed to trap a particle in three dimensions. Such objectives are readily available from various manufacturers.[4] A schematic of a typical objective is shown in Figs. 8.3a and 8.3b. It consists of a series of lenses carefully oriented and tightly packed into a tubular brass housing, which is encapsulated by the objective barrel and, as is the case with most oil-immersion objectives, is equipped with a spring-loaded retractable nosecone assembly to protect the front lens elements and the specimen from collision damage.

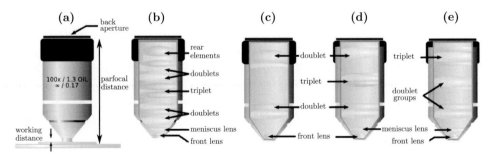

Figure 8.3 Objectives: (a) Oil-immersion infinity-corrected $100\times$ apochromat objective showing the definitions of working distance, parfocal distance and back aperture. (b) Cutaway diagram showing the lens system inside a typical $100\times$ apochromat objective. (c)–(e) Cutaway diagrams showing the increasing complexity of the lens system inside a $10\times$ objective for higher degrees of correction: (c) achromat; (d) fluorite (semi-apochromat); (e) apochromat.

---

[4] The major manufacturers of microscope objectives are Nikon, Olympus, Zeiss and Leica.

The specifics of any given objective are imprinted or engraved on the external portion of the barrel. A list of abbreviations for the specifications commonly found on microscope objectives is shown in Table 8.3. In the following, we discuss the objective characteristics that are most important for optical manipulation applications:

- **Numerical aperture.** For optical manipulation, the single most important number characterising an objective is its numerical aperture [Eq. (2.22)]. This is a critical value that indicates the light acceptance angle, which in turn determines the light-gathering power, the resolving power and the depth of field of the objective. Although the numerical aperture of objectives working in air is naturally limited to about 0.95, the numerical aperture can be dramatically increased by designing the objective to be used with an immersion medium, such as water (typically 1.20) or oil (typically 1.30).[5]
- **Immersion medium.** Although most objectives are designed to image specimens, with air as the medium between the objective and the cover glass, in order to attain higher numerical apertures, many objectives are designed to image the specimen through another medium that reduces refractive index differences between glass and the imaging medium. High-resolution plan apochromat objectives can achieve numerical apertures up to 1.40 when the immersion medium is a special oil with a refractive index of about 1.51. Other common immersion media are water (refractive index 1.33) and glycerol (refractive index 1.47). Objectives designed for special immersion media usually have a colour-coded ring inscribed around the circumference of the objective barrel, as listed in Table 8.4. Some important properties of immersion oils are described in Box 8.1.

> **Suggestion: Check your immersion oil.** The advantages of oil-immersion objectives are severely compromised if the wrong immersion oil is used. It is advisable to employ only the oil intended by the objective manufacturer and not to mix immersion oils between manufacturers to avoid unpleasant artefacts such as crystallisation or phase separation. Also, always check the expiry date of your immersion oil.

- **Working distance and coverslip thickness.** The working distance is the distance between the objective front lens and the top of the cover glass when the specimen is in focus, as shown in Fig. 8.3a. It is often inscribed on the barrel as a distance in millimetres. In most instances, the working distance of an objective decreases as the magnification increases. The working distance should be matched to the employed coverslip thickness [Table 8.2], which is typically 0.17 mm in most optical manipulation applications, i.e., a coverslip No. 1.5,[6] which is compatible with the working distances of immersion objectives

---

[5] Some objectives specifically designed for transmitted light fluorescence and dark-field imaging are equipped with an internal iris diaphragm that allows adjustment of the effective numerical aperture. Objectives designed for total internal reflection microscopy have numerical aperture higher than that of the medium where the focus is generated (typically water, $n_m = 1.33$) in order to produce evanescent illumination.

[6] There is often some variation in thickness within a batch of coverslips. For this reason, some of the more advanced objectives have a correction collar for adjustment of the internal lens elements to compensate for this variation. Optical correction for spherical aberration is produced by rotating the collar, which causes two of the lens element groups in the objective to move either closer together or farther apart.

Table 8.3 Objective abbreviations

Achro, Achromat	Achromatic aberration correction
Fluor, Fl, (Neo)fluar, Fluotar	Fluorite aberration correction
Apo	Apochromatic aberration correction
Plan, Pl, Achroplan, Plano	Flat field optical correction
EF, Acroplan	Extended field (field of view less than plan)
N, NPL	Normal field of view plan
Plan Apo	Apochromatic and flat field correction
UPLAN	Olympus universal plan[a]
LU	Nikon luminous universal[a]
L, LL, LD, LWD	Long working distance
ELWD, SLWD, ULWD	Extra-long, super-long, ultra-long working distance
Corr, W/Corr, CR	Correction collar
I, Iris, W/Iris	Adjustable numerical aperture (with iris diaphragm)
Oil, Oel	Oil immersion
Water, WI, Wasser	Water immersion
HI	Homogeneous immersion
Gly	Glycerin immersion
DIC, NIC	Differential or Nomarski interference contrast
CF, CFI	Chrome-free, chrome-free infinity-corrected (Nikon)
ICS	Infinity colour-corrected system (Zeiss)
RMS	Royal Microscopical Society objective thread size
M25	Metric 25 mm objective thread
M32	Metric 32 mm objective thread
Phase, PHACO, PC	Phase contrast
Ph 1, 2, 3, etc.	Phase condenser annulus 1, 2, 3, etc.
DL, DM	Phase contrast: dark low, dark medium
PL, PLL	Phase contrast: positive low, positive low low
PM, PH	Phase contrast: positive medium, high contrast[b]
NL, NM, NH	Phase contrast: negative low, medium, high contrast[c]
P, Po, Pol, SF	Strain-free, low birefringence, for polarised light
U, UV, Universal	UV-transmitting ($\gtrsim 340$ nm) for epifluorescence
M	Metallographic (no coverslip)
NC, NCG	No coverslip
EPI	Oblique or epi illumination
TL BBD, HD, B/D	Bright or dark-field (hell, dunkel)
D	Dark-field
H	For use with a heating stage
U, UT	For use with a universal stage
DI, MI, TI	Interferometry, noncontact, multiple beam (Tolanski)

[a] Bright-field, dark-field, DIC and polarised light.
[b] Regions with higher refractive index appear darker.
[c] Regions with higher refractive index appear lighter.

Table 8.4 Immersion medium colour codes

Immersion medium	Colour code
Oil	Black
Water	White
Glycerol	Orange
Special	Red

> **Box 8.1** **Immersion oil properties**
>
> Usually, the refractive index of an immersion oil is specified for one of the spectral lines given in the following table:
>
> | Fraunhofer symbol | Element | Wavelength |
> |---|---|---|
> | e | Mercury | 546.1 nm |
> | D | Sodium | 589.3 nm |
> | d | Helium | 587.6 nm |
> | F' | Cadmium | 480.0 nm |
> | F | Hydrogen | 486.1 nm |
> | C' | Cadmium | 643.8 nm |
> | C | Hydrogen | 656.3 nm |
>
> DIN specification for immersion oil is a refractive index of $n_e = 1.5150 \pm 0.0004$ (at $23\,^\circ$C) and a dispersion given by the Abbé V-number $v_e = 44 \pm 5$ (at $23\,^\circ$C), where $v_e = \frac{n_e - 1}{n_F - n_C}$.
>
> The viscosity of an immersion oil is also an important parameter. Low-viscosity oils are less apt to retain small bubbles and may be preferred when the distance from the cover glass to the objective is very small. Higher-viscosity oils fill larger gaps more satisfactorily and are also reusable in that a slide may be removed and replaced with another to make contact with the oil drop that remains on the objective lens. The kinematic viscosity of some representative immersion oils is given in the table below in the cgs unit of centistokes (cSt), where $1\,\text{cSt} \equiv 10^{-6}\,\text{m}^2\text{s}$:
>
> | Oil | Viscosity |
> |---|---|
> | Cargille type A (low viscosity) | 150 cSt |
> | Cargille type B (medium viscosity) | 1250 cSt |

Table 8.5 Objective magnification colour codes

Magnification	Colour code
0.5×	Not assigned
1×, 1.25×, 1.5×	Black
2×, 2.5×	Brown
4×, 5×	Red
10×	Yellow
16×, 20×	Green
25×, 32×	Turquoise
40×, 50×	Light blue
60×, 63×	Cobalt blue
100×, 150×, 250×	White

- **Trasmission efficiency.** Most objectives are designed to work within the visible range of optical wavelengths. Typically, their transmission efficiency is about 60% within this range for high-numerical-aperture immersion objectives. However, it can become much less at longer wavelengths in the near infrared for some objectives. Because these wavelengths are often employed in optical manipulation applications, e.g., to prevent photodamage to biological samples, one should pay attention to this issue in order not to lose too much laser power.
- **Magnification.** There exist commercial objectives with magnifications ranging from 0.5× to 250×. The most commonly employed for three-dimensional optical trapping applications are 100× oil-immersion objectives and 60× water-immersion objectives. We must remark, however, that the magnification is not the most important parameter for optical trapping applications. To help with the rapid identification of the magnification, microscope manufacturers label their objectives with colour codes, which are given in Table 8.5.
- **Optical aberration corrections.** Objectives are corrected to reduce their chromatic and spherical aberrations [Box 2.2]. Depending on the objective corrections, the internal system of lenses becomes more and more complex (and the objective correspondingly more and more expensive).
  - The simplest objectives are *achromatic objectives* with the least amount of correction [Fig. 8.3b]. Achromats are corrected for axial chromatic aberrations at two wavelengths, typically blue (486 nm) and red (656 nm), and for spherical aberrations in the green (546 nm). This limited correction can lead to substantial artefacts when specimens are examined and imaged with colour microscopy; for example, if the focus is chosen in the green region of the spectrum, images will have a reddish-magenta halo (residual colour). Nevertheless, these objectives can be a good compromise for optical tweezers application when only one wavelength is employed and there is no specific need for imaging the sample at high quality.

Table 8.6 Objective aberration corrections			
Objective type	Spherical aberration	Chromatic aberration	Field curvature
Achromat	1 colour	2 colours	Uncorrected
Plan achromat	1 colour	2 colours	Corrected
Fluorite	2–3 colours	2–3 colours	Uncorrected
Plan fluorite[a]	3–4 colours	2–4 colours	Corrected
Apochromat	3–4 colours	4–5 colours	Uncorrected
Plan apochromat	3–4 colours	4–5 colours	Corrected

[a] Plan fluorite is sometimes referred to as plan semi-apochromat.

- The next level of correction is found in *fluorite objectives*[7] or *semi-apochromat objectives* [Fig. 8.3c]. The lenses of fluorites are produced from advanced glass formulations that contain materials such as fluorspar or newer synthetic substitutes, which allow greatly improved correction of optical aberrations. Fluorites are corrected chromatically for red and blue light and spherically for two or three colours. The superior correction of fluorites compared to achromats enables these objectives to be made with a higher numerical aperture.
- The highest level of correction is found in *apochromatic objectives* [Fig. 8.3d]. Apochromats are corrected chromatically for three colours (red, green and blue), almost eliminating chromatic aberrations, and are corrected spherically for either two or three wavelengths. Although the lens design is similar to that for fluorites, they employ more advanced materials. Apochromats are the best choice for colour photomicrography in white light. Because of their high level of correction, apochromat objectives usually have, for a given magnification, higher numerical apertures than achromats or fluorites. The best high-performance fluorite and apochromat objectives are corrected for four (dark blue, blue, green, and red) or more colours chromatically and four colours spherically.

All three types of objectives suffer from pronounced field curvature and project images that are curved rather than flat, an artefact that increases in severity with higher magnification. To overcome this inherent condition, optical designers have produced flat-field corrected objectives, which yield images that are in common focus throughout the viewfield. Objectives that have flat-field correction and low distortion are called *plan achromats*, *plan fluorites*, or *plan apochromats*, depending upon their degree of residual aberration. An overview of the main objective characteristics is given in Table 8.6.

- **Parfocal distance.** The parfocal distance is measured from the nosepiece objective mounting hole to the point of focus on the specimen, as shown in Fig. 8.3a. This is another specification that can often vary by manufacturer. The major manufacturers

---

[7] They are named after the mineral fluorite, which was originally used in the manufacture of their lenses.

produce objectives that have a 45 mm or a 60 mm parfocal distance [Table 8.1]. Most manufacturers also make their objective nosepieces parcentric, meaning that when a specimen is centred in the field of view for one objective, it remains centred when another objective is brought into use.

- **Mechanical tube length.** This is the length of the microscope body tube between the nosepiece opening, where the objective is mounted, and the top edge of the observation tubes, where the oculars (eyepieces) are inserted. The tube length is usually inscribed on the objective in millimetres (typically between 160 and 200 mm depending on the manufacturer, as shown in Table 8.1) for fixed tube lengths or with the infinity symbol (∞) for infinity-corrected tube lengths.
- **Objective screw threads.** Historically, the mounting threads on almost all objectives conformed to the Royal Microscopical Society (RMS) standard in order to enable universal compatibility. This standard is still used in the production of infinity-corrected objectives by the manufacturer Olympus. Zeiss, Nikon and Leica have broken from the standard with the introduction of new infinity-corrected objectives that have a wider mounting thread size, making these objectives usable only on their own microscopes [Table 8.1].
- **Specialised optical properties.** Microscope objectives often have design parameters that optimise performance under certain conditions. For example, there are special objectives designed for polarised illumination signified by the abbreviations P, Po, POL, or SF (strain-free and/or having all barrel engravings painted red), phase contrast (PH and/or green barrel engravings), differential interference contrast (DIC), and many other abbreviations for additional applications. A list of several abbreviations, often manufacturer specific, is presented in Table 8.3.

8.2.2 Illumination schemes

Proper illumination of the specimen is crucial for obtaining the best-quality imaging in a microscope, and impacts both the ease of operation of an optical tweezers experiment and the efficacy of subsequent analysis by, e.g., particle detection, counting and tracking algorithms. Here, we review the basics of achieving suitable illumination in the microscope and more advanced techniques that can be used to improve the contrast, particularly of biological samples.

Bright-field microscopy

Bright-field illumination is the simplest and most common scheme used in transmitted light microscopy: the illumination light passes through the sample to be collected by the objective and imaged onto a camera. This is the illumination scheme used in the inverted microscope we built [Figs. 8.2g and 8.2h]. Contrast in the image is produced by refraction, which is induced by local changes in refractive index and absorption in the specimen. Control of the illumination can be achieved by regulation of two diaphragms, shown in Fig. 8.4: the *condenser diaphragm*, which is placed at the condenser aperture to adjust the

8.2 Inverted microscope construction

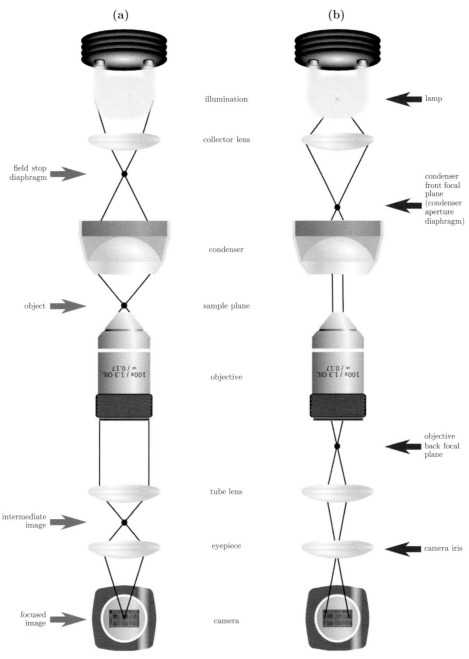

Figure 8.4 Köhler illumination: Location (a) of the field planes, conjugate with the field stop aperture, and (b) of the aperture planes, conjugate with the lamp filament.

effective numerical aperture of the condenser lens and controls, in combination with the objective, the resolution of the system; and the *field stop diaphragm*, which controls the width of the bundle of rays reaching the condenser and is used to eliminate excess light scattering that may degrade the contrast in the image.

Köhler illumination

The Köhler illumination scheme, first introduced by August Köhler in 1893, produces optimum illumination of the sample, achieving high contrast by illuminating the sample evenly. The principle is illustrated by the ray paths shown in Fig. 8.4. Fig. 8.4a shows the *field planes*, i.e., the planes that are conjugate with the field stop diaphragm: the object in the sample plane, an intermediate image between the tube lens and the eyepiece (or the lens which focuses the light onto the camera), and the camera sensor where the (magnified) image is formed. Fig. 8.4b shows the *aperture planes*, i.e., the planes that are conjugate with the filament of the illumination lamp: the front focal plane of the condenser, the back focal plane of the objective and, in this case, the iris of the camera. This arrangement ensures that the structure of the filament is not imaged onto the sample or the camera, so that it is prevented from causing significant variation in intensity across the image.

The microscope can be configured for Köhler illumination following these steps:

1. observe a specimen (a sample of microspheres stuck to a coverslip is useful for this purpose) under simple bright-field illumination with both field diaphragm and condenser aperture diaphragm fully open;
2. close down the field stop diaphragm and translate the condenser lens (this can be done by using the focusing control knob in a commercial microscope) until the edges of the diaphragm are in focus;
3. adjust the transverse position of the condenser to achieve maximum intensity through the closed-down field diaphragm;
4. open the field stop diaphragm;
5. adjust the aperture diaphragm to illuminate the sample and achieve the best balance between contrast and resolution.

We note that following these steps should properly configure the microscope for Köhler illumination for the objective with which the procedure is carried out, whereas changing the objective will require re-adjustment of the condenser aperture and field diaphragms for optimum illumination.

Phase contrast

The phase contrast technique, invented by Frits Zernike in 1934,[8] is best applied to thin specimens that have little absorption and small refractive index difference from the surrounding medium and, hence, have poor visibility in standard bright-field microscopy. Its

---

[8] Zernike was awarded the Nobel Prize in Physics in 1953 'for his demonstration of the phase contrast method, especially for his invention of the phase contrast microscope'.

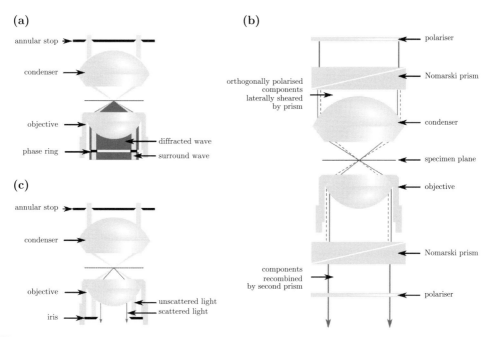

Figure 8.5 Contrast enhancement techniques: (a) phase contrast; (b) dark-field illumination; (c) differential interference contrast (DIC).

principle of operation is illustrated in Fig. 8.5a. Phase contrast microscopy requires the addition of an *annular stop* at the condenser aperture and the use of a corresponding *phase contrast objective*, denoted by the legend Ph1, Ph2 or Ph3 on the objective barrel [Table 8.3], where the number refers to the diameter of the phase stop annulus. Under conditions of Köhler illumination, the specimen is illuminated by plane waves emanating from the condenser annular stop. Waves that do not interact with the specimen (sometimes called *surround waves*) are brought to a focus in the back focal plane of the objective, which is conjugate with the condenser aperture, as shown in Fig. 8.4b. For a weakly scattering specimen, i.e., one with a low refractive index contrast, the scattered waves (sometimes called *diffracted waves*) have a small amplitude compared to that of the surround waves but, crucially, acquire a small phase shift in their interaction with the sample; for example, for a low-contrast biological sample with a thickness of ≈ 5 μm, this phase shift is about $\pi/2$ for light in the green part of the optical spectrum. The diffracted waves occupy the whole extent of the pupil, as shown in Fig. 8.5a. The phase and amplitude of the surround waves, which would otherwise overwhelm the small-amplitude diffracted waves, are controlled by a phase ring in the back focal plane of the objective, which acts predominantly on the surround waves by advancing their phase by $\pi/2$ and by attenuating them. Eventually, the surround and diffracted waves interfere at the image plane and the contrast of the specimen over the background illumination is greatly enhanced compared to that with bright-field illumination.

Dark-field microscopy

Dark-field microscopy also uses an annular condenser stop to illuminate the specimen at oblique angles. In the dark-field configuration, though, unscattered waves are either simply not collected by a low-numerical-aperture objective or, as shown in Fig. 8.5b, stopped by an adjustable iris inside a specialised high-NA objective. The effect is to increase contrast, as only light that has been scattered by the specimen is collected, resulting in a bright image on a dark background.

Differential interference contrast

Differential interference contrast (DIC), sometimes called Nomarski interference contrast after its inventor, Georges Nomarski, increases contrast by converting gradients in optical path difference into intensity modulation. In DIC, initially linearly polarised light is separated into two orthogonally polarised beams with a small displacement by a modified Wollaston (Nomarski) prism, as illustrated in Fig. 8.5c. If the specimen presents gradients in the optical path experienced by the two beams become of gradients in thickness and/or refractive index, then the two beams acquire a relative phase shift. The lateral shear between the two beams is removed by a second Nomarski prism and an analyser (polariser) projects the polarisation of the two beams onto the same axis. Gradients in optical path in the sample are then converted into gradients in intensity in the interfering beam, resulting in improved contrast, although as one side of a feature in the specimen appears brighter than the opposite side, the features often present a shadowed, pseudo three-dimensional appearance. DIC microscopy requires the use of a specialised condenser and prism system. Unlike phase contrast and dark-field objectives, the objective is not modified internally, but is designed to be used with an external Nomarski prism.

8.3 Sample preparation

We are now at the point where we need to prepare a sample with some trappable objects. The range of trappable objects is quite large, as we have seen in Fig. 1.3. However, to start it is simplest to use some synthetic microparticle in aqueous solution. Particles with very good characteristics, e.g., sphericity and size variance, are widely available from commercial providers.[9] The ideal size to start with lies in the intermediate size regime, i.e., from about 0.5 to 2 μm radius, where optical forces are maximised, as we have seen in Fig. 5.1. It is also possible to use some simple living organisms, e.g., yeast cells, which can be obtained easily and safely from a bakery. At first, one can simply place a droplet of the solution on top of a glass slide placed on the microscope sample stage and then slightly raise the objective (or lower the sample) in order to have the focal spot inside the droplet and be able to observe

---

[9] Some providers of microparticles are Duke Scientific, Bangs Labs, Microparticles.de, Polysciences and Kisker Biotech.

Table 8.7 Microparticle material properties

Material	Refractive index	Density
Melamin resin (MF)	1.68	$1.51\,\text{g cm}^{-3}$
Polystyrene (PS)	1.59	$1.05\,\text{g cm}^{-3}$
Poly(methyl methacrylate) (PMMA)	1.48	$1.19\,\text{g cm}^{-3}$
Silica (SiO$_2$)	1.42 to 1.46	$1.8\,\text{g cm}^{-3}$ to $2.0\,\text{g cm}^{-3}$

the objects in the suspension. Of course, we will soon need a more professional sample. In fact, the sample prepared as just described is not stable; e.g., it is subject to evaporation. In the following, we provide some recipes for how to prepare a sealed sample. These should be treated as starting guidelines to perfect one's own sample. Indeed, it is safe to say that there are at least as many approaches to sample preparation as researchers, because the specific sample to be employed strongly depends on the experiment to be conducted and also on the preferences of the experimentalist preparing it.

The sample preparation starts by suspending microparticles in a solution (often an aqueous solution) at an appropriate concentration. A series of typical microparticle parameters is given in Table 8.7. Microparticles are usually supplied by manufacturers at high concentration. The number density, i.e., the number of particles per millilitre of liquid, can be found as

$$n = \frac{w \times 10^{12}}{\frac{4}{3}\pi a^3 \rho}, \tag{8.1}$$

where w is the particle concentration in mg/ml, ρ is the particle material density in g/cm^3 and a is the particle radius in μm. For example, $a = 1$ μm radius polystyrene spheres may be sold at 2.5% solids (w/v) or $w = 25$ mg/ml, i.e., 25 mg of solid material per millilitre of solution, giving a number density of 4.6×10^{10} particles/ml. A solution at workable concentration typically requires diluting this concentration by several orders of magnitude, e.g., in the ratio of one part microparticle solution to $\approx 10^6$ parts solution.[10]

We now need a sample cell to host our solution on top of the microscope sample stage. A possible preparation process is shown in Fig. 8.6. The process starts with a carefully cleaned microscope glass slide [Fig. 8.6a], on top of which two stripes of parafilm (about 100 μm thick) are placed [Fig. 8.6b]. A thin coverslip is then placed on top of the parafilm [Fig. 8.6c] and this sample is heated so that the parafilm sticks to the glass surfaces. Finally, after removal of the excess parafilm [Fig. 8.6d], the chamber is filled with sample solution using a pipette, sealed using vacuum grease [Fig. 8.6e], placed upside down on the sample stage and gently fixed with some plastic screws [Fig. 8.6g]. It is now possible to observe the sample and to apply techniques such as digital video microscopy [Section 9.1] to study the Brownian motion of particles.

---

[10] This is best achieved by subsequent dilutions of the commercial solution. For example, it is possible to make a mother solution by diluting 100 to 1000 times the commercial solution and then dilute the mother solution further to create the solutions to be employed in the actual experiments.

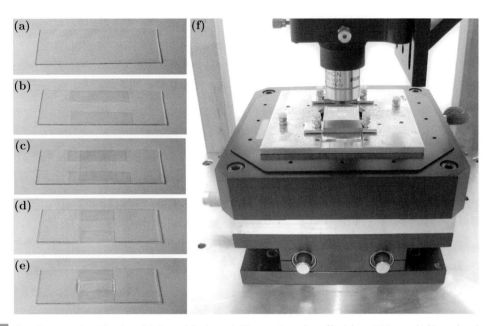

Figure 8.6 Sample preparation: (a) a glass slide is carefully cleaned; (b) two stripes of parafilm (about 100 μm thick) are placed on the glass; (c) a coverslip is placed on top of the parafilm, and the sample is heated in order for the parafilm to stick to the glass surfaces (not shown); (d) the excess parafilm is removed; (e) the chamber is filled with sample solution (e.g., microparticles suspended in an aqueous solution) and is sealed using vacuum grease; (g) the chamber is placed upside down on the sample stage and is gently fixed with some plastic screws.

Starting from the recipe explained, several variations are possible in order to build a sample cell. In particular, it is possible to use microscope slides having a well. Standard 76 mm × 26 mm slides with a 15 mm-diameter recess in one face can be obtained readily from suppliers. The process involves the following steps: fill or slightly overfill the recess with 80 μl of liquid (a calibrated pipette is useful for this purpose); place a coverslip on the slide adjacent to the recess and then slide it over while maintaining downward pressure; finally, fix the coverslip in place, using epoxy or even nail polish. Another alternative is to use adhesive spacers, such as Invitrogen SecureSeal. A convenient size is the 9 mm-diameter spacer available from Fisher Scientific (Product Code 10634453), which may be stuck to a plain slide, filled with a small volume of the sample solution and sealed using a coverslip.

> **Troubleshooting: My particles get stuck to the coverslip and/or between themselves.**
> This is most likely due to electrostatic interactions between the particles and/or the coverslip. When working with an aqueous solution, you can perform one or more of the following operations:
>
> - use deionised water to increase the screening length and, therefore, the electrostatic repulsion between particles;

- use a hydrophilic coverslip to increase the electrostatic repulsion between the coverslip and the particles (the coverslip can be made hydrophilic by plasma cleaning it or by leaving it for several minutes in a 1 mM NaOH solution);
- add a small amount of surfactant to the solution, e.g., sodium dodecyl sulphate (SDS) or Triton-X.

Troubleshooting: I took the sample out and put it back. Now the image on the camera looks terrible. If you have an oil-immersion objective, the act of removing and replacing the slide may have created small bubbles in the oil so that there is no longer a continuous medium between the objective and coverslip. Clean both the objective and the coverslip, and start again with fresh immersion oil.

8.4 Optical beam alignment

We are now ready to transform our inverted microscope into an optical tweezers by coupling a laser beam to the set-up we realised in the previous sections. The first crucial step is to choose an adequate laser, which depends strongly on the foreseen application. The most important considerations to be kept in mind are the following (a more in-depth overview of lasers is given in Subsection 8.4.1):

- **Laser power.** Typically, a laser with a power of about 10 to 100 mW is more than sufficient to generate a single optical trap capable of manipulating microscopic particles. As a rule of thumb, one can expect the optical stiffness to have a value similar to the ones reported in Fig. 5.1. Note that, because optical losses occur at the objective (which can be as large as 50%) and because of the necessary overfilling, the laser power at the focus, which is the laser power that ultimately determines the stiffness of the optical tweezers, might be significantly lower than the original one.
- **Wavelength.** In many cases, the wavelength of the laser light plays a crucial role. In particular, it is important to consider the potential for photodamage to the specimen, because the power density in even a low-power optical beam brought to a diffraction-limited spot can exceed 10^6 W/cm^2. For example, for biological applications it is important to choose a wavelength that minimises absorption (to prevent local heating) and photodamage (to maintain the biological samples' viability). Because most biological samples are in an aqueous medium, it is useful to keep in mind the absorption coefficient of water, shown in Fig. 8.7a.[11] Low photodamage is typically achieved for wavelengths in the near infrared,

---

[11] The absorption coefficient of a material, e.g., α_w for water shown in Fig. 8.7a, is expressed in cm^{-1}. According to Beer's law, the intensity of an electromagnetic wave penetrating a material falls off exponentially with distance from the surface, i.e., $I(d) = I_0 e^{-\alpha_w d}$, where $I(d)$ is the light transmitted at a distance d (expressed in cm) from the surface and I_0 is the incident intensity.

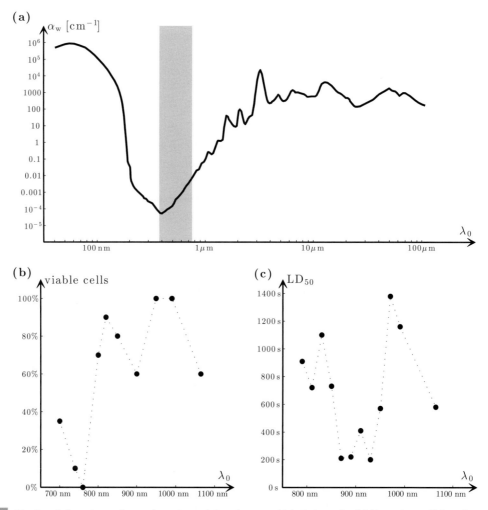

Figure 8.7 Wavelength dependence of water absorption and photodamage to biological samples: (a) Absorption coefficient of water α_w as a function of the vacuum wavelength λ_0. The shaded area represents the visible portion of the spectrum. (b) and (c) show the photodamage to biological material held in optical tweezers as a function of wavelength: (b) percentage of Chinese hamster ovary cells that remain viable after 5 minutes of trapping at $88\,\mathrm{mW}$ power (Liang et al., 1996); (c) time after which the rotation rate of the flagella of *E. coli* bacteria held in an optical tweezers at $100\,\mathrm{mW}$ trapping power decreases to 50% the initial rate (Neuman et al., 1999).

as is shown for the case of Chinese hamster ovary cells in Fig. 8.7b and for the case of *E. coli* bacteria in Fig. 8.7c. Furthermore, one should consider the transmission properties of the objective lens at the wavelengths being used, as the transmission across the near infrared is considerably less than 100% and, for some objectives, drops dramatically for wavelengths above $\approx 850\,\mathrm{nm}$.[12]

---

[12] It is advisable to check carefully the transmission characteristics supplied by the manufacturer of an objective not only at the trapping wavelength but also at any other wavelengths that may be used, e.g., for acquisition of

- **Beam stability and quality.** Depending on the application at hand, the stability and quality of the beam can play a crucial role. This is particularly true in applications that require high reproducibility and quantitative measurements, e.g., measurement of nanoscopic forces and torques. For more qualitative applications, e.g., the optical trapping and delivery of cells or vesicles, it might be possible to relax the requirements on beam stability and quality.

> **Suggestion: Start with a visible laser.** When building your first optical tweezers, we advise you to use a laser emitting in the visible portion of the spectrum with relatively low power. For example, HeNe lasers at 633 nm (red) or solid state lasers at 532 nm (green) with output power about 50 mW are ideal.

Once the laser has been chosen, one has to firmly fix it onto the optical table, as shown in Fig. 8.8a. In our case, we use a solid state laser with wavelength 1064 nm. To increase the stability of the laser beam, we prefer not to change the laser output power during our experiments. Therefore, we opt for an alternative way of controlling the laser power: we place along the beam path a linear polariser, a half-wave plate and a second beam polariser; in this way, we can tune the power of the beam without altering the laser setting by rotating the half-wave plate. The next step is to direct the laser through the objective, making use of a series of mirrors, as shown in Fig. 8.8b. We first use mirror M_4 to align the beam parallel to the optical table.[13] Then we proceed to direct the laser beam through the objective using mirrors M_5 and DM_6. In doing this, we apply the basic principle of beam steering, which is illustrated in detail in Fig. 8.9 and, in a nutshell, permits one to use two steerable mirrors to get a laser beam to go along any straight line. We also note at this point that DM_6 is in fact a dichroic mirror, which permits one to split the light of the laser beam, which goes to the objective, and the light of the lamp, which goes to the camera. We also add a filter in the light path leading to the camera in order to remove any laser light that might saturate the camera.

> **Suggestion: Centre your beams.** Optical beams should always go through the centres of optical elements, e.g., lenses and mirrors. This is because these optical elements are usually optimised for a beam going through their centres (e.g., using a lens off axis can result in the introduction of aberrations that can severely affect the qualify of focusing) and also because this makes any future correction to the beam path easier.

> **Suggestion: Use dielectric mirrors.** It is better to use a dielectric mirror instead of a metallic mirror in order to reduce losses. In fact, whereas losses of about 5% can be expected at metallic mirrors, at dielectric mirrors these losses can be reduced to about 0.5% or less.

Raman or fluorescence spectroscopy signals. Also, the range of wavelengths over which aberrations are well corrected is important, as an objective designed for, e.g., fluorescence microscopy in the visible spectrum may have relatively poor aberration correction in the near infrared.

[13] It is also usually useful to align the beam such that it goes straight along a series of holes in the optical table.

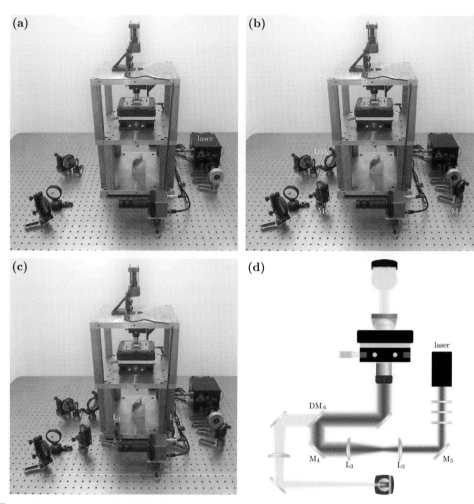

Figure 8.8 Alignment of the laser beam to generate an optical tweezers: (a) The laser head is firmly clamped on the optical table and the laser beam is prepared by using two polarisers and a half-wave plate to control the beam power without changing the laser power setting (a filter is also added in the camera path to prevent the laser light from reaching the camera); (b) a series of mirrors (M_4, M_5 and DM_6) are added to guide the beam through the objective in order to generate the optical trap; (c) a telescope (constituted by lenses L_1 and L_2) is added to increase the beam waist size and overfill the objective back aperture. (d) Corresponding schematic of the optical tweezers set-up.

Finally, as shown in Fig. 8.8c, we need to add a telescope in order to create a beam with an appropriate size to overfill the objective back aperture and to generate a strong optical trap. The beam waist can be straightforwardly measured by fitting a Gaussian profile to an image of the beam profile acquired by a digital camera (some neutral density filters might be necessary to prevent the camera from saturating).[14]

---

[14] Alternatively, if the beam is larger than your camera chip, it is possible to use a knife-edge and a photodiode to measure the transmitted power while cutting the laser beam transversally along a direction (x). The transmitted

8.4 Optical beam alignment

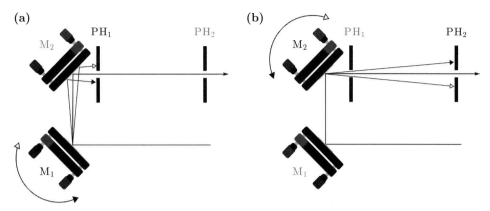

Figure 8.9 Beam alignment technique. In order to get a laser beam to go along any line (indicated by the two pinholes PH_1 PH_2), it is possible to use two manually steerable mirrors (M_1 and M_2) and follow the procedure illustrated here: (a) with M_1, centre the beam on PH_1; (b) with M_2, centre the beam on PH_2; repeat the previous two steps iteratively. The convergence of this procedure is greatly improved by placing PH_1 as close as possible to M_2 and PH_2 as far away as possible from M_2.

Finally, we can align the beam so that the focus is on the glass–air interface. This is best done using a sample without any liquid in order to increase the amount of reflected light that goes back to the camera, but can also be done with a sample filled with a solution, even though the intensity of the back-scattered light will be lower. It might also be useful to remove the filter in front of the camera. In practice, one has to approach the objective to the sample slowly, until the back-scattered pattern size becomes minimised, as shown in Fig. 8.10. The resulting pattern should be as symmetric as possible. Furthermore, the pattern should remain symmetric and with the same centre as the relative distance between focus and interface is changed.

> **Troubleshooting: I don't get a symmetric back-scattered light pattern.** The lack of symmetry in the back-scattered light pattern is most likely due to misalignment of the beam. You should check that the beam is perfectly centred and aligned with the objective axis.

power is the integral of the cut laser beam intensity profile, i.e.,

$$P(x) = \frac{P_{\text{beam}}}{2} \left[\text{erf}\left(\frac{\sqrt{2}x}{w_0}\right) + 1 \right],$$

where P_{beam} is the total beam power, w_0 is the beam waist and

$$\text{erf}(\xi) = \frac{2}{\sqrt{\pi}} \int_{-\infty}^{\xi} e^{-t^2} dt$$

is the *error function*. Thus, by fitting the measured power as a function of the knife-edge position, it is possible to measure the laser beam waist.

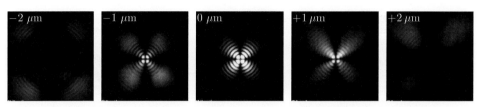

Figure 8.10 Back-scattered light patterns from a focused beam: patterns generated at a glass–air interface as a function of the focus–interface distance (z). It can be seen how the pattern becomes smaller as the focus–interface distance decreases. The cross is due to the presence of polarisation elements; often, one can observe rings instead.

> **Troubleshooting: As I move the objective, the back-scattered light pattern changes its centre.** This is most likely due to misalignment between the beam and objective axes, i.e., the beam is tilted. Correct the vertical alignment of the beam.

8.4.1 Lasers

Laser safety

Correct safety procedures when working with lasers are of paramount importance. In fact, lasers are potentially harmful devices, which, depending on their power and wavelength, can cause eyesight impairment and other physical injuries. We will not discuss laser safety procedures in depth here, as this book should not be a substitute for rigorous laser safety training. However, we urge, all users to undergo suitable safety training before starting to work with lasers and to observe all local safety regulations, including, in particular, the use of appropriate eye safety equipment.

For reference only, we provide in Table 8.8 a summary description of the hazards by laser safety class according to the IEC (International Electrotechnical Commission) 60825-1 and ANSI (American National Standards Institute) Z136.1-2007 Laser classification schemes for lasers and laser systems operating in the range from 0.18 μm to 1 mm. The classification of a laser depends on the wavelength, output power, beam divergence and user's exposure time.

Technical characteristics

Several different lasers are available. A list of the main kinds that have been used in optical trapping experiments is given in Table 8.9. In the following, we discuss the laser characteristics that are most important for optical manipulation applications:

- **Beam quality.** The quality of the laser beam is critical to achieve the tightly focused (as close to diffraction-limited as possible) spot required for optical trapping. The quality of the laser beam is often expressed as the parameter M^2, which is the ratio between the *beam parameter product* of the laser beam and that of a diffraction-limited beam. The beam parameter product is the product of the beam waist and the beam divergence

Table 8.8 Laser safety classification scheme

Laser class	Description of hazard
Class 1	Any wavelength. Safe under reasonable foreseeable conditions due to low emission power (e.g., < 0.39 mW for red laser) or due to total enclosure.
Class 1M	Any wavelength. Geometrical spread of beam reduces naked eye exposure to maximum permissible exposure (MPE) level, but beam is hazardous with viewing aids.
Class 2	Visible lasers only, $\lambda_0 = 400$ to 700 nm. Low power. Blink reflex affords adequate protection.
Class 2M	Visible lasers only. Moderate power where beam divergence reduces the hazard, but may be hazardous if viewing aids or optics are used.
Class 3R	Any wavelength. Moderate power: accessible emission limit is $5\times$ Class 1 limit for invisible lasers or $5\times$ Class 2 limit for visible. Risk of injury is relatively low, but prevent direct exposure to the beam.
Class 3B	Any wavelength. Direct viewing is hazardous. Upper limit can be harmful to skin and pose scatter hazard.
Class 4	Any wavelength. No upper power limit. Harmful to eyes and skin. Possible scattered light hazard.

Table 8.9 Lasers

Laser	Typical wavelength	Typical power
Diode lasers	Visible – near infrared	< 250 mW
Fibre lasers	Near infrared (e.g., 1064 nm, 1070 nm)	1 W to 10 W
Diode-pumped solid state (DPSS) lasers, e.g., Nd:YAG, Nd:YLF	Near infrared 1064 nm, 1047 nm, 1053 nm	1 W to 10 W
Frequency-doubled DPSS lasers, e.g., frequency-doubled Nd:YAG	Visible 532 nm	1 W to 5 W
Helium–neon (HeNe) lasers	633 nm	< 100 mW

half-angle, which for a diffraction-limited beam is λ_0/π. A Gaussian beam has $M^2 = 1$ and, for optical tweezers applications, a laser beam with M^2 as close as possible to this diffraction-limited performance is preferable. Typically, a spatial mode TEM_{00} will also be specified.

- **Pointing stability.** For optical tweezers applications, good pointing stability is necessary to keep the position of the optical trap steady. Fluctuations in beam-pointing direction can arise from, e.g., mechanical vibrations of optical elements in the laser resonator or thermal effects in the laser gain medium. Several different quantities are used to express the pointing stability, so care should be exercised when trying to interpret this parameter, paying attention in particular to the conditions under which it has been measured and to whether the quoted stability refers to the average direction of the beam or to the root-mean-square (rms) fluctuations in direction. One commonly used measure is the change in beam direction with a change in temperature, usually specified in µrad/°C.

- **Power stability and noise.** In optical trapping experiments, fluctuations in laser power lead to fluctuations in the strength of the optical trap. In laser data sheets, there are usually specified the *power stability*, i.e., the drift in average laser power measured over an extended period of time, and the *noise*, i.e., fluctuations around the average of the laser power within a specified bandwidth, typically from a few hertz to (tens of) megahertz. These quantities are often normalised by the average value during the measurement and quoted as either rms or peak-to-peak deviations from the average, expressed as a percentage.
- **Frequency stability.** In general, the frequency stability depends on the type of laser. It can range from several hundred to a few megahertz for a free-running laser. Active stabilisation through an external cavity and locking to atomic lines yields improvements of several orders of magnitude, as the linewidth of lasers locked to atomic transitions can range in the hundreds of kilohertz. Although the frequency stability of the laser is crucial for laser cooling of atoms, as we will see in Chapter 24, it is not very important for most mesoscopic optical tweezers applications.

In dealing with infrared lasers, a means of visualising the beam path is necessary. A laser viewing card can be used for this purpose. The card has a photosensitive area that glows with visible light in the spot illuminated by the laser beam, thus enabling the user to follow the beam path. Cards made with several materials with different sensitivity to different parts of the spectrum are available.

8.4.2 Lenses

Lenses are used in the optical train to manipulate the beam diameter and divergence. Factors that should be taken into account when selecting lenses include the following:

- **Material.** Optical tweezers typically use a laser with a wavelength in the visible to near-infrared portion of the spectrum. These lenses are typically made of N-BK7, a kind of glass which has a high transmission in this spectral range. Its refractive index is $n_d = 1.5167$ and its dispersion is $v_d = \frac{n_d - 1}{n_F - n_C} = 64$ (measured at the wavelength of the sodium D-line [Box 8.1]).[15]
- **Lens shape.** Cross-sections showing the shape of different common lens types are shown in Fig. 8.11. As most lenses readily available from commercial manufacturers have spherical surfaces, they introduce a small amount of spherical aberration, which can be minimised by turning the curved side towards the side where the beam is collimated,[16]

---

[15] Note, however, that although its Abbé V-number is high, indicating relatively low dispersion, at the wavelength $\lambda_0 = 1064$ nm, which is commonly used in optical tweezers, the refractive index is $n_{1064} = 1.5066$.
[16] Or towards the side where the beam has lower divergence, in case the beam is not collimated on either side.

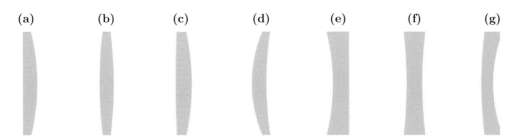

Figure 8.11 Lens shapes: (a) plano-convex; (b) bi-convex; (c) best form; (d) positive meniscus; (e) plano-concave; (f) biconcave; (g) negative meniscus. The lenses shown here have a focal length of (a)–(d) $f = +100$ mm (converging or positive lenses) or (e)–(g) $f = -100$ mm (diverging or negative lenses). The focal lengths are calculated assuming lenses made of N-BK7 glass and with a diameter of 25.4 mm (1 inch).

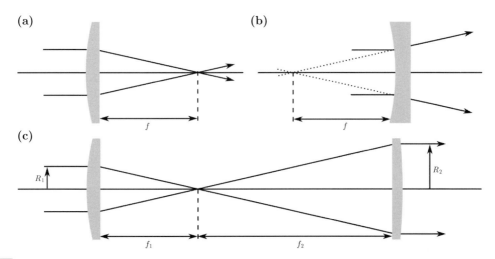

Figure 8.12 The action of lenses illustrated with ray diagrams. (a) A plano-convex (positive) lens focuses an incoming collimated beam. (b) A plano-concave (negative) lens makes an incoming collimated beam divergent. (c) A telescope expands a beam. Note that the curved side of each plano-convex lens is turned towards the side where the beam is collimated (parallel rays) to minimise spherical aberration.

as shown in Fig. 8.12. Table 8.10 describes the most common uses of the various kinds of lens.
- **Surface quality.** The surface quality of a lens is measured by a quantity known as scratch-dig. A scratch is a mark of the polished or coated surface. A dig is a small rough spot or pit on the polished or coated optical surface. Scratch-dig is reported as a pair of numbers, such as '60–40', where the first number indicates the maximum width of a scratch in micrometres and the second number is the maximum diameter for a dig in hundredths of a millimetre. A scratch-dig of 60-40 is regarded as low quality. Lenses for manipulating the laser beam for an optical tweezers should have a scratch-dig of 40-20 or better.
- **Anti-reflection coating.** A small fraction of the intensity incident on the air–glass interface at the surface of a lens will be reflected. This fraction, which can be calculated from

Table 8.10 Lens shapes	
Plano-convex	Best suited for infinite conjugate applications, e.g., focusing a collimated beam or collimating a diverging beam.
Biconvex	Most suitable for finite conjugate imaging where the ratio of the image and object distances is between 0.2 and 5.
Best form	Designed to minimise spherical aberration while still using spherical surfaces to form the lens. They provide the best possible performance from a spherical lens for collimating and focusing beams.
Positive meniscus	Most commonly paired with another positive lens to increase the numerical aperture of the system.
Plano-concave	Used to diverge a collimated beam from a virtual focus. Commonly used in Galilean beam expanders.
Biconcave	Used to increase the divergence of a converging beam.
Negative meniscus	Used in combination with another lens to decrease the numerical aperture of the system.

the Fresnel reflection coefficients [Eqs. (2.5) and (2.7)], for most glass types is about 4% (at near-normal incidence). To reduce this loss, lenses are often coated with an anti-reflection (AR) coating, which can reduce the reflection coefficient at each surface of the lens to less than 0.25%. AR coatings can be either specialised for a particular laser line or broadband to cover a portion of the spectrum (typically up to several hundred nanometres).

8.4.3 Mirrors

Mirrors are used to control the beam direction in the optical train. The surface of a mirror is made highly reflective by either a metallic or a multi-layer dielectric coating. Metallic mirrors are commonly coated with silver (reflection coefficient $R > 97.5\%$ in the visible to near-infrared), aluminium ($R > 90\%$ in the visible and ultraviolet) or gold ($R > 96\%$ in the near-infrared to mid-infrared). Dielectric mirrors are available with high-reflection (HR) coatings with reflectivities in excess of 99% both for particular laser lines and broadband. It should be noted, however, that the reflection coefficient is a function of polarisation and of angle of incidence, and performance may degrade for, e.g., angles of incidence in excess of 45°. As for lenses, the surface quality of mirrors is specified by scratch-dig, with very high quality scratch-dig 10-5 dielectric mirrors being readily available. Metallic mirrors tend to be of lower surface quality.

8.4.4 Filters

Neutral density (ND) filters are used to attenuate the intensity of an optical beam. ND filters may be either absorptive or reflective; i.e., they reduce the intensity either by absorbing or reflecting power. Absorptive ND filters are usually made of an absorbing glass, whereas reflective ND filters are constituted by a glass substrate with a metallic (reflective) coating.

The degree of attenuation is specified by the optical density (OD) of the filter, defined as

$$\text{OD} = -\log_{10}(T), \tag{8.2}$$

where T is the transmission coefficient. For example, an ND filter with OD $=0.3$ transmits approximately 50% of the incident intensity. The optical density will usually be specified at a particular test wavelength, e.g., that of the helium–neon laser (633 nm), and it is advisable to verify the value of the OD at the actual wavelength being used.

Wavelength-selective filters can be used to separate or to selectively stop light of a particular wavelength (or range of wavelengths). An *edge filter* transmits wavelengths above (below) a threshold wavelength and rejects wavelengths below (above) it. A *longpass filter* transmits wavelengths longer than the threshold (known in this case as the *cut-on wavelength*), whereas a *shortpass filter* transmits wavelengths shorter than the threshold (*cut-off wavelength*). A *bandpass filter* transmits wavelengths in a specified range (the *passband*) around a centre wavelength, whereas a *band-stop filter* performs the inverse operation and stops wavelengths in a particular range (*stop-band*). A *notch filter* stops transmission within a much narrower range of wavelengths, typically within tens of nanometres around the centre wavelength, with very high transmission ($> 90\%$) outside this range. Notch filters are useful in selectively blocking laser light while allowing other wavelengths to pass, e.g., in order to stop scattered laser light from reaching the imaging camera in an optical tweezers experiment. High-quality filters are made from a dielectric multi-layer stack, which operates as a Fabry-Perot etalon to transmit wavelengths only within a certain range, often with an additional broadband absorber to block out-of-band transmission. Such filters usually have a preferred orientation in the optical beam (indicated on the filter itself) such that the absorber follows the dielectric multilayer to minimise heating effects from absorption. Coloured glass filters are a lower-cost alternative to edge and bandpass filters.

Dichroic mirrors are used to combine or separate beams of different wavelengths as wavelength-selective filters do, but in doing so introduce minimum wavefront distortion into both the transmitted and reflected beams. These are, therefore, best used when the wavefront quality of both beams is important, e.g., in combining the optical tweezers trapping beam with a probe beam of a different wavelength where both beams must subsequently pass though the microscope objective to be focused on a diffraction-limited spot.

8.4.5 Polarisation control

The output beam from most laser systems is a TEM$_{00}$ (Gaussian) beam with linear polarisation. Control of the state of polarisation of the beam is important both for manipulating the beam, e.g., splitting a single beam into multiple beams, and for preparing it for use with polarisation-sensitive devices, e.g., a spatial light modulator. Control of polarisation is performed using *wave plates*. Wave plates are thin slices of a transparent birefringent material such as quartz, which is cut so that the components of the electric field vector in the directions of the two crystal axes (known as the ordinary and extra-ordinary directions) acquire different phase shifts. When housed in a rotatable mount, wave plates constitute a powerful toolbox for continuous control of the state of polarisation of a laser beam.

A *half-wave plate* (HWP) introduces a phase difference of π between the ordinarily and extra-ordinarily polarised components of the beam. For an input beam that is linearly polarised, the net effect is a rotation of the direction of linear polarisation by twice the angle that the input polarisation direction makes with the HWP extra-ordinary axis.

A *quarter-wave plate* (QWP) introduces a phase difference of $\pi/2$ between the ordinarily and extra-ordinarily polarised components of the beam. For an input beam that is linearly polarised, the output beam is elliptically polarised, with ellipticity and handedness that depend on the orientation of the QWP. In particular, an input polarisation that is linear at an angle of $45°$ to the QWP axes produces a circularly polarised output. For an input beam that is circularly polarised, the output beam is linearly polarised at an angle of $45°$ to the QWP axes.

A *polarising beam splitter* (PBS) is capable of separating a beam into two linearly polarised components. *Plate polarising beam splitters* have the beam splitter coating applied to the front face. A *PBS cube* is made from a pair of right-angle prisms cemented together with the coating applied to the hypoteneuse face of one. Both plate and cube PBS reflect light that is *s*-polarised with respect to the beam-splitting face and transmit light that is *p*-polarised.

8.5 Optical trapping and manipulation

At this point, everything is ready for prime time. We can therefore place our diluted sample of microparticles on the sample stage and proceed to trap our first particle. First of all, we need to place the sample on the stage and adjust the position of the objective and sample holder such that we can visualise the plane just above the glass–solution interface. The image recorded by the camera should show the sedimented particles and look similar to the one shown in Fig. 8.13a.[17] In order to get the image in focus, it might be necessary to adjust the lens in front of the camera. At this point, it should be sufficient to move the laser focus a few micrometres above the interface (by moving the sample stage down) and approach some of the particles. If a particle is free to move and not stuck to the glass, you will see the particle jump into the laser beam and become optically trapped, as shown in Fig. 8.13a. To double-check that you have actually optically trapped the particle, you can now try to move it around. You can first move the sample stage further down so that the optically trapped particle gets moved vertically above the glass–solution interface; as shown in Fig. 8.13b, you can see that the non-optically trapped particles on the sample cell bottom go out of focus, while the optically trapped particle remains in focus. Then you can also move the sample stage horizontally; again, as shown in Fig. 8.13c, the image of the optically trapped particle remains at the same spot within the image captured by the camera, while the background particles sedimented on the coverslip appear to be displaced.

---

[17] In most cases the particles have higher density than the solution. If this is not the case, the particles are to be found near the solution–glass interface at the top of the sample cell.

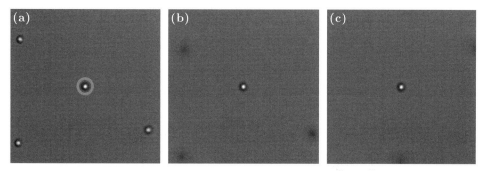

Figure 8.13 Optically trapped particle: (a) Optically trapped particle (circled particle). As the sample stage is moved (b) vertically and then (c) horizontally, the optically trapped particle appears to remain at the same spot within the image captured by the camera, whereas the background particles sedimented on the coverslip appear defocused when the sample is moved vertically in (b) and displaced when it is moved horizontally in (c).

> **Troubleshooting: I can see the particles in focus, but I cannot trap them.** If the problem persists after you have checked that your laser beam is effectively switched on and unobstructed, it is most likely due to the fact that the camera imaging plane and the laser focusing plane are significantly different. This can be due to chromatic aberrations (especially with simpler objectives) or to a misalignment of the camera lens. To solve this problem, check that you can effectively focus your laser beam on the coverslip while observing the sedimented particles at the same time (in order to ensure that the two focal planes are the same). This alignment can be achieved by focusing the laser on the coverslip and then moving the camera lens in order to get the sedimented particles in focus. This might not be possible if the difference between the two planes is too large.

> **Troubleshooting: My particle is pushed away from the trap.** Probably there is too much scattering force. Try to reduce the laser power and/or to increase the overfilling. You can also use particles with lower refractive index contrast or of larger dimensions (to increase the effective gravity). It is also possible that you are trying to trap too far away from the coverslip, where spherical aberrations have degraded the focal spot; in this case, try to trap closer to the coverslip.

8.5.1 Steerable optical tweezers

The optical tweezers we have built so far has only one fixed trap. As we have seen in Figs. 8.13b and 8.13c, it is possible to move an object relative to its surroundings by moving the sample. However, sometimes it is more advantageous to be able to move the trap over the field of view, i.e., to have steerable optical tweezers. To create a movable trap that also is stable, i.e., with the same amount of trapping power, regardless of the movement of the beam, whenever moved the laser beam has to (1) pivot around the entrance aperture

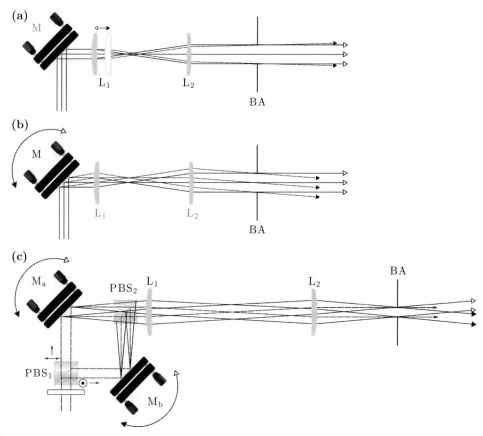

Figure 8.14 Trap steering. The position of an optical trap can be changed within the sample (a) along the vertical z-direction by displacing one of the lenses in a telescopic arrangement and (b) in the horizontal xy-plane by conjugating the plane of a steerable mirror to the objective back aperture (BA). (c) It is also possible to construct a dual optical tweezers where a beam is split into two and recombined using two polarising beam splitters (PBS_1 and PBS_2) in order to create two independently steerable optical traps; the choice of polarising beam splitters is to avoid losing optical power.

of the microscope objective and (2) retain the same degree of overfilling of the microscope entrance aperture.

Beam steering along the vertical z-direction can easily be achieved by moving one of the lenses in the telescope in order to change the curvature of the trapping beam, as shown in Fig. 8.14a. Beam steering in the horizontal xy-plane can be achieved by tilting the beam at a plane conjugated to the back aperture of the objective, as shown in Fig. 8.14b. In this way, the beam entering the objective is slightly tilted, resulting in a focus in a slightly offset position. Instead of a gimbal-mounted mirror, it is possible to use a galvometre-mounted mirror or an acousto-optic deflector to move the trap in a very controllable way. In this way it is also possible to generate multiple optical traps by time-sharing a single optical beam.

An interesting approach to a dual-beam steerable optical tweezers was presented by Fällman and Axner (1997). The schematic is shown in Fig. 8.14c. This set-up makes use of a couple of polarising beam splitters to split a beam into two orthogonally polarised beams and then recombine them before the objective entrance. Beam steering is achieved by using two gimbal mirrors.

More advanced techniques to control the position and characteristics of optical tweezers can be implemented using wavefront engineering and spatial light modulators, as we will explain in detail in Chapter 11.

8.6 Alternative set-ups

The optical tweezers set-up we presented in this chapter is but one of infinitely many possible implementations. Several alternative approaches have been presented in the literature, based both on homemade and commercial microscopes. In particular, there are the set-ups proposed by Smith et al. (1999), Bechhoefer and Wilson (2002), Mellish and Wilson (2002), Appleyard et al. (2007), Lee et al. (2007) and Mathew et al. (2009).

Problems

8.1 Build an optical tweezers with a near-infrared laser source. Trap several different particles such as latex beads of different sizes, yeast cells, and gold nanoparticles. Find the minimum trapping power needed to have a stable trap. What does this tells you about the trapping efficiency?

8.2 How can you build an optical tweezers set-up with multiple traps? How many particles can you trap? Does the number of particles you can trap depend on their size? Why?

8.3 Design a steerable optical tweezers using a gimbal mirror or a pair of galvo-mirrors. How can you use such a set-up to move different particles in controlled patterns such as a line, a circle or a Lissajous figure?

8.4 Design an optical tweezers set-up capable of using non-Gaussian laser beams, such as Laguerre–Gaussian beams and cylindrical vector beams. What kinds of experiments can be performed with such a set-up that cannot be performed with a standard optical tweezers?

References

Appleyard, D. C., Vandermeulen, K. Y., Lee, H., and Lang, M. J. 2007. Optical trapping for undergraduates. *Am. J. Phys.*, **75**, 5–14.

Bechhoefer, J., and Wilson, S. 2002. Faster, cheaper, safer optical tweezers for the undergraduate laboratory. *Am. J. Phys.*, **70**, 393–400.

Fällman, E., and Axner, O. 1997. Design for fully steerable dual-trap optical tweezers. *Appl. Opt.*, **36**, 2107–13.

Lee, W. M., Reece, P. J., Marchington, R. F., Metzger, N. K., and Dholakia, K. 2007. Construction and calibration of an optical trap on a fluorescence optical microscope. *Nature Prot.*, **2**, 3225–38.

Liang, H., Vu, K. T., Trang, T. C., et al. 1996. Wavelength dependence of cell cloning efficiency after optical trapping. *Biophys. J.*, **70**, 1529–33.

Mathew, M., Santos, S. I. C. O., Zalvidea, D., and Loza-Alvarez, P. 2009. Multimodal optical workstation for simultaneous linear, nonlinear microscopy and nanomanipulation: Upgrading a commercial confocal inverted microscope. *Rev. Sci. Instrumen.*, **80**, 073701.

Mellish, A. S., and Wilson, A. C. 2002. A simple laser cooling and trapping apparatus for undergraduate laboratories. *Am. J. Phys.*, **70**, 965–71.

Neuman, K. C., Chadd, E. H., Liou, G. F., Bergman, K., and Block, S. M. 1999. Characterization of photodamage to *Escherichia coli* in optical traps. *Biophys. J.*, **77**, 2856–63.

Smith, S. P., Bhalotra, S. R., Brody, A. L., et al. 1999. Inexpensive optical tweezers for undergraduate laboratories. *Am. J. Phys.*, **67**, 26–35.

9 Data acquisition and optical tweezers calibration*

The optical tweezers we built in the previous chapter can trap and manipulate small objects. However, to perform more complex experiments, where, e.g., optical tweezers are used as sensitive force transducers to exert and measure forces ranging from tens of femtonewtons to hundreds of piconewtons, it is necessary to track the motion of the trapped objects and to measure the corresponding optical forces. In this chapter, we will discuss how to do this using a trapped spherical particle. The position of *several* Brownian particles can be tracked by *digital video microscopy* with nanometre and millisecond resolution. It is also possible to employ *interferometric techniques* for more precise (down to a fraction of a nanometre) and faster (up to several tens of megahertz) measurements on a *single* Brownian particle. Knowing the trajectory of an optically trapped particle, it is possible to calibrate the optical trap, i.e., to obtain the value of the *trap stiffness*, using one of various alternative calibration procedures such as *mean squared displacement analysis* [Fig. 9.1a], *autocorrelation function analysis* [Fig. 9.1b] and *power spectrum analysis* [Fig. 9.1c], which we will also explore in this chapter.

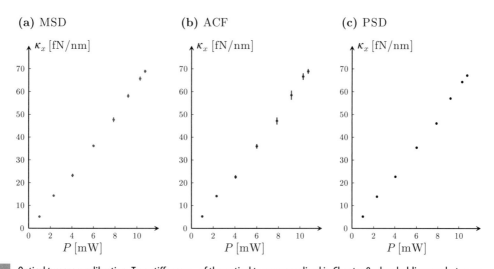

Figure 9.1 Optical tweezers calibration. Trap stiffness κ_x of the optical tweezers realised in Chapter 8 when holding a polystyrene spherical particle with radius $a = 1.03\,\mu\text{m}$ as a function of the trapping laser power P at the sample, obtained with (a) mean squared displacement analysis, (b) autocorrelation function analysis and (c) power spectrum analysis. The resulting κ_x is linear in P and the three measurements are in very good agreement.

* This chapter was written together with Giuseppe Pesce.

9.1 Digital video microscopy

As a typical optical tweezers set-up comes already equipped with a digital camera, the most straightforward means to measure the motion of a Brownian particle is to record a video of its position and then to track the position of the particle frame by frame. This technique, known as *digital video microscopy*, has found widespread application in several fields and, in particular, in colloidal studies. It is especially well suited to study systems where multiple particles are present, but it is relatively slow, being limited by the camera frame rate, which typically goes up to a few thousands frames per second. Here, we adapt the methods proposed by Crocker and Grier (1996) and by Sbalzarini and Koumoutsakos (2005).

Each frame of a digital video is a digital image. A digital image is a two-dimensional matrix of pixels associated with a colour value. In the case of *greyscale images*, each pixel is associated with a single number representing the light *intensity* between 0 (black) and $2^N - 1$ (white), where N is the number of bits of the encoding. The value of N depends on the sensitivity of the sensor and the precision of the analogue-to-digital converter (ADC). Often $N = 8$, corresponding to only 256 shades of grey, which may not be sufficient for accurate digital video microscopy; thus, it is better to use a camera with $N = 10$, 12 or 16. In the case of *colour images*, more than one value is stored for each pixel. For example, in the often-employed *RGB colour model*, three values are stored, corresponding to the intensity of the red, green and blue light at the pixel. It is possible to covert a colour image to greyscale, e.g., by selecting only one of its colour channels or by calculating its total intensity.

In the following, we will consider a greyscale image $I_t(m_x, m_y)$, where each pixel can assume an integer value between 0 and 255, the subscript t represents the acquisition time, the integer coordinate $m_x = 1, \ldots, M_x$ is the pixel row index and $m_y = 1, \ldots, M_y$ is the pixel column index. Fig. 9.2a shows a bright-field greyscale image of a ternary mixture of silica spheres (with radii $a = 1.55$, 2.31 and 3.16 µm) freely diffusing at the bottom of a sample cell. The particles appear as bright circles whose size depends on the radius.

The simplest method of tracking the positions of the particles is by *thresholding*. Because the pixel value at the particle is larger than the one of the background, it is possible to fix a threshold N_{th} and convert the greyscale image into a black and white image such that pixels whose value is lower than N_{th} are set to black (0) and pixels whose value is higher than N_{th} are set to white (255), i.e.,

$$I_t^{th}(m_x, m_y) = \begin{cases} 0 & I_t(m_x, m_y) < N_{th} \\ 255 & I_t(m_x, m_y) \geq N_{th}. \end{cases} \quad (9.1)$$

Following the thresholding operation, it is possible to apply some morphological filters, such as *dilation* and *erosion filters*, in order to eliminate some common causes of noise such as salt-and-pepper noise, i.e., sparsely occurring white and black pixels. The resulting black and white image is shown in Fig. 9.2b. Each white region potentially corresponds to a particle. The size of the region can be used to assess the size of the particles and also to identify false detections, e.g., by excluding regions whose size lies outside a given interval.

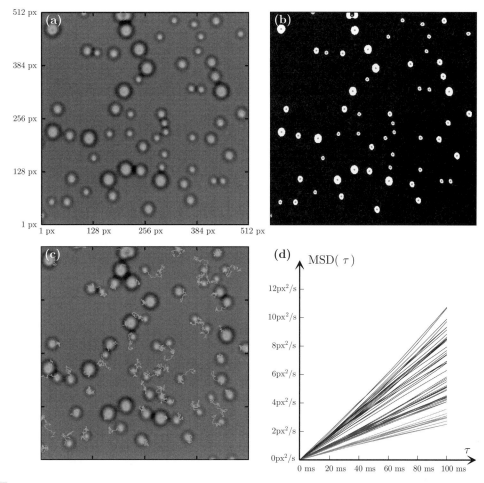

Figure 9.2 Digital video microscopy. (a) Original and (b) thresholded digital image of a ternary mixture of silica spheres (radii $a = 1.55, 2.31$ and 3.16 μm) freely diffusing on the bottom of a sample cell. The circles in (a) and the symbols in (b) represent the measured position and size of the particles. (c) Particle trajectories relative to an acquisition time of 10 s and (d) corresponding mean squared displacements (MSDs). Note in (d) that the MSDs loosely cluster around three values, corresponding to the three particle sizes.

The positions of the particles can then be calculated as the centroids of these regions (dots in Fig. 9.2b). Thanks to the averaging in the centroid calculation, this technique permits one to achieve sub-pixel resolution, typically down to about a tenth of the pixel size (about 10 nm) in the x- and y-directions. The area of the regions can be used to estimate the radii of the corresponding particles (circles in Fig. 9.2a).

A more advanced particle detection technique, known as *feature point detection*, makes use of the fact that the particle intensity profiles on the image are in first approximation Gaussian, as can be seen from the image of the optically trapped particle shown in Fig. 9.3a. This technique consists of the following steps:

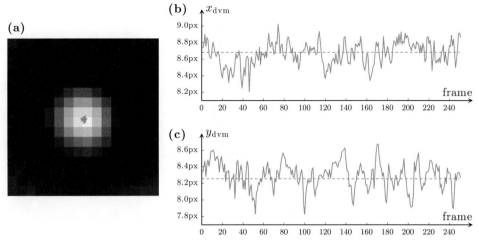

Figure 9.3 Optical trapped particle tracked by digital video microscopy. (a) Image (16 pixels × 16 pixels) of an optically trapped polystyrene sphere ($a = 1.03$ μm) and corresponding detected trajectory (solid line), which is completely contained within a single pixel. Detected position along the (b) x- and (c) y-directions (solid lines) and average particle position (dashed lines). The frame rate is 2500 frames/s.

1. Normalisation of the pixel values between 0 and 1 as

$$I_t^n(m_x, m_y) = \frac{I_t(m_x, m_y) - I_{\min}}{I_{\max} - I_{\min}}, \tag{9.2}$$

where $I_{\min}$ and $I_{\max}$ are the global (i.e., across all the frames of the movie rather than within each frame individually) minimum and maximum of intensity.

2. Image enhancement by convolution with a Gaussian kernel, i.e.,

$$I_t^c(m_x, m_y) = \sum_{\Delta m_x = -w}^{w} \sum_{\Delta m_y = -w}^{w} I_t^n(m_x + \Delta m_x, m_y + \Delta m_y) \, K(\Delta m_x, \Delta m_y), \tag{9.3}$$

where

$$K(\Delta m_x, \Delta m_y) = \left[\sum_{\Delta m_x = -w}^{w} e^{-\frac{\Delta m_x^2}{4w^2}} \right]^{-2} e^{-\frac{\Delta m_x^2 + \Delta m_y^2}{4w^2}}, \tag{9.4}$$

with w a positive parameter representing the size of the filter in pixels. By setting w larger than a single point's apparent radius but smaller than the smallest inter-point separation, this filter can be used to reduce long-wavelength modulations of the background intensity due to non-uniform sensitivity among the camera pixels or uneven illumination.[1]

3. A preliminary estimation of the particles' positions is then obtained by finding local intensity maxima (possibly above a certain threshold) in $I_t^c(m_x, m_y)$. A pixel $[\tilde{x}_P, \tilde{y}_P]$ is

---

[1] By setting w much smaller than a single point's apparent radius (typically $w = 1$), this filter can also be used to reduce the discretisation noise from the digital camera.

taken as the approximate location of a particle if no other pixel within a distance of w is brighter.

4. Under the assumption that $[\tilde{x}_P, \tilde{y}_P]$ is near the true geometric centre $[x_P, y_P]$ of the particle, an approximation to the offset is given by the distance to the brightness-weighted centroid in $I_t^c(m_x, m_y)$,

$$\begin{cases} \Delta \tilde{x}_P = \dfrac{1}{m_0(P)} \displaystyle\sum_{\Delta m_x^2 + \Delta m_y^2 < w^2} \Delta m_x \, I_t^c(m_x + \Delta m_x, m_y + \Delta m_y), \\ \Delta \tilde{y}_P = \dfrac{1}{m_0(P)} \displaystyle\sum_{\Delta m_x^2 + \Delta m_y^2 < w^2} \Delta m_y \, I_t^c(m_x + \Delta m_x, m_y + \Delta m_y), \end{cases} \quad (9.5)$$

where

$$m_0(P) = \sum_{\Delta m_x^2 + \Delta m_y^2 < w^2} I_t^c(m_x + \Delta m_x, m_y + \Delta m_y). \quad (9.6)$$

The location estimate is, then, refined as $[\tilde{x}_P + \Delta \tilde{x}_P, \tilde{y}_P + \Delta \tilde{y}_P]$. If either $|\Delta \tilde{x}_P|$ or $|\Delta \tilde{y}_P|$ are larger than 0.5 pixel, the candidate location is accordingly moved by one pixel and the refinement is re-calculated.

We have applied this algorithm to the detection of the optically trapped particle shown in Fig. 9.3a. The corresponding detected positions along the x and y direction are shown in Figs. 9.3b and 9.3c, respectively; the sub-pixel precision of this detection technique can be appreciated, as the maximum excursion of the particle from its equilibrium position (dashed lines) is less than a pixel in both directions.[2]

To achieve the best results, independent of the detection technique employed, often some preliminary steps are required to prepare the images, such as removal of salt-and-pepper noise by median filtering, subtraction of a (fixed) background image, or normalisation of the image intensity. It is also important to optimise the illumination intensity and contrast, e.g., using the techniques described in Subsection 8.2.2. Moreover, using a coherent source of illumination (*holographic video microscopy*), it is possible to get better contrast, especially along the z-direction.

As particle detection is applied to each frame, it delivers a series of sets of positions, each corresponding to the particles detected in the frame acquired at time t. At this point, it is necessary to link positions corresponding to the same physical particle in subsequent frames into trajectories. The basic idea of the linking algorithm is that a particle at time t is identified with a particle at time $t + \Delta t$, where Δt is the time difference between the two frames, if the two particles are distant by less than a certain value. In the case of freely diffusing Brownian particles, this value can be set to a multiple of $\sqrt{4D\Delta t}$ in order to account for the Brownian diffusion of the particle frame to frame. This algorithm can be extended so that each linking step may consider several frames to account for particle occlusion. By performing this linking, it is finally possible to obtain the particle trajectories

---

[2] This technique achieved not only sub-pixel *precision*, but also sub-pixel *accuracy*. This is demonstrated in Figs. 9.8a and 9.8b, where the particle trajectories detected by digital video microscopy are compared with the ones detected by interferometry.

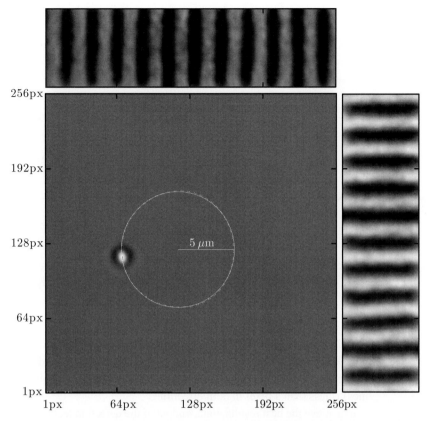

Figure 9.4 Microscope calibration. The calibration of the system can be achieved either by imaging a regular structure or by tracking a particle stuck to the bottom of the sample (dashed line) as it is controllably moved by the stage. In our case we used a homemade grating with a period of 2 μm and we moved the particle along a 5 μm circle (solid line). The resulting calibration factors are approximatively 90 nm/px in both directions with both techniques.

shown by the solid lines in Fig. 9.2c, which can be then used for various kinds of statistical analysis, e.g., the calculation of the MSDs shown in Fig. 9.2d.

Finally, it is important to be able to convert the particle position measurements expressed in pixels into actual physical units of length. This requires calibration of the microscope. The easiest way of doing this is by imaging a regular object, e.g., a microfabricated grating. Such gratings can be microfabricated by standard lithography techniques, but it is also possible to acquire them commercially. An alternative approach is to track a particle stuck on the bottom of the sample as this is controllably moved by the stage. These two approaches are illustrated in Fig. 9.4.

Exercise 9.1.1 Simulate the image of a particle as acquired by a digital camera by pixelating a Gaussian intensity distribution. Use the thresholding and feature point detection algorithms to recover the centre of the Gaussian distribution from its pixelated version. How much is the error with the thresholding algorithm? And with the point feature detection algorithm? [Hint: You can adapt the code `dvmgauss` from the book website.]

Exercise 9.1.2 Detect the particle positions in the video recording of the diffusion of three different kinds of particles at the bottom of a sample given on the book website using the thresholding and feature point detection algorithms. How do the algorithm parameters influence the detection? What differences do you notice? [Hint: You can use the code `dvmtest` from the book website.]

Exercise 9.1.3 Using the particle positions obtained in the previous exercise, obtain the particle trajectories by running the linking algorithm and calculate the MSDs of the particles. [Hint: You can use the code `dvmtest` from the book website.]

Exercise 9.1.4 How do the thresholding and point feature detection algorithms perform in detecting the image of a particle illuminated by a plane wave (coherent illumination). Study the case when this light is collected by a condenser lens and projected onto a screen and show that it can be used to detect not only the lateral position, but also the axial position of the particle (holographic video microscopy). [Hint: Use Mie scattering theory.]

9.1.1 Digital cameras

We now present some of the basic properties and settings of digital cameras, whose understanding is crucial to achieve the best results for digital video microscopy:

- **CCD vs. CMOS sensors.** Almost all digital cameras use either a *charge-coupled device* (CCD) or a *complementary metal oxide semiconductor* (CMOS) sensor to capture images. CCD sensors are silicon chips containing arrays of photosensitive sites. Charge packets proportional to the impinging light intensity are generated at the photosensitive sites and then moved around the chip before being converted to a voltage by a capacitor.[3] This approach limits the readout speed and leads to the phenomenon known as blooming, i.e., the smearing out of bright spots due to charge spills to neighbouring pixels, but also improves the sensitivity and pixel-to-pixel uniformity of the analogue-to-digital conversion (as each charge packet is converted by the same ADC). In CMOS sensors, the charge from the photosensitive pixel is converted to a voltage at the pixel site. Thus, each site is essentially a photodiode. This explains the speed of CMOS sensors, but also their lower sensitivity as well as high fixed-pattern noise due to fabrication inconsistencies in the multiple charge-to-voltage conversion circuits. Thus, CMOS sensors are faster, but have a lower signal-to-noise ratio and lower overall image quality than CCD sensors.
- **Interlaced vs. progressive scan.** Some CCD cameras use interlaced scanning across the sensor, whereby the odd rows and the even rows are read out alternately and then integrated to produce the full frame. In digital video microscopy, if the particles move

---

[3] In fact, the term 'charge-coupled device' refers to the shift register used to move the charge packets on the chip from the photosensitive sites to the analogue-to-digital converter (ADC), akin to the notion of a bucket brigade.

significantly between the capture of the odd and even rows, this lead to ghosting or blurring effects, which can be mitigated by deinterlacing.

- **Area of interest.** The area of interest allows a subset of the camera sensor array to be read out. This is useful for reducing the field of view (FOV) in order to decrease the amount of data transferred, thereby increasing the possible frame rate.
- **Binning/subsampling.** The value of adjacent pixels can be averaged together to form larger effective pixels (*binning*) or only a subset of pixels can be read out (*subsampling*) when one needs the entire FOV, but not the camera's full resolution. Akin to reducing the area of interest, binning and subsampling increase the acquisition speed by decreasing the amount of data transferred.
- **Gain.** The gain controls the amplification of the signal from the camera sensor, including any associated background noise. Gain can be before or after the ADC. Gain before the ADC can be useful for taking full advantage of the bit-depth of the camera in low light conditions, although it is almost always the case that careful illumination of the sample is more desirable. In general, gain should be used only after optimising the exposure setting and then only after setting the exposure time to its maximum for a given frame rate. Gain after the ADC is not true gain, but rather digital gain and, in fact, may lose some information in the process.
- **Gamma.** The gamma controls the greyscale reproduced on the image. An image gamma of unity indicates that the camera sensor is precisely reproducing the object greyscale (linear response). A gamma setting much greater than unity results in a silhouetted image in black and white.

9.2 Interferometry

An optically trapped particle scatters the trapping beam so that the light field in the forward direction is the superposition of the incoming and scattered light. This basic observation has been harnessed in the development of interferometric position detection techniques. Because interferometric techniques make use of photodetectors, which are inherently less noisy and faster than digital cameras, it is possible to measure the three-dimensional position of a particle located near the focal region with high precision (down to a fraction of a nanometre) and at very high speed (up to several megahertz). The trapping and detection operations can be made independent by illuminating the particle with an auxiliary beam weak enough not to generate significant optical forces. This is particularly useful in experiments where the position or intensity of the trapping beam needs to be changed during the experiment.

A typical schematic of a forward interferometric position detection scheme is shown in Fig. 9.5a. The trapping beam is focused by the objective, whereas the condenser collects the interference pattern arising from the interference between the incoming and scattered fields and a photodetector located in the condenser back-focal plane records the resulting signals. The concrete implementation we realised in our set-up is shown in Fig. 9.5b.

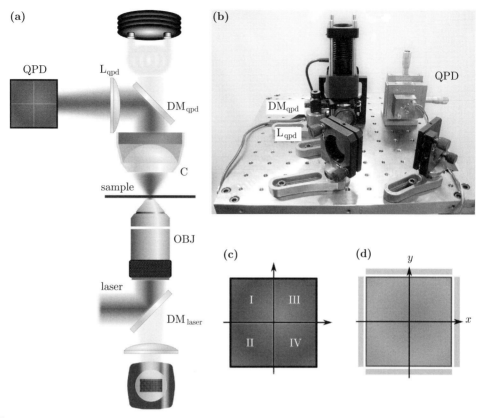

Figure 9.5 Interferometric position detection set-up. (a) Schematic of a forward interferometric position detection set-up and (b) corresponding implementation in our optical tweezers using an InGaAs quadrant photodetector (QPD) with a small area to increase signal sensitivity at $\lambda_0 = 1064$ nm and to decrease the detector noise level. (c) QPDs work by measuring the intensity difference between the light impinging on the left (top) and right (bottom) sides of the detection plane [Eq. (9.11)]. (d) Position sensing detectors (PSDs) measure the position of the centroid of the impinging light intensity distribution [Eq. (9.39)].

Suppose that the detection beam is focused by an aplanatic objective generating a convergent spherical wave, as shown in Fig. 5.8. We will assume the objective to have numerical aperture $\mathrm{NA_o}$, focal length f_o and aperture radius $R_o = f_o \, \mathrm{NA_o}/n_\mathrm{m}$, where n_m is the medium refractive index. The wave propagates to a diffraction-limited axial image and generates a high-intensity spot at the focal point **O**. Let **C** be a vector which describes the position of the particle relative to the focal point. A condenser lens collects the incident light and forward scattering onto a forward position detector, which is located at a plane conjugated to the back focal plane of the condenser lens. We will assume the condenser to be an aplanatic lens with numerical aperture $\mathrm{NA_c}$, focal length f_c and aperture radius $R_c = f_c \, \mathrm{NA_c}/n_\mathrm{m}$. In order to quantify the signals associated with the particle position, we need to know the field at the back focal plane. We will assume x_d and y_d to be the coordinates on the back focal plane and, since the condenser is aplanatic, the coordinates on the entrance pupil of the condenser are $x_c = x_d$, $y_c = y_d$, $z_c = \sqrt{f_c^2 - x_d^2 - y_d^2}$. The scattered field on the back focal plane is

$$\mathbf{E}_{s,\text{BFP}}(x_d, y_d) = \sqrt{\frac{n_m}{n_d}} \frac{1}{\sqrt{\cos\theta}} \{[\mathbf{E}_s(x_c, y_c, z_c) \cdot \hat{\boldsymbol{\theta}}]\hat{\boldsymbol{\rho}} + [\mathbf{E}_s(x_c, y_c, z_c) \cdot \hat{\boldsymbol{\varphi}}]\hat{\boldsymbol{\varphi}}\}, \qquad (9.7)$$

where $\mathbf{E}_s(x_c, y_c, z_c)$ is the scattered field on the entrance pupil of the condenser, $\hat{\boldsymbol{\theta}}$, $\hat{\boldsymbol{\varphi}}$ and $\hat{\boldsymbol{\rho}}$ are the radial, azimuthal and polar unit vectors defined in Section 4.5, the prefactor $\sqrt{n_m/(n_d \cos\theta)}$ is needed to conserve the electromagnetic wave energy and is analogous to the one for the focusing of a beam [Eq. (4.40)], and n_d is the refractive index in the back focal plane of the condenser (typically, $n_d = 1$).[4] The incoming field on the back focal plane is

$$\mathbf{E}_{i,\text{BFP}}(x_d, y_d) = M^{-1} e^{i\pi} e^{ik_0 n_m (f_c - f_o)} \mathbf{E}_i(-x_i, -y_i), \qquad (9.8)$$

where $M = f_c/f_o$ is the magnification factor, $x_i = M^{-1} x_d$ and $y_i = M^{-1} y_d$ are the coordinates on the entrance pupil of the objective, the phase factor $e^{i\pi}$ is the Gouy phase shift [Fig. 4.6c] and the phase factor $e^{ik_0 n_m (f_c - f_o)}$ takes into account the phase accumulated during propagation. Therefore, the total field at the back focal plane of the condenser is given by

$$\mathbf{E}_{\text{FS}}(x_d, y_d) = \mathbf{E}_{s,\text{BFP}}(x_d, y_d) + \mathbf{E}_{i,\text{BFP}}(x_d, y_d), \qquad (9.9)$$

and the intensity of the light impinging on a photodetector placed in a plane conjugated to the back focal plane is given by

$$I_{\text{FS}}(x_d, y_d) = \begin{cases} \dfrac{\varepsilon_0 c}{2} |\mathbf{E}_{\text{FS}}(x_d, y_d)|^2 & \sqrt{x_d^2 + y_d^2} \leq R_c \\ 0 & \sqrt{x_d^2 + y_d^2} > R_c. \end{cases} \qquad (9.10)$$

The intensity on the back focal plane for particles of various radii, i.e., $a = 400$, 600 and $1000\,\text{nm}$, is presented in Fig. 9.6a as the particle is displaced along the transverse x-direction. For all sizes of particles we consider, the forward-scattering field essentially has a single-lobe angular intensity distribution, which moves as the particle moves along the x-direction. Thus, by tracking the movement of this intensity distribution, it is possible to measure the particle position in the transverse xy-plane. We remark that the two crucial parameters of any position detection system are (1) the *displacement sensitivity*, i.e., the signal as a function of the particle displacement, typically expressed in volts per metre; and (2) the *linear response range* of the position detection system. Both parameters depend on the intensity distribution that reaches the detector.

---

[4] Some numerically convenient simplifications can be made when $f_c \gg |\mathbf{C}|$ and $f_c \gg \lambda_0$, as is usually the case. In particular, in the calculation of the scattered field in the T-matrix method, it can be convenient to translate the incoming field instead of the particle and then, in order to take into account the difference in the propagation distances, multiply the scattered field by the phase factor $\exp(-ik_0 n_m \mathbf{C} \cdot \hat{\mathbf{r}})$, where $\hat{\mathbf{r}}$ is the unit vector associated with $\mathbf{r} = [x_c, y_c, z_c]$.

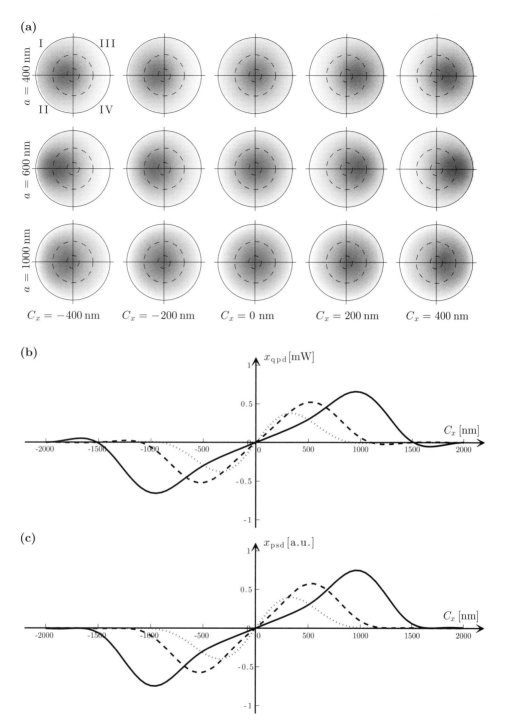

Figure 9.6 Transverse forward scattering and transverse position detection. (a) Images of the field at the back focal plane of the condenser for various particle radii a and particle transverse displacement $\mathbf{C} = [C_x, 0, 0]$. The circles delimit the areas of the field that is captured by the condenser lens with $\mathrm{NA}_o = 0.20, 0.65$ and 1.20 (solid line) from the centre outwards. Signal from a (b) QPD and (c) PSD as a function of C_x for the case of $a = 400$ nm (dotted line), $a = 600$ nm (dashed line) and $a = 1000$ nm (solid line) for an incident beam with power 1 mW and detection $\mathrm{NA}_c = 1.20$.

Two types of photodetectors are typically used as position sensors. The *quadrant photodetector* (QPD) works by measuring the intensity difference between the the left and right (and between the top and bottom) sides of the detection plane [Fig. 9.5c]. The response of a QPD is given by

$$\begin{cases} x_{\text{qpd}} = \iint_{x_{\text{d}}>0} I_{\text{FS}}(x_{\text{d}}, y_{\text{d}}) \, dx_{\text{d}} \, dy_{\text{d}} - \iint_{x_{\text{d}}<0} I_{\text{FS}}(x_{\text{d}}, y_{\text{d}}) \, dx_{\text{d}} \, dy_{\text{d}}, \\ y_{\text{qpd}} = \iint_{y_{\text{d}}>0} I_{\text{FS}}(x_{\text{d}}, y_{\text{d}}) \, dx_{\text{d}} \, dy_{\text{d}} - \iint_{y_{\text{d}}<0} I_{\text{FS}}(x_{\text{d}}, y_{\text{d}}) \, dx_{\text{d}} \, dy_{\text{d}}, \\ z_{\text{qpd}} = \iint I_{\text{FS}}(x_{\text{d}}, y_{\text{d}}) \, dx_{\text{d}} \, dy_{\text{d}}. \end{cases} \quad (9.11)$$

Fig. 9.6b shows the response of a QPD to the lateral displacement of the particle. It can be seen that approximately up to a particle radius the detector signal is proportional to the particle displacement. Furthermore, the signal absolute value grows together with the particle size and the corresponding scattering. The *position sensing detector* (PSD) measures the position of the centroid of the collected intensity distribution [Fig. 9.5d], giving a more adequate response for non-Gaussian profiles. The response of a PSD is given by the first moments of the intensity distribution, i.e.,

$$\begin{cases} x_{\text{psd}} = \iint x_{\text{d}} \, I_{\text{FS}}(x_{\text{d}}, y_{\text{d}}) \, dx_{\text{d}} \, dy_{\text{d}} \\ y_{\text{psd}} = \iint y_{\text{d}} \, I_{\text{FS}}(x_{\text{d}}, y_{\text{d}}) \, dx_{\text{d}} \, dy_{\text{d}} \\ z_{\text{psd}} = \iint I_{\text{FS}}(x_{\text{d}}, y_{\text{d}}) \, dx_{\text{d}} \, dy_{\text{d}}. \end{cases} \quad (9.12)$$

Fig. 9.6c shows the response of a PSD to the lateral displacement of the particle. It can be seen that this response is very similar to the one of a QPD. Both QPD and PSD fare well when assessed for sensitivity and linear range along the transverse direction. This is generally true for the forward-scattering detection scheme, but usually it is not true when non-Gaussian intensity profiles are considered, e.g., in the case of backward-scattering position detection.

Fig. 9.7a shows the light intensity on the condenser back focal plane as the particles are displaced along the longitudinal z-direction. In this case, the size of the spot changes as a consequence of the change of relative phase between incoming and scattered waves because of the Gouy phase shift inherent in focused beams. The response of a QPD photodetector (z_{qpd}) is shown in Fig. 9.7b. The grey lines show the total intensity collected by a photodetector with $\text{NA}_c = 1.20$. It can be seen that under almost all conditions, the total amount of collected light is close to the total power of the incoming beam, so that the QPD signal is fairly insensitive to the particle position. Furthermore, the signal does not appear to be very linear around $z_{\text{d}} = 0$. To overcome these difficulties, it is possible to reduce the numerical aperture of the detection system. For example, for $\text{NA}_c = 0.65$, we

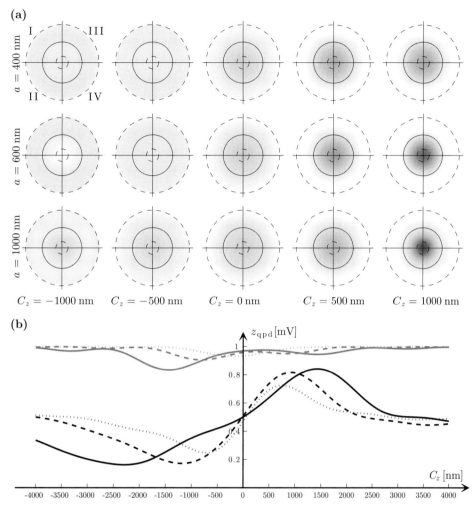

Figure 9.7 Longitudinal forward scattering and longitudinal position detection. (a) Images of the field at the back focal plane of the condenser for various particle radii a and particle longitudinal displacement $\mathbf{C} = [0, 0, C_z]$. The circles delimit the areas of the field that is captured by the condenser lens with $\mathrm{NA}_o = 0.20, 0.65$ (solid line) and 1.20 from the centre outwards. (b) Signal from a QPD as a function of C_z for the case of $a = 400$ nm (dotted line), $a = 600$ nm (dashed line) and $a = 1000$ nm (solid line) for an incident beam with power 1 mW and detection $\mathrm{NA}_c = 0.65$ (black lines) and 1.20 (grey lines).

obtain the black lines shown in Fig. 9.7b, which present good sensitivity and linear range. The numerical aperture of the detection system can be changed by changing the condenser lens, but also by placing an opportunely sized iris in front of the photodetector.

In Fig. 9.8, we present the position detection results obtained with the set-up we built in Chapter 8 and the detection apparatus shown in Fig. 9.5b. We use a sensor based on an InGaAs QPD with a small area to increase signal sensitivity at $\lambda_0 = 1064$ nm and to decrease the detector noise level. The two transverse signals and the longitudinal signal are

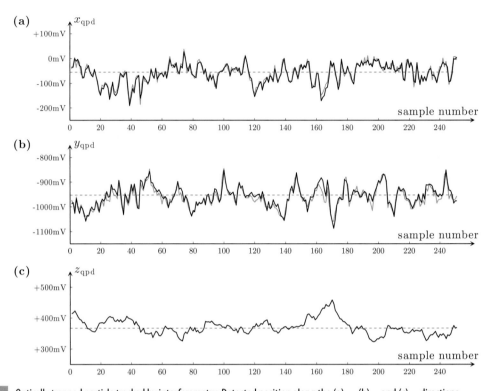

Figure 9.8 Optically trapped particle tracked by interferometry. Detected position along the (a) x, (b) y and (c) z-directions (solid lines) and average particle position (dashed lines) of an optically trapped polystyrene sphere ($a = 1.03$ μm) using forward interferometric detection with QPD. The fact that the average is not null is mainly due to the amplification stages in the QPD acquisition card. The lighter lines in (a) and (b) show the results of a simultaneous position detection using digital video microscopy. These are the same data shown in Fig. 9.3 rescaled and shifted so that their variance and average are equal to those of the interferometric data. The sampling rate is 2500 frames/s.

plotted by the black solid lines in Figs. 9.8a, 9.8b and 9.8c, respectively. The average signal is shown by the dashed lines. The reader might be puzzled by the fact that the average signal for x_{qpd} and y_{qpd} is not null. This is because of two facts. On one hand, some offset is introduced by the fact that the four quadrants of a QPD are not perfectly equal, but have slightly different responses. On the other hand, following the QPD sensor, there is a series of analogue amplifiers that condition the signal and introduce some additional offset. In any case, in normal situations this offset is not a problem and, if needed, can easily be removed when the signals are processed. To check the quality of this position detection scheme, we have compared these results with the results of simultaneous position detection using digital video microscopy. Thus, the digital video microscopy data have been rescaled and shifted so that their variance and average are equal to those of the interferometric data. The resulting time series are shown by the lighter lines in Figs. 9.8a and 9.8b (these data are in fact the same as those reported in Fig. 9.3). It can be seen how the two independent techniques deliver essentially the same result.

The forward-scattering position detection scheme is not always possible. In a number of experiments, geometrical constraints may prevent access to the forward-scattering light, forcing one to make use of the backward-scattered light instead. This occurs, for example, in biophysical applications where one of the two faces of a sample holder needs to be coated with some specific non-transparent material and in plasmonics applications where a plasmon wave needs to be excited from one of the faces of the sample holder. Furthermore, the backward mode of operation makes it easier to combine the optical trap with other techniques such as atomic force microscopy, which requires access to one side of the holder. A special feature of backward-scattering detection compared to the forward-scattering case is that the same lens is used both for trapping and for collecting the scattering light, as can be seen in schematics of typical backward-scattering position detection set-ups shown in Figs. 9.9a and 9.9b. In this case, a beam splitter is necessary in order to separate the incoming beam from the backward-scattered light. The resulting field at the detector is

$$\mathbf{E}_{\mathrm{BS}}(x_\mathrm{d}, y_\mathrm{d}) = \sqrt{\frac{n_\mathrm{m}}{n_\mathrm{d}}} \frac{1}{\sqrt{\cos\theta}} \left\{ -\left[\mathbf{E}_\mathrm{s}(x_\mathrm{o}, y_\mathrm{o}, z_\mathrm{o}) \cdot \hat{\boldsymbol{\theta}}\right] \hat{\boldsymbol{\rho}} + \left[\mathbf{E}_\mathrm{s}(x_\mathrm{o}, y_\mathrm{o}, z_\mathrm{o}) \cdot \hat{\boldsymbol{\varphi}}\right] \hat{\boldsymbol{\varphi}} \right\},$$
(9.13)

where $x_\mathrm{c} = x_\mathrm{d}$, $y_\mathrm{c} = y_\mathrm{d}$, $z_\mathrm{c} = \sqrt{f_\mathrm{c}^2 - x_\mathrm{d}^2 - y_\mathrm{d}^2}$ are the coordinates on the pupil of the objective and, importantly, there is no interference with the incoming beam.[5] We will not go into the details of the calculation of the signals because the analysis is essentially the same as for the forward-scattering detection scheme. The light intensities at the detector are presented in Fig. 9.9c for various particle sizes and transverse particle displacements. Whereas the forward-scattered field essentially has a single-lobe angular intensity distribution for all probe sizes, the backward-scattered field distribution becomes more and more complex as the particle size increases. The responses of a QPD to the particles in the backward-scattering scheme are shown in Fig. 9.9d, where it can be noticed that the intensity of the light reaching the detector is significantly lower that in the forward-scattering case.[6] Comparing the results obtained for particles with $a = 400$ nm and $a = 600$ nm, we see that the sign of the response sensitivity changes. Whereas the spot of a 400 nm particle moves to the left for $C_x > 0$ and to the right for $C_x < 0$, the intensity pattern produced by the 600 nm probe moves in the opposite direction. For the 1000 nm particle, both sides of the detector receive almost the same intensity from the two-lobe backward-scattering intensity distribution, producing a loss of sensitivity. Therefore, the use of backward-scattering detection presents a number of difficulties that are absent from forward-scattering detection. Particular attention should be paid to the probe size, as there exist specific sizes for which the probe displacement cannot be detected. Some of these problems might be mitigated by using the appropriate detection numerical aperture and choosing either QPD or PSD.[7]

---

[5] This is only partly true, because in any real set-up there is almost always some stray reflection of the incoming beam that interferes with the backward-scattered light.

[6] This can easily be understood with a back-of-the-envelope calculation by considering that the reflectivity of glass ($n_\mathrm{p} = 1.55$) when immersed in water ($n_\mathrm{m} = 1.33$) is just 0.6%.

[7] An in-depth discussion of this phenomenon can be found in Volpe et al. (2007).

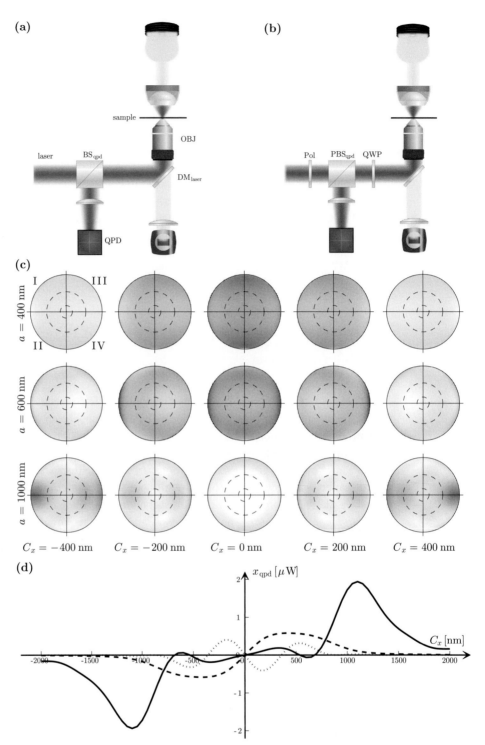

Figure 9.9 Backward scattering position detection. (a) Schematic of a backward position detection set-up. (b) losses can be reduced using a polarising beam splitter (PBS_{qpd}) in combination with a linear polariser (Pol) and a quarter-wave plate (QWP). (c) Images of objective back-focal-plane fields for various particle radii a and transverse displacement $\mathbf{C} = [C_x, 0, 0]$. The circles delimit the field captured by the condenser lens with $NA_o = 0.20, 0.65$ and 1.20 (solid line) from the centre outwards. (d) Signal from a QPD as a function of C_x for $a = 400$ nm (dotted line), 600 nm (dashed line) and 1000 nm (solid line) for an incident beam with power 1 mW and detection $NA_o = 1.20$.

Finally, we should remark that signals obtained with interferometric techniques are not all in physical units of length, but typically in volts. Therefore, a calibration of the voltage-to-length conversion coefficient need to be performed. This calibration factor can be obtained in several alternative ways. In particular, it is possible to scan an optically trapped particle across a (much weaker) detection beam, in which case one need to have at least two beams. Luckily, it is also possible to perform this calibration self-consistently using only the position signal of an optically trapped particle acquired by a photodetector, as we will see in detail in Sections 9.6, 9.7 and 9.8.

Exercise 9.2.1 Study numerically the responses of a QPD and PSD in forward scattering as a function of the particle and medium parameters, the incoming beam parameters and the detection numerical aperture. In particular, focus your attention on the sensitivity and linear range of the detection system. [Hint: You can adapt the code `fs` from the book website.]

Exercise 9.2.2 Repeat the previous study in backward scattering. Note how the sensitivity and linear range are affected. In particular, note how the slope of the detection signal changes as a function of the parameters. For which parameters is the sensitivity slope null? [Hint: You can adapt the code `bs` from the book website and use the work by Volpe et al. (2007) as guidance.]

Exercise 9.2.3 Study how QPD and PSD can be used to detect a particle position using Hermite–Gaussian, Laguerre–Gaussian and cylindrical vector beams. [Hint: Pay attention to the role played by the Gouy phase shift. You can find some guidance in Garbin et al. (2009).]

Exercise 9.2.4 Using a polarising beam splitter and the appropriate polarisation optics, it is possible to perform backward-scattering position detection without losing any trapping power or (almost) any scattered light. Suggest how this may be done. [Hint: What happens when a linearly polarised light beam passes through a quarter-wave plate and is reflected by a mirror back through the same quarter-wave plate?]

9.2.1 Photodetectors

A photodiode is a semiconductor device based on a p–n junction that produces a small current (the *photocurrent*, I_{PD}) when illuminated. As seen above, for position-tracking applications, either segmented photodiodes (commonly a four-segment QPD) or lateral-effect photodiodes (PSD) can be used. In selecting a photodiode, several parameters describing the performance of the device should be considered, in particular

- **Responsivity.** The responsivity $R(\lambda_0)$ of a photodiode is the ratio of the generated photocurrent I_{PD} to the incident light power P at a given wavelength λ_0 (units of ampere per watt):

$$R(\lambda_0) = \frac{I_{PD}}{P}. \tag{9.14}$$

Table 9.1 Photodiode materials

Material	Approximate range	Approximate peak
Silicon (Si)	400 to 1000 nm	800 to 900 nm
Germanium (Ge)	900 to 1600 nm	1500 to 1600 nm
Indium gallium arsenide (InGaAs)	900 to 1700 nm	1300 to 1600 nm

The approximate range of wavelengths over which photodiodes of different materials respond is shown in Table 9.1.

- **Quantum efficiency.** The quantum efficiency measures the fraction of the incident photons that contribute to the photocurrent. The photocurrent I_{PD} arising from incident power P for a photodiode with quantum efficiency η is

$$I_{PD} = \eta e_- \frac{P}{hf}, \tag{9.15}$$

where e_- is the charge of the electron, h is the Planck constant and f is the frequency of the incident light. The quantum efficiency is usually expressed as a percentage and can vary significantly with wavelength.

- **Dark current.** The dark current is the current flowing through the photodiode when no light is incident. The magnitude of the dark current depends on factors such as temperature, bias voltage, photodiode material and active area. It is expressed in units of current (typically picoamperes to nanoamperes) at a particular bias voltage.

- **Junction capacitance.** The capacitance of the p–n junction determines the speed of the photodiode response and, therefore, the photodiode bandwidth. The capacitance of a junction tends to decrease with increasing reverse bias voltage (as a consequence of the increased width of the depletion zone), which thereby increases the response speed. It is expressed in units of capacitance (typically less than 1 nF).

- **Bandwidth.** The bandwidth is determined by the photodiode junction capacitance C and the load resistance R_L through which the photocurrent flows, as

$$f_{BW} = \frac{1}{2\pi R_L C}. \tag{9.16}$$

- **Noise equivalent power (NEP).** The NEP is a measure of the minimum detectable optical power. The NEP is the input optical power that produces a signal with power spectral density equal to that of the noise from an unilluminated photodiode (i.e., a signal-to-noise ratio of unity) in a bandwidth of 1 Hz. It is expressed in units of power per square-root frequency, e.g., $W/\sqrt{Hz}$. Alternatively, the minimum detectable power may be expressed using the *specific detectivity* $D^* = \sqrt{A}/\text{NEP}$, where A is the detector active area.

- **Mode of operation (photoconductive or photovoltaic).** A photodiode may be operated either with a reverse bias voltage (photoconductive operation) or with zero bias (photovoltaic operation). Generally, the reverse bias voltage increases the extent of the depletion region, hence reducing the capacitance of the junction and permitting very high speed operation. However, this occurs at the expense of an increased dark current. In photovoltaic operation, the sensitivity and linearity of the photodiode response are better, and dark current is minimal arising only from Johnson noise.

9.2.2 Acquisition hardware

The tracking signal generated by the photodetector needs to be acquired and stored in a computer for analysis. This requires a device to convert the analogue position tracking signal into a digital form. This can be achieved straightforwardly with a digital oscilloscope, or using a computer-mounted data acquisition (DAQ) board. A range of DAQ boards from manufacturers such as National Instruments and LabJack are available. These DAQ boards can connect to the controlling computer via USB, ethernet, WiFi, PCI or PXI bus. Interfacing and controlling the DAQ board from the computer can be performed by the manufacturer's software (e.g., National Instruments LabView or LabJack LJM), although support for other software (e.g., MatLab, Python, C/C++, DAQFactory) is often also offered. When selecting a data acquisition board, the most important specifications that need to be considered include

- **Number of (analogue) input channels.** The acquisition board needs enough analogue channels to record as many tracking signals as required by the application at hand.
- **Resolution.** The analogue tracking signal is converted into a digital value by an analogue-to-digital converter (ADC). The number of bits N_{bits} allocated to storing the digital value sets the number of levels into which the signal is digitised as $2^{N_{\text{bits}}}$; e.g., a 12-bit device converts an input signal into one with $2^{12} = 4096$ discrete values.
- **Bandwidth.** The bandwidth is the range of frequencies that are passed unattentuated by the analogue signal input circuit *before* the input of the ADC. The bandwidth is usually measured to the 3 dB point, where the amplitude of the signal has decreased by a factor of $\sqrt{2}$; i.e., the power of the signal is reduced by half. The bandwidth of an acquisition device should be several times the highest frequency of interest in the measured signal. Units are those of frequency (typically hundreds of kilohertz to megahertz).
- **Sample rate.** The sample rate is the frequency at which the analogue signal is digitalised by the ADC (note that this is not the same as the input channel bandwidth). Units are samples per second, typically from 10 kilosample/s to over 1 megasample/s.
- **Analogue input range.** This is the range of signal voltage amplitudes that can be acquired.
- **Single-ended vs. differential inputs.** In describing analogue inputs, 'single-ended' and 'differential' refer to the reference for the signal voltage. A single-ended input is usually referenced towards ground, while a differential signal is referenced to a secondary input.

9.3 Calibration techniques: An overview

Once the trajectory of a Brownian particle has been measured either by digital video microscopy or by interferometry, as described in the previous sections, it is possible to use it to quantitatively study the optical potential. In calibrating an optical trap, the main objective is to determine the trap stiffness κ_x. A common secondary objective is to determine the conversion factor S_x from measurement units (e.g., pixels or volts) to physical units of length (metres) in order to measure absolute displacements in the trap and hence the forces that cause them. In the following, we will present the various techniques that permit one to preform this calibration.

Most commonly, *passive* calibration techniques are employed, where the trajectory of the optically trapped particle is measured within a fixed optical trap. These techniques include *potential analysis* [Section 9.4], the *equipartition method* [Section 9.5], *mean squared displacement analysis* [Section 9.6], *autocorrelation function analysis* [Section 9.7] and *power spectrum analysis* [Section 9.8]. There are also some *active* calibration techniques, where the effect on the optically trapped particle of a known force is measured, typically by applying a fluid flow [Section 9.9]. Ideally, all techniques should deliver the same result, as shown in Fig. 9.1 for the case of three techniques applied to our data. In fact, employing more than one calibration technique and confirming their consistency is a good check of the quality of acquired experimental data. An additional check is given by the fact that the trap stiffness should be proportional to the optical power, which is also satisfied in the data shown in Fig. 9.1.

In the following, we will consider a spherical particle of radius a immersed in a fluid of viscosity η at absolute temperature T with friction coefficient

$$\gamma = 6\pi \eta a \tag{9.17}$$

and bulk diffusion coefficient

$$D = \frac{k_B T}{\gamma}, \tag{9.18}$$

where k_B is the Boltzmann constant. Even though the motion of the particle can be recorded in three dimensions, we will mainly focus on only one axis, the x-axis transverse to the beam propagation. Nevertheless, all the methods we describe can be applied to motion in the other two dimensions.

9.4 Potential analysis

The trajectory $x(t)$ of a Brownian particle can be described by the Langevin equation [Eq. (7.10)]

$$\frac{d}{dt}x(t) = -\frac{1}{\gamma}\frac{d}{dx}U(x) + \sqrt{2D}W_x(t), \tag{9.19}$$

where $U(x)$ is the potential and $W_x(t)$ is a white noise, whose properties are described in Box 7.3. Because the Brownian particle is in thermal equilibrium with the heat bath constituted of the fluid molecules, its probability distribution follows the Maxwell–Boltzmann distribution given by Eq. (7.19),

$$\rho(x) = \rho_0 \exp\left[-\frac{U(x)}{k_B T}\right], \tag{9.20}$$

where ρ_0 is a normalisation factor.[8] By solving Eq. (9.20) for $U(x)$, we obtain

$$U(x) = -k_B T \log[\rho(x)] + U_0, \tag{9.21}$$

where U_0 is a constant. Differently from the methods presented in the following sections, this method does not assume that the optical tweezers potential is harmonic, and therefore can be used to verify the hypothesis that the optical potential is harmonic or to probe more complex potentials.

In practice, a series of *independent*[9] samples x_l are acquired at times t_l during the measurement time T_s. The positions x_l are then sorted into a series of equally spaced bins along the x-axis, so that c_m is the count of the positions falling within the mth bin, i.e., the number of x_l such that $\left(m - \frac{1}{2}\right)\Delta x < x_l < \left(m + \frac{1}{2}\right)\Delta x$, where typically $\Delta x \approx 2$ to 20 nm, depending on the precision of the measurement. The resulting histogram $\{c_m\}$ is an approximation to the position probability distribution $\rho(x)$. This procedure can be repeated several times to obtain an average value and a standard deviation for the counts. It is finally possible to calculate the potential from $\{c_m\}$ as $U_m = -k_B T \log(c_m)$. The probability distributions shown in Fig. 9.10a correspond to the motion of a particle optically trapped at three different laser powers: the symbols represent the averages and the error bars the standard deviations from five experimental runs. The symbols in Fig. 9.10b show the corresponding potentials.

The determination of the potential by this method is subject to some systematic errors, as it can be smeared out by the presence of uncorrelated noise, e.g., low-frequency mechanical vibrations of the set-up and detection errors. Also, if the acquired data are correlated, the presence of low-pass frequency filters in the acquisition system can alter the appearance of the potential.

Exercise 9.4.1 Apply the potential measurement technique to the reconstruction of the potential where a particle is moving starting from the recording of a trajectory. [Hint: You can simulate a trajectory in a generic potential [Section 7.4] or, alternatively, you can use the experimental trajectory of an optically trapped particle `trajectory` from the book website. For the analysis you can adapt the code `calpot` from the book website.]

Exercise 9.4.2 Study how the presence of noise affects the potential measurement technique. Consider, in particular, the effect of low-frequency vibrations. [Hint: You can simulate a trajectory with noise (e.g., low-frequency vibrations) [Section 7.4] and analyse it with the code `calpot` from the book website.]

---

[8] The normalisation factor $\rho_0 = Z^{-1}$, where Z is the partition function, is needed so that $\int \rho(x)dx = 1$.
[9] More precisely, one needs to have *enough* independent samples. The particle position need to be sampled over a measurement time much longer than the characteristic time scale τ_{ot} of the motion of the particle in the optical potential, i.e., $T_s \gg \tau_{ot}$. This sets some bounds on the minimum acquisition time, but permits one also to decrease the acquisition frequency, as the samples do not need to be correlated. Furthermore, the samples do not need to be taken at regular time intervals.

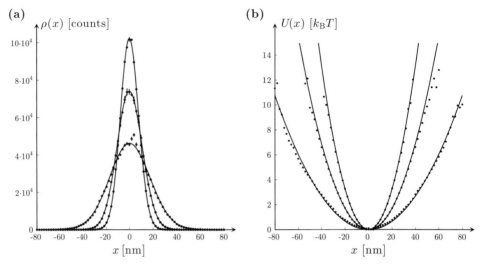

Figure 9.10 Potential and equipartition analysis. (a) Position histograms and (b) potentials corresponding to an optically trapped polystyrene sphere ($a = 1.03\,\mu\text{m}$). The symbols represent the potential method analysis of three series of experimental data corresponding to 2.3, 6.0 and 10.8 mW laser power ($\lambda_0 = 1064$ nm) at the sample. The solid lines correspond to the position distribution and the potential of an optical trap with the stiffness obtained from the equipartition analysis [Eq. (9.28)] of the experimental data ($\kappa_x^{(\text{ex})} = 13.9 \pm 0.3, 35.7 \pm 0.7$ and 68.0 ± 0.3 fN/nm). These data demonstrate that the trap's potential is harmonic up to $10\,k_\text{B}T$ at least.

9.5 Equipartition method

The potential associated with an optical trap is to a very good first approximation harmonic, i.e.,

$$U(x) = \frac{1}{2}\kappa_x \left[x - x_{\text{eq}}\right]^2, \tag{9.22}$$

where κ_x is the trap stiffness and x_{eq} is the equilibrium position. Therefore, the equation of motion of an optically trapped Brownian particle is [Eqs. (7.23)]

$$\frac{dx(t)}{dt} = -\kappa_x \left[x(t) - x_{\text{eq}}\right] + \sqrt{2D}\, W_x(t), \tag{9.23}$$

and, from Eq. (9.20), the particle probability distribution is Gaussian,

$$\rho(x) = \rho_0 \exp\left[-\frac{\kappa_x \left(x - x_{\text{eq}}\right)^2}{2k_\text{B}T}\right]. \tag{9.24}$$

Therefore, once the harmonic hypothesis has been verified following the procedure described in the previous section, it is possible to use the equipartition theorem [Box 7.3], which states that

$$\langle U(x) \rangle = \frac{1}{2}\kappa_x \left\langle (x - x_{\text{eq}})^2 \right\rangle = \frac{1}{2}k_B T, \tag{9.25}$$

where instead of the ensemble average one can employ a time average because of the ergodicity of the system.

In practice, because the probability distribution of particle position is Gaussian, given a series of *independent*9 samples x_l with $l = 1, \ldots, L$ acquired at times t_l during the measurement time T_s, we can estimate the equilibrium position as

$$x_{\text{eq}}^{(\text{ex})} = \frac{1}{L} \sum_{l=1}^{L} x_l, \qquad (9.26)$$

the position variance as

$$\sigma_x^{2,(\text{ex})} = \frac{1}{L} \sum_{l=1}^{L} \left(x_l - x_{\text{eq}}^{(\text{ex})}\right)^2 \qquad (9.27)$$

and the trap stiffness as

$$\kappa_x^{(\text{ex})} = \frac{k_B T}{\sigma_x^{2,(\text{ex})}}. \qquad (9.28)$$

More precise estimates of $\kappa_x^{(\text{ex})}$ and $x_{\text{eq}}^{(\text{ex})}$ can be obtained by repeating the experiments and averaging the results. This procedure permits one also to obtain the standard deviations. The solid lines in Figs. 9.10a and 9.10b show, respectively, the Gaussian probability distributions and the harmonic potentials estimated from the experimental data corresponding to various optical trap powers.

As in the determination of the potential described in the previous section, the experimental data points need to sample the probability distribution, so they do not need to be acquired with a fixed time step, as long as a sufficient number of independent points are acquired. Also, as for the potential analysis method, the estimation of $\kappa_x^{(\text{ex})}$ with the equipartition method is prone to systematic errors due to the presence of uncorrelated noise.

Exercise 9.5.1 Estimate the stiffness of an optical trap from the recording of the trajectory of an optically trapped particle using the equipartition potential method. [Hint: You can simulate a trajectory in a harmonic potential [Section 7.4] or, alternatively, you can use the experimental trajectory of an optically trapped particle `trajectory` from the book website. For the analysis you can adapt the code `caleq` from the book website.]

Exercise 9.5.2 Study how the presence of noise affects the estimation of the stiffness of an optical trap with the equipartition method. [Hint: You can simulate a trajectory with noise (e.g., low-frequency vibrations) [Section 7.4] and analyse it with the code `caleq` from the book website.]

Exercise 9.5.3 Study how the presence of a low-pass filter in the acquisition system affects the estimation of the stiffness of an optical trap with the equipartition method. [Hint: You can simulate a trajectory [Section 7.4], low-pass filter it and analyse it with the code `caleq` from the book website.]

Exercise 9.5.4 Show that Eqs. (9.26), (9.27) and (9.28) are optimal for the estimation of $x_{\text{eq}}^{(\text{ex})}$, $\sigma_x^{2(\text{ex})}$ and $\kappa_x^{(\text{ex})}$ in the sense that they optimise the maximum likelihood estimation of the potential parameters (under the assumption that this is harmonic). [Hint: Use the fact that independent particle positions in the optical trap constitute a uncorrelated Gaussian process.]

9.6 Mean squared displacement analysis

A more precise characterisation of the optical trap can be obtained from the *mean squared displacement* (MSD) of the optically trapped particle. The MSD quantifies how a particle moves from its initial position: for ballistic motion the MSD is proportional to t^2, for diffusive motion it is proportional to t and for a trapped particle it saturates to a constant value. For the case of an optically trapped particle the MSD is given by Eq. (7.28), i.e.,

$$\text{MSD}_x(\tau) = \overline{[x(t+\tau) - x(t)]^2} = 2\frac{k_\text{B} T}{\kappa_x}\left[1 - e^{-\frac{|\tau|}{\tau_{\text{ot},x}}}\right], \tag{9.29}$$

where κ_x is the trap stiffness and $\tau_{\text{ot},x} = \gamma/\kappa_x$ is the trap characteristic time. As discussed in Subsection 7.4.2, $\text{MSD}_x(\tau)$ features a transition from linear growth corresponding to free diffusion behaviour at short time scales ($\tau \ll \tau_{\text{ot},x}$) to a plateau due to confinement at long time scales ($\tau \gg \tau_{\text{ot},x}$).

In practice, differently from the potential analysis and the equipartition method, now we need to acquire a time series of *correlated* particle positions at regular time intervals. Thus, we obtain the samples $x_j = x(t_j)$, where $j = 1, \ldots, N$, at the sampling times $t_j = j\Delta t$. The MSD can be calculated from this trajectory at discrete time lags $\tau_k = k\Delta t$ as

$$\text{MSD}_{x,k} = \frac{1}{N-k}\sum_{j=1}^{N-k}\left[x_{j+k} - x_j\right]^2. \tag{9.30}$$

The experimental MSD given by Eq. (9.30) can be fitted by least squares to the theoretical MSD given by Eq. (9.29) in order to obtain $\kappa_x^{(\text{ex})}$. By repeating this analysis on various experimental series, it is possible to obtain an average value and a standard deviation for $\kappa_x^{(\text{ex})}$, as well as for the $\text{MSD}_{x,k}$. The symbols in Fig. 9.11 show the experimental MSDs for an optically trapped particle with three different optical powers: the stronger the trap, the sooner the plateau is reached. The solid lines are the theoretical MSDs [Eq. (9.29)] corresponding to the fitted parameters.

Often the trajectory obtained from the position detection system is not naturally given in units of length. For example, in the case of digital video microscopy the trajectory is naturally given in pixels and in the case of interferometric detection in voltage. Therefore, it can be useful to determine the conversion factors to units of length by fitting the value of

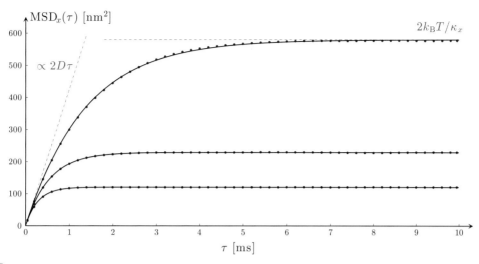

Figure 9.11 Mean squared displacement analysis. MSDs corresponding to a polystyrene sphere ($a = 1.03\ \mu\text{m}$) optically trapped with 2.3, 6.0 and 10.8 mW laser power ($\lambda_0 = 1064$ nm) at the sample. The symbols and error bars are the experimental MSDs [Eqs. (9.30)] and the solid lines are the theoretical MSDs [Eq. (9.29)] corresponding to the fitted parameters. The fitted values of the stiffness are $\kappa_x^{(\text{ex})} = 14.3 \pm 0.3, 36.2 \pm 0.6$ and 68.9 ± 0.6 fN/nm and of the unit conversion factor [Eq. (9.31)] are $S_x^{(\text{ex})} = 478, 191$ and 106 nm/V. At short time scales the MSD is diffusive and at long time scales, it reaches a plateau due to the trap confinement – the stronger the trap, the smaller the value of the plateau.

the experimental friction coefficient $\gamma^{(\text{ex})}$. Then the conversion factor to units of length is given by

$$S_x^{(\text{ex})} = \sqrt{\frac{\gamma}{\gamma^{(\text{ex})}}}, \tag{9.31}$$

where γ is the friction coefficient given by Eq. (9.17).[10] This procedure can serve also as a consistency check for the analysis procedure.

Exercise 9.6.1 Estimate the stiffness of an optical trap from the recording of its trajectory using MSD analysis. [Hint: You can use the experimental trajectory of an optically trapped particle `trajectory` from the book website or you can simulate the trajectory of an optically trapped particle [Section 7.4]. For the analysis you can adapt the code `calmsd` from the book website.]

---

[10] We note that, although $\gamma^{(\text{ex})}$ can be determined with high accuracy and precision, the accuracy with which γ is known is limited by actual knowledge of the particle and medium. Microspheres are commercially available with radius known to within 1% accuracy and similar accuracy of the spherical shape. However, much less accuracy might be attainable on the dynamic viscosity of the fluid where the experiments take place, especially in biophysical experiments.

9.7 Autocorrelation analysis

Another method often employed in characterising an optical tweezers is analysis of the *autocorrelation function* (ACF) of the particle position. This method is particularly suitable when one wishes to deconvolve a more complex time-domain dynamics of the trapped particle from its tracking signal, e.g., non-conservative effects and dynamics of non-spherical particles. The position ACF provides a measure of the time it takes for the particle to 'forget' its initial position. The time scale $\tau_{\text{ot},x}$ characterises how a particle falls into the trap: as the stiffness increases, the particle undergoes a stronger restoring force and $\tau_{\text{ot},x}$ decreases. The position ACF is given by Eq. (7.27), i.e.,

$$C_x(\tau) = \frac{k_B T}{\kappa_x} e^{-\frac{|\tau|}{\tau_{\text{ot},x}}}, \qquad (9.32)$$

where κ_x is the trap stiffness and $\tau_{\text{ot},x} = \gamma/\kappa_x$ is the trap characteristic time.

In practice, as in the case of the MSD, we need a time series of *correlated* particle positions at regular time intervals, i.e., the sample positions $x_j = x(t_j)$, where $j = 1, \ldots, N$, at the sampling times $t_j = j\Delta t$. The ACF can be calculated from this trajectory at discrete time lags $\tau_k = k\Delta t$ as

$$C_{x,k} = \frac{1}{N-k} \sum_{j=1}^{N-k} x_{j+k} x_j. \qquad (9.33)$$

The experimental ACF given by Eq. (9.33) can be fitted by least squares to the theoretical ACF given by Eq. (9.32) in order to obtain an estimate of κ_x. By repeating this analysis on various experimental series, it is possible to obtain an average value and a standard deviation for $\kappa_x^{(\text{ex})}$, as well as for the $C_{x,k}$. The symbols in Fig. 9.12 show the experimental ACFs for an optically trapped particle with three different optical powers: the stronger the trap, the faster the ACF decays. The solid lines are the theoretical ACFs [Eq. (9.32)] corresponding to the fitted parameters.

Also, in the case of the ACF analysis, it can be useful to determine the conversion factor from the acquisition system units to units of length by also fitting $\gamma^{(\text{ex})}$ and using Eq. (9.46), as in Section 9.6, to which we refer for a discussion of related issues.

Exercise 9.7.1 Estimate the stiffness of an optical trap from the recording of its trajectory using ACF analysis. [Hint: You can use the experimental trajectory of an optically trapped particle `trajectory` from the book website or you can simulate the trajectory of an optically trapped particle [Section 7.4]. For the analysis you can adapt the code `calacf` from the book website.]

9.7.1 Crosstalk analysis and reduction

The crosstalk between the various position signals can be characterised using the cross-correlation function (CCF). For example, given the transverse QPD signals x_{qpd} and y_{qpd}

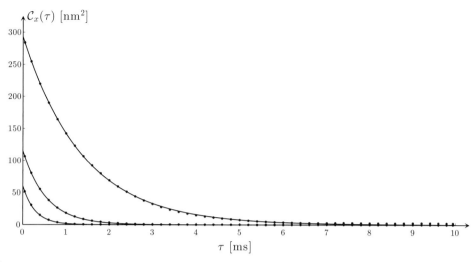

Figure 9.12 Autocorrelation analysis. Position autocorrelation functions (ACFs) corresponding to a polystyrene sphere ($a = 1.03$ μm) optically trapped with 2.3, 6.0 and 10.8 mW laser power ($\lambda_0 = 1064$ nm) at the sample. The symbols and error bars are the experimental ACFs [Eqs. (9.33)] and the solid lines are the theoretical ACFs [Eq. (9.32)] corresponding to the fitted parameters. The fitted values of the stiffness are $\kappa_x^{(\text{ex})} = 14.1 \pm 0.4$, 36.0 ± 1.0 and 68.9 ± 1.0 fN/nm and those of the unit conversion factor [Eq. (9.31)] are $S_x^{(\text{ex})} = 480$, 192 and 107 nm/V. As can be clearly seen, the stronger the trap, the faster the decay of the ACF.

as a function of time, i.e., $x_{\text{qpd},j}$ and $y_{\text{qpd},j}$, where $j = 1, \ldots, N$, acquired at the sampling times $t_j = j \Delta t$, the CCF can be calculated at discrete time lags $\tau_k = k \Delta t$ as

$$C_{xy,\text{qpd},k} = \frac{1}{N-k} \sum_{j=1}^{N-k} x_{\text{qpd},j+k}\, y_{\text{qpd},j}. \qquad (9.34)$$

The black solid line in Fig. 9.13a shows the CCF we obtain from our data. The fact that it is not negligible when compared to the ACF of the x (grey solid line) and y (grey dashed line) signals tells us that there is some crosstalk between the two channels. We can see this even more clearly from the fact that the joint probability distribution of the signals x_{qpd} and y_{qpd} plotted in the inset in Fig. 9.13a is an inclined elliptical Gaussian distribution whose major axes do not coincide with x_{qpd} and y_{qpd}.

To reduce the crosstalk to a minimum, it is possible to decorrelate the x_{qpd} and y_{qpd} by applying a *principal component analysis transform* (PCA transform), which is an orthogonal transformation of the coordinates, i.e., a rotation.[11] The new coordinate axes (y'_{qpd} and y'_{qpd}) now coincide with the major axes of the elliptical distribution, as can be seen in the inset in Fig. 9.13b. If we now calculate the CCF of the signals expressed in the coordinates y'_{qpd} and y'_{qpd}, we obtain a virtually null function, as shown by the black solid line in Fig. 9.13b, which is hardly distinguishable from the lag time axis.

---

[11] In fact, the PCA transform is optimal among all possible orthogonal transforms to minimise the correlation between two signals.

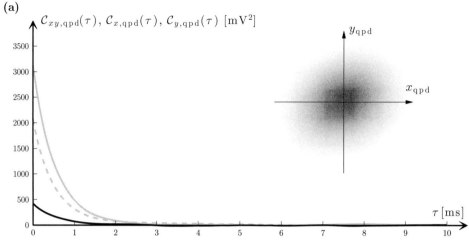

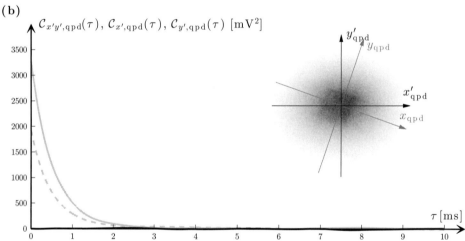

Figure 9.13 Cross-correlation function and crosstalk reduction. (a) Cross-correlation function (CCF) between the QPD signals x_{qpd} and y_{qpd} (black solid line), autocorrelation function (ACF) of x_{qpd} (grey solid line) and ACF of y_{qpd} (dashed solid line). Inset: the probability distribution of x_{qpd} and y_{qpd} is an inclined elliptical Gaussian distribution. (b) CCF between the decorrelated signals x'_{qpd} and y'_{qpd} (black solid line) is close to zero and coincides with the x-axis in the graph. ACF of x'_{qpd} (grey solid line) and ACF of y'_{qpd} (dashed solid line). Inset: the major axes of the probability distribution of x'_{qpd} and y'_{qpd} coincide with the coordinate axes.

Now that we have successfully decorrelated our signals, it is time to understand the reason that there was a correlation in the first place. Here we list some common causes of correlation:

- In general, an optical trap is not rotationally symmetric, but elliptic about the beam axis (z-axis), e.g., because the laser beam is linearly polarised and the focal spot has an elliptical intensity profile [Chapter 4]. Therefore, the equation of motion decouples in some orthogonal coordinates (x', y') that coincide with the major and minor axes of the ellipse, so that

$$U(x', y') = \frac{1}{2}\kappa_{x'}x'^2 + \frac{1}{2}\kappa_{y'}y'^2. \tag{9.35}$$

If $\kappa_{x'} = \kappa_{y'}$, then the harmonic trapping potential is perfectly rotationally symmetric about the beam axis, and the particle's equations of motion decouple no matter which pair of Cartesian coordinates are used in the transverse plane orthogonal to the beam axis. This condition is in fact approximately met in most experiments, at least for particles larger than the wavelength of light or for circular polarisation of the trapping beam.[12] In our case, the grey lines in Fig. 9.13b show that the correlation functions for the decorrelated signals (and in fact also of the original signals) along x'_{qpd} and y'_{qpd} are approximately proportional to each other. This means that the two signals have nearly the same characteristic time and the trap is nearly rotationally symmetric, permitting us to exclude the possibility that the optical trap is significantly asymmetric.

- With an asymmetric trap excluded, the simplest explanation for the correlation between x_{qpd} and y_{qpd} is a difference in sensitivity of the four quadrants of the QPD. Note that such asymmetry need not be a property of the diode itself. All four quadrants could be identical, but have a nonlinear relationship between input light intensity and output voltage. In that case, less than perfect centring of the laser beam on the diode will cause different amounts of light to fall on different quadrants and, hence, make them respond with different sensitivity to the small changes in light that correspond to the movement of the bead.

- Another explanation, which does not exclude the previous one, could be a small asymmetry in the spot of light scattered by the bead onto the photodiode. This would cause different amounts of light to shift between quadrants for identical shifts of the bead in the x and y-directions. If, furthermore, some of that asymmetrically scattered light falls beyond the edge of the QPD, then a shift of the bead in the x-direction will change the y-signal and hence register as a correlated change in y.

Exercise 9.7.2 Use the PCA transform to decorrelate some correlated signals. [Hint: You can use the experimental trajectory of an optically trapped particle `trajectoryxy` from the book website or you can simulate the trajectory of an optically trapped particle in an elliptical potential. For the analysis you can adapt the code `calpca` from the book website.]

Exercise 9.7.3 Study how different light sensitivities for the four quadrants of a QPD can introduce correlation of the acquired signals.

9.8 Power spectrum analysis

We now consider the *power spectrum* analysis of the trajectory of an optically trapped particle. Power spectrum analysis is widely regarded as the most reliable method, at least

---

[12] In this regard, it is always important to align the QPD with the polarisation axes of the beam, so to minimise crosstalk.

with spherical particles. In fact, it has the advantage of working in the frequency domain, which permits one to remove common causes of noise relatively easily; such as slow mechanical drifts at low frequency and sinusoidal noise, e.g., noise due to the mains electricity network, which appear as peaks in the *power spectral density* (PSD). In this section, we will follow the detailed analysis of the power spectrum of an optically trapped particle realised by Berg-Sørensen and Flyvbjerg (2004).

We will start from the overdamped Langevin equation of motion of an optically trapped particle given in Eq. (7.23), which in one dimension can be recast as

$$\frac{dx(t)}{dt} + 2\pi f_{c,x} x(t) = \sqrt{2D} W_x(t), \qquad (9.36)$$

where D is the diffusion coefficient, $W_x(t)$ is a white noise, whose properties are described in Box 7.3, and we have introduced the corner frequency

$$f_{c,x} = \frac{\kappa_x}{2\pi\gamma}, \qquad (9.37)$$

where κ_x is the trap stiffness, $\gamma = 6\pi\eta a$ the particle friction coefficient, η the medium viscosity and a the particle radius. The Fourier transform of Eq. (9.36) is

$$2\pi(f_{c,x} - if)\check{X}(f) = \sqrt{2D}\check{W}_x(f), \qquad (9.38)$$

where $\check{X}(f)$ and $\check{W}_x(f)$ are the Fourier transforms of $x(t)$ and $W_x(t)$, respectively. Taking the square modulus of both sides of Eq. (9.38) gives the PSD of the particle motion,

$$P_x(f) = |\check{X}(f)|^2 = \frac{D/(2\pi^2)}{f_{c,x}^2 + f^2}, \qquad (9.39)$$

where we have used the fact that $|\check{W}_x(f)|^2 \equiv 1$ because white noise is uncorrelated in time and has unitary power.

In practice, the trajectory $x(t)$ is sampled with frequency f_s for a time T_s. Thus, we obtain the samples $x_j = x(t_j)$, where $j = 1, \ldots, N$, at the sampling times $t_j = j\Delta t$, where $\Delta t = 1/f_s$ is the sampling time step. The finite difference equation that describes x_j is given by Eq. (7.24), which in the present case becomes

$$x_{j+1} = (1 - 2\pi f_{c,x}\Delta t)\, x_j + \sqrt{2D\Delta t}\, w_{j,x}, \qquad (9.40)$$

where $w_{j,x}$ are independent Gaussian random numbers with zero mean and unitary variance. From this series of data we calculate the discrete Fourier transform

$$\check{X}_k = \Delta t \sum_{j=1}^{N} e^{i2\pi f_k t_j} x_j = \Delta t \sum_{j=1}^{N} e^{i2\pi jk/N} x_j, \qquad (9.41)$$

where $f_k = k/T_s$ and, usually, $k = -N/2 + 1, \ldots, N/2$. This discrete Fourier transform is a good approximation to the continuous one for $|f_k| \ll f_s$.[13] Taking the discrete Fourier

---

[13] Incidentally, this also implies that the approximation is good only for frequencies smaller than the *Nyquist frequency* $f_{\text{Nyq}} = f_s/2$.

transform of Eq. (9.40), we obtain

$$e^{i2\pi k/N} \check{X}_k = (1 - 2\pi f_{c,x} \Delta t) \check{X}_k + \sqrt{2D\Delta t}\, \check{W}_{k,x}. \tag{9.42}$$

Eq. (9.42) can be simplified using the approximation $e^{i2\pi k/N} \approx 1 + i2\pi k/N$, valid for $k \ll N$, or equivalently $|f_k| \ll f_s$, and observing that $k/N = f_k \Delta t$, so that the resulting experimental power spectrum is

$$P_k = \frac{|\check{X}_k|^2}{\sqrt{T_s}} = \frac{D/(2\pi^2 \Delta T_s)}{f_{c,x}^2 + f_k^2} |\check{W}_{k,x}|^2. \tag{9.43}$$

Because $w_{j,x}$ are independent Gaussian random numbers, $\text{Re}\{\check{W}_{k,x}\}$ and $\text{Im}\{\check{W}_{k,x}\}$ are also independent Gaussian random numbers. Thus, $|\check{W}_{k,x}|^2$ are independent non-negative random numbers with *exponential* distribution, whose expected values are

$$\langle P_k \rangle = \frac{D/(2\pi^2 \Delta T_s)}{f_{c,x}^2 + f_k^2} \tag{9.44}$$

and whose standard deviations are

$$\sigma(P_k) = \langle P_k \rangle. \tag{9.45}$$

In order for a least squares fitting of the experimentally obtained power spectrum [Eq. (9.43)] to its theoretical expression [Eq. (9.39)] to be performed, each data point needs be drawn from a Gaussian distribution and different data points need to be statistically independent. The second condition is satisfied by P_k in Eq. (9.43), but the first is not, because P_k is exponentially distributed. The solution is to perform data compression, which results in a smaller data set with less noise and, because of the central limit theorem, in normally distributed data. Data compression can be performed by repeating the experiment n_r times and averaging the resulting values of P_k. Alternatively, it is possible to perform data compression by *blocking*. Blocking replaces a block of n_b consecutive data points with a single new data point with coordinates that are simply block averages.[14] When n_r and/or n_b are sufficiently large, the resulting data points have a Gaussian distribution by virtue of the central limit theorem and are thus amenable to least squares fitting. It is therefore possible to fit the value of $f_{c,x}$. By repeating this analysis on various experimental data series, it is possible to obtain an average value and a standard deviation for $f_{c,x}$, as well as for the P_k. The symbols in Fig. 9.14 show the experimental PSDs for a particle optically trapped with three different optical powers. The solid lines are the theoretical PSDs [Eq. (9.39)] corresponding to the fitted parameters.

Also in the case of PSD analysis, it can be useful to experimentally determine the conversion factor from the acquisition system units to units of length. We refer to Section 9.6 for a more detailed discussion of the related issues. In the case of PSD analysis, this is most naturally done by fitting $D^{(\text{ex})}$ and by using the following expression of the conversion

---

[14] One of the advantages of blocking is that the resulting data points do not need to be equidistributed along the frequency axis, which can be particularly useful for data display with logarithmic axes.

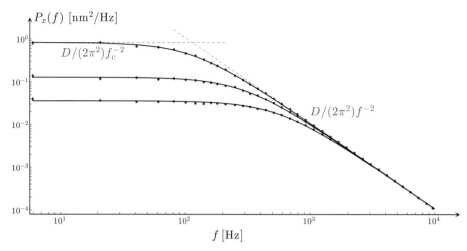

Figure 9.14 Power spectrum analysis. Power spectral densities (PSDs) corresponding to a polystyrene sphere ($a = 1.03$ μm) optically trapped with $2.3, 6.0$ and 10.8 mW laser power ($\lambda_0 = 1064$ nm) at the sample. The symbols and error bars are the experimental PSDs [Eqs. (9.43)] and the solid lines are the theoretical PSDs [Eq. (9.39)] corresponding to the fitted parameters. The fitted values of the corner frequency are $f_{c,x} = 114 \pm 2, 290 \pm 3$ and 550 ± 5 Hz (corresponding to $\kappa_x^{(\text{ex})} = 13.9 \pm 0.2, 35.4 \pm 0.4$ and 67.1 ± 0.6 fN/nm) and those of the unit conversion factor [Eq. (9.31)] are $S_x^{(\text{ex})} = 483, 192$ and 107 nm/V.

factor:

$$S_x^{(\text{ex})} = \sqrt{\frac{D^{(\text{ex})}}{D}}, \tag{9.46}$$

where D is the diffusion coefficient given by Eq. (9.18).

Exercise 9.8.1 Estimate the stiffness of an optical trap from the recording of its trajectory using PSD analysis. [Hint: You can use the experimental trajectory of an optically trapped particle `trajectory` from the book website or you can simulate the trajectory of an optically trapped particle. For the analysis you can adapt the code `calpsd` from the book website.]

Exercise 9.8.2 Study how the fact that the photodetector can act as a low pass filter for the signal may alter the calibration procedure using PSD. [Hint: You can follow the procedure explained in Berg-Sørensen and Flyvbjerg (2004).]

Exercise 9.8.3 Study how the presence of anti-aliasing filters can alter the calibration procedure using the PSD. [Hint: You can follow the procedure explained in Berg-Sørensen and Flyvbjerg (2004).]

9.8.1 Analytical least square fitting

It is possible to fit the experimental PSD given in Eq. (9.43) analytically, following the approach of Berg-Sørensen and Flyvbjerg (2004). After blocking (n_b) and averaging (n_r), we obtain the experimental values of the PSD $P_k^{(\text{ex})}$, which have a Gaussian distribution. Therefore, we can fit the PSD by minimising

$$\chi^2 = \sum_k \left[\frac{P_k^{(\text{ex})} - P_k^{(\text{fit})}}{P_k^{(\text{fit})}/\sqrt{n_b}} \right]^2 = n_r n_b \sum_k \left[\frac{P_k^{(\text{ex})}}{P_k^{(\text{fit})}} - 1 \right]^2, \qquad (9.47)$$

which can be done analytically. Because the theoretical PSD can be written as

$$P_k^{(\text{fit})} = \frac{1}{a + b f_k^2}, \qquad (9.48)$$

with a and b positive parameters to be fitted, χ^2 is a quadratic function of a and b and its minimisation gives

$$\begin{cases} f_c^{(\text{fit})} = \sqrt{\dfrac{a}{b}} = \sqrt{\dfrac{S_{0,1} S_{2,2} - S_{1,1} S_{1,2}}{S_{1,1} S_{0,2} - S_{0,1} S_{1,2}}}, \\[6pt] \dfrac{D^{(\text{fit})} T_s}{2\pi^2} = \dfrac{1}{b} = \dfrac{S_{0,2} S_{2,2} - S_{1,2}^2}{S_{1,1} S_{0,2} - S_{0,1} S_{1,2}}, \\[6pt] \dfrac{\chi^2_{\min}}{n_b} = S_{0,0} - \dfrac{S_{0,1}^2 S_{2,2} + S_{1,1}^2 S_{0,2} - 2 S_{0,1} S_{1,1} S_{1,2}}{S_{0,2} S_{2,2} - S_{1,2}^2}, \end{cases} \qquad (9.49)$$

where

$$S_{p,q} = \sum_k f_k^{2p} P_k^{(\text{ex})q}. \qquad (9.50)$$

From this fitting, it is possible to calculate the error using propagation of errors. The resulting formulas are

$$\begin{cases} \dfrac{\sigma(f_c^{(\text{fit})})}{f_c^{(\text{fit})}} = \dfrac{s_{f_c}(x_{\min}, x_{\max})}{\sqrt{\pi f_c^{(\text{fit})} T_s}}, \\[8pt] \dfrac{\sigma(D^{(\text{fit})})}{D^{(\text{fit})}} = \sqrt{\dfrac{1 + \pi/2}{\pi f_c^{(\text{fit})} T_s}} s_D(x_{\min}, x_{\max}), \end{cases} \qquad (9.51)$$

where $x_{\min} = f_{\min}/f_c^{(\text{fit})}$ and $x_{\max} = f_{\max}/f_c^{(\text{fit})}$,

$$s_{f_c}(x_1, x_2) = \sqrt{\frac{\pi}{u(x_1, x_2) - v(x_1, x_2)}},$$

$$s_D(x_1, x_2) = \sqrt{\frac{u(x_1, x_2)}{(1 + \pi/2)(x_2 - x_1)}},$$

$$u(x_1, x_2) = \frac{2x_2}{1 + x_2^2} - \frac{2x_1}{1 + x_1^2} + 2 \arctan\left[\frac{x_2 - x_1}{1 + x_1 x_2}\right]$$

and
$$v(x_1, x_2) = \frac{4}{x_2 - x_1} \arctan^2\left[\frac{x_2 - x_1}{1 + x_1 x_2}\right].$$

The function s_{f_c} is normalised so that $s_{f_c}(0, \infty) = 1$. Thus, $s_{f_c}(x_{\min}, x_{\max}) \geq 1$, because maximum precision is achieved only by fitting to the whole spectrum. It is important to choose a range of frequency where systematic errors do not alter the shape of the Lorentzian.

9.8.2 Hydrodynamic corrections

When a rigid body moves through a dense fluid such as water, the friction between the body and fluid depends on the body's past motion, because that determines the fluid's present motion. This phenomenon is known as the *hydrodynamic memory effect*. For a sphere performing linear harmonic motion $x(t)$ with angular frequency $\omega = 2\pi f$ in an incompressible fluid and at vanishing Reynolds number, the Navier–Stokes equations can be solved analytically, obtaining a force

$$F_{\text{fric}} = -\gamma\left[1 + \frac{R}{\delta}\right]\frac{dx(t)}{dt} - \left[3\pi\rho a^2\delta + \frac{2}{3}\pi\rho a^3\right]\frac{d^2 x(t)}{dt^2}, \quad (9.52)$$

where γ is the Stokes friction coefficient, a is the radius of the particle, δ is the penetration depth, which characterises the exponential decrease of the fluid's velocity field as a function of the distance from the oscillating sphere, and ρ is the density of the fluid. The term containing $\frac{dx(t)}{dt}$ is a friction term that dissipates energy and the term containing $\frac{d^2 x(t)}{dt^2}$ is an elastic term related to the inertia of the fluid moved by the particle. The penetration depth δ is frequency-dependent,

$$\delta(f) = \sqrt{\frac{\nu}{\pi f}} = a\sqrt{\frac{f_\nu}{f}}, \quad (9.53)$$

where ν is the kinematic friction coefficient of the fluid and $f_\nu = \nu/(\pi a^2)$. Typically, $\delta(f)$ will be quite large when compared to a for the frequency range in which we are interested.

Because Fourier decomposition describes any trajectory as a sum of linear oscillatory motions, the friction in Eq. (9.52) also appears in the frequency representation of the generalised Langevin equation describing the Brownian motion of a harmonically trapped sphere in an incompressible fluid,

$$\check{X}(f) = \frac{2k_B T \, \text{Re}\{\gamma_{\text{Stokes}}(f)\}}{m(-i2\pi f)^2 + \gamma_{\text{Stokes}}(f)(-i2\pi f) + \kappa} \check{W}(f), \quad (9.54)$$

where

$$\gamma_{\text{Stokes}}(f) = \frac{\check{F}_{\text{fric}}(f)}{i2\pi f \check{X}(f)} = \gamma\left[1 + (1-i)\frac{a}{\delta} - i\frac{2a^2}{9\delta^2}\right], \quad (9.55)$$

m is the mass of the sphere, κ the spring constant of the harmonic trapping force and $k_B T$ the thermal energy. This equation becomes the Einstein–Ornstein–Uhlenbeck theory in the limit of $f \to 0$.

Therefore, the hydrodynamically corrected value of the PSD is given by

$$P_{\text{hydro}}(f) = \frac{D/(2\pi)^2 \left[1 + (f/f_v)^2\right]}{\left[f_c - f^{3/2}/f_v^{1/2} - f^2/f_m\right]^2 + \left[f + f^{3/2}/f_v^{1/2}\right]^2}, \qquad (9.56)$$

where $f_m = \gamma/(2\pi \tilde{m})$ and $\tilde{m} = m + 2\pi \rho a^3/3 \approx 3m/2$ for normal microspheres.[15] This power spectrum contains the same two fitting parameters, f_c and D, as the Lorentzian of the Einstein–Ornstein–Uhlenbeck theory, but differs significantly from it, except at low frequencies. The radius a of the bead now also occurs in f_m and f_v, and not only through f_c, but it is not a parameter we must fit, because it is known to 1% uncertainty and occurs only in terms that are so small that this small uncertainty of a has negligible effect on $P_{\text{hydro}}(f)$.

The frictional force in Eq. (9.52) was derived by Stokes under the assumption that the oscillating sphere is infinitely deep inside the fluid volume. For optimal designs, the lens that focuses the light into an optical trap typically has a short focal length. So a microsphere caught in such a trap is typically near a microscope coverslip. Consequently, the hydrodynamic interaction between the microsphere and the essentially infinite surface of the coverslip must be accounted for. Faxén (1922) has done this for a sphere moving parallel to an infinite plane with constant velocity in an incompressible fluid bounded by the plane and asymptotically at rest under conditions of vanishing Reynolds number. Solving perturbatively in a/l, where l is the distance from the sphere's centre to the plane, Faxén found

$$\gamma_{\text{Faxen}}(a/l) = \frac{\gamma}{1 - \frac{9}{16}\frac{a}{l} + \frac{1}{8}\left(\frac{a}{l}\right)^3 - \frac{45}{256}\left(\frac{a}{l}\right)^4 - \frac{1}{16}\left(\frac{a}{l}\right)^5 + \ldots}. \qquad (9.57)$$

There is no second-order term in the denominator, so this formula remains good to within 1% for $l > 3a$, if one ignores all but the first-order term. This first-order result was first obtained by Lorentz. If his first-order calculation is repeated for a sphere undergoing linear oscillation parallel to a plane, one finds a friction formula that has Stokes' and Lorentz's formulas as limiting cases:

$$\gamma(f, a/l) = \gamma_{\text{Stokes}}(f) \left\{ 1 + \frac{9}{16}\frac{a}{l}\left[1 - \frac{1-i}{3}\frac{a}{l} + \frac{2i}{9}\left(\frac{a}{l}\right)^2 - \frac{4}{3}\left(1 - e^{-(1-i)(2l-a)/\delta}\right)\right]\right\}. \qquad (9.58)$$

The effect of the infinite plane is to increase friction, but less so at larger frequencies where δ is smaller.

By replacing $\gamma_{\text{Stokes}}(f)$ with $\gamma(f, a/l)$ in Eqs. (9.52) and (9.54), one obtains a power spectrum that accounts for all relevant physics of the bead in the trap, with the expected

---

[15] This simple relation between f_m and f_v might tempt one to eliminate one of these frequencies in favour of the other. However, they parameterise different physics. f_v parameterises the flow pattern established around a sphere undergoing linear harmonic oscillations in an incompressible fluid; this pattern is completely unrelated to the mass of the sphere, which in fact does not need to have any mass. f_m parameterises the time it takes for friction to dissipate the kinetic energy of the sphere and the fluid it entrains; it crucially depends on the mass of the sphere. By keeping both parameters in formulas, the physical origin of various terms is kept clear.

value

$$P_{\text{hydro}}(f,a/l) = \frac{D/(2\pi)^2 \text{Re}\left\{\frac{\gamma(f,a/l)}{\gamma}\right\}}{\left[f_c - f\text{Im}\left\{\frac{\gamma(f,a/l)}{\gamma}\right\} - f^2/f_m\right]^2 + \left[f\text{Re}\left\{\frac{\gamma(f,a/l)}{\gamma}\right\}\right]^2}, \quad (9.59)$$

where

$$\text{Re}\left\{\frac{\gamma(f,a/l)}{\gamma}\right\} = 1 + \sqrt{\frac{f}{f_\nu}} - \frac{3}{16}\frac{a}{l} + \frac{3}{4}\frac{a}{l}\exp\left\{-\frac{2l}{a}\sqrt{\frac{f}{f_\nu}}\right\}\cos\left\{-\frac{2l}{a}\sqrt{\frac{f}{f_\nu}}\right\}$$

and

$$\text{Im}\left\{\frac{\gamma(f,a/l)}{\gamma}\right\} = -\sqrt{\frac{f}{f_\nu}} + \frac{3}{4}\frac{a}{l}\exp\left\{-\frac{2l}{a}\sqrt{\frac{f}{f_\nu}}\right\}\sin\left\{-\frac{2l}{a}\sqrt{\frac{f}{f_\nu}}\right\}.$$

We remark that this physical power spectrum differs from the recorded power spectrum because the data acquisition system samples the signal at discrete times and contains filters.

9.8.3 Noise tests

The PSD can be used effectively to detect and quantify the presence of noise in the set-up. In particular, it is possible to perform two basic tests, which are described in the following.

The *dark spectrum* is the PSD recorded when the photodetector is kept in total darkness and is a measurement of the equipment's electronic noise. The dark spectrum shown in Fig. 9.15a is flat, except for a spike at 50 Hz from the power supply[16] and at a few higher frequencies. All values are a factor of 10^3 to 10^5 below that of the PSD of the optically trapped particles; hence, this noise may contribute about 1% to 10% to the spectra, because amplitudes, not spectra, add up. However, in addition to being small, these spikes are so narrow and few that (statistically) they do not matter in the calibration procedure.

The *light spectrum* is the PSD recorded with the laser light impinging directly on the photodiode while no microsphere is in the trap. The light spectrum is often a good approximation to the noise because of the mechanical instabilities of the set-up, e.g., caused by the limited pointing stability of the laser beam and of the optics it passes through on its way to the photodiode.[17] For example, the spectrum shown in Fig. 9.15b features a significant low-frequency noise that seems to fall off as f^{-2}, which is what one would observe if the direction of the laser was doing a slow random walk. The presence of mechanical vibrations in the set-up, usually due to the presence of some underdamped mechanical resonances, can also be identified from the light spectrum as they manifest themselves as peaks in the light spectrum; furthermore, it is possible to mitigate them by studying how the light spectrum varies as a function of changes introduced in the set-up.

---

[16] Clearly, this spike can be expected to be at 60 Hz in most of the Americas and in some Asian countries.

[17] Strictly speaking, however, the light spectrum only provides a lower bound on all mechanical noise. For example, mechanical vibrations may be transmitted to the fluid volume (and therefore to an optically trapped particle), but not to the light spectrum, because the fluid is transparent.

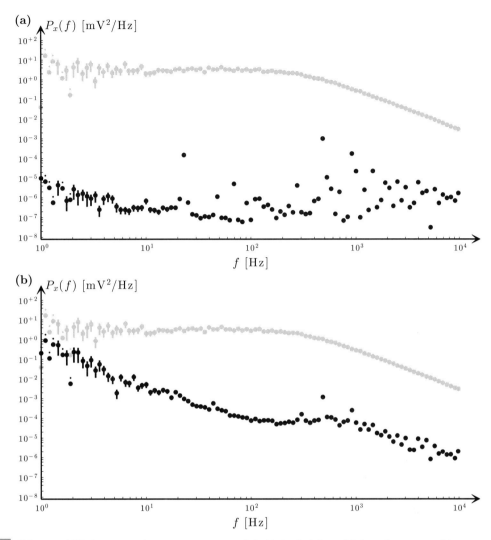

Figure 9.15 Noise tests. (a) Dark spectrum: the power spectrum recorded with the diode in total darkness is a measure of the electronic noise level. The spike at 50 Hz is caused by the power supply. All values are a factor of 10^3 to 10^5 below the optically trapped particle spectrum (grey spectrum). (b) Light spectrum: the power spectrum recorded with the trap's laser light impinging directly onto the photodiode with no microsphere in the trap is a measure of the stability of the set-up. The low-frequency noise is most likely caused by the presence of mechanical instabilities in the set-up.

9.9 Drag force method

Until now, we have only considered passive calibration methods. However, it is also possible to infer the spring constant κ_x by measuring the displacement of the particle in the optical trap resulting from the application of a known force. In the case of a spherical bead surrounded by a fluid medium such as water, this can be accomplished by monitoring the position of the trapped bead while the fluid flows past it at a known velocity v_{fluid}. The force on the bead due to the fluid flow is then given by

$$F_{\text{drag}} = -\gamma \, v_{\text{fluid}}. \tag{9.60}$$

The fluid flow can be produced by effectively flowing the fluid in a microfuidic chamber or, alternatively, by moving the sample stage at a velocity v_{stage} while keeping the trap fixed, in which case $v_{\text{fluid}} = v_{\text{stage}}$.

A more precise measurement can be obtained by trapping a bead in an optical trap without flow and then oscillating the cell holding the liquid that surrounds the trapped bead at a fixed frequency and amplitude. If the position of the cell as a function of time is given by

$$x_{\text{stage}}(t) = A \sin\left(2\pi f_{\text{stage}} t\right), \tag{9.61}$$

then the velocity of the cell, and therefore of the liquid surrounding the bead, is

$$v_{\text{stage}}(t) = 2\pi f_{\text{stage}} A \cos\left(2\pi f_{\text{stage}} t\right) \tag{9.62}$$

and the force due to viscous drag on the trapped bead is

$$F_{\text{drag}}(t) = -2\pi f_{\text{stage}} A \gamma \cos\left(2\pi f_{\text{stage}} t\right). \tag{9.63}$$

The resulting equation of motion is

$$\left[\frac{dx(t)}{dt} - v_{\text{stage}}(t)\right] + 2\pi f_c x(t) = \sqrt{2D} w_x(t). \tag{9.64}$$

Therefore, the PSD is

$$P(f) = \frac{D/\pi^2}{f^2 + f_c^2} + \frac{A^2}{2\left(1 + \frac{f_c}{f_{\text{stage}}}\right)} \delta(f - f_{\text{stage}}). \tag{9.65}$$

This PSD consists of the familiar Lorentzian first term, which originates from the Brownian motion of the bead in the harmonic trapping potential, plus a delta-function spike second term at the frequency with which the stage is driven. An example of such a PSD is shown in Fig. 9.16. By fitting the experimentally obtained PSD to Eq. (9.65), it is possible to determine at the same time the calibration factor from measurement units to metres, the diffusivity of the particle and the stiffness of the trap. In particular, the calibration factor for distances is

$$S_x^{(\text{ex})} = \sqrt{\frac{W_{\text{stage}}}{W_{\text{ex}}}}, \tag{9.66}$$

where W_{ex} is the experimentally determined power in the spike, measured in measurement units squared (typically volts squared), and W_{stage} is the same quantity measured in units of

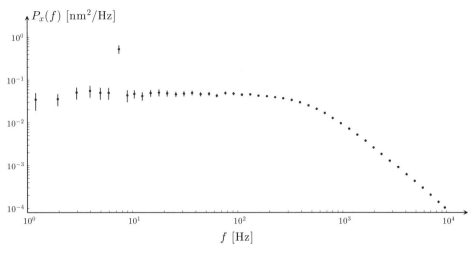

Figure 9.16 Oscillating optical tweezers. Power spectrum density of a microscopic particle held in an oscillating optical tweezers. It consists of a thermal background caused by Brownian motion plus a spike at f_{stage}. The stage is moving sinusoidally with frequency $f_{\text{stage}} = 8$ Hz and amplitude $A = 150$ nm along the x-direction.

length squared, which can be straightforwardly determined from Eq. (9.65) to be

$$\frac{A^2}{2\left(1 + \frac{f_c}{f_{\text{stage}}}\right)}. \tag{9.67}$$

Further details on this technique can be found in Tolić-Nørrelykke et al. (2006).

Problems

9.1 *Allan variance and trap stability.* The Allan variance is a measure of frequency stability commonly employed for clocks, oscillators and amplifiers. Given a time series of N samples and a total measurement time of $T_{\text{msr}} = N/f_{\text{msr}}$, where f_{msr} is the acquisition frequency, consider N_{int} time intervals of length $T_{\text{int}} = T_{\text{msr}}/N_{\text{int}}$, and denote $\bar{x}_i$ as the mean over the ith interval, then the Allan variance is defined as

$$\sigma^2_{\text{Allan},x}(T_{\text{int}}) = \frac{1}{2(N_{\text{int}} - 1)} \sum_{i=1}^{N_{\text{int}}-1} [\bar{x}_{i+1} - \bar{x}_i]^2.$$

First, show analytically that the standard error on the positional fluctuations of a trapped particle is

$$\frac{1}{\sqrt{N_{\text{int}}}} \sqrt{\overline{x^2}} \approx \sqrt{\frac{2k_B T \gamma}{\kappa_x^2 T_{\text{int}}}}$$

and check it on some experimental (or simulated) particle trajectory. This thermal limit cannot be beaten by any measurement that does not oversample a desired signal.

Now, following Czerwinski et al. (2009), calculate the square-root Allan variance (SRAV) for the trajectory of a trapped particle; show that its absolute maximum is found at the optical trap characteristic time τ_{ot} and that the SRAV overlaps with the thermal limit for $T_{int} > \tau_{ot}$. This is an indication of a very stable set-up. At what time does the SRAV reach a local minimum? What does this tell you about the optimum time span for your measurements? Investigate how the Allan variance changes when the optical trap is weaker or stronger.

9.2 *Enhanced position detection.* Study the possible use of multiple laser beams and detectors to track the position of a particle. Can such approaches increase the sensitivity or linear range of the detection?

9.3 *Non-spherical particles.* Study the possibility of performing interferometric detection of non-spherical optically trapped particles. Propose some configurations where it is possible to detect all degrees of freedom, i.e., the position of the particle centre of mass and the orientation of the particle within the optical trap.

9.4 *Cloaking.* Study for what kinds of particles the response of a photodetector can be made to show no sensitivity to the particle position, for example, following Volpe et al. (2007). Can this be interpreted as an effective cloaking of the particle?

9.5 *Viscoelastic media.* Explore how passive and active calibration methods can be combined in order to reconstruct the viscoelastic response of a medium from the position data of an optically trapped particle. You can, for example, follow the approach proposed by Buosciolo et al. (2004).

References

Berg-Sørensen, K., and Flyvbjerg, H. 2004. Power spectrum analysis for optical tweezers. *Rev. Sci. Instrumen.*, **75**, 594–612.

Buosciolo, A., Pesce, G., and Sasso, A. 2004. New calibration method for position detector for simultaneous measurements of force constants and local viscosity in optical tweezers. *Opt. Commun.*, **230**, 357–68.

Crocker, J. C., and Grier, D. G. 1996. Methods of digital video microscopy for colloidal studies. *J. Colloid Interface Sci.*, **179**, 298–310.

Czerwinski, F., Richardson, A. C., and Oddershede, L. B. 2009. Quantifying noise in optical tweezers by Allan variance. *Opt. Express*, **17**, 13 255–69.

Faxén, H. 1922. Der Widerstand gegen die Bewegung einer starren Kugel in einer zähen Flüssigkeit, die zwischen zwei parallelen ebenen Wänden eingeschlossen ist. *Ann. Physik*, **373**, 89–119.

Garbin, V., Volpe, G., Ferrari, E., Versluis, M., Cojoc, D., and Petrov, D. 2009. Mie scattering distinguishes the topological charge of an optical vortex: A homage to Gustav Mie. *New J. Phys.*, **11**, 013046.

Sbalzarini, I. F., and Koumoutsakos, P. 2005. Feature point tracking and trajectory analysis for video imaging in cell biology. *J. Struct. Biol.*, **151**, 182–95.

Tolić-Nørrelykke, S. F., Schäffer, E., Howard, J., et al. 2006. Calibration of optical tweezers with positional detection in the back focal plane. *Rev. Sci. Instrumen.*, **77**, 103101.

Volpe, G., Kozyreff, G., and Petrov, D. 2007. Backscattering position detection for photonic force microscopy. *J. Appl. Phys.*, **102**, 084701.

10 Photonic force microscope

One of the most successful applications of optical tweezers has been the measurement of small forces: because the optical restoring force is elastic, at least for most standard optical tweezers configurations, it is possible to measure the force acting on an optically trapped particle by multiplying its measured displacement from its equilibrium position and the optical trap stiffness. As shown in Fig. 10.1, the forces measurable by such a *photonic force microscope* (PFM) are typically in the range from tens of piconewtons down to a few femtonewtons. Because these forces are comparable to most forces acting between biomolecules, the PFM has been extremely successful at measuring such forces. In this chapter, we will review how the PFM can be employed to measure torques as well as forces, how the presence of surfaces or other particles affects the force measurement process and how great the influence of non-conservative effects is. Finally, we will describe how the optical force can be directly measured from the scattering associated with the trapping process.

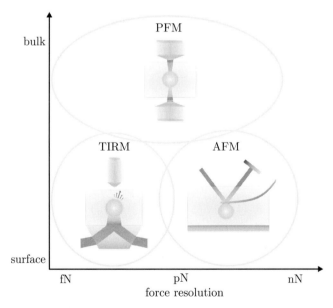

Figure 10.1 Force measurement techniques at the nanoscale. Main techniques for measuring forces at microscopic and nanoscopic length scales, classified according to their force resolution and the working conditions (surface/bulk) for which they are best suited: atomic force microscopy (AFM), photonic force microscopy (PFM) and total internal reflection microscopy (TIRM). Reprinted figure from Brettschneider et al., *Phys. Rev. E* **83**, 041113. Copyright (2011) by the American Physical Society.

10.1 Scanning probe techniques

Various techniques have been developed to probe the mechanical properties of microsystems. In the early 1990s, several kinds of scanning probe microscopy were already established. Binnig et al. (1982) invented the scanning tunnelling microscope (STM), which made it possible to resolve crystallographic structures and organic molecules at the atomic level, and later Binnig et al. (1986) invented the atomic force microscope (AFM). These instruments have been successfully employed to study biological and nanofabricated structures, overcoming the diffraction limit of traditional optical microscopes. Furthermore, they developed from pure imaging tools into more general techniques for the manipulation and measurement of nanoscopic objects such as single atoms and molecules. However, all these techniques require a macroscopic mechanical device to guide the probe and present some limitations when dealing with samples in a liquid environment. The advent in 1986 of the first three-dimensional optical trap paved the way towards new kinds of probes. In 1993, Ghislain and co-workers devised a new kind of scanning force microscopy using an optically trapped microparticle as a probe (Ghislain and Webb, 1993; Ghislain et al., 1994). This technique was later called *photonic force microscopy* (PFM) by Florin et al. (1997). The PFM provides the capability of measuring forces in the range from femtonewtons to piconewtons. Because these values are well below those that can be reached with techniques that are based on microfabricated mechanical cantilevers, the PFM ideally complements the AFM, as shown in Fig. 10.1.

A typical PFM comprises an *optical trap* that holds a probe – a dielectric or metallic particle, which randomly moves due to *Brownian motion* in the potential well formed by the optical trap – and a *position sensing system*. We have analysed the thermal motion of an optically trapped particle in Chapter 7, and we have seen how the three-dimensional particle position can be recorded using various devices, e.g., quadrant photodiodes, position sensing detectors or digital cameras, in Sections 9.1 and 9.2.

For a small displacement of the probe from the centre of an optical trap, the restoring force is proportional to the displacement itself. Hence, an optical trap acts like a Hookean spring with a fixed stiffness, which can be characterised by the various methods we have discussed in Chapter 9. In particular, autocorrelation analysis [Section 9.7] and power spectrum analysis [Section 9.8] are considered the most reliable, allowing one to determine the trap stiffness from the position fluctuations of the probe relying only on the knowledge of the temperature and the viscosity of the surrounding medium.

A constant and homogeneous external force $\mathbf{F}_{\text{ext}}$ acting on the probe results in a shift in the probe equilibrium position proportional to the force itself, as shown in Fig. 10.2 for a force acting along the x-direction. Therefore, the values of the trap stiffness being known, it is possible to retrieve the magnitude and direction of the force by measuring the average probe displacement, i.e.,

$$\begin{cases} F_{\text{ext},x} = \kappa_x \Delta x_{\text{eq}}, \\ F_{\text{ext},y} = \kappa_y \Delta y_{\text{eq}}, \\ F_{\text{ext},z} = \kappa_z \Delta z_{\text{eq}}, \end{cases} \quad (10.1)$$

Box 10.1 Total internal reflection microscopy

Total internal reflection microscopy (TIRM) is a sensitive non-invasive technique for measuring the interaction potential between a colloidal particle and a wall with femtonewton resolution. The equilibrium distribution of the particle–wall separation distance $z(t)$ is sampled by monitoring the intensity $I(t)$ scattered by the Brownian particle under evanescent illumination. A sketch of a typical TIRM geometry with a single colloidal particle in front of a planar transparent wall is shown below (a). The particle is illuminated by the evanescent field created by the total internal reflection of a laser beam at the glass-fluid interface while undergoing Brownian motion. The scattered intensity $I(t)$ is measured using a photomultiplier (PMT) (b). The forces acting on the particle are due to gravity $G_{\rm eff}$ and electrostatic interactions $F_{\rm el}$ (arrows). Knowing the relationship between the scattered intensity and distance, i.e., $I(z)$ (c), the intensity time series $I(t)$ can be converted into the particle trajectory $z(t)$ (d). More details on this technique can be found in Prieve (1999), Volpe et al. (2009) and Walz (1997).

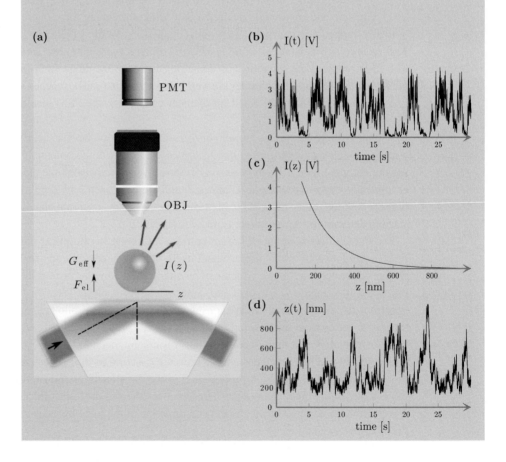

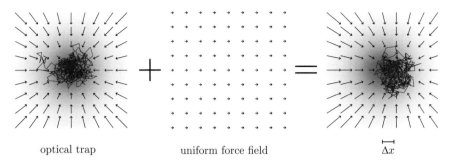

Figure 10.2 Photonic force microscope. An optical tweezers produces a force field that is harmonic, so that an optically trapped particle experiences an elastic restoring force with stiffness κ_x (force field on the left). As a uniform force $F_{\text{ext},x}$ acts on the optically trapped particle (force field in the centre), it displaces the particle probability distribution by Δx, so that one can measure $F_{\text{ext},x} = \kappa_x \Delta x$ (force field on the right).

where κ_x, κ_y and κ_z are the stiffnesses of the trap, and Δx_{eq}, Δy_{eq} and Δy_{eq} are the average particle displacements from the equilibrium position.

Exercise 10.1.1 Show that the Langevin equations of an optically trapped particle under the action of an external force $\mathbf{F}_{\text{ext}}$ are (in three dimensions)

$$\begin{cases} \dfrac{dx(t)}{dt} = -\dfrac{F_{\text{ext},x}}{\gamma} - \dfrac{\kappa_x}{\gamma} x(t) + \sqrt{2D}\, W_x(t), \\ \dfrac{dy(t)}{dt} = -\dfrac{F_{\text{ext},y}}{\gamma} - \dfrac{\kappa_y}{\gamma} y(t) + \sqrt{2D}\, W_y(t), \\ \dfrac{dz(t)}{dt} = -\dfrac{F_{\text{ext},z}}{\gamma} - \dfrac{\kappa_z}{\gamma} z(t) + \sqrt{2D}\, W_z(t), \end{cases} \qquad (10.2)$$

where x and y represent the position of the particle in the plane perpendicular to the beam propagation direction, z represents the position of the particle along the propagation direction, κ_x, κ_y and κ_z are the stiffnesses of the trap, γ is the particle friction coefficient, and $W_x(t)$, $W_y(t)$ and $W_z(t)$ are independent white noises. Study them numerically and demonstrate the validity of Eqs. (10.1). [Hint: Start from the Langevin equations of an optically trapped particle given in Eqs. (7.23).]

Exercise 10.1.2 Show analytically that the autocorrelation function and power spectrum of an optically trapped particle are not affected by a constant force. [Hint: Assume a harmonic trapping potential.]

Exercise 10.1.3 Study how the error in the force measurement with the PFM technique depends on the stiffness of the optical trap as well as on the acquisition time for each measurement. Note that the signal-to-noise ratio (SNR) does not depend on the stiffness of the optical trap, but only on the acquisition time. [Hint: You can perform this study both numerically and analytically.]

Exercise 10.1.4 In an optical trap, restoring optical forces are elastic only for relatively small displacements of the probe (typically within the radius of the particle). Study numerically what happens for larger displacements. In particular, how are Eqs. (10.1) altered?

Exercise 10.1.5 Optical tweezers can also be used as *force transducers*. For small displacements of the probe, the applied force is elastic, whereas for relatively large displacements of the particle (typically similar to the radius) the restoring optical force becomes locally constant. Show how this fact can be used to produce a force clamp optical tweezers that permits one to apply a constant force using an optically trapped particle. [Hint: Study the work by Lang et al. (2002), which developed this technique and applied it to single-molecule studies.]

10.2 Photonic torque microscope

As we have seen in the previous section, the PFM has been employed successfully to measure forces in the piconewton and femtonewton range; it works by measuring the displacement of an optically trapped probe under the action of a homogeneous force field [Fig. 10.2]. Another kind of mechanical perturbation is a torque acting on the optically trapped particle. As shown in Fig. 10.3, the presence of a rotational force field centred on the trap axis does not produce any shift in the equilibrium position of an optically trapped particle, and therefore it is not possible to apply the PFM technique as described in the previous section. Nevertheless, the rotational force field induces a cross-correlation between the movement along the x- and y-directions, which can be experimentally measured and used to determine the value of the rotational force field. In the rest of this section, we will explain this technique following the original work by Volpe and Petrov (2006) and Volpe et al. (2007).

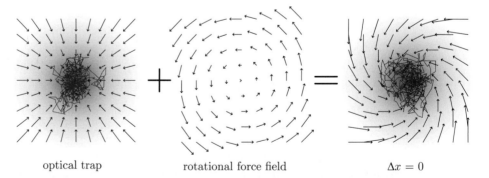

Figure 10.3 Photonic force microscope with rotational force fields. If a rotational force field is superimposed on the harmonic trapping potential generated by an optical tweezers, no shift of the optically trapped particle occurs. Therefore, more sophisticated statistical analysis techniques are necessary to detect the presence of the rotational force field than to detect the one presented in Fig. 10.2.

Consider a sphere of mass m and radius a suspended in a liquid medium and confined within a harmonic potential well, where it moves randomly because of thermal excitation. In the presence of a rotational force field in the xy-plane, the sphere rotates around the z-axis with a constant angular velocity $\mathbf{\Omega} = \Omega \hat{\mathbf{z}}$, whose value results from a balance between the torque applied to the sphere and the drag torque, i.e.,

$$\mathbf{T}_{\text{drag}} = \mathbf{r} \times \mathbf{F}_{\text{drag}} = \gamma \mathbf{r} \times \mathbf{v} = \gamma \mathbf{r} \times (\mathbf{r} \times \mathbf{\Omega}), \tag{10.3}$$

where $\mathbf{r}$ is the sphere position, $\mathbf{v}$ is its velocity, $\gamma = 6\pi \eta a$ is its friction coefficient and η is the medium viscosity. Hence, the time average of the torque exerted on the particle is

$$\overline{\mathbf{T}} = \gamma \overline{\mathbf{r} \times (\mathbf{r} \times \mathbf{\Omega})} = \gamma \Omega \sigma_{xy}^2 \hat{\mathbf{z}}, \tag{10.4}$$

where σ_{xy}^2 is the variance of the particle position in the plane orthogonal to the torque. Adding the rotational force field to the Langevin equations for the Brownian motion of an optically trapped particle in the xy-plane [Eq. (7.23)], we obtain

$$\begin{cases} \dfrac{dx(t)}{dt} = -\dfrac{\kappa}{\gamma} x(t) - \Omega y(t) + \sqrt{2D}\, W_x(t) \\ \dfrac{dy(t)}{dt} = -\dfrac{\kappa}{\gamma} y(t) + \Omega x(t) + \sqrt{2D}\, W_y(t), \end{cases} \tag{10.5}$$

where κ is the stiffness of the optical trap in the xy-plane, D is the particle diffusion coefficient, and $W_x(t)$ and $W_y(t)$ are independent white noises. The terms $-\Omega y(t)$ and $+\Omega x(t)$ introduce a coupling between the equations, which becomes apparent in the fact that the cross-correlation is non-zero. The resulting autocorrelation functions (ACFs) and cross-correlation function (CCF) are

$$C_x(\tau) = C_y(\tau) = \frac{k_B T}{\kappa} e^{-\frac{|\tau|}{\tau_{\text{ot}}}} \cos(\Omega \tau) \tag{10.6}$$

and

$$C_{xy}(\tau) = \frac{k_B T}{\kappa} e^{-\frac{|\tau|}{\tau_{\text{ot}}}} \sin(\Omega \tau), \tag{10.7}$$

where $\tau_{\text{ot}} = \gamma/\kappa$ is the trap characteristic time, k_B is the Boltzmann constant and T is the absolute temperature. The actual shape of the correlation functions depends on the value of $\tau_{\text{ot}}\Omega$. For $\tau_{\text{ot}}\Omega < 1$ [Fig. 10.4a], the restoring force induced by the potential well is dominant and, at the limit $\tau_{\text{ot}}\Omega \ll 1$, the ACF is reduced to the expression for the behaviour of a Brownian particle in a harmonic potential given by Eq. (9.32), whereas the cross-correlation is negligible; this is the case of the standard PFM, which we have studied in the previous section. For $\tau_{\text{ot}}\Omega > 1$ [Fig. 10.4c], the rotational effect is dominant and, at the limit $\tau_{\text{ot}}\Omega \gg 1$, we get the sinusoidal ACFs and CCFs typical of a rotating particle; in this case, the torque can be measured straightforwardly from the particle rotation rate, which can be assessed either by digital video microscopy or by measuring the modulation frequency of the light scattered by the particle as it rotates. Analysis of the ACFs and CCFs

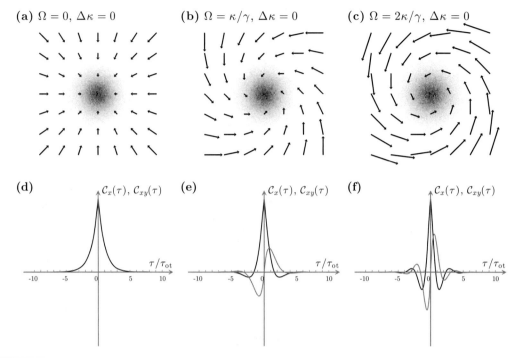

Figure 10.4 Photonic force microscope in a rotationally symmetric potential. (a)–(c) Photonic torque microscope with a rotationally symmetric harmonic potential, i.e., $\kappa_x = \kappa_y$ or, equivalently, $\Delta\kappa = 0$, for various values of Ω. The shaded areas show the probability distribution function of the probe particle position in the corresponding force field (arrows). When $\Omega = 0$, the probe movement can be separated along two orthogonal directions. When the (absolute) value of Ω increases, the force field lines bend and the probe movements along the x- and y-directions are not independent any more. (d)–(f) Corresponding autocorrelation functions $C_x(\tau)$ (black solid line) and cross-correlation functions $C_{xy}(\tau)$ (grey solid line) [Eqs. (10.6) and (10.7)]. Note that $C_y(\tau) = C_x(\tau)$ and $C_{yx}(\tau) = C_{xy}(-\tau)$.

given by Eqs. (10.6) and (10.7) bridges these two cases, providing new insights into the intermediate situation, such as that shown in Fig. 10.4b.

Exercise 10.2.1 Simulate the Brownian motion of an optically trapped particle in a rotational force field [Eqs. (10.5)]. Observe how the trajectories, the ACFs and the CCFs change as a function of the trap sitffness κ, the friction coefficient γ and the angular velocity Ω. [Hint: You can adapt the code `torque` from the book website.]

Exercise 10.2.2 Demonstrate that the motion of a Brownian particle near an equilibrium position, $\mathbf{r}_{eq} = \mathbf{0}$, in an external force field, $\mathbf{F}_{ext}(\mathbf{r})$, is

$$\frac{d\mathbf{r}(t)}{dt} = \gamma^{-1}\mathbf{J_0}\mathbf{r}(t) + \sqrt{2D}\mathbf{W}(t), \tag{10.8}$$

where $\mathbf{r}(t) = [x(t), y(t)]^T$ is the particle position, $\mathbf{W}(t) = [W_x(t), W_y(t)]^T$ is a vector of independent white noises and $\mathbf{J_0}$ is the Jacobian of $\mathbf{F}_{ext}(\mathbf{r})$ calculated at the equilibrium point. Some examples of force fields are shown in Figs. 10.4a–c and 10.5a–c.

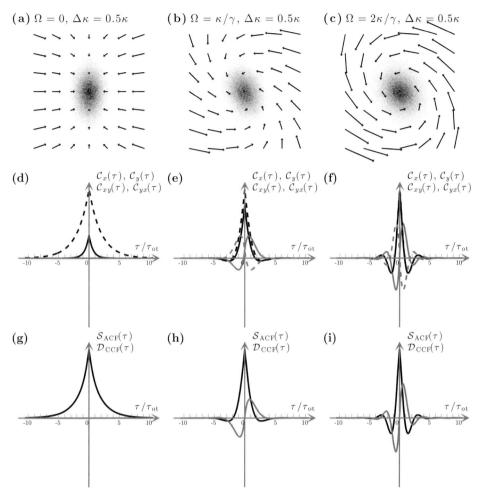

Figure 10.5 Photonic force microscope in a non-rotationally-symmetric potential. (a)–(c) Photonic torque microscope with a non-rotationally-symmetric harmonic potential, i.e., $\kappa_x \neq \kappa_y$ or, equivalently, $\Delta\kappa \neq 0$, and for various values of Ω. The shaded areas show the probability distribution function of the probe particle position in the corresponding force field (arrows). As the value of $\Delta\kappa$ increases, the probability density function becomes more and more elliptical, until for $\Delta\kappa \geq \kappa$ the probe is confined only along the x-direction and the confinement along the y-direction is lost (not shown). For values of $\Omega \geq \kappa/\gamma$, the rotational component of the force field becomes dominant over the conservative one. The presence of a rotational component masks the asymmetry in the conservative one, since the probability density function assumes a more rotationally symmetric shape. (d)–(f) Corresponding autocorrelation and cross-correlation functions [Eqs. (10.16), (10.17), (10.18) and (10.19)]: $\mathcal{C}_x$ (black solid line), $\mathcal{C}_y$ (black dashed line), $\mathcal{C}_{xy}$ (grey solid line), and $\mathcal{C}_{yx}$ (grey dashed line). (g)–(i) Corresponding sum (black line) and difference (grey line) [Eqs. (10.23) and (10.24)], which are independent from the choice of the reference system.

Show that $\mathbf{J}_0$ can be written as

$$\mathbf{J}_0 = \mathbf{J}_c + \mathbf{J}_{nc}, \qquad (10.9)$$

where

$$\mathbf{J}_c = \begin{bmatrix} \frac{\partial f_x(0)}{\partial x} & \frac{1}{2}\left(\frac{\partial f_x(0)}{\partial y} + \frac{\partial f_y(0)}{\partial x}\right) \\ \frac{1}{2}\left(\frac{\partial f_y(0)}{\partial x} + \frac{\partial f_x(0)}{\partial y}\right) & \frac{\partial f_y(0)}{\partial y} \end{bmatrix} \qquad (10.10)$$

and

$$\mathbf{J}_{nc} = \begin{bmatrix} 0 & \frac{1}{2}\left(\frac{\partial f_x(0)}{\partial y} - \frac{\partial f_y(0)}{\partial x}\right) \\ \frac{1}{2}\left(\frac{\partial f_y(0)}{\partial x} - \frac{\partial f_x(0)}{\partial y}\right) & 0 \end{bmatrix}. \qquad (10.11)$$

Show that $\mathbf{J}_c$ is the conservative component of the force field, and that $\mathbf{J}_{nc}$ is the rotational component. Furthermore, note that, if the coordinate system is chosen so that $\frac{\partial f_x(0)}{\partial y} = -\frac{\partial_x f_y(0)}{\partial x}$,

$$\mathbf{J}_0 = \begin{bmatrix} -\kappa_x & \gamma\Omega \\ -\gamma\Omega & -\kappa_y \end{bmatrix}, \qquad (10.12)$$

where $\kappa_x = -\frac{\partial f_x(\bar{r})}{\partial x}$, $\kappa_y = -\frac{\partial f_y(\bar{r})}{\partial y}$ and $\Omega = \gamma^{-1}\frac{\partial f_x(\bar{r})}{\partial y} = -\gamma^{-1}\frac{\partial f_y(\bar{r})}{\partial x}$ is the value of the constant angular velocity of the probe rotation around the z axis due to the presence of the rotational force field. [Hint: Use the Helmholtz decomposition [Box 5.2]. Note that each linear conservative force field comes from a quadratic potential $U(x, y) = ax^2 + bxy + cy^2$ and a symmetric Jacobian can be always associated with such a potential in an appropriate frame of reference.]

Exercise 10.2.3 Demonstrate that the conditions for the stability of the Brownian motion described by Eqs. (10.8) are

$$\begin{cases} \mathrm{Det}\,(\mathbf{J}_0) = \kappa^2 - \Delta\kappa^2 + \gamma^2\Omega^2 > 0, \\ \mathrm{Tr}\,(\mathbf{J}_0) = -2\kappa < 0, \end{cases} \qquad (10.13)$$

where $\kappa = (\kappa_x + \kappa_y)/2$ and $\Delta\kappa = (\kappa_x - \kappa_y)/2$. Reproduce the stability diagram shown in Fig. 10.6.

Exercise 10.2.4 Show that the solution of Eq. (10.8) in a coordinate system where the conservative and rotational components of the force field are readily separable, i.e., where the Jacobian is given by Eq. (10.12), is given by

$$\mathbf{r}(t) = \sqrt{2D}\int_{-\infty}^{t} \mathbf{K}(t)\mathbf{K}^{-1}(s)\mathbf{W}(s)ds, \qquad (10.14)$$

where

$$\mathbf{K}(t) = \gamma^{-1}\begin{bmatrix} \Delta\kappa + \sqrt{\Delta\kappa^2 - \gamma^2\Omega^2} & \gamma\Omega \\ \gamma\Omega & \Delta\kappa - \sqrt{\Delta\kappa^2 - \gamma^2\Omega^2} \end{bmatrix}\begin{bmatrix} e^{\lambda_+ t} & 0 \\ 0 & e^{\lambda_- t} \end{bmatrix} \qquad (10.15)$$

is the Wronskian of the system and $\lambda_\pm = -\phi \pm \sqrt{\Delta\phi^2 - \Omega^2}$ are its eigenvalues.

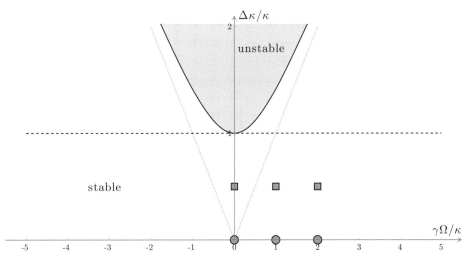

Figure 10.6 Stability diagram. Stability diagram of a particle in a potential well as a function of the parameters $\gamma\Omega/\kappa$ and $\Delta\kappa/\kappa$. The parameters corresponding to the white region are stable, i.e., satisfy the stability conditions given in Eq. (10.13). The dashed line represents $\Delta\kappa = \kappa$ and the dotted line represents the $\Delta\kappa = \gamma|\Omega|$ curves. The standard PFM, discussed in Section 10.1, corresponds to $\Delta\kappa = 0$ and $\Omega = 0$. When a rotational term is added, i.e., $\Omega \neq 0$ and $\Delta\kappa = 0$, the system remains stable. When there is no rotational contribution to the force field ($\Omega = 0$), the equilibrium point becomes unstable as soon as $\Delta\kappa \geq \kappa$, as this implicates that $\kappa_y < 0$ and, therefore, the probe is not confined along the y-direction. In the presence of a rotational component ($\Omega \neq 0$), the stability region becomes larger, as the equilibrium point now becomes unstable only for $\Delta\kappa \geq \sqrt{\kappa^2 + \gamma^2\Omega^2}$. The circles represent the parameters that are further investigated in Fig. 10.4 and the squares those that are investigated in Fig. 10.5.

Exercise 10.2.5 Using Eq. (10.14), show that the ACFs and CCFs of the probe motion near an equilibrium position are

$$C_x(\tau) = \frac{k_B T}{\kappa} e^{-\frac{|\tau|}{\tau_{\text{ot}}}}$$
$$\times \left[\left(\frac{\gamma^2 \Omega^2 - \alpha^2 \Delta\kappa^2}{\gamma^2 \Omega^2 - \Delta\kappa^2} - \alpha^2 \frac{\Delta\kappa}{\kappa} \right) \mathcal{C}(\tau) - \alpha^2 \frac{\Delta\kappa}{\kappa} \left(1 - \frac{\Delta\kappa}{\kappa} \right) \mathcal{S}(|\tau|) \right], \quad (10.16)$$

$$C_y(\tau) = \frac{k_B T}{\kappa} e^{-\frac{|\tau|}{\tau_{\text{ot}}}}$$
$$\times \left[\left(\frac{\gamma^2 \Omega^2 - \alpha^2 \Delta\kappa^2}{\gamma^2 \Omega^2 - \Delta\kappa^2} + \alpha^2 \frac{\Delta\kappa}{\kappa} \right) \mathcal{C}(\tau) + \alpha^2 \frac{\Delta\kappa}{\kappa} \left(1 + \frac{\Delta\kappa}{\kappa} \right) \mathcal{S}(|\tau|) \right], \quad (10.17)$$

$$C_{xy}(\tau) = \frac{k_B T}{\kappa} e^{-\frac{|\tau|}{\tau_{\text{ot}}}} \frac{\gamma\Omega}{\kappa} \left[+\mathcal{S}(\tau) + \alpha^2 \frac{\Delta\kappa}{\kappa} (\mathcal{C}(\tau) + \mathcal{S}(|\tau|)) \right], \quad (10.18)$$

$$C_{yx}(\tau) = \frac{k_B T}{\kappa} e^{-\frac{|\tau|}{\tau_{\text{ot}}}} \frac{\Omega}{\kappa} \left[-\mathcal{S}(\tau) + \alpha^2 \frac{\Delta\kappa}{\kappa} (\mathcal{C}(\tau) + \mathcal{S}(|\tau|)) \right], \quad (10.19)$$

where

$$\alpha^2 = \frac{\kappa^2}{\kappa^2 + (\gamma^2\Omega^2 - \Delta\kappa^2)} \qquad (10.20)$$

is a dimensionless positive parameter,

$$C(t) = \begin{cases} \cos\left(\gamma^{-1}\sqrt{|\gamma^2\Omega^2 - \Delta\kappa^2|}t\right) & \gamma^2\Omega^2 > \Delta\kappa^2 \\ 1 & \gamma^2\Omega^2 = \Delta\kappa^2 \\ \cosh\left(\gamma^{-1}\sqrt{|\gamma^2\Omega^2 - \Delta\kappa^2|}t\right) & \gamma^2\Omega^2 < \Delta\kappa^2 \end{cases} \qquad (10.21)$$

and

$$S(t) = \begin{cases} \dfrac{\kappa \sin\left(\gamma^{-1}\sqrt{|\gamma^2\Omega^2 - \Delta\kappa^2|}t\right)}{\sqrt{|\gamma^2\Omega^2 - \Delta\kappa^2|}} & \gamma^2\Omega^2 > \Delta\kappa^2, \\ \dfrac{\kappa}{\gamma}t & \gamma^2\Omega^2 = \Delta\kappa^2, \\ \dfrac{\kappa \sinh\left(\gamma^{-1}\sqrt{|\gamma^2\Omega^2 - \Delta\kappa^2|}t\right)}{\sqrt{|\gamma^2\Omega^2 - \Delta\kappa^2|}} & \gamma^2\Omega^2 < \Delta\kappa^2. \end{cases} \qquad (10.22)$$

In Figs. 10.4d–f and 10.5d–f, these functions are plotted for different ratios of the conservative and rotational components of the force field.

Exercise 10.2.6 The expression for the ACFs and CCFs in Eqs. (10.16), (10.17), (10.18) and (10.19) were obtained in a specific coordinate system, where the conservative and rotational components of the force field can readily be identified. However, typically the experimental time series of the probe position are acquired in a rotated coordinate system. Show that, if the rotated coordinate system is such that $\mathbf{r}'(t) = \mathbf{R}\mathbf{r}$, where $\mathbf{r}'(t) = [x'(t), y'(t)]^{\mathrm{T}}$, $\mathbf{r}(t) = [x(t), y(t)]^{\mathrm{T}}$ and

$$\mathbf{R} = \begin{bmatrix} \cos\theta & -\sin\theta \\ \sin\theta & \cos\theta \end{bmatrix},$$

the ACFs and CCFs are linear combinations of Eqs. (10.16)-(10.19); i.e.,

$$C_{x'}(\tau) = \cos^2\theta\, C_x(\tau) - \cos\theta\sin\theta\, C_{xy}(\tau) - \sin\theta\cos\theta\, C_{yx}(\tau) + \sin^2\theta\, C_{yy}(\tau),$$

$$C_{y'}(\tau) = \sin^2\theta\, C_x(\tau) + \sin\theta\cos\theta\, C_{xy}(\tau) + \cos\theta\sin\theta\, C_{yx}(\tau) + \cos^2\theta\, C_{yy}(\tau),$$

$$C_{x'y'}(\tau) = \cos\theta\sin\theta\, C_x(\tau) + \cos^2\theta\, C_{xy}(\tau) - \sin^2\theta\, C_{yx}(\tau) - \sin\theta\cos\theta\, C_{yy}(\tau),$$

$$C_{y'x'}(\tau) = \sin\theta\cos\theta\, C_x(\tau) - \sin^2\theta\, C_{yy}(\tau) + \cos^2\theta\, C_{yx}(\tau) - \cos\theta\sin\theta\, C_{yy}(\tau).$$

Note that, although these ACFs and CCFs depend on θ, the difference of the CCFs,

$$\mathcal{D}_{\mathrm{CCF}}(\tau) = C_{x'y'}(\tau) - C_{y'x'}(\tau) = \frac{k_B T}{\kappa} e^{-\frac{|\tau|}{\tau_{\mathrm{ot}}}} \frac{\gamma\Omega}{\kappa} S(\tau), \qquad (10.23)$$

and the sum of the ACFs,

$$S_{\text{ACF}}(\tau) = C_{x'}(\tau) + C_{y'}(\tau)$$
$$= 2D\frac{e^{-\phi|\tau|}}{\phi}\left[\left(\frac{\Omega^2 - \alpha^2\Delta\phi^2}{\Omega^2 - \Delta\phi^2}\right)C(\tau) + \alpha^2\frac{\Delta\phi^2}{\phi^2}S(|\tau|)\right]. \quad (10.24)$$

are invariant. These functions are shown for some parameter values in Fig. 10.5.

Exercise 10.2.7 Once the probe position time series $\mathbf{r}'(t) = [x'(t), y'(t)]^{\text{T}}$ has been acquired in a generic coordinate system, it needs to be statistically analysed in order to reconstruct all the parameters of the force field, i.e., κ, $\Delta\kappa$, Ω and θ. Show that the data analysis can be performed in three steps:

1. evaluation of the parameters κ, $\Delta\kappa$ and Ω by a fitting to $\mathcal{D}_{\text{CCF}}(\tau)$ [Eq. (10.23)] and $\mathcal{S}_{\text{ACF}}(\tau)$ [Eq. (10.24)];
2. estimation of the the rotation angle θ by solving

$$\begin{cases} \cos(2\theta) = \dfrac{\kappa\,\mathcal{D}_{\text{ACF}}(0) - \gamma\Omega\,\mathcal{S}_{\text{CCF}}(0)}{\kappa - 2k_{\text{B}}T\alpha^2\frac{\Delta\kappa}{\kappa}\left(\frac{\gamma\Omega}{\kappa}\right)^2}, \\[2ex] \sin(2\theta) = \dfrac{\gamma\Omega\,\mathcal{D}_{\text{ACF}}(0) - \kappa\,\mathcal{S}_{\text{CCF}}(0)}{\kappa - 2k_{\text{B}}T\alpha^2\frac{\Delta\kappa}{\kappa}\left(\frac{\gamma\Omega}{\kappa}\right)^2}, \end{cases} \quad (10.25)$$

where $\mathcal{D}_{\text{ACF}}(\tau) = C_{x'}(\tau) - C_{y'}(\tau)$ and $\mathcal{S}_{\text{CCF}}(\tau) = C_{x'y'}(\tau) + C_{y'x'}(\tau)$;[1]
3. reconstruction of the total force field, possibly subtracting the trapping force field to retrieve the external force field under investigation.

Apply this analysis method to some numerically simulated data.

10.3 Force measurement near surfaces

The forces acting on a microscopic object immersed in a fluid medium can be assessed either by studying the underlying potential or by studying their effect on the object's trajectory. The first approach – to which we shall refer as the *equilibrium distribution method* – requires sampling the equilibrium distribution of the particle position. Accordingly, it can be only applied under conditions where the investigated system is at (or close to) thermodynamic equilibrium with a heat bath. The second method – to which we shall refer as the *drift method* – does not require the object to be at (or even close to) thermal equilibrium. Therefore, it can be applied also to systems that are intrinsically out of equilibrium, e.g., molecular machines, transport through pores and DNA stretching. The latter method, however, requires the recording of the object trajectory with high sampling

---

[1] If $\Delta\kappa = 0$, there is radial symmetry and any orientation can be used. If $\Omega = 0$, the orientation of the coordinate system coincide with the axes of the position distribution function ellipsoid so that a convenient alternative means to determine their directions is to apply a principal component analysis (PCA) decomposition.

rates, which can be technologically more challenging, in particular when combined with a high spatial resolution. In this section, following the work of Volpe et al. (2010) and Brettschneider et al. (2011), we will review the challenges that arise when such methods are applied in the presence of diffusion gradients, which typically occur when Brownian particles diffuse near surfaces or in the presence of other particles.

10.3.1 Equilibrium distribution method

A microscopic object in contact with a thermal heat bath at constant temperature does not come to rest, but keeps jiggling around because of the presence of thermal agitation. As we have seen in Section 7.3, when the particle is subjected to an external potential $U(z)$, this leads to the Boltzmann position distribution given by Eq. (7.19),

$$\rho(z) = \exp\left(-\frac{U(z)}{k_B T}\right),$$

where k_B is the Boltzmann constant and T is the temperature of the heat bath. Therefore, it is possible to sample the steady-state position probability distribution $\rho(z)$ by measuring a large number of uncorrelated object positions, as shown in Fig. 10.7a. The equilibrium potential can then be derived as $U(z) = -k_B T \ln \rho(z)$, as shown in Fig. 10.7b, and the force as

$$F(z) = -\frac{dU(z)}{dz} = \frac{k_B T}{\rho(z)} \frac{d\rho(z)}{dz}, \tag{10.26}$$

as shown in Fig. 10.7c. Because of the exponential dependence of the probability distribution on the potential depth, in typical experiments only potential minima within less than $\approx 5\,k_B T$ are explored by the particle.

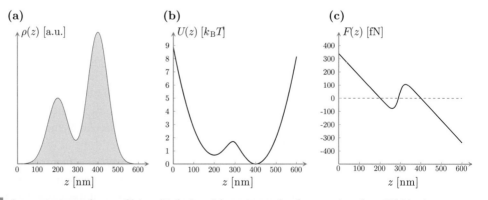

Figure 10.7 Force measurement from equilibrium distributions. Schematic procedure for measuring a force $F(z)$ by the equilibrium distribution method. From (a) the measured equilibrium probability distribution $\rho(z)$ of a particle, one obtains (b) the potential energy distribution $U(z) = -k_B T \ln \rho(z)$, which then gives (c) the force $F(z) = -\frac{d}{dz}U(z)$ [Eq. (10.26)].

10.3.2 Drift method

Because, for a microscopic body suspended in a liquid medium, viscous forces prevail over inertial effects by several orders of magnitude, a constant force F applied to a microscopic particle results in a constant drift velocity $v = F/\gamma$, where γ is the object's friction coefficient. Because $v = \Delta z/\Delta t$ can be retrieved from the measured particle displacement Δz within time Δt, the force can be measured as

$$F = \gamma \frac{\Delta z}{\Delta t}. \tag{10.27}$$

For large forces, i.e., forces whose effects are much larger than those of the thermal noise, this leads to a univocal result. However, when the *drift* force amplitude is comparable to the effect of the thermal noise, the measured particle displacement Δz and, thus, the drift force vary between identical experiments, leading to a statistical *distribution* of the measured values [Fig. 10.8],

$$F(z) = \gamma \left\langle \frac{\Delta z_j(z)}{\Delta t} \right\rangle, \tag{10.28}$$

where $\Delta z_j(z)$ denotes the jth experimental value of the particle's displacement after time Δt.

Although Eq. (10.28) is the key to measuring forces under nonequilibrium conditions, it is only valid in situations where the diffusion coefficient $D = k_{\rm B} T/\gamma$ of the object to which the force is applied is constant. When D becomes position-dependent, Eq. (10.28) must be corrected by an additional term,

$$F(z) = \gamma(z) \left\langle \frac{\Delta z_j(z)}{\Delta t} \right\rangle - \underbrace{\gamma(z) \frac{d D(z)}{dz}}_{\text{spurious force}}. \tag{10.29}$$

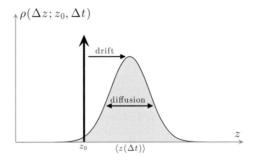

Figure 10.8 Force measurement from drift velocity. Schematic view of the propagator $\rho(\Delta z; z_0, \Delta t)$, which gives the distribution of the particle position increments Δz from its initial position z_0 after a time step Δt. The distribution is Gaussian for sufficiently small Δt with standard deviation $\sqrt{2 D(z_0) \Delta t}$. The force can be estimated from the measured drift according to Eq. (10.29).

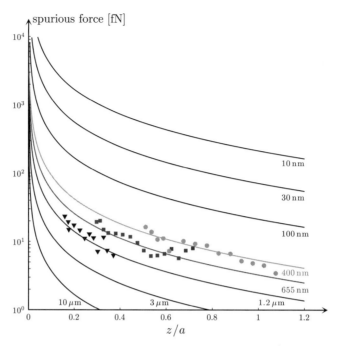

Figure 10.9 Spurious force. Distance dependence of the theoretically calculated spurious force $\gamma \frac{d}{dz} D$ for various particle radii a (solid lines). Experimentally measured spurious forces are shown for $a = 400$ nm (circles), $a = 655$ nm (squares) and $a = 1180$ nm (triangles) (Brettschneider et al., 2011).

This spurious force is related to the spurious drift that we discussed in Section 7.6 and, in particular, in Eq. (7.37).[2] The spurious force depends only on the particle radius, and is independent of the particle's material and density. In particular, as can be seen in Fig. 10.9, it increases for particles closer to a wall and for smaller particles because of the higher values of the diffusion coefficient, reaching values on the order of several piconewtons for particles with radius $a = 10$ nm, i.e., the size of a macromolecule.

Exercise 10.3.1 Simulate the motion of a Brownian particle in a diffusion gradient (e.g., the gradient produced by the presence of a planar wall as described in Section 7.6) and

---

[2] A qualitative physical understanding of the correction term in Eq. (10.29) can be gained by considering the effect of a diffusion gradient on a Brownian particle initially localised at position z_{initial} at time t_0. In the simplest picture, the particle diffusion results in a dichotomous movement either to the left or to the right with the same probability; therefore after time Δt the particle is displaced to $z_{\text{initial}} \pm \sqrt{2D\Delta t}$. In a more realistic picture, the final particle position has a continuous probability distribution. In both cases, assuming D constant, the final particle position distribution $\rho(\Delta z; z_0, \Delta t)$ is symmetric. In the presence of a diffusion gradient, the value of D is different at the initial and final positions and, therefore, the evaluation of the displacement is not univocal. Assuming $D = D(z_{\text{intial}})$, $\rho(\Delta z; z_0, \Delta t)$ is symmetric, as in the constant diffusion coefficient case. However, it could be argued that D should be averaged over the particle displacement; assuming thus that $D = D(z_{\text{middle}})$, $\rho(\Delta z; z_0, \Delta t)$ becomes asymmetric, because the particle displaces further when moving towards increasing diffusion. Finally, assuming that $D = D(z_{\text{final}})$, $\rho(\Delta z; z_0, \Delta t)$ becomes even more asymmetric. The *spurious drift* and the related *spurious force* account for such asymmetry.

evaluate the forces acting on the particle using both the equilibrium distribution method and the drift method. In particular, evaluate the spurious force and how it scales with particle size and particle-wall separation. [Hint: You can use the code `driffusion` from the book website.]

10.4 Relevance of non-conservative effects

To use optical tweezers to assess the mechanical properties of microscopic systems, it is crucial to have an accurately calibrated optical probe. As we have seen in the Chapter 9 and in the previous sections of this chapter, there are several straightforward methods of experimentally measuring the trap parameters – i.e., the trap stiffness and the conversion factor between measurement units and length – and, therefore, the force exerted by the optical tweezers on an object. An implicit assumption of most of these calibration methods is that, for small displacements of the probe from the centre of an optical trap, the restoring force is proportional to the displacement. Hence, an optical trap is assumed to act on the probe like a Hookean spring with a fixed stiffness. This condition implies that the force field produced by the optical forces must be conservative, excluding the possibility of a rotational component. This is actually true to a great extent in the plane perpendicular to the beam propagation direction for a standard optical trap generated by a Gaussian beam. However, this is not true in a plane parallel to the beam propagation.

Back in 1992 Ashkin already pointed out that, in principle, scattering forces in optical tweezers do not conserve mechanical energy, and that this could have some measurable consequences (Ashkin, 1992). In particular, this non-conservative force would produce a dependence of the axial equilibrium position of a trapped microsphere on its transverse position in the trapping beam; such prediction was confirmed by Merenda et al. (2006). Later, Roichman et al. (2008) directly investigated the non-conservative component and discussed the implications that this might have for optical tweezers-based experiments making use of thermal fluctuations in the calibration procedure.

A technique to evaluate the relative weight of the non-conservative component of the optical forces was introduced by Pesce et al. (2009) extending the concept of photonic torque microscope, which we have seen in Section 10.2. The basic idea of this technique is to study the cross-correlation function between the radial (r) and axial (z) position of the particle in the optical trap. Such cross-correlation emerges in the presence of non-conservative forces. Using this technique, Pesce et al. (2009) analysed various optically trapped particles under different trapping conditions, concluding that non-conservative effects are effectively negligible and do not affect the standard calibration procedure, unless for extremely low-power trapping, far away from the trapping regimes usually used in experiments. In particular, for medium-high laser powers (from a few tens of milliwatts at the sample upwards), the effect is undetectable even for long acquisition times. This is because the deviation from a conservative force field is larger far away from the trap centre, which is explored more often in a weaker trap.

10.5 Direct force measurement

Most of the force measurement methods presented in Chapter 9 and in the previous sections of this chapter require the determination of the stiffness of the optical tweezers and of the position of the optically trapped particle. Often, they rely on data analysis techniques that assume the optical trap to be harmonic, which is predicated on various prerequisites, e.g., the use of non-aberrated Gaussian beams and spherical particles. Furthermore, the value of the stiffness is typically valid only for a given configuration of the optical set-up and, thus, a recalibration is required whenever the parameters of the optical set-up are changed. An alternative and more direct approach is to measure the momentum change of the optical beam as it traps a particle. This approach has the advantage of being based on first principles (Newton's second and third laws). Thus, it is largely independent of the experimental conditions; it allows the use of non-spherical particles and trapping beams with arbitrary intensity profiles, as well as the measurement of forces in homogeneous buffers with unknown viscosity and/or refractive index; and it does not require continued recalibration. This is particularly useful for experiments in living cells because no *in situ* calibration is needed. This technique was originally developed by Smith et al. (2003) using a low-NA set-up with counter-propagating laser beams and has been recently extended to the case of a single-beam optical tweezers by Farré and Montes-Usategui (2010).

The basic principle is illustrated in Fig. 10.10. Using the angular spectrum representation [Eq. (4.10)], the forward scattering of the electromagnetic field in the sample plane $\mathbf{E}_f(x, y, z)$ can be decomposed into a set of plane waves as

$$\mathbf{E}_f(x, y, z) = \int\!\!\!\int_{-\infty}^{+\infty} \check{\mathbf{E}}_f(k_{m,x}, k_{m,y}; 0)\, e^{i[k_{m,x}x + k_{m,y}y + k_{m,z}z]}\, dk_{m,x}\, dk_{m,y}, \qquad (10.30)$$

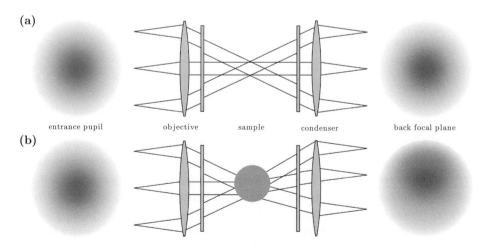

Figure 10.10 Direct force measurement. The force is proportional to the change of momentum of the beam. Thus, it can be measured as the difference between the transverse momenta of the beam (a) without particle and (b) with particle.

where $\check{\mathbf{E}}_f(k_{m,x}, k_{m,y}; 0)$ is the Fourier transform of the field $\mathbf{E}_f(x, y, 0)$, and $k_{m,x}$, $k_{m,y}$ and $k_{m,z}$ are the components of the wavevector. Because in a plane wave all photons have the same momentum, $\mathbf{p}_m = \hbar \mathbf{k}_m$, where $\hbar$ is the reduced Planck constant and $\mathbf{k}_m = [k_{m,x}, k_{m,y}, k_{m,z}]$ is the wavevector,[3] $\check{\mathbf{E}}_f(k_{m,x}, k_{m,y}; 0)$ also represents the momentum spectrum of $\mathbf{E}_f(x, y, 0)$; i.e., $\check{\mathbf{E}}_f(k_{m,x}, k_{m,y}; 0)$ is related to the number of photons in $\mathbf{E}_f(x, y, 0)$ having transverse momentum components $p_{m,x} = \hbar k_{m,x}$ and $p_{m,y} = \hbar k_{m,y}$. When this light is collected by a condenser lens, the intensity pattern $I_{FS}(x_d, y_d)$ collected on the back focal plane corresponds to the momentum distribution of the light in the sample plane, where x_d and y_d are the coordinates on the detector plane. If, furthermore, the condenser satisfies Abbe's sine condition [Eq. (4.39)],[4] a linear relation holds between x_d (y_d) and $p_{m,x}$ ($p_{m,y}$),

$$x_d = f_c \frac{k_{m,x}}{k_0} = f_c \frac{p_{m,x}}{p_0} \quad \text{and} \quad y_d = f_c \frac{k_{m,y}}{k_0} = f_c \frac{p_{m,y}}{p_0}, \quad (10.31)$$

where f_c is the focal length of the condenser, k_0 is the wavevector in vacuum and p_0 is the momentum of a photon in vacuum. Therefore, assuming it is possible to collect all light, the total momentum of the beam in the transverse plane is

$$\begin{cases} P_{\text{opt},x} = \iint p_{m,x}(x_d, y_d) \frac{I_{FS}(x_d, y_d)}{u} dx_d dy_d \\ \qquad = \frac{1}{f_c c} \iint x_d I_{FS}(x_d, y_d) dx_d dy_d, \\ P_{\text{opt},y} = \iint p_{m,y}(x_d, y_d) \frac{I_{FS}(x_d, y_d)}{u} dx_d dy_d \\ \qquad = \frac{1}{f_c c} \iint y_d I_{FS}(x_d, y_d) dx_d dy_d, \end{cases} \quad (10.32)$$

where $u = c p_0$ is the energy of a photon and c is the speed of light in vacuum. As we have seen in Section 9.2, it is possible to use a position-sensing detector [Eq. (9.39)] to measure the signal

$$\begin{cases} S_{\text{psd},x} = \eta_{\text{psd}} \iint \frac{x_d}{R_{\text{psd}}} I_{FS}(x_d, y_d) dx_d dy_d, \\ S_{\text{psd},y} = \eta_{\text{psd}} \iint \frac{y_d}{R_{\text{psd}}} I_{FS}(x_d, y_d) dx_d dy_d, \end{cases} \quad (10.33)$$

where η_{psd} is the detector efficiency and R_{psd} is the detector radius. The force is proportional to the change of momentum of the beam. Thus, it can be measured as the difference between the transverse momenta of the beam without particle [Fig. 10.10a] and with particle [Fig. 10.10b]. In fact, often the procedure is further simplified because the trapping

---

[3] See also Box 2.1 for a discussion of the momentum of light in a medium. For the sake of this technique, the definition of the momentum of light is largely irrelevant, as the measurement is performed in air, where the issue does not arise.

[4] This, in particular, implies that its principal surface must be spherical. This is the same process we have seen when focusing a beam in Section 4.5, just in reverse.

laser profile is generally centre-symmetric and the transverse momenta of the beam without trapped particle are zero. We finally obtain

$$\begin{cases} F_{\text{opt},x} = \dfrac{R_{\text{psd}}}{\eta_{\text{psd}}\, f_c\, c} S_{\text{psd},x}, \\ F_{\text{opt},y} = \dfrac{R_{\text{psd}}}{\eta_{\text{psd}}\, f_c\, c} S_{\text{psd},y}, \end{cases} \qquad (10.34)$$

where the prefactor $\frac{R_{\text{psd}}}{\eta_{\text{psd}} f_c c}$ is a constant that depends only on the set-up macroscopic configuration and not on the details of the sample or of the beam. Although in principle this constant can be determined from macroscopic measurements of the optical set-up, it is often convenient to determine it from a calibration experiment under very controllable conditions, where the optical forces are measured with the methods explained in Chapter 9 and in the previous sections of this chapter.

In principle, the practical implementation of this technique requires the collection of all light scattered over a 4π solid angle. This can be realised using a counter-propagating optical tweezers set-up, as shown in Fig. 10.11a. The basic set-up requires two objectives, which are used both to focus the two counter-propagating beams and to collect the scattered light, and two position-sensing detectors, which are placed on the back focal planes of the objectives. In order to collect as much scattered light as possible, it is useful not to overfill the objectives, so that the effective numerical aperture of the trapping beams is significantly smaller than the numerical aperture of the condenser.

This technique can also be employed in single-beam optical traps, as shown in Fig. 10.11b, even though its implementation becomes significantly more challenging. In order to collect almost the whole forward-scattered light,[5] a high-NA condenser lens (e.g., a NA $=$ 1.40 oil-immersion condenser lens) must be used; furthermore, some precautions must be taken, i.e., the numerical aperture of the condenser must be chosen to be higher than the refractive index of the medium used to suspend the particle and the trapping position must be close to the upper cover-slip of the suspension chamber for the light to quickly reach the buffer–glass interface [Fig. 10.11c]. As a result of these restrictions on the particle's axial position, it becomes necessary to use aberration-free objectives with long working distances (water-immersion typically providing better performance than oil-immersion lenses).

Exercise 10.5.1 Simulate the force detection technique described in this section. Verify the relation between scattered light and force. Compare its performance with that of techniques based on the determination of the trap stiffness and particle position from the measurement of forward-scattered light with a quadrant photodetector and a position sensing detector, explained in Chapter 9.

---

[5] The light travelling in the backward direction usually corresponds to a small fraction of the whole scattered light and, thus, the information about the momentum change is mainly concentrated in the forward-scattered light. The error associated with neglecting the backward-scattered light is typically on the order of a few per cent or less, which is acceptable in most biophysical experiments.

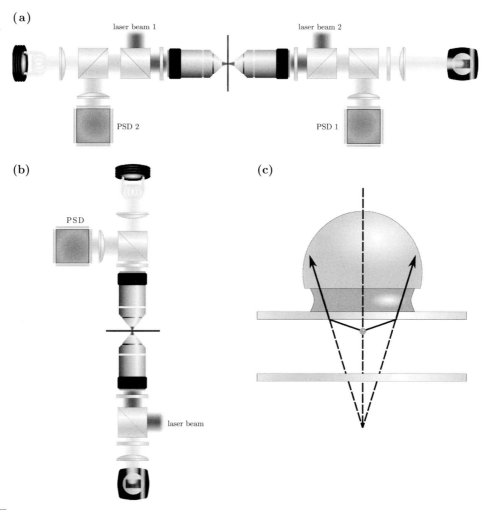

Figure 10.11 Set-ups for direct force measurement. (a) Almost all scattered light can be collected using a counter-propagating optical tweezers set-up. (b) In single-beam optical traps, this technique becomes significantly more challenging as, in order to collect almost the whole forward-scattered light, a high-NA condenser lens must be used. (c) Close-up of the objective acceptance angle for the single-beam high-NA configuration.

Exercise 10.5.2 Show that the change of propagation direction at the interface does not affect the lateral momentum of the beam. [Hint: Use the fact that the transverse components of the wavevector are preserved by Snell's law.]

Exercise 10.5.3 Inhomogeneities in the medium surrounding the particle might change the angular distribution or intensity of the scattered light, and therefore the momentum of the beam. How does this affect the accuracy of the force detection? How can their effect be minimised?

Exercise 10.5.4 How can this technique be extended to measure the force along the optical axis?

Exercise 10.5.5 The approach to the direct measurement of the optical force acting on a particle described in this section can be extended to measure also the transfer of spin angular momentum. Devise a set-up capable of doing this. [Hint: Make use of polarising beams splitters. Read Bishop et al. (2004).]

Exercise 10.5.6 How can this technique be extended further to measure the transfer of orbital angular momentum? [Hint: Make use of a spatial light modulator.]

Problems

10.1 *Angular fluctuations of nanowires.* A dielectric nanowire, i.e., an elongated cylinder with a transverse size much smaller and an axial length much larger than the trapping wavelength, in an optical tweezers aligns with the optical axis. The tracking signals on a quadrant photodiode are the result of both translational and off-axis angular thermal fluctuations in the trap. Under what assumptions can you calibrate optical forces on the nanowire with a quadrant photodiode? How are the power spectrum and correlation functions of the tracking signals modified by the angular fluctuations?

10.2 *Non-conservative effects on cylinders.* Consider an optically trapped cylinder. During its thermal dynamics the cylinder is set to an angle with respect to its equilibrium position and a transverse component of the radiation force occurs because of its non-spherical shape. Show that this force is non-conservative. What is the effect of this non-conservative transverse component on the cylinder dynamics? What effect do you expect to see on the correlation functions?

10.3 *Intrinsic rotations.* Consider an optically trapped asymmetric particle rotating about the trap axis. Calculate the power spectrum and the auto- and cross-correlation functions of the signals from a quadrant photodiode. How do the signals change when the particle features symmetries?

10.4 *Position clamping.* Consider a spherical particle trapped in an optical tweezers. Suppose a feedback (with proportional gain G_P and delay time τ_{loop}) is applied to the focus position depending on the particle position. Write down the Langevin equation describing this position-clamped optical tweezers. Calculate its power spectrum and derive its effective stiffness in terms of the unclamped stiffness. How can you realise this experimentally?

10.5 *Dual trap dumb-bell assay.* In high-resolution optical tweezers experiments, often a dumb-bell configuration is employed [see Chapter 13], consisting of two trapped beads connected by an elastic tether. Neglecting inertial and hydrodynamic coupling effects, write down the set of linear coupled Langevin equations describing the motion of the beads. Derive the expression for their power spectra. How can you use power spectrum analysis to extract the tether spring constant in terms of the traps' and beads' parameters? What changes if you include the hydrodynamic coupling between the spheres?

10.6 *Force clamping in a double trap.* Consider a dual trap dumb-bell configuration, as in the previous problem. Assume one of the traps is force-clamped with a feedback loop. How are the Langevin equations and power spectra modified by the force clamp? How can you extract the tether spring constant from the trap and feedback loop parameters?

10.7 *Shifted rotational force field.* Consider a rotational force field with a broken symmetry, i.e., with its centre shifted from the trap equilibrium position by $x_{\rm shift}$. How do the equations in Section 10.2 need to be modified?

References

Ashkin, A. 1992. Forces of a single-beam gradient laser trap on a dielectric sphere in the ray optics regime. *Biophys. J.*, **61**, 569–82.

Binnig, G., Rohrer, H., Gerber, C., and Weibel, E. 1982. Surface studies by scanning tunneling microscope. *Phys. Rev. Lett.*, **49**, 57–60.

Binnig, G., Quate, C. F., and Gerber, C. 1986. Atomic force microscope. *Phys. Rev. Lett.*, **56**, 930–33.

Bishop, A. I., Nieminen, T. A., Heckenberg, N. R., and Rubinsztein-Dunlop, H. 2004. Optical microrheology using rotating laser-trapped particles. *Phys. Rev. Lett.*, **92**, 198104.

Brettschneider, T., Volpe, G., Helden, L., Wehr, J., and Bechinger, C. 2011. Force measurement in the presence of Brownian noise: Equilibrium-distribution method versus drift method. *Phys. Rev. E*, **83**, 041113.

Farré, A., and Montes-Usategui, M. 2010. A force detection technique for single-beam optical traps based on direct measurement of light momentum changes. *Opt. Express*, **18**, 11 955–68.

Florin, E.-L., Pralle, A., Hörber, J. K. H., and Stelzer, E. H. K. 1997. Photonic force microscope based on optical tweezers and two-photon excitation for biological applications. *J. Struct. Biol.*, **119**, 202–11.

Ghislain, L. P., and Webb, W. W. 1993. Scanning-force microscope based on an optical trap. *Opt. Lett.*, **18**, 1678–80.

Ghislain, L. P., Switz, N. A., and Webb, W. W. 1994. Measurement of small forces using an optical trap. *Rev. Sci. Instrumen.*, **69**, 2762–8.

Lang, M. J., Asbury, C. L., Shaevitz, J. W., and Block, S. M. 2002. An automated two-dimensional optical force clamp for single molecule studies. *Biophys. J.*, **83**, 491–501.

Merenda, F., Boer, G., Rohner, J., Delacrétaz, G., and Salathé, R.-P. 2006. Escape trajectories of single-beam optically trapped micro-particles in a transverse fluid flow. *Opt. Express*, **14**, 1685–99.

Pesce, G., Volpe, G., Luca, A. C. De, Rusciano, G., and Volpe, G. 2009. Quantitative assessment of non-conservative radiation forces in an optical trap. *Europhys. Lett.*, **86**, 38002.

Prieve, D. C. 1999. Measurement of colloidal forces with TIRM. *Adv. Colloid Interface Sci.*, **82**, 93–125.

Roichman, Y., Sun, B., Stolarski, A., and Grier, D. G. 2008. Influence of nonconservative optical forces on the dynamics of optically trapped colloidal spheres: The fountain of probability. *Phys. Rev. Lett.*, **101**, 128301.

Smith, S. B., Cui, Y., and Bustamante, C. 2003. Optical-trap force transducer that operates by direct measurement of light momentum. *Methods Enzymol.*, **361**, 134–62.

Volpe, G., and Petrov, D. 2006. Torque detection using Brownian fluctuations. *Phys. Rev. Lett.*, **97**, 210603.

Volpe, G., Volpe, G., and Petrov, D. 2007. Brownian motion in a nonhomogeneous force field and photonic force microscope. *Phys. Rev. E*, **76**, 061118.

Volpe, G., Brettschneider, T., Helden, L., and Bechinger, C. 2009. Novel perspectives for the application of total internal reflection microscopy. *Opt. Express*, **17**, 23 975–85.

Volpe, G., Helden, L., Brettschneider, T., Wehr, J., and Bechinger, C. 2010. Influence of noise on force measurements. *Phys. Rev. Lett.*, **104**, 170602.

Walz, J. Y. 1997. Measuring particle interactions with total internal reflection microscopy. *Curr. Opin. Colloid Interface Sci.*, **2**, 600–606.

11 Wavefront engineering and holographic optical tweezers*

The most basic optical tweezers set-up, e.g., the one we built in Chapter 8, produces a *single* optical trap by focusing a *single* optical beam. By using more than one beam, or by splitting a single beam, it is possible to generate multiple traps. This procedure, however, leads to rather complex, and sometimes messy, set-ups. Additionally, to steer these traps it is necessary to move some mechanical components of the optical set-up. Finally, using non-Gaussian beams demands the use of specialised optical components, e.g., axicons to generate Bessel beams or holographic masks to generate higher-order Laguerre–Gaussian beams. To overcome these difficulties, it is possible to employ *holographic optical tweezers* (HOTs). HOTs use a computer-controlled diffractive optical element (DOE) to split a single collimated laser beam into several separate beams, each of which is focused into an optical tweezers. These optical traps can be made dynamic and displaced in three dimensions by projecting a sequence of holograms and, furthermore, non-Gaussian beam profiles can be straightforwardly encoded in the holographic mask, such as the Laguerre–Gaussian beams employed in the optical trapping experiment shown in Fig. 11.1. In this chapter, we explain how to design and operate HOTs.

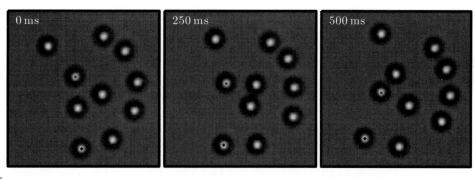

Figure 11.1 Rotating particles in Laguerre–Gaussian beams. Particles (polystyrene, 1 μm radius) trapped in two concentric Laguerre–Gaussian beams with opposite helicity: the particles in the outer ring are pushed clockwise by a Laguerre–Gaussian beam with helicity $+40$, while the ones in the inner ring are pushed anticlockwise (at a faster angular speed) by a Laguerre–Gaussian beam with helicity -10. The particles are confined in a quasi-two-dimensional space by the scattering force that pushes them against the upper coverslip of the sample chamber.

---

* This chapter was written together with Giuseppe Pesce.

11.1 Basic working principle

The basic working principle of HOTs is shown in Fig. 11.2. A DOE is employed to shape the profile of an incoming optical beam, which for simplicity we assume to have uniform intensity and, more importantly, uniform phase. The DOE is positioned at the front focal plane of the Fourier lens (to which we will refer from now on as the *DOE plane*), which collects the first-order diffracted beam, so that the complex amplitude in the back focal

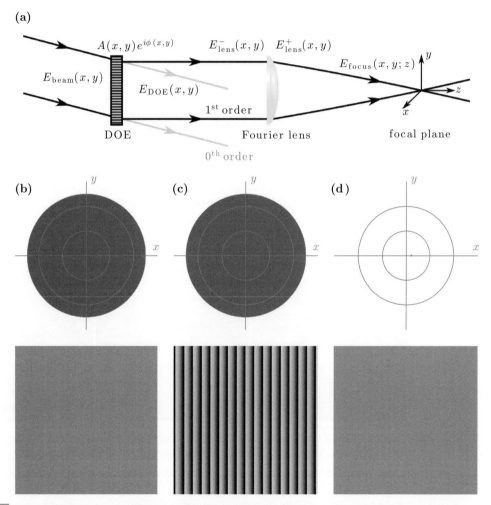

Figure 11.2 HOT working principle. (a) A diffractive optical element (DOE) alters the beam profile of an incoming optical beam (with uniform intensity and phase). The DOE is positioned in the front focal plane of the Fourier lens (the DOE plane), which collects the first-order diffracted beam, so that the the complex amplitude in the back focal plane of the Fourier lens (the focal plane) is the Fourier transform of the complex amplitude in the DOE plane. Intensity (top) and phase (bottom) profiles of (b) the incoming beam, (c) the beam after the DOE and (d) the focused beam at the focal plane.

> **Box 11.1** **Fresnel diffraction integral**
>
> Given the electromagnetic field $E(x, y; 0)$ defined in the plane $z = 0$, the field $E(x, y; z)$ in the plane $z = $ constant can be calculated using the *Fresnel diffraction integral*, i.e.,
>
> $$E(x, y; z) = \frac{e^{ik_0 z}}{i\lambda_0 z} \int\!\!\!\int_{-\infty}^{+\infty} E(x', y'; 0) e^{i\frac{k_0}{2z}[(x-x')^2 + (y-y')^2]} dx' dy',$$
>
> where k_0 and λ_0 are the vacuum wavenumber and wavelength. This integral is the convolution
>
> $$E(x, y, z) = h(x, y, z) * E(x, y, 0),$$
>
> where
>
> $$h(x, y, z) = \frac{e^{ik_0 z}}{i\lambda_0 z} e^{i\frac{k_0}{2z}[x^2 + y^2]}$$
>
> is the *impulse response* of free space propagation. Taking the two-dimensional Fourier transform, we obtain its expression in the frequency domain,
>
> $$\check{E}(f_x, f_y; z) = \check{H}(f_x, f_y; z)\check{E}(f_x, f_y; 0),$$
>
> where $f_x = \lambda_0^{-1} x/z$ and $f_y = \lambda_0^{-1} y/z$ are the spatial frequencies, $\check{E}(f_x, f_y; z)$ is the Fourier transform of $E(x, y; z)$, $\check{E}(f_x, f_y; 0)$ is the Fourier transform of $E(x, y; 0)$ and
>
> $$\check{H}(f_x, f_y; z) = e^{ik_0 z} e^{-i\pi \lambda z [f_x^2 + f_y^2]}$$
>
> is the Fourier transform of the impulse response multiplied by $4\pi^2$ (this factor can also be 1 or 2π depending on the convention used to define the Fourier transform).

plane of the Fourier lens (from now on the *focal plane*) is the Fourier transform of the complex amplitude at the DOE plane.

We will now make these considerations more quantitative. We consider an incoming paraxial beam with transverse profile $E_{\text{beam}}(x, y)$. The effect of the DOE is to alter the beam profile by a complex factor $A(x, y)e^{i\phi(x,y)}$, where $A(x, y)$ is a (real) intensity attenuation factor and $\phi(x, y)$ is a phase factor. Therefore, the electric field just after the DOE is

$$E_{\text{DOE}}(x, y) = A(x, y)e^{i\phi(x,y)} E_{\text{beam}}(x, y). \tag{11.1}$$

We can now propagate $E_{\text{DOE}}(x, y)$ from the DOE plane up to the entrance pupil of the lens using the Fresnel diffraction integral [Box 11.1], obtaining

$$E_{\text{lens}}^{-}(x, y) = \frac{e^{ik_0 f}}{i\lambda_0 f} e^{i\frac{k_0}{2f}[x^2 + y^2]} \int\!\!\!\int_{-\infty}^{+\infty} E_{\text{DOE}}(x', y') e^{i\frac{k_0}{2f}[x'^2 + y'^2]} e^{-i\frac{2\pi}{\lambda_0 f}[xx' + yy']} dx' dy',$$

$$\tag{11.2}$$

where k_0 is the vacuum wavenumber, λ_0 is the vacuum wavelength, f is the focal distance of the lens and the superscript '$-$' indicates that we consider the beam just *before* the lens.

Within the thin lens approximation and assuming the diameter of the lens to be significantly larger than the size of the beam, the effect of the Fourier lens is to add a quadratic phase factor to the beam profile, i.e.,

$$E^+_{\text{lens}}(x, y) = E^-_{\text{lens}}(x, y) e^{-i\frac{k_0}{2f}[x^2+y^2]}, \tag{11.3}$$

where the superscript '+' indicates that we consider the beam just *after* the lens,[1] so that

$$E^+_{\text{lens}}(x, y) = \frac{e^{ik_0 f}}{i\lambda_0 f} \int\!\!\!\int_{-\infty}^{+\infty} E_{\text{DOE}}(x', y') e^{i\frac{k_0}{2f}[x'^2+y'^2]} e^{-i\frac{2\pi}{\lambda_0 f}[xx'+yy']} dx' dy'. \tag{11.4}$$

At this point, we apply the Fresnel diffraction formula to propagate $E^+_{\text{lens}}(x, y)$ from the lens exit pupil to its focal plane, obtaining

$$E_{\text{focus}}(x, y) = \frac{e^{ik_0 f}}{i\lambda_0 f} e^{i\frac{k_0}{2f}[x^2+y^2]} \int\!\!\!\int_{-\infty}^{+\infty} E^+_{\text{lens}}(x', y') e^{i\frac{k_0}{2f}[x'^2+y'^2]} e^{-i\frac{2\pi}{\lambda_0 f}[xx'+yy']} dx' dy'. \tag{11.5}$$

By substituting Eq. (11.3) into Eq. (11.5), we obtain

$$E_{\text{focus}}(x, y) = \frac{e^{ik_0 f}}{i\lambda_0 f} e^{i\frac{k_0}{2f}[x^2+y^2]} 4\pi^2 \check{E}^-_{\text{lens}}(f_x, f_y), \tag{11.6}$$

where $f_x = \lambda_0^{-1} x/f$, $f_y = \lambda_0^{-1} y/f$ and

$$\check{E}^-_{\text{lens}}(f_x, f_y) = \frac{1}{4\pi^2} \int\!\!\!\int_{-\infty}^{+\infty} E^-_{\text{lens}}(x, y) e^{-i2\pi[f_x x + f_y y]} dx dy \tag{11.7}$$

is the Fourier transform of $E^-_{\text{lens}}(x, y)$.

Finally, we can substitute the Fourier transform of Eq. (11.2) [Box 11.1], i.e.,

$$\check{E}^-_{\text{lens}}(f_x, f_y) = e^{ik_0 z} e^{-i\pi\lambda z[f_x^2 + f_y^2]} \check{E}_{\text{DOE}}(f_x, f_y), \tag{11.8}$$

into Eq. (11.6) to derive the expression relating the field at the DOE plane and at the focal plane,

$$E_{\text{focus}}(x, y) = \frac{e^{2ik_0 f}}{i\lambda_0 f} 4\pi^2 \check{E}_{\text{DOE}}(f_x, f_y). \tag{11.9}$$

As anticipated at the beginning of this section, Eq. (11.9) shows that the field in the focal plane of the lens is the Fourier transform of the field at the DOE plane. This means that the field at a certain position in the focal plane is proportional to the amplitude of a certain spatial frequency in the DOE plane.

---

[1] To be precise, a constant phase factor of the form $e^{ik_0 n_1 \Delta l}$ should also be considered, where n_1 is the lens refractive index and Δl is the lens central thickness. However, this becomes relevant only if interferometric measurement is considered, so that it can be safely ignored in our case.

Exercise 11.1.1 Show that, if (as is often the case) the input beam can be modelled with a plane wave, i.e., $E_{\text{beam}}(x, y) \equiv E_0$, the field in the focal plane is determined solely by the profile imposed by the DOE. What happens if the incoming field is a Gaussian beam? And if it is a Laguerre–Gaussian beam?

Exercise 11.1.2 Show that the effect of a blazed grating on the DOE, e.g., along the x-direction, is to deflect the impinging beam (assume it to have a constant phase and amplitude) by an angle

$$\alpha_x \approx \tan \alpha_x = \frac{\lambda_0}{\Lambda_x}, \qquad (11.10)$$

where Λ_x is the periodicity of the grating. This is typically a small angle, at most about $1°$. How much is it for the case when $\lambda_0 = 1064$ nm and the pixel size of the DOE $d = 12$ μm? Show that such a blazed grating produces a displacement of the intensity spot at the front focal plane equal to

$$\Delta x = f \frac{\lambda_0}{\Lambda_x}. \qquad (11.11)$$

Repeat the same analysis for a blazed grating along the y-direction and show that the sum of gratings along the x- and y-directions results in a grating that produces a diagonal shift equal to the combination of their shifts; i.e., the effects of gratings are additive.

Exercise 11.1.3 Show that the relation between the field at the DOE plane and the field at planes different from (but close to) the focal plane is

$$E_{\text{focus}}(x, y; z) = \frac{e^{ik_0[2f+z]} e^{ik_0[x^2+y^2]}}{i\lambda_0 f} 4\pi^2 \check{E}_{\text{DOE},z}(f_x, f_y), \qquad (11.12)$$

where $\check{E}_{\text{DOE},z}(f_x, f_y)$ is the Fourier transform of

$$E_{\text{DOE}}(x, y) e^{-i \frac{\pi z}{\lambda_0 f^2}[x^2+y^2]}. \qquad (11.13)$$

Note that the term $e^{-i \frac{\pi z}{\lambda_0 f^2}[x^2+y^2]}$ represents the phase of a *Fresnel lens*, whose role is to alter the curvature of the incoming beam (see also the following exercise).

Exercise 11.1.4 Show that the effect of imposing a Fresnel lens on the DOE is to alter the curvature of the incoming beam (assume it to have a constant phase and amplitude) and that, if the Fresnel lens focal length is f_F, the intensity spot at the front focal plane is displaced along the z-direction by

$$\Delta z = \frac{f^2}{f_F}. \qquad (11.14)$$

Exercise 11.1.5 Show that imposing on the DOE a phase profile corresponding to the sum of a grating and a Fresnel lens results in a displacement of the focal spot in the lateral and axial directions, corresponding respectively to the effect of the grating and of the Fresnel lens.

11.2 Computer-generated holograms

In optical trapping and manipulation applications, point traps, higher-order beams and continuous intensity distributions for the desired trapping pattern are typically employed. In principle, given a desired intensity distribution in the front focal plane, it is sufficient to take its inverse Fourier transform to determine the appropriate hologram to place on the DOE. However, the result of this simple operation is typically a hologram that modulates both the phase and the amplitude of the input beam. On one hand, employing such a hologram would require modulating both the phase and the amplitude of the incoming laser beam. On the other hand, amplitude modulation would remove power from the beam, leading to significantly reduced efficiency. Therefore, it is usually more convenient to employ a phase-only hologram.

In the following, we will consider the incoming beam to be a uniform plane wave, i.e., $E_{\text{beam}}(x, y) \equiv E_0$. The DOE is typically a two-dimensional pixellated device whose pixels have area d^2, row index $m_x = 1, \ldots, M_x$ and column index $m_y = 1, \ldots, M_y$. The complex amplitude imposed on the incoming beam by pixel $[m_x, m_y]$ is $A_{m_x,m_y} = A \exp\{i\phi_{m_x,m_y}\}$, where A is a (real) attenuation constant and ϕ_{m_x,m_y} is a phase shift. Therefore, the electric field after the DOE is $E_{m_x,m_y} = A \exp\{i\phi_{m_x,m_y}\}E_0$. From Eq. (11.12), we can now calculate the electric field in the trap volume around the focal plane of the objective as

$$E_{\text{focus}}(x, y, z) = \frac{e^{i\frac{2\pi}{\lambda_0}[2f+z]} e^{i\frac{\pi}{\lambda_0 z}[x^2+y^2]}}{i\lambda_0 f} d^2 \sum_{m_x=1}^{M_x} \sum_{m_y=1}^{M_y} |E_0|^2 A^2 e^{i[\phi_{m_x,m_y} - \Delta_{m_x,m_y}(x,y,z)]}, \quad (11.15)$$

where

$$\Delta_{m_x,m_y}(x, y, z) = \frac{2\pi}{\lambda_0 f}[x_{m_x,m_y} x + y_{m_x,m_y} y] + \frac{\pi z}{\lambda_0 f^2}[x^2_{m_x,m_y} + y^2_{m_x,m_y}] \quad (11.16)$$

and $[x_{m_x,m_y}, y_{m_x,m_y}]$ are the pixels' coordinates.

We can now consider how much laser power flows through a diffraction-limited area around the position $[x, y, z]$ within the trapping volume. The power, i.e., the total time-averaged energy flux, just after the DOE is

$$P_t = \frac{c\varepsilon_0^2}{2} M_x M_y d^2 |E_0|^2 A^2. \quad (11.17)$$

The energy flux flowing through a diffraction-limited spot with area $f^2 \lambda_0^2/(M_x M_y d^2)$, around position $[x, y, z]$ is

$$P(x, y, z) = \frac{c\varepsilon_0}{2} \frac{f^2 \lambda_0^2}{M_x M_y d^2} \frac{d^4}{\lambda_0^2 f^2} |E_0|^2 A^2 \left| \sum_{m_x=1}^{M_x} \sum_{m_y=1}^{M_y} e^{i(\phi_{m_x,m_y} - \Delta_{m_x,m_y}(x,y,z))} \right|^2. \quad (11.18)$$

Therefore, we can introduce the dimensionless variable

$$V(x, y, z) = \frac{1}{M_x M_y} \sum_{m_x=1}^{M_x} \sum_{m_y=1}^{M_y} e^{i(\phi_{m_x,m_y} - \Delta_{m_x,m_y}(x,y,z))}, \quad (11.19)$$

which is characterised by the property that

$$|V(x, y, z)|^2 = \frac{P(x, y, z)}{P_t} \qquad (11.20)$$

and for $z = 0$ corresponds to the discrete Fourier transform of ϕ_{m_x,m_y} evaluated at the spatial frequencies $[\lambda_0^{-1} x/f, \lambda_0^{-1} y/f]$.

In the following subsections, our task will be to optimise the values of ϕ_{m_x,m_y} imposed on the DOE in order to maximise the modulus of $V(x, y, z)$ at some trapping sites, which are located at $[x_{\text{ot},n}, y_{\text{ot},n}, z_{\text{ot},n}]$ for $n = 1, \ldots, N$. We will focus our attention on standard Gaussian optical traps, whereas we consider higher-order beams in Section 11.3 and continuous optical potentials in Section 11.4.

11.2.1 Single steerable trap

The simplest case is a single trap, i.e., $N = 1$, placed at $[x_{\text{ot},1}, y_{\text{ot},1}, z_{\text{ot},1}]$. The optimal phase modulation is immediately found to be

$$\phi^S_{m_x,m_y}(x_{\text{ot},1}, y_{\text{ot},1}, z_{\text{ot},1}) = \Delta_{m_x,m_y}(x_{\text{ot},1}, y_{\text{ot},1}, z_{\text{ot},1}), \qquad (11.21)$$

for which $V(x_{\text{ot},1}, y_{\text{ot},1}, z_{\text{ot},1}) = 1$; i.e., all the diffracted power is deflected into the trap. Thus,

$$\phi^S_{m_x,m_y} = \underbrace{\frac{2\pi}{\lambda_0 f} \left[x_{m_x,m_y} x_{\text{ot},1} + y_{m_x,m_y} y_{\text{ot},1} \right]}_{\text{Diffraction grating}} + \underbrace{\frac{\pi z_{\text{ot},1}}{\lambda_0 f^2} \left[x^2_{m_x,m_y} + y^2_{m_x,m_y} \right]}_{\text{Fresnel lens}}, \qquad (11.22)$$

where the first term gives rise to a *blazed diffraction grating* phase pattern shifting the spot laterally [Figs. 11.3a, 11.3b and 11.3c], and the second term generates a phase retardation that is comparable to that of a *Fresnel lens* [Figs. 11.3d and 11.3e], shifting the spot along the axial direction. A combination of the two produces a shift of the trap in three dimensions [Fig. 11.3f]. Because the generation of the holographic mask given by Eq. (11.22) is computationally very fast, this is a very effective way to generate a single optical trap that can be moved in three dimensions in real time.

Exercise 11.2.1 Simulate the intensity profile in the focal plane generated by a phase profile such as the one in Eq. (11.22). What happens when the parameters of the grating and of the Fresnel lens are changed? In particular, study the case of a grating made of alternating rows of pixels with phase $0°$ and $180°$ and note that the resulting intensity profile produces two symmetric spots. Why is this so? [Hint: You can adapt the program `slm` from the book website.]

Exercise 11.2.2 Show that any optical field in the focal plane can be shifted laterally by adding a diffraction grating phase profile to the hologram.

Exercise 11.2.3 Show that any optical field in the focal plane can be shifted axially by adding a Fresnel lens phase profile to the hologram.

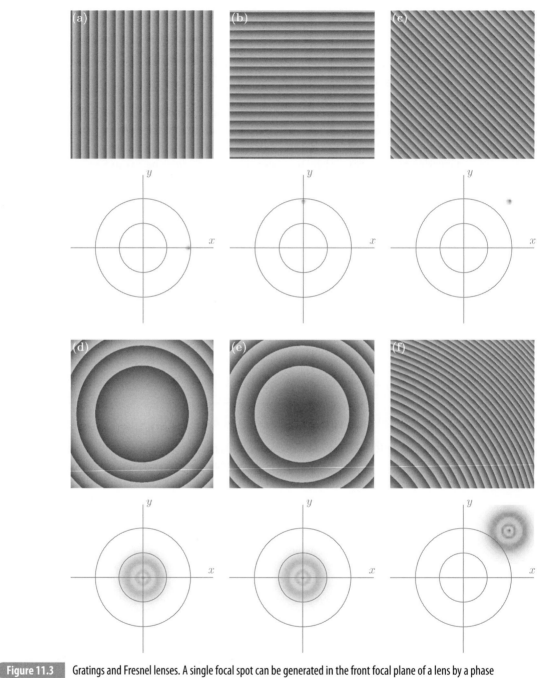

Figure 11.3 Gratings and Fresnel lenses. A single focal spot can be generated in the front focal plane of a lens by a phase modulation corresponding to a grating and/or a Fresnel lens: (a) grating along x; (b) grating along y; (c) combination of gratings along x and y; (d) positive Fresnel lens; (e) negative Fresnel lens; (f) combination of the grating in (c) and Fresnel lens in (d). Both the phase modulation (top) and the focal spot intensity (bottom) are shown.

11.2.2 Random mask encoding

To generate multiple traps, i.e., $N > 1$, one of the fastest, but also least efficient, algorithms is the *random mask encoding* algorithm. For every pixel of the SLM, a phase shift is determined as if the hologram was made to generate only one of the N traps, i.e.,

$$\phi^{RM}_{m_x,m_y} = \phi^{S}_{m_x,m_y}\left(x_{ot,n(m_x,m_y)}, y_{ot,n(m_x,m_y)}, z_{ot,n(m_x,m_y)}\right), \quad (11.23)$$

where $\phi^{S}_{m_x,m_y}$ is given by Eq. (11.22) and $n(m_x, m_y) \in [1, \ldots, N]$ is chosen randomly for each pixel. This technique is very fast and permits one to achieve good uniformity amongst traps. Nevertheless, the overall efficiency can be very low when N is large, because on the average only $M_x M_y/N$ pixels interfere constructively to generate each trap.

In order to quantify the performance of this algorithm and to benchmark it against the other algorithms we present in the following subsections, we test all algorithms against a standard task of computing a 1000×1000 hologram for the generation of $N = 100$ traps arranged on a 10×10 square lattice located in the focal plane ($z = 0$). Their performance is quantified by four parameters, *efficiency*,

$$I_{tot} = \sum_{n=1}^{N} I_n, \quad (11.24)$$

average intensity,

$$\langle I \rangle = \frac{I_{tot}}{N}, \quad (11.25)$$

uniformity,

$$u = 1 - \frac{\max[I_n] - \min[I_n]}{\max[I_n] + \min[I_n]} \quad (11.26)$$

and *percentage standard error*,

$$\sigma = \frac{\sqrt{\sum_{n=1}^{N}(I_n - \langle I \rangle)^2}}{N\langle I \rangle} \times 100, \quad (11.27)$$

where I_n is the intensity of the nth trap. As shown in Table 11.1, the random mask encoding obtains an efficiency $I_{tot} = 0.01 \approx 1/N$, which is quite low, as each trap has only an intensity $\langle I \rangle = 0.0001 \approx 1/M^2$, whereas the uniformity is quite good at $u = 0.67$.

Exercise 11.2.4 Simulate the intensity profile in the focal plane generated by a phase profile such as the one in Eq. (11.23). Study how the efficiency [Eq. (11.24)] and the uniformity [Eq. (11.26)] of the trap array change as a function of its parameters. [Hint: You can adapt the program slm from the book website.]

Exercise 11.2.5 Show how the random mask encoding algorithm can be quite efficient at generating some helper tweezers on top of a complex light structure obtained via a precalculated hologram using temporarily a fraction of the SLM pixels. Simulate the generation and use of helper tweezers numerically.

Table 11.1 Comparison of algorithms to generate holograms

Algorithm[a]	I_{tot}[b]	u[c]	σ[d]	T_{comp}[e]
Random mask encoding	0.01	0.67	0.14	0.3 s
Superposition of gratings and lenses	0.29	0.007	2.59	4.6 s
Superposition of gratings and lenses with additional random phase	0.84	0.18	0.37	4.6 s
Gerchberg–Saxton algorithm	0.93	0.33	0.27	316 s
Adaptive-additive algorithm	0.93	0.99	0.0062	348 s

[a] All algorithms are tested against a standard task of computing a 1000 × 1000 hologram for the generation of $N = 100$ traps arranged on a 10×10 square lattice located in the focal plane ($z = 0$).
[b] Efficiency given by Eq. (11.24).
[c] Uniformity given by Eq. (11.26).
[d] Percentage standard error given by Eq. (11.27).
[e] Computational time on a standard laptop computer.

Exercise 11.2.6 Show that the random mask encoding algorithm can be adapted to the generation of arrays of weighted traps by randomly assigning the available SLM pixels to the individual traps proportionally to the ratio of the square roots of the desired trap intensities. Study numerically the performance of the resulting algorithm. [Hint: You can adapt the program `slm` from the book website.]

11.2.3 Superposition of gratings and lenses

Another straightforward algorithm, which achieves a better efficiency than the random mask encoding at a just slightly higher computational cost, is the *superposition of gratings and lenses* algorithm. In this algorithm, the phase of each pixel is chosen to be equal to the argument of the complex sum of single-trap holograms given by Eq. (11.22), i.e.,

$$\phi_{m_x,m_y}^{\text{SGL}} = \arg\left\{\sum_{n=1}^{N} e^{i\phi_{m_x,m_y}^{\text{S}}(x_{\text{ot},n},y_{\text{ot},n},z_{\text{ot},n})}\right\}. \quad (11.28)$$

This algorithm typically has good efficiency at the cost of poor uniformity. For example, in the benchmark test we present in Table 11.1, we obtained $I_{\text{tot}} = 0.29$ and $u = 0.007$. A typical problem that arises using this algorithm is that, when highly symmetrical trap geometries are sought, such as the one of our benchmark, a consistent part of the energy is diverted to unwanted ghost traps.[2]

---

[2] If the precise location of the traps is not crucial, it might be useful to add some small amount of noise to the location of each trap in order to improve the array uniformity.

Exercise 11.2.7 Simulate the intensity profile in the focal plane generated by a hologram obtained from the superposition of gratings and lenses [Eq. (11.28)]. Study how the efficiency [Eq. (11.24)] and the uniformity [Eq. (11.26)] of the trap array change as a function of its parameters. [Hint: You can adapt the program `slm` from the book website.]

Exercise 11.2.8 Show that the superposition of gratings and lenses [Eq. (11.28)] results from the maximisation of

$$\sum_{n=1}^{N} \text{Re}\left\{V(x_{\text{ot},n}, y_{\text{ot},n}, z_{\text{ot},n})\right\},$$

with respect to the arguments ϕ_{m_x,m_y}, i.e., it maximises the sum of the projections of the $V(x_{\text{ot},n}, y_{\text{ot},n}, z_{\text{ot},n})$ [Eq. (11.19)] along the real axis of the complex plane. [Hint: Impose the vanishing of the first derivative with respect to ϕ_{m_x,m_y} to find the stationary solutions and, then, identify the local maxima imposing that the Hessian matrix is negative definite.]

Exercise 11.2.9 Show that the efficiency of the superposition of gratings and lenses algorithm can be improved by adding a random phase to each single-trap hologram, i.e.,

$$\phi_{m_x,m_y}^{\text{SGL}} = \arg\left\{\sum_{n=1}^{N} e^{i\phi_{m_x,m_y}^{\text{S}}(x_{\text{ot},n},y_{\text{ot},n},z_{\text{ot},n}+\theta_n)}\right\}, \tag{11.29}$$

where θ_n are random phases. This algorithm is known as *random superposition of gratings and lenses*. Show that this algorithm corresponds to maximising the sum of the amplitudes of $V(x_{\text{ot},n}, y_{\text{ot},n}, z_{\text{ot},n})$ [Eq. (11.19)] projected on a randomly chosen direction in the complex plane, i.e., to maximising

$$\sum_{n=1}^{N} \text{Re}\left\{V(x_{\text{ot},n}, y_{\text{ot},n}, z_{\text{ot},n})e^{-i\theta_n}\right\}.$$

Exercise 11.2.10 Show that an array of weighted traps can be generated by weighting the terms of the sum in Eq. (11.28) by the square roots of the desired trap intensities. Verify the result numerically. [Hint: You can adapt the program `slm` from the book website.]

11.2.4 Gerchberg–Saxton algorithm

The *Gerchberg–Saxton algorithm* was originally developed by Gerchberg and Saxton (1972) for crystallographic applications. The basic idea is illustrated in Fig. 11.4: it is an iterative algorithm that permits one to find a phase distribution that turns a given input intensity distribution arriving at the DOE plane into a desired intensity distribution in the trapping plane by propagating the complex amplitude back and forth between these two planes and replacing at each step the intensity on the trapping plane with the target intensity and that on the DOE plane with the laser's actual intensity profile. It typically converges after a few tens of iterations; for example, in the benchmark test we present in Table 11.1, we obtained

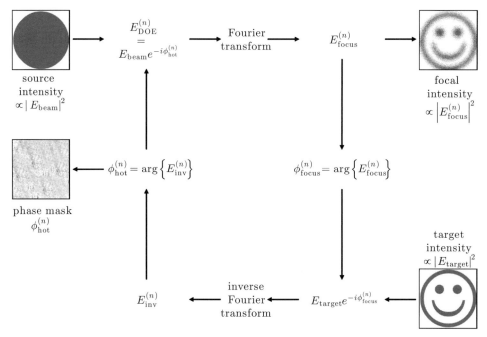

Figure 11.4 The Gerchberg–Saxton algorithm. Given the incoming beam electric field E_{beam}, the Gerchberg–Saxton algorithm permits one to obtain a phase distribution for the hologram at the DOE plane ϕ_{hot} that permits one to approximate the target intensity distribution at the focal plane. At the nth iteration of the algorithm, the amplitude of the field at the DOE, $E_{\text{DOE}}^{(n)}$, is given by E_{beam} and its phase by $\phi_{\text{hot}}^{(n)}$. The focal field $E_{\text{focus}}^{(n)}$ is given by the Fourier transform of $E_{\text{DOE}}^{(n)}$. Next, the amplitude of the focal field is substituted with the target intensity amplitude and the resulting field is inverse Fourier transformed, obtaining $E_{\text{inv}}^{(n)}$. Finally, the hologram phase is updated to the phase of $E_{\text{inv}}^{(n)}$. If the focal intensity and the target intensity differ more than an acceptable error, the cycle is then repeated (until a certain maximum number of iterations is reached).

$I_{\text{tot}} = 0.93$ and $u = 0.33$ after 30 iterations. This algorithm is ideally suited to deal with continuous intensity distributions, as we will see in more detail in Section 11.4.

Exercise 11.2.11 Implement the Gerchberg–Saxton algorithm and study its performance.

Exercise 11.2.12 Noting that, when the target intensity is an array of point traps, it is unnecessary to calculate the field complex amplitude in points whose amplitude will be replaced by zero before back-propagation, propose an alternative (more efficient) implementation of the Gerchberg–Saxton algorithm where the field is computed only at the trap locations.

Exercise 11.2.13 Show that the uniformity of the resulting trap array can be improved by aiming at a slightly modified target intensity distribution. [Hint: At each iteration, adjust the target intensity profile in order to reduce deviations from the average intensity. This algorithm is known as the *weighted Gerchberg–Saxton algorithm*.]

Exercise 11.2.14 Extend the Gerchberg–Saxton algorithm to three-dimensional trapping geometries by considering multiple planes for forward propagation and obtaining the back-propagated field as the complex sum of the corrected and back-propagated fields from the target planes. Then, proceed to generalise the Gerchberg–Saxton algorithm to full three-dimensional shaping.

11.2.5 Adaptive–additive algorithm

An alternative iterative approach is the *adaptive-additive algorithm*, which was first applied to the generation of holographic optical trap arrays by Dufresne et al. (2001). The basic idea of this algorithm is presented in Fig. 11.5: starting with an arbitrary guess of the phase profile and an initial input wavefront, the Fourier transform of this wavefront is the starting estimate for the output electric field; the resulting error in the focal plane is reduced by mixing a proportion, a, of the desired amplitude into the field in the focal plane; inverse-Fourier transforming the resulting focal field yields the corresponding field in the input plane; the amplitude in the input plane is replaced with the actual amplitude of the laser profile; finally, the algorithm is iterated. The main advantage of this algorithm in comparison to the Gerchberg–Saxton algorithm is that it permits one to achieve better

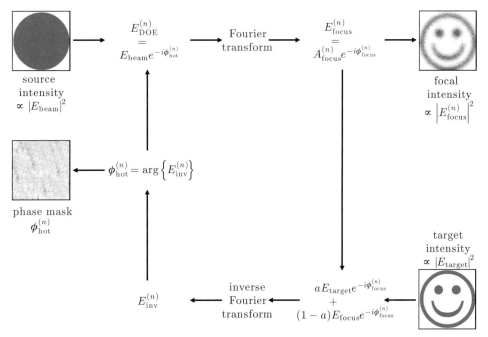

Figure 11.5 The adaptive–additive algorithm. The algorithm is quite similar to the Gerchberg–Saxton algorithm illustrated in Fig. 11.4. At the nth iteration, the amplitude of the field at the DOE, $E_{\text{DOE}}^{(n)}$, is given by E_{beam} and its phase by $\phi_{\text{hot}}^{(n)}$. The focal field $E_{\text{focus}}^{(n)}$ is given by the Fourier transform of $E_{\text{DOE}}^{(n)}$. At this point, the amplitude of the focal field is substituted with a weighted sum of the target amplitude and the focal field amplitude itself. The resulting field is inverse Fourier transformed, obtaining $E_{\text{inv}}^{(n)}$. Finally, the hologram phase is updated to the phase of $E_{\text{inv}}^{(n)}$. If the focal intensity and the target intensity differ more than an acceptable error, the cycle is then repeated (until a certain maximum number of iterations is reached).

uniformity over all the traps in the array. In the benchmark test we present in Table 11.1, we obtained $I_{\text{tot}} = 0.93$ and $u = 0.99$ after 30 iterations.

Exercise 11.2.15 Implement the adaptive-additive algorithm and study its performance.

11.2.6 Direct search algorithms

The holograms obtained with the previous algorithms can be refined by *direct search algorithms* where a *gain function* is defined and optimised. For example, we can choose a gain function obtained as a linear combination of the average intensity [Eq. (11.25)] and percentage standard error metrics [Eq. (11.27)], i.e.,

$$F_{\text{gain}} = \langle I \rangle - w\sigma, \qquad (11.30)$$

which is a function of all ϕ_{m_x,m_y} where w is a weighting factor. Then, starting from a good initial guess for the hologram, obtained using one of the algorithms explained above, a pixel is picked at random and cycled through all its possible grey levels while looking for an increase of F_{gain}. In this way, it is possible to improve the uniformity and efficiency of the hologram and to adjust the relative importance of the two by tuning w (typically $0 < w \leq 1$). Further improvements can be obtained by allowing for moves that temporarily decrease the gain (as in Monte Carlo algorithms) in a process known as *simulated annealing*.

Exercise 11.2.16 Any given optical field propagating through an optical system can be expressed as a composition of modes in an arbitrary orthogonal representation (e.g., orthogonal plane waves). Show that optimal focusing is achieved if all the modes meet at a selected point in space with the same phase and suggest a method to achieve this using a DOE and a direct search algorithm. [Hint: You can follow the method proposed by Čižmár et al. (2010).]

Exercise 11.2.17 How can the DOE be used in order to correct aberrations present in the optical system in order to achieve a more tight focus and, thus, more efficient optical traps? [Hint: For example, try adding (modulo 2π) Zernike polynomials to the original hologram.]

11.3 Higher-order beams and orbital angular momentum

Until now, we have only considered how to realise arrays of traps generated by Gaussian beams. Another powerful area that can be accessed using wavefront engineering is that of non-Gaussian beams. Here, we describe the techniques that permit one to generate, in particular, Hermite–Gaussian beams, Laguerre–Gaussian beams and non-diffracting beams, which we have described in detail in Subsections 4.4.2, 4.4.3 and 4.4.4, respectively. These beams can be straightforwardly generated starting from a uniform phase profile by

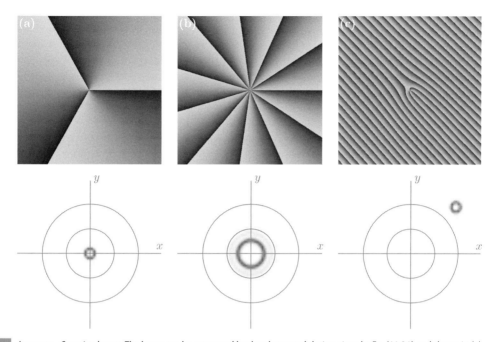

Figure 11.6 Laguerre–Gaussian beam. The beam can be generated by the phase modulation given by Eq. (11.31) and shown in (a) and (b) for beams of azimuthal order $l = 3$ and 11. Note that there are exactly $l\ 2\pi$ phase jumps along the angular direction. In (c), a grating is added to the phase mask in (a) to alter the direction of the 1st-order Laguerre–Gaussian beam away from the 0th-order unmodulated light. Both the phase modulation (top) and the focal spot intensity (bottom) are shown.

projecting onto the DOE their own characteristic phase profile, which we have seen in Figs. 4.7, 4.8 and 4.9, respectively. For example, the phase profile of a Laguerre–Gaussian beam of orders p and l at the beam waist w_0 can be obtained from the expression of the beam in Eq. (4.28) as

$$\phi(\rho, \varphi) = il\varphi + \pi \Theta \left\{ L_p^l \left(2\frac{\rho^2}{w_0^2} \right) \right\}, \qquad (11.31)$$

where $\Theta \{\cdot\}$ is the Heaviside step function. The azimuthal order causes $l\ 2\pi$ phase jumps along the angular direction. The radial order p leads to $p + 1$ radial regions, whose borders correspond to the p zeros of the corresponding generalised Laguerre function. It is typically convenient to shift the generated Laguerre–Gaussian beam away from the unmodulated 0th-order beam by adding a blazed grating to the phase modulation mask, as shown in Fig. 11.6 and explained in Section 11.5.

Exercise 11.3.1 Simulate the generation of Laguerre–Gaussian beams using a phase mask. [Hint: You can adapt the program `slmlg` from the book website.]

Exercise 11.3.2 Show that a Hermite–Gaussian beam of orders m_x and m_y can be generated using the phase modulation

$$\phi(x, y) = \Theta \left\{ H_{m_x}\left(\sqrt{2}\frac{x}{w_0}\right) H_{m_y}\left(\sqrt{2}\frac{y}{w_0}\right) \right\}. \tag{11.32}$$

Simulate its generation using an appropriate phase mask. [Hint: Start from Eq. (4.25). For the simulation, you can adapt the program `slmlg` from the book website.]

Exercise 11.3.3 How can Bessel beams and non-diffracting beams be generated using a phase mask? [Hint: Peruse Subsection 4.4.4 and Rose et al. (2012).]

Exercise 11.3.4 How can multiple Laguerre–Gaussian beams be generated in order to produce the counter-rotating optical traps shown in Fig. 11.1?

11.4 Continuous optical potentials

Wavefront engineering can also be employed to generate continuous optical potentials. In fact, the Gerchberg–Saxton and adaptive-additive algorithms, which we have seen in Subsections 11.2.4 and 11.2.5, respectively, are ideally suited to obtain phase profiles capable of generating continuous intensity distributions. The main difference regards the definition of the criterion employed to define the quality of the generated phase profile. In order to evaluate the convergence and to be able to impose a stop criterion, the resulting modulus of the complex amplitude of the scalar field at the focal plane is compared to the desired one,

$$C = \iint \left[A_{\text{focus}}(x, y) - A_{\text{target}}(x, y) \right]^2 dx dy, \tag{11.33}$$

where $A_{\text{focus}}(x, y)$ is the electric field intensity at the focal plane and $A_{\text{target}}(x, y)$ is the target intensity. The loop is iterated until the stop criterion is met, i.e., either a maximum number of iterations or a threshold error is reached.

Finally, we note that the same algorithm can be also applied to obtain a continuous distribution in multiple planes.[3] The algorithm, which generalises the ones presented in Figs. 11.4 and 11.5, is as follows:

1. propagate the field at the DOE to the multiple planes;
2. calculate the total error as the sum of the error calculated at each plane per Eq. (11.33);
3. inverse-Fourier transform these fields, multiply them by the inverse lens factor and add all contributions;
4. replace the amplitude at the DOE plane with the imposed one;
5. iterate this loop until the stop criterion is met.

---

[3] A generalisation to full three-dimensional shaping is currently far too slow for interactive use, taking days to calculate the desired phase modulation (Shabtay, 2003; Whyte and Courtial, 2005). Nevertheless, as described in Box 11.2, future improvement of GPU performance holds the promise that this will be soon feasible within reasonable time scales.

> **Box 11.2** **GPU computation of holograms**
>
> Unlike the central processing unit (CPU), which consists of a few cores optimised for sequential serial processing, the graphics processing unit (GPU) is designed for parallel processing of simultaneous operations on thousands of cores. To exploit the capabilities of the GPU and enable parallelisation of computation, NVIDIA has introduced the complete unified device architecture (CUDA). By optimising the code as explained, e.g., by Bianchi and Di Leonardo (2010), it is possible to achieve the calculation of holograms hundreds of times faster than on the CPU. Furthermore, Haist et al. (2006) cite other advantages of using the GPU, including increased performance per unit cost, ease of upgrading graphics boards and the expectation that the rate of increase in GPU performance (driven by the demand for ever more realistic computer game graphics) will continue to outstrip that in CPU performance. Potential drawbacks include the complexity of optimisation of code, which may also then need to be re-optimised for different GPUs, the consequent lack of flexibility and possibly even a reduction in speed if large amounts of data are transferred to the GPU, negating its advantage.

Exercise 11.4.1 Write a code to generate continuous optical potentials using the Gerchberg–Saxton algorithm both on a single plane and on multiple planes.

Exercise 11.4.2 Show that, when the Gerchberg–Saxton algorithm is used to generate a continuous optical potential, the error as defined by Eq. (11.33) diminishes or stays the same for each iteration. [Hint: Read the article by Fienup (1982).]

Exercise 11.4.3 Write a code to generate continuous optical potentials using the adaptive-additive algorithm both in a single plane and in multiple planes.

11.5 Set-up implementation

The configurations most commonly employed to actually build a holographic optical tweezers (HOT) are illustrated in Fig. 11.7. In the scheme presented in Fig. 11.7a, starting from the configuration described in Section 11.1 and illustrated in Fig. 11.2a, the DOE plane is optically conjugated to the back focal plane of the optical tweezers objective, so that the complex amplitude at the trapping plane, i.e., the front focal plane of the objective, is the Fourier transform of the complex amplitude in the DOE plane. Therefore, optical conjugation is achieved by positioning the DOE in the input plane of a telescope whose role is to project the hologram onto the input pupil of the objective lens. To calculate the effective displacement of the optical trap in the objective front focal plane, one has to consider the optical train after the DOE, which typically consists of a telescope and the microscope objective. Because the telescope magnifies the DOE field by a factor s, which is equal to the ratio of the focal lengths of its lenses, the grating period imaged by this telescope is equal to $s\Lambda$ and the angle of the 1st-order diffraction on the back focal plane of the microscope objective is scaled by a factor of $1/s$ to α/s. Therefore, the displacement

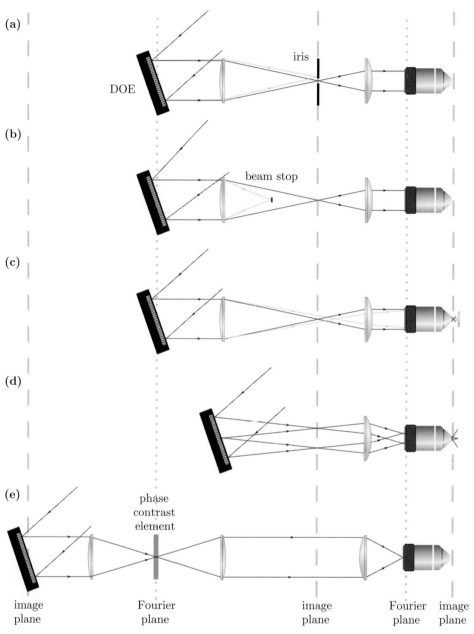

Figure 11.7 HOT configurations. Holographic optical tweezers (HOTs) can be configured in different ways. (a) The spatial light modulator (SLM) can be positioned exactly in the Fourier plane and the 1st-order diffracted beam can be deflected away from the 0th-order beam and spatially selected using an iris, or (b) close to the Fourier plane to allow the 0th-order beam to be removed with a spatial filter (beam stop). (c) A counter-propagating optical trapping configuration can be generated by placing a mirror behind the sample plane, allowing for the use of objectives with lower numerical apertures and longer working distances. Furthermore, the SLM can also be placed (d) in a Fresnel plane or (e) in a plane conjugated to the image plane, e.g., to use the phase contrast approach discussed in Subsection 11.6.2.

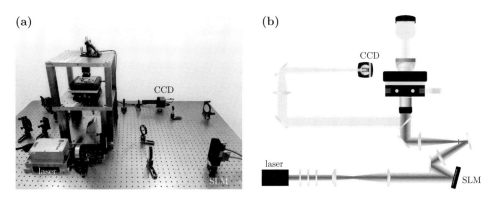

Figure 11.8 HOT set-up. (a) Holographic optical tweezers (HOT) set-up and (b) corresponding schematic.

of the beam in the front focal plane of the objective is

$$u = f_{\text{OBJ}} \frac{\lambda_0}{s \Lambda}, \tag{11.34}$$

where f_{OBJ} is the objective focal length. Analogous considerations apply to the effect of a Fresnel lens. We note that although Eq. (11.9) provides a first approximation to the field in the focal plane of the objective, some deviations arise because of the high numerical aperture of the objective and the exact focal field can be calculated using the focusing theory explained in Section 4.5.

The HOT set-up we will build is shown in Fig. 11.8. Starting from the optical tweezers we realised in Chapter 8, we removed the laser and build a new optical train with a 4f-configuration in order to add a DOE, which in our case is a reflective spatial light modulator (SLM). One important technical detail is that the SLM is arranged so that the 0th-order beam, i.e., the beam simply reflected by the SLM, is reflected in a direction different from that of the optical train of the optical trap; i.e., it is made to point to the side and downwards (for laser safety). Consequentially, we need to project a diffraction grating onto the SLM in order to generate a 1st-order beam aligned along the optical train of the optical trap for the following alignment procedure; an iris placed in the intermediate focal plane can help in removing the residual components of the 0th-order, as shown in Fig. 11.7a.[4] Then we can proceed to align the system as discussed in Section 8.4. Finally, we can place our sample on the stage and start our holographic optical trapping adventure on the lines of what we did in Section 8.5, but without the need for moving the sample. After a particle has been trapped, it is possible to move the trap vertically by adding a Fresnel lens to the computer-generated hologram: the optically trapped particle will get out of focus, whereas the non-optically-trapped particles on the sample cell bottom remain in focus. It is also possible to move the trap laterally by adding a blazed grating: the image of the optically trapped particle will move within the field of view of the camera, whereas the background

---

[4] An alternative approach to removing the 0th-order beam is to use a slightly convergent beam and to place a beam stop before the intermediate focus, as shown in Fig. 11.7b. In fact, as the focus of the incoming beam lies before the intermediate focal plane, the 0th-order beam can easily be removed by a spatial filter, e.g., a glass slide with a microfabricated reflecting or absorbing dot.

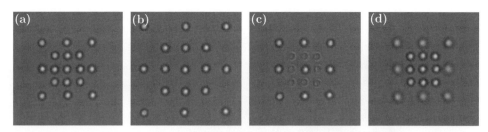

Figure 11.9 Holographically optically trapped particles. An array of holographic optical traps is used (a) to trap 17 particles, (b) to move them laterally in the focal plane and (c), (d) to move them vertically in the direction perpendicular to the focal plane.

particles sedimented on the coverslip will appear to remain in the same position. Of course, at this point, the interesting part of using HOTs is to create multiple reconfigurable traps, as shown in Fig. 11.9, or non-Gaussian traps, as shown in Fig. 11.1.

Some other configurations are also possible. For example, a counter-propagating beam trap can be generated in a set-up where the sample cell has a reflecting surface: two beams with different curvatures, and thus different axial focal positions, can be generated by the SLM and employed as two counter-propagating beams, as schematically shown in Fig. 11.7c. Such systems can employ lower NA than conventional three-dimensional traps and thus feature a longer working range.

One is not restricted to using the SLM in the Fourier plane. Indeed, it is also possible to use it in an intermediate Fresnel or image plane. Positioning the SLM in an intermediate Fresnel plane, as shown in Fig. 11.7d, simplifies the hologram design in that the SLM aperture can be divided into different trap regions so that moving the position of one of the traps slightly is simply accomplished by translating the corresponding section of the SLM pattern, which is computationally quick. Positioning the SLM in an image plane, as shown in Fig. 11.7d, permits one to directly image the image projected onto the SLM to the optical trap plane; this is computationally very efficient, even though it can lead to greater losses than phase-only modulation.

In all configurations, the calculation of the optimum hologram for a large number of optical traps or for multiple-plane continuous optical potentials is a computationally intensive task. These limitations become particularly evident in applications where real-time and interactive optical manipulation (see, e.g., Subsection 12.6) is required, as even optimised algorithms may not perform fast enough and thus may be limited to running pre-calculated series of holograms to steer the optical traps along pre-defined paths. A possible approach to overcoming this problem is shifting the intensive parts of the computation to the graphics processing unit (GPU) in order to exploit its parallel architecture, as explained in Box 11.2.

11.5.1 Spatial light modulators

A spatial light modulator (SLM) is a device that functions as a DOE by control of either the amplitude or the phase of an optical beam in a plane transverse to the direction of propagation. Amplitude modulators (which include digital micromirror devices) obviously have a lower diffraction efficiency for transformation of the input beam into the desired pattern for optical trapping, because a significant proportion of the input beam is simply blocked or directed away from the path of the trapping beam. Phase modulators, which are available in both transmissive and reflective forms, use a pixellated liquid crystal display to imprint a spatially varying pattern that ideally alters only the phase via the change in refractive index associated with the reorientation of the liquid crystal under an applied voltage. In a transmissive SLM, the beam passes once through the liquid crystal layer (and its supporting substrate). In a reflective SLM (commonly known as liquid crystal on silicon, or LCoS), the beam passes through the liquid crystal, reflects from the substrate and then passes through the liquid crystal again. This means that a reflective SLM can employ a thinner liquid crystal layer to achieve the same phase retardation, which can result in an increase in the switching speed (which varies as the square of the layer thickness). Liquid crystal-based technologies have the advantage of being relatively advanced, being lower in cost, and having shorter switching times.[5]

Two distinct classes of liquid crystal SLMs based on the type of liquid crystal used should be considered. Ferroelectric liquid crystal SLMs can switch between only two phase retardation levels (0 and π), which limits diffraction efficiency, but are capable of fast switching speeds of up to 1 kHz. The more common nematic liquid crystal SLMs approximate continuous phase variation with a large number (typically 256) of discrete retardation levels from 0 up to the maximum dynamic range of the device – most often 2π, although modulators which produce a greater than 2π maximum retardance are available. Nematic liquid crystal SLMs typically switch at video frame rate, i.e., 60 Hz.

Other factors to consider when selecting a suitable spatial light modulator for optical tweezers applications may include

- **Pixel size and number.** These will affect the resolution of, for example, positioning of a trap in the transverse plane. Currently LCoS SLMs with a pixel pitch as small as 6.4 μm and up to 1920 × 1080 pixels are available.
- **Pixel filling factor.** This refers to the area of the display taken up by the liquid pixel crystals: a higher filling factor reduces the inactive region between pixels, thereby increasing the efficiency of conversion into the phase modulated beam. For most devices a filling factor over 80% is specified.
- **Wavelength range.** Ideally the SLM will be anti-reflection coated at the laser wavelength to be employed and capable of at least 2π retardation. Models for wavelength ranges from the blue end of the visible spectrum to telecom (near-infrared) wavelengths are available.
- **Phase resolution.** Greyscale resolution is typically 8-bit (256 levels), although SLMs with 12-bit resolution are already available.
- **Polarisation sensitivity.** Standard LCoS SLM devices will apply the phase modulation to one direction of linear polarisation only; however, polarisation-independent SLMs have recently been introduced.

---

[5] Manufacturers of liquid crystal spatial light modulators currently available include Boulder Nonlinear Systems (USA) (now part of Meadowlark Optics), HoloEye Photonics (Germany), Hamamatsu (Japan) and Jenoptik (Germany).

- **Damage threshold.** When using high-power or pulsed lasers care should be taken not to exceed the maximum power density that would avoid damage to the liquid crystal layer.

11.6 Alternative approaches

Several alternative approaches are available to generate large arrays of optical traps. For example, as discussed in Subsections 4.4.4 and 12.2, using multi-beam interference (discrete beams) is simple, produces high-quality optical lattices over extended three-dimensional volumes and can tolerate high beam powers. However, such approaches are limited to (quasi-)symmetric patterns. Alternatively, galvo-mirrors or piezoelectrics have served as the basis for designs involving scanning laser tweezers, building on the approach described in Subsection 8.5.1. Another (DOE-based) strategy that allows flexible generation of trap arrays uses the *generalised phase-contrast* (GPC) method. In addition, arrays covering large areas have now been produced using evanescent waves. In the remainder of this section, we will discuss the time-shared optical trap approach [Subsection 11.6.1] and the GPC approach [Subsection 11.6.2]; evanescent trapping will be discussed in Section 12.4.

11.6.1 Time-shared optical traps

Multiple time-shared optical traps can be obtained using several beam steering techniques and a 4f configuration, such as the one described in Subsection 8.5.1. The laser needs to be repositioned on a time scale short enough so that the trapped particles experience only a time-averaged potential. Two of the most commonly employed devices are acousto-optic deflectors (AODs) and galvo-mirrors. AODs create a blazed grating by generating acoustic standing waves within a crystal and thus creating a periodic alteration of the crystal's refracting index so that a light beam can be deflected in one direction; using two such crystals, it is possible to deflect a beam in two directions (x and y). The main advantage of AODs over SLMs, is that they can be scanned at hundreds of kilohertz.[6] Galvo-driven mirrors can be controlled by an electro-optical modulator, which can yield smoothly varying intensity modulations in a continuous optical potential. In applications where a smooth potential is required, such as in the creation of a smooth ring trap (Blickle et al., 2007), galvo-mirrors might be preferred over either HOTs or AODs because they provide a much higher throughput of the incident light than either AODs or Fourier-plane HOTs; however, inertia limits the scan speed of any macroscopic mirror to a fraction of what is available via AODs. Unlike the SLM-based techniques, systems based on AODs

---

[6] Some technical issues need to be accounted for when working with AODs. In particular, ghost traps might appear in analogue AOD systems (as the beam is often sequentially repositioned in x and then in y, so the generation of two traps along the diagonal of a square yields an unintended spot at one of the other corners). Furthermore, as the AOD efficiency falls off as a function of the deflection angle, for applications requiring uniform arrays, one must compensate, either by spending more time at peripheral traps or by increasing the power sent to those traps. Finally, AOD-generated arrays can be thought of as being made up of incoherent light (different beams do not interfere, being present only one at a time).

and galvo-mirrors cannot normally do mode conversion or aberration correction, and they cannot generate three-dimensional arrays of traps.

In all time-shared optical trapping techniques, in order for each trapped particle to feel only the time-averaged potential, the maximal time that the laser can spend away from any one trap needs to be much smaller than the characteristic time it takes the particle to diffuse across the trap, i.e., the time scale $\tau_c = 1/f_c$, where f_c is the corner frequency in the power spectrum [Section 9.8]. In fact, while the trap is not at the site, the particle diffuses away from its nominal trapping site by a characteristic distance $d = \sqrt{6Dt_{\text{off}}}$, where D is the particle diffusion coefficient and t_{off} is the time the laser is off at the site. For example, for a 1 µm radius particle in water and $t_{\text{off}} = 1$ ms, we can expect an average diffusion of $d = 36$ nm. Clearly, the smaller the particle, the shorter τ_c becomes; also, the less viscous the medium, the shorter τ_c.[7] This, together with the requirement that the laser spends sufficient time at each trap site to produce the time-averaged power required for the desired trap strength and the fact that, although trap strength depends only on the time-average power, sample damage due to two-photon absorption and local heating contain a dependence upon the peak power, limits both the type of arrays that can be constructed and the accuracy to which the spheres can be positioned using time-shared trapping.

11.6.2 Generalised phase contrast

Another alternative approach positions the SLM in the image plane of the traps, as shown in Fig. 11.7e. This method, known as the *generalised phase-contrast* (GPC) technique, was developed by Mogensen and Glückstad (2000), building on the phase contrast imaging method developed by Frits Zernike, which we discussed in Subsection 8.2.2. In this configuration, the SLM acts as a phase-contrast element where a phase pattern displayed on the SLM produces the same intensity distribution in the sample plane and, thus, complicated distributions of traps can be produced without any need for sophisticated computing. As opposed to a simple intensity mask, which creates the desired pattern by blocking the unwanted light, phase contrast is optically efficient, ensuring that all of the laser light is directed into the optical traps; in fact, this set-up can reach a much higher modulation efficiency (up to 90%) without speckle noise or ghost traps. Because, in the GPC approach, the SLM is conjugate to the trapping plane, no computations are required to convert the phase-only modulation of the SLM into an intensity modulation in the image plane; instead there is a direct, one-to-one correspondence between the phase pattern displayed on the SLM and the intensity pattern created in the trapping plane. In the Fourier plane, a small π-phase filter shifts the focused light coming from the SLM so that at the image plane it will interfere with a plane-wave component. The result is a system that only requires the user to write the desired two-dimensional patterns on the SLM. The downside to this

---

[7] This practically means that when particles are being trapped in air, as we will discuss in Chapter 19, the corner frequency can reach value of several kilohertz even for large (several micrometre) radius particles. This is an issue that, in fact, has limited the application of time-shared techniques to the trapping of aerosols, leaving the use of HOTs as a preferable choice.

is that xy-positions are limited to pixel positions, meaning that ultra-high-precision trap positioning is not possible to the degree it is with the other SLM-based techniques.

Extension of the GPC method to three dimensions requires the use of counter-propagating beam traps, rather than optical tweezers, and, therefore, three-dimensional control is, in some sense, more involved. For this reason, Dam et al. (2007) have developed an automated alignment protocol for users interested in three-dimensional control. Nevertheless, it is not possible to place traps behind each other controllably with this method, but there are now many impressive demonstrations of three-dimensional manipulation using the GPC technique, e.g., in some of the applications described in Chapter 16. The version of GPC using counter-propagating beam traps can also use low-NA optics, which can have a large field of view and a large Rayleigh range. So, although Fourier-plane holographic optical traps can provide only a small range of axial displacements, limited by spherical aberration, the low-NA GPC trap arrays are sometimes called *optical elevators* because of the wide range over which the traps can be displaced along the optical axis, as the working distance can be up to 1 cm, which is roughly 100 times that of a conventional optical tweezers set-up.

Problems

11.1 *Genetic algorithms.* Develop a program that uses genetic algorithms to generate holograms to produce arrays of optical traps and optical trapping potentials. Compare the performance of these algorithms with that obtained using the algorithms described in the text.

11.2 Calculate the phase and intensity structure of two optical vortices with the same vorticity but located at opposite positions within a Gaussian profile. Study (experimentally or by simulations) the evolution when the positions get closer to the optical axis until merging. Does the total vorticity of the beam change as vortices move towards the higher intensity region? What happens if the two vortices possess opposite charges?

11.3 *Fractional optical vortices.* Laguerre–Gaussian beams possess a phase structure that yields a perfect helicoidal phase front only for integer l. Following the work of Berry (2004), investigate theoretically the phase structure of beams that possess fractional l values and show that they have a complex phase structure comprising many vortices at different locations within the beam cross-section. Design an SLM-based set-up and the corresponding holograms needed to study these fractional optical vortices. Compare your experimental (or simulated) results for different fractional values of l with those of Leach et al. (2004) and Lee et al. (2004).

11.4 *Interferometric synthesis of cylindrical vector beams.* Design a set-up that can generate cylindrical vector beams interferometrically with one SLM that works either in transmission or in reflection. Consider first a design that has a 4f configuration and describe the holographic pattern you need to transfer to the SLM to generate radially and azimuthally polarised beams. Then consider a design that includes a Sagnac interferometer. What are the advantages and disadvantages of the two configurations? How can you synthesise fractional cylindrical vector beams? Suppose you use a reflection SLM that needs to be used at small reflection angles. How can you modify the Sagnac interferometer to avoid right angle reflection at the SLM? Compare your designs with those of Maurer et al. (2007) and Jones et al. (2009).

References

Berry, M. V. 2004. Optical vortices evolving from helicoidal integer and fractional phase steps. *J. Opt. A Pure Appl. Opt.*, **6**, 259–68.

Bianchi, S., and Di Leonardo, R. 2010. Real-time optical micro-manipulation using optimized holograms generated on the GPU. *Comp. Phys. Commun.*, **181**, 1444–8.

Blickle, V., Speck, T., Lutz, C., Seifert, U., and Bechinger, C. 2007. The Einstein relation generalized to non-equilibrium. *Phys. Rev. Lett.*, **98**, 210601.

Čižmár, T., Mazilu, M., and Dholakia, K. 2010. *In situ* wavefront correction and its application to micromanipulation. *Nature Photon.*, **4**, 388–94.

Dam, J. S., Rodrigo, P. J., Perch-Nielsen, I. R., Alonzo, C. A., and Glückstad, J. 2007. Computerized "drag-and-drop" alignment of GPC-based optical micromanipulation system. *Opt. Express*, **15**, 1923–31.

Dufresne, E. R., Spalding, G. C., Dearing, M. T., Sheets, S. A., and Grier, D. G. 2001. Computer-generated holographic optical tweezer arrays. *Rev. Sci. Instrumen.*, **72**, 1810–16.

Fienup, J. R. 1982. Phase retrieval algorithms: A comparison. *Appl. Opt.*, **21**, 2758–69.

Gerchberg, R. W., and Saxton, W. O. 1972. A practical algorithm for the determination of the phase from image and diffraction plane pictures. *Optik*, **35**, 237–46.

Haist, T., Reicherter, M., Wu, M., and Seifert, L. 2006. Using graphics boards to compute holograms. *Comput. Sci. Eng.*, **8**, 8–13.

Jones, P. H., Rashid, M., Makita, M., and Maragò, O. M. 2009. Sagnac interferometer method for synthesis of fractional polarization vortices. *Opt. Lett.*, **34**, 2560–62.

Leach, J., Yao, E., and Padgett, M. J. 2004. Observation of the vortex structure of a non-integer vortex beam. *New J. Phys.*, **6**, 71.

Lee, W. M., Yuan, X.-C., and Dholakia, K. 2004. Experimental observation of optical vortex evolution in a Gaussian beam with an embedded fractional phase step. *Opt. Commun.*, **239**, 129–35.

Maurer, C., Jesacher, A., Fürhapter, S., Bernet, S., and Ritsch-Marte, M. 2007. Tailoring of arbitrary optical vector beams. *New J. Phys.*, **9**, 78.

Mogensen, P. C., and Glückstad, J. 2000. Dynamic array generation and pattern formation for optical tweezers. *Opt. Commun.*, **175**, 75–81.

Rose, P., Boguslawski, M., and Denz, C. 2012. Nonlinear lattice structures based on families of complex nondiffracting beams. *New J. Phys.*, **14**, 033018.

Shabtay, G. 2003. Three-dimensional beam forming and Ewald's surfaces. *Opt. Commun.*, **226**, 33–7.

Whyte, G., and Courtial, J. 2005. Experimental demonstration of holographic three-dimensional light shaping using a Gerchberg–Saxton algorithm. *New J. Phys.*, **7**, 117.

12 Advanced techniques

As we have seen in the previous chapters, a single strongly focused laser beam is a powerful tool for trapping and manipulating microscopic particles. Furthermore, we have seen in Chapter 11 that multiple and dynamic optical tweezers can be generated using diffractive optics. However, this does not exhaust the potential of optical trapping and manipulation. Indeed, as optical forces have been applied to the solution of problems in different fields, they have been adapted to and hybridised with the techniques available and needed in these fields. An example of such an advanced hybrid application is shown in Fig. 12.1, where a feedback optical trap is used to hold a nanoparticle in a bow-tie nanoaperture fabricated on the tip of a tapered metallised optical fibre. In this chapter, we will survey more advanced techniques, including the hybridisation of optical tweezers and spectroscopic techniques, the generation of extended optical force landscapes, the use of optical fibre traps, evanescent wave traps and feedback traps, and the exploitation of haptic devices to interface the microworld and the user.

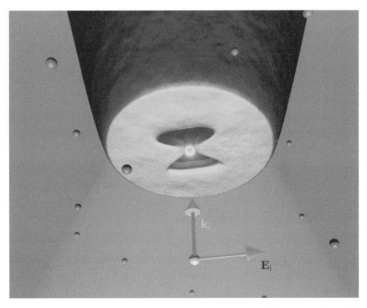

Figure 12.1 Self-induced back action optical trap. Artist's depiction of a SIBA optical trap: a feedback optical trap holds a dielectric nanoparticle in place in a bow-tie nanoaperture fabricated on the tip of a tapered metallised optical fibre. Reprinted by permission from Macmillan Publishers Ltd: Berthelot et al., *Nature Nanotech.* **9**, 295–9, copyright 2014.

12.1 Spectroscopic optical tweezers

Spectroscopic optical tweezers (SOTs) are obtained by enhancing optical tweezers with spectroscopic functionalities. SOTs allow one to study the chemical and physical properties of a single optically trapped microparticle or nanoparticle by probing its electronic (photoluminescence), vibrational (Raman) and nonlinear (e.g., two-photon photoluminescence) properties *in situ*. A typical SOT apparatus incorporates two optical beams, one to trap the particle and one to probe its properties. Trapping is often accomplished with a near-infrared laser to minimise photodamage to biomaterials, as we have seen in Section 8.4, whereas visible light is often used for excitation. Such a two-beam arrangement, shown in Fig. 12.2a, offers the versatility to trap, manipulate and excite specific zones of the object under investigation by displacing the trapping and excitation beams independently. In some cases, though, it is possible to use a single beam for both tasks, which may have the advantage of simplicity and increased stability. This simpler configuration is shown in Fig. 12.2b and has been widely exploited since the 1980s (Petrov, 2007). A more complex and versatile version uses two separate objectives to focus the beams for trapping and excitation, as shown in Fig. 12.2c. In all cases, one should always keep in mind that all components along

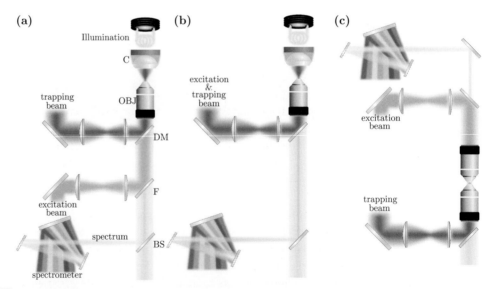

Figure 12.2 Basic configurations of spectroscopic optical tweezers. (a) Set-up integrating optical tweezers and spectroscopy. The trapping and excitation beams are combined using a dichroic mirror (DM) and focused through a high-numerical-aperture objective lens (OBJ). For scattering spectroscopy, the sample can also be excited using a halogen lamp focused by a dark-field condenser (C) from the top (Illumination); in this case, the signal is collected through the trapping objective lens. Notch/edge filters (F) are used to cut out the elastic scattering at the excitation/trapping wavelengths. A beam splitter (BS) divides the imaging light from the spectral signal. A grating spectrometer equipped with a CCD camera or an avalanche photodiode acquires the spectroscopic signal. (b) Simpler configuration based on a single beam and single objective. (c) More complex and versatile configuration using two independent objectives for the trapping and excitation beams.

the optical path can potentially contribute to the spectroscopic signal, and therefore great care should be exercised to minimise spurious signals. In particular, the use of thin (80 μm, number 0 [Table 8.1]) quartz or CaF_2 cover-slips can decrease background fluorescence signals from those for typical 170 μm (number 1 or 1.5) glass cover-slips. Furthermore, it might be useful to avoid oil-immersion objectives and prefer water-immersion ones instead, in order to avoid the background signals coming from the immersion oil. In this section, we will give a brief introduction to the technical details of SOT set-ups. The applications of SOTs will be discussed in more detail in Chapter 15.

12.1.1 Fluorescence tweezers

Fluorescence is caused by the absorption of radiation at one wavelength followed by nearly immediate re-radiation at a longer wavelength. Fluorescence microscopy is arguably the most widely used technique for obtaining information on the position or conformation of a molecule. This is particularly true in the life sciences, where recent advances have enabled spatial resolution well below the conventional optical diffraction limit (Hell, 2008) and localisation of fluorophores to as little as one nanometre accuracy (Yildiz et al., 2003). Therefore, the combination of fluorescence microscopy with optical tweezers is an attractive proposition, as it may permit simultaneous determination of the position, mechanical properties and biochemical state of a molecule.

An example of a fluorescence optical tweezers is shown in Fig. 12.3a. This experimental apparatus was realised by van Dijk et al. (2004) to use with fluorophores (Cy3 succinimidyl ester, Alexa555-maleimide and carboxytetramethylrhodamine succinimidyl ester) that absorb in the visible with a peak at around $\approx$ 560 nm and emit in the range from about 530 to 700 nm. The trapping is performed using a wavelength in the near infrared (850 nm) and the fluorescence excitation using a visible wavelength (532 nm). Dichroic mirrors are used to combine the laser beams for trapping and fluorescence excitation and filters are used to isolate the fluorescence emission (see Subsection 8.4.4 for details on these optical elements). The 'NIR reject filter' is used to stop any scattered light from the trapping laser beam in the imaging path, the '610/75 bandpass filter' allows fluorescence light (from 610 nm to 675 nm) to pass and finally, the '532 notch filter' stops any remaining excitation light. Although van Dijk et al. (2004) found no evidence of multiphoton absorption of the 850 nm trapping light, the trapping laser was found to affect the rate of photobleaching, i.e., the rate at which fluorophores enter a non-fluorescent state as a result of oxidation. The mechanism proposed was a two-step process involving absorption of a green and a near-infrared photon followed by ionisation of the dye molecule. Different dyes, however, exhibit different photobleaching rates, making the choice of a suitable dye a crucial aspect in fluorescence tweezers applications.

For fluorescence microscopy applications requiring localisation of emission down to the nanometre scale, the stability of the apparatus is a significant issue. Furthermore, position fluctuations of the optical trap are affected by inherent laser pointing instabilities, as we have seen in Subsection 8.4.1. The mechanical and thermal stability of the apparatus can be improved by following the suggestions provided in Section 8.1. For example, using an actively vibration-isolated optical table and elastomeric vibration isolators to mount the microscope, Capitanio et al. (2005, 2007) were able to achieve 1 nm accuracy in fluorescence

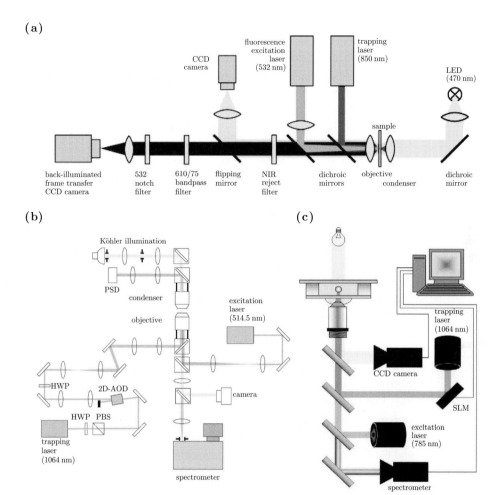

Figure 12.3 Concrete examples of spectroscopic optical tweezers. (a) Fluorescence optical tweezers. The near-infrared (850 nm) laser beam is used for optical trapping and the green (532 nm) laser beam for fluorescence excitation. A number of optical filters are used along the imaging path to isolate the fluorescence. Reprinted with permission from van Dijk et al., *J. Phys. Chem. B* **108**, 6479–84. Copyright (2004) American Chemical Society. (b) Two-beam photoluminescence optical tweezers. The 1064 nm laser beam (steerable in two dimensions by acousto-optic deflectors) is used for optical trapping and two-photon photoluminescence, and the 514.5 nm laser beam for direct absorption excitation. Spectra are recorded using a grating spectrometer with a CCD array. Reprinted with permission from Wang et al., *Nano Lett.* **11**, 4149–53. Copyright (2011) American Chemical Society. (c) Two-beam Raman optical tweezers set-up. The 1064 nm laser beam and the spatial light modulator are used to generate a set of steerable optical traps, whereas the 785 nm laser beam is used to excite the Raman signal. Reprinted from Creely et al., *Opt. Express* **13**, 6105–10. Copyright (2005) The Optical Society.

imaging. Even higher stability was achieved by Abbondanzieri et al. (2005) by enclosing the apparatus in a sealed box filled with helium, which has a refractive index even closer to unity than air, at atmospheric pressure. Intensity fluctuations of the laser arising from optical feedback can be nearly eliminated using an optical (Faraday) isolator in the beam path close to the laser source.

12.1.2 Photoluminescence tweezers

Photoluminescence is a process in which a material goes into a higher electronic state by absorbing a photon and, at a later stage, relaxes and returns to a lower energy level by releasing energy in the form of photons. Therefore, photoluminescence spectroscopy is an optical method of probing the electronic properties of materials. Photoluminescence spectroscopy has been successfully integrated with optical tweezers. For example, Wang et al. (2011) used the set-up shown in Fig. 12.3b to investigate nonlinear photoexcitation in optically trapped InP nanowires. Optical trapping is performed by the 1064 nm laser beam, which can be steered in two dimensions using acousto-optic deflectors. The trapping beam is also used to excite second harmonic generation at $\lambda_{SHG} = 532$ nm, together with band-edge photoluminescence emission at $\lambda_{PL} = 890$ nm due to two- and three-photon absorption. Direct absorption photoluminescence was excited by a separate laser beam at 514.5 nm. From the red shift between the two-photon absorption photoluminescence and the direct absorption photoluminescence, Wang et al. (2011) were able to probe band-filling at the single nanowire level. Furthermore, by implementing a holographic optical tweezers with a spatial light modulator, as explained in Chapter 11, Wang et al. (2013) were able to move a trapped nanowire relative to the focus of the excitation beam, thereby mapping structural inhomogeneities and enabling sorting of nanowires with specific characteristics.

12.1.3 Raman tweezers

Raman spectroscopy is based on inelastic scattering of photons by molecular vibrations and it is therefore able to probe the molecular properties of a sample. It is one of the most powerful analytical techniques in physics, chemistry, biology and corresponding interdisciplinary branches of science. By combining Raman spectroscopy with a confocal optical microscope, one can achieve chemical characterisation on a scale of femtolitre volumes within the sample to be probed. Furthermore, Raman spectroscopy can be combined usefully with optical tweezers to probe the chemical and physical properties of a trapped particle through its vibrational fingerprint.

Raman tweezers were first demonstrated by probing the Raman spectrum of an emulsion of 1-bromonaphthlene or iodobenzene microdroplets in water (Lankers et al., 1994). In this experiment, both trapping and excitation were performed by the same 514.5 nm laser beam. Later, Xie et al. (2002) demonstrated a single-beam Raman tweezers suitable for use with biological material. In these experiments, the power of the trapping and excitation laser beam (785 nm) was switched between low power for long-term trapping and high power for spectroscopy during short periods in order to avoid photodamage to the trapped specimen from prolonged illumination.

Greater functionality in Raman tweezers can be achieved by an apparatus with separate beams for tweezing and spectroscopy that can be moved relative to each other. For example, the Raman tweezers apparatus used by Creely et al. (2005) for spatially resolved Raman spectroscopy on single optically trapped cells is shown in Fig. 12.2c. Optical trapping is performed by an infrared laser (1064 nm) multiplexed and controlled by a spatial light modulator, as we have seen in Chapter 11. Excitation is performed by a separate near-infrared laser (785 nm). Differing spatial distributions of proteins and lipids were imaged by plotting the intensity of the relevant Raman bands as an optically trapped cell was scanned through the Raman excitation beam using the steerable trapping beams.

12.2 Optical potentials

Going beyond a single or a discrete set of optical traps, one can consider extended optical force potentials. Several techniques have been proposed to produce such extended optical potentials, typically generated using interference of several laser beams, which can produce both regular and random light patterns, although trapping is achieved against a surface to compensate for the presence of scattering forces, which frequently exceed axial gradient forces in this configuration. Extended three-dimensional trapping landscapes have also been achieved, e.g., using spatial light modulators and acousto-optic modulators.

12.2.1 Periodic and quasi-periodic potentials

The use of several interfering beams can produce complex optical landscapes where particles can be trapped in two dimensions (against a surface or at a liquid–liquid or liquid–air interface). In this way, up to thousands of colloidal particles have been trapped and arranged in crystalline (e.g., lattice with square or hexagonal symmetry) and also quasicrystalline structures (Mikhael et al., 2007), as shown in Fig. 12.4. Because of the two-dimensional confinement, it is possible to use a relatively low optical power (on the order of about $0.1 \, \text{mW}/\mu\text{m}^2$ or even less). Careful control of the various interfering beams is necessary to ensure their correct interference.

Alternative techniques for obtaining extended optical landscapes make use of holographic optical tweezers (HOTs) or acousto-optic deflectors (AODs). As we have seen in Chapter 11, HOT techniques can be used to produce several discrete optical traps or a continuous optical landscape. Their main advantage is the possibility of generating multiple reconfigurable optical traps and therefore of manipulating several particles in three dimensions. However, because typically a high-NA objective is required, the field of view is intrinsically limited to about $100 \, \mu\text{m} \times 100 \, \mu\text{m}$ and the number of trapping sites is severely limited by the hologram resolution and the available laser power (typically up to several tens of trapping sites). AODs can be used to generate arbitrary dynamical optical landscapes by deflection of a single incident laser beam from a travelling acoustic wave inside a transparent crystal. By changing the frequency and amplitude of this wave, the incoming laser beam can be modulated with respect to its angle and intensity at frequencies up to about 50 kHz. As in the case of standard optical tweezers, AOD-enhanced techniques require an overfilled

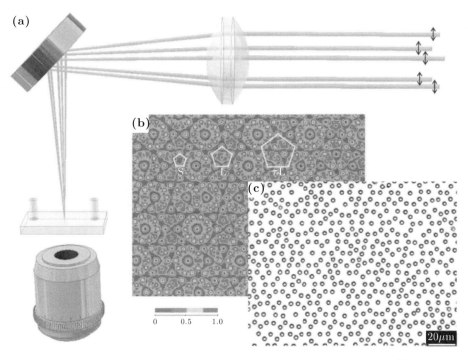

Figure 12.4 Experimental realisation of colloidal quasicrystals. (a) Five linearly polarised (polarisation as indicated by arrows) parallel laser beams forming a regular pentagon are focused into a thin sample cell. (b) Experimentally determined intensity distribution of the interference pattern, which acts as a substrate potential for the colloids. The pattern displays a decagonal symmetry and the predominating motifs are pentagons (indicated in white) with sides of different lengths related by the golden ratio $\tau = S/L$. Here $S = 5.64$ μm and $L = 9.13$ μm. The shades of grey of the intensity field reflect the variation in potential well depth. (c) Configuration of colloidal particles at a density of 0.0264 particles/μm exposed to a decagonal substrate interference pattern. Reprinted by permission from Macmillan Publishers Ltd: Mikhael et al., *Nature* **454**, 501–4, copyright 2008.

high-NA objective, so the field of view is intrinsically limited to about 100 μm × 100 μm, and differently from HOTs, only permit manipulation of particles in two dimensions, i.e., in the focal plane.

12.2.2 Random potentials and speckle tweezers

Pseudo-random optical potentials can be generated using spatial light modulators or acousto-optic deflectors, as discussed in the previous subsection. An alternative approach, proposed by Volpe et al. (2014a), relies on the use of speckle light fields. Speckle light fields with the required statistical properties are routinely generated over large areas using various processes, such as scattering of a laser beam from a rough surface, multiple scattering in an optically complex medium, or mode mixing in a multimode fibre, as shown in Figs. 12.5a, 12.5b and 12.5c, respectively. In general, the motion of a Brownian particle in a static speckle field is the result of random thermal forces and deterministic optical forces (Volpe et al., 2014b). Optical gradient forces are the dominant deterministic forces acting

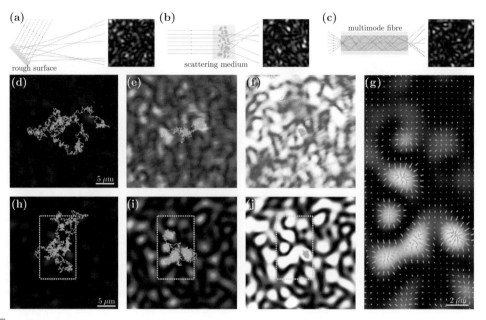

Figure 12.5 Speckle optical tweezers. Speckle light fields result from the interference of several optical waves with random phase. They can be generated by several techniques, e.g., (a) reflection from a rough surface, (b) transmission through a highly scattering medium (e.g., a biological tissue) and (c) mode mixing in a multimode optical fibre; the laser beam is sketched as a series of rays that interfere and generate the speckle. (d)–(f) The experimental trajectories (solid lines) of a silica bead ($a = 1.03 \pm 0.03$ μm, $n_p = 1.42$) in water ($n_m = 1.33$) under the action of the optical forces produced by a speckle show the progressive particle confinement as the average speckle intensity increases ($\langle I \rangle = 0.12$ μW/μm² in (d), $\langle I \rangle = 1.43$ μW/μm² in (e) and $\langle I \rangle = 5.77$ μW/μm² in (f)). The backgrounds are the corresponding images of the speckle patterns generated by mode mixing in a multimode optical fibre. (g) Calculated optical force field (arrows) exerted on a silica bead in a simulated speckle pattern (background). (h)–(j) Corresponding simulated trajectories (solid lines) of silica particles moving in speckle fields of the same average intensity as in (d)–(f). The dashed lines delimit the area corresponding to the force field distribution in (g). The average modulus of the calculated force exerted by the speckle field is (h) $\langle F \rangle = 0.14$ fN, (i) $\langle F \rangle = 1.82$ fN, and (j) $\langle F \rangle = 7.3$ fN. All trajectories are recorded or simulated for 420 s. Figures (d)–(j) reprinted from Volpe et al., *Opt. Express* **22**, 18 159–67. Copyright (2014) The Optical Society.

on dielectric particles whose size is comparable to or smaller than the average speckle grain, and they attract particles with high refractive index towards the intensity maxima of the optical field. As a particle moves in the speckle field, the optical force acting on it changes both in magnitude and in direction with a characteristic time scale that to first approximation is inversely proportional to the average speckle intensity, as we will see in Section 20.6 [Eq. (20.14)]. Let us consider first the simplest case, e.g., the motion of an isolated silica bead (radius $a = 1.03 \pm 0.03$ μm, refractive index $n_p = 1.42$) in a static speckle pattern. As shown by the trajectory (solid line) in Fig. 12.5d, when the average speckle intensity is relatively low ($\langle I \rangle = 0.12$ μW/μm²), the particle is virtually freely diffusing. As the intensity increases (Fig. 12.5e, $\langle I \rangle = 1.43$ μW/μm²), the particle becomes metastably trapped in the speckle grains, although it can still jump from one grain to the next from time to time. Finally, for even higher intensities (Fig. 12.5f, $\langle I \rangle = 5.77\,\mu\text{W}/\mu\text{m}^2$), the particle remains trapped in one of the speckle grains virtually forever. Further insight into the underlying physics can be gained by calculating the optical force field acting on a silica particle moving in a simulated speckle pattern [Fig. 12.5g]. As the particle size is significantly larger than the light wavelength, this calculation can be performed using the ray optics approach described in Chapter 2, whereas the particle motion can be simulated with the Brownian dynamics simulations described in Chapter 7. The features of the resulting simulated trajectories [Figs. 12.5h, 12.5i and 12.5j] are in very good agreement with the experimental data shown in Figs. 12.5d, 12.5e and 12.5f.

12.3 Counter-propagating traps and optical fibre traps

Ashkin (1970) reported the stable trapping of 2.68 μm-diameter latex spheres between a pair of weakly diverging, counter-propagating Gaussian laser beams, as illustrated in Fig. 12.6a. As we have seen in Section 2.4, the main advantage of counter-propagating optical traps is that they permit the to use of low-numerical-aperture objectives, which allows a large field of view and a long working distance. The main disadvantage is that they require the use of two optical beams focused by two independent objectives. This problem was partly overcome by the *optical mirror trap*, shown in Fig. 12.6b (Pitzek et al., 2009; Thalhammer et al., 2011), which used co-propagating beams with longitudinally displaced foci generated using a spatial light modulator and reflected from a mirror to realise a stable trapping geometry. Perhaps the most straightforward implementation of this trapping scheme, however, is to use the weakly diverging beams emitted by two optical fibres, as first demonstrated by Constable et al. (1993).

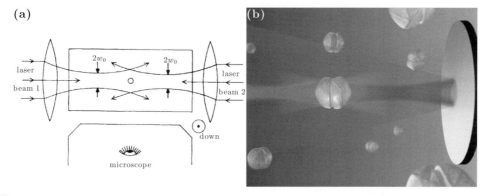

Figure 12.6 Counter-propagating optical traps. (a) Ashkin's 1970 realisation of an optical trap. Reprinted figure from Ashkin, *Phys. Rev. Lett.* **24**, 156–9. Copyright (1970) by the American Physical Society. (b) Optical mirror trap. Reproduced with permission from Thalhammer et al., *J. Opt.* **13**, 044024. Copyright (2011) IOP Publishing. Reproduced by permission of IOP Publishing. All rights reserved.

Figure 12.7a illustrates a typical dual fibre trap geometry: two beams diverging from waists located at the outputs of two optical fibres trap a particle at the mid-point between the fibres. In this configuration, confinement in the axial direction is produced by the scattering forces ($F_{\text{scat},1}$ and $F_{\text{scat},2}$) from the counter-propagating beams and in the transverse direction by gradient forces (F_{grad}). The optical fibre trap has the potential benefits of a large particle capture volume compared to the conventional optical tweezers, the possibility of trapping relatively large particles and the increased space around the trapping region permitting greater optical or mechanical access. A fibre trap can be constructed using relatively inexpensive methods (Piñón et al., 2013), although alignment of the beams is critical to the stability of the resulting trap, as translationally or rotationally misaligned fibres can give rise to off-axis trapping or circulatory motion within the trap (Constable et al., 1993; Sidick et al., 1997). Optical fibres also have the benefit of easy integration into microfluidic chips to realise optofluidic devices (Monat et al., 2007; Psaltis et al., 2006), as we will see in Chapter 16.

12.3.1 Optical stretcher

A powerful development has been the application of the optical fibre trap technique to soft dielectric particles, i.e., particles that can be deformed by optical forces. Guck et al. (2000) demonstrated that, although the force on the particle's centre of mass – which is the

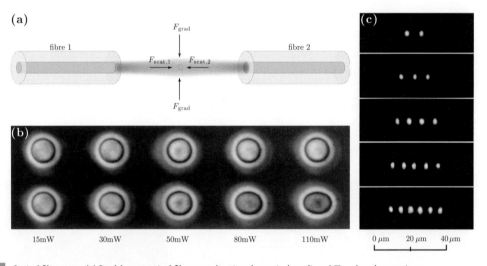

Figure 12.7 Optical fibre traps. (a) Dual-beam optical fibre trap showing the optical gradient (F_{grad}) and scattering ($F_{\text{scat},1}$, $F_{\text{scat},2}$) forces that act on the trapped particle. (b) Optical stretching of an osmotically swollen red blood cell (RBC) in an optical fibre trap. As the laser power is increased (text below the images), the RBC becomes elongated along the trap axis. The top row shows the RBC trapped at 5 mW in each beam; the bottom row was obtained by increasing the power to the value indicated below the image and then reducing it again to 5 mW to demonstrate the reversibility of the process. Reprinted from Guck et al., *Biophys. J.* **81**, 767–84, copyright (2001), with permission from Elsevier. (c) Longitudinal optical binding of an increasing number of particles in an optical fibre trap. Reprinted figure from Gordon et al., *Phys. Rev. B* **77**, 245125. Copyright (2008) by the American Physical Society.

result of the integrated optical stress on the particle surface – is responsible for trapping, the distribution of the stress can have significant consequences for the resulting shape of a relatively soft object such as a biological cell. For an initially spherical object held between counter-propagating Gaussian laser beams, the optical stress distribution on the surface can be approximated as $\sigma(\alpha) = \sigma_0 \cos^2(\alpha)$, where α is the angular coordinate on the cell surface measured from the trap axis and σ_0 the peak value of the optical stress. The method was applied to human red blood cells, osmotically swollen so that they adopted a spherical shape. In this situation, the outward-directed stress is maximal on the axis of the trap, resulting in increased elongation of the red blood cell along the trap axis as the laser power is increased, as shown in Fig. 12.7b, where the overall shape is a result of the balance between optical and elastic membrane forces (Guck et al., 2001). From the observed deformation, Guck et al. (2001) were able to calculate the mechanical properties of the red blood cell membrane, i.e., the product of the Young's modulus E and membrane thickness h, obtaining $Eh = (3.9 \pm 1.4) \times 10^{-5}$ N/m. Several refinements of the optical stretching model have been put forward, e.g., the inclusion of internal reflections in ray-optics models (Ekpenyong et al., 2009), the calculation of optical stresses from generalised Lorentz–Mie theory (Boyde et al., 2012) and the inclusion of the properties of cytoskeletal components in the mechanical response of the cell (Ananthakrishnan et al., 2006). The last development used either a thick-shell model of the actin cortex or a three-layer model including actin cortex, interior polymer assembly and cell nucleus. The thick shell model was found to describe deformations only for a localised stress distribution, whereas for a broad stress distribution the three-layer model accounting for interior structural elements and their coupling to the transiently cross-linked actin cortex better fitted the resulting shell shape. In particular, the three-layer model permitted a quantifiable difference in the mechanical properties of healthy and malignant fibrobasts to be identified, namely a decrease in the cortical shear modulus.

12.3.2 Longitudinal optical binding

Another interesting phenomenon that occurs in the optical fibre trap geometry is the spontaneous self-organisation of a number of colloidal particles into arrays aligned along the trap axis (Singer et al., 2003), shown in Fig. 12.7c. The explanation of this self-organising phenomenon is similar to that discussed in Section 3.10 for optical binding. Considerable insight can be gained by treating the particles as point scatterers to obtain a self-consistent solution for the total field distribution that determines the inter-particle spacing in the array in the experiment of Singer et al. (2003). The spacing between particles was observed to increase with particle size up to a diameter of 2 μm, where bistable behaviour between arrays with large inter-particle spacing and with the particles just touching was observed. Gordon et al. (2008) used a generalised multipole technique to calculate the scattered field distribution from an array of particles in a fibre trap, followed by a Maxwell stress tensor analysis to determine the optical force on the particles. Hydrodynamic forces arising from particle motion were accounted for through the Oseen tensor [Section 17.1], resulting in an optohydrodynamic theory capable of predicting array equilibrium spacings and dynamics. Interestingly, in this case it was observed that the particle equilibrium spacing was a

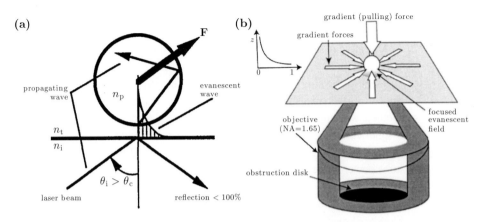

Figure 12.8 Evanescent wave trapping. (a) Optical forces acting on a microparticle from unidirectional evanescent wave illumination. Reprinted from Kawata and Sugiura, *Opt. Lett.* **17**, 772–4. Copyright (1992) The Optical Society. (b) Stable evanescent wave optical tweezers obtained by focusing an evanescent wave. Reprinted from Gu et al., *Appl. Phys. Lett.* **84**, 4236–8. Copyright (2004) American Institute of Physics.

function of the number of particles in the array (similar results were reported by Tatarkova et al. (2002)) and that the inter-particle spacing was not uniform, becoming larger towards the ends of the array. Because axial trapping in the fibre trap relies on scattering forces, the inhomogeneous spacing was explained as arising from the balance of scattering forces from the beam emitted from the fibre and the forward Mie scattering of other particles in the array.

12.4 Evanescent wave traps

A beam incident on an interface between two media with refractive indices n_i and n_t with $n_i > n_t$ from the side of the higher refractive index (n_i), undergoes total internal reflection if incident at an angle greater than the total internal reflection angle $\theta_c = \arcsin(\frac{n_t}{n_i})$ [Eq. (2.12)]. In this situation, an evanescent field penetrates a short distance into the low-refractive-index side. The presence of a particle (refractive index n_p) close to the interface alters the boundary conditions to allow a wave to cross the interface and propagate inside the particle. This wave then exerts a scattering force on the particle centre of mass as illustrated in Fig. 12.8a and first demonstrated experimentally by Kawata and Sugiura (1992).

12.4.1 Evanescent tweezers

The configuration proposed by Kawata and Sugiura (1992) and shown in Fig. 12.8a permits one to apply an optical pushing force on a particle, but not to stably trap a particle at a given position. An actual optical tweezers based on evanescent fields was demonstrated by Gu et al. (2004) using an annular beam created by a circular obstruction strongly focused by a

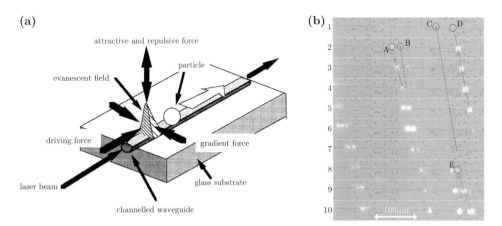

Figure 12.9 Optical waveguide forces. (a) Forces on a microparticle from illumination by the evanescent field of a channel waveguide. Reprinted from Kawata and Tani, *Opt. Lett.* **21**, 1768–70. Copyright (1996) The Optical Society. (b) Propulsion of particles along a sub-wavelength diameter tapered optical fibre. Reprinted from Brambilla et al., *Opt. Lett.* **32**, 3041–3. Copyright (2007) The Optical Society.

high-NA TIRF objective lens (NA = 1.65), as shown in Fig. 12.8b. We have discussed this kind of evanescent focusing in detail in Subsection 4.7.2. In a nutshell, when the full aperture of the objective is illuminated, a conventional optical tweezers capable of trapping in three dimensions is formed. By blocking the central part of the beam, it is possible to allow to pass only those rays coming from near the edge of the aperture and incident at the interface to the suspending medium at an angle greater than the critical angle. The resulting strongly focused evanescent field has strong gradients of intensity in the transverse direction that confine the particle laterally near the optical axis, albeit with reduced trapping efficiency compared to conventional optical tweezers. In the axial direction, an intensity gradient exists in the rapidly decaying evanescent field which acts to hold the particle against the interface. One advantage of this method is the greatly reduced size of the trapping volume in the axial direction, because the evanescent field decays in intensity very quickly, e.g., by 50% over a distance of approximately 60 nm in the work by Gu et al. (2004).

12.4.2 Waveguides

An alternative means of generating an evanescent field capable of pushing particles is to use an optical waveguide, i.e., a region of increased refractive index, compared with the surrounding medium (called the cladding), capable of guiding light. An optical waveguide with sub-wavelength dimensions supports the fundamental mode, although a significant fraction of the optical power is guided in an evanescent field outside the waveguide core. The gradient of the evanescent field traps a particle against the waveguide, while radiation pressure propels it along the waveguide. Optical manipulation of particles using optical waveguides has been realised using, e.g., a channel waveguide (Kawata and Tani, 1996) [Fig. 12.9a] and an optical fibre tapered to a waist with micrometric or sub-micrometric diameter (Brambilla et al., 2007) [Fig. 12.9b]. Sub-wavelength optical fibres can be fabricated from conventional optical fibres by a heat-and-pull technique capable of tapering the fibre to a waist of a few hundred nanometres (Brambilla et al., 2004). Such a tapered fibre is sometimes referred to as a biconical taper to distinguish it from a fibre that is pulled until it fractures, leaving a sharp tip, which can also be used for optical trapping applications (Xin et al., 2012). In the tapering region, the diameter of the original core glass of the waveguiding structure is reduced to zero, leaving a waveguide created by the cladding glass material and the surrounding medium. The evanescent field of the optical nanofibre in the region of the taper waist has been used to demonstrate optical trapping and manipulation of species as diverse as laser-cooled atoms (Sagué et al., 2007), colloidal particles (Skelton et al., 2012) and bacteria (Xin et al., 2013). Tapered optical fibres have the advantage of easy integration into optical systems using standard optical components, although their fragile nature may be a drawback in some applications. In this regard, the extra complication of fabrication of channel waveguides may be offset by the more robust waveguiding structure. Channel waveguides have been used to manipulate and transport a range of micro- and nanoscale objects including colloidal gold nanoparticles (Ng et al., 2000), cells (Gaugiran et al., 2005) and semiconductor nanowires (Néel et al., 2009).

12.4.3 Optical binding

The optical forces pushing the particles in the evanescent trapping configuration shown in Fig. 12.8a can be overcome using counter-propagating beams. This was first demonstrated by Garcés-Chávez et al. (2005), who used counter-propagating surface waves to balance the radiation pressure on the particles to be trapped, with the addition of transverse structure created by a Ronchi ruling which projects a pattern of linear fringes onto the interface. The intensity variation in the fringes provided an additional gradient force in the transverse direction to localise particles within them. Both colloidal particles and red blood cells were observed to be propelled along the fringes when illuminated by a single beam. In counter-propagating beams, they were trapped by the balanced radiation pressure and observed to self-organise to maintain a fixed separation as a result of an optical binding interaction similar to the one described in Section 3.10.

The counter-propagating evanescent field geometry was used by Mellor et al. (2006) to make a detailed study of the structures that formed as a result of optical binding. A variety of structures with hexagonal or rectangular symmetries, one example of which is shown in Fig. 12.10a, were observed depending on the particle size (or effective size as measured taking into account the Debye length, which changes with the concentration of electrolyte) and polarisation state of the laser beams, which controls the presence of interference fringes that give additional structure to the optical field. The size-dependence of optical forces under illumination from counter-propagating beams was exploited by Čižmár et al. (2006) to sensitively separate particles according to their size. This experiment used two beams with an adjustable phase delay and incident on the surface with a range of angles around the critical angle, such that a standing wave formed at the surface from both evanescent and propagating waves. The optical potential depth experienced by particles in a standing wave is a function of size: some particles are strongly confined, whereas others are almost unaffected. Moving the interference pattern by adjusting the phase between the

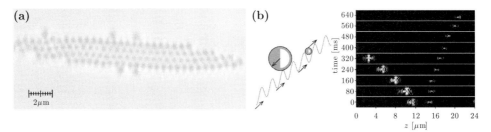

Figure 12.10 Evanescent optical binding. (a) An example of a particle structure (hexagonal lattice) formed by 520 nm diameter polystyrene spheres in counter-propagating evanescent fields. Reprinted from Mellor et al., *Opt. Express* **14**, 10079–88. Copyright (2006) The Optical Society. (b) Separation of particles according to size using a combination of evanescent and propagating fields: 350 nm-diameter particles are moved to the right by a moving interference pattern between two evanescent fields, whereas 750 nm-diameter particles, which are insensitive to the spatial modulation of intensity in the standing wave, are moved to the left by unbalanced radiation pressure arising from the unequal intensities of the two beams. Reprinted figure from Čižmár et al., *Phys. Rev. B* **74**, 035105. Copyright (2006) by the American Physical Society.

beams allowed the trapped particles to be translated. Furthermore, by introducing a small imbalance in intensity between the two beams, it was possible to propel the untrapped particles in the opposite direction. This is illustrated in Fig. 12.10b where smaller (350 nm-diameter) particles are transported to the right by the moving interference pattern, whereas larger (750 nm-diameter) particles, which are not trapped in the standing wave, are pushed to the left by the unequal radiation pressure from the two beams.

12.4.4 Plasmonic traps

In Section 3.9, we introduced the concept of *localised* surface plasmons in the context of metallic nanoparticles and the resonant behaviour in optical properties that arise as a result of exciting oscillations in the charge densities of these particles. The enhanced electromagnetic field associated with exciting a *propagating* surface plasmon at a metallic–dielectric interface (a surface plasmon polariton) can also be be exploited for optical trapping applications in order to enhance optical forces.

The first experiment that demonstrated the existence of plasmon radiation forces was performed by Volpe et al. (2006) with the set-up shown in Fig. 12.11a. In this experiment, a microscope coverslip with a 40 nm-thick gold coating provides the interface at which the surface plasmon is excited and constitutes the upper surface of the sample chamber. Excitation of the surface plasmon was performed by a helium–neon laser beam ($\lambda_0 = 633$ nm) coupled by total internal reflection via a prism so that it is incident at the plasmon resonance angle and with transverse magnetic (TM) polarisation (the *Kretschmann geometry*). A separate optical tweezers (generated by the $\lambda_0 = 532$ nm laser beam) was used to quantify the force exerted on dielectric particles near the gold surface by photonic force microscopy, as explained in Chapter 10. The surface plasmon enhanced evanescent wave was thus observed to exert a force with a component that draws a dielectric particle towards the surface and

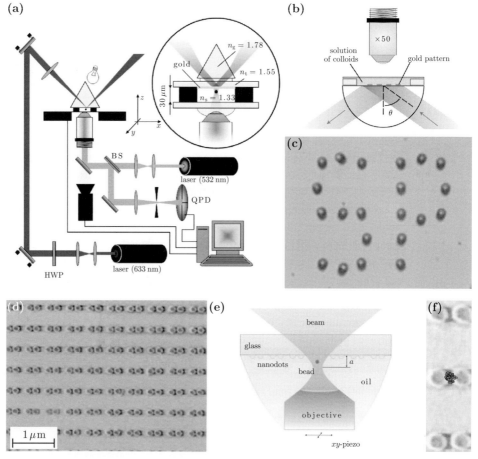

Figure 12.11 Plasmonic traps. (a) Set-up with which the first measurement of plasmon radiation forces was performed. Reprinted figure from Volpe et al., *Phys. Rev. Lett.* **96**, 238101. Copyright (2006) by the American Physical Society. (b) Kretschmann geometry to excite surface plasmons. (c) Patterned plasmonic optical traps filled with 4.88 µm-diameter polystyrene microparticles in the shape of the letters 'SP'. Figures (b) and (c) are reprinted by permission from Macmillan Publishers Ltd: Righini et al., *Nature Phys.* **3**, 477–80, copyright 2007. (d) Plasmonic nanoantenna structure consisting of pairs of nanodots of height 90 nm, diameter 134 nm and separated by 200 nm. (e) Experimental set-up to observe trapping in the nanoantenna structure. The $\lambda_0 = 1064$ nm laser beam excites a localised surface plasmon in the nanoantennas, which concentrates the field in the small gap between the nanodots. (f) Trajectory of a trapped 200 nm-diameter particle superimposed on an image of nanoantennas. The position fluctuations have a root-mean-square deviation of 18 nm. Figures (d)–(f) are reprinted by permission from Macmillan Publishers Ltd: Grigorenko et al., *Nature Photon.* **2**, 365–70, copyright 2008.

a component that propels the particle parallel to the surface. Resonant excitation of the surface plasmon resulted in a force around 40 times greater than non-resonant excitation, which could be obtained by altering either the incidence angle or the polarisation state of the incident beam to transverse electric (TE).

The surface plasmon polariton excited at a flat gold film generates a uniform force that pushes a particle along but does not localise it. Localisation, and therefore trapping, can be achieved by patterning the substrate by fabricating some metallic structures on a dielectric interface. This is illustrated in the experiments of Righini et al. (2007), shown in Fig. 12.11b, where a Kretschmann geometry illumination of a surface (incident angle $\theta = 68°$) patterned with gold pads of diameter 4.8 μm and thickness 40 nm results in an array of traps that can be filled with dielectric particles of similar size, here 4.88 μm-diameter polystyrene spheres, as shown in Fig. 12.11c. The dimensions of the metallic pads for surface plasmon trapping are significant: as the size of the pads is reduced to approach the wavelength of the surface plasmon, the pad will only support a localised surface plasmon, rather than a propagating surface plasmon polariton.

To trap sub-micrometric and nanometric particles, the field enhancement of coupled plasmonic antennas can be exploited. A coupled plasmonic nanoantenna consists of two metallic nanostructures with a small (nanoscale) dielectric gap. When illuminated, the field confined in the dielectric gap is enhanced, producing an intense *hot spot* where nanoparticles can be trapped. Such a plasmonic nanoantenna trap was demonstrated by Grigorenko et al. (2008) using the apparatus illustrated in Figs. 12.11d and 12.11e. The gold nanodots used to form the nanoantenna were fabricated on a regular lattice in a double-pillar structure with a 200 nm gap between them. The dimensions of the nanodots were chosen so that the localised plasmon resonance of the double-pillar structure could be excited by the infrared ($\lambda_0 = 1064$ nm) laser beam used as a conventional optical tweezer in the same apparatus. In this case, the localised surface plasmon could be excited by a propagating beam, and evanescent wave coupling by the Kretschmann geometry as used for the surface plasmon polariton trap was not needed. Fig. 12.11f shows the trajectory of a 200 nm particle in the gap of the nanoantenna (superimposed on an electron micrograph of the structure); the root-mean-square deviation of the position fluctuations was just 18 nm, a factor of 10 smaller than in the conventional optical tweezers in the same apparatus, demonstrating the strong confinement possible for nanostructures using a plasmonic optical trap.

12.5 Feedback traps

Feedback optical traps are the class of traps where the motion or position of the particle itself influences the strength of the trap. A paradigmatic example of feedback trapping is the *self-induced back action* (SIBA) trap, demonstrated by Juan et al. (2009); see Fig. 12.12. In the SIBA trap, the wavelength of the trapping laser beam is chosen to be slightly longer than the transmission cut-off wavelength of a nanoaperture in a metallic film. Because the transmission of the aperture is sensitive to local refractive index changes, the presence of a dielectric particle as small as 50 nm in the aperture modifies its transmission properties. The electric field distribution within the aperture has maxima at the aperture edge along the direction of polarisation of the trapping laser, and it is at this location that the particle is trapped. Indeed, the presence of the particle near the edge concentrates the electric field further, thereby resulting in an enhancement of the optical trapping of the particle induced

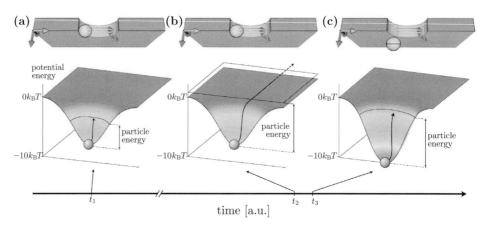

Figure 12.12 Self-induced back action traps. Schematic of SIBA trapping set-up employing a 310 nm aperture in a 100 nm Au film to trap 50 nm radius polystyrene spheres in water: as the particle is about to escape the nanohole, the optical potential becomes deeper and pulls it back. Reprinted by permission from Macmillan Publishers Ltd: Juan et al., *Nature Photon.* **5**, 349–56, copyright 2011.

by the presence of the particle itself. If the particle moves out of the aperture, the resulting change in transmission means there is a decrease in optical momentum flux through the aperture, and hence a restoring force acting on the particle directing it back towards the aperture (Juan et al., 2011). One of the great advantages of this technique is that the active role the particle itself plays in the trapping mechanism means that extremely low power can be used to trap particles tens of nanometres in size, which would otherwise experience very weak confinement in a conventional optical tweezers. As shown in Fig. 12.1, this technique has been extended by Berthelot et al. (2014) to apertures fabricated at the end of a tapered metallised optical fibre, which is used to carry the trapping light to the aperture and also allows the trap to be repositioned by moving the fibre tip.

12.6 Haptic optical tweezers

We cannot directly feel and experience the forces typically associated with optical tweezers, as they are far too small. It is, however, possible to use haptic feedback devices to create the illusion of substance and force within the virtual world.[1] Haptic optical tweezers permit the operator to sense the microscopic world with its low inertia and high viscosity. By doing so, they will permit more complex micromanipulation tasks to be performed with higher dexterity than is currently possible.

Haptic optical tweezers are a natural extension of various techniques developed for the interactive control of optical tweezers via a variety of advanced user interfaces including

---

[1] The word 'haptic' derives from the ancient Greek word ἅπτω, meaning 'I touch'.

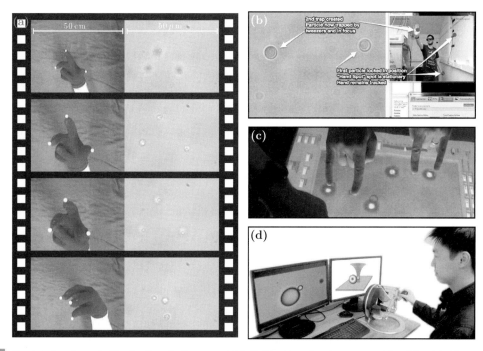

Figure 12.13 User interfaces for controlling haptic optical tweezers. (a) Optical tracking of markers on a user's hand. Reprinted from Whyte et al., *Opt. Express* **14**, 12 497–502. Copyright (2006) The Optical Society. (b) Optical tracking of hand positions using a Microsoft Kinect game controller. Reproduced with permission from McDonald et al., *J. Opt.* **15**, 035708. Copyright (2013) IOP Publishing. Reproduced by permission of IOP Publishing. All rights reserved. (c) Multi-user touch-screen table-top controller. Reprinted from Grieve et al., *Opt. Express* **17**, 3595–602. Copyright (2009) The Optical Society. (d) Haptic optical tweezers utilising the Novint Falcon interface. Reprinted from Pacoret and Régnier, *Rev. Sci. Instrumen.* **84**, 081301. Copyright (2013) American Institute of Physics.

joystick control (Gibson et al., 2007), tracking of points on a user's hand (Whyte et al., 2006) [Fig. 12.13a], tracking of the user's whole hand with a Microsoft Kinect game controller (McDonald et al., 2013) [Fig. 12.13b], touch-screen table-top control (Grieve et al., 2009) [Fig. 12.13c] and control using an iPad interface (Bowman et al., 2011). The actual movement of the optical traps can be achieved though a variety of different techniques, which can either displace the sample (e.g., mechanised or piezoelectric stages) or alter the focused laser beams (e.g., spatial light modulators, galvanometer mirrors and acousto-optic modulators). Haptic optical tweezers have been realised with a number of different commercial controllers that feed back information about the microscopic environment where the optical tweezers operate through the application of pressure or motion to the user via a specialised controller.[2]

---

[2] Manufacturers of haptic devices that have been used in optical tweezers include SensAble Technologies (now part of Geomagic, USA), Force Dimension (Switzerland) and Novint (USA).

The *haptic coupling loop*, which includes both the control and feedback processes, consists of the following steps:

1. positional control commands from the user interface are scaled down and used to direct the movement of an optically trapped microscopic particle;
2. the displacement of the particle is measured and used to calculate the scaled-up force that the user should experience through the haptic controller.

The three main factors that must be considered in the design of an effective haptic control system are (Pacoret and Régnier, 2013)

- the number of *degrees of freedom* of both measurement and movement, i.e., the number of microscopic actuators that can be sensed and controlled in parallel;
- the *bandwidth* of both detection and actuation, which limits the transparency of control due to latency in transmission;
- the *feedback intensity*, i.e., the amplitude of the feedback force.

Such bidirectional interaction between the user and an otherwise unreachable microscopic environment was first demonstrated by Arai et al. (2000), who detected particle position using light scattered by the particle on a quadrant photodiode, as explained in Section 9.2, and steered the particle using a combination of galvanometer mirrors and piezo-stage movement; the force acting on the particle was fed back to the user via a SensAble Technologies PHANTOM haptic device. Later, Pacoret et al. (2009) demonstrated haptic control of optical tweezers; in Fig. 12.13d, their haptic controller can be seen at work.

The motion of an optically trapped particle can be achieved using a piezoelectric stage, such as the one we used in the realisation of our optical tweezers set-up in Chapter 8. For example, this was the strategy employed by Basdogan et al. (2007). Because the piezoelectric stage moves the entire sample, rather than the optical trap, its response time is typically long because of the mass of the components and it is not suited for high-bandwidth control; however, an advantage for some applications is that the optically trapped particles remain aligned with the measurement system. Higher-speed actuation can be achieved by beam steering using, e.g., galvanometer mirrors, acousto-optic deflectors or spatial light modulators, as we have seen in Chapter 11. The first two methods typically have a higher bandwidth than a user's force perception (approximately 1 kHz), but only permit actuation in two dimensions, whereas spatial light modulators permit three-dimensional control but bandwidth is limited to lower frequencies (Onda and Arai, 2012). These methods also have the advantage of permitting simultaneous actuation of multiple optical traps.

Haptic control also requires the position of the particle to be measured with high bandwidth. As seen in Chapter 9, numerous particle detection and position measurement schemes for optical tweezers have been developed. When the system best suited for haptic control is considered, factors such as the dimensionality of the position measurement, speed of acquisition and signal processing should be considered. Back-focal-plane interferometry [Section 9.2] permits three-dimensional position measurement at rates exceeding tens or even hundreds of kilohertz; however, the effective workspace over which the position can be measured may be limited to less than a few micrometres due to increasing misalignment with the detection system. Digital video microscopy using fast CMOS cameras [Section 9.1] can achieve very high frame rates, up to several thousands of frames per second for a reduced region of interest, which is adequate for the highly damped motion of microscopic particles in an aqueous environment. For example, Cheng et al. (2013) have achieved real-time three-dimensional particle tracking at rates up to 10 000 frames per second using a CMOS camera and field programmable gate array (FPGA).

References

Abbondanzieri, E. A., Greenleaf, W. J., Shaevitz, J. W., Landick, R., and Block, S. M. 2005. Direct observation of base-pair stepping by RNA polymerase. *Nature*, **438**, 460–5.

Ananthakrishnan, R., Guck, J., Wottawah, F., et al. 2006. Quantifying the contribution of actin networks to the elastic strength of fibroblasts. *J. Theor. Biol.*, **242**, 502–16.

Arai, F., Ogawa, M., and Fukuda, T. 2000. Indirect manipulation and bilateral control of the microbe by the laser manipulated microtools. Pages 665–670 of *Proc. 2000 IEEE/RSJ International Conference on Intelligent Robots and Systems, 2000*, vol. 1.

Ashkin, A. 1970. Acceleration and trapping of particles by radiation pressure. *Phys. Rev. Lett.*, **24**, 156–9.

Basdogan, C., Kiraz, A., Bukusoglu, I., Varol, A., and Doğanay, S. 2007. Haptic guidance for improved task performance in steering microparticles with optical tweezers. *Opt. Express*, **15**, 11 616–21.

Berthelot, J., Aćimović, S. S., Juan, M. L., et al. 2014. Three-dimensional manipulation with scanning near-field optical nanotweezers. *Nature Nanotech.*, **9**, 295–9.

Bowman, R. W., Gibson, G., Carberry, D., et al. 2011. iTweezers: Optical micromanipulation controlled by an Apple iPad. *J. Opt.*, **13**, 044002.

Boyde, L., Ekpenyong, A., Whyte, G., and Guck, J. 2012. Comparison of stresses on homogeneous spheroids in the optical stretcher computed with geometrical optics and generalized Lorenz-Mie theory. *Appl. Opt.*, **51**, 7934–44.

Brambilla, G., Finazzi, V., and Richardson, D. 2004. Ultra-low-loss optical fiber nanotapers. *Opt. Express*, **12**, 2258–63.

Brambilla, G., Murugan, G. S., Wilkinson, J. S., and Richardson, D. J. 2007. Optical manipulation of microspheres along a subwavelength optical wire. *Opt. Lett.*, **32**, 3041–3.

Capitanio, M., Cicchi, R., and Pavone, F. S. 2005. Position control and optical manipulation for nanotechnology applications. *Eur. Phys. J. B*, **46**, 1–8.

Capitanio, M., Maggi, D., Vanzi, F., and Pavone, F. S. 2007. FIONA in the trap: The advantages of combining optical tweezers and fluorescence. *J. Opt. A Pure Appl. Opt.*, **9**, S157–S163.

Cheng, P., Jhiang, S. M., and Menq, C.-H. 2013. Real-time visual sensing system achieving high-speed 3D particle tracking with nanometer resolution. *Appl. Opt.*, **52**, 7530–9.

Čižmár, T., Šiler, M., Šerý, M., et al. 2006. Optical sorting and detection of submicrometer objects in a motional standing wave. *Phys. Rev. B*, **74**, 035105.

Constable, A., Kim, J., Mervis, J., Zarinetchi, F., and Prentiss, M. 1993. Demonstration of a fiber-optical light-force trap. *Opt. Lett.*, **18**, 1867–9.

Creely, C., Volpe, G., Singh, G. P., Soler, M., and Petrov, D. 2005. Raman imaging of floating cells. *Opt. Express*, **13**, 6105–10.

Ekpenyong, A. E., Posey, C. L., Chaput, J. L., et al. 2009. Determination of cell elasticity through hybrid ray optics and continuum mechanics modeling of cell deformation in the optical stretcher. *Appl. Opt.*, **48**, 6344–54.

Garcés-Chávez, V., Dholakia, K., and Spalding, G. C. 2005. Extended-area optically induced organization of microparticles on a surface. *Appl. Phys. Lett.*, **86**, 031106.

Gaugiran, S., Gétin, S., Fedeli, J., et al. 2005. Optical manipulation of microparticles and cells on silicon nitride waveguides. *Opt. Express*, **13**, 6956–63.

Gibson, G., Barron, L., Beck, F., Whyte, G., and Padgett, M. 2007. Optically controlled grippers for manipulating micron-sized particles. *New J. Phys.*, **9**, 14.

Gordon, R., Kawano, M., Blakely, J. T., and Sinton, D. 2008. Optohydrodynamic theory of particles in a dual-beam optical trap. *Phys. Rev. B*, **77**, 245125.

Grieve, J. A., Ulcinas, A., Subramanian, S., et al. 2009. Hands-on with optical tweezers: A multitouch interface for holographic optical trapping. *Opt. Express*, **17**, 3595–602.

Grigorenko, A. N., Roberts, N. W., Dickinson, M. R., and Zhang, Y. 2008. Nanometric optical tweezers based on nanostructured substrates. *Nature Photon.*, **2**, 365–70.

Gu, M., Haumonte, J.-B., Micheau, Y., Chon, J. W. M., and Gan, X. 2004. Laser trapping and manipulation under focused evanescent wave illumination. *Appl. Phys. Lett.*, **84**, 4236–8.

Guck, J., Ananthakrishnan, R., Moon, T. J., Cunningham, C. C., and Käs, J. 2000. Optical deformability of soft biological dielectrics. *Phys. Rev. Lett.*, **84**, 5451–4.

Guck, J., Ananthakrishnan, R., Mahmood, H., et al. 2001. The optical stretcher: A novel laser tool to micromanipulate cells. *Biophys. J.*, **81**, 767–84.

Hell, S. W. 2008. Towards fluorescence nanoscopy. *Nature Biotechnol.*, **21**, 1347–55.

Juan, M. L., Gordon, R., Pang, Y., Eftekhari, F., and Quidant, R. 2009. Self-induced back-action optical trapping of dielectric nanoparticles. *Nature Phys.*, **5**, 915–19.

Juan, M. L., Righini, M., and Quidant, R. 2011. Plasmon nano-optical tweezers. *Nature Photon.*, **5**, 349–56.

Kawata, S., and Sugiura, T. 1992. Movement of micrometer-sized particles in the evanescent field of a laser beam. *Opt. Lett.*, **17**, 772–4.

Kawata, S., and Tani, T. 1996. Optically driven Mie particles in an evanescent field along a channeled waveguide. *Opt. Lett.*, **21**, 1768–70.

Lankers, M., Popp, J., and Kiefer, W. 1994. Raman and fluorescence spectra of single optically trapped microdroplets in emulsions. *Appl. Spectrosc.*, **48**, 1166–8.

McDonald, C., McPherson, M., McDougall, C., and McGloin, D. 2013. HoloHands: Games console interface for controlling holographic optical manipulation. *J. Opt.*, **15**, 035708.

Mellor, C. D., Fennerty, T. A., and Bain, C. D. 2006. Polarization effects in optically bound particle arrays. *Opt. Express*, **14**, 10 079–88.

Mikhael, J., Roth, J., Helden, L., and Bechinger, C. 2008. Archimedean-like tiling on decagonal quasicrystalline surfaces. *Nature*, **454**, 501–4.

Monat, C., Domachuk, P., and Eggleton, B. J. 2007. Integrated optofluidics: A new river of light. *Nature Photon.*, **1**, 106–14.

Néel, D., Gétin, S., Ferret, P., et al. 2009. Optical transport of semiconductor nanowires on silicon nitride waveguides. *Appl. Phys. Lett.*, **94**, 253115.

Ng, L. N., Zervas, M. N., Wilkinson, J. S., and Luff, B. J. 2000. Manipulation of colloidal gold nanoparticles in the evanescent field of a channel waveguide. *Appl. Phys. Lett.*, **76**, 1993–5.

Onda, K., and Arai, F. 2012. Multi-beam bilateral teleoperation of holographic optical tweezers. *Opt. Express*, **20**, 3633–41.

Pacoret, C., and Régnier, S. 2013. Invited Article: A review of haptic optical tweezers for an interactive microworld exploration. *Rev. Sci. Instrumen.*, **84**, 081301.

Pacoret, C., Bowman, R., Gibson, G., et al. 2009. Touching the microworld with force-feedback optical tweezers. *Opt. Express*, **17**, 10259–64.

Petrov, D. V. 2007. Raman spectroscopy of optically trapped particles. *J. Opt. A Pure Appl. Opt.*, **9**, S139–S156.

Piñón, T. M., Castelli, A. R., Hirst, L. S., and Sharping, J. E. 2013. Fiber-optic trap-on-a-chip platform for probing low refractive index contrast biomaterials. *Appl. Opt.*, **52**, 2340–5.

Pitzek, M., Steiger, R., Thalhammer, G., Bernet, S., and Ritsch-Marte, M. 2009. Optical mirror trap with a large field of view. *Opt. Express*, **17**, 19 414–23.

Psaltis, D., Quake, S. R., and Yang, C. 2006. Developing optofluidic technology through the fusion of microfluidics and optics. *Nature*, **442**, 381–6.

Righini, M., Zelenina, A. S., Girard, C., and Quidant, R. 2007. Parallel and selective trapping in a patterned plasmonic landscape. *Nature Phys.*, **3**, 477–80.

Sagué, G., Vetsch, E., Alt, W., Meschede, D., and Rauschenbeutel, A. 2007. Cold-atom physics using ultrathin optical fibers: Light-induced dipole forces and surface interactions. *Phys. Rev. Lett.*, **99**, 163602.

Sidick, E., Collins, S. D., and Knoesen, A. 1997. Trapping forces in a multiple-beam fiber-optic trap. *Appl. Opt.*, **36**, 6423–33.

Singer, W., Frick, M., Bernet, S., and Ritsch-Marte, M. 2003. Self-organized array of regularly spaced microbeads in a fiber-optical trap. *J. Opt. Soc. Am. B*, **20**, 1568–74.

Skelton, S. E., Sergides, M., Patel, R., et al. 2012. Evanescent wave optical trapping and transport of micro- and nanoparticles on tapered optical fibers. *J. Quant. Spectrosc. Rad. Transfer*, **113**, 2512–20.

Tatarkova, S. A., Carruthers, A. E., and Dholakia, K. 2002. One-dimensional optically bound arrays of microscopic particles. *Phys. Rev. Lett.*, **89**, 283901.

Thalhammer, G., Steiger, R., Bernet, S., and Ritsch-Marte, M. 2011. Optical macro-tweezers: Trapping of highly motile micro-organisms. *J. Opt.*, **13**, 044024.

van Dijk, M. A., Kapitein, L. C., van Mameren, J., Schmidt, C. F., and Peterman, E. J. G. 2004. Combining optical trapping and single-molecule fluorescence spectroscopy: Enhanced photobleaching of fluorophores. *J. Phys. Chem. B*, **108**, 6479–84.

Volpe, G., Quidant, R., Badenes, G., and Petrov, D. 2006. Surface plasmon radiation forces. *Phys. Rev. Lett.*, **96**, 238101.

Volpe, G., Volpe, G., and Gigan, S. 2014a. Brownian motion in a speckle light field: Tunable anomalous diffusion and selective optical manipulation. *Scientific Reports*, **4**, 3936.

Volpe, G., Kurz, L., Callegari, A., Volpe, G., and Gigan, S. 2014b. Speckle optical tweezers: Micromanipulation with random light fields. *Opt. Express*, **22**, 18 159–67.

Wang, F., Reece, P. J., Paiman, S., et al. 2011. Nonlinear optical processes in optically trapped InP nanowires. *Nano Lett.*, **11**, 4149–53.

Wang, F., Toe, W. J., Lee, W. M., et al. 2013. Resolving stable axial trapping points of nanowires in an optical tweezers using photoluminescence mapping. *Nano Lett.*, **13**, 1185–91.

Whyte, G., Gibson, G., Leach, J., et al. 2006. An optical trapped microhand for manipulating micron-sized objects. *Opt. Express*, **14**, 12 497–502.

Xie, C., Dinno, M. A., and Li, Y. 2002. Near-infrared Raman spectroscopy of single optically trapped biological cells. *Opt. Lett.*, **27**, 249–51.

Xin, H., Xu, R., and Li, B. 2012. Optical trapping, driving, and arrangement of particles using a tapered fibre probe. *Scientific Reports*, **2**, 818.

Xin, H., Cheng, C., and Li, B. 2013. Trapping and delivery of *Escherichia coli* in a microfluidic channel using an optical nanofiber. *Nanoscale*, **5**, 6720–4.

Yildiz, A., Forkey, J. N., McKinney, S. A., et al. 2003. Myosin V walks hand-over-hand: Single fluorophore imaging with 1.5-nm localization. *Science*, **300**, 2061–5.

PART III

APPLICATIONS

'Would you tell me, please, which way I ought to go from here?'
'That depends a good deal on where you want to get to', said the Cat.
'I don't much care where . . .' said Alice.
'Then it doesn't matter which way you go', said the Cat.
'. . . so long as I get *somewhere*', Alice added as an explanation.
'Oh, you're sure to do that', said the Cat, 'if you only walk long enough.'

Alice's Adventures in Wonderland – Lewis Carroll

13 Single-molecule biophysics

The application of optical tweezers methods to the study of biological systems has been one of the outstanding triumphs of the technique. Optical tweezers are particularly suited to the study of the physical properties of single biological molecules because of the overlap of the characteristic length (from a fraction of a nanometre to a fraction of a micrometre) and force (from several femtonewtons to hundreds of piconewtons) scales between biomolecules and optical traps. In many experiments biomolecules are not trapped directly, but are attached to the surface of a microsphere that acts as a handle for the optical tweezers. Several powerful variations on the standard optical tweezers technique have been devised to enable the study of single biological molecules, including the two-bead assay [Fig. 13.1], active position-clamping and angular optical trapping (the *optical torque wrench*). In this chapter, we illustrate the use of optical tweezers for single-molecule biophysical studies focusing on two particular examples: the mechanics of DNA and the dynamics of motor proteins.

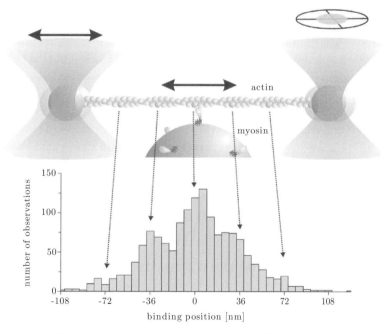

Figure 13.1 Single-molecule two-bead assay using a dual optical tweezers. Scanning one of the beads confined in optical tweezers pulls the actin past the myosin motor protein, which preferentially binds at sites separated by the 36 nm helical repeat distance of actin. Reprinted by permission from Macmillan Publishers Ltd: Veigel et al., *Nature Cell Biol.* **4**, 59–65, copyright 2001.

13.1 DNA mechanics: Stretching

The mechanical properties of DNA can, under conditions of small extension, be described by the so-called *worm-like chain model* (WLC model), which considers a polymer as a line that bends smoothly under the influence of random thermal fluctuations. In this model, the correlation between the orientations of two segments of the polymer decays exponentially with a characteristic *flexural persistence length* d.[1] A molecule whose persistence length is small compared to its *contour length* L, the molecule's length at its maximum possible extension, tends to adopt a compact random-coil structure and, as a result, the average end-to-end distance of the molecule is much shorter than L. For such a molecule, there are thus many different conformations with the same end-to-end length. Therefore, stretching the molecule by increasing the end-to-end distance reduces the number of available configurations and, hence, is entropically unfavourable – a phenomenon known as *entropic elasticity*. This is indeed the case for a (sufficiently long) DNA molecule. In the WLC model the force–extension relationship (Marko and Siggia, 1995) has the form

$$F(x) = \frac{k_B T}{d} \left[\frac{1}{4\left(1 - \frac{x}{L}\right)^2} - \frac{1}{4} + \frac{x}{L} \right], \tag{13.1}$$

where F is the force applied to the molecule and x is its corresponding extension. Early works on the elasticity of DNA (Smith et al., 1992; Bustamante et al., 1994) used magnetic beads to stretch double-stranded DNA (dsDNA) with a tension in the range from 0.01 to 10 pN and measured the resulting force–extension curves, which could be interpreted as being consistent with the WLC model with a persistence length $d \approx 50$ nm, equivalent to approximately 150 base pairs in 150 mM Na$^+$ buffer, where the DNA has a rise of 0.34 nm per base pair (Bustamante et al., 2003).

Evidence for the *intrinsic elasticity* of the molecule, however, only appears when it is overstretched beyond its normal contour length at high applied force. To probe this elastic response, Smith et al. (1996) held a single DNA molecule (48.5 kbp, $L = 16.4$ μm) between two latex beads, one immobilised on a micropipette and the other held in an optical tweezers. In this experiment, the optical tweezers was formed by two counter-propagating beams brought to a common focus by relatively low-numerical-aperture (NA $= 0.85$) objective lenses, which permitted the bead to be held away from the surfaces of the experimental cell, as shown in Fig. 13.2a. The molecule was stretched by moving the micropipette and the extension was determined from the separation of the beads. The force acting on the molecule was measured by the displacement of the bead in the calibrated optical tweezers, as determined by the transmitted laser light on position-sensitive detectors [Section 10.5]. The enthalpic elasticity of the dsDNA molecule was evident from the discrepancy in the

---

[1] The flexural persistence length is simply related to the bending rigidity of the polymer κ as $\kappa = k_B T d$, where $k_B T$ is the thermal energy, k_B is the Boltzmann constant and T is the absolute temperature.

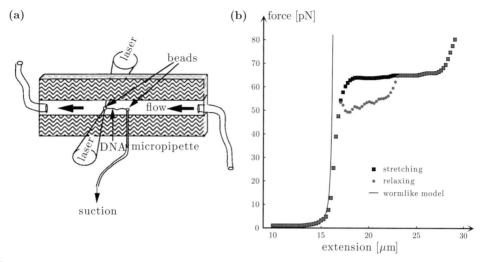

Figure 13.2 Probing the mechanical properties of single DNA molecules. (a) Optical tweezers single-molecule assay for measuring the force–extension curve of DNA and (b) measured force–extension curve for λ-phage DNA. Figure (a) is reprinted from Baumann et al., *Proc. Natl. Acad. Sci. U.S.A.* **94**, 6185–91. Copyright (1997) National Academy of Sciences, U.S.A.

measured force–extension curve compared to the expectation from an inextensible WLC model at extensions where the separation of the beads approaches the contour length of the molecule, as shown in Fig. 13.2b. The Young's modulus measured in this way was $Y = (3.5 \pm 0.3) \times 10^8$ Pa. The elastic modulus measured by this technique (and assuming a linear approximation for the change in length ΔL in response to the force F) was $K = FL/\Delta L = 1100$ pN. The elastic modulus can be related to the Young's modulus and, hence, to the bending rigidity and persistence length of the molecule. Modelling dsDNA as a rod with circular cross-section, the resulting persistence length is $d \approx 66$ nm, or 193 base pairs.[2] At higher applied force (65 pN) the DNA molecule underwent a transition into an overstretched form (as shown by the plateau in Fig. 13.2b), increasing considerably in length for an increase of only 2 pN in applied force (suggestive of a highly cooperative process) before elastic behaviour recovered. The fully overstretched molecule showed a contour length of 28 µm, 1.7 times the length of B-form dsDNA, and thus a rise of 5.8 nm per base pair. Another feature evident in the force–extension curves was hysteresis during relaxation after overstretching. A possible explanation for this was the fraying of one strand of DNA in the region of a 'nick' in the molecule. The existence of single-stranded regions along the molecule accounts for the lower measured force in the return part of the stretching cycle. The elasticity of single-stranded DNA (ssDNA) was thus probed with the same optical tweezers assay technique and observed to be significantly different from that of dsDNA. For small extensions the behaviour could be described by a modified freely jointed chain

---

[2] It has been suggested that the discrepancy between this value and that of $d \approx 50$ nm previously reported arises from the presence of permanent bends along the dsDNA making a 'static' contribution to the persistence length, apparently reducing the persistence length of a molecule with no curvature.

(FJC) model, where orientationally independent (Kuhn) segments could stretch as well as align under applied force.

Because the WLC model of Marko and Siggia (1995) includes only the entropic contribution to the elasticity of the DNA molecule, this theory was modified by Wang et al. (1997) to include the enthalpic contribution to elasticity in the form of an elastic modulus, K, as

$$F(x) = \frac{k_B T}{d} \left[\frac{1}{4\left(1 - \frac{x}{L} + \frac{F}{K}\right)^2} - \frac{1}{4} + \frac{x}{L} - \frac{F}{K} \right]. \tag{13.2}$$

The force–extension curve of DNA molecules with contour length as short as $L = 1314$ nm was probed using a position-clamping optical tweezers to achieve the high forces required for extensions approaching the contour length. In the position-clamping mode a closed-loop feedback circuit is operated so that position information about the trapped particle is used to control the intensity of the trapping laser beam. The optical tweezers then maintains a constant particle position (away from the position of zero optical force) by changing the optical force in response to the particle movement. In the experimental assay, a single DNA molecule was bound between a microscope coverglass and a bead held in the optical trap. The microscope stage was driven to stretch the DNA between the coverglass and the bead, pulling the bead away from the centre of the trap in open-loop mode, thereby increasing the optical force acting on the bead. When the bead reached the position clamp set-point, the feedback loop was activated; this increased the laser intensity to hold the position of the particle constant while the optical force increased. The switch between open- and closed-loop operation permits the measurement to be optimised for sensitivity in the low-force regime and dynamic range in the high-force regime during the stretching procedure. During the experiment the maximum force applied was lower than that applied in Smith et al. (1996) that caused the transition to the overstretched state. Data for the force–extension curve were well-fitted by the modified WLC model of Eq. (13.2) over the complete range of applied force. In a 10 mMNa$^+$ buffer the fit returned parameters for the DNA molecule of persistence length $d = 47.4 \pm 1.0$ nm, elastic modulus $K = 1008 \pm 38$ pN and contour length $L = 1343 \pm 5$ nm.

13.2 DNA mechanics: Thermal fluctuations

As shown in the previous section, using optical tweezers as a force transducer provides a very sensitive tool for determining the force–extension curve and, thus, the mechanical and elastic properties of DNA. The resolution of this technique, though, is limited by the thermal fluctuations of the optically trapped beads that are attached to the DNA and used as handles. Meiners and Quake (2000) devised a method of overcoming this limitation by using the correlations introduced between the thermal fluctuations of beads in separate optical traps by a DNA molecule that is held between them, as shown in Fig. 13.3a. In the

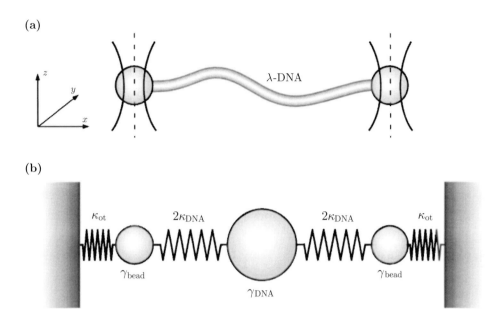

Figure 13.3 Probing DNA thermal fluctuations. (a) Experimental configuration to observe thermal force fluctuations in a single DNA molecule and (b) corresponding model. Reprinted figure from Meiners and Quake, *Phys. Rev. Lett.* **84**, 5014–17. Copyright (2000) by the American Physical Society.

absence of the linking DNA molecule, the thermal motion of the optically trapped beads can be quantified to a good first approximation by their position autocorrelation functions [Section 9.7], i.e.,

$$\overline{x_i(t+\tau)x_i(\tau)} = \frac{k_B T}{\kappa_{ot}} e^{-\tau/\tau_{ot}}, \tag{13.3}$$

where $i = 1, 2$ is an index indicating the bead, and κ_{ot} and τ_{ot} are the spring constant and relaxation time of the beads in the optical traps. Their cross-correlation function exhibits a pronounced anti-correlation at short times because of their hydrodynamic coupling [Section 17.1]. When they are joined by an elastic polymer, the cross-correlation of the position fluctuations of the beads attached to the polymer ends is described by a stretched exponential,

$$\overline{x_1(t+\tau)x_2(\tau)} = \frac{k_B T}{\kappa_c} e^{-(\tau/\tau_c)^\nu}, \tag{13.4}$$

where κ_c and τ_c are a spring constant and a relaxation time, and ν is the stretch exponent. The system can be modelled as shown in Fig. 13.3b, where the beads, which have hydrodynamic friction coefficient γ_{bead}, are confined in optical traps with spring constant κ_{ot}, and the DNA is modelled as a sphere with friction coefficient γ_{DNA} and two springs of stiffness $2\kappa_{DNA}$. The DNA stiffness and relaxation time can then be obtained from the autocorrelation and cross-correlation functions as

$$\kappa_{\text{DNA}} = \kappa_c \frac{\gamma_{\text{bead}}}{\tau_{\text{ot}}(\kappa_{\text{ot}} + \kappa_c)}, \quad (13.5)$$

$$\tau_{\text{DNA}} = \tau_c \frac{\kappa_{\text{ot}}(1 - \tau_{\text{ot}}/\tau_c) - \kappa_c(1 + \tau_{\text{ot}}/\tau_c)}{(\kappa_{\text{ot}} + \kappa_c)(1 - \tau_{\text{ot}}/\tau_c)}. \quad (13.6)$$

In the limit of a stiff optical trap with short relaxation time compared to that of the extended DNA, i.e., $\kappa_{\text{ot}} \gg \kappa_{\text{DNA}}$ and $\tau_{\text{ot}} \ll \tau_{\text{DNA}}$, these reduce to $\kappa_c = \kappa_{\text{DNA}}$ and $\tau_c = \tau_{\text{DNA}}$. At low extensions of the DNA molecule ($< 0.82L$) the cross-correlation data fitted well with a stretch exponent $\nu = 1$; i.e., the DNA relaxes as a simple exponential. Above this extension, a stretch exponent of $\nu > 1$ was required. The measured DNA stiffnesses for both longitudinal and transverse extensions increased rapidly up to the maximum extension tested of $0.92L$. The behaviour of the DNA stiffness with extension was found to be in good agreement with the WLC model, assuming a persistence length of $d = 53$ nm and a contour length of $L = 16.1 \pm 0.2$ µm.

13.3 DNA mechanics: Torsional properties

The response of DNA to torsion (twisting) can be approximated by a torsional potential that is harmonic with the amount of twist θ,

$$E_{\text{torsion}} = \frac{1}{2} \frac{G}{L} \theta^2, \quad (13.7)$$

where G is the torsional modulus and L the contour length. Optical tweezers provide a means of measuring the torsional modulus of DNA by exerting a controlled torque on the molecule. To apply torque to the DNA molecule, an optical trap capable of producing controlled rotations of the trapped particle is required. Rotations of an optically trapped particle can arise either as a result of the properties of the trapping beam, or as a consequence of shape or optical anisotropy of the particle. Oroszi et al. (2006) used a disc-shaped particle formed by squeezing 1 µm-diameter polystyrene microspheres. The preferred orientation direction of the flat particle in the optical trap was controlled by the direction of polarisation of the trapping light and the confinement about this direction by the ellipticity of the polarisation. These two parameters could be changed in the experiment by rotations of a half-wave and quarter-wave plate, respectively. A mixture of λ-DNA in buffer solution and polystyrene disk suspension was incubated for several hours, resulting in DNA attachment to the disks and to a plastic-coated substrate, until eventually some fraction of disks were connected to the substrate by a single DNA molecule. The orientation of the disk tethered to the substrate by the single molecule in this way can then be determined by the balance of optical torque and the torsional strain of the molecule, as shown in Fig. 13.4a. If the torsional stiffness of the optical trap is κ_T and that of the molecule is κ_M, then, assuming that the response of both systems is within the linear regime, the equilibrium orientation will be attained for

$$\kappa_M \alpha_M = \kappa_T \alpha_T, \quad (13.8)$$

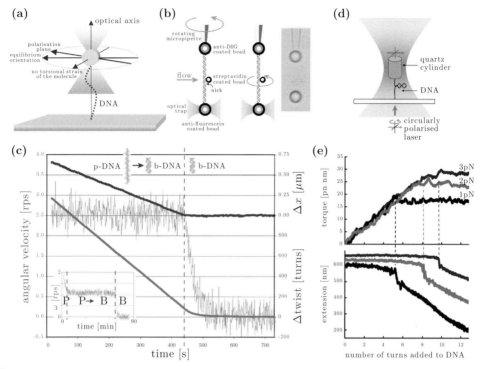

Figure 13.4 Twisting DNA. (a) Torque is applied to dsDNA by the rotation of an optically trapped disc-shaped particle whose orientational angle is controlled by the polarisation direction of the trapping light. Reprinted figure from Oroszi et al., *Phys. Rev. Lett.* **97**, 058301. Copyright (2006) by the American Physical Society. (b) Experimental system for overwinding a DNA molecule by rotating a micropipette while maintaining the DNA under tension using the optical trap. (c) Unwinding of overwound DNA with accompanying change in length and structural transformation. Figures (b) and (c) are reprinted by permission from Macmillan Publishers Ltd: Bryant et al., *Nature* **424**, 338–41, copyright 2003. (d) Torque applied to DNA by rotation of a nanofabricated birefringent quartz cylinder. Reprinted by permission from Macmillan Publishers Ltd: Deufel et al., *Nature Methods* **4**, 223–5, copyright 2007. (e) Buckling of DNA on overwinding indicative of plectoneme formation. Reprinted figure from Forth et al., *Phys. Rev. Lett.* **100**, 148301. Copyright (2008) by the American Physical Society.

where the angles α_M and α_T are measured from the orientations of zero molecular and optical torsion, respectively. Furthermore, the combined system will have an effective torsional stiffness of

$$\kappa_{\text{eff}} = \kappa_M + \kappa_T. \tag{13.9}$$

Because the angle α_T is not easily directly measurable, the change in equilibrium angular orientation was instead measured as the direction of polarisation, i.e., the direction of zero optical torque, was rotated. From the changes in angle of polarisation and equilibrium position and the effective torsional stiffness of the combined system (measured from the orientational fluctuations about the equilibrium angle) the torsional stiffness of the molecule, k_M, and hence the torsional modulus, $G = k_M L$, could be determined. A further difficulty lies in the fact that the molecular torsional stiffness is three orders of magnitude smaller than the optical torque. To enable measurement, the DNA molecule was thus 'pre-stressed' (while still remaining within the linear regime) by adding several complete twists before starting a measurement. An effective torsional modulus, $G_{\text{eff}} = L\Delta\tau/\Delta\theta$, where $\Delta\tau$ is the change in torque and $\Delta\theta$ the change in twist of the molecule, was thus measured. In the model of Moroz and Nelson (1997), the effective torsional modulus is dependent upon the relative extension of the molecule. A fit of the Moroz–Nelson model to the data at relative extensions of the molecule of 0.50 and 0.75 produced a value for the torsional modulus of $420 \pm 43\,\text{pNnm}^2$.

The torsional properties of DNA measured by this technique can be compared to the results obtained by Bryant et al. (2003) for the torsional constant in a different experimental geometry where an optical trap was used to hold the DNA molecule under tension. The experimental configuration is shown in Fig. 13.4b. The ends of the DNA molecule were attached to microbeads that in turn were held by a micropipette and an optical trap. Here, the optical trap was used to maintain constant tension on the molecule by operating in a force feedback mode (Smith et al., 2003). The upper part of the molecule between the microbead held in the micropipette and a small 'rotor' bead attached at a point on the molecule was twisted by rotation of the micropipette while the rotor bead was held in place by a fluid flow. When the fluid flow was released, this segment of the molecule untwisted, as revealed by the motion of the rotor bead in video microscopy. For a small twist, where the linear relationship between torque and angle was maintained, a torsional modulus of $410 \pm 30\,\text{pNnm}^2$ was measured. An alternative measurement made by observing the amplitude of thermal angular fluctuations and applying equipartition of energy produced a torsional modulus of $440 \pm 40\,\text{pNnm}^2$, in good agreement with the dynamic measurement and also with the results of Oroszi et al. (2006). Furthermore, if the DNA molecule was highly overwound while held at tension, a transition from the usual (B-DNA) form to an over-extended high-helicity form (P-DNA) occurred. As the overwound molecule relaxed, P-DNA converted back to B-DNA. The signature of this conversion was unwinding at constant angular velocity (constant torque) accompanied by reduction in length of the molecule, as shown in Fig. 13.4c. When the re-conversion was complete, the B-DNA decelerated as it relaxed towards zero torsion. From the rotational speed during P-DNA to B-DNA conversion a critical torque (for the conditions of the experiment) was deduced as $\tau_{\text{crit}} = 34 \pm 2\,\text{pNnm}$.

An alternative means of generating a torque on an optically trapped particle is to exploit the optical anisotropy of a birefringent material. La Porta and Wang (2004) demonstrated angular trapping and controlled rotations of microscopic quartz particles arising from the transfer of optical spin angular momentum to the particle; they termed this technique an *optical torque wrench*. Deufel et al. (2007) produced nanofabricated quartz cylinders (diameter $0.53 \pm 0.05\,\mu\text{m}$, height $1.1 \pm 0.1\,\mu\text{m}$) with their crystal extraordinary axis perpendicular to the cylinder axis:[3] when held in an optical tweezers their elongated shape forced alignment of the cylinder axis with the trapping laser axis and the optical anisotropy

---

[3] The positive birefringence of quartz leads to confinement in two of the three Euler angles, in contrast to calcite, which exhibits negative birefringence and may also be optically rotated (Friese et al., 1998).

in the transverse plane forced alignment of the extraordinary optical axis with the laser electric field direction, as shown in Fig. 13.4d. The optical torque wrench realised in this manner can be operated in either a passive or an active mode. In the passive mode, the direction of the trapping laser linear polarisation is rotated rapidly, so that the nanofabricated cylinder cannot rotate fast enough to follow the polarisation direction, resulting in a small constant torque on the particle (Inman et al., 2010). In the active mode, capable of applying higher torques, a constant torque is maintained via feedback on the cylinder orientation to control the input polarisation angle in order to maintain a fixed torque (La Porta and Wang, 2004). The ability to exert a high torque while maintaining the DNA molecule under tension revealed the same B-DNA to supercoiled P-DNS (scP-DNA) transition observed by Bryant et al. (2003). Deufel et al. (2007) also functionalised one end of their nanofabricated quartz cylinders so that a length of dsDNA was tethered between the cylinder and a glass coverslip. The dsDNA was held at a constant tension of 10 pNnm and a positive supercoil added by rotating the cylinder. For small amounts of supercoiling the extension of the DNA molecule when held at moderate tension actually increased. The measured torque then increased with applied twist until a critical value of supercoiling was reached where a plateau at a critical value of torque was observed, indicative of the phase transition from B-DNA to scP-DNA. The value for this critical torque was approximately 33 pNnm, in good agreement with the measurements of Bryant et al. (2003). The surprising initial increase in length was accounted for by a twist-stretch coupling in the molecule. Gore et al. (2006) proposed that this behaviour could arise from a DNA structure that, rather than being an isotropic rod, consisted of an inner elastic rod surrounded by a stiff outer wire. Twist–stretch coupling then would occur as an extension of the inner core when under tension its diameter decreases, allowing the outer wire to wrap more turns over the length of the molecule. The energy of the molecule when subject to tension and torsion must then be adapted by the addition of a term describing twist–stretch coupling of the form $g\theta\frac{x}{L}$, where g is the twist–stretch coupling constant. A measurement of the coupling constant by Gore et al. (2006) using a technique similar to that of Bryant et al. (2003) yielded the value $g = -90 \pm 20$ pNnm. Under more moderate tension and torque, the DNA molecule can absorb the extra twist produced by overwinding by buckling into supercoiled loops or plectonemes (Forth et al., 2013). By applying the optical torque wrench method under tension of 1 pNnm to 3 pNnm, Forth et al. (2008) observed a plateau in the torque on the molecule accompanied by a sharp drop in the DNA extension on the addition of between 5 and 10 turns to the DNA, as shown in Fig. 13.4e. As the DNA was overwound further, the torque on the molecule remained constant while the extension decreased, indicative of the formation of additional loops in the plectoneme.

13.4 Motor proteins

The dynamic range and sensitivity of optical tweezers for both displacement and force measurements also make them an ideal tool for studying the dynamics of molecular motors. An important example of a molecular motor is the myosin molecule. This motor is driven

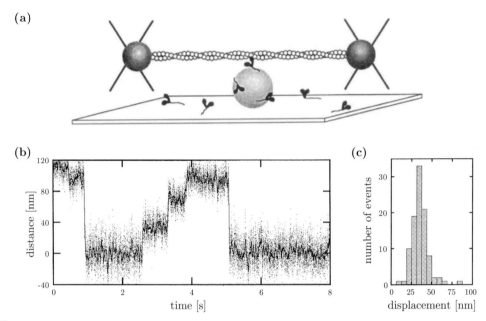

Figure 13.5 Probing the mechanics of molecular motors. (a) Two-bead assay for measuring step size and force generation by a molecular motor. Reprinted by permission from Macmillan Publishers Ltd: Finer et al., *Nature* **368**, 113–19, copyright 1994. (b) Sample data from the motility assay of myosin-V demonstrating resolution of individual steps of the molecular motor. (c) Histogram of myosin-V step sizes, showing a mean step size of about 36 nm. Figures (b) and (c) are reprinted by permission from Macmillan Publishers Ltd: Mehta et al., *Nature* **400**, 590–93, copyright 1999.

by chemical energy released from the hydrolysis of adenosine triphosphate (ATP), which is converted into the relative sliding of actin and myosin filaments that drives muscle contraction. The mechanics of this molecular motor were probed by Finer et al. (1994) using a motility assay based on a dual optical tweezers and illustrated in Fig. 13.5a. Silica beads were attached to a coverslip and coated at low density with skeletal muscle heavy meromyosin (HMM). An actin filament was held under a tension of about 2 pN between two polystyrene beads in separate optical traps. The traps could be steered at low speed over a large distance by motorised mirrors and one trap could be re-positioned at high speed over short distances by an acousto-optic modulator (AOM). For high force measurements, a feedback loop was implemented using the monitored position of the trapped bead to adjust the trap location (rather than the trap intensity as in the experiments of Wang et al. (1997)) using the AOM. To probe the molecular motor dynamics, the actin filament was brought close to the silica bead so that the filament could interact with HMM molecules present on the bead surface. Initially, under conditions of low load on the motor (i.e., relatively low stiffness of the optical traps), when the actin and myosin were allowed to interact, the position of the bead in the optical tweezers connected to the filament was observed to undergo rapid transient movements parallel to the direction of the filament. The average size of the displacements was measured to be 12 nm and, at high concentration of ATP, its duration was 7 ms. These displacements were ascribed to the 'stepping' of myosin along the actin filament during the ATP-driven motor cycle: myosin binds to actin, undergoes a conformational change-driven 'power stroke', and then dissociates from actin. Reducing the concentration of ATP had no effect on the size of the myosin step, but increased its duration. In order to measure the force exerted by a single molecule during the power stroke, the feedback loop was closed to increase the trap stiffness by two orders of magnitude. In this case, when actin and myosin were allowed to interact, force transients with a magnitude of 3.4 ± 1.2 pN were observed. The force distributions were again unchanged at lower ATP concentration.

A similar two-bead motility assay technique was used by Mehta et al. (1999a) to study the myosin-V molecular motor. Similar to skeletal muscle myosin (myosin-II), myosin-V is an actin-based motor fuelled by the hydrolysis of ATP. Myosin-V, however, performs a different task in a different environment: it is associated with organelle transport in regions that are sparsely populated with both actin filaments and myosin and, thus, it is expected to have mechanical properties different from those of the class of myosin motors that drive high-speed motility in muscle contraction. Following a procedure similar to that used in the myosin-II experiments described previously, an actin filament was held at tension between two polystyrene beads and brought towards a silica sphere immobilised on a surface and coated at low density with myosin-V. The low density of myosin-V revealed the processive nature of the actin–myosin interaction: at each interaction a single motor goes through a number of cycles to advance along a filament, rather than motility being the result of interaction with a number of co-localised non-processive motors. The steps produced by the motor cycles of myosin-V are clearly seen in the experimental data of the displacements of one of the optically trapped beads attached to the end of the actin filament, an example of which is shown in Fig. 13.5b. In these experiments, the myosin motor typically took between three and five steps before stalling, corresponding to a load required to stall the motor of 3.0 ± 0.3 pN. In interpreting data such as the ones shown in Fig. 13.5b, one must account for the stiffnesses of all components of the assay, including the linkages from the actin filament and the myosin molecule to the respective substrates as well as the optical trap stiffness. To this end, the position of the optical trap was oscillated, pulling the bead with it. Because of the finite length of the filament between the motor and the bead, the motion of the bead did not follow the trap over its full range of motion and was instead 'clipped' at some finite amplitude, corresponding to the restraint imposed by the length of the tethered filament. During the motor stepping, the displacement of the bead before clipping, and hence the length of the tether, decreased by an average of about 36 nm, as shown in the histogram of displacement in Fig. 13.5c; this can thus be inferred to be the step size of the myosin-V motor. Further detailed study of the motion of myosin-V by Veigel et al. (2001) revealed that the stepping occurred in two distinct stages. The myosin-V protein itself is a double-headed structure, which suggests that it may move along actin in a 'walking' (processing) manner. Veigel et al. (2001) were able to study the kinetics and dynamics of myosin procession using the now-familiar two-bead optical tweezers assay technique. A single-headed form of myosin was observed to bind to actin, producing an initial step of 16 nm followed some time later by a second 5 nm step. The net working stroke is thus 21 nm, rather shorter than the 36 nm step measured for double-headed myosin by Mehta et al. (1999a). Runs of several steps by the double-headed form of myosin-V were also observed in this experiment with a step size (not counting the first step of the run) of 34.5 ± 0.6 nm. However, the sizes of the first step of a run and of events when only one step occurred were both rather shorter at 26.2 ± 2.3 nm and 25.7 ± 0.6 nm, respectively. These data suggested that the working stroke of myosin ranges is between 21 and 25 nm and that the longer step size is a consequence of the processional motion of the double-headed myosin consisting of the working stroke followed by a forward-biased diffusion of the unbound head to a preferred binding site on the filament. The actin filament itself is a helical structure with a pseudorepeat distance of 36 nm. The twist of the actin gives rise to a periodic energy profile for the myosin-V molecule, such that after the working stroke the unbound head is positioned in an energetically unfavourable region, resulting in biased diffusion of the head towards a binding site. The presence of periodic binding sites was revealed in the two-bead assay by scanning an actin filament past a single-headed myosin and observing a myosin–actin binding event from the reduction in Brownian motion of a bead, as shown in the upper part of Fig. 13.1. A histogram of the intervals between binding events, reproduced in the lower part of Fig. 13.1, shows binding events occurring at multiples of 36 nm, indicating that there are indeed preferred binding sites on the filment with the same spacing as the helical pseudorepeat distance.

13.5 Further reading

There is an extensive literature on the application of optical tweezers to single biological molecules; without aiming at being exhaustive, here we highlight a few pertinent articles on topics related to those discussed in this chapter. Wang et al. (1998) used optical tweezers to measure the speed at which RNA polymerase moves along the DNA molecule during transcription while under an applied load in order to determine a force–velocity curve, finding little variation in transcriptional velocity up to the force required to stall movement of around 25 pN. Wuite et al. (2000) measured the force generated by T7 DNA polymerase during replication, finding that polymerisation stalled at 34 ± 8 pN; crucial to this experiment were the differing elasticities of double-stranded and single-stranded DNA, described in the WLC model by persistence lengths of 0.7 nm and 53 nm, respectively. For further studies of the myosin motor, the reader is referred, e.g., to Molloy et al. (1995) on the force produced by a single myosin head and to Fujita et al. (2012), who elucidated the contribution to force generation during the stepping cycle arising from a lever-arm swing followed by a Brownian 'search-and-catch' mechanism. Svoboda et al. (1993) used optical tweezers to study the motor protein kinesin, which also exhibited a stepping behaviour with a mean step size of 8 nm, close to the tubulin monomer repeat distance; in these experiments, kinesin was reported to transport silica beads out of a trap with a maximum force of 5 pN, which differs from the experiments of Kuo and Sheetz (1993), which found an isometric force of 1.9 ± 0.4 pN; possible sources for this discrepancy are discussed in Svoboda and Block (1994). Additional reviews of optical tweezers methods applied to single-molecule biophysics can be found in Mehta et al. (1999b) and Perkins (2009).

References

Baumann, C. G., Smith, S. B., Bloomfield, V. A., and Bustamante, C. 1997. Ionic effects on the elasticity of single DNA molecules. *Proc. Natl. Acad. Sci. U.S.A.*, **94**, 6185–90.

Bryant, Z., Stone, M. D., Gore, J., et al. 2003. Structural transitions and elasticity from torque measurements on DNA. *Nature*, **424**, 338–41.

Bustamante, C., Marko, J. F., Siggia, E. D., and Smith, S. 1994. Entropic elasticity of lambda-phage DNA. *Science*, **265**, 1599–1600.

Bustamante, C., Bryant, Z., and Smith, S. B. 2003. Ten years of tension: Single-molecule DNA mechanics. *Nature*, **421**, 423–7.

Deufel, C., Forth, S., Simmons, C. R., Dejgosha, S., and Wang, M. D. 2007. Nanofabricated quartz cylinders for angular trapping: DNA supercoiling torque detection. *Nature Methods*, **4**, 223–5.

Finer, J. T., Simmons, R. M., and Spudich, J. A. 1994. Single myosin molecule mechanics: Piconewton forces and nanometre steps. *Nature*, **368**, 113–19.

Forth, S., Deufel, C., Sheinin, M. Y., et al. 2008. Abrupt buckling transition observed during the plectoneme formation of individual DNA molecules. *Phys. Rev. Lett.*, **100**, 148301.

Forth, S., Sheinin, M. Y., Inman, J., and Wang, M. D. 2013. Torque measurement at the single molecule level. *Annu. Rev. Biophys.*, **42**, 583–604.

Friese, M. E. J., Nieminen, T. A., Heckenberg, N. R., and Rubinsztein-Dunlop, H. 1998. Optical alignment and spinning of laser-trapped microscopic particles. *Nature*, **394**, 348–50.

Fujita, K., Iwaki, M., Iwane, A. H., Marcucci, L., and Yanagida, T. 2012. Switching of myosin-V motion between the lever-arm swing and Brownian search-and-catch. *Nature Commun.*, **3**, 956.

Gore, J., Bryant, Z., Nöllmann, M., et al. 2006. DNA overwinds when stretched. *Nature*, **442**, 836–9.

Inman, J., Forth, S., and Wang, M. D. 2010. Passive torque wrench and angular position detection using a single-beam optical trap. *Opt. Lett.*, **35**, 2949–51.

Kuo, S. C., and Sheetz, M. P. 1993. Force of single kinesin molecules measured with optical tweezers. *Science*, **260**, 232–4.

La Porta, A., and Wang, M. D. 2004. Optical torque wrench: Angular trapping, rotation, and torque detection of quartz microparticles. *Phys. Rev. Lett.*, **92**, 190801.

Marko, J. F., and Siggia, E. D. 1995. Stretching DNA. *Macromolecules*, **28**, 8759–70.

Mehta, A. D., Rock, R. S., Rief, M., Spudich, J. A., Mooseker, M. S., and Cheney, R. E. 1999a. Myosin-V is a processive actin-based motor. *Nature*, **400**, 590–3.

Mehta, A. D., Rief, M., Spudich, J. A., Smith, D. A., and Simmons, R. M. 1999b. Single-molecule biomechanics with optical methods. *Science*, **283**, 1689–95.

Meiners, J.-C., and Quake, Stephen R. 2000. Femtonewton force spectroscopy of single extended DNA molecules. *Phys. Rev. Lett.*, **84**, 5014–17.

Molloy, J. E., Burns, J. E., Kendrick-Jones, J., Tregear, R. T., and White, D. C. S. 1995. Movement and force produced by a single myosin head. *Nature*, **378**, 209–12.

Moroz, J. D., and Nelson, P. 1997. Torsional directed walks, entropic elasticity, and DNA twist stiffness. *Proc. Natl. Acad. Sci. U.S.A.*, **94**, 14 418–22.

Oroszi, L., Galajda, P., Kirei, H., Bottka, S., and Ormos, P. 2006. Direct measurement of torque in an optical trap and its application to double-strand DNA. *Phys. Rev. Lett.*, **97**, 058301.

Perkins, T. T. 2009. Optical traps for single molecule biophysics: A primer. *Laser Photon. Rev.*, **3**, 203–20.

Smith, S. B., Finzi, L., and Bustamante, C. 1992. Direct mechanical measurements of the elasticity of single DNA molecules by using magnetic beads. *Science*, **258**, 1122–6.

Smith, S. B., Cui, Y., and Bustamante, C. 1996. Overstretching B-DNA: The elastic response of individual double-stranded and single-stranded DNA molecules. *Science*, **271**, 795–9.

Smith, S. B., Cui, Y., and Bustamante, C. 2003. Optical-trap force transducer that operates by direct measurement of light momentum. *Methods Enzymology*, **361**, 134–62.

Svoboda, K., and Block, S. M. 1994. Force and velocity measured for single kinesin molecules. *Cell*, **77**, 773–84.

Svoboda, K., Schmidt, C. F., Schnapp, B. J., and Block, S. M. 1993. Direct observation of kinesin stepping by optical trapping interferometry. *Nature*, **365**, 721–7.

Veigel, C., Wang, F., Bartoo, M. L., Sellers, J. R., and Molloy, J. E. 2001. The gated gait of the processive molecular motor, myosin V. *Nature Cell Biol.*, **4**, 59–65.

Wang, M. D., Yin, H., Landick, R., Gelles, J., and Block, S. M. 1997. Stretching DNA with optical tweezers. *Biophys. J.*, **72**, 1335–46.

Wang, M. D., Schnitzer, M. J., Yin, H., Landick, R., Gelles, J., and Block, S. M. 1998. Force and velocity measured for single molecules of RNA polymerase. *Science*, **282**, 1902–7.

Wuite, G. J. L., Smith, S. B., Young, M., Keller, D., and Bustamante, C. 2000. Single-molecule studies of the effect of template tension on T7 DNA polymerase activity. *Nature*, **404**, 103–6.

14 Cell biology

As with single biological molecules, discussed in Chapter 13, optical tweezers have also played a significant role in advancing our understanding of complex cellular processes. To this end, mechanical interactions between cells, or between cells and their environments, have been probed using optical tweezers as force transducers, utilising their capability as a powerful tool to probe forces that are relevant to cell biology, such as adhesion forces, in a minimally invasive way. Optical tweezers have also been coupled with other techniques from cell biology to aid in the elucidation of signalling pathways that regulate processes such as binding or growth. Furthermore, optical tweezers have been employed as a powerful means to alter and control the behaviour of cells, such as the rate and direction of growth of neurons. In this chapter, we exemplify the application of optical tweezers to problems in cell biology, reviewing in detail three examples: measurement of cellular adhesion forces, probing the structures by which bacteria bind to surfaces, and guiding the growth of neurons [Fig. 14.1].

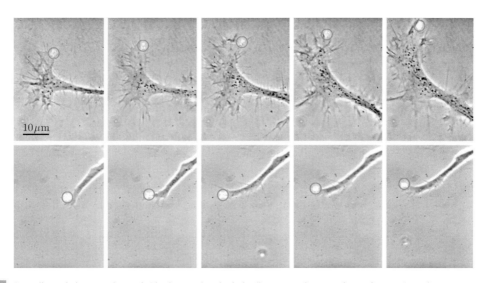

Figure 14.1 Optically guided neuronal growth: Flat (top row) and tubular (bottom row) neuronal growth cones. In each experiment, the position of the laser spot is highlighted with a circle. Frames are shown at time intervals of 10 minutes (top) and 5 minutes (bottom). Reprinted from Ehrlicher et al., *Proc. Natl. Acad. Sci. U.S.A.* **99**, 16 024–8. Copyright (2002) National Academy of Sciences, U.S.A.

14.1 Cellular adhesion forces

Cellular adhesion plays a vital role in cellular functions such as migration and endocytosis. Adhesion occurs as a consequence of a specific, non-covalent binding between transmembrane receptors, such as integrins (Evans and Calderwood, 2007), and either counter-receptors on other cells or extracellular ligands. Cellular bonds are dynamic, and their association and dissociation rates may be altered by environmental factors such as mechanical stress. Such regulation is important in cellular dynamics such as spreading or migration, where, for example, the exertion of force by a cell on a substrate depends on the linkage of the cytoskeleton to the extracellular matrix via the integrin complex. Optical tweezers have been employed to probe these forces either by trapping the cells themselves or by trapping microscopic colloidal probe particles, as we will see in the following two examples.

Thoumine et al. (2000) used optical tweezers to probe the interaction between integrins and the extracellular matrix protein fibronectin in fibroblasts. The association (k_{on}) and dissociation (k_{off}) rates of receptors and ligands determine the time evolution of the probability $p_1(t)$ that a single receptor is bound as

$$p_1(t) = \frac{k_{on}}{k_{on} + k_{off}} \left[1 - e^{-(k_{on}+k_{off})t}\right]. \tag{14.1}$$

Assuming that adhesion bonds form independently, the probability, $p_n(t)$, that n out of a total population of N receptors are bound at time t is given by a binomial distribution. Therefore, if N_a is the number of bonds necessary for adhesion, the probability that the cell is attached to the surface evolves as

$$p_a(t) = \sum_{n=N_a}^{N} p_n(t) = 1 - \sum_{n=0}^{N_a-1} p_n(t). \tag{14.2}$$

Fibroblasts were held in optical tweezers and brought into contact with a fibronectin-coated coverslip for a set time. The microscope stage was then quickly moved so that the trap location was separated from the cell and, thus, an optical force F_{pull} acted to detach the cell from the surface. The adhesion probability as a function of cell-to-surface contact time for low ($F_{pull} \leq 13$ pN) and high ($F_{pull} \geq 28$ pN) force was then fitted to the theoretical time evolution of the adhesion probability given by Eq. (14.2) to retrieve the association and dissociation rates. The time evolution of adhesion probability for pulling with low force was well fitted with $N_a = 1$ bonds giving rates of $k_{on} = 1.64 \times 10^{-6}$ s^{-1} and $k_{off} = 0.06$ s^{-1} and, for high force, with $N_a = 2$ with rates $k_{on} = 2.50 \times 10^{-6}$ s^{-1} and $k_{off} = 0.21$ s^{-1}. The transition from requiring one to two bonds to describe the adhesion probability between $F_{pull} = 13$ and 28 pN suggested that the strength of a single receptor–ligand bond lies within this range.

Jiang et al. (2003) used optical tweezers to probe the interaction of the cytoplasmic tail of integrins with the cytoskeleton. Silica beads (0.64 μm diameter) were coated with a

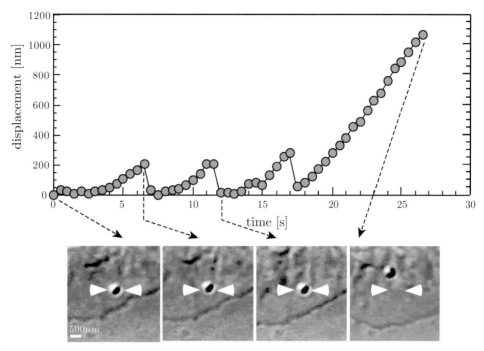

Figure 14.2 Measurement of the strength of the cytoskeleton–integrin bond. The small (0.64 μm-diameter) bead in the lower series of snapshots is held in a calibrated optical trap, whose location is shown by the arrows, near the leading edge of a motile lamellipodium. As the bead is pulled backwards at a constant speed of 60 nm/s, the load on the bond increases until the bond ruptures and the bead returns to the trap centre, as shown in the graph of bead displacement versus time. Reprinted by permission from Macmillan Publishers Ltd: Jiang et al., *Nature* **424**, 334–7, copyright 2003.

trimer of the fibronectin–integrin binding domain, held in optical tweezers and brought into contact with the leading edge of the lamellipodia of a motile fibroblast. The concentration of fibronectin trimer on the surface was sufficiently low so that the probability of the binding being a result of a single trimer was in excess of 80%. The beads were observed to move steadily backwards, away from the leading edge, at a speed of 60 nm/s, thereby pulling the bead away from equilibrium in the optical tweezers and increasing the load on the link between the bead and the cytoskeleton. When the link was broken, the bead was pulled back to the centre of the trap by the optical trapping force, where it was free to form another bond and again be pulled out of the optical trap. This is illustrated in Fig. 14.2, where the bead can be seen to undergo three pulling and breaking events before being pulled out of the optical trap altogether. A histogram of the force required to break the link showed a strong peak at a force of 2 pN. Because, as we have seen previously, the fibronectin–integrin bond strength was measured by Thoumine et al. (2000) to be considerably higher than this, Jiang et al. (2003) inferred that it was the connection between the cytoplasmic tail of the integrin and the actin cytoskeleton that was being broken.

14.2 Adhesion and structure of bacterial pili

Bacteria adhere to tissues during colonisation and infection processes. Bacterial adhesion is mediated by *pili*, which are hair-like structures on the surface of the bacterium. For example, the P pilus expressed by the bacterium *Escherichia coli* consists of a long helical rod with a short flexible fibrillum tip; the P pilus fimbrial adhesins bind to disaccharides present on the surface of epithelial cells in the urinary tract, providing a route to colonisation and infection.

Fällman et al. (2004) built an optical tweezers force transducer system to measure the strength of *E. coli* adhesion. The bacteria were attached to the surface of large (9.6 μm-diameter) carboxyl polystyrene beads that were immobilised on a glass coverslip. Smaller (3 μm-diameter) polystyrene beads functionalised with galabiose, which binds specifically to the P pilus adhesin, were captured in an optical tweezers and brought close to the bacterium so that the pilus could bind [Fig. 14.3a]. Force was then applied to the bond by moving the large bead using a piezo stage at constant speed to pull the bacterium away from the optical tweezers [Fig. 14.3b]. The displacement of the small bead in the optical tweezers was measured using the forward-scattered light from a probe laser beam on a position-sensitive detector, as we have seen in Section 9.2. As the small bead was moved out of the trap, the force on the P pilus–galabiose bond increased until the bond broke and the bead returned to the centre of the optical trap. For a calibrated optical tweezers (stiffness $\kappa_x = 10$ pN/μm), the displacement of the bead provides a measurement of the force applied to the bond, as we have seen in Section 10.1. An example of the evolution of

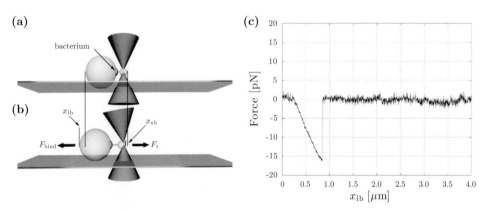

Figure 14.3 Measurement of bacterial adhesion forces. (a) The optically trapped bead coated with receptor molecules is brought close to a bacterium stuck on the surface of the large bead. (b) The bond is loaded by moving the piezo stage by a distance x_{lb} and the resulting displacement of the small bead, x_{sb}, is measured by the deflection of the probe beam. (c) The force applied to the bond as a function of time is determined from the movement of the small bead and the measured trap stiffness. This measurement suggests that a single bond was formed and that it was broken at an applied force of approximately 16 pN. Reprinted from Fällman et al., *Biosens. Bioelectron.* **19**, 1429–37, copyright 2004, with permission from Elsevier.

the force that needs to be applied in order to pull the bacterium out of the trap is shown in Fig. 14.3c, where the piezo stage is withdrawn from the trapped bead at a constant speed of 0.5 μm/s. The fact that this force–displacement curve increases linearly until it falls back to zero suggests that a single bond was formed and then broken, and that the typical bond rupture force lies in the range from 10 to 15 pN.

Using a similar experimental procedure, Jass et al. (2004) were able to measure the mechanical properties of the P pilus itself. The force required to extend the pilus, of (unstretched) length l_0, has a complicated form and three distinct regimes can be identified depending on the pilus (stretched) length l,

$$F_{\text{pilus}}(l) = \begin{cases} \kappa_{\text{pilus}}(l - l_0) & l_0 \leq l < l_1, \\ F_u & l_1 \leq l < l_2, \\ F_{\text{III}}(l) & l_2 \leq l. \end{cases} \quad (14.3)$$

In the first regime ($l_0 \leq l \leq l_1$), the applied force produces an elastic stretching of the pilus with elastic constant κ_{pilus}. In the second regime ($l_1 \leq l \leq l_2$), the helical structure of the pilus starts to unfold in a sequential manner, resulting in extension at constant force (the unfolding force F_u). In the third regime ($l_2 \leq l$), the entire rod of the pilus has been unfolded and the force–extension curve has a more complicated form, denoted $F_{\text{III}}(l)$. The force–extension curves reported by Jass et al. (2004) showed the general form expected from Eq. (14.3), although interpretation in terms of the elastic modulus and unfolding force of an individual pilus requires some care because of the likelihood of multiple pili contributing to the binding. In the first regime, corresponding to extension of the pilus in its elastic regime, abrupt changes in applied force due to successive detachment of individual pili (rather than structural changes) were sometimes observed. For normal pili, linear force–extension behaviour was observed for displacements of the large bead to which the bacteria were attached by around 1 μm (taking into account the finite stiffness of the optical trap, this corresponds to pilus extensions that are somewhat smaller) before a transition to the second regime corresponding to unfolding of the pilus rod; the fractional elongation of a pilus before entering the second regime was $17.5 \pm 2.5\%$. In the second regime, extension at a constant force at values that were multiples of $F_u = 27 \pm 2$ pN were observed, which could be understood as the simultaneous unfolding of multiple pili, where F_u is the force required to unfold a single pilus. This interpretation was reinforced by the observation of discrete steps in the applied force where the force changed by F_u, indicative of successive detachment of pili from the bead. In the third regime, the origin of the complicated behaviour of the force–extension curves was not elucidated from these measurements alone, and could possibly arise from stretching of the interaction between adjacent subunits or from stretching of the subunits themselves.

14.3 Directed neuronal growth

The growth and motility of cells necessarily involve extension and distortion of the cell membrane. Although the membrane itself is believed to be passive in its response to deformation of the cytoskeleton, the mechanical properties of the membrane certainly play a role in the deformability, motility and growth of the cell. For this reason, the mechanical properties of cell membranes have been the subject of extensive studies. In this section, we exemplify the role of optical tweezers in this kind of studies by reviewing how they have been employed to explore and control the growth processes of neuronal cells.

Dai and Sheetz (1995) used optical tweezers to study the mechanics of the membrane of a neuronal growth cone by pulling the membrane into elongated tethers (see also Section 18.2). Spheres (0.5 µm-diameter) coated with immunoglobulin G (IgG) were trapped in an optical tweezers and held against the surface of the growth cone of chick dorsal root ganglias (DRGs), clusters of sensory neurons that carry signal to the spinal cord. Under all experimental conditions, approximately 55% to 65% of spheres were observed to bind irreversibly to the cell surface. The sphere was then pulled away from the surface by moving the optical trap at a constant velocity, which resulted either in the formation of an elongated membrane tether or in the bead being pulled out of the trap and remaining on the surface. On release from the optical tweezers, the tethers were observed to retract, indicating the presence of a force pulling the membrane back. By measuring the displacement of the bead in the calibrated optical tweezers [Section 10.1], the force acting on it during the pulling out of the tether was found. This was repeated for different velocities of trap movement and the force was extrapolated to zero velocity to find the critical force required for tether growth, $F_0 = 6.6 \pm 0.3$ pN. The probability of tether formation on pulling, the critical force and the rate of increase of force with pulling velocity were observed to be affected when treatments that decreased membrane–cytoskeleton interaction were applied. Specifically, treatment with cytochalasin B and D, which inhibit actin dynamics and cause the cytoskeleton to separate from the membrane, and variation in the concentration of organic solvents (DMSO and ethanol), while having no effect on the probability of attachment, all resulted in an increase in the probability of tether formation. These treatments decreased both the critical force for tether formation and the gradient of the force–velocity graph. Neither ATP depletion nor treatment with nocodazole, which disrupts microtubules, caused a change in the probability of tether formation or in the critical force, although the gradient of the force–velocity graphs was also decreased. The conclusions that could be drawn from quantifying the force required to extend a tether in this way were that the membrane interacts with the cytoskeleton through reversible weak bonds and that the force needed to extend the membrane is comparable to the force produced by two myosin molecules [Section 13.4].

Ehrlicher et al. (2002) used an optical tweezers positioned near the leading edge of a neuronal growth cone to influence the direction of growth. An example is shown in Fig. 14.1 for both flat (top row) and tube-like (bottom row) growth cones. In both sets of pictures, the direction of growth of the cone is clearly guided towards the laser spot (circle). Furthermore, the rate of growth in the optically guided case was estimated to increase by a factor of five from that in the unguided case. The mechanism at work behind the directed growth was suggested to be the action of a weak optical dipole force on actin monomers and oligomers (fragments of actin filaments). An actin monomer has a size of approximately 3 nm, so the depth of the optical potential well for typical experimental parameters was much less than the thermal energy. Furthermore, larger (not diffraction-limited) laser spots (or even scanning spots to address a larger area of the lamellipodium) were used, reducing the potential well depth even further. Therefore, the mechanism at work was explained in these terms: although not large enough to confine actin monomers, the optical force should have been sufficient to bias the diffusion of monomers and fragments towards the focus, resulting in an optically induced increase in concentration of actin and in an increased rate of polymerisation leading to an extension of the lamellipodium in the direction of the laser.

A similar technique was employed by Mohanty et al. (2005), who used a line optical tweezers formed by shaping the beam with a cylindrical lens to guide the growth cone. The focused beam had an elliptical shape, 1 μm × 40 μm, and the intensity profile was made asymmetric along the long axis by tilting the beam with respect to the cylindrical lens. The lamellipodia were observed to extend along the elliptical focus towards the higher-intensity end at a rate significantly higher (32 ± 6 μm/h) than that to unilluminated neurons (1 ± 1 μm/h). For a line tweezer with a symmetric intensity profile, no enhancement in growth rate was observed. The induced extension of the growth cone was found to be permanent after approximately 20 minutes of illumination. However, contradictory results were reported by Carnegie et al. (2008), who found that an asymmetric line optical tweezers produced no discernible difference in average growth rate whether the asymmetric intensity gradient was forward-biased, which was expected to favour actin diffusion towards the edge of the growth cone (growth rate 65 ± 11 μm/h), or reverse-biased, which was expected to favour diffusion away from the leading edge (growth rate 78 ± 9 μm/h). Carnegie et al. (2008) also reported, unlike Mohanty et al. (2005), that a line tweezers with a symmetric intensity profile showed a significantly higher growth rate (111 ± 11 μm/h) and that the guided growth was not permanent, as after 20 to 30 minutes growth stalled and the cells started to retract.

Therefore, given the experimental evidence currently available, the mechanism behind optical guiding of neuronal growth is not clear-cut. Ehrlicher et al. (2007) suggest four possible mechanisms:

1. filipodial asymmetries, where the trapping laser may bias and fix the distribution of fillipodia at the edge of the lamellipodia, favouring growth in that direction;
2. inhibition of actin retrograde flow, where the optical tweezers may prevent the actin network from moving inwards, increasing the apparent resistance to the extension of the lamellipodia;
3. enhanced actin polymerisation, where the action of the laser tweezers on the cell membrane may increase the probability that actin monomers polymerise and thus push the membrane further forward; and
4. laser-induced heating, where the localised heating produced by the focused laser beam may enhance the rate of actin polymerisation into filaments, thereby enhancing the growth rate.

To address one of these mechanisms, Stevenson et al. (2006) made measurements of the heating effect of optical traps at two different near-infrared wavelengths, which had both been used for successful guided growth of neurons, and found the small increase in temperature insufficient to significantly increase the rate of actin polymerisation. Ebbesen and Bruus (2012), however, proposed that the temperature gradients that were produced in the range of experiments so far performed may promote neuronal growth by an alternative mechanism: transient receptor potential (TRP) channels, which regulate the flux of calcium

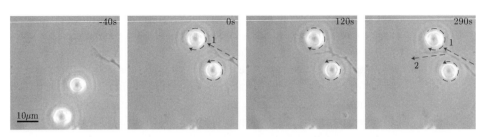

Figure 14.4 Directed growth of neurons. Two counter-rotating optically trapped vaterite particles (rotation direction shown by curved arrows, rotation rate about 1 Hz) are positioned so that the induced microfluidic flow pattern directs the growth of an axon between the spheres. Reprinted by permission from Macmillan Publishers Ltd: Wu et al., *Nature Photon.* **6**, 62–7, copyright 2012.

(Ca^{2+}) ions, may be activated by an increase in temperature, giving rise to influx of Ca^{2+} and increased growth rate.

The direction of growth also depends on fluid shear forces, which can be controlled using optical tweezers. Wu et al. (2012) used birefringent vaterite particles, which can be optically trapped and set into rotation by circularly polarised light. In the low-Reynolds-number regime applicable to these experiments [Box 7.2], the induced flow around the spinning sphere, corrected for the proximity of a fixed coverslip on which the axon is growing, gave rise to a shear stress on an object on the coverslip that has a peak value of 1.7×10^{-3} N/m^2 at a distance approximately 4 μm away from the rotation axis, and directed azimuthally in the same sense as the sphere's rotation. Positioning a spinning vaterite sphere adjacent to a growing axon caused the direction of growth to be deflected by up to 30° in the direction of the induced flow in a significant number of cases. Reversing the direction of rotation of the sphere caused the growth of the axon to be deflected in the opposite direction. A pair of counter-rotating particles were also used to control the microfluidic flow pattern and steer the growth of an axon between the two spheres in the direction of the flow as shown in Fig. 14.4. Because neuronal growth cones respond to their environments (e.g., to the size, shape and elasticity of the substrate), a probable explanation for the observed control of growth direction is that the growth cones are responding to the fluid shear force, even though the cellular signalling pathway by which turning is controlled is not elucidated.

14.4 Further reading

As is the case for single-biological-molecule studies, which we have discussed in Chapter 13, there is a vast literature on applications of optical tweezers to cell biology, stretching back to the seminal experiments on the trapping of bacteria performed by Ashkin and Dziedzic (1987), on the manipulation of organelles within cells also performed by Ashkin and Dziedzic (1989) and on the torsional compliance of flagella in tethered *Escherichia coli* and a motile *Streptococcus* performed by Block et al. (1989). For example, the structure and adhesion properties of the bacterial pilus have been further probed in a series of optical tweezers experiments reported in papers including Andersson et al. (2006a, 2006b) and Bjornham and Axner (2010). Koch and Rohrbach (2012) used an acousto-optic modulator to adapt the shape of the trapping potential to the shape of an elongated 200 nm thin helical bacterium (*Spiroplasma*) and were able to measure its shape deformations at 800 Hz by exploiting local phase differences in coherently scattered trapping light. The force exerted by filopodia and lamellipodia of neuronal growth cones was measured by Cojoc et al. (2007), who found that a single filopodium exerted a force on a substrate of up to 3 pN (requiring actin polymerisation), whereas lamellipodia exerted a force of up to 20 pN (requiring microtubule polymerisation). An automated system for the guidance of motile cells and direction of neuronal growth over a two-dimensional substrate using an acousto-optic deflector to steer the beam was demonstrated by Stuhrmann et al. (2005), and one employing a spatial light modulator (SLM) was developed by Carnegie et al. (2009); both systems are capable of real-time detection of the shape of the extending growth cone, with the SLM system also permitting more complex beam shaping, such as line traps or Bessel beams. Guidance of neuron growth in three dimensions was demonstrated by Graves et al. (2009). Comprehensive reviews of optical tweezers methods applied to biology can be found in the articles by Svoboda and Block (1994), Ou-Yang and Wei (2010) and Stevenson et al. (2010).

References

Andersson, M., Fällman, E., Uhlin, B. E., and Axner, O. 2006a. A sticky chain model of the elongation and unfolding of *Escherichia coli* P pili under stress. *Biophys. J.*, **90**, 1521–34.

Andersson, M., Fällman, E., Uhlin, B. E., and Axner, O. 2006b. Dynamic force spectroscopy of *E. coli* P pili. *Biophys. J.*, **91**, 2717–25.

Ashkin, A., and Dziedzic, J. M. 1987. Optical trapping and manipulation of viruses and bacteria. *Science*, **235**, 1517–20.

Ashkin, A, and Dziedzic, J M. 1989. Internal cell manipulation using infrared laser traps. *Proc. Natl. Acad. Sci. U.S.A.*, **86**, 7914–18.

Bjornham, O., and Axner, O. 2010. Catch-bond behavior of bacteria binding by slip bonds. *Biophys. J.*, **99**, 1331–41.

Block, S. M., Blair, D. F., and Berg, H. C. 1989. Compliance of bacterial flagella measured with optical tweezers. *Nature*, **338**, 514–18.

Carnegie, D. J., Stevenson, D. J., Mazilu, M., Gunn-Moore, F., and Dholakia, K. 2008. Guided neuronal growth using optical line traps. *Opt. Express*, **16**, 10 507–17.

Carnegie, D. J., Čižmár, T., Baumgartl, J., Gunn-Moore, F. J., and Dholakia, K. 2009. Automated laser guidance of neuronal growth cones using a spatial light modulator. *J. Biophoton.*, **2**, 682–92.

Cojoc, D., Difato, F., Ferrari, E., et al. 2007. Properties of the force exerted by filopodia and lamellipodia and the involvement of cytoskeletal components. *PLoS ONE*, **2**, e1072.

Dai, J., and Sheetz, M. P. 1995. Mechanical properties of neuronal growth cone membranes studied by tether formation with laser optical tweezers. *Biophys. J.*, **68**, 988–96.

Ebbesen, C. L., and Bruus, H. 2012. Analysis of laser-induced heating in optical neuronal guidance. *J. Neurosci. Methods*, **209**, 168–77.

Ehrlicher, A., Betz, T., Stuhrmann, B., et al. 2002. Guiding neuronal growth with light. *Proc. Natl. Acad. Sci. U.S.A.*, **99**, 16 024 –8.

Ehrlicher, A., Betz, T., Stuhrmann, B., et al. 2007. Optical neuronal guidance. *Methods Cell Biol.*, **83**, 495–520.

Evans, E. A., and Calderwood, D. A. 2007. Forces and bond dynamics in cell adhesion. *Science*, **316**, 1148–53.

Fällman, E., Schedin, S., Jass, J., et al. 2004. Optical tweezers based force measurement system for quantitating binding interactions: System design and application for the study of bacterial adhesion. *Biosens. Bioel.*, **19**, 1429–37.

Graves, C. E., McAllister, R. G., Rosoff, W. J., and Urbach, J. S. 2009. Optical neuronal guidance in three-dimensional matrices. *J. Neurosci. Methods*, **179**, 278–83.

Jass, J., Schedin, S., Fällman, E., et al. 2004. Physical properties of *Escherichia coli* P pili measured by optical tweezers. *Biophys. J.*, **87**, 4271–83.

Jiang, G., Giannonea, G., Critchley, D. R., Fukumoto, E., and Sheetz, M. P. 2003. Two-piconewton slip bond between fibronectin and the cytoskeleton depends on talin. *Nature*, **424**, 334–7.

Koch, M., and Rohrbach, A. 2012. Object-adapted optical trapping and shape-tracking of energy-switching helical bacteria. *Nature Photon.*, **6**, 680–6.

Mohanty, S. K., Sharma, M., Panicker, M. M., and Gupta, P. K. 2005. Controlled induction, enhancement, and guidance of neuronal growth cones by use of line optical tweezers. *Opt. Lett.*, **30**, 2596–8.

Ou-Yang, H. D., and Wei, M.-T. 2010. Complex fluids: Probing mechanical properties of biological systems with optical tweezers. *Ann. Rev. Phys. Chem.*, **61**, 421–40.

Stevenson, D. J., Lake, T. K., Agate, B., et al. 2006. Optically guided neuronal growth at near infrared wavelengths. *Opt. Express*, **14**, 9786–93.

Stevenson, D. J., Gunn-Moore, F., and Dholakia, K. 2010. Light forces the pace: Optical manipulation for biophotonics. *J. Biomed. Opt.*, **15**, 041503.

Stuhrmann, B., Gögler, M., Betz, T., et al. 2005. Automated tracking and laser micromanipulation of motile cells. *Rev. Sci. Instrum.*, **76**, 035105.

Svoboda, K., and Block, S. M. 1994. Biological applications of optical forces. *Annu. Rev. Biophys. Biomol. Struct.*, **23**, 247–85.

Thoumine, O., Kocian, P., Kottelat, A., and Meister, J.-J. 2000. Short-term binding of fibroblasts to fibronectin: Optical tweezers experiments and probabilistic analysis. *Eur. Biophys. J.*, **29**, 398–408.

Wu, T., Nieminen, T. A., Mohanty, S., et al. 2012. A photon-driven micromotor can direct nerve fibre growth. *Nature Photon.*, **6**, 62–7.

15 Spectroscopy

Optical spectroscopy is one of the most powerful analytical techniques for the characterisation of materials and biological samples. Raman spectroscopy, for example, is routinely used for chemical and physical measurements in materials science, geology, microelectronics and biology. It is possible to integrate spectroscopic functionalities into optical tweezers set-ups, as we have seen in Section 12.1. The resulting *spectroscopic optical tweezers* (SOTs) permit one to manipulate and analyse a wide range of microscopic and nanoscopic particles. In particular, SOTs allow one to study biological entities found in suspension, e.g., living cells, viruses, bacteria and organelles, in their most natural environment, i.e., without the need to fix them on a substrate, as shown in Fig. 15.1 for the case of normal and thalassemic red blood cells. Applications at the nanoscale include, e.g., the possibility of identifying and probing the properties of a single nanostructure *in situ*. In this chapter, we will give a brief account of some applications that combine optical trapping and manipulation with spectroscopic techniques.

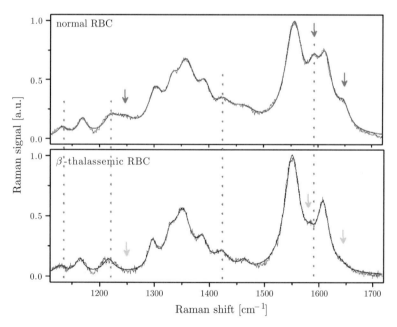

Figure 15.1 Raman spectra of optically trapped red blood cells. Comparison between the spectra of optically trapped normal and β-thalassemic RBCs. The arrows indicate the spectral features affected by intensity changes, whereas the dashed lines highlight the observed energy shifts. Reprinted from De Luca et al., *Opt. Express* **16**, 7943–57. Copyright (2008) The Optical Society.

15.1 Absorption and photoluminescence spectroscopy

The interaction between light and matter is largely determined by the energy levels of the internal degrees of freedom of the material system. For example, the motion of electrons determines the electronic energy levels in an atom and the motion of atoms in a molecule the vibrational and rotational energy levels of the molecule (Demtröder, 2003). The energy spectrum of a molecule can be represented by a *Jabłoński diagram*, as shown in Fig. 15.2a. The electronic energy levels are shown as curves (in bold) in terms of the interatomic coordinates. The equilibrium positions of the atoms correspond to the energy minima of the curves. The vibrational states for each electronic level (thin lines) are obtained by the quantisation of the atomic motion about its equilibrium configuration. This yields a ladder of (approximately harmonic) energy levels that has equal spacing near the equilibrium. Optical transitions involve different electronic states and are indicated by thick solid vertical lines in the Jabłoński diagram, whereas vibrational or rotational transitions within the same electronic state are shown with thin solid vertical lines.

Optical absorption occurs when a photon excites a molecule from its ground state to an excited state. If the transition involves different electronic states, the wavelength of the absorption process generally falls in the UV spectral region (approximately from 200 to 400 nm) or visible range (approximately 400 to 800 nm); therefore, spectroscopy in the UV and visible spectral regions typically probes the electronic structure of a system. Photons can also excite molecules to higher vibrational/rotational levels. In this case, the energy

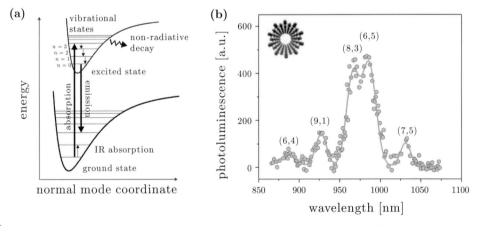

Figure 15.2 Jabłoński diagram and photoluminescence spectrum. (a) Jabłońsky diagram of the electronic (bold solid lines) and vibrational/rotational (thin horizontal lines) energy levels of a molecule. The allowed transitions are indicated with arrows: thick solid vertical lines for radiative, i.e., dipole-allowed, transitions; thin solid vertical lines for vibrational or rotational transitions within the same electronic state; and squiggly lines for non-radiative transitions.
(b) Photoluminescence spectrum of a single-walled nanotube bundle confined by spectroscopic optical tweezers. Nanotube families in the bundle emit at different wavelengths according to their chiral numbers (shown in brackets). Adapted from Maragò et al., *Nature Nanotech.* **8**, 807–19 (2013).

of the absorbed photon is much smaller than that for electronic absorption and the typical wavelength falls in the infrared (approximately 1 to 10 μm) spectral region; therefore, infrared absorption spectroscopy probes the vibrational/rotational structure of molecules. *Optical emission* is a process that involves the transition from an excited level to a lower energy level of the molecular system with consequent emission of a photon. Optical emission can be *spontaneous* or *stimulated*. Both optical absorption and emission processes involve a single photon. Fig. 15.2a depicts a typical fluorescence or photoluminescence process: an incident photon with energy $\hbar\omega_L$ excites the molecule from the ground to an electronic excited state; from there, the system relaxes through the vibrational sub-structure to the lowest vibrational energy level within a few picoseconds; finally, spontaneous emission occurs as the system relaxes into a vibrational level of the electronic ground state within some nanoseconds with consequent emission of a fluorescence photon with energy $\hbar\omega_{fluo} < \hbar\omega_L$. Fluorescence and photoluminescence are therefore two-step processes, occurring on the same time scale as spontaneous emission. Another class of phenomena crucial in optical spectroscopy is scattering processes. Such scattering processes can be *elastic* or *inelastic* according to whether the energy of the scattered photon is equal, *Rayleigh scattering*, or different, *Raman scattering*, to the energy of the incident photon.

SOTs allow one to study the chemical properties of a single particle by probing its vibrational (Raman), electronic (fluorescence/photoluminescence), plasmonic (extinction, thermal) or nonlinear (e.g., two-photons photoluminescence) properties *in situ*. Using SOTs, it is possible, e.g., to select systems with specific physico-chemical properties, such as single-walled carbon nanotubes with certain chiral indices (Rodgers et al., 2008), out of an ensemble of particles with different properties. A typical SOT set-up incorporates two optical beams, as discussed in Section 12.1: one to trap the particle and one to excite it. In some cases, it is possible to use a single beam for both tasks. The single-beam configuration is, indeed, simpler and more stable, whereas the two-beam arrangement offers more versatility to trap, manipulate and excite specific zones of the object, by displacing the trapping and excitation beams independently (Creely et al., 2005). Trapping is often accomplished with a near-infrared laser to minimise photodamage of biomaterials (Xie et al., 2002), whereas visible light is often used for excitation of the particles (Reece et al., 2009; Wang et al., 2011).

Photoluminescence spectroscopy is an optical method of probing the electronic and structural properties of materials, which has been successfully integrated into optical tweezers. As an example, Reece et al. (2009) investigated the structural properties of single InP nanowires in a liquid environment by combining a 1064 nm optical tweezers with a 514.5 nm photoluminescence laser beam. Based on the energy maximum in the photoluminescence emission, the spectra of individually trapped nanowires allowed differentiation of particles with different structures (zinc-blende, wurtzite and mixed phases). Moreover, by implementing an holographic optical tweezers [Chapter 11], Wang et al. (2013) were able to scan the excitation spot along the trapped nanowires, mapping structural inhomogeneities and enabling sorting of specific particles prior to their incorporation into devices. Two-beam SOTs were also used to investigate nonlinear photoexcitation in optically trapped InP nanowires (Wang et al., 2011). Under strong ($\sim$100 MW/cm^2) excitation at 1064 nm, second harmonic generation at 532 nm was observed from individual nanowires, together with band-edge photoluminescence emission at 890 nm, due to two- and three-photon absorption. From the red shift between the two-photon absorption photoluminescence and the direct absorption photoluminescence (excited at 514.5 nm), it was possible to probe band-filling at the single nanowire level. Optical manipulation of semiconductor nanowires with such techniques offers an attractive route for the development of novel devices with engineered electronic properties and for component-wise assembly of nanophotonic devices (Huang et al., 2005).

Perovskite alkaline niobate nanowires have attracted much attention for their interesting nonlinear optical response and their possible use as mechano-optical probes (Dutto et al., 2011). Nakayama et al. (2007) studied polarisation-dependent second harmonic generation from optically trapped nanowires with SOTs, showing waveguiding, i.e., enabling the second harmonic generation signal to propagate at the nanowire apex, thus acting as a nanoscopic light source for photonic force microscopy.

Photoluminescence spectroscopy is also one of the most important tools for characterisation of single-walled carbon nanotubes (O'Connell et al., 2002). Photoluminescence spectra allow determination of the chiral indices, as well as providing information on bundling via the study of exciton energy transfer (Tan et al., 2007) and interaction with the local environment (as dielectric screening shifts the photoluminescence (Hertel et al., 2005)). SOTs allow one to perform single bundle analysis in solution. Fig. 15.2b shows the photoluminescence of a single bundle dispersed in a water/taurodeoxycholate solution (Bonaccorso et al., 2010), optically trapped and probed in a single-beam SOT ($\lambda_0 = 633$ nm). Such a characterisation of the bundle permits one to determine its chirality (Maragò et al., 2013).

15.2 Raman spectroscopy

When light is scattered from a molecule, most photons are elastically diffused; i.e., the scattered photons have the same energy of the incident ones. However, a small fraction of the light (approximately 1 in 10^7 photons) is scattered at energies slightly different from, and usually lower than, the incident light energy. The process leading to this inelastic scattering is referred to as the *linear Raman effect*. Raman spectroscopy relies upon the detection of inelastically scattered light from a sample as a result of illumination of the sample by a laser beam. The energy shift provides information on the vibrational and rotational energies of molecular bonds and on the chemical species forming those bonds. Raman spectra can be considered, therefore, as the chemical fingerprints of molecules and of their interaction with the environment. In the context of biological studies, this enables chemical identification of specific components in cells, tissues and organisms. In carbon-based nanotechnology, Raman spectroscopy is routinely applied to retrieve information on the diameter, electronic type (metallic or semiconductor) and number of walls of carbon nanotubes (Jorio et al., 2001; Hartschuh et al., 2003), and to investigate the properties of graphene samples (Ferrari et al., 2007; Ferrari and Basko, 2013).

The Raman effect can be described quite accurately by treating both the material system (e.g., molecule, crystal lattice) and electromagnetic field classically. In fact, the appearance of inelastic scattering at wavelengths different from the incident one is correctly predicted. In Chapter 3, we have seen that, when a molecule interacts with light, an electric dipole moment, p, is induced. We now consider a generalisation including nonlinear terms depending on higher powers of the incident electric field, so that the induced dipole moment components are written as

$$p_i = \sum_j \alpha_{ij}\mathcal{E}_j + \frac{1}{2}\sum_{jk}\beta_{ijk}\mathcal{E}_j\mathcal{E}_k + \frac{1}{6}\sum_{jkl}\gamma_{ijkl}\mathcal{E}_j\mathcal{E}_k\mathcal{E}_l, \qquad (15.1)$$

where $\mathbb{A} = \{\alpha_{ij}\}$ is the linear polarisability, and $\mathbb{B} = \{\beta_{ijk}\}$ and $\mathbb{C} = \{\gamma_{ijkl}\}$ are respectively the first and second hyper-polarisability tensors of the molecule. These are dependent on the positions of the nuclei and, thus, on the molecule or lattice vibrations. The linear term is associated with Rayleigh and Raman scattering. The second-order term is related to the hyper-Rayleigh and hyper-Raman scattering. The third-order term is associated with the second hyper-Rayleigh and second hyper-Raman scattering. This describes all four-wave mixing processes, including the coherent anti-Stokes Raman scattering (CARS), which is the most important and is described in the following section. We now limit our analysis to the first linear term in Eq. (15.1). For small vibrations, the dependence of the linear polarisability on the nuclei's positions is accounted for by expanding $\mathbb{A}$ around the equilibrium position and keeping only the linear term in the normal mode coordinates, i.e.,

$$\mathbb{A} \approx \mathbb{A}_0 + \sum_n \mathbb{A}'_n q_n,$$

where $\mathbb{A}_0$ is the linear polarisability tensor at the equilibrium positions, q_n are the normal modes' coordinates, $\mathbb{A}'_n = |\partial\alpha_{ij}/\partial q_n|_0$ are the derivatives of the polarisability calculated at the equilibrium positions, and the sum is taken over all possible vibrational modes. Considering only the nth vibrational mode with a harmonic response $q_n = q_{n0}\cos(\omega_n t)$ with amplitude q_{n0} and frequency ω_n, the polarisability is $\mathbb{A}_n \approx \mathbb{A}_0 + \mathbb{A}'_n q_{n0}\cos(\omega_n t)$. If we excite the system with an incident field $\mathcal{E} = \mathcal{E}_0\cos(\omega_L t)$, the induced dipole, and hence the optical response, related to the nth vibrational mode contains three different frequency components,

$$p = \mathbb{A}_0\mathcal{E}_0\cos(\omega_L t) + \frac{1}{2}\mathbb{A}'_n q_{n0}\mathcal{E}_0[\cos(\omega_L + \omega_n)t + \cos(\omega_L - \omega_n)t].$$

The first term corresponds to Rayleigh scattering: it oscillates at ω_L, it occurs at the same frequency as that of the driving field and it does not depend on the molecular vibrations but only on the linear polarisability at equilibrium. The mixed terms correspond to the anti-Stokes and Stokes Raman scattering: they oscillate at $\omega_L + \omega_n$ and $\omega_L - \omega_n$ and depend on the tensor $\mathbb{A}'_n$, which must have some non-zero elements for Raman scattering to occur. An example of a typical Raman spectrum is shown in Fig. 15.3a for carbon tetrachloride molecules (Gucciardi et al., 2007), where the Stokes and anti-Stokes Raman emission peaks are clearly visible.

Although the classical description is capable of predicting the appearance of Raman peaks, it fails to explain the intensity difference between Stokes and anti-Stokes emission or its dependence on temperature. To this end, it is necessary to treat the material system quantum-mechanically, adopting a semi-classical approach to the problem where

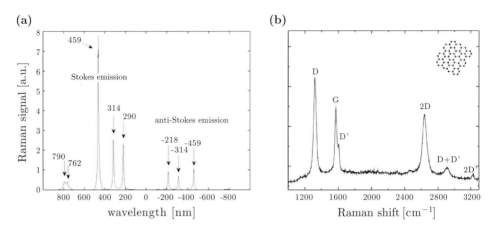

Figure 15.3 Raman spectra of carbon tetrachloride and graphene. (a) Stokes and anti-Stokes Raman spectra of carbon tetrachloride molecules (CCl_4) at room temperature excited by an argon ion laser at 514.5 nm. Reprinted from Gucciardi et al. (2007) with kind permission from Springer Science and Business Media. (b) Raman spectrum of an optically trapped graphene flake with 633 nm trapping and excitation wavelength. Besides the G and 2D peaks, typical of carbon structures, it features significant peaks at D, D' and the combination mode D+D'. Although the D peak is generally associated with structural defects, in this case it is related to the edges of the trapped sub-micrometre flakes (Ferrari and Basko, 2013). Reprinted with permission from Maragò et al., *ACS Nano* **4**, 7515–7523. Copyright (2010) American Chemical Society.

the radiation field is still described classically. A good review of the topic can be found in Long (2002); here we only provide a brief general description. The interaction between the quantum system and the classical field, at frequency ω_L, is treated in a perturbation framework, where the classical induced dipole is replaced with the electric dipole matrix elements associated with a transition of the molecule from a lower energy initial to a higher energy final state. Vibrations of the nth mode are characterised by an energy ladder spaced by $\hbar\omega_n$ [Fig. 15.2a]. The perturbation is kept at first order with only a dipole interaction, whereas higher-order multipoles are neglected. By expanding the wavefunctions associated with the initial and final states in terms of the external perturbation, it is possible to identify the occurring Rayleigh scattering and Raman processes. A level diagram sketch of the different processes is depicted in Fig. 15.4. In all processes, the radiation field induces two transitions: the absorption of a photon from the initial state to a higher-energy virtual state and the emission of a photon that brings the system to its final state. In the Rayleigh scattering process [Fig. 15.4a], the initial and final states are the same and the scattered photon has the same frequency as the incident one, $\omega_R = \omega_L$. When the energy of the final state is higher than the energy of the initial state, a Stokes process occurs and the scattered photon has a frequency lower than the incident driving frequency; i.e., $\omega_S = \omega_L - \omega_n$ [Fig. 15.4b]. Conversely, when the energy of the final state is lower than that of the initial state, an anti-Stokes photon is emitted with frequency higher than that of the incident one, $\omega_{AS} = \omega_L + \omega_n$ [Fig. 15.4c]. If the energy of the virtual state occurs in proximity to an allowed electronic transition, the Raman process is resonant. The quantisation of the molecular system yields the correct selection rules and transition amplitudes for the

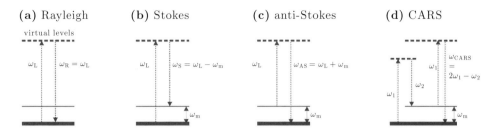

Figure 15.4 Energy level schemes for different scattering processes. (a) In Rayleigh scattering, the emitted photon has the same frequency as the incident one. (b) For Stokes emission, the scattered photon has a frequency lower than the incident one. (c) In anti-Stokes processes, the scattered photon has a frequency higher than the one of the incident light. (d) In coherent anti-Stokes Raman scattering, a four-wave mixing process takes place and resonantly excites the emission of anti-Stokes photons.

Raman processes. An important result is the correct intensity ratio between the Stokes and anti-Stokes emissions and its temperature dependence, i.e.,

$$\frac{I_S}{I_{AS}} = \frac{(\omega_L - \omega_n)^4}{(\omega_L + \omega_n)^4} e^{-\hbar \omega_n / k_B T}.$$

Raman spectroscopy is able to detect and analyse extremely small molecular-sized objects with high resolution. When combined with optical confocal microscopy, Raman spectroscopy allows analysis and chemical imaging with spatial resolution limited only by diffraction. Raman tweezers, realised by coupling a Raman spectrometer with optical tweezers [Section 12.1], have allowed an even greater degree of analytical capability. Raman tweezers use optical tweezers to suspend and manipulate micro/nanoparticles without direct contact, so that the molecular Raman spectra may be recorded while in their most natural state. Thus, the spectra collected are more indicative of the true nature of the molecule under study and, therefore, of more significance.

Raman tweezers were first introduced for the investigation of biological materials and, for example, they were shown to be able to discriminate between living and dead yeast cells (Xie et al., 2002). Their ability to trap and analyse individual nanostructures was further demonstrated on 40 nm polystyrene beads (Ajito and Torimitsu, 2002). Raman imaging of floating cells (Creely et al., 2005) and real-time detection of hyperosmotic stress in yeast cells (Singh et al., 2005) were then demonstrated. A Raman tweezers system has been employed to examine the oxygenation capability (i.e., the functionality) of β-thalassemic red blood cells (De Luca et al., 2008), as shown in Fig. 15.1, where the Raman spectrum of an optically trapped healthy red blood cell is compared with that from a heterozygous β-thalassemic cell. Raman bands characteristic of oxygenated Hb (oxyHb) are strongly depressed for the β-thalassemic cell (indicated by the arrows). Moreover, numerous Raman bands, affected by oxygenation condition, are energy-shifted (dashed lines). Both experimental outcomes clearly demonstrate a lower efficiency of β-thalassemic cells in carrying out their natural role, namely oxygen transportation from lungs to the whole organism. These results demonstrate that Raman tweezers have enormous potential for the monitoring of blood diseases and their response to drug therapies.

The potential of Raman tweezers as a tool for analysis and manipulation of nanostructures in liquid was demonstrated on carbon nanotubes (Tan et al., 2004). Rodgers et al. (2008) showed that Raman tweezers can be used to selectively trap and aggregate single-walled carbon nanotubes with specific chiralities. This was done by focusing a 633 nm beam (10 mW) on a solution containing dispersed nanotubes and mapping the increase of radial breathing modes related to nanotubes of specific chiralities as a function of time. Raman tweezers are also ideally suited to trap, manipulate and sort individual graphene flakes in solution (Maragò et al., 2010), sorting them as a function of shape and number of layers, with prospects for accurately positioning them within devices with controlled properties. In fact, Raman spectroscopy allows one to extract structural and electronic information on individual flakes (Ferrari et al., 2007; Ferrari and Basko, 2013), as shown in Fig. 15.3b for the Raman spectrum of a trapped flake measured at 633 nm.

15.3 Coherent anti-Stokes Raman spectroscopy

If the incident laser intensity is sufficiently high, the nonlinear response of the molecular system related to the higher-order terms in Eq. (15.1) becomes important. Coherent anti-Stokes Raman scattering (CARS) is a nonlinear Raman scattering process that involves a four-wave mixing process linked to the third-order susceptibility of a material system (Maker and Terhune, 1965). In CARS, the system is coherently excited through the beating of two incoming electromagnetic waves having frequencies ω_1 and ω_2, and then mixed with the incoming field at ω_1, resulting in a coherent output signal at the anti-Stokes frequency $\omega_{CARS} = 2\omega_1 - \omega_2$, as shown in Fig. 15.4d. When the frequency difference $\omega_1 - \omega_2$ coincides with a Raman-active level of the molecule at ω_m, a CARS signal is resonantly created.

The intensity of the CARS signal is proportional to the square of the intensity of the input beam, I_1, and linearly dependent on the intensity of the second beam, I_2. Moreover, the signal will be maximal when momentum conservation is satisfied, i.e., the phase-matching condition $\Delta \mathbf{k} = \mathbf{k}_{CARS} - 2\mathbf{k}_1 + \mathbf{k}_2 = 0$ is satisfied. Because CARS is generated only in a small volume at the crossing of three beams, high spatial resolution can be achieved by crossing the beams at different angles, provided that the phase matching condition is always fulfilled.

Two examples of the integration and use of CARS with optical tweezers can be found in Chan et al. (2005) and Shi and Liu (2007).

15.4 Rayleigh spectroscopy and surface-enhanced Raman spectroscopy

Rayleigh spectroscopy measures the spectral dependence of the elastic light scattering cross-section and can be used to probe plasmon resonances in metal nanoparticles (Maier, 2007). Metal nanoparticles are interesting as optically resonant nanoantennas, capable of spatially confining and enhancing the local electromagnetic field by orders of magnitude. In particular, nanoparticle dimers, trimers and fractal aggregates with novel functionalities and higher field enhancement capabilities can, in principle, be created using optical forces. By combining Rayleigh scattering with optical tweezers, Prodan et al. (2003) showed plasmon hybridisation, due to the close encounter between a trapped silver nanoparticle and an immobilised one (Prikulis et al., 2004). The plasmon resonance energy shift (Aizpurua et al., 2005) can be used as a parameter to study quantitatively the interaction potential between colloidal nanoparticles in an optical tweezers (Tong et al., 2011) and reconstruct the inter-particle potential energy landscape as a function of distance, allowing tuning of the optical interaction between the nanoparticles in the dimer. Optical forces were also shown to be strongly affected by near-field coupling amongst nanoparticles simultaneously trapped in a single optical trap (Ohlinger et al., 2011). The coupling was found to strengthen the nanoparticles' interaction with the trapping light, causing a gradual shift of the plasmon resonance towards the laser wavelength. This resulted in thermal destabilisation of the system because of the enhanced light absorption and consequent overheating of the water layer around the nanoparticles.

Surface-enhanced Raman spectroscopy (SERS) takes advantage of the local field enhancement offered by optically resonant metal nanoparticles to amplify the Raman signal (Maier, 2007) and in principle allows high-sensitivity label-free identification of molecular species (Kneipp et al., 2006). Optically coupled metal nanoparticles are amongst the most efficient substrates for SERS of molecular adsorbates (Aizpurua et al., 2005). Optical tweezers were proven to be an effective tool to create SERS-active metal nanocolloid aggregates (Svedberg et al., 2006). SOTs therefore have a great potential for ultrasensitive, label-free molecular recognition in liquids via SERS (Bjerneld et al., 2003). Svedberg et al. (2006) used optical forces to bring two silver nanoparticles into near-field contact in a liquid solution containing thiophenol (10 μM) and created a SERS-active dimer capable of strongly enhancing the Raman signal compared to that for a single trapped silver nanoparticle. Repulsive optical forces can also be used to deposit silver nanoparticles on glass coated with 3-aminopropyltrimethoxysilane, to form SERS-active aggregates. Bjerneld et al. (2003) used this substrate to detect Rhodamine 6G molecules in solution at 0.1 μM concentration.

Biomolecules, e.g., proteins or nucleic acids, find their natural functional environment in liquid. The rapid, ultrasensitive, label-free detection of pathology biomarkers in body fluids is a field in which plasmonic nanosensors can find revolutionary applications (Willets and Van Duyne, 2007). Two different concepts of SERS-based nanosensors for the detection of biomolecules in liquid were recently demonstrated using SOTs. Rao et al. (2010) anchored a double-stranded deoxyribonucleic acid (DNA) molecule, tagged with biotin and dioxydenine at each end, between two optically trapped ($\lambda_{trap} = 1064$ nm) polystyrene beads coated with streptavidin and anti-dioxydenine. The DNA was thus suspended in a solution containing SERS-active gold nanocolloids and excited with a second laser beam ($\lambda_{exc} = 785$ nm) that allowed recovery of the enhanced signal of three vibrational bands. Messina et al. (2011) showed that gold nanocolloidal aggregates optically trapped in single-beam SOTs (785 nm) enable SERS detection of molecules adsorbed on their surface, as shown in Fig. 15.5.

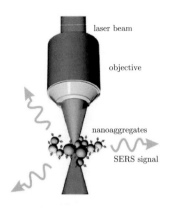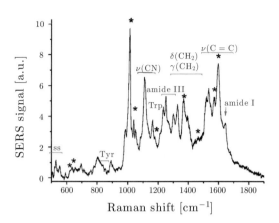

Figure 15.5 Surface-enhanced Raman scattering: SERS of bovine serum albumin (BSA) proteins performed in liquid by optically trapping gold colloidal aggregates on which the protein is adsorbed. The enhanced BSA peaks are indicated: SS, disulfide bridges; Tyr, tyrosine; Trp, tryptophan, amide bands, stretching modes. The asterisks indicate the SERS signal from pyridine. Reprinted with permission from Messina et al., *J. Phys. Chem. C* **115**, 5115–22. Copyright (2011) American Chemical Society.

A photonic force microscope can benefit from the ability to fabricate SERS-active nanometric probes, paving the way for local enhanced spectroscopy of biological surfaces. A route to accomplishing this is to design special SERS-active probes consisting of metal colloids (silver and gold) tightly bound to micrometric silica beads (Bálint et al., 2009) or to nanowires, with reduced thermal fluctuations compared to those of individual metal nanoparticles (Irrera et al., 2011). Bálint et al. (2009) showed that by optically manipulating and exciting such silica hybrid probes in close contact with the surface of cells incubated in emodin (concentration 2 µM) within a single-beam optical tweezers ($\lambda_{\text{laser}} = 785$ nm), it was possible to detect the SERS fingerprint of membrane emodin molecules.

SERS optical tweezers couple high molecular sensitivity with contactless, label-free, three-dimensional capability of operation in liquids. SERS-active probes can be highly specific because functionalised probes allow the selective interaction with specific sample sites. Thus, they represent a promising tool for the development of next-generation biosensors capable of detecting biomolecules and investigating biological samples in their natural environment.

15.5 Further reading

Any advanced textbook on spectroscopy is a very good starting point to acquire a working knowledge of spectroscopic optical tweezers. A classic general textbook is Demtröder (2003); another introductory textbook focused on Raman spectroscopy is Ferraro et al. (2003); and an excellent textbook on the principles of SERS and plasmonics is Le Ru and Etchegoin (2008). A collection of advanced papers on plasmon-enhanced spectroscopies and the *nanoantenna* concept can be found in Lamy de la Chapelle and Pucci (2013). An excellent review of Raman tweezers is Petrov (2007). Linear and non-linear microspectroscopy (Raman, hyper-Raman, hyper-Rayleigh and two-photon luminescence) have been combined with optical tweezers by Fontes et al. (2005). Raman tweezers have found wide application in the identification of bacterial spores (Chan et al., 2004), biological cells (Xie et al., 2004), viruses inside trapped cells (Hamden et al., 2005), and trapped mitochondria (Tang et al., 2007). *Photonic jets* (Lecler et al., 2005) created beyond a trapped dielectric particle have been employed for near-field Raman spectroscopy (Kasim et al., 2008) below the diffraction limit to resolve PMOS transistors and gold nanopatterns. A method of increasing the signal-to-noise ratio in Raman tweezers by phase-sensitive detection was demonstrated by Rusciano et al. (2006). The combination of optical tweezers and optical injection of nanoparticles inside living cells was recently demonstrated (McDougall et al., 2009). Photoporation (Stevenson et al., 2006), i.e., the process of creating a transient pore on a cell membrane with a focussed laser beam, was used for the targeted delivery of 100 nm gold nanoparticles into a specific region of the interior of an individual mammalian cell (McDougall et al., 2009). This provides a novel all-optical methodology for integrating nanobiosensors within specific intracellular regions.

References

Aizpurua, J., Bryant, G. W., Richter, L. J., et al. 2005. Optical properties of coupled metallic nanorods for field-enhanced spectroscopy. *Phys. Rev. B*, **71**, 235420.

Ajito, K., and Torimitsu, K. 2002. Single nanoparticle trapping using a Raman tweezers microscope. *Appl. Spectrosc.*, **56**, 541–4.

Bálint, S., Kreuzer, M. P., Rao, S., et al. 2009. Simple route for preparing optically trappable probes for surface-enahnced Raman scattering. *J. Phys. Chem. C*, **113**, 17 724–9.

Bjerneld, E. J., Svedberg, F., and Käll, M. 2003. Laser-induced growth and deposition of noble-metal nanoparticles for surface-enhanced Raman scattering. *Nano Lett.*, **3**, 593–6.

Bonaccorso, F., Sun, Z., Hasan, T., and Ferrari, A. C. 2010. Density gradient ultracentrifugation of nanotubes: Interplay of bundling and surfactants encapsulation. *J. Phys. Chem. C*, **114**, 17267.

Chan, J. W., Esposito, A. P., Talley, C. E., et al. 2004. Reagentless identification of single bacterial spores in aqueous solution by confocal laser tweezers Raman spectroscopy. *Anal. Chem.*, **76**, 599–603.

Chan, J. W., Winhold, H., Lane, S. M., and Huser, T. 2005. Optical trapping and coherent anti-Stokes Raman scattering (CARS) spectroscopy of submicron-size particles. *IEEE J. Sel. Top. Quant. Electron.*, **11**, 858–63.

Creely, C., Volpe, G., Singh, G., Soler, M., and Petrov, D. 2005. Raman imaging of floating cells. *Opt. Express*, **13**, 6105–10.

De Luca, A. C., Rusciano, G., Ciancia, R., et al. 2008. Spectroscopical and mechanical characterization of normal and thalassemic red blood cells by Raman tweezers. *Opt. Express*, **16**, 7943–57.

Demtröder, W. 2003. *Laser spectroscopy*. Heidelberg, Germany: Springer Verlag.

Dutto, F., Raillon, C., Schenk, K., and Radenovic, A. 2011. Nonlinear optical response in single alkaline niobate nanowires. *Nano Lett.*, **11**, 2517–21.

Ferrari, A. C., and Basko, D. M. 2013. Raman spectroscopy as a versatile tool for studying the properties of graphene. *Nature Nanotechnol.*, **8**, 235–46.

Ferrari, A. C., Meyer, J. C., Scardaci, V., et al. 2007. Raman spectrum of graphene and graphene layers. *Phys. Rev. Lett.*, **97**, 187401.

Ferraro, J. R., Nakamoto, K., and W., Brown C. 2003. *Introductory Raman spectroscopy*. Waltham, MA: Academic Press.

Fontes, A., Ajito, K., Neves, A. A. R., et al. 2005. Raman, hyper-Raman, hyper-Rayleigh, two-photon luminescence and morphology-dependent resonance modes in a single optical tweezers system. *Phys. Rev. E*, **72**, 012903.

Gucciardi, P. G., Trusso, S., Vasi, C., Patanè, S., and Allegrini, M. 2007. Near-field Raman spectroscopy and imaging. Pages 287–329 of: *Applied scanning probe methods V: Scanning probe microscopy techniques*. Heidelberg, Germany: Springer Verlag.

Hamden, K. E., Bryan, B. A., Ford, P. W., et al. 2005. Spectroscopic analysis of Kaposi's sarcoma-associated herpesvirus infected cells by Raman tweezers. *J. Virology Methods*, **129**, 145–51.

Hartschuh, A., Pedrosa, H. N., Novotny, L., and Krauss, T. D. 2003. Simultaneous fluorescence and Raman scattering from single carbon nanotubes. *Science*, **301**, 1354–6.

Hertel, T., Hagen, A., Talalaev, V., et al. 2005. Spectroscopy of single-and double-wall carbon nanotubes in different environments. *Nano Lett.*, **5**, 511–14.

Huang, Y., Duan, X., and Lieber, C. M. 2005. Nanowires for integrated multicolor nanophotonics. *Small*, **1**, 142–7.

Irrera, A., Artoni, P., Saija, R., et al. 2011. Size-scaling in optical trapping of silicon nanowires. *Nano Lett.*, **11**, 4879–84.

Jorio, A., Saito, R., Hafner, J. H., et al. 2001. Structural (n,m) determination of isolated single-wall carbon nanotubes by resonant Raman scattering. *Phys. Rev. Lett.*, **86**, 1118–21.

Kasim, J., Ting, Y., Meng, Y. Y., et al. 2008. Near-field Raman imaging using optically trapped dielectric microsphere. *Opt. Express*, **16**, 7976–84.

Kneipp, K., Moskovits, M., and Kneipp, H. 2006. *Surface-enhanced Raman scattering*. Berlin: Springer Verlag.

Lamy de la Chapelle, M., and Pucci, A. 2013. *Nanoantenna – Plasmon-enhanced spectroscopies for biotechnological applications*. Singapore: Pan Stanford Publishing.

Le Ru, E., and Etchegoin, P. 2008. *Principles of surface-enhanced Raman spectroscopy*. Amsterdam: Elsevier.

Lecler, S., Takakura, Y., and Meyrueis, P. 2005. Properties of a three-dimensional photonic jet. *Opt. Lett.*, **30**, 2641–3.

Long, D. A. 2002. *The Raman effect: A unified treatment of the theory of Raman scattering by molecules*. New York: Wiley.

Maier, S. A. 2007. *Plasmonics: Fundamentals and applications*. New York: Springer Verlag.

Maker, P. D., and Terhune, R. W. 1965. Study of optical effects due to an induced polarization third order in the electric field strength. *Phys. Rev.*, **137**, A801–A818.

Maragò, O. M., Jones, P. H., Gucciardi, P. G., Volpe, G., and Ferrari, A. C. 2013. Optical trapping and manipulation of nanostructures. *Nature Nanotech.*, **8**, 807–19.

Maragò, Onofrio M., Bonaccorso, F., Saija, R., et al. 2010. Brownian motion of graphene. *ACS Nano*, **4**, 7515–23.

McDougall, C., Stevenson, D. J., Brown, C. T. A., Gunn-Moore, F., and Dholakia, K. 2009. Targeted optical injection of gold nanoparticles into single mammalian cells. *J. Biophoton.*, **2**, 736–43.

Messina, E., Cavallaro, E., Cacciola, A., et al. 2011. Manipulation and Raman spectroscopy with optically trapped metal nanoparticles obtained by pulsed laser ablation in liquids. *J. Phys. Chem. C*, **115**, 5115–22.

Nakayama, Y., Pauzauskie, P. J., Radenovic, A., et al. 2007. Tunable nanowire nonlinear optical probe. *Nature*, **447**, 1098–1101.

O'Connell, M. J., Bachilo, S. M., Huffman, C. B., et al. 2002. Band gap fluorescence from individual single-walled carbon nanotubes. *Science*, **297**, 593–6.

Ohlinger, A., Nedev, S., Lutich, A. A., and Feldman, J. 2011. Optothermal escape of plasmonically coupled silver nanoparticles from a three-dimensional optical trap. *Nano Lett.*, **11**, 1770–4.

Petrov, D. V. 2007. Raman spectroscopy of optically trapped particles. *J. Opt. A Pure Appl. Opt.*, **9**, S139–S156.

Prikulis, J., Svedberg, F., Käll, M., et al. 2004. Optical spectroscopy of single trapped metal nanoparticles in solution. *Nano Lett.*, **4**, 115–18.

Prodan, E., Radloff, C., Halas, N. J., and Nordlander, P. 2003. A hybridization model for the plasmon response of complex nanostructures. *Science*, **302**, 419–22.

Rao, S., Raj, S., Balint, S., et al. 2010. Single DNA molecule detection in an optical trap using surface-enhanced Raman scattering. *Appl. Phys. Lett.*, **96**, 213701.

Reece, P. J., Paiman, S., Abdul-Nabi, O., et al. 2009. Combined optical trapping and microphotoluminescence of single InP nanowires. *Appl. Phys. Lett.*, **95**, 101109.

Rodgers, T., Shoji, S., Sekkat, Z., and Kawata, S. 2008. Selective aggregation of single-walled carbon nanotubes using the large optical field gradient of a focused laser beam. *Phys. Rev. Lett.*, **101**, 127402.

Rusciano, G., De Luca, A. C., Sasso, A., and Pesce, G. 2006. Phase-sensitive detection in Raman tweezers. *Appl. Phys. Lett.*, **89**, 261116.

Shi, K., Li, P., and Liu, Z. 2007. Broadband coherent anti-Stokes Raman scattering spectroscopy in supercontinuum optical trap. *Appl. Phys. Lett.*, **90**, 141116.

Singh, G. P., Creely, C. M., Volpe, G., Grötsch, H., and Petrov, D. 2005. Real-time detection of hyperosmotic stress response in optically trapped single yeast cells using Raman microspectroscopy. *Anal. Chem.*, **77**, 2564–8.

Stevenson, D., Agate, B., Tsampoula, X., et al. 2006. Femtosecond optical transfection of cells: Viability and efficiency. *Opt. Express*, **14**, 7125–33.

Svedberg, F., Li, Z., Xu, H., and Käll, M. 2006. Creating hot nanoparticle pairs for surface-enhanced Raman spectroscopy through optical manipulation. *Nano Lett.*, **6**, 2639–41.

Tan, P. H., Rozhin, A. G., Hasan, T., et al. 2007. Photoluminescence spectroscopy of carbon nanotube bundles: Evidence for exciton energy transfer. *Phys. Rev. Lett.*, **99**, 137402.

Tan, S., Lopez, H. A., Cai, C. W., and Zhang, Y. 2004. Optical trapping of single-walled carbon nanotubes. *Nano Lett.*, **4**, 1415–19.

Tang, H., Yao, H., Wang, G., et al. 2007. NIR Raman spectroscopic investigation of single mitochondria trapped by optical tweezers. *Opt. Express*, **15**, 12 708–16.

Tong, L., Miljković, V. D., Johansson, P., and Käll, M. 2011. Plasmon hybridization reveals the interaction between individual colloidal gold nanoparticles confined in an optical potential well. *Nano Lett.*, **11**, 4505–8.

Wang, F., Reece, P. J., Paiman, S., et al. 2011. Nonlinear optical processes in optically trapped InP nanowires. *Nano Lett.*, **11**, 4149–53.

Wang, F., Toe, W. J., Lee, W. M., et al. 2013. Resolving stable axial trapping points of nanowires in an optical tweezers using photoluminescence mapping. *Nano Lett.*, **13**, 1185–91.

Willets, K. A., and Van Duyne, R. P. 2007. Localized surface plasmon resonance spectroscopy and sensing. *Annu. Rev. Phys. Chem.*, **58**, 267–97.

Xie, C., Dinno, M. A., and Li, Y. 2002. Near-infrared Raman spectroscopy of single optically trapped biological cells. *Opt. Lett.*, **27**, 249–51.

Xie, C., Goodman, C., Dinno, M., and Li, Y.-Q. 2004. Real-time Raman spectroscopy of optically trapped living cells and organelles. *Opt. Express*, **12**, 6208–14.

16 Optofluidics and lab-on-a-chip

Optofluidics is the integration of photonics with microfluidics. The goal is to construct and provide platforms that enable enhanced optical sensing and manipulation of different types of samples that are of interest to interdisciplinary science. The broader vision is the *lab-on-a-chip* concept shown in Fig. 16.1, i.e., the scaling down of an entire laboratory to fit on a chip with biological, chemical, physical and optical sensing capabilities. In this context, optical forces find their perfect place, affording the possibility to trap, manipulate, sort and characterise samples without mechanical contact. In this chapter, we discuss specific applications, such as optical sorting of particles and cells by size or refractive index using radiation pressure or optical potentials, the monolithic integration of fibres, cavities and waveguides for optomechanical probing and the realisation of micromachines and microrobots.

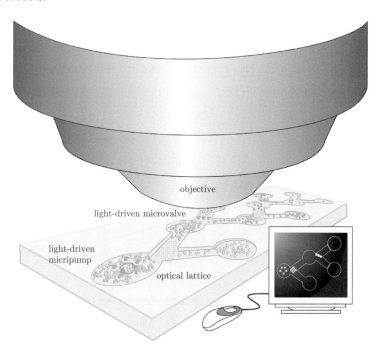

Figure 16.1 Light-driven lab-on-a-chip concept. Pictorial view. The integration of microfluidics with optical tweezers makes it possible to sort and assemble particles as well as to exploit and control light-driven pumps, valves and mixers, all monitored by a computer-controlled interface. Further analysis is achieved by applying micro-spectroscopy techniques for Raman and fluorescence investigation in an all-optical micro-laboratory. Reprinted by permission from Macmillan Publishers Ltd: Glückstad, *Nature Mater.* **3**, 9–10, copyright 2004.

16.1 Optical sorting

The isolation or separation of particles and cells is a crucial function in optofluidic experiments (Hunt and Wilkinson, 2008). Samples can be selected and delivered to a sensing or imaging location using a microfluidic system (Fan and White, 2011). A simple way to achieve sorting exploits the controlled propulsion by scattering forces. Buican et al. (1987) first used this technique to sort Chinese hamster ovary cells. Later, the combination of radiation pressure with an opposing fluid flow led to the realisation of *optical chromatography* (Imasaka, 1998), i.e., particle separation in a microfluidic channel by radiation pressure (Hart et al., 2007). Optical separation of particles based on size and refractive index was achieved (Terray et al., 2005), including spores of *Bacillus anthracis* and *Bacillus thuringiensis* (Hart et al., 2006). A recent development of radiation pressure sorting is its application to chiral structures such as liquid crystal microdroplets (Tkachenko and Brasselet, 2014) or chiral solid microresonators (Cipparrone et al., 2011; Hernández et al., 2013). Optomechanical chiral forces (Tkachenko and Brasselet, 2013; Donato et al., 2014) arise from the polarisation-dependent interaction of light with chiral particles. Tkachenko and Brasselet (2014) demonstrated chiral optical sorting of cholesteric liquid crystal droplets using the radiation pressure of counter-propagating beams with opposite circular polarisation crossing a microfluidic channel.

The sorting capabilities of optical forces can be greatly enhanced by using extended optical landscapes (Glückstad, 2004). A three-dimensional optical lattice formed by the interference of five laser beams was used by MacDonald et al. (2003) to sort a mixed sample of particles flowing in a microfluidic system, as shown in Fig. 16.2. In this configuration, some particles are strongly deflected from their original trajectories, while others pass

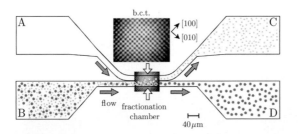
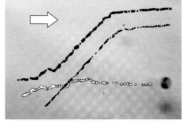

Figure 16.2 Microfluidic sorting in a three-dimensional optical lattice. Two particle species with different properties (e.g., size, refractive index) flow from B to D in a microfluidic circuit. The optomechanical interaction with an optical lattice in the fractionation chamber pushes one species into the upper stream of the laminar flow from chamber A, sorting and collecting them into chamber C. The picture on the right shows the optical sorting by size of drug delivery agents (microcapsules): the black tracks show the deflection of 2 μm-diameter protein microcapsules flowing at 20 μm/s from left to right across the optical landscape; a 4 μm-diameter microcapsule (white track) of the same material is nearly unaffected by the interaction with the optical lattice. Reprinted by permission from Macmillan Publishers Ltd: MacDonald et al., *Nature* **426**, 421–4, copyright 2003.

through the light field unaffected, depending on the optomechanical interaction with the optical lattice. In particular, they showed that some protein microcapsules can be separated by size and delivered with an efficiency of 96% even for concentrated solutions thanks to the 45° angular deflection obtained. Static optical sorting, i.e., without the need for fluid flow, has also been achieved using extended interference patterns close to a surface (Čižmár et al., 2006).

Arrays of holographic optical tweezers can be used to create periodically modulated optical potentials where particles can be directionally locked (Ladavac et al., 2004). Such ratchet potentials are able to produce continuous and reconfigurable fractionation at the mesoscale that depends on the particles' properties, such as size and refractive index, with part-per-thousand resolution (Xiao and Grier, 2010).

Recently, Volpe et al. (2014) have used speckle fields to create random optical potentials for particle sorting based on size and refractive index (see also the *speckle tweezers* discussed in Subsection 12.2.2). When extended static and time-varying speckle fields are employed in a microfluidic flow, the statistical interaction between particles and random optical potentials allows one to achieve optical sieving, guiding and sorting. Fig. 16.3 shows

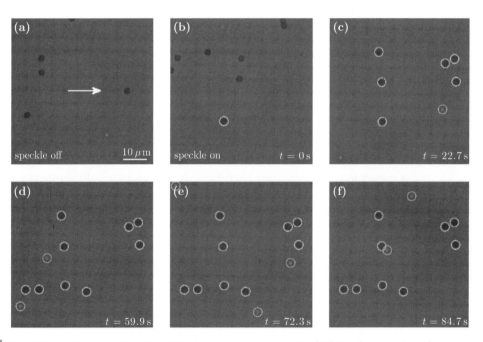

Figure 16.3 Microfluidic sorting in a speckle pattern. Optical sieving using a static speckle field. Time-lapse snapshots of the flow (flow speed 3.01 ± 0.12 μm/s in the direction of the arrow) of two types of particles with similar radii, $a \approx 1$ μm, but different refractive indices, in a speckle field: the brighter particles are made of silica, $n_p = 1.42$, whereas the darker ones are made of melamine, $n_p = 1.68$. The speckle landscape traps the particles with higher refractive index while letting the ones with lower refractive index flow away. Reprinted from Volpe et al., *Opt. Express* **22**, 18159–66. Copyright (2014) The Optical Society.

images demonstrating optical sieving by static speckle fields of particles with similar size (about 1 μm radius) but different refractive index: the particles with higher refractive index ($n_p = 1.68$) are trapped, while the ones with lower refractive index ($n_p = 1.42$) flow through the speckle potentials almost unperturbed.

Finally, as we will see in more detail in Chapter 22, the exploitation of plasmon-enhanced fields in metal nanostructures and platforms has recently paved the way for the realisation of plasmonic tweezers, frequency-dependent plasmonic sorting (Ploschner et al., 2012) and plasmonic optofluidics (Escobedo et al., 2012; Patra et al., 2014).

16.2 Monolithic integration

The lab-on-a-chip concept aims at the miniaturisation and integration on a single substrate of several functionalities for chemical, physical and biological analysis (Harrison et al., 1993). Microfluidic channels are used to mix, sort, analyse and deliver small amount of fluids (from the microlitre down to the femtolitre range), where particles and cells might also be suspended. A key element of the lab-on-a-chip concept is the monolithic integration of the different optical elements needed for trapping, manipulation and characterisation of different samples. In fact, full integration of microdevices will help in the construction of automated, cost-effective, compact and portable microfluidic systems. In this context, femtosecond laser microstructuring (Osellame et al., 2011) permits the monolithic integration in glass of both microchannels for microfluidics and optical waveguides for delivering and collecting light for spectroscopic detection. Femtosecond laser microstructuring is a direct maskless fabrication technique that enables the creation of photonic elements, e.g., waveguides, splitters or interferometers, by simply moving the sample glass substrate with respect to the femtosecond laser focus. It enables a three-dimensional architecture to be built at arbitrary depths inside the glass, allowing more compact geometries and greater freedom in lab-on-a-chip design.

An example of the potential of femtosecond microstructuring design is the monolithic optical stretcher shown in Figs. 16.4a and 16.4b. The cells flowing in the microchannel are trapped and stretched by the counter-propagating light delivered by optical fibres through the microfabricated waveguides (see also the *optical stretcher* discussed in Subsection 12.3.1). Both microchannel and waveguides are fabricated using femtosecond laser micromachining. For the microchannel, laser irradiation of the fused silica structure is followed by wet chemical etching. The practical realisation shown in Fig. 16.4b demonstrates the compact and robust design, which has been experimentally tested by optically trapping and stretching both red and white blood cells (Bellini et al., 2010, 2012).

Another example of integration has been demonstrated by Bragheri et al. (2012). They used femtosecond laser fabrication to build a fully integrated fluorescence-activated cell sorter with single-cell sensitivity. Two optical waveguides are built crossing a microfluidic splitter: the first is used to excite and collect fluorescence from cells in the fluid flow, while

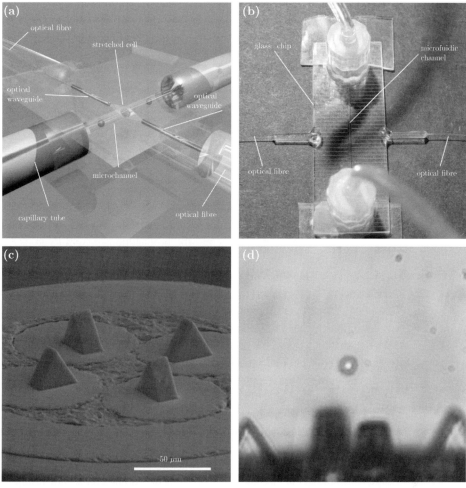

Figure 16.4 Fibre tweezers integrated into microfluidic devices. (a) Rendering of a monolithic optical stretcher fabricated by femtosecond laser micromachining: the cells flow in the capillary tubes of the microfluidic circuit and are trapped and stretched in the spot of the dual-beam fibre trap created by the optical waveguides. Reprinted from Bellini et al., *Opt. Express* **18**, 4679–88. Copyright (2010) The Optical Society. (b) Picture showing a monolithic optical stretcher. Reprinted from Bellini et al., *Biomed. Opt. Express* **3**, 2658–68. Copyright (2012) The Optical Society. (c) SEM image of the end face of a fibre-tweezers. The prisms completely cover the fibres' core and are used to reflect light to the trapping spot on the axis of the bundle of fibres. (d) A red blood cell is trapped in hypotonic solution by a fibre-tweezers. Reprinted by permission from Macmillan Publishers Ltd: Liberale et al., *Sci. Rep.* **3**, 1258, copyright 2013.

the second, which is displaced along the microchannel, delivers a light beam that is used to push selected cells away from the main flow into a separated collecting chamber.

A different fully integrated system was obtained by sculpting elements directly on optical fibres (Liberale et al., 2007). Liberale et al. (2013) further demonstrated miniaturised fibre-based optical tweezers by fabricating micro-prism beam deflectors using two-photon lithography at the tip of bundled fibres, as shown in Fig. 16.4c. The light guided in the fibres is reflected by the micro-prisms towards the bundle axis, creating a trapping spot, where cells can be trapped and analysed Fig. 16.4d. The fibre-bundle microtweezers has been used to trap red blood cells and collect fluorescence and Raman spectra of cancer cells towards the goal of a *lab-on-a-fibre* concept that opens perspectives for diagnosis and screening purposes *in vivo*.

16.3 Photonic crystal cavities

Photonic crystals are periodic structures made of materials with different refractive indices that are capable of strongly affecting the propagation of light (Sakoda, 2005). They are properly designed in one, two, and three dimensions so that only specific wavelength bands can propagate. In particular, the design and fabrication of photonic crystal slab structures, shown in Fig. 16.5, can enable very efficient spatial confinement of electromagnetic fields by combining total internal reflection and multiple scattering. This cavity-like resonant near-field confinement can be exploited for optical trapping with photonic crystals (Rahmani and Chaumet, 2006). As in the case of plasmonic tweezers, which we will discuss in detail in Chapter 22, this photonic arrangement yields feedback optical forces that are based on the interplay between the trapped particle and the trapping resonant field (Barth and Benson, 2006) – an example of self-induced back action [Section 12.5]. Photonic crystal structures are well suited for integration into optofluidic and lab-on-a-chip applications because of their increased stability in providing near-field light for trapping, manipulation and sensing.

An example of a photonic crystal structure fabricated on a silicon-on-insulator wafer is shown in Fig. 16.5a. Descharmes et al. (2013) used this hollow two-dimensional photonic crystal design to demonstrate single-particle trapping, manipulation and characterisation. The photonic platform is embedded in an optofluidic chip and it is composed of two distinct resonant cavities with resonances separated by 10 nm, as shown in Fig. 16.5b. Thus, dielectric nanoparticles in the fluid flow are selectively trapped in either cavity depending on the cavity excitation wavelength [Figs. 16.5c and 16.5d] with just a fraction of a milliwatt of power.

Photonic crystal patterns can also be used for the controlled assembly of microparticles (Renaut et al., 2012), whereas extended photonic crystal slabs can be arranged to yield precise templated self-assembled colloidal patterns controlled resonantly by light (Jaquay et al., 2013). The enhanced field spatial confinement of photonic crystal resonators allows trapping, manipulation and sensing at the nanoscale (Mandal et al., 2010), e.g., enabling the successful handling of individual quantum dots and Wilson disease proteins (Chen et al., 2012). Because this enhanced performance is obtained with silicon structures, it entails negligible heating effects when compared to plasmonic tweezers.

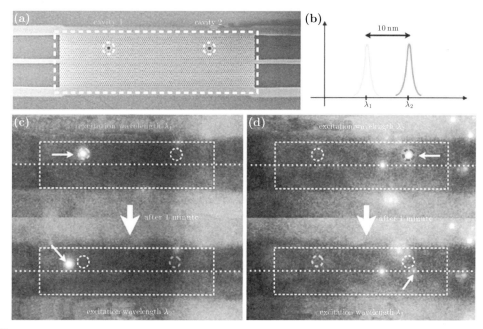

Figure 16.5 Selective optical trapping with a photonic crystal cavity. (a) SEM image of a hollow photonic crystal structure with two distinct nanocavities resonant at two different wavelengths, $\lambda_1 = 1568$ nm and $\lambda_2 = 1578$ nm, as schematically illustrated in (b). (c) Demonstration of selective particle trapping when cavity 1 is excited and of particle release from the trap when excitation is switched to cavity 2. (d) Trapping when cavity 2 is excited and release when switching to cavity 1. Reproduced by permission from Descharmes et al., *Lab Chip* **13**, 3268–74. Copyright (2013) The Royal Society of Chemistry.

16.4 Micromachines

The development of non-contact techniques to enable controlled manipulation, orientation and rotation of micro- and nanoparticles is of crucial importance for their incorporation as active elements in next generation micro- and nanomachines (Palima and Glückstad, 2013). These require a range of components for the conversion of energy into motion. In particular, devices capable of continuous rotational motion are of critical importance for the operation of micromachines in applications such as microscale fluid pumping in microfluidic chips (Maruo and Inoue, 2006). Unidirectional rotational motion on the microscale has been demonstrated in systems such as catalytic microrotors (Wang et al., 2009) and molecular rotors (Koumura et al., 1999). However, the use of optical trapping methods opens up the possibility of assembling and driving nanostructured components while simultaneously monitoring their operation and performance. The need to produce controlled rotations in addition to translations has led to a number of different techniques for rotating microparticles. These include methods that rely on properties such as particle

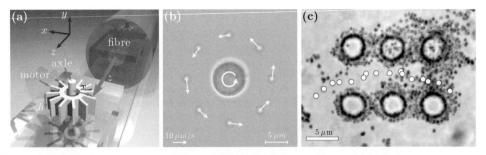

Figure 16.6 Light-driven micromachines. (a) Rendering of an optical microrotor with the integration of an optical fibre to supply the driving optical force. Reproduced with permission from Metzger et al., *J. Opt.* **13**, 044018. Copyright (2011) IOP Publishing. Reproduced by permission of IOP Publishing. All rights reserved. (b) Optical microrotor based on a vaterite particle (4 μm radius) set into rotation by transfer of spin angular momentum. The flow field (arrows) around the rotor is measured by multipoint holographic optical velocimetry. Reprinted figure from Di Leonardo et al., *Phys. Rev. Lett.* **96**, 134502. Copyright (2006) by the American Physical Society. (c) Particles (800 nm diameter) trapped and rotated in holographic optical vortices. The resulting device operates as an optomechanical micropump. The white circles identify the trajectory of a single sphere as it moves through the pump 25 μm to the left in 7 s. Reprinted from Ladavac and Grier, *Opt. Express* **12**, 1144–9. Copyright (2004) The Optical Society.

birefringence and consequent transfer of spin angular momentum from circularly polarised beams (Bishop et al., 2004), form birefringence (Neale et al., 2005), anisotropic scattering (*windmill effect*) from the particle shape (Galajda and Ormos, 2001; Jones et al., 2009) and transfer of optical orbital angular momentum from Laguerre–Gaussion beams (Asavei et al., 2009; Padgett and Bowman, 2011). The other crucial aspect of micromachines is microfabrication. Laser fabrication (Juodkazis et al., 2008) can realise controlled designs of microstructures with sub-diffraction resolution (Palima et al., 2012); for example, two-photon polymerisation can be used to fabricate complex-shaped microscopic objects that can be trapped, rotated and propelled by laser light. An example of system integration is shown in Fig. 16.6a, where light is delivered through an optical fibre to actuate a laser-sculpted microrotor (Metzger et al., 2011).

Another key application in microfluidics is the demonstration of micropumps and mixers. A successful implementation, shown in Fig. 16.6b, exploited birefringent vaterite microparticles (Bishop et al., 2004), which can spin, creating a rotational fluid flow in their proximity. Di Leonardo et al. (2006) used holographic optical velocimetry to measure the flow field around the rotating vaterite particle and realise an all-optical controlled microfluidic pump (Leach et al., 2006), whereas Volpe et al. (2008) extended this technique to the characterisation of singular points in fluid flows. Alternatively, Fig. 16.6c shows the demonstration of an optomechanical micropump based on particles trapped and rotated in holographic optical vortices (Ladavac and Grier, 2004); the fluid drag by the trapped particles creates a stream flow that guides a larger particle within the pump region.

Holographic optical trapping and manipulation can also be used effectively in lab-on-a-chip devices (Padgett and Di Leonardo, 2011). Interestingly, it is also possible to exploit low-numerical-aperture lenses in a counter-propagating configuration (Rodrigo et al., 2005).

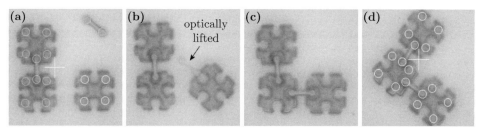

Figure 16.7 Microassembly of reconfigurable microenvironments. Optical assembly of two-photon polymerised microblock components that can be used as reconfigurable microenvironments. (a) A cluster of four traps is selected to rotate a microblock in two dimensions. (b) Three-dimensional optical manipulation of dumb-bell microstructures with a pair of traps fitting a spherical lobe of a dumb-bell to a hole of a complementary microblock. (c) Successful assembly of all the microcomponents. (d) Grouping of all traps to collectively manipulate the entire assembly. Reprinted from Rodrigo et al., *Opt. Express* **17**, 6578–83. Copyright (2009) The Optical Society.

In fact, holographic micromanipulation has developed towards the concept of microrobotics (Palima and Glückstad, 2013). Robotic microgrippers can be realised with optical forces as the actuation mechanism (Gibson et al., 2007), whereas amplification of optical forces at the microscale can be obtained using microlevers (Lin et al., 2011). As shown in Fig. 16.7, optical assembly of microblocks constructed by two-photon polymerisation is obtained with great precision and can serve as a reconfigurable micro-environment for chemical and biological assays (Rodrigo et al., 2009).

16.5 Further reading

Optofluidics in liquid crystals has also been at the centre of much research work (Gleeson et al., 2006). This includes the manipulation of colloids and defects in liquid crystals (Trivedi et al., 2011), as well as the controlled rotation of nematic (Juodkazis et al., 2003) or chiral (Yang et al., 2008) liquid crystal droplets. Zhang et al. (2004) demonstrated microfluidic optophoresis based on the reduction of cell velocity by an optical tweezers aligned across a microfluidic channel, which was able to identify cancerous and non-cancerous cells by their different times of flight. Photophoretic frequency-dependent chiral sorting of single-walled carbon nanotubes has been theoretically explored by Smith et al. (2014). Optical conveyors (Ruffner and Grier, 2012), i.e., active tractor beams that can selectively move objects, have been shown by Ruffner and Grier (2014) to be able to transport dielectric latex nanoparticles through water for about 60 μm. A long-range polarisation-dependent tractor beam has also been realised by Shvedov et al. (2014) to trap and manipulate gold-coated glass spheres over tens of centimetres. Finally, anti-resonant reflecting liquid-core optical waveguides have been used by Kühn et al. (2009) as an optofluidic trap for microparticles and bacteria.

References

Asavei, T., Loke, V. L. Y., Barbieri, M., et al. 2009. Optical angular momentum transfer to microrotors fabricated by two-photon photopolymerization. *New J. Phys.*, **11**, 093021.

Barth, M., and Benson, O. 2006. Manipulation of dielectric particles using photonic crystal cavities. *Appl. Phys. Lett.*, **89**, 253114.

Bellini, N., Vishnubhatla, K. C., Bragheri, F., et al. 2010. Femtosecond laser fabricated monolithic chip for optical trapping and stretching of single cells. *Opt. Express*, **18**, 4679–88.

Bellini, N., Bragheri, F., Cristiani, I., et al. 2012. Validation and perspectives of a femtosecond laser fabricated monolithic optical stretcher. *Biomed. Opt. Express*, **3**, 2658–68.

Bishop, A. I., Nieminen, T. A., Heckenberg, N. R., and Rubinsztein-Dunlop, H. 2004. Optical microrheology using rotating laser-trapped particles. *Phys. Rev. Lett.*, **92**, 198104.

Bragheri, F., Minzioni, P., Vazquez, R. M., et al. 2012. Optofluidic integrated cell sorter fabricated by femtosecond lasers. *Lab Chip*, **12**, 3779–84.

Buican, T. N., Smyth, M. J., Crissman, H. A., et al. 1987. Automated single-cell manipulation and sorting by light trapping. *Appl. Opt.*, **26**, 5311–16.

Chen, Y.-F., Serey, X., Sarkar, R., Chen, P., and Erickson, D. 2012. Controlled photonic manipulation of proteins and other nanomaterials. *Nano Lett.*, **12**, 1633–7.

Cipparrone, G., Mazzulla, A., Pane, A., Hernandez, R. J., and Bartolino, R. 2011. Chiral self-assembled solid microspheres: A novel multifunctional microphotonic device. *Adv. Materials*, **23**, 5773–8.

Čižmár, T., Šiler, M., Šerý, M., et al. 2006. Optical sorting and detection of submicrometer objects in a motional standing wave. *Phys. Rev. B*, **74**, 035105.

Descharmes, N., Dharanipathy, U. P., Diao, Z., Tonin, M., and Houdré, R. 2013. Single particle detection, manipulation and analysis with resonant optical trapping in photonic crystals. *Lab Chip*, **13**, 3268–74.

Di Leonardo, R., Leach, J., Mushfique, H., et al. 2006. Multipoint holographic optical velocimetry in microfluidic systems. *Phys. Rev. Lett.*, **96**, 134502.

Donato, M. G., Hernandez, J., Mazzulla, A., et al. 2014. Polarization-dependent optomechanics mediated by chiral microresonators. *Nature Commun.*, **5**, 3656.

Escobedo, C., Brolo, A. G., Gordon, R., and Sinton, D. 2012. Optofluidic concentration: Plasmonic nanostructure as concentrator and sensor. *Nano Lett.*, **12**, 1592–6.

Fan, X., and White, I. M. 2011. Optofluidic microsystems for chemical and biological analysis. *Nature Photon.*, **5**, 591–7.

Galajda, P., and Ormos, P. 2001. Complex micromachines produced and driven by light. *Appl. Phys. Lett.*, **78**, 249–51.

Gibson, G., Barron, L., Beck, F., Whyte, G., and Padgett, M. 2007. Optically controlled grippers for manipulating micron-sized particles. *New J. Phys.*, **9**, 14.

Gleeson, H. F., Wood, T. A., and Dickinson, M. 2006. Laser manipulation in liquid crystals: An approach to microfluidics and micromachines. *Phil. Trans. Royal Soc. A: Math. Phys. Eng. Sci.*, **364**, 2789–805.

Glückstad, J. 2004. Sorting particles with light. *Nature Mater.*, **3**, 9–10.

Harrison, D. J., Fluri, K., Seiler, Z. H., et al. 1993. Micromachining a miniaturized capillary electrophoresis-based chemical analysis system on a chip. *Science*, **261**, 895–7.

Hart, S. J., Terray, A., Leski, T. A., Arnold, J., and Stroud, R. 2006. Discovery of a significant optical chromatographic difference between spores of *Bacillus anthracis* and its close relative, *Bacillus thuringiensis. Anal. Chem.*, **78**, 3221–5.

Hart, S. J., Terray, A., Arnold, J., and Leski, T. A. 2007. Sample concentration using optical chromatography. *Opt. Express*, **15**, 2724–31.

Hernández, R. J., Mazzulla, A., Pane, A., Volke-Sepúlveda, K., and Cipparrone, G. 2013. Attractive-repulsive dynamics on light-responsive chiral microparticles induced by polarized tweezers. *Lab Chip*, **13**, 459–67.

Hunt, H. C., and Wilkinson, J. S. 2008. Optofluidic integration for microanalysis. *Microfluid. Nanofluid.*, **4**, 53–79.

Imasaka, T. 1998. Optical chromatography. A new tool for separation of particles. *Analusis*, **26**, 53.

Jaquay, E., Martínez, L. J., Mejia, C. A., and Povinelli, M. L. 2013. Light-assisted, templated self-assembly using a photonic-crystal slab. *Nano Lett.*, **13**, 2290–94.

Jones, P. H., Palmisano, F., Bonaccorso, F., et al. 2009. Rotation detection in light-driven nanorotors. *ACS Nano*, **3**, 3077–84.

Juodkazis, S., Matsuo, S., Murazawa, N., Hasegawa, I., and Misawa, H. 2003. High-efficiency optical transfer of torque to a nematic liquid crystal droplet. *Appl. Phys. Lett.*, **82**, 4657–9.

Juodkazis, S., Mizeikis, V., Matsuo, S., Ueno, K., and Misawa, H. 2008. Three-dimensional micro- and nano-structuring of materials by tightly focused laser radiation. *Bull. Chem. Soc. Japan*, **81**, 411–48.

Koumura, N., Zijlstra, R. W. J., van Delden, R. A., Harada, N., and Feringa, B. L. 1999. Light-driven monodirectional molecular rotor. *Nature*, **401**, 152–5.

Kühn, S., Measor, P., Lunt, E. J., et al. 2009. Loss-based optical trap for on-chip particle analysis. *Lab Chip*, **9**, 2212–16.

Ladavac, K., and Grier, D. 2004. Microoptomechanical pumps assembled and driven by holographic optical vortex arrays. *Opt. Express*, **12**, 1144–9.

Ladavac, K., Kasza, K., and Grier, D. G. 2004. Sorting mesoscopic objects with periodic potential landscapes: Optical fractionation. *Phys. Rev. E*, **70**, 010901.

Leach, J., Mushfique, H., di Leonardo, R., Padgett, M., and Cooper, J. 2006. An optically driven pump for microfluidics. *Lab Chip*, **6**, 735–9.

Liberale, C., Minzioni, P., Bragheri, F., et al. 2007. Miniaturized all-fibre probe for three-dimensional optical trapping and manipulation. *Nature Photon.*, **1**, 723–7.

Liberale, C., Cojoc, G., Bragheri, F., et al. 2013. Integrated microfluidic device for single-cell trapping and spectroscopy. *Sci. Rep.*, **3**, 1258.

Lin, C.-L., Lee, Y.-H., Lin, C.-T., et al. 2011. Multiplying optical tweezers force using a micro-lever. *Opt. Express*, **19**, 20 604–9.

MacDonald, M. P., Spalding, G. C., and Dholakia, K. 2003. Microfluidic sorting in an optical lattice. *Nature*, **426**, 421–4.

Mandal, S., Serey, X., and Erickson, D. 2010. Nanomanipulation using silicon photonic crystal resonators. *Nano Lett.*, **10**, 99–104.

Maruo, S., and Inoue, H. 2006. Optically driven micropump produced by three-dimensional two-photon microfabrication. *Appl. Phys. Lett.*, **89**, 144101.

Metzger, N. K., Mazilu, M., Kelemen, L., Ormos, P., and Dholakia, K. 2011. Observation and simulation of an optically driven micromotor. *J. Opt.*, **13**, 044018.

Neale, S. L., MacDonald, M. P., Dholakia, K., and Krauss, T. F. 2005. All-optical control of microfluidic components using form birefringence. *Nature Mater.*, **4**, 530–33.

Osellame, R., Hoekstra, H. J. W. M., Cerullo, G., and Pollnau, M. 2011. Femtosecond laser microstructuring: An enabling tool for optofluidic lab-on-chips. *Laser Photon. Rev.*, **5**, 442–63.

Padgett, M., and Bowman, R. 2011. Tweezers with a twist. *Nature Photon.*, **5**, 343–8.

Padgett, M., and Di Leonardo, R. 2011. Holographic optical tweezers and their relevance to lab on chip devices. *Lab Chip*, **11**, 1196–205.

Palima, D., and Glückstad, J. 2013. Gearing up for optical microrobotics: Micromanipulation and actuation of synthetic microstructures by optical forces. *Laser Photon. Rev.*, **7**, 478–94.

Palima, D., Bañas, A. R., Vizsnyiczai, G., et al. 2012. Wave-guided optical waveguides. *Opt. Express*, **20**, 2004–14.

Patra, P. P., Chikkaraddy, R., Tripathi, R. P. N., Dasgupta, A., and Kumar, G. V. P. 2014. Plasmofluidic single-molecule surface-enhanced Raman scattering from dynamic assembly of plasmonic nanoparticles. *Nature Commun.*, **5**, 4357.

Ploschner, M., Cizmar, T., Mazilu, M., Di Falco, A., and Dholakia, K. 2012. Bidirectional optical sorting of gold nanoparticles. *Nano Lett.*, **12**, 1923–7.

Rahmani, A., and Chaumet, P. C. 2006. Optical trapping near a photonic crystal. *Opt. Express*, **14**, 6353–8.

Renaut, C., Dellinger, J., Cluzel, B., et al. 2012. Assembly of microparticles by optical trapping with a photonic crystal nanocavity. *Appl. Phys. Lett.*, **100**, 101103.

Rodrigo, P. J., Daria, V. R., and Glückstad, J. 2005. Four-dimensional optical manipulation of colloidal particles. *Appl. Phys. Lett.*, **86**, 074103.

Rodrigo, P. J., Kelemen, L., Palima, D., et al. 2009. Optical microassembly platform for constructing reconfigurable microenvironments for biomedical studies. *Opt. Express*, **17**, 6578–83.

Ruffner, D. B., and Grier, D. G. 2012. Optical conveyors: A class of active tractor beams. *Phys. Rev. Lett.*, **109**, 163903.

Ruffner, D. B., and Grier, D. G. 2014. Universal, strong and long-ranged trapping by optical conveyors. *Opt. Express*, **22**, 26 834–43.

Sakoda, K. 2005. *Optical properties of photonic crystals*, Vol. 80. Heidelberg, Germany: Springer Verlag.

Shvedov, V., Davoyan, A. R., Hnatovsky, C., Engheta, N., and Krolikowski, W. 2014. A long-range polarization-controlled optical tractor beam. *Nature Photon.*, **8**, 846–50.

Smith, D., Woods, C. J., Seddon, A., and Hoerber, H. 2014. Photophoretic separation of single-walled carbon nanotubes: A novel approach to selective chiral sorting. *Phys. Chem. Chem. Phys.*, **16**, 5221–8.

Terray, A., Arnold, J., and Hart, S. J. 2005. Enhanced optical chromatography in a PDMS microfluidic system. *Opt. Express*, **13**, 10 406–15.

Tkachenko, G., and Brasselet, E. 2013. Spin controlled optical radiation pressure. *Phys. Rev. Lett.*, **111**, 033605.

Tkachenko, G., and Brasselet, E. 2014. Optofluidic sorting of material chirality by chiral light. *Nature Commun.*, **5**, 3577.

Trivedi, R. P., Engström, D., and Smalyukh, I. I. 2011. Optical manipulation of colloids and defect structures in anisotropic liquid crystal fluids. *J. Opt.*, **13**, 044001.

Volpe, G., Volpe, G., and Petrov, D. 2008. Singular-point characterization in microscopic flows. *Phys. Rev. E*, **77**, 037301.

Volpe, G., Kurz, L., Callegari, A., Volpe, G., and Gigan, S. 2014. Speckle optical tweezers: Micromanipulation with random light fields. *Opt. Express*, **22**, 18 159–66.

Wang, Y., Fei, S., Byun, Y.-M., et al. 2009. Dynamic interactions between fast microscale rotors. *J. Am. Chem. Soc.*, **131**, 9926–7.

Xiao, K., and Grier, D. G. 2010. Multidimensional optical fractionation of colloidal particles with holographic verification. *Phys. Rev. Lett.*, **104**, 028302.

Yang, Y., Brimicombe, P. D., Roberts, N. W., et al. 2008. Continuously rotating chiral liquid crystal droplets in a linearly polarized laser trap. *Opt. Express*, **16**, 6877–82.

Zhang, H., Tu, E., Hagen, N. D., et al. 2004. Time-of-flight optophoresis analysis of live whole cells in microfluidic channels. *Biomed. Microdev.*, **6**, 11–21.

17 Colloid science

A colloid is a mesoscopic particle, whose size typically ranges from about one nanometre to several micrometres. Often colloidal particles are dispersed in a host medium, typically a liquid or a gas. Frenkel (2002) usefully defines a colloid by its behaviour, as the dynamics of colloids is significantly affected by their Brownian motion, which we have studied in Chapter 7. This definition encompasses nearly all of the objects to which the methods of optical tweezers have been applied, going from aerosols [Chapter 19] to nanoparticles [Chapter 23]. The interactions between particles in a colloidal suspension depend on the suspending solvent, as in the case of *hydrodynamic interactions*, *electrostatic forces* and *depletion forces*. These interactions can give rise to complex behaviours, as in the case of the hydrodynamic interactions between particles shown in Fig. 17.1. Optical tweezers provide an ideal tool for probing these colloidal interactions. In this chapter, we review experiments that use optical traps to measure and exploit the coupling between colloidal particles in suspension, taking as examples experiments aimed at determining the hydrodynamic, electrostatic and depletion interactions.

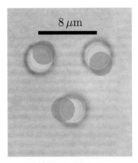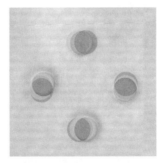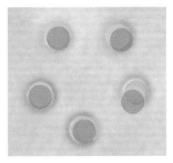

Figure 17.1 Hydrodynamic synchronisation of colloids. Regular arrays of driven colloidal particles synchronised into steady collective dynamical states by their hydrodynamic coupling interactions. The shaded circles show the particles' positions at the initial, 20th and 40th frames of a video recording. The particles in the three-particle system perform in-phase oscillations, the ones in the four-particle system perform oscillations in anti-phase with their neighbours, and the particles in the five-particle system feature phase locking between neighbours. The frame rate is between 200 and 300 frames per second, depending on the case. Reproduced by permission from Damet et al., *Soft Matter* **8**, 8672–8. Copyright (2012) The Royal Society of Chemistry.

17.1 Hydrodynamic interactions

In the seminal experiment of Meiners and Quake (1999), two colloidal spheres were held in independent optical tweezers and their thermal fluctuations tracked. In such a situation, the Langevin equations that describe the motion of the two particles are

$$\frac{d\mathbf{r}_n}{dt} = \sum_{m=1,2} \mathbb{H}_{nm}(\mathbf{r}_n - \mathbf{r}_m)[-\kappa \mathbf{r}_m + \mathbf{f}_m(t))], \tag{17.1}$$

where $\mathbf{r}_n$ with $n = 1, 2$ are the positions of the two particles, κ is the trap stiffness (assumed to be equal in all directions and for both traps), $\mathbf{f}_m(t)$ are randomly fluctuating forces such that $\langle \mathbf{f}_m(t) \rangle = 0$ and $\langle \mathbf{f}_n(t)\mathbf{f}_m(t+\tau) \rangle = 2\mathbb{H}_{nm}^{-1} k_B T \delta(\tau)$ [Subsection 7.2.2] and $\mathbb{H}_{nm}(\mathbf{r})$ is the hydrodynamic mobility tensor given by the *Oseen tensor*. The Oseen tensor represents the leading term in the multipole expansion of the Stokes flow produced by a given distribution of force and is given by

$$\begin{cases} \mathbb{H}_{nn}(\mathbf{r}) = \dfrac{\mathbb{I}}{\gamma} \\ \mathbb{H}_{nm}(\mathbf{r}) = \dfrac{3a}{4r}\dfrac{1}{\gamma}\left[\mathbb{I} + \dfrac{\mathbf{r} \otimes \mathbf{r}}{r^2}\right], \end{cases} \tag{17.2}$$

where $\gamma = 6\pi \eta a$ is the viscous drag coefficient for a sphere of radius a [Eq. (7.8)], $\mathbb{I}$ is the unit tensor and the symbol $\otimes$ represents a tensor (dyadic) product [Box 5.1]. Meiners and Quake (1999) showed that the position fluctuations exhibited a pronounced anti-correlation at short times, as shown in Fig. 17.2a. This behaviour is straightforwardly explained by considering the normal modes of the system (Metzger et al., 2007), which for two particles are those describing the centre-of-mass motion, $\mathbf{r}_+ = (\mathbf{r}_1 + \mathbf{r}_2)/2$, and the relative motion, $\mathbf{r}_- = (\mathbf{r}_1 - \mathbf{r}_2)$. In this coordinate system, the correlations of the modes decay with different time constants, with that corresponding to centre-of-mass motion decaying more quickly. Physically, this arises because the movement of one particle perturbs the surrounding fluid, which then transports the second particle along the same direction; i.e., the second particle follows the wake of the first. Fluctuations in the centre-of-mass coordinate are thus damped more quickly than those in the relative motion, where fluid between the particles must be displaced.

Hydrodynamic coupling between optically trapped colloidal particles gives rise to a number of interesting emergent phenomena in out-of-equilibrium systems. Sokolov et al. (2011) observed spontaneous particle pairing in an experimental system consisting of two colloidal particles driven around a ring-shaped optical potential created by an optical vortex beam. The signature of the pairing interaction is a strongly peaked histogram of angular separations, $\Delta\theta$, between the particles. The mechanism behind such pairing arises from the hydrodynamic interaction between the particles, which are both driven around the ring by the azimuthal component of the scattering force of the optical vortex beam. The symmetry of the hydrodynamic coupling is broken by the curvature of the ring: the force exerted by one particle on the other has a component in the tangential direction, which is symmetric under particle exchange, and also one in the radial direction, which is antisymmetric. This

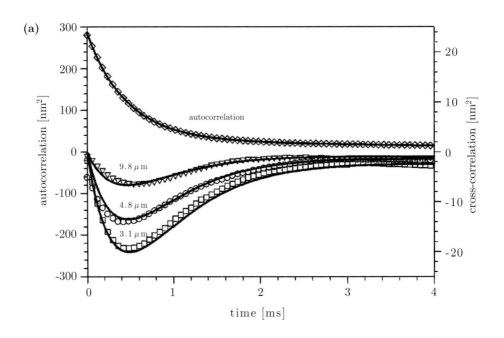
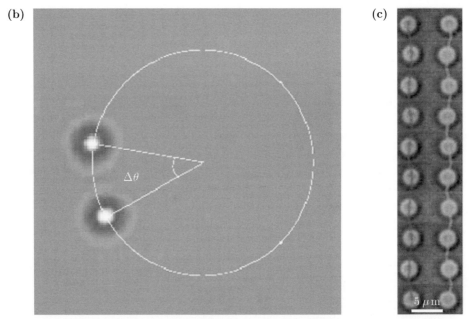

Figure 17.2 Hydrodynamic interactions between trapped colloidal particles. (a) Cross-correlation in the position fluctuations of two trapped colloidal particles for three different particle separations. The lines through the data points are the expected form of the cross-correlation using a theory based on the hydrodynamic coupling described in the text. Reprinted figure from Meiners and Quake, *Phys. Rev. Lett.* **82**, 2211–14. Copyright (1999) by the American Physical Society. (b) Pairing of particles driven around the circumference of a ring-shaped vortex beam trap. Reprinted figure from Sokolov et al., *Phys. Rev. Lett.* **107**, 158302. Copyright (2011) by the American Physical Society. (c) Hydrodynamic synchronisation of a pair of optically trapped oscillators. Reprinted from Kotar et al., *Proc. Natl. Acad. Sci. U.S.A.* **107**, 7669–73. Copyright (2010) National Academy of Sciences, U.S.A.

has the effect of shifting the leading (following) particle's motion to a larger (smaller) average radius; thus, for constant tangential force, the trailing particle's angular velocity is increased and the particles form a bound pair, as shown in Fig. 17.2b.

Kotar et al. (2010) observed the spontaneous synchronisation of a pair of driven colloidal oscillators arising as a result of their hydrodynamic interaction. In their experiment, two colloidal spheres in separate optical traps were driven by being subject to a 'geometric switch' protocol: the position of each trap was switched between two locations along the axis joining the particles, with the timing of the switch triggered by the particle moving to within a small pre-defined distance of the current equilibrium position. The hydrodynamic coupling between the particles caused them to synchronise and oscillate in antiphase, as shown in Fig. 17.2c, although sychronisation was lost by decreasing the coupling (increasing the separation) between the particles or by making the trap frequencies different (detuning the oscillators). Later work by Bruot et al. (2012) showed that the synchronisation state depended crucially on the form of the potential and the amount of noise (thermal fluctuations in force). More complicated structures subject to a similar geometric switch protocol show a rich variety of emergent behaviours in their collective dynamics. Damet et al. (2012) trapped a number of particles, N, around the circumference of a ring on the vertices of a regular polygon and subjected them to a drive tangential to the ring, as shown in Fig. 17.1. For $N = 3$ particles, the dominant mode was synchronised in-phase oscillation of all three particles. For $N = 4$ particles, there was again a single dominant mode, but in this case adjacent particles oscillated in anti-phase and this behaviour was observed also for higher even values of N. For $N = 5$ and higher odd values of N, a different behaviour emerged: nearest neighbour particles oscillated with a fixed phase delay, which could be either positive or negative, thus describing a wave propagating around the ring of particles in either a clockwise or anti-clockwise direction.

17.2 Electrostatic interactions

In the experiments of Section 17.1, electrostatic interactions between colloidal particles were assumed to be negligible. However, generally, when colloidal spheres are dispersed in water, ionic groups bonded to their surfaces dissociate and give rise to a screened electrostatic interaction. The interaction between such charged colloidal spheres was formulated in the 1940s and is known as the Derjaguin–Landau–Verwey–Overbeek theory or, more briefly, as the *DLVO theory* (Derjaguin and Landau, 1941; Verwey, 1947). If the separations between spheres are large enough so that the Debye–Hückel linearisation approximation can be made for the electrostatic potential of a system of ions in an electrolyte, then the DLVO theory gives the potential between colloidal spheres of radius a in an electrolyte of permittivity ε_m as

$$U_{\text{DLVO}}(r) = \frac{Z^2 e_-^2}{\varepsilon_m} \left[\frac{e^{a/l_D}}{1 + a/l_D} \right]^2 \frac{e^{-r/l_D}}{r}, \qquad (17.3)$$

where r is the centre-to-centre inter-particle distance, Z is the sphere's effective surface charge (which can be considerably smaller than the fully dissociated surface charge), e_- is the electron elementary charge and l_D is the Debye–Hückel radius. This last quantity characterises the distance at which the charge on the sphere is screened by counter-ions in solution. Crocker and Grier (1994) were able to make a quantitative measurement of the electrostatic interaction potential for a pair of sub-micrometre colloidal polystyrene sulphate spheres. The method involved positioning two spheres at a fixed separation in a pair of optical traps using the experimental set-up illustrated in Fig. 17.3a, releasing them and tracking their subsequent motion (blinking optical tweezers). By finding an intensity-weighted centroid of the particle image (digital video microscopy), an uncertainty in the position of the spheres as little as 50 nm was achieved. Because the probability of finding the particles with separation x is related to the interaction potential through the Boltzmann distribution, $\rho(x) = \exp(-U(x)/k_B T)$ [Section 9.4], the analysis of several thousand pair trajectories permitted the reconstruction of the potential as a function of inter-particle separation, as shown in Fig. 17.3b. The results showed excellent quantitative agreement with the DLVO theory of Eq. (17.3), for fitted parameters of effective charge $Z = (1991 \pm 150)e^-$ and a Debye–Hückel screening length of $l_D = 161 \pm 10$ nm.

Although excellent agreement with DVLO theory was recovered for large particle separations, several contemporaneous experiments seemed to show an attractive potential at small separations, particularly in thin sample cells. Although several explanations for this curious 'like-charge attraction' were suggested, including electrostatic interactions with the charges carried by the walls and the resulting interaction between the counter-ion clouds of the spheres and the walls, the problem was eventually resolved by detailed consideration of imaging artefacts that occur when digital video microscopy is performed on particles that are separated by less than twice their diameter (Baumgartl and Bechinger, 2005). At such small separations, an overlap in the images of the two particles causes a systematic deviation in the measured position compared to the true position: at very small separations, i.e., less than 1.15 diameters, the particles are actually closer than they appear, whereas for separations in the range from 1.5 to 1.8 diameters they are further apart than the locations of the intensity weighted centroids would suggest. After correcting for this imaging artefact, Baumgartl et al. (2006) demonstrated that this apparent like-charge attraction disappeared and the interaction was described by a screened Coulomb potential at all separations, as shown in Fig. 17.3c.

17.3 Depletion interactions

Unlike the electrostatic interaction discussed in Section 17.2, where the solvent screens the Coulomb interaction between charged colloidal particles, the depletion interaction is purely a solvent effect. In the Asakura–Oosawa theory of the depletion interaction, the surrounding solvent molecules are excluded from a thin shell surrounding a colloidal particle. This *depletion zone* is represented by the grey line surrounding the large spheres in Fig. 17.4a, inside which the centres of the surrounding small spheres cannot penetrate.

17.3 Depletion interactions

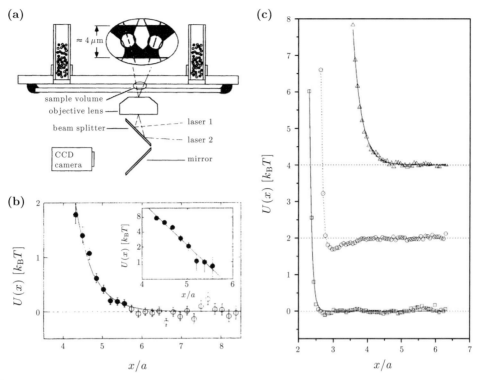

Figure 17.3 Electrostatic interactions between trapped colloidal particles. (a) Dual blinking optical tweezers experimental set-up for measuring the electrostatic interaction between two colloidal particles. (b) Measured interaction potential for a pair of spheres with radius $a = 0.326$ μm as a function of their separation x. For larger separations, the interaction potential is well fitted by the potential predicted by the DVLO theory with fitted parameters $Z^* = 1991\ e^-$ and $l_D = 161$ nm. The inset is a semilogarithmic plot whose gradient is used to determine the screening length. Figures (a) and (b) are reprinted from Crocker and Grier, *Phys. Rev. Lett.* **73**, 352–5. Copyright (1994) by the American Physical Society. (c) Measurement of pairwise electrostatic interaction potentials showing the importance of correcting the particles' positions obtained from digital video microscopy at small particle separations. The upper trace shows that the interaction potential (vertically shifted by $4k_B T$) for large separations of the particles is fitted well by a screened repulsive Coulomb interaction. When the particles are allowed to sample small separations (less than approximately twice their diameter), an apparently attractive potential appears, as shown in the middle curve (vertically shifted by $2k_B T$). After correction for artefacts arising from imaging particles at small separations, the interaction potential is found to be repulsive at all separations and in agreement with the DLVO theory, as shown by the lower trace. Reproduced by permission from Baumgartl et al., *Soft Matter* **2**, 631–5. Copyright (2006) The Royal Society of Chemistry.

When the two large spheres are in close proximity, their depletion zones overlap, and thus the total volume from which the small spheres (or solvent molecules) are excluded decreases and the total volume available to the large spheres increases, thereby increasing their entropy. Therefore, the depletion interaction causes an entropy-driven attraction between the large spheres. The strength of this interaction was probed in experiments by Crocker et al. (1999) using a pair of colloidal polyethylmethacrylate (PMMA) spheres (diameter

$2a_L = 1100 \pm 15$ nm) confined in a line optical tweezers formed by rapidly scanning the focus of a laser beam and suspended in a solvent containing a volume fraction ϕ_S of smaller polystyrene spheres (diameter $2a_S = 83$ nm). Similarly to the experiments that measured the electrostatic interaction described in the previous section, the depletion interaction potential is measured by digital video microscopy [Section 9.1] and the particle trajectories are used to find the probability $\rho(r)$ of finding the particles at separation r. The probability is related to the system's Helmholtz free energy as found by the Asakura–Oosawa theory:

$$U_{AO}(r) = \frac{k_B T \tilde{\phi}_S}{(2\tilde{a}_S)^3} [2\tilde{a}_S + 2a_L - r]^2 \left[2\tilde{a}_S + 2a_L + \frac{r}{2}\right], \quad (17.4)$$

where the small sphere radius and volume fraction have been modified as $\tilde{a}_S = a_S + \delta a_S$ and $\tilde{\phi}_S = \phi_S(1 + \delta a_S/a_S)^3$ to account for the range of electrostatic interactions δa_S [Section 17.2].

The interaction energy between pairs of large spheres measured as a function of the small sphere volume fraction is shown in Fig. 17.4c. For $\phi_S = 0$, the weak attraction is ascribed to van der Waals interactions. At a small volume fraction (e.g., $\phi_S = 0.04$ and 0.07), the free energy is fitted by the Asakura–Oosawa theory, accounting for the electrostatic interaction over a distance $a_S = 7 \pm 3$ nm. Features not accounted for in the Asakura–Oosawa theory are the repulsive potential at separations $r \approx 1.2$ μm, which appears at moderate volume fraction ($\phi_S = 0.15, 0.21$ and 0.26), and the oscillatory nature of the potential at high volume fraction ($\phi_S = 0.34$ and 0.42). The explanation for these features lies in the behaviour of the small spheres in the region between the large spheres, where they tend to form a layered shell structure, as artistically shown in Fig. 17.4b.

Ohshima et al. (1997) were able to determine the depletion interaction between colloidal polystyrene spheres and a plane glass surface arising from the solvent containing polyethylene oxide (PEO), which, for the molecular weight used in the experiments, has a radius of gyration of $r_g = 0.0677$ μm. As in the case of the depletion interaction between two large colloidal spheres, there is an entropy-driven force between the polystyrene spheres and the planar surface because of the overlap of their respective depletion regions and the consequent change in excluded volume for the polymer molecules. A focused laser beam incident on a polystyrene sphere at the surface and directed normal to the surface was used to first trap the sphere in two dimensions; then, by a gradual increase in the laser power, the optical force on the sphere was increased until it was sufficient to overcome the depletion force attracting the particle to the surface. Only if the concentration of polymer was sufficiently high was there an appreciable interaction between sphere and surface, i.e., a free energy change exceeding $k_B T$. The force required to overcome the depletion interaction was found to be proportional to the polymer concentration only up to a critical value. Above this concentration, it was suggested that the polymer molecules become entangled, and thus the characteristic length scale is no longer the radius of gyration, as in the dilute solution, but the characteristic length scale of smaller units termed 'blobs', resulting in a decrease in the free energy gain from the depletion interaction.

The entropic force between a colloidal sphere suspended in a solution of small spheres and a planar surface was exploited by Dinsmore et al. (1996), who demonstrated that a passive structured surface could trap particles as a result of the depletion interaction and

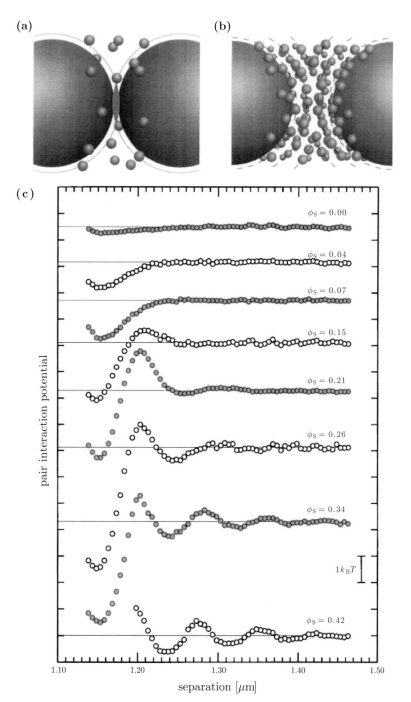

Figure 17.4 Depletion interactions between colloidal particles. (a) The large spheres are surrounded by a *depletion region* from which the smaller spheres are excluded, shown by the grey line. When the large spheres are close together, the total excluded volume decreases and, thus, the volume available to the small spheres increases. (b) With a higher volume fraction, the small spheres form a layered shell structure around the larger spheres. (c) Measured depletion interaction free energy as a function of small sphere volume fraction ϕ_S. Reprinted figures from Crocker et al., *Phys. Rev. Lett.* **82**, 4352–5. Copyright (1999) by the American Physical Society.

the free energy barrier to the sphere's motion created by a step. Optical tweezers were used to position a 0.460 µm-diameter sphere close to the edge of a glass coverslip. The suspension medium also contained a high volume fraction of small spheres (diameter 0.083 µm) and 0.01 M NaCl to screen the electrostatic interaction to a distance of approximately 5 nm. After the particle's release from the optical trap, its motion was used to reconstruct the potential in which it moved as a consequence of the change in Helmholtz free energy occurring across the step, demonstrating an energy barrier of approximately $2k_BT$.

17.4 Further reading

Optical tweezers have proved an invaluable tool for probing the hydrodynamic interactions between particles. A particularly interesting set of experiments reported in Di Leonardo et al. (2008, 2009) has probed the interaction in two-dimensional fluid films, finding that the reduced dimensionality leads to stronger longer-range coupling, decaying logarithmically with distance. Furthermore, the interaction between pairs of rod-shaped particles (Di Leonardo et al., 2011) was then observed to make a transition from behaving as the one between two-dimensional point particles for small separations to behaving as the one between three-dimensional point particles for large separations. Interactions between aerosol particles have also been quantified in a detailed study by Koehler et al. (2011), who compared blinking optical tweezers with a direct force measurement technique based on the displacement of the particle from its equilibrium position, finding good agreement between the two methods and DLVO theory for electrostatic interactions. The study of synchronised states of driven oscillators and rotors in various geometries has also been a highly productive subject for investigation in recent years (Cicuta et al., 2012; Di Leonardo et al., 2012; Lhermerout et al., 2012; Koumakis and Di Leonardo, 2013; Arzola et al., 2014). Similar partial synchronisation has also been observed in oscillators driven by thermal fluctuations by detecting coincidences in the hopping rate from one potential well to another for a pair of particles in closely spaced double-well optical potentials (Curran et al., 2012). Finally, an in-depth classical review of the application of optical tweezers methods to colloidal systems can be found in Grier (1997).

References

Arzola, A. V., Jákl, P., Chvátal, L., and Zemánek, P. 2014. Rotation, oscillation and hydrodynamic synchronization of optically trapped oblate spheroidal microparticles. *Opt. Express*, **22**, 16 207–21.

Baumgartl, J., and Bechinger, C. 2005. On the limits of digital video microscopy. *Europhys. Lett.*, **71**, 487–93.

Baumgartl, J., Arauz-Lara, J. L., and Bechinger, C. 2006. Like-charge attraction in confinement: Myth or truth? *Soft Matter*, **2**, 631–5.

Bruot, N., Kotar, J., de Lillo, F., Cosentino Lagomarsino, M., and Cicuta, P. 2012. Driving potential and noise level determine the synchronization state of hydrodynamically coupled oscillators. *Phys. Rev. Lett.*, **109**, 164103.

Cicuta, G. M., Onofri, E., Cosentino Lagomarsino, M., and Cicuta, P. 2012. Patterns of synchronization in the hydrodynamic coupling of active colloids. *Phys. Rev. E*, **85**, 016203.

Crocker, J. C., and Grier, D. G. 1994. Microscopic measurement of the pair interaction potential of charge-stabilized colloid. *Phys. Rev. Lett.*, **73**, 352–5.

Crocker, J. C., Matteo, J. A., Dinsmore, A. D., and Yodh, A. G. 1999. Entropic attraction and repulsion in binary colloids probed with a line optical tweezer. *Phys. Rev. Lett.*, **82**, 4352–5.

Curran, A., Lee, M. P., Padgett, M. J., Cooper, J. M., and Di Leonardo, R. 2012. Partial synchronization of stochastic oscillators through hydrodynamic coupling. *Phys. Rev. Lett.*, **108**, 240601.

Damet, L., Cicuta, G. M., Kotar, J., Cosentino Lagomarsino, M., and Cicuta, P. 2012. Hydrodynamically synchronized states in active colloidal arrays. *Soft Matter*, **8**, 8672–8.

Derjaguin, B., and Landau, L. 1941. Theory of the stability of strongly charged lyophobic sols and of the adhesion of strongly charged particles in solutions of electrolytes. *Acta Physicochim. U. R. S. S.*, **14**, 633–62.

Di Leonardo, R., Keen, S., Ianni, F., et al. 2008. Hydrodynamic interactions in two dimensions. *Phys. Rev. E*, **78**, 031406.

Di Leonardo, R., Ianni, F., Saglimbeni, F., et al. 2009. Optical trapping studies of colloidal interactions in liquid films. *Colloids Surf. A*, **343**, 133–6.

Di Leonardo, R., Cammarota, E., Bolognesi, G., Schäfer, H., and Steinhart, M. 2011. Three-dimensional to two-dimensional crossover in the hydrodynamic interactions between micron-scale rods. *Phys. Rev. Lett.*, **107**, 044501.

Di Leonardo, R., Búzás, A., Kelemen, L., et al. 2012. Hydrodynamic synchronization of light driven microrotors. *Phys. Rev. Lett.*, **109**, 034104.

Dinsmore, A. D., Yodh, A. G., and Pine, D. J. 1996. Entropic control of particle motion using passive surface microstructures. *Nature*, **383**, 239–42.

Frenkel, D. 2002. Soft condensed matter. *Physica A*, **313**, 1–31.

Grier, D. G. 1997. Optical tweezers in colloid and interface science. *Curr. Opin. Colloid Interface Sci.*, **2**, 264–70.

Koehler, T. P., Brotherton, C. M., and Grillet, A. M. 2011. Comparison of interparticle force measurement techniques using optical trapping. *Colloids Surf. A*, **384**, 282–8.

Kotar, J., Leoni, M., Bassetti, B., Cosentino Lagomarsino, M., and Cicuta, P. 2010. Hydrodynamic synchronization of colloidal oscillators. *Proc. Natl. Acad. Sci. U.S.A.*, **107**, 7669–73.

Koumakis, N., and Di Leonardo, R. 2013. Stochastic hydrodynamic synchronization in rotating energy landscapes. *Phys. Rev. Lett.*, **110**, 174103.

Lhermerout, R., Bruot, N., Cicuta, G. M., Kotar, J., and Cicuta, P. 2012. Collective synchronization states in arrays of driven colloidal oscillators. *New J. Phys.*, **14**, 105023.

Meiners, J.-C., and Quake, S. R. 1999. Direct measurement of hydrodynamic cross correlations between two particles in an external potential. *Phys. Rev. Lett.*, **82**, 2211–14.

Metzger, N. K., Marchington, R. F., Mazilu, M., et al. 2007. Measurement of the restoring forces acting on two optically bound particles from normal mode correlations. *Phys. Rev. Lett.*, **98**, 068102.

Ohshima, Y. N., Sakagami, H., Okumoto, K., et al. 1997. Direct measurement of infinitesimal depletion force in a colloid–polymer mixture by laser radiation pressure. *Phys. Rev. Lett.*, **78**, 3963–6.

Sokolov, Y., Frydel, D., Grier, D. G., Diamant, H., and Roichman, Y. 2011. Hydrodynamic pair attractions between driven colloidal particles. *Phys. Rev. Lett.*, **107**, 158302.

Verwey, E. J. W. 1947. Theory of the stability of lyophobic colloids. *J. Phys. Chem.*, **51**, 631–6.

18 Microchemistry

Optical tweezers can be used to manipulate droplets, vesicles and vesicle membranes, as shown in the example of a sucrose-filled giant unilamellar vesicle (GUV) in Fig. 18.1. In fact, just as with solid particles, it is possible to transport and release droplets and vesicles. Differently from solid particles, though, droplets and vesicles can be brought together and made to coalesce into larger droplets and vesicles in a process in which their contents get mixed. By doing so, it is possible to use droplets and vesicles as movable reaction vessels containing tiny amounts of desired reagents for combinatorial chemistry studies using ultrasmall (in the femtolitre range) volumes. These mechanical manipulations can then be coupled to spectroscopic probing techniques to measure and understand mixing and coalescence processes as they happen, in real time. In this chapter, we will review progress in experiments regarding the fusion of droplets and vesicles, the spectroscopic study of the resulting chemical mixing processes and also the action of optical trapping on the encapsulating vesicle membrane itself.

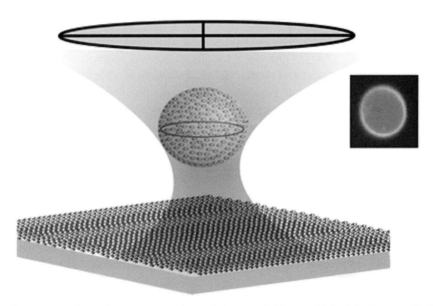

Figure 18.1   Optically trapped vesicle: artistic representation of an optical tweezers holding a vesicle loaded with sucrose. The inset shows a confocal microscopy image of an optically trapped giant unilamellar vesicle (GUV whose membrane is labeled with a red dye and whose lumen is labeled with a green dye. Reprinted with permission from Bendix and Oddershede, *Nano Lett.* **11**, 5431–7. Copyright (2011) American Chemical Society.

18.1 Liquid droplets

The measurement of reactant concentration in aqueous droplets opens the way towards using them as microchemical reaction vessels, where reactants can diffuse freely within a confined space, mimicking the environment in which biochemical processes occur in a cell. As we will see in Chapter 19, aqueous droplets can be optically trapped and manipulated as aerosols. Furthermore, optical tweezers can also be used to bring droplets together in order to coagulate them and mix their contents. This coagulation process is, in itself, an important process, as the size distribution of aerosols regulates such properties as their light-scattering behaviour and chemical activity.

Buajarern et al. (2006) used cavity-enhanced Raman scattering (CERS) to study the coagulation and mixing of aerosols. The CERS spectrum of an aqueous aerosol droplet contains a background spectrum arising from inelastic scattering and from the excitation of the OH stretching vibrational modes of water with superimposed discrete lines arising from stimulated Raman scattering at the wavelengths of the droplet *whispering gallery modes* (WGMs), which are related to the droplet size. This provides a tool for interrogation of droplets during the coagulation process. An example is shown in Fig. 18.2, where two droplets in separate holographic optical tweezers are brought together until they fuse. Evident in the CERS spectrum are the WGMs of the two original droplets and the modes of the final coagulated droplet. From these modes, the radii and hence the volumes of the droplets may be accurately determined. In particular, Buajarern et al. (2006) found that droplet volume is conserved in this process.

The mixing of aerosol droplets of different compositions was also monitored via the CERS spectra. When an aqueous sodium chloride droplet is mixed with an ethanol droplet, the CH band of the ethanol molecule appears in the spectrum of the resulting droplet. Further analysis of the calculated Mie spectra of a homogeneously mixed droplet versus a partly unmixed droplet consisting of an ethanol shell and an ethanol/sodium chloride/water core revealed that the locations of the WGMs were consistent with the droplet being fully mixed. Conversely, coagulation of an aqueous sodium chloride droplet with a decane droplet gives rise to a CERS spectrum containing the CH and OH stretching vibrations, but lacking the stimulated Raman mode structure. Because decane and water are immiscible, the coagulation produced a droplet with separated phases and a loss of spherical symmetry, thereby reducing the quality factor of the WGMs.

Micrometre-sized aqueous droplets also make ideal reaction vessels for femtolitre volumes containing small numbers of molecules. They are particularly advantageous for single-molecule studies, as the molecule can be confined within the detection volume of a microscope without the need for surface attachment. In contrast to the aerosol trap described previously, Reiner et al. (2006) optically trapped aqueous droplets (refractive index, $n_p = 1.33$) dispersed in a suspending fluorocarbon, FC-77 ($n_m = 1.29$). Droplets were prepared containing a dilute solution of a number of different dyes. An avalanche photodiode was used to collect the fluorescence emitted on excitation by a green laser. Discrete steps in the emission corresponding to photobleaching of individual dye molecules were evident, revealing droplets containing as few as three, two or one molecule.

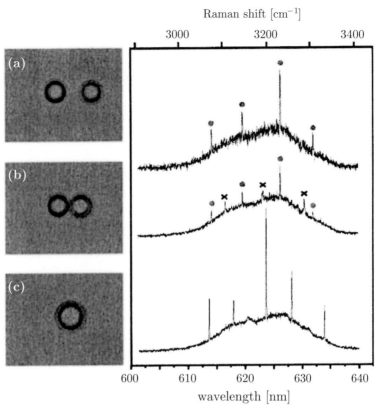

Figure 18.2 Coagulation of optically trapped aerosol droplets: simultaneous manipulation, coagulation and spectroscopic sizing of optically trapped aqueous aerosol droplets. Two droplets in separate holographic optical tweezers are brought together until they fuse. Evident in the CERS spectrum are the WGMs of the two droplets, identified with circles or crosses in spectra (a) and (b), and the modes of the final coagulated droplet in part (c). Reprinted from Buajarern et al., J. Chem. Phys. **125**, 114506. Copyright (2006) American Institute of Physics.

18.2 Vesicle and membrane manipulation

A vesicle is a (quasi-)spherical membrane enclosing a liquid reservoir (Vasdekis et al., 2013). The membrane material may be a lipid (in which case the vesicle is referred to as a *liposome*), a surfactant (*niosome*) or a block copolymer (*polymerosome*). The contents of the vesicle are protected by the membrane so that they can be transported to a targeted location and potentially released on demand. Both transport of the vesicle and release of the cargo may be performed using optical manipulation tools. Stable optical trapping of a vesicle can, however, be challenging, as frequently the encapsulated material and external medium have a similar refractive index. Ichikawa and Yoshikawa (2001) were able to optically trap unilamellar vesicles of phosphatidyl–glycerol up to 10 μm in diameter (*giant unilamellar vesicles*, or GUVs) filled with 100 mM D-glucose solution. The small refractive index difference ($\Delta n = 0.0021$) between the contents and the 50 nM NaCl solution suspending

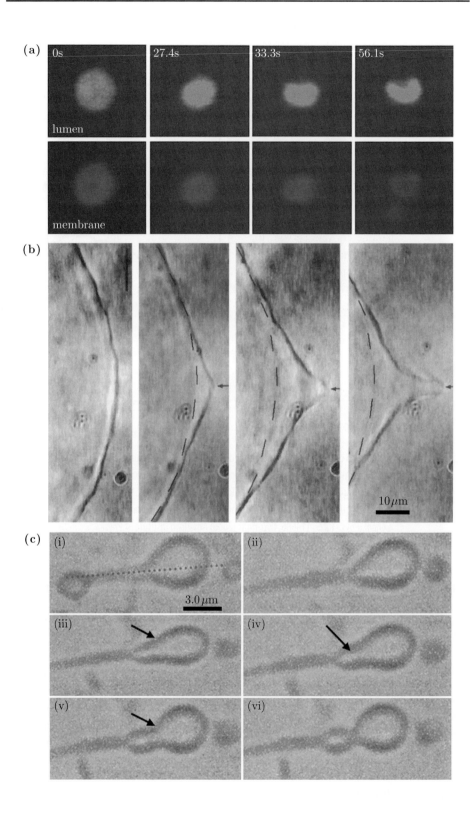

medium enabled optical trapping with a very low trapping efficiency of $Q = 0.01$, but still an order of magnitude greater than that achieved for D-glucose-containing vesicles suspended in the same concentration glucose solution. Bendix and Oddershede (2011) were able to optically trap vesicles with a diameter as small as 50 nm containing 1 M sucrose solution. Water-containing vesicles were also trapped, but in this case the optical tweezers acted on the membrane material only. Furthermore, small aqueous-core vesicles were deformed by the action of the trapping laser light as shown in Fig. 18.1a.

Bar-Ziv et al. (1998) used optical forces to deform the membrane of a vesicle. The strong electric field gradient of an optical tweezers applied to a point on the surface of a large vesicle results in the lipid membrane material being pulled towards the trap and hence in a laser-induced surface tension in the membrane. For typical optical tweezers parameters this surface tension is on the order of a few piconewtons per micrometre, which is several orders of magnitude higher than the equilibrium surface tension of a large vesicle, but several orders of magnitude lower than its stretching modulus. Optical tweezers are therefore capable of exciting shape transitions (bending modes rather than stretching modes) in the membrane. In this configuration, membrane material flows into the optical trap, but at distances more than a few times the trap size this flow is hindered by viscous effects from the surrounding water, and therefore the tension in the membrane decreases to zero. This is illustrated in Fig. 18.1b, where a point on the circumference of a large flaccid vesicle is slowly pulled outwards against the restoring elastic force by optical tweezers. In this experiment the maximum force applied by the optical tweezers was calculated from the extremal shape of the deformed membrane before it was pulled out of the trap, using *a priori* knowledge of the bending modulus. Continuous tweezing of a point on the membrane can lead to irreversible loss of membrane material via detachment of lipid fragments from the region of the laser trap (Bar-Ziv et al., 1995a). This loss of surface area eventually leads to an increase in membrane tension and internal pressure so that an initially flaccid vesicle can become spherical. Such laser-induced pressurisation was observed to lead to spontaneous expulsion of enclosed inner vesicles from an oligovesicular structure.

Spyratou et al. (2009) used a line optical tweezers to induce a budding transition in multilammellar liposomes. In their experiment, the trapping laser beam was transformed into an elliptical cross-section using cylindrical lenses, resulting in a focal line of dimensions

Figure 18.3 Vesicle membrane manipulation by optical tweezers. (a) A 1 μm-diameter vesicle with aqueous core held in optical tweezers is deformed at high power. The shape of the vesicle is revealed by the distribution of fluorescence from dyes contained in the vesicle interior (upper panels) and membrane (lower panels). Reprinted from Bendix and Oddershede, *Nano Lett.* **11**, 5431–7. Copyright (2011) American Chemical Society. (b) A part of the membrane of a giant flaccid vesicle (which extends beyond the region of the image) is pulled by optical tweezers applied at the point indicated by the arrow. Reprinted from Bar-Ziv et al., *Biophys. J.* **75**, 294–320, copyright (1998), with permission from Elsevier. (c) A line tweezers, whose location is shown by the dotted line in frame (i), is used to manipulate a vesicle. In frames (ii), (iii), (iv) and (v), the vesicle can be seen to undergo a budding transition as a result of the laser-induced surface tension. In frame (vi), two separate vesicles are visible. Reprinted from Spyratou et al. (2009), *Colloids Surf. A* **349**, 35–42, copyright (2009), with permission from Elsevier.

1 μm × 10 μm in the trapping plane. As shown in Fig. 18.1c, the additional optically induced tension in the membrane of the trapped vesicle causes it to minimise its projected area at constant volume until no further decrease is possible and the vesicle undergoes a budding transition into two spherical daughter liposomes with different radii. It should be noted, however, that other mechanisms arising from the heating induced by the trapping laser and the resulting changed permeability of the membrane may also play a role in this transition. At long times and for sufficiently large vesicles (diameter > 3 μm), irreversible shape changes – from spherical to ellipsoidal – of the trapped vesicles were observed. This was a result of the decreasing vesicle inner volume due to evaporation from the suspending medium and the resulting osmotic efflux from the vesicle. The laser-induced tension was then sufficient to transform the remaining flaccid membrane material into a filament. At short times no hysteresis in the membrane deformation was observed, meaning that measurement of the membrane shear modulus was possible from the known applied optical force and the resulting deformation, and it was found to be $\mu = 1.0 \pm 0.2$ μN/m.

18.3 Vesicle fusion

From having established that a whole vesicle can be optically trapped and that the membrane material can also be optically manipulated, it is clear that micrometre-sized vesicles present potential as microreaction vessels in a manner similar to that of the aqueous droplets considered in Section 18.1. Although aqueous droplets present some advantages in this regard, such as, e.g., spontaneous fusion, encapsulation within a vesicle more closely replicates the partitioning between the interior of a cell and the outside world created by the cell membrane. Kulin et al. (2003) demonstrated the controlled manipulation of giant multilamellar vesicles using optical tweezers and an additional pulsed ultraviolet (UV) laser to initiate membrane fusion when two vesicles were brought into contact, shown in Fig. 18.4. In the majority of these experiments, it was observed that the resulting fused vesicle was spherical and that the volume of the vesicles was conserved during the fusion process. The implication of this is that the membranes are likely multilamellar, as unilamellar vesicles would conserve the surface area of the vesicles and the resulting vesicle would not be spherical without the uptake of additional fluid from the surroundings, prohibited by the local nature of the membrane disruption induced by the UV laser. Mixing of reagents throughout the volume of the fused vesicle was demonstrated by fusing one liposome containing the dye sulforhodamine B with another containing only buffer solution. The fluorescence from the resulting vesicle was observed to be uniformly emitted throughout the volume and of lower intensity, demonstrating that the dye had diffused throughout the vesicle. The observed mixing time was at the limit of the temporal resolution of the video imaging technique used. Reaction between reagents contained in separate vesicles was demonstrated by fusing two liposomes containing the dye Fluo-3 and calcium ions. Fluo-3 chelates ions in solution and fluoresces with a considerably higher intensity with calcium ions because of their higher affinity. The fluorescence was not only observed throughout the volume of the fused vesicle, but also at a higher intensity because of the reaction between ions and dye.

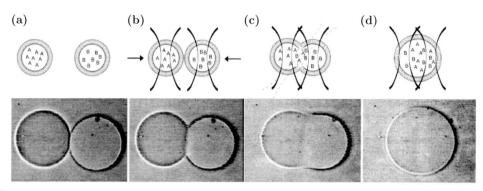

Figure 18.4 Controlled vesicle fusion in optical tweezers. The top row shows a schematic of the fusion and mixing process: (a) chemical reagents A and B are contained in two separate liposomes; (b) the liposomes are brought into contact using optical tweezers; (c) disruption of the membrane in the region of contact is initiated by a pulsed UV laser; (d) a single large liposome is formed in which the reagents can mix. The bottom row reproduces photographs from the experiment demonstrating controlled vesicle fusion at times (a) $t = 0$ ms, (b) $t = 132$ ms, (c) $t = 264$ ms and (d) $t = 528$ ms. Reprinted with permission from Kulin et al., *Langmuir* **19**, 8206–10. Copyright (2003) American Chemical Society.

18.4 Further reading

Further studies using CERS as a tool for sizing optically trapped aerosol droplets have been reported by Hopkins et al. (2004) and for monitoring droplet coagulation and mixing by Power et al. (2012). Extensive work on the properties of membranes manipulated with optical tweezers has enabled the study of diverse interesting phenomena such as instabilities and 'pearling' states (spherical volumes connected by thin tubes) on a tubular membrane that result from an increase in local surface tension from the action of optical tweezers (Bar-Ziv and Moses, 1994), the unbinding of membrane bilayers over a region that extends away from the point at which they are 'pinched' together by optical tweezers (Bar-Ziv et al., 1995b) and the spontaneous expulsion of smaller vesicles from inside larger vesicles during optical tweezing (Moroz et al., 1997). Shitamichi et al. (2009) used a dual optical tweezers to hold and stretch liposomes and measure the membrane surface tension and bending rigidity from the observed deformation from a known applied force. Brown et al. (2011) were able to study the frequency response of vesicle membranes by driving oscillations using multiple optical traps around the vesicle equator.

References

Bar-Ziv, R., and Moses, E. 1994. Instability and "pearling" states produced in tubular membranes by competition of curvature and tension. *Phys. Rev. Lett.*, **73**, 1392–5.

Bar-Ziv, R., Frisch, T., and Moses, E. 1995a. Entropic expulsion in vesicles. *Phys. Rev. Lett.*, **75**, 3481–4.

Bar-Ziv, R., Menes, R., Moses, E., and Safran, S. A. 1995b. Local unbinding of pinched membranes. *Phys. Rev. Lett.*, **75**, 3356–9.

Bar-Ziv, R., Moses, E., and Nelson, P. 1998. Dynamic excitations in membranes induced by optical tweezers. *Biophys. J.*, **75**, 294–320.

Bendix, P. M., and Oddershede, L. B. 2011. Expanding the optical trapping range of lipid vesicles to the nanoscale. *Nano Lett.*, **11**, 5431–7.

Brown, A. T., Kotar, J., and Cicuta, P. 2011. Active rheology of phospholipid vesicles. *Phys. Rev. E*, **84**, 021930.

Buajarern, J., Mitchem, L., Ward, A. D., et al. 2006. Controlling and characterizing the coagulation of liquid aerosol droplets. *J. Chem. Phys.*, **125**, 114506.

Hopkins, R. J., Mitchem, L., Ward, A. D., and Reid, J. P. 2004. Control and characterisation of a single aerosol droplet in a single-beam gradient-force optical trap. *Phys. Chem. Chem. Phys.*, **6**, 4924–7.

Ichikawa, M., and Yoshikawa, K. 2001. Optical transport of a single cell-sized liposome. *Appl. Phys. Lett.*, **79**, 4598–4600.

Kulin, S., Kishore, R., Helmerson, K., and Locascio, L. 2003. Optical manipulation and fusion of liposomes as microreactors. *Langmuir*, **19**, 8206–10.

Moroz, J. D., Nelson, P., Bar-Ziv, R., and Moses, E. 1997. Spontaneous expulsion of giant lipid vesicles induced by laser tweezers. *Phys. Rev. Lett.*, **78**, 386–9.

Power, R., Reid, J. P., Anand, S., et al. 2012. Observation of the binary coalescence and equilibration of micrometer-sized droplets of aqueous aerosol in a single-beam gradient-force optical trap. *J. Phys. Chem. A*, **116**, 8873–84.

Reiner, J. E., Crawford, A. M., Kishore, R. B., et al. 2006. Optically trapped aqueous droplets for single molecule studies. *Appl. Phys. Lett.*, **89**, 013904.

Shitamichi, Y., Ichikawa, M., and Kimura, Y. 2009. Mechanical properties of a giant liposome studied using optical tweezers. *Chem. Phys. Lett.*, **479**, 274–8.

Spyratou, E., Mourelatou, E. A., Georgopoulos, A., et al. 2009. Line optical tweezers: A tool to induce transformations in stained liposomes and to estimate shear modulus. *Colloids Surf. A*, **349**, 35–42.

Vasdekis, A. E., Scott, E. A., Roke, S., Hubbell, J. A., and Psaltis, D. 2013. Vesicle photonics. *Annu. Rev. Mater. Res.*, **43**, 283–305.

19 Aerosol science

Aerosols are microscopic solid or liquid particles dispersed in a gas. Their study has great significance in fields such as combustion science and atmospheric chemistry. Holding such particles in optical tweezers permits one to measure properties that contribute to the chemical and physical state of the aerosol and hence to its action on the environment, such as nucleation rates, mass and heat transfer and particle composition, size and mixing state. Optical trapping and manipulation of aerosols poses unique challenges, mainly due to the very low viscosity of the suspending medium, which leads to important inertial effects, but also permits one to explore some new phenomena, such as photophoretic forces, as shown in Fig. 19.1. In this chapter, we review experiments on both solid and liquid airborne particles, discussing the features that differentiate the manipulation and interrogation of optically trapped aerosols from optical manipulation in a liquid medium.

Figure 19.1 Photophoretic optical trap: Trapping of an agglomerate of carbon nanoparticles at the intensity minimum of an *optical bottle*, i.e., the spherically aberrated focus of a laser beam. Because photophoretic forces are several order of magnitude strong than optical forces, trapping is achieved with just 5 µW laser power. Reprinted from Shvedov et al., *Opt. Express* **19**, 17 350–6. Copyright (2011) The Optical Society.

19.1 Optical tweezers in the gas phase

Although Ashkin and Dziedzic (1975) were able to levitate liquid droplets optically in 1975, i.e., to hold them in a slightly focused laser beam pointing upwards and producing scattering forces sufficient to balance gravity, the first proper optical tweezers for airborne particles was not demonstrated until much later by Omori et al. (1997). In their experiment, micrometre-sized *solid* glass spheres were projected from a glass surface by driving the surface with a piezoelectric transducer; a relatively low axial trapping efficiency ($Q_z = 0.01$) was reported for 2.5 µm-radius spheres trapped using 40 mW of laser power. Magome et al. (2003) reported the first optically trapped (rather than optically levitated) *liquid* aerosols: they nucleated water droplets from a supersaturated vapour and were able to achieve a much higher axial efficiency ($Q_z = 0.46$ was reported for a 5.7 µm-radius water droplet using only 5 mW of laser power), which can be explained at least partly by the fact that they used a higher-NA oil-immersion objective lens. More recent experiments have used a nebuliser to introduce aerosol particles into the experimental chamber; this technique works well both for aqueous droplets (McGloin et al., 2008) and for solid particles, which are delivered suspended in a liquid, e.g., ethanol, that then rapidly evaporates (Summers et al., 2008). In a humidity-controlled environment the size distribution of aqueous droplets in the aerosol reaches an equilibrium, and stable trapping over several hours is possible (McGloin et al., 2008). An experimental scheme for optical tweezing of nebulised aerosols is shown in Fig. 19.2a.

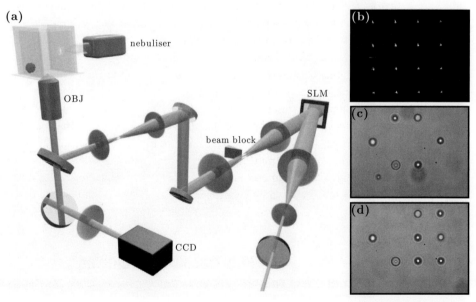

Figure 19.2 Aerosol optical tweezers: (a) Schematic of an experimental set-up to optically trap aerosols, including a nebuliser to produce aerosol droplets, an SLM to generate multiple optical traps and a CCD camera to track the position of the droplets. (b) A 4 × 4 array of optical traps, whose trapping sites are populated by aerosol droplets over time as shown in (c) and (d). Reprinted from Burnham and McGloin, *Opt. Express* **14**, 4175–81. Copyright (2006) The Optical Society.

This particular scheme also includes a spatial light modulator (SLM) in the beam path to produce multiple holographic optical traps, as we have seen in Chapter 11. Fig. 19.2b shows the holographically generated array of trapping sites, and Figs. 19.2c and 19.2d show aqueous aerosol droplets populating a fraction of these sites over time.

There are several important differences between optically tweezing particles in a liquid and in a gas. Most of these differences are related to the dissimilar viscosities of the media. In fact, whereas Brownian motion in a liquid medium is typically overdamped, i.e., at relatively long time scales (from a fraction of a millisecond upwards), inertial effects can be neglected as we have seen in Chapter 7, in a gaseous medium it is underdamped and inertial effects become important. For example, in the overdamped case the power spectral density (PSD) of the fluctuations of an optically particle is given by Eq. (9.39), which has a Lorentzian form and is independent of the particle mass; i.e.,

$$P_{\text{over}}(f) = \frac{D/(2\pi^2)}{f_c^2 + f^2}, \tag{19.1}$$

where $D = k_B T/\gamma$ is the diffusion coefficient, k_B is the Boltzmann constant, T is the absolute temperature, γ is the friction coefficient, $f_c = \kappa/(2\pi\gamma)$ is the corner frequency and κ is the trap stiffness. In the underdamped regime the PSD instead becomes

$$P_{\text{under}}(f) = \frac{D/(2\pi^2)}{(f_c - 2\pi m/\gamma f^2)^2 + f^2}, \tag{19.2}$$

which acquires a peak at $f_m = (2\pi)^{-1}\sqrt{\kappa/m}$ and at high frequency decays as f^{-4}, rather than as f^{-2}. A first practical consequence of the underdamped regime is that traps cannot be filled by actively seeking particles; instead they need to be passively filled by slow-moving particles, as in the experiment shown in Fig. 19.2. This is due to the fact that the trajectory the droplet must take to enter the trap passes through a zone where the scattering force is significant and, because of the lack of viscous damping from the suspending medium, a particle on such a trajectory is pushed through the trapping region; therefore, only droplets near the plane of the laser focus have a high probability of entering the trap and remaining there. Furthermore, it is observed that the size of the aerosols that may be stably trapped is proportional to the laser power and that, at higher laser power, the lack of dissipation of the scattering force from the suspending medium leads to trap instability and particle loss. Another interesting consequence of the underdamped regime, known as *parametric resonance*, was observed by Di Leonardo et al. (2007) as an increase in the amplitude of the PSD at f_m when they modulated the laser power at $2f_m$.

19.2 Trapping and guiding

As is the case for conventional optical tweezers in liquid media, the ability to control the position of trapped aerosols is highly desirable. In the holographic aerosol optical tweezers demonstrated by Burnham and McGloin (2006) and illustrated in Fig. 19.2a, the individual aerosols were manipulated over the 45 μm × 60 μm field of view by dynamically changing the kinoform displayed on the SLM. Longer-range optical guiding of aerosols has been achieved using a non-diffracting Bessel beam [Subsection 4.4.4]. Summers et al. (2006) have investigated guiding of liquid aerosols including water, ethanol and dodecane and made a quantitative comparison of the distance over which the droplets are guided in both Gaussian and Bessel beams in the vertical direction (antiparallel to gravity). As may be expected, guiding over significantly larger distances was possible using the Bessel beam, e.g., up to 2.75 mm for dodecane aerosols and 1.2 mm for water (the longer distance for dodecane was attributed to its higher refractive index giving rise to a greater optical scattering force); these distances were several times those achieved using the Gaussian beam. Carruthers et al. (2010) have also demonstrated guiding of aerosols over millimetric distances in an experiment that employed two counter-propagating Bessel beams, which, differently from the previous experiment were oriented in the horizontal plane; evidence of optical binding [Section 3.10] was also observed in this experiment, as several droplets formed stable chain structures that underwent collective motion with fixed separation distances.

19.3 Photophoretic trapping and guiding

Many atmospheric aerosols are not transparent, a particular example being black carbon arising from combustion processes. These airborne particles are not amenable to conventional optical trapping, because of the presence of overwhelming scattering optical forces due to light absorption. Nevertheless, an alternative scheme for trapping and guiding absorbing aerosols based on radiometric forces (photophoresis) has been successfully implemented by Shvedov et al. (2009). Photophoresis occurs when an absorbing particle in a gaseous medium is non-uniformly heated by incident light. The integrated effect of gas molecules rebounding asymmetrically from the surface of the particle gives rise to a force on the particle centre of mass. For air at room temperature, such a photophoretic force can exceed radiation pressure force by several orders of magnitude. In their experiment, Shvedov et al. (2009) realised a photopheretic trap for absorbing carbon nanoclusters using counter-propagating Laguerre–Gaussian beams [Subsection 4.4.3]. The positive photophoretic force[1] for these particles repels them from the high-intensity part of the beam, so transverse confinement is achieved in the dark core of the Laguerre–Gaussian beams. An estimate of the photophoretic force for micrometre-sized carbon revealed that it exceeded the radiation pressure by at least four orders of magnitude. Longitudinal confinement is achieved by counter-propagating beams that have a small separation between their respective beam waists. Clusters with sizes ranging from less than a micrometre to over 10 μm were trapped, and guiding of particles up to 2 mm along the axis of the trap was achieved by altering the relative powers of the counter-propagating beams. A photograph of absorbing particles held in the photophoretic trap is shown in Fig. 19.3. Remarkably, this trap operates in the open air rather than in a sealed chamber.

---

[1] Thermodiffusive phenomena, of which photophoresis is an example, are labeled *positive* when particles move from a hot to a cold region and *negative* when the reverse is true. Typically, the heavier/larger species in a mixture exhibit positive thermophoretic behaviour, whereas the lighter/smaller species exhibit negative behaviour.

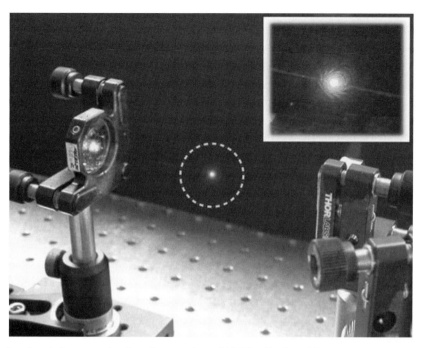

Figure 19.3 Photophoretic optical trap: trapping of a carbon nanocluster (highlighted by the circle) in counter-propagating Laguerre–Gaussian beams. The inset shows a close-up of the trapped particle. Reprinted from Shvedov et al. (2010b) with kind permission from Springer Science and Business Media.

Three-dimensional photophoretic traping has also been realised using a single-beam trap (Shvedov et al., 2011). This technique relies on the spherical aberration that is unavoidably introduced when a Gaussian beam is focused using a spherical lens. For the correct degree of aperture filling, the aberrated focus produced by the lens was shown to contain a series of intensity minima along the axis surrounded in all directions by high-intensity shells; these focused beams have been termed *optical bottles*. These intensity minima thus form robust photophoretic traps for absorbing carbon particles, as shown in Fig. 19.1. Indeed, the strength of the aberration-based trap is such that particles could be trapped in open air and the focusing objective subjected to large-scale movements.

19.4 Further reading

Rudd et al. (2008) performed optical trapping of aerosol droplets up to 20 μm in size in a fibre trap, also observing optical binding effects similar to the ones described above. The ability of a Bessel beam to trap and guide aerosol particles of varying size has been extensively studied by Meresman et al. (2009) and Preston et al. (2014). Notably, Preston et al. (2014) highlighted the significance of whispering gallery modes in determining the optical force and the equilibrium position (or retention distance) for aerosols of differing sizes when they are optically trapped against a counter-propagating gas jet, leading to a size-dependent spatial separation or optical chromatography. Bessel beam traps were also used by Cotterell et al. (2014) to monitor the hygroscopic growth of aerosols as a function of relative humidity. Desyatnikov et al. (2009) developed a model to describe the photophoretic trapping force, obtaining results in good agreement with the experiments described in the previous section. Shvedov et al. (2010a) were also able to trap carbon-coated hollow glass spheres rather than irregularly shaped clusters, which permitted a higher degree of control and repeatability of measurements, thus enabling precise (sub-millimetre) deposition of the spheres on a target from a distance of 0.5 m. Furthermore, multiple particles were trapped in air in both a regular array formed by a diffractive mask (Shvedov et al., 2010b) and in a speckle pattern (Shvedov et al., 2010c).

References

Ashkin, A., and Dziedzic, J. M. 1975. Optical levitation of liquid drops by radiation pressure. *Science*, **187**, 1073–5.

Burnham, D. R., and McGloin, D. 2006. Holographic optical trapping of aerosol droplets. *Opt. Express*, **14**, 4175–81.

Carruthers, A. E., Reid, J. P., and Orr-Ewing, A. J. 2010. Longitudinal optical trapping and sizing of aerosol droplets. *Opt. Express*, **18**, 14 238–44.

Cotterell, M. I., Mason, B. J., Carruthers, A. E., et al. 2014. Measurements of the evaporation and hygroscopic response of single fine-mode aerosol particles using a Bessel beam optical trap. *Phys. Chem. Chem. Phys.*, **16**, 2118–28.

Desyatnikov, A. S., Shvedov, V. G., Rode, A. V., Krolikowski, W., and Kivshar, Y. S. 2009. Photophoretic manipulation of absorbing aerosol particles with vortex beams: Theory versus experiment. *Opt. Express*, **17**, 8201–11.

Di Leonardo, R., Ruocco, G., Leach, J., et al. 2007. Parametric resonance of optically trapped aerosols. *Phys. Rev. Lett.*, **99**, 010601.

Magome, N., Kohira, M. I., Hayata, E., Mukai, S., and Yoshikawa, K. 2003. Optical trapping of a growing water droplet in air. *J. Phys. Chem. B*, **107**, 3988–90.

McGloin, D., Burnham, D. R., Summers, M. D., et al. 2008. Optical manipulation of airborne particles: Techniques and applications. *Faraday Discuss.*, **137**, 335–50.

Meresman, H., Wills, J. B., Summers, M., McGloin, D., and Reid, J. P. 2009. Manipulation and characterisation of accumulation and coarse mode aerosol particles using a Bessel beam trap. *Phys. Chem. Chem. Phys.*, **11**, 11 333–9.

Omori, R., Kobayashi, T., and Suzuki, A. 1997. Observation of a single-beam gradient-force optical trap for dielectric particles in air. *Opt. Lett.*, **22**, 816–18.

Preston, T. C., Mason, B. J., Reid, J. P., Luckhaus, D., and Signorell, R. 2014. Size-dependent position of a single aerosol droplet in a Bessel beam trap. *J. Opt.*, **16**, 025702.

Rudd, D., Lopez-Mariscal, C., Summers, M., et al. 2008. Fiber based optical trapping of aerosols. *Opt. Express*, **16**, 14 550–60.

Shvedov, V. G., Desyatnikov, A. S., Rode, A. V., Krolikowski, W., and Kivshar, Y. S. 2009. Optical guiding of absorbing nanoclusters in air. *Opt. Express*, **17**, 5743–57.

Shvedov, V. G., Rode, A. V., Izdebskaya, Y. V., et al. 2010a. Giant optical manipulation. *Phys. Rev. Lett.*, **105**, 118103.

Shvedov, V. G., Desyatnikov, A. S., Rode, A. V., et al. 2010b. Optical vortex beams for trapping and transport of particles in air. *Appl. Phys. A*, **100**, 327–31.

Shvedov, V. G., Rode, A. V., Izdebskaya, Y. V., et al. 2010c. Selective trapping of multiple particles by volume speckle field. *Opt. Express*, **18**, 3137–42.

Shvedov, V. G., Hnatovsky, C., Rode, A. V., and Krolikowski, W. 2011. Robust trapping and manipulation of airborne particles with a bottle beam. *Opt. Express*, **19**, 17 350–56.

Summers, M. D., Reid, J. P., and McGloin, D. 2006. Optical guiding of aerosol droplets. *Opt. Express*, **14**, 6373–80.

Summers, M. D., Burnham, D. R., and McGloin, D. 2008. Trapping solid aerosols with optical tweezers: A comparison between gas and liquid phase optical traps. *Opt. Express*, **16**, 7739–47.

20 Statistical physics

The behaviour of optically trapped particles is the result of the interplay between a natural well-defined noisy background and a finely controllable deterministic force field, as shown in Fig. 20.1 for the simplest case of a particle held in a harmonic optical trap. In fact, on one hand, as we have seen in Chapter 7, colloidal particles are constantly moving because of the presence of Brownian motion, which introduces a well-defined noisy background. On the other hand, as we have seen all though this book, it is possible to use optical forces to introduce deterministic perturbations acting on the particles in a very controllable way. Therefore, optically trapped particles can be a very powerful tool for studying statistical physics phenomena whose dynamics is driven by both random and deterministic forces, ranging from biomolecules and nanodevices to financial markets and human organisations. In this chapter, we will discuss as examples how optically trapped particles have been employed to study Kramers rates, stochastic resonance, spurious drift, crystal formation and anomalous diffusion.

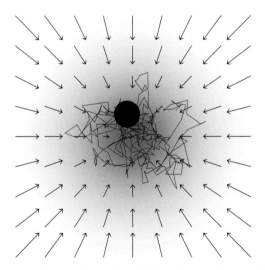

Figure 20.1 Interplay of random and deterministic forces: The motion (solid line) of an optically trapped Brownian particle (sphere) is a dynamic system whose behaviour results from the interplay of random forces due to the presence of Brownian motion and deterministic forces due to the presence of optical forces (arrows).

20.1 Colloids as a model system for statistical physics

Molecules and atoms are continuously in motion because of their finite temperature. However, it is difficult to observe this motion directly in experiments because of their small size, much smaller than the wavelength of light, and because of their rapid motion, on time scales of attoseconds. Colloidal particles also undergo continuous motion because of bombardment by molecules of the surrounding fluid and have the advantage of being directly observable with a conventional microscope using digital video microscopy, as we have seen in Section 9.1. In fact, their size is comparable to the wavelengths of visible light, so that one can easily gather information on the motion of single particles in a system. Furthermore, the characteristic time scale of the motion of colloidal particles, being on the order of milliseconds, is also more easily accessible than that of atoms. These characteristics make them an ideal model system for statistical physics (Babič et al., 2005).

We can quantify the difference between atoms and colloids by considering that the time τ it takes a particle to diffuse over a distance comparable to its radius a in a medium of viscosity η in three dimensions is

$$\tau = \frac{a^2}{6D}, \tag{20.1}$$

where $D = \frac{k_B T}{\gamma}$ is the Stokes–Einstein diffusion coefficient given by Eq. (7.20), k_B is the Boltzmann constant, T is the absolute temperature, and γ is the friction coefficient, which for a spherical particle is $\gamma = 6\pi \eta a$. For atoms $\tau \approx 1$ ps, whereas for colloids $\tau \approx 1$ s.

We must remark, however, that much of the interesting behaviour of colloids is related to the fact that they are, in many respects, not like atoms (Frenkel, 2002). In fact, on one hand, colloids are the computer simulator's dream, because many of them can be represented quite well by models – such as the hard-sphere and Yukawa models – that are far too simple to represent molecular systems. On the other hand, colloids are also the simulator's nightmare, or at least challenge, because, at a closer look, simple models do not always work, as we have already seen in Chapter 17.

20.2 Kramers rates

Activated escape from a metastable state underlies many physical, chemical and biological processes, such as diffusion in solids, switching in superconducting junctions, chemical reactions and protein folding. Kramers (1940) presented the first quantitative calculation of the rates of thermally driven transitions, which have come to be known as *Kramers rates*. A mesoscopic particle suspended in a liquid and confined within a bistable potential well, e.g., a double optical trap, provides an ideal representation of Kramers' ideas: the particle moves at random within one of the wells until a large fluctuation propels it over the energy barrier between the two wells. These jumps can be used to address quantitatively the problem of transition rates, as long as the confining potential can be accurately determined. Such a particle can be described by the Langevin equation

$$\frac{dx}{dt} = -\frac{1}{\gamma}\frac{dU(x)}{dx} + \sqrt{2D}W(t), \tag{20.2}$$

where x is the particle position; γ is the friction coefficient; $U(x)$ is the bistable potential,[1] taken to possess two minima x_+ and x_- associated with the two stable states, separated by a maximum corresponding to an intermediate unstable state x_s (barrier); $W(t)$ is a random Gaussian white noise of zero mean and variance equal to $2D$; and $D = k_B T/\gamma$ is the particle diffusion coefficient. The variable x is confined, most of the time, around x_+ or x_-, but every now and then there are noise-driven abrupt transitions from x_+ to x_- (or vice versa) across the unstable state x_s. The kinetics of these transitions is determined by the noise strength D and the potential barrier $\Delta U_\pm$, defined by

$$\Delta U_\pm = U(x_s) - U(x_\pm). \tag{20.3}$$

In the limit where $k_B T \ll \Delta U_\pm$, the Kramers rates are given by

$$r_\pm = \frac{1}{\tau_\pm} = \frac{\omega_\pm \omega_s}{2\pi\gamma} \exp\left\{-\frac{\Delta U_\pm}{k_B T}\right\}, \tag{20.4}$$

where $\tau_\pm$ are the mean transition times, and $\omega_\pm$ and ω_s characterise, respectively, the curvatures of the potential at the minimum from which the system escapes and at the unstable point.

McCann et al. (1999) reported a detailed experimental analysis of the Brownian dynamics of a sub-micrometre-sized dielectric particle confined in a double-well optical trap measuring the associated Kramers rates. They produced a double potential well by focusing two parallel laser beams through a single objective lens. Each beam created a stable three-dimensional trap for a spherical silica particle of diameter $2a = 0.3$ μm. When the two stable particle positions, $\mathbf{r}_+$ and $\mathbf{r}_-$, were displaced by 0.25 to 0.45 μm, Kramers transitions occurred through a saddle point at $\mathbf{r}_s$, as depicted in Fig. 20.2a. An example of an experimentally recorded trajectory $\mathbf{r}(t)$ is shown in Fig. 20.2b. The probability density $\rho(\mathbf{r})$ was found from the time series $\mathbf{r}(t)$, typically using 10^6 to 10^7 frames corresponding to time intervals much longer than the mean inter-well transition times. Because, as explained in Section 7.3, the spatial probability density is [Eq. (7.19)]

$$\rho(\mathbf{r}) = \rho_0 \exp\left\{-\frac{U(\mathbf{r})}{k_B T}\right\}, \tag{20.5}$$

where $U(\mathbf{r})$ is the potential energy as a function of particle position and ρ_0 is a normalisation constant, it is possible to obtain the potential directly from measurements as

$$U(\mathbf{r}) = -k_B T \ln \rho(\mathbf{r}), \tag{20.6}$$

which is shown in Fig. 20.2c. This experiment obtained excellent agreement between the predictions of Kramers' theory [Eq. (20.4)] and the measured transition rates, without any adjustable or free parameters, over a substantial range of barrier heights.

---

[1] It is helpful to consider a bistable potential that can be expressed analytically in a particularly simple form as $U(x) = ax^4/4 - bx^2/2$, where a and b are constant.

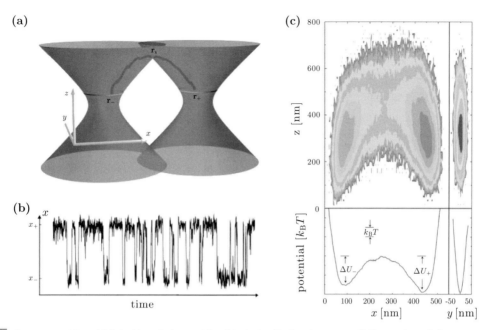

Figure 20.2 Kramers transitions: (a) A double optical potential well is obtained by focusing two parallel beams through the same objective. The dark line illustrates the path of a particle in an inter-well transition between the equilibrium positions $\mathbf{r}_+$ and $\mathbf{r}_-$ (which are at the level of the rings around the beams and displaced above the focal plane because of scattering forces) through the saddle position at $\mathbf{r}_s$. (b) Projection of particle trajectory, acquired by digital video microscopy, along the x-axis perpendicular to the beams, where x_+ and x_- are the equilibrium positions along the x-axis. The sampling interval is 5 ms and the total duration of the record is approximately 8 s. The particle spends most of its time in the vicinity of the stable equilibrium points, with infrequent transitions between them. (c) Potential obtained from 4×10^6 camera frames containing 94 000 inter-well transitions, represented as two-dimensional contour plots (top) and as one-dimensional graphs (bottom). Reprinted by permission from Macmillan Publishers Ltd: McCann et al., *Nature* **402**, 785–7, copyright 1999.

20.3 Stochastic resonance

Another phenomenon that can be observed and studied using a double-well potential is *stochastic resonance*, whereby an appropriate dose of noise can significantly enhance the synchronisation of a system's response to a periodic signal (Gammaitoni et al., 1998). Since the discovery of stochastic resonance by Benzi et al. (1981), its working principle has been applied in a wide variety of systems, ranging from electric circuits to neuronal networks, and it is commonly invoked when noise and nonlinearity concur to determine an improvement of a system's response to a weak signal.[2]

---

[2] And also climatic change as stochastic resonance was intially proposed by Benzi et al. (1982) to explain the periodic recurrence of ice ages.

Starting from Eq. (20.2), one considers a bistable dynamical system subjected simultaneously to noise and to a weak periodic forcing, i.e.,

$$\frac{dx}{dt} = -\frac{1}{\gamma}\frac{\partial U(x)}{\partial x} + \sqrt{2D}W(t) + \epsilon h(x)\cos(\omega_0 t + \varphi), \qquad (20.7)$$

where ϵ, ω_0 and ϕ are respectively the amplitude, frequency and phase of the periodic forcing. The forcing contribution can be cast in a form similar to the potential term by introducing the generalised time-dependent potential

$$\tilde{U}(x) = U(x) - \epsilon g(x)\cos(\omega_0 t + \varphi), \qquad (20.8)$$

with $\frac{d}{dx}g(x) = h(x)$. Differently from the Kramers transition case studied in the previous section, the barrier $\Delta W_\pm$ is now periodically modulated in time, leading to situations where states $x_\pm$ are found at the bottom of wells that are successively less shallow and more shallow than those in the forcing-free system. One is thus led to expect that the transitions will be facilitated during part of this cycle, provided the periodicity of the forcing matches somehow[3] the Kramers time in Eq. (20.4). In this way, the transitions across the barrier can be synchronised to follow, on the average, the periodicity of the external forcing. Furthermore, for given ω_0 and ϵ, the response of the systems goes through a sharp maximum for an intermediate (finite) value of the noise, as shown in Fig. 20.3a. This enhancement of the system's response to a (weak) periodic signal is the signature of stochastic resonance.

Experimentally, Simon and Libchaber (1992) were the first to use a double optical trap to study the synchronisation of inter-well transitions by periodic forcing and, in doing so, performed one of the earliest works where an optically trapped colloid was used to study a statistical physics phenomenon. To observe stochastic resonance, the depths of the potential wells were modulated periodically by modulating the intensity of the two laser beams. When they measured the distribution of times the dielectric sphere stayed in one well before it was kicked into the other one in the absence of modulation, they obtained an exponentially decaying distribution. In the presence of periodic forcing, they observed a sequence of peaks at odd multiples of the half driving period, with exponentially decaying peak height, as shown in Figs. 20.3b and 20.3c; in fact, in the case when the dwell time equals half the period, optimal synchronisation occurs, leading to concentration of the escape within the first period.

20.4 Spurious drift in diffusion gradients

Diffusion gradients emerge naturally when a Brownian particle is in a complex or crowded environment. For example, the diffusion coefficient decreases when a particle is close to a wall because of hydrodynamic interactions, as we have seen in Section 7.6. The consequences of the presence of diffusion gradients have recently been explored by several

---

[3] More precisely, the period of the driving should equal approximately twice the noise-induced escape time.

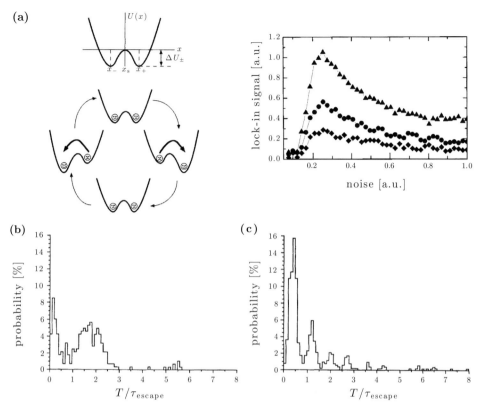

Figure 20.3 Stochastic resonance. (a) Stochastic resonance in a symmetric double well potential. Sketch of the double-well potential $U(x) = \frac{1}{4}bx^4 - \frac{1}{2}ax^2$: the minima are located at $x_\pm = \pm\sqrt{a/b}$ and are separated by a potential barrier $\Delta U_\pm = a^2/(4b)$, which is located at $x_s = 0$. The cartoon shows how, in the presence of periodic driving, the double-well potential is tilted back and forth, thereby successively raising and lowering the potential barriers of the right and the left well, respectively, in an antisymmetric manner: a suitable dose of noise (i.e., when the period of the driving equals approximately twice the noise-induced escape time) will make the 'sad face' happy by allowing synchronised hopping to the globally stable state. The graph on the right shows one of the principal signatures of stochastic resonance, the fact that the amplitude of the periodic component of the response of a bistable system simultaneously to noise and to a weak periodic forcing reaches a sharp maximum as a function of the noise intensity. Reprinted figure from Gammaitoni et al., *Rev. Mod. Phys.* **70**, 223–87. Copyright (1988) by the American Physical Society. (b, c) Experimental escape-time distributions of a particle in the double-well potential generated by a two-beam optical trap. The time is measured in units of the mean escape time τ_{escape} (from one potential minimum to the other). The period of the forcing was chosen as (b) 3.08 and (c) 0.76 times the escape time τ_{escape}. Whereas in (c) the peaks are clearly located at odd multiples of half the forcing period, the second peak in (b) is clearly shifted to the left. Reprinted figure from Simon and Libchaber, *Phys. Rev. Lett.* **68**, 3375–8. Copyright (1992) by the American Physical Society.

experimental techniques making use of optical forces. For example, it has been found that the motion of Brownian particles is affected by the presence of hydrodynamic memory, where the motion of the fluid surrounding the particle affects the future motion of the particle itself (Franosch et al., 2011). Also, spurious forces due to the presence of diffusion

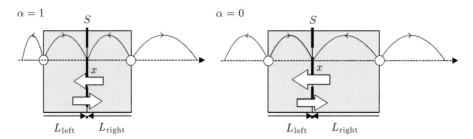

Figure 20.4 Spurious drift without flux. The spurious drift, given by Eq. (20.11) with $\alpha = 1$, permits one to preserve the relation between the external forces acting on the particle and the Maxwell–Boltzmann probability distribution. For example, in the absence of external forces, the Maxwell–Boltzmann probability distribution is uniform and the particle flux though any surface S must be zero (here we consider S placed perpendicular to the diffusion coefficient gradient at coordinate x). During a time interval Δt, all the particles crossing S from the left (right) are half of those included in the volume SL_{right} (SL_{left}), where L_{right} (L_{left}) is the right (left) step terminating at x taken by a walker during that time interval. Therefore, the initial particle distribution is indeed an equilibrium distribution only if the position-dependent step $\sqrt{2D\Delta t}$ in the diffusion gradient is evaluated at the final position of the step ($\alpha = 1$, left cartoon). If $\alpha < 1$, there is a negative flux leading to an accumulation of particles in the direction opposite to the diffusion gradient (in particular for $\alpha = 0$, right cartoon). Reproduced with permission from Lançon et al., *Europhys. Lett.* **54**, 28–34. Copyright (2001) IOP Publishing. Reproduced by permission of IOP Publishing. All rights reserved.

gradients have recently been measured (Volpe et al., 2010; Brettschneider et al., 2011), indicating the need to account for their presence in force measurement, as discussed in Section 10.3. To gain an intuitive understanding of how the spurious drift emerges, we will present a simple model system proposed by Lançon et al. (2001) and illustrated in Fig. 20.4. We will first consider how a particle moves in a (simplified) time step lasting Δt in the presence of a constant diffusion coefficient D, i.e.,

$$x(t + \Delta t) = x(t) \pm \sqrt{2D\Delta t}, \tag{20.9}$$

where the '$\pm$' sign is used because the Brownian walker can move in either direction. If there is a diffusion gradient, i.e., $D = D(x)$, the length of the position-dependent step $\sqrt{2D(x)\Delta t}$ is not uniquely determined, as it depends on the value of $D(x)$ chosen along the step length; in fact, $D(x)$ can be evaluated at the initial point, at some point in the middle, or at the arrival point. Thus, the diffusion coefficient can be written as

$$D(x + \alpha \Delta x) = D(x) + \alpha \frac{dD(x)}{dx} \Delta x, \tag{20.10}$$

where $\alpha = 0$, $1/2$ or 1 depending on the point chosen. The position increment for a Brownian walker with a position-dependent diffusion coefficient is obtained by combining Eqs. (20.10) and (20.9), i.e.,

$$x(t + \Delta t) = x(t) \pm \frac{1}{\gamma}\sqrt{2D(x(t))\Delta t} + \alpha \frac{dD(x)}{dx}\Delta t. \tag{20.11}$$

An ensemble average of Eq. (20.11) gives the average step length

$$\langle x(t) - x(0) \rangle = \alpha \frac{dD(x)}{dx} \Delta t. \tag{20.12}$$

In the case of systems in equilibrium with a heat bath $\alpha = 1$, because this value permits one to preserve the relation between the external forces acting on the particle and the Maxwell–Boltzmann probability distribution, as we have seen in Section 7.6, and thus a spurious drift is present, which must be taken into account in force measurement, as discussed in Section 10.3.

20.5 Colloidal crystals and quasicrystals

Colloidal crystals provide a model system for studying problems in soft-matter physics, ranging from crystallisation to colloidal transport. Traditionally, colloidal crystals are prepared by gravity sedimentation, resulting in thick (millimetres) three-dimensional polycrystalline samples contained within a fluid. An alternative approach makes use of extended optical potentials. As we have seen in Subsection 12.2.1, these optical potentials can be generated by the interference of multiple coherent laser beams by using holographic techniques or acousto-optic modulators. This approach is particularly suited to work as a model system because its parameters (e.g., particle size and material, illumination light) are easily controllable and the dynamics of the colloids is easily accessible by standard optical microscopy techniques. Typically, these structures are two-dimensional, as the scattering radiation force of the laser beams is employed to push the particles against a surface, thus confining them in a quasi-two-dimensional space. In particular, the interference of multiple beams was employed in the seminal work by Burns et al. (1990) to generate optically bound crystalline and quasicrystalline structures. On the same line several works followed. In particular, Mikhael et al. (2008) investigated the phase behaviour of a colloidal monolayer interacting with a quasicrystalline decagonal substrate created by interfering five laser beams, using the set-up shown in Fig. 12.4, and found a new phase showing both crystalline and quasicrystalline structural properties. Roichman and Grier (2005) used holographic optical tweezers to demonstrate the assembly of two-dimensional and three-dimensional dielectric quasicrystals, including structures with specifically engineered defects, as shown in Fig. 20.5. Other studies have observed, e.g., giant diffusion induced by an oscillating periodic potential (Lee and Grier, 2006), and have demonstrated guiding and sorting of particles using either moving periodic potentials or static periodic potentials in microfluidic flows (MacDonald et al., 2003; Xiao and Grier, 2010), as we have seen in Section 16.1.

20.6 Random potentials and anomalous diffusion

Beyond periodic or quasiperiodic potentials, random potentials also play a major role in many phenomena, from the motion of molecules undergoing anomalous diffusion within

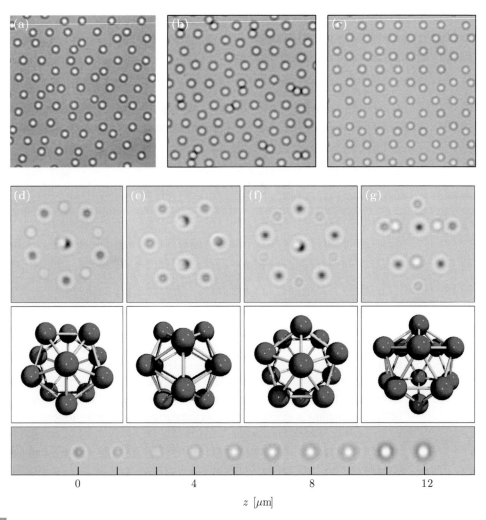

Figure 20.5 Holographically assembled quasicrystals. (a) five-fold, (b) seven-fold and (c) eight-fold colloidal quasicrystals organised using holographic optical tweezers. (d) five-fold axis, (e) two-fold axis, (f) five-fold axis and (g) midplane view of a rolling colloidal icosahedron realised using holographic optical tweezers. In all cases the particles are 1.53 μm-diameter silica spheres immersed in water. Bottom row: a sequence of images of a trapped particle moving through a focal plane used for colibration of the digital video microscopy. Reprinted from Roichman and Grier, *Opt. Express* **13**, 5434–9. Copyright (2005) The Optical Society.

the cytoplasm of a cell to the Brownian motion of stars within galaxies. As we have seen in Subsection 12.2.2, a powerful model system to study these phenomena is the motion of a Brownian particle in a random optical potential generated by a speckle pattern, which was originally proposed by Volpe et al. (2014). Even though speckle light patterns can be generated by different processes (e,g., scattering of a laser on a rough surface, multiple scattering in an optically complex medium or mode-mixing in a multimode fibre, as shown in Figs. 12.5a, 12.5b and 12.5c, respectively), they are always the result of the interference of a large number of waves propagating along different directions and with a random phase distribution. Therefore, despite their random appearance, they share some universal statistical properties. In particular, a speckle pattern has a negative exponential intensity distribution, and the normalised spatial autocorrelation function $C_i(\Delta r)$ can be approximated by a Gaussian, i.e.,

$$C_i(\Delta \mathbf{r}) = \frac{\langle I(\mathbf{r} + \Delta \mathbf{r}) I(\mathbf{r}) \rangle}{\langle I(\mathbf{r})^2 \rangle} \approx e^{-\frac{|\Delta \mathbf{r}|^2}{2\sigma^2}}, \qquad (20.13)$$

where $I(\mathbf{r})$ is the speckle pattern intensity as a function of the position $\mathbf{r}$ and the standard deviation $\sigma \approx d/3$ is proportional to the average speckle grain size d. Eq. (20.13) is shown by the dashed line in Fig. 20.6a.

The motion of a Brownian particle in a static speckle field is the result of random thermal forces and deterministic optical forces. As a particle moves in a speckle field, the optical force acting on it changes both in magnitude and in direction with a characteristic time scale $\tau_s = \frac{L}{\bar{v}}$, where L is the correlation length of the optical force field and $\bar{v}$ is the average particle drift speed. Thus, the optical force field correlation function is (Volpe et al., 2014)

$$C_f(\Delta \mathbf{r}) = \frac{\kappa^2}{\sigma^2} \langle I \rangle^2 \left(2 - \frac{|\Delta \mathbf{r}|^2}{\sigma^2}\right) e^{-\frac{|\Delta \mathbf{r}|^2}{2\sigma^2}}, \qquad (20.14)$$

where $\kappa = \frac{1}{4}\text{Re}\{\alpha\}$, α is the particle polarisability and $L = \sqrt{2}\sigma$. Eq. (20.14) is shown by the solid line (theory) and the circles (simulations) in Fig. 20.6a. Because the particle motion is overdamped, the average particle drift speed is $\bar{v} = \langle F \rangle / \gamma$, where γ is the particle friction coefficient and $\langle F \rangle = \kappa/\sigma$ is the average force. Thus, one obtains

$$\tau_s \approx \sqrt{2} \frac{\sigma^2 \gamma}{\kappa \langle I \rangle}. \qquad (20.15)$$

The motion in a static speckle pattern has been illustrated in Fig. 12.5. When the optical forces are relatively low, the particle is virtually freely diffusing. As the forces increase, first a subdiffusive behaviour emerges where the particle is metastably trapped in the speckle grains, although it can still move between them. Finally, for even stronger forces, the particle remains trapped in one of the speckle grains for a very long time. These observations can be interpreted in terms of τ_s: for relatively strong forces, τ_s is quite low, which means that the particle, on the average, experiences a restoring force towards an equilibrium position quite soon in its motion, having little possibility of escaping a speckle grain; for relatively weak forces, instead, τ_s is much higher, which means that the particle has time to diffuse away from a speckle grain before actually experiencing the influence of the optical forces exerted by it. These qualitative considerations can be made more precise by calculating the mean squared displacement $\text{MSD}(\tau)$ of the particle motion. As shown in Fig. 20.6b, for weak optical forces and high τ_s, the mean squared displacement is substantially linear in τ, i.e., $\text{MSD}(\tau) \approx 4D\tau$, where D is the Stokes–Einstein diffusion coefficient. As the forces increase and τ_s decreases, there is a transition towards a subdiffusive regime characterised by $\text{MSD}(\tau) \propto \tau^\beta$ with $\beta < 1$. For very large τ, the motion again is diffusive, i.e., $\beta = 1$, albeit with an effective diffusion coefficient $D_{\text{eff}} < D$. Interestingly, in many naturally occurring anomalous diffusion processes, such as the subdiffusion of molecules within living cells,

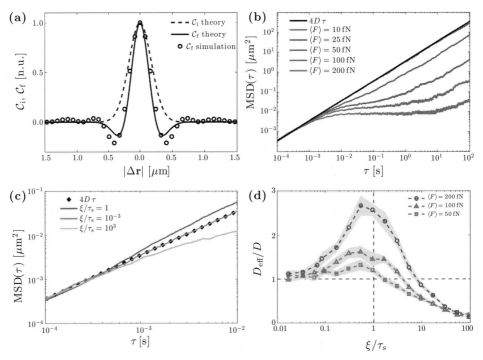

Figure 20.6 Anomalous diffusion in a random potential. (a) Theoretical [Eq. (20.14)] and simulated normalised autocorrelation function of the force field produced by the optical forces associated with a speckle pattern. The dashed line represents the theoretical normalised autocorrelation function of the speckle pattern intensity. (b) Subdiffusive mean squared displacements (MSD) for a Brownian particle moving in a static speckle as a function of the average force $\langle F \rangle$ and its deviation from Einstein's free diffusion law (black line). (c) Superdiffusive MSD for a Brownian particle moving in a speckle pattern that varies on a time scale ξ; τ_s [Eq. (20.15)] is the characteristic time scale of the motion of the particle due to the optical forces associated with the speckle pattern. The dots represent Einstein's free diffusion law. (d) The effective diffusion of the motion at long time scales as a function of ξ/τ_s shows a transition from subdiffusion ($D_{\text{eff}} < D$) to superdiffusion ($D_{\text{eff}} > D$). The maximum value of the superdiffusion appears for $\xi \approx \tau_s$ ($\tau \approx 22.5$ ms for $\langle F \rangle = 50$ fN, $\tau \approx 11.2$ ms for $\langle F \rangle = 100$ fN and $\tau \approx 5.6$ ms for $\langle F \rangle = 200$ fN). Every mean point is averaged over 500 particle trajectories 100 s long, whose initial position is randomly chosen within the speckle field. The grey shaded areas represent one standard deviation around the average values. In all cases, the particle is a 250 nm radius polystyrene ($n_p = 1.59$) sphere immersed in water at 300 K. Adapted from Volpe et al., *Sci. Rep.*, **4**, 3936 (2014).

the distribution of waiting times is expected to correspond to a random walker continually caught in potential wells whose depths are exponentially distributed. In fact, although periodic potentials are characterised by only a few potential depths, the distribution of potential depths (e.g., the distribution of intensities) in a speckle pattern follows a negative exponential distribution.

Employing a time-varying speckle pattern that changes over a time scale ξ similar to τ_s, it is also possible to control and tune anomalous diffusion continuously from subdiffusion to superdiffusion. A time-varying speckle pattern can result from a time-varying environment, but can also be produced in a more controllable way by modulating spatially or spectrally the laser that generates it. As can be seen in Fig. 20.6c, different diffusive regimes emerge depending on the value of the ratio ξ/τ_s, allowing one to tune the diffusive behaviour of the particle just by relying on external control of the speckle pattern time scale ξ. For $\xi/\tau_s \gg 1$, the speckle pattern motion is adiabatic, so that the particle can reach its equilibrium distribution in the optical potential before the speckle field changes. This leads to subdiffusive behaviour, i.e., $\text{MSD}(\tau) \propto \tau^\beta$ with $\beta < 1$, as in a static speckle pattern. For $\xi/\tau_s \ll 1$, the particle cannot follow the fast variation of the speckle pattern, so the average optical force on the particle is zero, leading to diffusive behaviour, i.e., $\text{MSD}(\tau) \approx 4D\tau$. For $\xi/\tau_s \approx 1$, the particle is subject to time-varying forces, which can induce superdiffusive behaviour, i.e., $\text{MSD}(\tau) \propto \tau^\beta$ with $\beta > 1$. Fig. 20.6d highlights the transition from subdiffusion to superdiffusion by plotting D_{eff}/D as a function of ξ/τ_s for various $\langle F \rangle$: if the resulting $D_{\text{eff}} > D$, the particle has undergone superdiffusion. In this way, the diffusive behaviour of a Brownian particle in a speckle field can be controlled by tuning the relevant adimensional parameter ξ/τ_s, thus providing a simple model system to study anomalous diffusion.

20.7 Further reading

Bistable optical traps have been employed in a wide range of experiments; for example, studying the potential generated by two orthogonally polarised optical beams, Stilgoe et al. (2011) showed that more than two stable trapping equilibrium positions could arise. Schmitt et al. (2006) compared stochastic resonance and stochastic activation, showing that, whereas in stochastic resonance the constructive role of noise is to optimise and synchronise the actual response when metastable states are periodically modulated, stochastic activation maximises the time-averaged escape rate over a time-modulated energy barrier. Interestingly, the case of a single optical trap has also been studied, giving insights into the behaviour of monostable dynamic systems; for example, Volpe et al. (2008) used a time-varying single-well optical potential to demonstrate the phenomenon named *stochastic resonant damping*, which is characterised by the counter-intuitive result that the variance of the particle position increases as the trap stiffness increases. Greinert et al. (2006) used an optically trapped particle to monitor the ageing process in a colloidal glass. More complex potentials have also been used, e.g., to demonstrate the properties of Brownian ratchets (Faucheux et al., 1995). Extended optical potentials have been widely used to study phenomenas such as crystallisation (Bechinger et al., 2001).

References

Babič, D., Schmitt, C., and Bechinger, C. 2005. Colloids as model systems for problems in statistical physics. *Chaos*, **15**, 026114.

Bechinger, C., Brunner, M., and Leiderer, P. 2001. Phase behavior of two-dimensional colloidal systems in the presence of periodic light fields. *Phys. Rev. Lett.*, **86**, 930–33.

Benzi, R., Sutera, A., and Vulpiani, A. 1981. The mechanism of stochastic resonance. *J. Phys. A Math. Gen.*, **14**, L453–L457.

Benzi, R., Parisi, G., Sutera, A., and Vulpiani, A. 1982. Stochastic resonance in climatic change. *Tellus*, **34**, 10–16.

Brettschneider, T., Volpe, G., Helden, L., Wehr, J., and Bechinger, C. 2011. Force measurement in the presence of Brownian noise: Equilibrium-distribution method versus drift method. *Phys. Rev. E*, **83**, 041113.

Burns, M. M., Fournier, J.-M., and Golovchenko, J. A. 1990. Optical matter: Crystallization and binding in intense optical fields. *Science*, **249**, 749–54.

Faucheux, L. P., Bourdieu, L. S., Kaplan, P. D., and Libchaber, A. J. 1995. Optical thermal ratchet. *Phys. Rev. Lett.*, **74**, 1504–7.

Franosch, T., Grimm, M., Belushkin, M., et al. 2011. Resonances arising from hydrodynamic memory in Brownian motion. *Nature*, **478**, 85–8.

Frenkel, D. 2002. Soft condensed matter. *Physica A*, **313**, 1–31.

Gammaitoni, L., Hänggi, P., Jung, P., and Marchesoni, F. 1998. Stochastic resonance. *Rev. Mod. Phys.*, **70**, 223–87.

Greinert, N., Wood, T., and Bartlett, P. 2006. Measurement of effective temperatures in an aging colloidal glass. *Phys. Rev. Lett.*, **97**, 265702.

Kramers, H. A. 1940. Brownian motion in a field of force and the diffusion model of chemical reactions. *Physica*, **7**, 284–304.

Lançon, P., Batrouni, G., Lobry, L., and Ostrowsky, N. 2001. Drift without flux: Brownian walker with a space-dependent diffusion coefficient. *Europhys. Lett.*, **54**, 28–34.

Lee, S.-H., and Grier, D. G. 2006. Giant colloidal diffusivity on corrugated optical vortices. *Phys. Rev. Lett.*, **96**, 190601.

MacDonald, M. P., Spalding, G. C., and Dholakia, K. 2003. Microfluidic sorting in an optical lattice. *Nature*, **426**, 421–4.

McCann, L. I., Dykman, M., and Golding, B. 1999. Thermally activated transitions in a bistable three-dimensional optical trap. *Nature*, **402**, 785–7.

Mikhael, J., Roth, J., Helden, L., and Bechinger, C. 2008. Archimedean-like tiling on decagonal quasicrystalline surfaces. *Nature*, **454**, 501–4.

Roichman, Y., and Grier, D. 2005. Holographic assembly of quasicrystalline photonic heterostructures. *Opt. Express*, **13**, 5434–9.

Schmitt, C., Dybiec, B., Hänggi, P., and Bechinger, C. 2006. Stochastic resonance vs. resonant activation. *Europhys. Lett.*, **74**, 937–43.

Simon, A., and Libchaber, A. 1992. Escape and synchronization of a Brownian particle. *Phys. Rev. Lett.*, **68**, 3375–8.

Stilgoe, A. B., Heckenberg, N. R., Nieminen, T. A., and Rubinsztein-Dunlop, H. 2011. Phase-transition-like properties of double-beam optical tweezers. *Phys. Rev. Lett.*, **107**, 248101.

Volpe, G., Perrone, S., Rubi, J. M., and Petrov, D. 2008. Stochastic resonant damping in a noisy monostable system: Theory and experiment. *Phys. Rev. E*, **77**, 051107.

Volpe, G., Helden, L., Brettschneider, T., Wehr, J., and Bechinger, C. 2010. Influence of noise on force measurements. *Phys. Rev. Lett.*, **104**, 170602.

Volpe, G., Volpe, G., and Gigan, S. 2014. Brownian motion in a speckle light field: Tunable anomalous diffusion and selective optical manipulation. *Sci. Rep.*, **4**, 3936.

Xiao, K., and Grier, D. G. 2010. Multidimensional optical fractionation of colloidal particles with holographic verification. *Phys. Rev. Lett.*, **104**, 028302.

21 Nanothermodynamics

Thermodynamics studies the relations between heat, energy (e.g., mechanical, electrical and chemical energy) and other quantities such as entropy and information. The first law of thermodynamics is a restatement of the principle of conservation of energy. The second law of thermodynamics states that heat can only spontaneously flow from a hotter to a cooler body or, equivalently, that the entropy of a closed system can only increase. For large systems, these laws can be expressed in terms of deterministic mathematical relations between the above-mentioned quantities. However, when dealing with microscopic systems, they unveil their true statistical nature: because of the presence of thermal noise, large fluctuations can occur leading, e.g., to an occasional decrease of entropy, as shown in Fig. 21.1. This chapter will review how optical tweezers have been instrumental in studying the thermodynamics of microscopic systems.

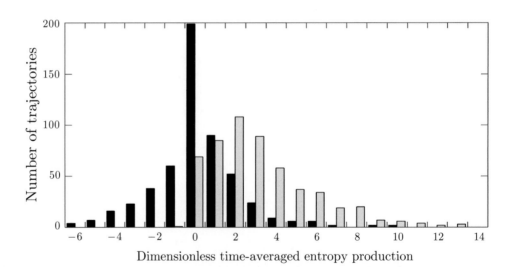

Figure 21.1 Violation of the second law for microscopic systems. Experimental confirmation of the fluctuation–dissipation theorem, which predicts measureable violations of the second law for small systems over short time scales. The histogram shows the dimensionless time-averaged entropy production from 540 experimental trajectories of a colloidal particle in an optical trap after initiation of stage translation: at short times (10 ms, black bars) the trajectories are distributed nearly symmetrically about zero with nearly equal numbers of trajectories consuming and producing entropy; at longer times (2000 ms, gray bars), the entropy-consuming trajectories occur less often and the mean entropy production shifts towards more positive numbers. Reprinted figure from Wang et al., *Phys. Rev. Lett.* **89**, 050601. Copyright (2002) by the American Physical Society.

21.1 Violation of the second law

The *second law of thermodynamics* states that for a closed system the entropy production rate is necessarily positive. One of the fundamental questions that must be addressed to understand the foundations of thermodynamics is how the second law of thermodynamics can be reconciled with reversible microscopic equations of motion. In fact, *Loschmidt's paradox* states that in a time-reversible system, for every phase-space trajectory there exists a time-reversed anti-trajectory; as the entropy production of a trajectory and its conjugate anti-trajectory are of identical magnitude but opposite sign, one cannot prove that entropy production is positive. As we will see in this section, the resolution of this paradox is that the law is strictly valid only for large systems and over long times.

Evans et al. (1993) introduced the *fluctuation theorem*, which provided a resolution of *Loschmidt's paradox* and a quantitative description of violations of the second law in finite systems. The fluctuation theorem relates the probability of observing a phase-space trajectory of duration t with entropy production $\Sigma_t = A$ to that of a trajectory with the same magnitude of entropy change, but where the entropy is consumed or absorbed; i.e.,

$$\frac{P(\Sigma_t = A)}{P(\Sigma_t = -A)} = e^A. \tag{21.1}$$

As the entropy production, Σ_t, is an extensive property, its magnitude scales with system size and observation time (or trajectory duration) t. Therefore, the fluctuation theorem entails that, as the system size gets larger or the trajectory duration becomes longer, entropy-consuming trajectories become unlikely, recovering the expected second law behaviour. Importantly, the fluctuation theorem predicts appreciable and measureable violations of the second law for small systems over short time scales.

Wang et al. (2002) quantitatively confirmed the predictions of the fluctuation theorem for transient systems (Evans and Searles, 1994) by experimentally studying the trajectory of a colloidal particle in an optical trap. In their experiment, the optical trap translated relative to the solvent at a velocity $\mathbf{v}_{ot}$, starting from time $t = 0$. From the particle's trajectory, $\mathbf{x}(t)$, and the optical forces acting on the particle, $\mathbf{F}_{ot}(t)$, it was possible to calculate Σ_t as

$$\Sigma_t = \frac{1}{k_B T} \int_0^t \mathbf{v}_{ot} \cdot \mathbf{F}_{ot}(s)\, ds, \tag{21.2}$$

where T is the temperature of the medium. Wang et al. (2002) showed that occurrence of anti-trajectories quantitatively followed the prediction of an integrated form of the fluctuation theorem. Fig. 21.1 shows the distribution of time-averaged entropy production, Σ_t/t, in 540 particle trajectories at times up to 2 s after the start of stage translation. For $t = 10^{-2}$ s, the trajectories are distributed nearly symmetrically around $\Sigma_t = 0$ with a nearly equal number of entropy-consuming and entropy-producing trajectories. At longer times, entropy-consuming anti-trajectories ($\Sigma_t < 0$) occur less often and the mean entropy production of the trajectories shifts towards more positive numbers, in accord with the fluctuation theorem. At even longer times, i.e., for t greater than a few seconds, these entropy-consuming trajectories cannot be observed, as predicted from the second law of thermodynamics. This experiment provided the first evidence that measurable violations of the second law of thermodynamics in small systems occur not only at simulation time scales (femtoseconds), but also at colloidal time and length scales (seconds).

21.2 The Jarzynski equality

As we saw in the previous section, the paradox at the core of the second law is captured by the following question: how can time-reversible microscopic equations of motion give rise to non-time-reversible macroscopic behaviour? Although, when applied to large systems, the behaviour of the second law emerges from the statistics of huge numbers, one can expect it to be enforced more leniently in systems with relatively few degrees of freedom. Following Jarzynski (2011), we can illustrate this point by considering a gas-and-piston set-up. A gas comprising $N = 10^{23}$ molecules begins in a state of thermal equilibrium inside a container enclosed by adiabatic walls. If the piston is rapidly pushed into the gas and then pulled back to its initial location, there will be a net increase in the internal energy of the gas because of the positive work W done on the gas; i.e.,

$$W > 0. \tag{21.3}$$

We must note at this point that on the microscopic scale there exist N-particle trajectories for which $W < 0$, but these are extremely improbable. However, for a gas of only a few particles, it would be possible to observe, at least once in a while, negative work, although we still expect Eq. (21.3) to hold on the average; i.e.,

$$\langle W \rangle > 0, \tag{21.4}$$

where the angular brackets represent an ensemble average, i.e., an average over many repetitions of the push–pull process with the tiny sample of gas re-equilibrated prior to each repetition. Therefore, when thermodynamics is applied to ever-smaller systems, the second law becomes increasingly blurred and although inequalities such as Eq. (21.3) remain true on the average, statistical fluctuations around the average become ever more important as fewer degrees of freedom come into play, as we have seen in the previous section and in Fig. 21.1.

The fluctuations themselves satisfy some useful laws. For example, Eq. (21.4) can be replaced by the equality (Jarzynski, 2011)

$$\left\langle e^{\frac{-W}{k_B T}} \right\rangle = 1, \tag{21.5}$$

where T is the temperature at which the gas is initially equilibrated and k_B is the Boltzmann constant. If we additionally assume that the piston is manipulated in a time-symmetric manner, e.g., pushed in at a constant speed and then pulled out at the same speed, then the statistical distribution of work values $\rho(W)$ satisfies the symmetry relation (Jarzynski, 2011)

$$\frac{\rho(+W)}{\rho(-W)} = e^{\frac{+W}{k_B T}}. \tag{21.6}$$

The validity of these results depends neither on the number of molecules in the gas nor on the rate at which the process is performed.

Along the lines of the simple example described above, Jarzynski (1997) proved an equality relating the irreversible work to the equilibrium free energy difference, ΔG. This result was particularly interesting because it stated that it was possible to obtain equilibrium thermodynamic parameters from processes carried out arbitrarily far from equilibrium. Liphardt et al. (2002) tested Jarzynski's equality by mechanically stretching a single molecule of RNA reversibly and irreversibly between two conformations. Application of this equality to the irreversible work trajectories recovers the ΔG profile of the stretching process to within $k_B T/2$ of its best independent estimate, i.e., the mean work of reversible stretching. This experiment provided the first example of the use of Jarzynski's equality to bridge the statistical mechanics of equilibrium and non-equilibrium systems.

21.3 Information-to-energy conversion

Leó Szilárd (1929) invented a feedback protocol in which a hypothetical intelligence, dubbed *Maxwell's demon*, pumps heat from an isothermal environment and transforms it into work. After a long and intense controversy, it was finally clarified that the demon's role does not contradict the second law of thermodynamics; in fact, it converts information into free energy. Toyabe et al. (2010) provided the first demonstration of this information-to-energy conversion by using an optically trapped particle and a feedback protocol. The basic idea is illustrated in Fig. 21.2. Let us consider a microscopic particle on a spiral-staircase-like potential, as shown in Fig. 21.2a. The height of each step is set to be comparable to the thermal energy $k_B T$, so that the particle can jump between steps stochastically. Downward jumps along the gradient are more frequent than upward jumps and, on the average, the particle falls down the stairs unless it is externally pushed up. We can now consider the feedback control shown in Fig. 21.2b: measure the particle's position at regular intervals and, if an upward jump is observed, place a block behind the particle to prevent subsequent downward jumps. If this procedure is repeated, the particle is expected to climb the stairs. Note that, in the ideal case, energy to place the block can be negligible; this implies that the particle can obtain free energy without any direct energy injection, making use only of information about the particle's position.

21.4 Micrometre-sized heat engine

There is a fundamental limitation on the miniaturisation of engines, as small engines are not simply rescaled versions of their larger counterparts. In particular, if the work performed during the engine duty cycle is comparable to the thermal energy, one can expect that the machine will operate in reverse over short time scales; i.e., heat energy from the surroundings will be converted into useful work allowing the engine to run backwards.

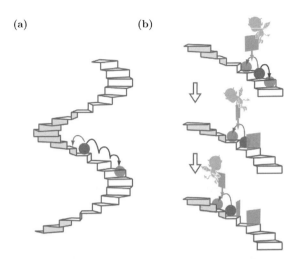

Figure 21.2 Experimental realisation of Maxwell's demon. (a) A microscopic particle is held on a spiral-staircase-like potential with a step height comparable to the thermal energy $k_B T$. Because of the presence of thermal fluctuations, the particle stochastically jumps between steps and, as the downward jumps along the gradient are more frequent than the upward ones, the particle on the average falls down the stairs. (b) When the feedback control is switched on, a block is placed behind the particle to prevent downward jumps whenever an upward jump has been observed. By repeating this cycle, the particle is expected to climb up the stairs without direct energy injection, using only the information gathered from the observation of its position. Reprinted by permission from Macmillan Publishers Ltd: Toyabe et al., *Nature Phys.* **6**, 988–92, copyright 2010.

The limitations associated with the miniaturisation of engines apply particularly to thermal engines. In fact, the original description of classical heat engines by Sadi Carnot in 1824 has largely shaped our understanding of work and heat exchange during macroscopic thermodynamic processes. It took almost two centuries before our technological prowess permitted us to study the limit of such engines on scales for which thermal fluctuations are important. Following a theoretical proposal by Schmiedl and Seifert (2008), Blickle and Bechinger (2012) realised experimentally a microscopic heat engine comprising a single colloidal particle subject to a time-dependent optical laser trap. The work associated with the system was shown to be a fluctuating quantity, which depended strongly on the cycle duration time t_{cycle}, which in turn determined the efficiency of the heat engine. Fig. 21.3 shows a comparison of a macroscopic Stirling cycle and its realisation in the microscopic experiment by Blickle and Bechinger (2012). The cycle started at time $t = 0$ with step (1), where the particle was trapped at room temperature $T_{\text{cold}} = 22\,°\text{C}$ (corresponding to the cold heat bath) in a shallow potential with $\kappa_{\min} = 281$ fN/μm. In analogy to a macroscopic engine (inset in Fig. 21.3), for $0 \leq t < t_{\text{cycle}}/2$ an isothermal compression was performed by linearly increasing the trap stiffness to $\kappa_{\max} = 1180$ fN/μm (step (1) $\to$ (2)). Next, the heating laser was suddenly (< 1 ms) turned on with a mechanical shutter, which led (relative to the sampling frequency) to an instantaneous temperature jump to $T_{\text{hot}} = 86\,°\text{C}$. Because the trapping potential remained constant during the step (2) $\to$ (3), this corresponded to an isochoric process. Next, the system was isothermally expanded during $t_{\text{cycle}}/2 \leq t < t_{\text{cycle}}$,

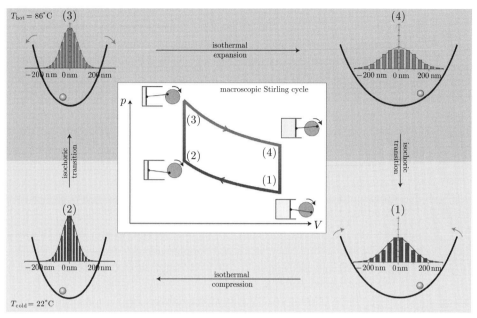

Figure 21.3 Microscopic realisation of a Stirling cycle: (1) → (2), isothermal increase of the stiffness of the optical trapping potential (isothermal compression) at $T_{\text{cold}} = 22\,°\text{C}$; (2) → (3), instantaneous temperature increase to $T_{\text{hot}} = 86\,°\text{C}$ at fixed optical potential (isochoric process); (3) → (4), isothermal decrease of trap stiffness (isothermal expansion) at T_{hot}; (4) → (1), instantaneous temperature decrease to T_{cold} at fixed optical potential (isochoric process). The histograms show the measured particle probability distributions of the corresponding stationary states together with Gaussian fits as solid lines. The width of the distribution at T_{hot} is 1.11 times greater than at T_{cold} for the same potential. This broadening is due to an increase of thermal energy. Inset: Stirling process in the pressure–volume diagram representation, where the enclosed area amounts to the work extracted by the machine. Reprinted by permission from Macmillan Publishers Ltd: Blickle and Bechinger, *Nature Phys.* **8**, 143–6, copyright 2012.

corresponding to (3) → (4), and the entire cycle of total duration t_{cycle} was completed with a temperature jump back to T_{cold} (step (4) → (1)).

21.5 Further reading

Good recent reviews of stochastic thermodynamics are Jarzynski (2011) and Seifert (2012). Apart from the Jarzynski equality, several other microscopic thermodynamic relations have been studied using optically trapped particle. Trepagnier et al. (2004) tested the Hatano and Sasa non-equilibrium steady-state (NESS) equality (Hatano and Sasa, 2001), which is a testable hypothesis deriving from a steady-state thermodynamic theory that encompasses non-equilibrium steady states and transitions between such states proposed by Oono and Paniconi (1998). Bérut et al. (2012) verified the Landauer principle, according to which the erasure of information is a dissipative process and a minimal quantity of heat, proportional to the thermal energy and called the Landauer bound, is necessarily produced when a classical bit of information is deleted (Landauer, 1961). Roldán et al. (2014) showed that an entropy decrease is a key ingredient of a Szilard engine (Parrondo, 2001) and Landauer's principle, and performed a direct measurement of the entropy change along symmetry-breaking transitions for a Brownian particle subject to a bistable potential realised through two optical traps.

References

Bérut, A., Arakelyan, A., Petrosyan, A., et al. 2012. Experimental verification of Landauer's principle linking information and thermodynamics. *Nature*, **483**, 187–9.

Blickle, V., and Bechinger, C. 2012. Realization of a micrometre-sized stochastic heat engine. *Nature Phys.*, **8**, 143–6.

Evans, D. J., and Searles, D. J. 1994. Equilibrium microstates which generate second law violating steady states. *Phys. Rev. E*, **50**, 1645–8.

Evans, D. J., Cohen, E. G. D., and Morriss, G. P. 1993. Probability of second law violations in shearing steady states. *Phys. Rev. Lett.*, **71**, 2401–4.

Hatano, T., and Sasa, S. 2001. Steady-state thermodynamics of Langevin systems. *Phys. Rev. Lett.*, **86**, 3463–6.

Jarzynski, C. 1997. Nonequilibrium equality for free energy differences. *Phys. Rev. Lett.*, **78**, 2690–93.

Jarzynski, C. 2011. Equalities and inequalities: Irreversibility and the second law of thermodynamics at the nanoscale. *Annu. Rev. Condens. Matter Phys.*, **2**, 329–51.

Landauer, R. 1961. Irreversibility and heat generation in the computing process. *IBM J. Res. Dev.*, **5**, 183–91.

Liphardt, J., Dumont, S., Smith, S. B., Tinoco, I., and Bustamante, C. 2002. Equilibrium information from nonequilibrium measurements in an experimental test of Jarzynski's equality. *Science*, **296**, 1832–5.

Oono, Y., and Paniconi, M. 1998. Steady state thermodynamics. *Prog. Theoret. Phys. Suppl.*, **130**, 29–44.

Parrondo, J. M. R. 2001. The Szilard engine revisited: Entropy, macroscopic randomness, and symmetry breaking phase transitions. *Chaos*, **11**, 725–33.

Roldán, É., Martinez, I. A., Parrondo, J. M. R., and Petrov, D. 2014. Universal features in the energetics of symmetry breaking. *Nature Phys.*, **10**, 457–61.

Schmiedl, T., and Seifert, U. 2008. Efficiency at maximum power: An analytically solvable model for stochastic heat engines. *Europhys. Lett.*, **81**, 20003.

Seifert, U. 2012. Stochastic thermodynamics, fluctuation theorems and molecular machines. *Rep. Prog. Phys.*, **75**, 126001.

Szilárd, L. 1929. On the decrease of entropy in a thermodynamic system by the intervention of intelligent beings. *Z. Phys.*, **53**, 840–56.

Toyabe, S., Sagawa, T., Ueda, M., Muneyuki, E., and Sano, M. 2010. Experimental demonstration of information-to-energy conversion and validation of the generalized Jarzynski equality. *Nature Phys.*, **6**, 988–92.

Trepagnier, E. H., Jarzynski, C., Ritort, F. 2004. Experimental test of Hatano and Sasa's nonequilibrium steady-state equality. *Proc. Natl. Acad. Sci. U.S.A.*, **101**, 15 038–41.

Wang, G. M., Sevick, E. M., Mittag, E., Searles, D. J, and Evans, D. J. 2002. Experimental demonstration of violations of the second law of thermodynamics for small systems and short time scales. *Phys. Rev. Lett.*, **89**, 050601.

22 Plasmonics

The exploitation of the plasmonic response of metals for optical trapping applications may be divided into two broad categories. The first is to use the localised surface plasmons (LSPs) supported by metallic nanoparticles, which we have discussed in Section 3.9, to enhance their mechanical reaction to the fields and thereby enable optical trapping. The second is to use the surface plasmon polaritons (SPPs) supported by nanostructures on a substrate to generate enhanced fields over a small volume, where particles can be more effectively trapped, as we have seen in Subsection 12.4.4. The distinction between SPPs and LSPs is illustrated in Fig. 22.1. SPPs are the propagating electromagnetic surface waves arising from the excitation of a collective oscillation of the free electrons in a thin metal film, i.e., a quasi two-dimensional metallic structure that is nanoscale in one dimension only; because a SPP is an evanescent mode, it must be excited by an evanescent electric field, which is typically achieved using the Kretschmann geometry. LSPs, instead, are associated with excitations of oscillations in the bound electrons of metallic nanoparticles, or nanovoids within a metallic substrate, and can be directly excited with a propagating field. In this chapter we review a number of optical trapping and manipulation experiments in which plasmonic force enhancement plays a crucial role.

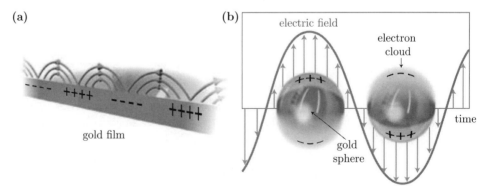

Figure 22.1 Plasmonic response of metal nanostructures. (a) Surface plasmon polaritons (SPPs) are evanescent propagating electromagnetic surface waves arising from the excitation of a collective oscillation of the free electrons in a thin metal film. (b) Localised surface plasmons (LSPs) are associated with excitations of oscillations in the bound electrons of a metallic nanoparticle. Reprinted by permission from Macmillan Publishers Ltd: Juan et al., *Nature Photon.* **5**, 349–56, copyright 2011.

22.1 Plasmonic nanoparticles

The interaction between electromagnetic radiation and conduction electrons in metal nanoparticles produces coherent localised plasmon oscillations, termed *LSPs*, with a resonant frequency dependent on the composition, size, geometry, dielectric environment and particle–particle distance (Maier, 2007). LSPs appear to be strongly dependent on metal nanostructure size and shape. As particle size increases, higher-order multipole modes appear and a description in terms of dipole contributions becomes inaccurate, as we have seen in Chapter 5. Furthermore, in the presence of aggregation, field enhancements significantly larger than for individual nanoparticles can be observed in the inter-junctions between adjacent particles, due to nearest-neighbour interactions. These inter-junction regions, characterised by highly intense and localised electromagnetic fields, are called *hot spots*.

The optical gradient force experienced by nanoparticles is typically very weak, because the dipolar polarisability given by Eqs. (3.15) and (3.25) scales with the particle volume. This volume scaling of the trapping force implies that, to confine nanoparticles against the destabilising effects of thermal fluctuations, a significantly higher optical power is required. The plasmonic nature of metallic nanoparticles can be exploited to enhance optical forces so that stable trapping can be achieved at a much lower power. Far from any plasmon resonances, the optical response of small (less than 100 nm) spherical metal nanoparticles is (mainly) the optical response of the free-electron plasma, yielding a large near-infrared polarisability, as we have seen in Section 3.9, and thus potentially a large optical trapping force when compared to dielectric particles of the same size. In particular, Svoboda and Block (1994) compared the optical trapping properties of 36.2 nm gold (plasmonic) nanospheres with 38 nm polystyrene (dielectric) ones, finding a maximum trapping force nearly seven times greater for the gold nanospheres as a result of the (seven times) greater polarisability at the 1064 nm trapping wavelength.

Metal nanoparticles are resonant systems and their optical properties (polarisability, cross-sections) are regulated by plasmon resonances that can be tuned by changing size, shape or aggregation. Pelton et al. (2006) demonstrated that the longitudinal surface plasmons of gold nanorods could be exploited for stable trapping and also alignment of the nanorods. The nanorods were synthesised with an aspect ratio producing a longitudinal surface plasmon resonance at 800 nm. With a trapping laser wavelength detuned to the long-wavelength side of the resonance, an increased residence time in the trapping region was observed (as monitored by the excitation of two-photon fluorescence by an additional probe laser beam coincident with the trap location), accompanied by a suppression of rotational diffusion, indicative of alignment. At wavelengths shorter than the plasmon resonance, the residence time in the trapping region decreased relative to free diffusion, indicating that the nanorods were repelled.

The plasmon resonance of a nanoparticle is also associated with an increase in its extinction cross-section and, hence, of its radiation pressure. The ability to stably trap a metallic nanoparticle, nanorod or nanowire depends on the details of the extinction

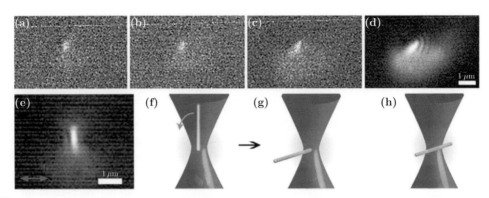

Figure 22.2 Trapping of plasmonic nanowires. (a)–(d) Dark-field images of an optically trapped gold nanowire evolving from (a) axially trapped to (d) trapped at one end. (e) Gold nanowire optically trapped in the middle. The arrow represents the direction of polarisation. (f)–(h) Illustration of the orientation of the nanowire in the traps: (f) corresponds to the orientation in (a), (g) to the one in (d) and (h) to the one in (e). Reprinted with permission from Yan et al., *ACS Nano* **7**, 8794–800. Copyright (2013) American Chemical Society.

spectrum arising from the geometry and composition of the nanostructures. For example, several experiments (Pauzauskie et al., 2006; Yan et al., 2012a) have shown that the presence of multiple longitudinal plasmon resonances in the extinction spectrum of silver nanowires renders them unstable to conventional single-beam optical tweezers, although they have been optically trapped by counter-propagating Bessel beams (Yan et al., 2012a). The same is not true, however, for gold nanowires, which can be trapped in three dimensions by a single-beam optical tweezers, as demonstrated by Yan et al. (2013). FDTD calculations reveal that the excitation of both longitudinal and transverse plasmonic resonances occurs, leading to a complex dependence of the stability of the trap on both the nanowire length and its orientation, but enabling optical trapping with either axial or transverse alignment. This is illustrated in Figs. 22.2a, 22.2b, 22.2c and 22.2d, which show dark-field images of an initially axially trapped gold nanowire reorienting until it is trapped at one end. Fig. 22.2e shows a gold nanowire trapped at the middle. Figs. 22.2f, 22.2g and 22.2h illustrate the orientation of the nanowire in these cases.

Even if the localised surface plasmon resonance of a single nanoparticle is far enough removed from the trapping wavelength so that it does not adversely affect optical trapping, the coupling between plasmons of more than one particle can shift the plasmon so that it becomes resonant with the trapping wavelength. This was observed by Ohlinger et al. (2011) by trapping 40 nm silver nanoparticles with a plasmon resonance at approximately 440 nm using a trapping laser of wavelength 808 nm. As more particles entered the trapping volume, the coupling between plasmons produced a red shift in the resonance of the coupled system relative to the single particle, as observed by the time evolution of the Rayleigh scattering spectrum from the trapping volume and the change in the colour of light scattered from blue to green to red, as shown in Fig. 22.3a. The change in the scattering spectrum is explained as due to the coupled plasmon resonance of a two-nanoparticle dimer, the wavelength of which shifts to the red as the nanoparticle separation in the dimer decreases, possibly as a result of an optical binding interaction [Section 3.10]. Measured scattered light intensities and

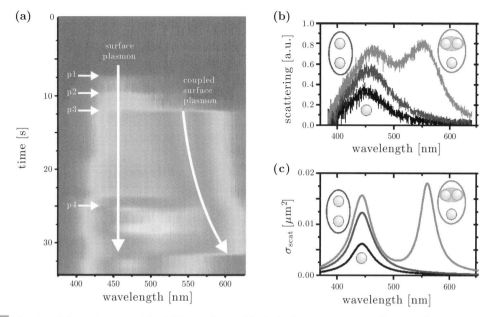

Figure 22.3 Trapping of plasmonic nanoparticles. (a) Time evolution of the Rayleigh scattering spectrum from optically trapped silver nanoparticles, showing the evolution of the wavelength of the coupled plasmon resonance. p1, p2, p3 and p4 indicate the presence of one, two, three and four nanoparticles in the trap, respectively. (b) Measured scattered intensity spectrum and (c) calculated scattering cross-section for one, two and three nanoparticles. Reprinted with permission from Ohlinger et al., *Nano Lett.* **11**, 1770–4. Copyright (2011) American Chemical Society.

calculated scattering cross-sections for a single nanoparticle, two nanoparticles and three nanoparticles (including a coupled dimerised pair) in the trap are shown in Figs. 22.3b and 22.3c. Eventually, as the coupled plasmon approaches the trapping wavelength, the particles are released from the trap as a result of the strong scattering and thermal effects.

As discussed previously, optical trapping of metal particles is often accompanied by significant heating, even at wavelengths far removed from the plasmon resonance. Seol et al. (2006) carried out a detailed investigation of the effect of heating on the optical trap spring constant by measuring it using three different methods: equipartition of energy [Section 9.5], power spectrum analysis [Section 9.8] and hydrodynamic drag [Section 9.9]. Although temperature appears explicitly in the spring constant determined by equipartition of energy, $\kappa_{eq} = k_B T / \langle x^2 \rangle$, the dependence of the power spectrum and drag-determined spring constants (κ_{ps} and κ_d) arises only through the temperature variation of fluid viscosity. Indeed, the three methods were found to produce consistent results in measurements of κ_{ps} and κ_d, which differed from κ_{eq}. Modelling the profile of the rise in temperature as a function of distance r around the nanoparticle as

$$\Delta T = \frac{P_{abs}}{4\pi r C}, \qquad (22.1)$$

where $P_{abs} = \sigma_{abs} I$ is the absorbed power and C the thermal conductivity of water, a very significant heating rate of $\Delta T = 266$ K per watt of laser power was calculated. It should be noted that boiling and bubble formation are suppressed around nanoparticles even at such elevated temperatures, because the nucleation of bubbles must overcome a surface pressure $P_{\text{surf}} = 2\Sigma/r$, where $\Sigma = 70\,\text{mN/m}$ is the surface tension of the air–water interface. For bubbles of a radius r comparable to that of the nanoparticle, P_{surf} is many times atmospheric pressure and the boiling point of water is raised very significantly (Bendix et al., 2010). When the temperature rise as a result of nanoparticle heating was taken into account, the spring constants measured by all three methods were brought into very good agreement.

22.2 Plasmonic substrates

As discussed in Subsection 12.4.4, the enhanced electromagnetic field associated with a SPP can also be exploited for optical trapping applications, as demonstrated by the experiments of Volpe et al. (2006). Excitation of a surface plasmon polariton can often be accompanied by local heating and give rise to effects such as thermophoresis of colloidal particles. Using a Kretschmann geometry for the excitation of the surface plasmon polariton in a 40 nm-thick gold layer on the surface of a prism using a 1064 nm laser beam, Garcés-Chávez et al. (2006) observed the formation of large-scale (more than 10 000 particles) predominantly hexagonally close-packed arrays of 5 μm silica spheres. The accumulation of particles is, in this case, mainly due to thermal gradient effects, because convection draws colloid-carrying fluid into the focal spot of the laser beam where the particles remain, as gravity prevents them from being transported upwards by the convective flow. As the number of particles increases, they form a close-packed array. The thermal gradient effects can be reduced by decreasing the thickness of the sample chamber, so that optical effects can be made to dominate. When this is done, particles are again drawn to the region of high-intensity electric field, but, rather than a close-packed array, they formed chains of particles parallel to the SPP wavevector, in a manner similar to that observed in optical binding in evanescent wave traps [Subsection 12.4.3]. The formation of the optically bound chains as the laser power was increased is shown in Fig. 22.4. The power density required to optically bind the particles was found to be lower than in a standard evanescent wave trap with a similar configuration, although not as low as the expected field enhancement of the

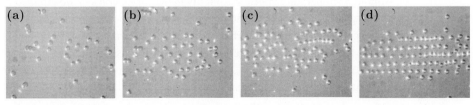

Figure 22.4 Optical binding induced by surface plasmons. Formation of particle chains in the enhanced electric field above a thin gold film supporting a SPP with increasing laser power (a) 20 mW, (b) 40 mW, (c) 50 mW and (d) 100 mW. Reprinted figure from Garcés-Chávez et al., *Phys. Rev. B* **73**, 085417. Copyright (1970) by the American Physical Society.

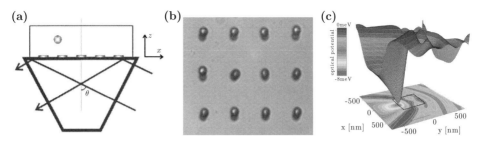

Figure 22.5 Plasmonic optical tweezers. (a) Excitation scheme for a patterned plasmonic substrate based on the Kretschmann configuration. Adapted from Quidant et al., *Opt. Lett.* **30**, 1009–11. Copyright (2005) The Optical Society. (b) Optical trapping of 4.8 μm-diameter polystyrene microspheres on a structured plasmonic substrate consisting of 4.8 μm-diameter, 40 nm-thick gold pads on a square array of side 25 μm. (c) Calculation of the optical potential energy for a 200 nm-diameter polystyrene sphere and a 0.45 μm-square gold pad. The arrow indicates the in-plane component of the exciting field wavevector, demonstrating that the potential minimum is in the forward direction from the gold pad. Figures (b) and (c) are reprinted by permission from Macmillan Publishers Ltd: Righini et al., *Nature Phys.* **3**, 477–80, copyright 2007.

surface plasmon might suggest, possibly as a result of reduced efficiency of coupling into the surface plasmon due to surface roughness. At still higher power densities, colloidal particles were driven away from the centre of the plasmon-excited region by transverse thermophoretic forces until these were balanced by the inward-directed convective force. In this region of equilibrium of thermal forces, optical forces were again seen to dominate and the particles adopted a linear chain formation.

The surface plasmon excited in a flat metal film naturally leads to a homogeneous potential that does not confine particles in the plane: the structure observed in the chain formation discussed earlier is the result of the modification of the local electric field by each particle. Transverse confinement could be achieved by patterning the optical field, but it is more common to pattern the metal surface and retain homogeneous illumination, as demonstrated by the excitation scheme shown in Fig. 22.5a (Quidant et al., 2005). With a properly designed substrate, patterning of the plasmonically enhanced near-field on a sub-optical wavelength scale is possible, leading to strong intensity gradients that are capable of confining dielectric particles (Quidant et al., 2004). This form of surface plasmon tweezers was experimentally demonstrated by Righini et al. (2007) using a pattern of 40 nm-thick gold pads on a glass substrate, illuminated under resonance conditions by a 785 nm laser beam. As can be seen in Fig. 22.5b, the enhanced field from each gold pad is capable of trapping a single colloidal polystyrene sphere. In this example, the spheres are of diameter 4.88 μm and the gold pads of diameter 4.8 μm. The intensity required to trap the array of several particles in the SP tweezers was found to be an order of magnitude less than that typically required to trap a single particle in conventional optical tweezers. Fig. 22.5b also shows that the particles are trapped towards one edge of the gold pads. This observation is borne out by calculations of the optical potential energy such as are shown in Figure 22.5c, which show a potential minimum in the forward direction, parallel to the in-plane wavevector. Subsequent work (Righini et al., 2008b) quantified the optical potential using the technique of photonic force microscopy [Chapter 10] and demonstrated

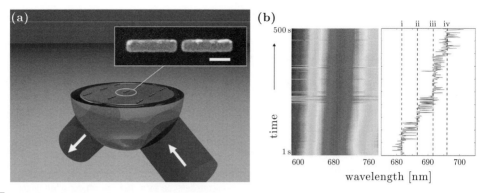

Figure 22.6 Optical traps based on plasmonic nanoantennas. (a) Set-up for evanescent wave excitation of a plasmonic nanoantenna (shown in inset; the scale bar is 200 nm). Adapted with permission from Righini et al., *Nano Lett.* **9**, 3387–91. Copyright (2009) American Chemical Society. (b) Rayleigh scattering spectra recorded sequentially, showing the shift in the resonance peak indicating the presence of a gold nanoparticle in the gap of a nanoantenna. The discrete steps in the value of the peak wavelength highlighted on the right correspond to one more nanoparticle entering or leaving the gap. Adapted with permission from Zhang et al., *Nano Lett.* **10**, 1006–11. Copyright (2010) American Chemical Society.

the controllability of the trap stiffness by varying the contributions of gradient and scattering forces, changing the incidence angle.

If the dimensions of the metallic surface features are reduced further so that they approach the wavelength of the surface plasmon, then the SPP is no longer supported. In this case the structures manifest a localised surface plasmon. Furthermore, if the metal pads are close together, they function as a plasmonic nanoantenna, with a hot spot of strongly enhanced field between the pads. Optical trapping of sub-micrometric dielectric particles in the enhanced field of such a gap nanoantenna was demonstrated by Grigorenko et al. (2008). Righini et al. (2009) further extended the technique to trap *Escherichia coli* bacteria in a nanoantenna formed by a pair of 500 nm-long, 50 nm-high gold wires with a 30 nm gap as illustrated in Fig. 22.6a. When covered with water, this structure presents a plasmon resonance at 800 nm and a field that is strongly confined to the gap between the wires with a 10-fold enhancement, in addition to hot spots at the distal ends of the gold rods. Rod-shaped *E. coli* bacteria were trapped and aligned by the plasmonic nanoantennas with an orientation parallel to the length of the rods.

Plasmonic nanoantenna traps have also been shown to be capable of trapping particles whose size lies well within the Rayleigh regime. Using a similar dipole nanoantenna excited by an evanescent field, Zhang et al. (2010) were able to demonstrate confinement of 20 nm gold nanoparticles in the enhanced field within the 25 nm gap between antenna arms of 80 nm. Here, evidence for trapping of a nanoparticle in the gap of the antenna was obtained by recording the Rayleigh scattering spectrum of the structure. The spectrum of the antenna had a resonance peak at a wavelength of approximately 690 nm, which was observed to shift to a longer wavelength of approximately 740 nm when a nanoparticle was trapped in the gap. Confinement of smaller nanoparticles (10 nm diameter) was possible after the antenna structure is optimised, i.e., by reducing the antenna gap. In this case, the shift in the Rayleigh scattering peak was smaller (5 nm) because of the smaller polarisability compared to that of a 20 nm nanoparticle, but discrete steps in the peak wavelength of the spectrum revealed the number of trapped nanoparticles, as shown in Fig. 22.6b.

Such flexibility in shaping the optical near-field via plasmonic structures opens up exciting possibilities for engineering the interaction between light and matter on the nanoscale, such as the use of plasmonic fractal structures (Volpe et al., 2011), which have been shown to produce a sub-diffraction-limit focal spot and stronger field confinement than a simple gap antenna of the same dimensions, or the control of the relative phases of interfering SPPs (Gjonaj et al., 2011) to shape a plasmonic focus at a desired location in the plane.

22.3 Plasmonic apertures

As mentioned previously, similarly to a nanoparticle or nanostructured antenna on a dielectric substrate, a nanoscale hole (or nanovoid) in a metallic substrate also possesses plasmonic resonance features. As demonstrated theoretically by Sainidou and García de Abajo (2008) for a gold nanorod in a nanovoid, the features of the spectrum of the nanohole are very sensitive to the particle within the gap, hole, cavity or void. The interaction between a particle and a nanohole can be exploited by properly engineering the spectrum of the nanohole so that the enhanced field within the hole is maximum when the particle is present. This is the essence of the self-induced back action (SIBA) trapping scheme demonstrated by Juan et al. (2009) by trapping 50 nm dielectric particles in a circular aperture [Subsection 12.5]. The SIBA trapping scheme is advantageous in that it does not rely on a resonance in the object to be trapped but on the combination of particle and nanoaperture, making it ideal for nanoscale dielectric or biological specimens. The presence of a higher-refractive-index nanoparticle in the aperture shifts the resonance to longer wavelengths, increasing the transmission and thus the flux of photon momentum through the aperture. The movement of a particle in the aperture is therefore correlated with a change in the transmitted light and hence momentum, which must be compensated for by a transfer of momentum to the particle itself, resulting in stable trapping at an equilibrium position within the aperture.

Despite the effectiveness of the SIBA trapping scheme for small dielectric particles with greatly reduced intensities, further optimisation of the aperture in which trapping takes place is necessary to maximise the effect and apply it to still smaller particles. Reducing the diameter of a circular nanoaperture leads to a rapid decay of the field enhancement and transmission intensity, and also shifts the resonance of the aperture to shorter wavelengths where absorption is increased. Changing the shape of the aperture, however, introduces features that can be exploited to optimise the SIBA trap. Chen et al. (2012) were able to trap dielectric particles as small as 22 nm using a rectangular nanopore created by micromachining a free-standing silicon membrane coated with a 100 nm gold layer, as shown in Fig. 22.7a. The geometry of the nanopore determines its resonance features, which can thus be engineered to be favourable for SIBA trapping. For incident polarisation

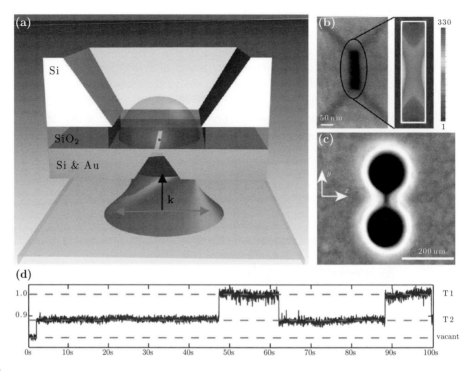

Figure 22.7 Optical traps based on plasmonic nanoapertures. (a) Set-up for illumination of a rectangular plasmonic nanoaperture in a gold film supported on a silicon membrane. (b) Calculation of the enhanced field distribution inside the aperture showing maxima at the mid-point of the long sides due to excitation of the Fabry–Pérot resonance at $\lambda_{SPP}/2$. (a) and (b) are adapted with permission from Chen et al., *Nano Lett.* **12**, 125–32. Copyright (2012) American Chemical Society. (c) Plasmonic double nanohole: when excited, the double nanohole shows an enhanced field strongly confined in the small gap between the two tips separating the holes. (d) Time-evolution of the intensity transmitted through the double nanohole shown in (c), showing two discrete transmission levels, T1 and T2, corresponding to the trapping of single BSA protein molecules in the native (N-state) and unfolded (F-state) states. Figures (c) and (d) are adapted with permission from Pang and Gordon, *Nano Lett.* **12**, 402–6. Copyright (2012) American Chemical Society.

parallel to the short sides, the rectangular nanopore presents a strongly confined surface plasmon gap mode along the short axis, with a Fabry–Pérot resonance in the direction of the long axis, which is half the SPP wavelength (Gordon et al., 2006). The resulting field in the aperture is thus maximal at the edges of the gap at the mid-points of the long sides, as shown in Fig. 22.7b, and for small enough particles is capable of trapping more than one particle in the aperture. Using just 2.5 mW of laser light at a wavelength of 1064 nm (detuned to the red of the 950 nm resonance of the nanoaperture), stable trapping of single and double nanoparticles with diameter 22 nm was observed. From the statistics of the distribution of intensity transmitted through the aperture, states corresponding to zero, one or two particles in the aperture could be resolved, along with the transition rates between these states, demonstrating that particles entered and left the aperture sequentially and also that the double-trapping state was more stable, arising from the interaction of the second particle with the field in the aperture as modified by the presence of the first particle, which could be explained as a form of optical binding. Chen et al. (2012) also define a useful figure of merit for comparing the efficiencies of nanoparticle and plasmonic-based traps as

$$\text{F.o.M.} = \frac{1}{gI_{\text{inc}}\alpha}, \qquad (22.2)$$

where I_{inc} is the incident intensity, g the enhancement factor in the aperture and α the particle polarisability. Using this definition, the F.o.M. of a conventional optical tweezers for an 18 nm-diameter gold particle (Hansen et al., 2005) is F.o.M. $\approx 40\,(\text{nmW})^{-1}$, whereas for the rectangular aperture SIBA trap using 22 nm dielectric particles this is increased to $10^4\,(\text{nmW})^{-1}$.

An alternative nanoaperture geometry that leads to strong field enhancement is the double nanohole (Kumar and Gordon, 2006) shown in Fig. 22.7c. In this structure, when the polarisation of the incident beam is perpendicular to the line joining the centres of the two holes, a large field enhancement occurs in the gap region. Compared with a single circular nanoaperture, because the field is concentrated in this small gap between the tips of the double-hole structure, the transmission is more sensitive to small changes of refractive index in this region. Using such a double nanohole, Pang and Gordon (2011) demonstrated trapping of dielectric particles as small as 12 nm. Compared with the rectangular aperture, the double-nanohole structure provides stronger confinement of the field in the short tip region between the holes rather than along the long side of the rectangular aperture. The same authors (Pang and Gordon, 2012) demonstrated the efficacy of the double-nanohole trap for biological material by trapping single molecules of protein bovine serum albumin (BSA), which has a hydrodynamic radius of 3.4 nm (in the N-state). An example of the intensity transmitted through the aperture is shown in Fig. 22.7d, which, as with other experiments using dielectric spheres, shows two discrete levels of intensity transmission (labelled T1 and T2). However, the authors hypothesise that the increased transmission state is not due to the presence of two BSA molecules in the aperture, but rather to an optically induced conformational change in the BSA molecule, unfolding it to the more extended, and more strongly polarisable, F-state. Supporting evidence for this hypothesis was gained by changing the pH of the buffer solution at the same optical power: at lower pH, where the BSA molecule already exists in the unfolded F-state, only the higher transmission level corresponding to trapping of this unfolded form was observed.

22.4 Further reading

Plasmonic optical trapping is one of the fastest-growing applications for optical forces. A number of excellent review articles covering both the trapping of metallic nanoparticles exhibiting LSPs and the use of SPPs for trapping have appeared in the last few years, including papers by Dienerowitz et al. (2008), Quidant and Girard (2008), Righini et al. (2008a), Juan et al. (2011), Shoji and Tsuboi (2014) and Urban et al. (2014).

In any optical trapping experiment that exploits plasmonic resonances, the issue of thermal effects must be addressed. Heating can be a significant factor and indeed

Ploschner et al. (2010) have suggested that it plays a role in the trapping of particles near a nanoantenna. To mitigate heating effects, Wang et al. (2011) introduced a design for plasmonic trapping using gold nanopillars on a gold film on top of a copper film supported on a silicon substrate. As these materials have a significantly higher thermal conductivity than the glass substrates used in previous experiments, an illumination intensity several orders of magnitude higher can be used before the same rise in temperature is reached.

Detailed studies of the optical trapping of both gold (Hansen et al., 2005; Hajizadeh and Reihani, 2010) and silver (Bosanac et al., 2008) nanoparticles have been conducted. In both cases, a maximum trapping force proportional to the third power of the particle radius was observed for diameters smaller than 100 nm, with a crossover to a lower exponent for larger radii. This size-scaling behaviour was interpreted by accounting for local heating of the surroundings (Seol et al., 2006; Kyrsting et al., 2011) and modelling the metallic nanoparticles as enclosed in a small steam bubble (Messina et al., 2011; Saija et al., 2009).

Non-spherical metallic nanoparticles, including gold nanorods (Pelton et al., 2006; Selhuber-Unkel et al., 2008), silver nanowires (Yan et al., 2012b) and aggregates of gold nanoparticles (Messina et al., 2011), can sustain plasmon resonances in a broad spectral region from the visible to the near infrared, which plays a crucial role in the enhancement of radiation forces and torques in optical tweezers. More specifically, elongated plasmonic nanostructures (such as nanowires and nanorods) are usually trapped with their axis parallel to the electric field vector of the trapping laser and orthogonal to the propagation axis (Borghese et al., 2007). This provides a means of controlling their orientation by rotating the laser polarisation (Jones et al., 2009). Tong et al. (2010) demonstrated that plasmonic nanostructures with lengths ranging from tens of nanometres to several micrometres could be aligned and rotated using the polarisation-dependent optical forces arising from their interaction with a single beam of linearly polarised near-infrared light. Ruijgrok et al. (2011) also studied the optically induced orientation of gold nanorods, finding that the resulting dynamics could be accounted for by allowing for a local heating of up to 70 K. Heating effects and trapping of non-spherical particles will be discussed further in the next chapter.

References

Bendix, P. M., Reihani, S. N. S., and Oddershede, L. B. 2010. Direct measurements of heating by electromagnetically trapped gold nanoparticles on supported lipid bilayers. *ACS Nano*, **4**, 2256–62.

Borghese, F., Denti, P., Saija, R., and Iatì, M. A. 2007. Optical trapping of nonspherical particles in the T-matrix formalism. *Opt. Express*, **15**, 11 984–98.

Bosanac, L., Aabo, T., Bendix, P. M., and Oddershede, L. B. 2008. Efficient optical trapping and visualization of silver nanoparticles. *Nano Lett.*, **8**, 1486–91.

Chen, C., Juan, M. L., Li, Y., et al. 2012. Enhanced optical trapping and arrangement of nano-objects in a plasmonic nanocavity. *Nano Lett.*, **12**, 125–32.

Dienerowitz, M., Mazilu, M., and Dholakia, K. 2008. Optical manipulation of nanoparticles: A review. *J. Nanophoton.*, **2**, 021875.

Garcés-Chávez, V., Quidant, R., Reece, P. J., et al. 2006. Extended organization of colloidal microparticles by surface plasmon polariton excitation. *Phys. Rev. B*, **73**, 085417.

Gjonaj, B., Aulbach, J., Johnson, P. M., et al. 2011. Active spatial control of plasmonic fields. *Nature Photon.*, **5**, 360–63.

Gordon, R., Kumar, L., Kiran Swaroop, L. K., and Brolo, A. G. 2006. Resonant light transmission through a nanohole in a metal film. *IEEE Trans. Nanotechnol.*, **5**, 291–4.

Grigorenko, A. N., Roberts, N. W., Dickinson, M. R., and Zhang, Y. 2008. Nanometric optical tweezers based on nanostructured substrates. *Nature Photon.*, **2**, 365–70.

Hajizadeh, F., and Reihani, S. N. S. 2010. Optimized optical trapping of gold nanoparticles. *Opt. Express*, **18**, 551–9.

Hansen, P. M., Bhatia, V. K., Harrit, N., and Oddershede, L. 2005. Expanding the optical trapping range of gold nanoparticles. *Nano Lett.*, **5**, 1937–42.

Jones, P. H., Palmisano, F., Bonaccorso, F., et al. 2009. Rotation detection in light-driven nanorotors. *ACS Nano*, **3**, 3077–84.

Juan, M. L., Gordon, R., Pang, Y., Eftekhari, F., and Quidant, R. 2009. Self-induced back-action optical trapping of dielectric nanoparticles. *Nature Phys.*, **5**, 915–19.

Juan, M. L., Righini, M., and Quidant, R. 2011. Plasmon nano-optical tweezers. *Nature Photon.*, **5**, 349–56.

Kumar, L. K. S., and Gordon, R. 2006. Overlapping double-hole nanostructure in a metal film for localized field enhancement. *IEEE J. Sel. Top. Quant. Electron.*, **12**, 1228–32.

Kyrsting, A., Bendix, P. M., Stamou, D. G., and Oddershede, L. B. 2011. Heat profiling of three-dimensionally optically trapped gold nanoparticles using vesicle cargo release. *Nano Lett.*, **11**, 888–92.

Maier, S. A. 2007. *Plasmonics: Fundamentals and applications*. New York: Springer Verlag.

Messina, E., Cavallaro, E., Cacciola, A., et al. 2011. Plasmon-enhanced optical trapping of gold nanoaggregates with selected optical properties. *ACS Nano*, **5**, 905–13.

Ohlinger, A., Nedev, S., Lutich, A. A., and Feldmann, J. 2011. Optothermal escape of plasmonically coupled silver nanoparticles from a three-dimensional optical trap. *Nano Lett.*, **11**, 1770–4.

Pang, Y., and Gordon, R. 2011. Optical trapping of 12 nm dielectric spheres using double-nanoholes in a gold film. *Nano Lett.*, **11**, 3763–7.

Pang, Y., and Gordon, R. 2012. Optical trapping of a single protein. *Nano Lett.*, **12**, 402–6.

Pauzauskie, P. J., Radenovic, A., Trepagnier, E., et al. 2006. Optical trapping and integration of semiconductor nanowire assemblies in water. *Nature Mater.*, **5**, 97–101.

Pelton, M., Liu, M., Kim, H. Y., et al. 2006. Optical trapping and alignment of single gold nanorods by using plasmon resonances. *Opt. Lett.*, **31**, 2075–7.

Ploschner, M., Mazilu, M., Krauss, T. F., and Dholakia, K. 2010. Optical forces near a nanoantenna. *J. Nanophoton.*, **4**, 041570.

Quidant, R., and Girard, C. 2008. Surface-plasmon-based optical manipulation. *Laser Photon. Rev.*, **2**, 47–57.

Quidant, R., Badenes, G., Cheylan, S., et al. 2004. Sub-wavelength patterning of the optical near-field. *Opt. Express*, **12**, 282–7.

Quidant, R., Petrov, D., and Badenes, G. 2005. Radiation forces on a Rayleigh dielectric sphere in a patterned optical near field. *Opt. Lett.*, **30**, 1009–11.

Righini, M., Zelenina, A. S., Girard, C., and Quidant, R. 2007. Parallel and selective trapping in a patterned plasmonic landscape. *Nature Phys.*, **3**, 477–80.

Righini, M., Girard, C., and Quidant, R. 2008a. Light-induced manipulation with surface plasmons. *J. Opt. A Pure Appl. Opt.*, **10**, 093001.

Righini, M., Volpe, G., Girard, C., Petrov, D., and Quidant, R. 2008b. Surface plasmon optical tweezers: Tunable optical manipulation in the femtonewton range. *Phys. Rev. Lett.*, **100**, 186804.

Righini, M., Ghenuche, P., Cherukulappurath, S., et al. 2009. Nano-optical trapping of Rayleigh particles and *Escherichia coli* bacteria with resonant optical antennas. *Nano Lett.*, **9**, 3387–91.

Ruijgrok, P. V., Verhart, N. R., Zijlstra, P., Tchebotareva, A. L., and Orrit, M. 2011. Brownian fluctuations and heating of an optically aligned gold nanorod. *Phys. Rev. Lett.*, **107**, 037401.

Saija, R., Denti, P., Borghese, F., Maragò, O. M., and Iatì, M. A. 2009. Optical trapping calculations for metal nanoparticles: Comparison with experimental data for Au and Ag spheres. *Opt. Express*, **17**, 10 231–41.

Sainidou, R., and García de Abajo, F. J. 2008. Optically tunable surfaces with trapped particles in microcavities. *Phys. Rev. Lett.*, **101**, 136802.

Selhuber-Unkel, C., Zins, I., Schubert, O., Sonnichsen, C., and Oddershede, L. B. 2008. Quantitative optical trapping of single gold nanorods. *Nano Lett.*, **8**, 2998–3003.

Seol, Y., Carpenter, A. E., and Perkins, T. T. 2006. Gold nanoparticles: Enhanced optical trapping and sensitivity coupled with significant heating. *Opt. Lett.*, **31**, 2429–31.

Shoji, T., and Tsuboi, Y. 2014. Plasmonic optical tweezers toward molecular manipulation: Tailoring plasmonic nanostructure, light source, and resonant trapping. *J. Phys. Chem. Lett.*, **5**, 2957–67.

Svoboda, K., and Block, S. M. 1994. Optical trapping of metallic Rayleigh particles. *Opt. Lett.*, **19**, 930–32.

Tong, L., Miljkovic, V. D., and Käll, M. 2010. Alignment, rotation, and spinning of single plasmonic nanoparticles and nanowires using polarization dependent optical forces. *Nano Lett.*, **10**, 268–73.

Urban, A. S., Carretero-Palacios, S., Lutich, A. A., et al. 2014. Optical trapping and manipulation of plasmonic nanoparticles: Fundamentals, applications, and perspectives. *Nanoscale*, **6**, 4458–74.

Volpe, G., Quidant, R., Badenes, G., and Petrov, D. 2006. Surface plasmon radiation forces. *Phys. Rev. Lett.*, **96**, 238101.

Volpe, G., Volpe, G., and Quidant, R. 2011. Fractal plasmonics: Subdiffraction focusing and broadband spectral response by a Sierpinski nanocarpet. *Opt. Express*, **19**, 3612–18.

Wang, K., Schonbrun, E., Steinvurzel, P., and Crozier, K. B. 2011. Trapping and rotating nanoparticles using a plasmonic nano-tweezer with an integrated heat sink. *Nature Commun.*, **2**, 469.

Yan, Z., Sweet, J., Jureller, J. E., et al. 2012a. Controlling the position and orientation of single silver nanowires on a surface using structured optical fields. *ACS Nano*, **6**, 8144–55.

Yan, Z., Jureller, J. E., Sweet, J., et al. 2012b. Three-dimensional optical trapping and manipulation of single silver nanowires. *Nano Lett.*, **12**, 5155–61.

Yan, Z., Pelton, M., Vigderman, L., Zubarev, E. R., and Scherer, N. F. 2013. Why single-beam optical tweezers trap gold nanowires in three dimensions. *ACS Nano*, **7**, 8794–800.

Zhang, W., Huang, L., Santschi, C., and Martin, O. J. F. 2010. Trapping and sensing 10 nm metal nanoparticles using plasmonic dipole antennas. *Nano Lett.*, **10**, 1006–11.

23 Nanostructures

Optical trapping and manipulation have flourished at the sub-nanometre scale, where light–matter mechanical coupling has enabled cooling of neutral atoms, ions and molecules, and at the micrometre scale, where the momentum transfer associated with the scattering of light allows the manipulation of microscopic particles and biological entities. However, it has been difficult to scale either of these techniques to the nanoscale, a size range of crucial importance for a wealth of potential applications based on technologically significant nanomaterials, such as quantum dots, nanowires, nanotubes, graphene and two-dimensional crystals, as shown in Fig. 23.1. Only recently have several novel approaches been suggested and realised to overcome such difficulties. In this chapter, we review the state of the art in trapping and manipulation of nanostructures with an emphasis on some of the most promising advances in controlled manipulation and assembly of individual and multiple nanostructures.

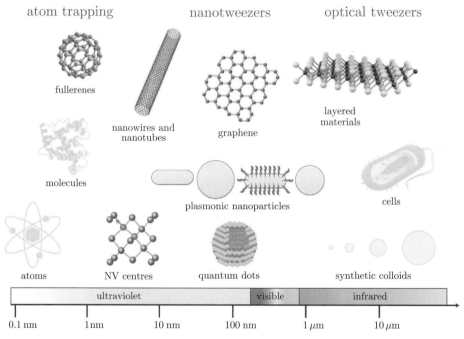

Figure 23.1 Size regimes of optical trapping: (From left to right) *atom trapping* (a few ångstroms to a few nanometres), *nanotweezers* (a few nanometres to a few hundred nanometres) and *optical tweezers* (from a fraction of a micrometre up). The horizontal scale bar shows the average object size and the corresponding light wavelength. Adapted from Maragò et al., *Nature Nanotechnol.* **8**, 807–19 (2013).

23.1 Metal nanoparticles

Optical forces acting on nanoparticles are generally in the femtonewton range as a result of the volumetric scaling of light-induced polarisability [Chapter 3]. Thus, they are often too weak to achieve stable optical trapping. Wright et al. (1993) explicitly evaluated the volume-scaling behaviour of the maximum trapping force for polystyrene nanospheres, showing a decrease by three orders of magnitude in the maximum trapping force as the sphere radius decreased from 100 nm to 10 nm. Therefore, to confine nanoparticles against the destabilising effects of thermal fluctuations, a significantly higher optical power is required: whereas a micrometre-sized polystyrene sphere can be stably trapped with a fraction of milliwatt in a standard optical tweezers set-up [Chapter 8], a 100 nm sphere requires 15 mW (Ashkin et al., 1986). This implies that for a 10 nm sphere about 1.5 W would be needed.

As anticipated in Chapter 22, metal nanoparticles can sustain localised surface plasmons with very high saturation intensities. These can yield enhanced optical forces leading to stable optical trapping with only a few milliwatts of power (Jones et al., 2009; Messina et al., 2011a,b). When light is tuned on the long-wavelength side far from any plasmon resonance, the response of individual metal nanoparticles to an incident light field is dictated by the free-electron plasma, resulting in increased dipolar polarisability in the near infrared. Thus, a tenfold enhancement of optical trapping forces is measured on individual metal nanospheres smaller than 100 nm with respect to polystyrene ones (Svoboda and Block, 1994; Hansen et al., 2005). Both gold nanoparticles with diameters from 9.5 to 254 nm (Hansen et al., 2005; Hajizadeh and Reihani, 2010) and silver nanoparticles with diameters from 20 to 275 nm (Bosanac et al., 2008) have been optically trapped in three dimensions. In both cases, a maximum trapping force proportional to the third power of the particle radius was observed for diameters smaller than 100 nm, with crossover to a lower exponent for larger radii. This size-scaling behaviour at large radii was interpreted by accounting for local heating of the surroundings (Seol et al., 2006; Kyrsting et al., 2011) and modelling the metallic nanoparticles as enclosed in a small steam bubble (Saija et al., 2009; Messina et al., 2011b).

Non-spherical metal nanoparticles, including gold nanorods (Pelton et al., 2006; Selhuber-Unkel et al., 2008), i.e., nanocylinders with an aspect ratio smaller than 10, silver nanowires (Yan et al., 2012) and gold nano-aggregates (Messina et al., 2011b) can sustain plasmon resonances in a broad spectral region in the visible and near infrared that play a crucial role in the enhancement of radiation forces and torques in optical trapping (Toussaint et al., 2007; Jones et al., 2009; Tong et al., 2010, 2011). More specifically, elongated plasmonic nanostructures, such as nanowires and nanorods, are usually trapped with their axes parallel to the electric field vector of the trapping laser and orthogonal to the propagation axis (Pelton et al., 2006; Selhuber-Unkel et al., 2008). The strength of this aligning torque is enhanced by tuning the laser close to the plasmon resonance (Pelton et al., 2006). This provides a means to control their orientation by rotating the laser polarisation (Jones et al., 2009). Plasmonic nanostructures with lengths from tens of nanometres to several micrometres were aligned and rotated using a single beam of linearly polarised near-infrared light (Tong et al., 2010). Dienerowitz et al. (2008) demonstrated experimentally the change in the sign of the gradient force by blue detuning the laser wavelength with respect to the metal nanoparticles' plasmon resonance, achieving particle confinement in the dark spot of an optical vortex beam. The frequency dependence of the plasmon-enhanced radiation force was also exploited in a system of two counter-propagating evanescent waves at different wavelengths to selectively guide metal nanoparticles of different sizes in opposite directions (Ploschner et al., 2012).

Resonant illumination of plasmonic nanoparticles gives rise to strong heating effects because of light absorption (Baffou and Quidant, 2013). Temperature increases of hundreds of kelvin were observed by trapping gold nanoparticles adjacent to fluorophore-containing lipid vesicles with permeability sensitive to temperature, as, when heated above the gel-transition temperature, fluorophores diffused out of the vesicle (Kyrsting et al., 2011). Further experiments exploited the differing longitudinal and transverse plasmon resonances of gold nanorods to control the local heating via the orientation of the nanorods with respect to the electric field vector of the trapping laser, suggesting that gold nanorods could be sensitive and switchable remote-controlled heat transducers for small volume samples (Ma et al., 2012).

Finally, optical trapping and manipulation of metal nanoparticles hold perspectives for several applications in nanofabrication, sensing, analytics, biology and medicine (Maragò et al., 2013; Urban et al., 2014).

23.2 Semiconductor nanostructures

Semiconductor nanostructures are of crucial importance in electronics and optoelectronics (Mitin et al., 1999). Technological progress towards miniaturisation has led to the development of crystal growth techniques and wet chemistry protocols that can build semiconductor structures layer by layer and engineer their morphological, electronic and optical properties atom by atom. The key element in the exploitation of *low-dimensional* nanostructures is *quantum confinement*, i.e., the fact that, when one or more dimensions of a semiconductor material are reduced to the nanoscale, the motion of electrons in the conduction band can be confined so much that quantum mechanics comes into play and the motion is quantised.

The length at which a quantum-mechanical description is needed is the *de Broglie wavelength* of the conduction electrons in the material, $\lambda_{dB} = h/\sqrt{\tilde{m}_e k_B T}$, where h is the Planck constant, $\tilde{m}_e$ is the electron effective mass in the crystal, k_B is the Boltzmann constant and T is the absolute temperature. Since the length scale at which the classical–quantum crossover occurs is related to λ_{dB}, it depends on both temperature and crystal properties through the effective mass. In normal circumstances, λ_{dB} is much smaller than the dimensions of the crystal and the electron kinetics can be described by classical physics. At room temperature and considering a typical semiconductor for which $\tilde{m}_e$ is 10% of the electron mass (Mitin et al., 1999), a length of about 10 nm or less is necessary to observe quantum confinement effects.

23.2 Semiconductor nanostructures

Table 23.1 Quantum confinement and density of states in low-dimensional nanostructures

	Bulk	Quantum well	Quantum wire	Quantum dot
ϵ	—	$\dfrac{\pi^2 \hbar^2 n^2}{2\tilde{m}_e L_z^2}$	$\dfrac{\pi^2 \hbar^2}{2\tilde{m}_e}\left(\dfrac{m^2}{L_y^2} + \dfrac{n^2}{L_z^2}\right)$	$\dfrac{\pi^2 \hbar^2}{2\tilde{m}_e}\left(\dfrac{l^2}{L_x^2} + \dfrac{m^2}{L_y^2} + \dfrac{n^2}{L_z^2}\right)$
E	$\dfrac{\hbar k^2}{2\tilde{m}_e}$	$\dfrac{\hbar^2(k_x^2 + k_y^2)}{2\tilde{m}_e} + \epsilon_n$	$\dfrac{\hbar^2 k_x^2}{2\tilde{m}_e} + \epsilon_{n,m}$	$\epsilon_{n,m,l}$
$g(E) \propto$	$\sqrt{E}$	$\Theta(E - \epsilon_n)$	$\dfrac{1}{\sqrt{E - \epsilon_{n,m}}}$	$\delta(E - \epsilon_{n,m,l})$

Confinement of the electron wavefunction in one, two and three dimensions yields very different effects on the optical response of materials. As it is directly related to the energy spectrum and number of accessible states, it can have great relevance also to optical forces. The relevant quantity that characterises quantum-confined systems is the density of states, $g(E) = dN/dE$, which is needed for the evaluation of the number of accessible states ΔN in an energy interval ΔE and is directly related to the absorption of light. For an unbound electron in bulk, $g(E)$ can be calculated from the energy of a free particle with an effective mass $\tilde{m}_e$ [Table 23.1]. First, we count the accessible states in k-space, $N(k)$, as the ratio between the volume of the sphere in k-space, $(4\pi/3)k^3$, and the minimum volume occupied by a single state, $(2\pi)^3/V$, where V is the bulk volume. Therefore, we get that $dN(k)/dk = Vk^2/(2\pi^2)$ and, because the relation between the wavevector and energy in bulk is $k = \sqrt{2\tilde{m}_e E/\hbar^2}$, where $\hbar = h/(2\pi)$, the density of accessible energy states is obtained as

$$g_{\text{bulk}}(E) = 2\frac{dN}{dk}\frac{dk}{dE} = \frac{V}{2\pi^2}\left(\frac{2\tilde{m}_e}{\hbar^2}\right)^{3/2}\sqrt{E}, \qquad (23.1)$$

where the factor of 2 accounts for the electron spin degeneracy.

For low-dimensional nanostructures, the density of states changes because of quantum confinement. If the motion is confined (hence quantised) in one direction and the electrons are free to move in the other two directions, the structure is called a *quantum well*. If the motion is confined in two dimensions, we have a *quantum wire*. If the electrons are confined in all three dimensions, we have a *quantum dot*, where the energy spectrum is fully quantised. As an example, we sketch the derivation of $g(E)$ for quantum wires. We will assume that the electron motion is confined in y and z, whereas it is free in x. Thus, the accessible states in the y and z directions are quantised with a bound energy (Mitin et al., 1999),

$$\epsilon_{n,m} = \frac{\pi^2 \hbar^2}{2\tilde{m}_e}\left(\frac{m^2}{L_y^2} + \frac{n^2}{L_z^2}\right), \qquad (23.2)$$

where L_y and L_z are the dimensions of the structure in the confined directions, and m and n are the quantum numbers associated with the energy states. We now write the total energy

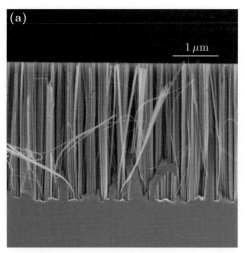
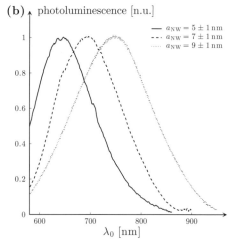

Figure 23.2 Photoluminescence of nanowires. (a) Cross-section of a SEM image of silicon nanowires obtained by the metal-assisted wet etching technique. Image credits: Alessia Irrera. (b) Photoluminescence spectra obtained by exciting silicon nanowire samples having different diameters with 488 nm light. The blue shift for smaller diameters indicates stronger quantum confinement. Reproduced with permission from Irrera et al., *Nanotechnol.* **23**, 075204. Copyright (2012) IOP Publishing. Reproduced by permission of IOP Publishing. All rights reserved.

in terms of the wavevector in the free direction k_x as

$$E = \epsilon_{n,m} + \frac{\hbar^2 k_x^2}{2\tilde{m}_e}. \tag{23.3}$$

Thus, by counting the number of states in the one-dimensional k-space $N(k_x) = k_x L_x/(2\pi)$, we get the density of energy states for one-dimensional nanostructures as

$$g(E) = \frac{L_x}{\pi} \sqrt{\frac{2\tilde{m}_e}{\hbar^2}} \frac{1}{\sqrt{E - \epsilon_{n,m}}}. \tag{23.4}$$

An example of quantum-confined system is shown in Fig. 23.2a. These are ultra-thin silicon nanowires synthesised by a wet etching process assisted by a metal thin film (Irrera et al., 2012). When the nanowires are optically excited, they exhibit photoluminescence emission at room temperature, as shown in Fig. 23.2b. The spectra corresponding to nanowires with decreasing diameter a_{NW} below 10 nm show a shift to higher energy (blue shift) as expected for increasing quantum confinement.

Amongst nanomaterials, a special place is held by those based on carbon. As shown in Fig. 23.3a, because of the flexibility of carbon bonding, carbon-based systems occur in many different structures with a wide variety of physical properties (Castro-Neto et al., 2009). In particular, graphene (Novoselov et al., 2004), where carbon atoms are arranged in a honeycomb structure, and carbon nanotubes (Iijima, 1991), which can be obtained by rolling graphene along a given direction and reconnecting the carbon bonds, are exemplar low-dimensional nanostructures whose mechanical, thermal, electronic and optical

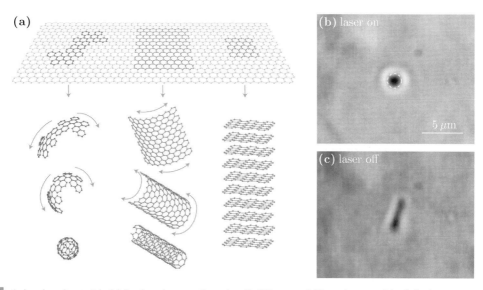

Figure 23.3 Carbon-based materials. (a) Graphene is a two-dimensional building material for carbon materials of all other dimensionalities: it can be wrapped up into (zero-dimensional) buckyballs, rolled into (one-dimensional) nanotubes or stacked into (three-dimensional) graphite. Reprinted by permission from Macmillan Publishers Ltd: Geim and Novoselov, *Nature Materi.* **6**, 183–91, copyright 2007. (b, c) Carbon nanotubes were the first example of carbon nanostructures to be optically trapped. Reprinted with permission from Maragò et al., *Nano Lett.* **8**, 3211–16. Copyright (2008) American Chemical Society.

properties are dramatically affected by the honeycomb atomic arrangement (Castro-Neto et al., 2009; Bonaccorso et al., 2010).

For optical trapping applications, the change in the spectral behaviour of low-dimensional nanostructures needs to be considered carefully. In particular, a laser wavelength that is too close to resonance might increase light absorption enough to increase scattering force and prevent stable trapping; thus it is often advisable to use trapping light shifted from the nanostructure resonances. Moreover, for non-spherical nanostructures such as nanotubes, nanowires, graphene or other two-dimensional crystals, the reduced symmetry means that optical torques are crucial in determining their alignment and orientation with respect to the beam propagation and polarisation directions (Borghese et al., 2008; Bareil and Sheng, 2010; Simpson and Hanna, 2011). For example, Fig. 23.3b shows how a nanotube bundle is trapped with its main axis along the light propagation direction, whereas it diffuses freely (in this case turning horizontally) because of thermal fluctuations when the trap is switched off [Fig. 23.3c]. Carbon nanotubes have also been also trapped with cylindrical vector beams (Donato et al., 2012) and organised through optical binding interactions (Maragò et al., 2013).

As we have seen in Chapter 10, optically trapped particles can be used to probe nanoscopic forces. This is also possible using nanostructures and an interferometric particle position detection technique. For spherical nanostructures, such as quantum dots, the detector signals are proportional to the centre-of-mass displacements (Jauffred et al., 2008), as is the case for larger particles. For non-spherical nanoparticles, the detector signals also contain angular information (Maragò et al., 2008a, 2010; Pollard et al., 2010); in fact, their length is the key parameter that regulates forces, torques and hydrodynamics (Irrera et al., 2011; Reece et al., 2011). The combination of sensitive position detection and low spring constant enables force resolution of a few femtonewtons, which far surpasses other scanning probe microscopy techniques (Neuman and Nagy, 2008). Furthermore, taking advantage of the low trapping stiffness (from 1 fN/nm to 1 pN/nm), it is possible to image soft structures (Dufresne and Grier, 1998; Rohrbach et al., 2004; Neuman and Nagy, 2008). Spatial resolution, however, may be limited by particle size and thermal fluctuations. The use of linear nanostructures as probes is an ideal option for increasing spatial resolution in microscopy (Phillips et al., 2011, 2012b; Olof et al., 2012), because the combination of their nanometric transverse size and micrometric length is key to enabling stable optical tweezing while maintaining high lateral resolution, even at very low laser power (Maragò et al., 2008b; Ikin et al., 2009; Pollard et al., 2010; Phillips et al., 2012a).

Finally, light-emitting or -guiding nanostructures, such as potassium niobate nanowires (Nakayama et al., 2007) or polymer nanofibres (Neves et al., 2010), offer tunable nanometric light sources that can be used to implement novel forms of subwavelength microscopy (Nakayama et al., 2007).

23.3 Optical force lithography and placement

The combination of laser manipulation and photopolymerisation allows one to build three-dimensional structures using nanoparticles and to place them on a substrate. For example, Ito et al. (2001) used an infrared trapping laser beam to collect near its focus nanoparticles suspended in solution and then applied an ultraviolet laser to induce photopolymerisation of a monomer, also present in solution, around the nanoparticles. It is also possible to use optical tweezers to trap and position single nanoparticles, e.g., to manipulate, assemble and fuse together different semiconductor nanowires (Pauzauskie et al., 2006), as shown in Fig. 23.4a. Controlled deposition of optically trapped In_2O_3 nanowires was realised by fast-scanning the trapping beam in order to rotate the nanowires and connect two branches of a circuit (Lee et al., 2009) [Fig. 23.4b]. Optical tweezers have also been used to trap and place gold nanoparticles on a substrate (Guffey and Scherer, 2010) with a positioning error of about 100 nm, largely due to Brownian fluctuations. This technique can be parallelised using, e.g., HOTs, as in Nedev et al. (2011), where HOTs were used to deposit gold nanoparticles on glass [Fig. 23.4c], and in Woerdemann et al. (2010), where they were used to organise zeolite crystals.

Another approach is optical-trapping-assisted nanopatterning (OTAN), where a micrometre-sized dielectric sphere is positioned close to a surface, where it acts as a near-field objective for pulsed laser processing (McLeod and Arnold, 2008). With this technique, patterns with features down to about 100 nm were generated. Surfaces with non-constant height were also patterned exploiting the non-diffracting properties of Bessel beams (Tsai et al., 2012).

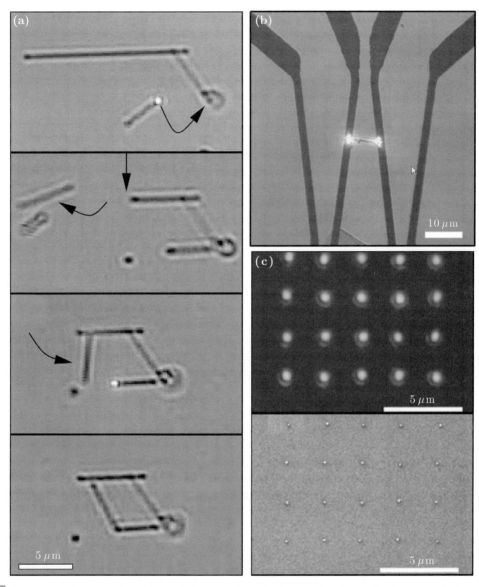

Figure 23.4 Optical force lithography. Manipulation and placement of nanostructures: (a) Assembly of a rhombus constructed from semiconductor CdS nanowires using HOTs. The assembly process entails nanowire translation, cutting and fusion with the substrate. Reprinted from Agarwal et al., *Opt. Express* **13**, 8906–12. Copyright (2005) The Optical Society. (b) Placement of a In_2O_3 nanowire by a scanning optical tweezers to connect two branches of a circuit. Reprinted from Lee et al., *Opt. Express* **17**, 17491–501. Copyright (2009) The Optical Society. (c) Deposition of gold nanoparticles on a glass substrate by HOTs: dark-field (top) and scanning electron microscope images (bottom) of a 5 × 5 pattern made of 80 nm particles. Reprinted with permission from Nedev et al., *Nano Lett.* **11**, 5066–70. Copyright (2011) American Chemical Society.

23.4 Prospects for nanotweezers

In the context of the ongoing trend towards miniaturisation of technology towards the nanoscale, the optical trapping and manipulation of nanostructures is opening up exciting possibilities for the assembly, characterisation and optical control of nanodevices and biomolecules. The realisation of these goals will require the development of new techniques for manipulating nanoparticles, beyond those that are currently available. Indeed, several barriers will need to be overcome. In particular, nanoparticles will need to be manipulated with sub-nanometre accuracy in order, e.g., to develop novel integrated devices. However, making a nanoscale trapping volume is just one side of the problem; being able to interrogate and control what is happening in the trap appears to be the unresolved challenge. Deleterious effects associated with heating of plasmonic particles must be mitigated via integration of schemes to dissipate heat. For optimal control and regulation of biomolecular interactions, specificity in single-molecule trapping will be required. It will be necessary to manipulate and assemble large numbers of particles in order to reach high-throughput, cost-efficient production, which could be achieved, e.g., by self-assembly of elementary building blocks. It will be necessary to develop autonomous nanodevices capable of their own locomotion and of exploring their environment. In this context, optical manipulation of individual nanoparticles will play a crucial role in the development and characterisation phase; however, more powerful, and largely new, parallel optical manipulation techniques could also play a major role in the production stage. Ideally, these new trapping schemes for nanostructures should be as flexible and widely applicable as optical tweezers have proved for microscopic materials.

23.5 Further reading

Nanotweezing techniques have been applied successfully to a variety of objects, e.g., metal nanoparticles (Svoboda and Block, 1994; Hansen et al., 2005; Bosanac et al., 2008; Hajizadeh and Reihani, 2010), plasmonic nanoparticles (Pelton et al., 2006; Toussaint et al., 2007; Dienerowitz et al., 2008; Selhuber-Unkel et al., 2008; Jones et al., 2009; Tong et al., 2010, 2011; Messina et al., 2011b; Ploschner et al., 2012), quantum dots (Jauffred et al., 2008; Chen et al., 2012), carbon nanotubes (Tan et al., 2004; Maragò et al., 2008a,b; Rodgers et al., 2008; Pauzauskie et al., 2009), graphene flakes (Maragò et al., 2010; Twombly et al., 2013), nanodiamonds (Geiselmann et al., 2013), polymer nanofibres (Neves et al., 2010) and semiconductor nanowires (Agarwal et al., 2005; Dutto et al., 2011; Pauzauskie et al., 2006; Nakayama et al., 2007; Reece et al., 2009, 2011; Irrera et al., 2011; Wang et al., 2013). Typically, these techniques rely either on special properties of the trapped objects themselves, e.g., force enhancement related to the plasmonic resonances supported by the trapped particles [Chapter 3] or highly anisotropic geometries, such as in nanotubes and nanowires; or on novel approaches to optical manipulation, e.g., by exploiting the field enhancement due to the surface plasmons supported by nanostructures on a substrate or by exploiting the feedback on the optical forces of the the trapped object [Chapter 22]. Optical manipulation has been used to build composite nano-assemblies (Agarwal et al., 2005; Pauzauskie et al., 2006; Ikin et al., 2009; Chen et al., 2012). Optical trapping of nanostructures has been developed to measure forces with femtonewton resolution [Chapter 10], enabling the study of interactions between nano-objects (Maragò et al., 2008b; Ikin et al., 2009; Pollard et al., 2010; Phillips et al., 2011, 2012a,b; Olof et al., 2012), and it has also been integrated with spectroscopic techniques [Chapter 15], such as Raman (Bjerneld et al., 2003; Tan et al., 2004; Svedberg et al., 2006; Rodgers et al., 2008; Bálint et al., 2009; Maragò et al., 2010; Rao et al., 2010; Messina et al., 2011a) and photoluminescence (Nakayama et al., 2007; Reece et al., 2009; Dutto et al., 2011; Wang et al., 2013, 2011; Geiselmann et al., 2013), paving the way to the selection and manipulation of nanoparticles after their individual characterisation. Finally the laser cooling of optically levitated nanoparticles towards their quantum-mechanical ground state has been investigated theoretically (Barker and Shneider, 2010; Chang et al., 2010; Romero-Isart et al., 2010) and experimentally (Gieseler et al., 2012).

References

Agarwal, R., Ladavac, K., Roichman, Y., et al. 2005. Manipulation and assembly of nanowires with holographic optical traps. *Opt. Express*, **13**, 8906–12.

Ashkin, A., Dziedzic, J.M., Bjorkholm, J.E., and Chu, S. 1986. Observation of a single-beam gradient optical trap for dielectric particles. *Opt. Lett.*, **11**, 288–90.

Baffou, G., and Quidant, R. 2013. Thermo-plasmonics: Using metallic nanostructures as nano-sources of heat. *Laser Photon. Rev.*, **7**, 171–87.

Bálint, S., Kreuzer, M. P., Rao, S., et al. 2009. Simple route for preparing optically trappable probes for surface-enahnced Raman scattering. *J. Phys. Chem. C*, **113**, 17 724–9.

Bareil, P. B., and Sheng, Y. 2010. Angular and position stability of a nanorod trapped in an optical tweezers. *Opt. Express*, **18**, 26 388–98.

Barker, P. F., and Shneider, M. N. 2010. Cavity cooling of an optically trapped nanoparticle. *Phys. Rev. A*, **81**, 023826.

Bjerneld, E. J., Svedberg, F., and Käll, M. 2003. Laser-induced growth and deposition of noble-metal nanoparticles for surface-enhanced Raman scattering. *Nano Lett.*, **3**, 593–6.

Bonaccorso, F., Hasan, T., Tan, P. H., et al. 2010. Density gradient ultracentrifugation of nanotubes: Interplay of bundling and surfactants encapsulation. *J. Phys. Chem C*, **114**, 17 267–85.

Borghese, F., Denti, P., Saija, R., Iatì, M. A., and Maragò, O. M. 2008. Radiation torque and force on optically trapped linear nanostructures. *Phys. Rev. Lett.*, **100**, 163903.

Bosanac, L., Aabo, T., Bendix, P. M., and Oddershede, L. B. 2008. Efficient optical trapping and visualization of silver nanoparticles. *Nano Lett.*, **8**, 1486–91.

Castro-Neto, A. H., Guinea, F., Peres, N. M. R., Novoselov, K. S., and Geim, A. K. 2009. The electronic properties of graphene. *Rev. Mod. Phys.*, **81**, 109–62.

Chang, D. E., Regal, C. A., Papp, S. B., et al. 2010. Cavity opto-mechanics using an optically levitated nanosphere. *Proc. Natl. Acad. Sci. U.S.A.*, **107**, 1005–10.

Chen, Y.-F., Serey, X., Sarkar, R., Chen, P., and Erickson, D. 2012. Controlled photonic manipulation of proteins and other nanomaterials. *Nano Lett.*, **12**, 1633–7.

Dienerowitz, M., Mazilu, M., Reece, P. J., Krauss, T. F., and Dholakia, K. 2008. Optical vortex trap for resonant confinement of metal nanoparticles. *Opt. Express*, **16**, 4991–9.

Donato, M. G., Vasi, S., Sayed, R., et al. 2012. Optical trapping of nanotubes with cylindrical vector beams. *Opt. Lett.*, **37**, 3381–3.

Dufresne, E. R., and Grier, D. G. 1998. Optical tweezer arrays and optical substrates created with diffractive optics. *Rev. Sci. Instrumen.*, **69**, 1974–7.

Dutto, F., Raillon, C., Schenk, K., and Radenovic, A. 2011. Nonlinear optical response in single alkaline niobate nanowires. *Nano Lett.*, **11**, 2517–21.

Geim, A. K., and Novoselov, K. S. 2007. The rise of graphene. *Nature Mater.*, **6**, 183–91.

Geiselmann, M., Juan, M. L., Renger, J., et al. 2013. Three-dimensional optical manipulation of a single electron spin. *Nature Nanotechnol.*, **8**, 175–9.

Gieseler, J., Deutsch, B., Quidant, R., and Novotny, L. 2012. Subkelvin parametric feedback cooling of a laser-trapped nanoparticle. *Phys. Rev. Lett.*, **109**, 103603.

Guffey, M. J., and Scherer, N. F. 2010. All-optical patterning of Au nanoparticles on surfaces using optical traps. *Nano Lett.*, **10**, 4302–8.

Hajizadeh, F., and Reihani, S. N. S. 2010. Optimized optical trapping of gold nanoparticles. *Opt. Express*, **18**, 551–9.

Hansen, P. M., Bhatia, V. K., Harrit, N., and Oddershede, L. 2005. Expanding the optical trapping range of gold nanoparticles. *Nano Lett.*, **5**, 1937–42.

Iijima, S. 1991. Helical microtubules of graphitic carbon. *Nature*, **354**, 56–8.

Ikin, L., Carberry, D. M., Gibson, G. M., Padgett, M. J., and Miles, M. J. 2009. Assembly and force measurement with SPM-like probes in holographic optical tweezers. *New J. Phys.*, **11**, 023012.

Irrera, A., Artoni, P., Saija, R., et al. 2011. Size-scaling in optical trapping of silicon nanowires. *Nano Lett.*, **11**, 4879–84.

Irrera, A., Artoni, P., Iacona, I., et al. 2012. Quantum confinement and electroluminescence in ultrathin silicon nanowires fabricated by a maskless etching technique. *Nanotechnol.*, **23**, 075204.

Ito, S., Yoshikawa, H., and Masuhara, H. 2001. Optical patterning and photochemical fixation of polymer nanoparticles on glass substrates. *Appl. Phys. Lett.*, **78**, 2566–8.

Jauffred, L., Richardson, A. C., and Oddershede, L. B. 2008. Three-dimensional optical control of individual quantum dots. *Nano Lett.*, **8**, 3376–80.

Jones, P. H., Palmisano, F., Bonaccorso, F., et al. 2009. Rotation detection in light-driven nanorotors. *ACS Nano*, **3**, 3077–84.

Kyrsting, A., Bendix, P. M., Stamou, D. G., and Oddershede, L. B. 2011. Heat profiling of three-dimensionally optically trapped gold nanoparticles using vesicle cargo release. *Nano Lett.*, **11**, 888–92.

Lee, S.-W., Jo, G., Lee, T., and Lee, Y.-G. 2009. Controlled assembly of In_2O_3 nanowires on electronic circuits using scanning optical tweezers. *Opt. Express*, **17**, 17 491–501.

Ma, H., Bendix, P. M., and Oddershede, L. B. 2012. Large-scale orientation dependent heating from a single irradiated gold nanorod. *Nano Lett.*, **12**, 3954–60.

Maragò, O. M., Jones, P. H., Bonaccorso, F., et al. 2008a. Femtonewton force sensing with optically trapped nanotubes. *Nano Lett.*, **8**, 3211–16.

Maragò, O. M., Bonaccorso, F., Saija, R., et al. 2010. Brownian motion of graphene. *ACS Nano*, **4**, 7515–23.

Maragò, O. M., Jones, P. H., Gucciardi, P. G., Volpe, G. V., and Ferrari, A. C. 2013. Optical trapping and manipulation of nanostructures. *Nature Nanotechnol.*, **8**, 807–19.

Maragò, O.M., Gucciardi, P. G., Bonaccorso, F., et al. 2008b. Optical trapping of carbon nanotubes. *Physica E*, **40**, 2347–51.

McLeod, E., and Arnold, C. B. 2008. Subwavelength direct-write nanopatterning using optically trapped microspheres. *Nature Nanotechnol.*, **3**, 413–17.

Messina, E., Cavallaro, E., Cacciola, A., et al. 2011a. Manipulation and Raman spectroscopy with optically trapped metal nanoparticles obtained by pulsed laser ablation in liquids. *J. Phys. Chem. C*, **115**, 5115–22.

Messina, E., Cavallaro, E., Cacciola, A., et al. 2011b. Plasmon-enhanced optical trapping of gold nanoaggregates with selected optical properties. *ACS Nano*, **5**, 905–13.

Mitin, V. V., Kochelap, V., and Stroscio, M. A. 1999. *Quantum heterostructures: Microelectronics and optoelectronics*. Cambridge, UK: Cambridge University Press.

Nakayama, Y., Pauzauskie, P. J., Radenovic, A., et al. 2007. Tunable nanowire nonlinear optical probe. *Nature*, **447**, 1098–1101.

Nedev, S., Urban, A. S., Lutich, A. A., and Feldmann, J. 2011. Optical force stamping lithography. *Nano Lett.*, **11**, 5066–70.

Neuman, K. C., and Nagy, A. 2008. Single-molecule force spectroscopy: Optical tweezers, magnetic tweezers and atomic force microscopy. *Nature Methods*, **5**, 491–505.

Neves, A. A. R., Camposeo, A., Pagliara, S., et al. 2010. Rotational dynamics of optically trapped nanofibers. *Opt. Express*, **18**, 822–30.

Novoselov, K. S., Geim, A. K., Morozov, S. V., et al. 2004. Electric field effect in atomically thin carbon films. *Science*, **306**, 666–9.

Olof, S. N., Grieve, J. A., Phillips, D. B., et al. 2012. Measuring nanoscale forces with living probes. *Nano Lett.*, **12**, 6018–23.

Pauzauskie, P. J., Radenovic, A., Trepagnier, E., et al. 2006. Optical trapping and integration of semiconductor nanowire assemblies in water. *Nature Mater.*, **5**, 97–101.

Pauzauskie, P. J., Jamshidi, A., Valley, J. K., Satcher, J. H., and Wu, M. C. 2009. Parallel trapping of multiwalled carbon nanotubes with optoelectronic tweezers. *Appl. Phys. Lett.*, **95**, 113104.

Pelton, M., Liu, M., Kim, H. Y., et al. 2006. Optical trapping and alignment of single gold nanorods by using plasmon resonances. *Opt. Lett.*, **31**, 2075–7.

Phillips, D. B., Grieve, J. A., Olof, S. N., et al. 2011. Surface imaging using holographic optical tweezers. *Nanotechnol.*, **22**, 285503.

Phillips, D. B., Simpson, S. H., Grieve, J. A., et al. 2012a. Force sensing with a shaped dielectric micro-tool. *Europhys. Lett.*, **99**, 58004.

Phillips, D. B., Gibson, G. M., Bowman, R., et al. 2012b. An optically actuated surface scanning probe. *Opt. Express*, **20**, 29679.

Ploschner, M., Cizmar, T., Mazilu, M., Di Falco, A., and Dholakia, K. 2012. Bidirectional optical sorting of gold nanoparticles. *Nano Lett.*, **12**, 1923–27.

Pollard, M. R., Botchway, S. W., Chichkov, B., et al. 2010. Optically trapped probes with nanometer-scale tips for femto-Newton force measurement. *New J. Phys.*, **12**, 113056.

Rao, S., Raj, S., Balint, S., et al. 2010. Single DNA molecule detection in an optical trap using surface-enhanced Raman scattering. *Appl. Phys. Lett.*, **96**, 213701.

Reece, P. J., Paiman, S., Abdul-Nabi, O., et al. 2009. Combined optical trapping and microphotoluminescence of single InP nanowires. *Appl. Phys. Lett.*, **95**, 101109.

Reece, P. J., Toe, W. J., Wang, F., et al. 2011. Characterization of semiconductor nanowires based on optical tweezers. *Nano Lett.*, **11**, 2375–81.

Rodgers, T., Shoji, S., Sekkat, Z., and Kawata, S. 2008. Selective aggregation of single-walled carbon nanotubes using the large optical field gradient of a focused laser beam. *Phys. Rev. Lett.*, **101**, 127402.

Rohrbach, A., Tischer, C., Neumayer, D., Florin, E.-L., and Stelzer, E. H. K. 2004. Trapping and tracking a local probe with a photonic force microscope. *Rev. Sci. Instrum.*, **75**, 2197–210.

Romero-Isart, O., Juan, M. L., Quidant, R., and Cirac, J. I. 2010. Toward quantum superposition of living organisms. *New J. Phys.*, **12**, 033015.

Saija, R., Denti, P., Borghese, F., Maragò, O. M., and Iatì, M. A. 2009. Optical trapping calculations for metal nanoparticles: Comparison with experimental data for Au and Ag spheres. *Opt. Express*, **17**, 10 231–41.

Selhuber-Unkel, C., Zins, I., Schubert, O., Sönnichsen, C., and Oddershede, L. B. 2008. Quantitative optical trapping of single gold nanorods. *Nano Lett.*, **8**, 2998–3003.

Seol, Y., Carpenter, A. E., and Perkins, T. T. 2006. Gold nanoparticles: Enhanced optical trapping and sensitivity coupled with significant heating. *Opt. Lett.*, **31**, 2429–31.

Simpson, S. H., and Hanna, S. 2011. Application of the discrete dipole approximation to optical trapping calculations of inhomogeneous and anisotropic particles. *Opt. Express*, **19**, 16 526–41.

Svedberg, F., Li, Z., Xu, H., and Käll, M. 2006. Creating hot nanoparticle pairs for surface-enhanced Raman spectroscopy through optical manipulation. *Nano Lett.*, **6**, 2639–41.

Svoboda, K., and Block, S. M. 1994. Optical trapping of metallic Rayleigh particles. *Opt. Lett.*, **19**, 930–32.

Tan, S., Lopez, H. A., Cai, C. W., and Zhang, Y. 2004. Optical trapping of single-walled carbon nanotubes. *Nano Lett.*, **4**, 1415–19.

Tong, L., Miljkovic, V. D., and Käll, M. 2010. Alignment, rotation, and spinning of single plasmonic nanoparticles and nanowires using polarization dependent optical forces. *Nano Lett.*, **10**, 268–73.

Tong, L., Miljković, V. D., Johansson, P., and Käll, M. 2011. Plasmon hybridization reveals the interaction between individual colloidal gold nanoparticles confined in an optical potential well. *Nano Lett.*, **11**, 4505–8.

Toussaint, K. C., Liu, M., Pelton, M., et al. 2007. Plasmon resonance-based optical trapping of single and multiple Au nanoparticles. *Opt. Express*, **15**, 12 017–29.

Tsai, Y.-C., Leitz, K.-H., Fardel, R., et al. 2012. Parallel optical trap assisted nanopatterning on rough surfaces. *Nanotechnol.*, **23**, 165304.

Twombly, C. W., Evans, J. S., and Smalyukh, I. I. 2013. Optical manipulation of self-aligned graphene flakes in liquid crystals. *Opt. Express*, **21**, 1324–34.

Urban, A. S., Carretero-Palacios, S., Lutich, A. A., et al. 2014. Optical trapping and manipulation of plasmonic nanoparticles: Fundamentals, applications, and perspectives. *Nanoscale*, **6**, 4458–74.

Wang, F., Reece, P. J., Paiman, S., et al. 2011. Nonlinear optical processes in optically trapped InP nanowires. *Nano Lett.*, **11**, 4149–53.

Wang, F., Toe, W. J., Lee, W. M., et al. 2013. Resolving stable axial trapping points of nanowires in an optical tweezers using photoluminescence mapping. *Nano Lett.*, **13**, 1185–91.

Woerdemann, M., Gläsener, S., Hörner, F., et al. 2010. Dynamic and reversible organization of zeolite L crystals induced by holographic optical tweezers. *Adv. Mater.*, **22**, 4176–9.

Wright, W. H., Sonek, G. J., and Berns, M. W. 1993. Radiation trapping forces on microspheres with optical tweezers. *Appl. Phys. Lett.*, **63**, 715–17.

Yan, Z., Jureller, J. E., Sweet, J., et al. 2012. Three-dimensional optical trapping and manipulation of single silver nanowires. *Nano Lett.*, **12**, 5155–61.

24 Laser cooling and trapping of atoms

In the last decades techniques for optical trapping at the atomic scale have flourished. Precise control of light–matter mechanical coupling has enabled trapping and cooling of neutral atoms, ions and molecules, reaching ever lower temperatures, as shown in Fig. 24.1. This tremendous progress has been recognised with the award of Nobel Prizes on three occasions: for laser cooling and trapping (1997), for the realisation of Bose–Einstein condensation (BEC) of neutral atoms (2001) and for the manipulation of individual quantum systems (2012). The purpose of this chapter is to present a brief overview of the fundamental concepts in this vast research field, including a description of Doppler and sub-Doppler cooling processes, atom-trapping techniques, Bose–Einstein condensation and optical lattices, with particular attention to techniques that make use of optical dipole forces as used in optical tweezers for microscopic particles.

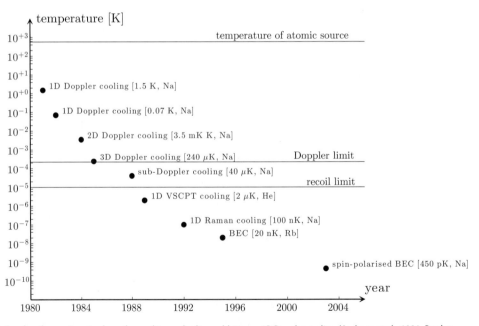

Figure 24.1 Landmark experiments along the road towards ultra-cold atoms: 1D Doppler cooling (Andreev et al., 1981; Prodan et al., 1982); 2D Doppler cooling (Balykin et al., 1985); 3D Doppler cooling (Chu et al., 1985); sub-Doppler cooling (Lett et al., 1988); 1D VSCPT cooling (Aspect et al., 1988); 1D Raman cooling (Kasevich and Chu, 1992); Bose–Einstein condensation (BEC) (Anderson et al., 1995); spin-polarised BEC (Leanhardt et al., 2003).

24.1 Laser cooling and optical molasses

The first proposals for laser cooling of atoms were made by Hänsch and Schawlow (1975), Wineland and Dehmelt (1975) and Letokhov et al. (1976). They proposed to exploit the Doppler effect to produce a velocity-dependent force on a collection of neutral atoms; this would permit them to reduce the rms (root-mean-square) velocity of the atoms and hence their effective temperature. The first Doppler coolings experiments in 1D were realised by Andreev et al. (1981) and Proden et al. (1982). The extension to 2D and 3D was then demonstrated by Bolytein et al. (1985) and Chu et al. (1985), respectively.

We have seen in Part I that the radiation force on a particle can be expressed in terms of the incident light intensity and radiation pressure cross-section given by Eq. (5.124). This expression, just like all other expressions for optical forces that we have encountered so far in this book, does not require the optical field to be quantised; this, in particular, means that a higher intensity yields a greater optical force. However, when dealing with the interaction between atoms and light, we need to describe the interaction in terms of energy quanta, i.e., photons, and to consider a quantum-mechanical description of the discrete atomic energy levels (Foot, 2005); differently from the classical case, this will yield, in particular, a saturation in the optical force when the light intensity is high.

For near-resonant light many atoms can, to a first approximation, be treated as two-level systems comprising a ground state, $|g\rangle$, and an excited state, $|e\rangle$, with an energy difference $E_0 = E_e - E_g = \hbar\omega_0$ and a natural lifetime of the excited state $\tau = 1/\Gamma$. The two-level system interacts with incoming radiation via the electric dipole Hamiltonian, $\hat{H}_d = -\hat{\mathbf{p}}_d \cdot \mathcal{E}_i$, and the strength of this interaction is described by the *Rabi frequency*, $\Omega_R = \langle g|\hat{H}_d|e\rangle/\hbar$. The Rabi frequency depends on the intensity of the electric field as $2\Omega_R^2/\Gamma^2 = I/I_{\text{sat}}$, where $I_{\text{sat}} = \pi hc\Gamma/3\lambda^3$ is the *saturation intensity* of the atomic transition. On illumination by near-resonant radiation, such a two-level atom undergoes processes of absorption and stimulated and spontaneous emission. Averaged over many cycles, these processes give rise to different mechanical effects on the centre of mass of an atom. When an atom absorbs a photon from an incident light beam with wavevector $k = \omega/c$, intensity I and detuning $\delta = \omega - \omega_0$ from the atomic transition, as shown in Fig. 24.2a, it receives a momentum kick of magnitude $\hbar k$ in the direction of propagation of the photon. If the excited atom then re-emits a photon by stimulated emission (arising from the same incident field), it receives a momentum kick of $\hbar k$ in the opposite direction, so that the net momentum transfer is zero. If, however, the excited atom decays by spontaneous emission, then emission occurs in a random direction (within the dipole emission pattern), with a corresponding momentum kick of $\hbar k$ in a direction that is uncorrelated with the direction of propagation of the incident beam. Over many cycles of absorption and spontaneous emission, the momentum transfer from emission averages to zero so that the resulting total scattering force, $F_{\text{scat}}(\delta, I)$, from an incident light beam on an atom at rest is related to the rate of photon scattering, $\mathcal{R}_{\text{scat}}(\delta, I)$, times the momentum transferred by a photon upon absorption; i.e.,

$$F_{\text{scat}}(\delta, I) = \hbar k \mathcal{R}_{\text{scat}}(\delta, I). \tag{24.1}$$

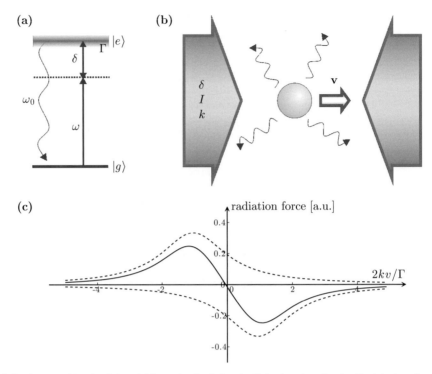

Figure 24.2 Optical molasses and two-level atom. (a) Energy levels of a two-level atomic system showing the detuning of an incident laser beam from resonance, $\delta = \omega - \omega_0$. (b) Configuration of laser beams for a one-dimensional optical molasses, i.e., a pair of counter-propagating beams detuned by δ from the atomic resonance. (c) Radiation force for a one-dimensional optical molasses as a function of atomic velocity. The linear region near $v = 0$ corresponds to viscous damping.

The scattering rate has a frequency dependence that follows a Lorentzian lineshape and includes the saturation of the atomic transition (Foot, 2005); i.e.,

$$\mathcal{R}_{\text{scat}}(\delta, I) = \frac{\Gamma}{2} \frac{I/I_{\text{sat}}}{1 + I/I_{\text{sat}} + 4\delta^2/\Gamma^2}. \tag{24.2}$$

Whereas in the classical picture the radiation force increases linearly with light intensity, the discrete energy levels in atomic systems yield a maximum value of the near-resonant radiation force in the limit of high intensity of $F_{\text{max}} = \hbar k \Gamma/2$. This arises through the saturation of populations in the ground and excited states, and thus the saturation of the scattering rate at high intensity, which tends towards a value of $\Gamma/2$. For an atom of mass m, this maximum radiation force yields an acceleration $a_{\text{max}} = F_{\text{max}}/m = \hbar k \Gamma/2m = v_{\text{r}}/2\tau$, where $v_{\text{r}} = \hbar k/m$ is the recoil velocity (the change in atomic velocity on absorption or emission). The maximum acceleration that can be exerted on an atom is thus given by the change in velocity upon absorption or emission of a photon in the typical time for the process to occur. For the practical case of sodium atoms with $m \approx 23$ a.m.u., a resonant wavelength of $\lambda = 589$ nm for the D2 line and an excited state lifetime $\tau \approx 16$ ns, the maximum acceleration is 10^5 times higher than the gravitational acceleration. For small intensities we recover the linear dependence of the radiation force with light intensity, i.e.,

$$F_{\text{scat}}(I \ll I_{\text{sat}}) \approx \frac{I}{c}\frac{\hbar\omega\Gamma}{2I_{\text{sat}}}\frac{1}{1+4\delta^2/\Gamma^2} = \frac{I}{c}\sigma_{\text{rad}}, \tag{24.3}$$

and, thus, for atomic species the classical picture can be obtained at low intensities with $\sigma_{\text{rad}} \approx \sigma_{\text{abs}}$.

We now consider an atom moving in one dimension at a velocity v and interacting with two counter-propagating light beams detuned by δ from the atomic resonance and with low intensity, as shown in Fig. 24.2b. This configuration is known as *optical molasses* because the atom experiences an optical force proportional to (and opposed to) its velocity. The atom is, in fact, subject to Brownian motion in velocity space as a consequence of the random kicks it receives from the photons from the laser beams. The Doppler effect shifts the effective detuning of the two beams as seen in the reference frame of the atom by a quantity $\mp kv$ so that the scattering rate and consequently the radiation force from each beam are now dependent on the atom velocity; i.e.,

$$F^\pm_{\text{scat}}(\delta \mp kv, I) \approx \pm\hbar k\frac{\Gamma}{2}\frac{I/I_{\text{sat}}}{1+4(\delta \mp kv)^2/\Gamma^2}. \tag{24.4}$$

The total radiation force on the atom is then the sum of the forces from the two counter-propagating beams, $F_{\text{mol}} = F^+_{\text{scat}} + F^-_{\text{scat}}$. For velocities much lower than the *capture velocity*, $v_c = \Gamma/2k$, the total force can be expanded in v to obtain a viscous force

$$F_{\text{mol}} \approx -2kv\frac{\partial F_{\text{scat}}}{\partial \omega}\bigg|_{v=0} = -\alpha(\delta, I)v \tag{24.5}$$

and we can define a damping coefficient as

$$\alpha(\delta, I) = 4\hbar k^2\frac{I}{I_{\text{sat}}}\frac{-2\delta/\Gamma}{[1+(2\delta/\Gamma)^2]^2}. \tag{24.6}$$

The dependence of the optical molasses force on atomic velocity is illustrated in Fig. 24.2c, where approximately linear dependence can be seen for the velocities within the capture range $|v| \ll v_c$. The damping term is positive for red detuning, i.e., $\delta < 0$, yielding a reduction of kinetic energy of the moving atom and thus an effective cooling of the atomic motion, whereas the damping is negative for blue detuning, i.e., $\delta > 0$, yielding an effective heating of the atomic motion. In summary, the Doppler cooling method exploits the Doppler shift to produce a velocity-dependent scattering rate such that faster-moving atoms experience a larger force opposing their direction of motion, thereby narrowing the spread of the distribution of atomic velocities.

We now focus on the cooling process ($\delta < 0$) and calculate the effective temperature reachable by the atoms as a balance between laser cooling and heating caused by the random fluctuations of the force and consequent atomic momentum diffusion due to photon scattering. In fact, the atom receives random kicks of $\hbar k$ each time it absorbs or re-emits a photon, diffusing in momentum space, much as we have seen a colloidal particle diffuses in water [Chapter 7]. Thus, laser cooling damps the energy E of the atoms at a rate $(dE/dt)_{\text{cool}} = vF_{\text{mol}} = -\alpha v^2 = -(2\alpha/m)E$ and this damping process occurs in a characteristic time $\tau_{\text{damp}} = m/2\alpha$ that falls in the range of milliseconds for a typical laser-cooled atomic species (e.g., caesium or rubidium). The heating process takes place at a rate $(dE/dt)_{\text{heat}} = D_p/m$, where D_p is the momentum diffusion constant. The equilibrium between these competing processes determines the mean squared velocity at equilibrium as $\overline{v^2} = D_p/m\alpha$ and, from equipartition of energy, we find the equilibrium temperature

$$T = \frac{m\overline{v^2}}{k_\text{B}} = \frac{D_p}{\alpha k_\text{B}}. \tag{24.7}$$

The momentum diffusion constant is given by an analysis of the random walk in momentum space. After $\mathcal{N}$ steps, considering the scattering from the two beams, we have that the mean squared momentum is $\overline{p^2} = 2\mathcal{N}(\hbar k)^2 = 2D_p t$. The kicks occur because of both absorption and spontaneous emission of photons that, in general, have an angular distribution that can be averaged to a factor ζ (Foot, 2005). Considering these contributions, we have that $\mathcal{N} = (1 + \zeta)\mathcal{R}_{\text{scat}} t$ and then, assuming isotropic angular redistribution of photons, we have that $\zeta = 1/3$ and the momentum diffusion is $D_p = (4/3)(\hbar k)^2 \mathcal{R}_{\text{scat}}$. For low laser intensities, the equilibrium temperature is only a function of the laser detuning, i.e.,

$$T = -\frac{\hbar\Gamma}{6k_\text{B}}\left(\frac{2\delta}{\Gamma} + \frac{\Gamma}{2\delta}\right), \tag{24.8}$$

and has a minimum for $\delta = -\Gamma/2$ corresponding to the Doppler cooling limit for one-dimensional optical molasses, $T_\text{D}^{(1\text{d})} = \hbar\Gamma/3k_\text{B}$. This shows that the limit of the Doppler cooling process is set only by the linewidth of the cooling transition. To damp the atomic motion in three orthogonal directions, six-beam optical molasses consisting of three counter-propagating pairs of beams is used. Here, we need to consider the contributions from all three pairs of beams and the factor taking into account the heating contributions from absorption and emission of photons becomes $(1 + 3\zeta) = 2$ for the three-dimensional configuration. Thus, we obtain the ultimate effective temperature for the Doppler cooling process in three dimensions, i.e., the *Doppler cooling limit*,

$$T_\text{D}^{3\text{d}} = \frac{\hbar\Gamma}{2k_\text{B}}. \tag{24.9}$$

As an example, for the caesium atom we have that $\Gamma = 2\pi \times 5.22\,\text{MHz}$, giving a Doppler limit of $T_\text{D}^{3\text{d}} = 125\,\mu\text{K}$.

The Doppler limit appears to be a fundamental limit on the temperatures that can be achieved by laser cooling. However, when temperatures much lower than the Doppler limit were reported (Lett et al., 1988), it was soon realised that a more sophisticated picture of the atom-light interaction was needed (Dalibard and Cohen-Tannoudji, 1989; Lett et al., 1989). In particular, the multi-level nature of real atoms used for laser cooling experiments and the polarisation gradients present in a standing wave formed by counter-propagating beams are of great significance. Such *sub-Doppler cooling* mechanisms are regulated by *optical pumping*. In Doppler cooling, the characteristic time of the process that leads to the Doppler limit is the excited level lifetime, $1/\Gamma$. In a sub-Doppler process, however, it is the optical pumping time between the different ground state sub-levels that is dependent on the intensity, polarisation and detuning of the laser light. This time is generally much longer than $1/\Gamma$, leading to much lower temperatures. An example of the working principles of sub-Doppler cooling is described in Fig. 24.3. The interference of counter-propagating

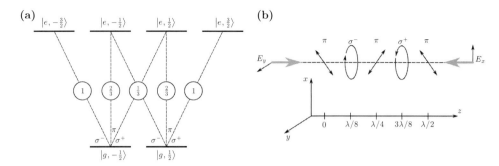

Figure 24.3 Sub-Doppler cooling: the combination of multi-level atoms with the polarisation gradients occurring in a standing wave is responsible for sub-Doppler cooling mechanisms. (a) Squares of the Clebsch–Gordan coefficients and polarisations necessary for an ideal $|J_g = 1/2\rangle \to |J_g = 3/2\rangle$ transition. (b) Polarisation gradients in a one-dimensional lin⊥lin standing wave. The polarisation changes from linear ($+45°$) to σ^- to linear ($-45°$) to σ^+ and back to linear ($+45°$) in half the laser wavelength.

beams with orthogonal linear polarisations (lin⊥lin configuration) results in a field with uniform intensity but spatially varying polarisation, which changes from linear to σ^+ to orthogonal linear to σ^- and back to linear in half the laser wavelength. A multi-level atom thus experiences a position-dependent *ac Stark shift*, or *light shift*, of the magnetic sub-levels, because the Clebsch–Gordan coefficient of the $|J_g = 1/2, m_J = 1/2\rangle \to |J_e = 3/2, m_J = 3/2\rangle$ transition excited at a location of pure σ^+ polarised light is three times greater than that of the $|J_g = 1/2, m_J = -1/2\rangle \to |J_e = 3/2, m_J = 1/2\rangle$ transition. The spatial correlation between the light shifts of the sub-levels and the optical pumping rate between sub-levels gives rise to the sub-Doppler (*Sisyphus*) cooling mechanism whereby atoms continually lose kinetic energy until they can no longer surmount the potential barrier and become localised in potential wells that are less than one optical wavelength in extent.

Another mechanism that plays a role in optical molasses is the *corkscrew cooling* mechanism. This takes place in the presence of a σ^+/σ^- counter-propagating beam configuration, where the light in the standing wave is always linear but its direction rotates over a distance of one wavelength. For a moving atom, this polarisation rotation results in a population imbalance between the magnetic sub-levels that yields a difference in the absorption between the two counter-propagating beams and hence a friction force slowing the atom down. The equilibrium temperature in sub-Doppler cooling processes is related to the Rabi frequency and detuning of the beams as

$$T_{\text{sub-D}} \propto \frac{\hbar \Omega_R^2}{k_B |\delta|} \propto \frac{I}{|\delta|}, \tag{24.10}$$

where the constant of proportionality is dependent on the beam configuration and level structure of the atom. In three dimensions (six-beam configuration), both Sisyphus and corkscrew cooling mechanisms play a role because both polarisation gradients are present (Hopkins and Durrant, 1997).

The minimum temperature that can be achieved by sub-Doppler mechanisms is determined by the quantisation of light and corresponds to the recoil of few photons. Because the atom recoils with a momentum $\hbar k$ for each photon emitted randomly, these motion fluctuations set the ultimate minimum temperature of sub-Doppler mechanisms, i.e., the *recoil limit*,

$$T_R = \frac{\hbar^2 k^2}{k_B m}. \tag{24.11}$$

For example, the recoil limit for caesium is $T_{R,Cs} = 0.2\ \mu K$.

To achieve a lower effective temperature, and thus a momentum with a spread that is much smaller than the photon recoil, it is necessary to drastically reduce the interaction between atoms and light at low velocity. This is obtained exploiting the creation of *dark states* between atomic sub-levels and inducing a velocity-dependent scattering rate, $\mathcal{R}_{scat}(v)$, that vanishes at zero velocity (Bardou et al., 2002). Sub-recoil temperatures have been achieved by two different methods. The first method, i.e., *velocity-selective coherent population trapping* (VSCPT), is based on a quantum interference effect involving the atomic sub-levels to create a dark state at zero velocity where the atomic population accumulates as the cooling process proceeds (Aspect et al., 1988). The second method, i.e., *Raman cooling*, exploits Raman pulses between the atomic sub-levels to obtain a velocity-selective excitation of the atoms that accumulate at zero velocity (Kasevich and Chu, 1992).

24.2 Atom trapping

In Chapter 3 we discussed optical trapping of small particles, i.e., the dipole approximation, within a classical framework. We showed that trapping is achieved by means of gradient forces. This classical picture is also valid in the case of atoms as soon as their interaction with light is regulated by dipole transitions and the classical expression for dynamic polarisability can be used. However, when the laser light is close to an atomic transition, a quantum description of the atom–light interaction is more appropriate to get a correct understanding of the atomic motion in the light field and of the quantum-mechanical origin of the dipole force. The expression for the dipole force exerted by a laser beam on an atom is (Foot, 2005)

$$\mathbf{F}_{dip} = -\frac{\hbar \delta}{4} \frac{\nabla \Omega_R^2}{\delta^2 + \Omega_R^2/2 + \Gamma^2/4}, \tag{24.12}$$

which can be expressed in terms of the laser intensity as

$$\mathbf{F}_{dip}(\delta, I) = -\frac{\hbar \delta}{2} \frac{\nabla (I/I_{sat})}{1 + I/I_{sat} + 4\delta^2/\Gamma^2}. \tag{24.13}$$

This expression should be compared with the scattering force given by Eq. (24.1). As in the classical picture, the dipole force is conservative and can easily be obtained as the gradient of the optical potential

$$U_{dip}(\delta, I) = \frac{\hbar \delta}{2} \ln \left(1 + \frac{I/I_{sat}}{1 + 4\delta^2/\Gamma^2}\right), \tag{24.14}$$

which, for large detuning, i.e., $|\delta| \gg \Gamma$, takes the simple form

$$U_{\text{dip}} \approx \frac{\hbar \Omega_R^2}{4\delta} = \frac{\hbar \Gamma^2 I}{8\delta I_{\text{sat}}}. \qquad (24.15)$$

This expression is practically useful because *dipole traps* are normally operated at large detuning to minimise the dissipation from the scattering force exerted by the trapping laser beams. In fact, for $|\delta| \gg \Gamma$, the scattering rate is $R_{\text{scat}} \approx \Gamma^3 I / (8\delta^2 I_{\text{sat}})$ and it is highly suppressed for large detuning, while a reasonable dipole trap depth is maintained by increasing the intensity. Therefore, the dipole force does not involve absorption of energy from the field and is purely related to the redistribution of momentum between the atom and the plane waves with different wavevectors composing the laser beam.[1]

For light with a frequency red-detuned from the atomic resonance, i.e., $\delta < 0$, the atoms are *strong field seekers*, i.e., they are attracted towards high-laser-intensity regions (e.g., the spot of a focused Gaussian beam), whereas for blue-detuned light, i.e., $\delta > 0$, the atoms become *weak field seekers* and the dipole force guides the atoms towards the low-intensity regions of the light field (e.g., the dark central region of a Laguerre–Gauss beam). These properties have been exploited in *atom optics* for the trapping, guiding, channeling, focusing and deposition of atoms (Meystre, 2001). For example, the transport of single atoms or Bose–Einstein condensates over long distances can be achieved by a red-detuned Gaussian beam focused with a simple lens, as, when the lens is mechanically translated, the atoms follow the trapping optical potential (Gustavson et al., 2001).

A dipole trap for atoms at room temperature is too weak in comparison to their thermal energy; thus the starting point of most experiments on cold atoms is a *magneto-optical trap* (MOT). The MOT was first experimentally demonstrated by Raab et al. (1987) following an earlier suggestion by Jean Dalibard (see footnote in Raab et al., 1987). It consists of six σ^+/σ^- counter-propagating beams and a spherical quadrupole magnetic field that increases linearly with distance from the trap centre, e.g., $\mathbf{B}_z = b_z \mathbf{z}$ for the axial direction of the trap. The principle of operation of a MOT is explained by considering a two-level atom with an excited state having magnetic sub-levels ($J > 0$). The magnetic field induces a position-dependent shift in the magnetic sub-levels' energy, i.e.,

$$\Delta U_{\text{mag}} = g_J \mu_B m_J B = g_J \mu_B m_J b_z z, \qquad (24.16)$$

where g_J is the Landé gyromagnetic (g-) factor and μ_B is the Bohr magneton. If the illuminating laser beams are red-detuned from the atomic resonance, there is a position (as well as velocity) dependence of the scattering rate: the further from the trap centre, the more closely the magnetic sub-levels are shifted into resonance and the greater the scattering force countering the atomic motion. Thus, an atom in a MOT is subject to a force (in one dimension and at low intensity) that can be approximated as the sum of the optical molasses viscous force [Eq. (24.5)] and an elastic force, $F_z \approx -\alpha v_z - \kappa_z z$, where the MOT

---

[1] In this way, we recover the picture described in Chapter 5, i.e., that optical trapping is a consequence of scattering processes in light beams that are superpositions of plane waves with different wavevectors (angular spectrum representation).

spring constant is (Cohen-Tannoudji and Guéry-Odelin, 2011)

$$\kappa_z = \mu_B b_z k \Omega_R^2 \frac{-\delta \Gamma}{\left(\delta^2 + \Gamma^2/4\right)^2}. \tag{24.17}$$

The standard six-beam MOT is easy to operate and robust, being fairly insensitive to unbalanced laser beam intensities, misaligned trapping beams and impure circular polarisations. It is indeed an efficient technique that can collect more than a billion atoms loaded directly from an atomic vapour or an atomic beam.

24.3 Optical dipole traps for cold atoms

The first demonstration of optical trapping of cold atoms was made by Chu et al. (1986). A cloud of sodium atoms, pre-cooled in an optical molasses to a temperature of 240 µK, was loaded into a dipole trap formed by a single strongly focused laser beam. As discussed previously, the detuning of the laser light from atomic resonance is important in cold atom traps to minimise the rate of heating due to spontaneous scattering. In these experiments, detunings of up to 1300 GHz to the red side of the D2 resonance (natural linewidth $\Gamma = 2\pi \times 9.8$ MHz, saturation intensity, $I_{sat} = 20$ mW) could be used with a laser power of 220 mW focused into a diffraction-limited spot of radius 10 µm, resulting in an optical potential well with a depth of up to 5 mK at the optimum detuning of 650 GHz. For these parameters, the atoms were confined within a cylindrical volume of around 10^3 µm^3 at a density of approximately 0.5 µm^{-3}; i.e., the trap contained about 500 atoms. Similarly to an optical tweezers for microscopic particles, the collection of trapped atoms could be moved by translating the focal spot of the trapping laser beam.

Frese et al. (2000) demonstrated the loading of a small (and deterministic) number of cold caesium atoms into a dipole trap from a MOT. Operating the MOT at a strong field gradient reduced the capture rate from the background gas, allowing the number of trapped atoms to be adjusted between 1 and 10 by adjusting the vapour pressure. Atoms collected in the MOT were transferred into an optical dipole trap formed by a single strongly focused Nd:YAG laser beam, the wavelength of which (1064 nm) was very far detuned to the red side of the caesium D2 transition (852 nm). With 2.5 W of laser power focused into a beam waist of 5 µm, the dipole trap depth was 16 mK and the photon scattering rate 190 s^{-1}. Although the dipole trap optical potential was shallow and conservative, and thus unable to capture atoms directly from the background vapour, operation of both the MOT and the dipole trap simultaneously cooled the atoms into the dipole trap, allowing a certain number of cold atoms to be transferred from the MOT to the dipole trap and back with close to 100% efficiency. The lifetime of atoms in the dipole trap was measured to be 51 ± 3 s.

Further progress towards trapping of single atoms was made by Schlosser et al. (2001) using a dipole trap with a very small volume. In their experiment, the waist of the laser beam forming the dipole trap was smaller than 1 µm, i.e., comparable to conventional optical tweezers. This was achieved using a nine-element objective lens with a working distance of 1 cm and NA = 0.7 mounted inside the vacuum chamber. As in the previously described experiments, the dipole trapping laser was far detuned to the red side of both the D1 and D2 transitions of the atomic species (in this case rubidium). The small focal spot used meant that a potential well depth of 1 mK could be achieved with relatively low (≈ 1 mW) laser power. More than one atom was prevented from occupying the dipole trap simultaneously by light-assisted atom–atom interactions; thus the occupation statistics of the trap was sub-Poissonian in that the occupation number fluctuated between 0 and 1 only.

24.4 The path to quantum degeneracy

A major quest in cold atoms research has been reaching quantum degeneracy. The path towards the realisation of extremely cold and dense atomic samples exhibiting quantum statistical phenomena has been long and complex (Cohen-Tannoudji and Guéry-Odelin, 2011). In fact, because of the photon recoil limit, laser cooling and magneto-optical trapping can yield atoms with temperatures in the microkelvin range and densities on the order of 10^{11} cm^{-3}, which are still five orders of magnitude too low compared to the phase space densities required for quantum degeneracy. *Magnetic trapping* and *forced evaporative cooling* have been the crucial techniques for overcoming these limitations. The typical protocol to reach quantum degeneracy is summarised in the following steps: (i) atoms are collected and laser-cooled in a MOT; (ii) the cloud of cold atoms is then loaded into a confining conservative potential, i.e., a magnetic or optical trap; and (iii) forced evaporative cooling is performed by lowering the trap depth.

Magnetic trapping is a powerful way to confine and manipulate neutral atoms. The magnetic confinement of neutral atoms is based on the interaction between an inhomogeneous magnetic field and the atomic magnetic dipole moment. Thus, the interaction energy is

$$U_{\mathrm{mag}} = -\boldsymbol{\mu} \cdot \mathbf{B} = m_J g_J \mu_{\mathrm{B}} B, \tag{24.18}$$

where B is the modulus of the magnetic field. A consequence of Earnshaw's theorem is that local field maxima are not allowed in free space (Wing, 1984) and it is only possible to achieve atom confinement using local magnetic field minima. Thus, atoms must be in a low-field seeking state, i.e., $m_J g_J > 0$.

The first and simplest magnetic trap for neutral atoms to be experimentally realised was a quadrupole trap (Migdall et al., 1985) consisting of two anti-Helmholtz coils. Although this simple configuration provides tight confinement, it has an intrinsic problem in that the local minimum at the centre of the trap corresponds to the region where $B = 0$. When an atom is close to this $B = 0$ region, it can undergo a so-called Majorana spin-flip, so that it ends up in an untrapped magnetic state where it feels a repulsive potential that ejects it from the trap. A way to avoid these Majorana losses is to use the repulsive optical force of a blue-detuned laser beam to plug the $B = 0$ hole of the magnetic trap. This solution was adopted in the first observation of Bose–Einstein condensation of sodium atoms by Davis et al. (1995).

A different approach is used in the time-averaged orbiting potential (TOP) trap (Petrich et al., 1995) that was used for the realisation of the first rubidium condensate

(Anderson et al., 1995). Its principle of operation relies on adding a rotating bias field to a quadrupole trap. If the rotation is faster than the atomic motion, but slower than the Larmor frequency, $\omega_L = \mu B/\hbar$, the atoms experience a time-averaged harmonic potential and the magnetic field at the centre of the trap is equal to the added bias field. In ultra-cold atoms experiments, magnetic traps with a tight confinement and long lifetime are generally needed. Many experiments use a variation of the Ioffe–Pritchard trap (Pritchard, 1983) in which the field is magnetostatic with an adjustable minimum value.

Evaporative cooling (Ketterle and Van Druten, 1996) is the other essential ingredient that enabled quantum degeneracy to be reached. This is applied to atomic samples trapped in a potential with a finite depth, such as in a magnetic or optical trap (Barrett et al., 2001). The technique relies on the controlled and selective removal of the higher-energy particles from the ensemble of trapped atoms. If the removed atoms escaping the trap have higher energy per particle than the average, then the remaining trapped particles have an average energy low than the initial energy. The remaining atoms re-thermalise through elastic collisions to a lower energy and hence to a lower temperature. An essential requirement for evaporative cooling is a long lifetime of the atomic sample compared to the collisional re-thermalisation. Evaporative cooling was first suggested by Hess (1986) and used by Masuhara et al. (1988) to cool spin-polarised atomic hydrogen. In magnetic traps, the sample of atoms is spin-polarised and forced evaporation is based on controlled transitions into untrapped magnetic states that are generally induced by tuning a radio-frequency field on the trapped–untrapped states resonance. At the end of the evaporative cooling stage, Bose–Einstein condensates of 10^6 to 10^7 atoms are typically created with temperature ranging between nanokelvin and picokelvin regimes (Leanhardt et al., 2003).

24.5 Bose–Einstein condensation

Bose–Einstein condensation (BEC) is a phase transition that results purely from the quantum statistics of a group of identical particles with integer spin, i.e., *bosons*. It was predicted by Einstein (1925) following the work of Bose (1924). In this phenomenon, a macroscopic number of particles accumulate in the ground state of the system below a critical temperature, T_0. This occurs at high densities and low temperatures, when the thermal de Broglie wavelength of the atoms becomes comparable to the inter-particle distance so that the atomic wavefunctions overlap and the phase space density of the gas is of the order of unity. The conditions for this quantum degeneracy to occur are very extreme in real systems, where interactions among the particles may lead to a strong depletion or complete destruction of the single state macroscopic occupation. However, BEC underlies the physics of several phenomena, ranging from superfluids and superconductors (Tilley and Tilley, 1990) to exciton-polaritons (Deng et al., 2010) to nuclear and subnuclear matter (Bohr and Mottelson, 1998).

The statistical mechanics of an ensemble of N non-interacting bosons is well described in basic textbooks (Foot, 2005). Here, we recall the most relevant results for the weakly interacting Bose gas in a confining potential that could be realised either by a magnetic trap or by an optical trap. The confining potential has a substantial effect on the physics of BEC. If we consider a harmonic confinement, with characteristic frequency ω_{ho}, the critical temperature at which condensation occurs is $T_0 = 0.94 N^{1/3} \hbar \omega_{\text{ho}} / k_B$, i.e., the condensation temperature depends only on the total number of particle and the harmonic oscillator energy. Moreover, the condensate fraction N_0/N has a dependence on temperature according to a third power law, i.e.,

$$\frac{N_0}{N} = 1 - \left(\frac{T}{T_0}\right)^3. \tag{24.19}$$

Interactions modify substantially the physics of Bose–Einstein condensation. For example, in the case of liquid helium (He-II), which was the first system where the occurrence of BEC was suggested (London, 1938), the condensate is highly depleted (about 10% condensate fraction) by the strong interactions among particles. However, dilute ultra-cold gases are weakly interacting systems, so interactions can be sufficiently well described by binary collisions only. Condensation occurs while the gas is still dilute and the de Broglie wavelength is one order of magnitude larger than the range of interatomic forces. Therefore, the interactions in these systems can be modelled by a contact potential. In particular, at very low temperature the two-body collisions can be described by a single parameter, i.e., the s-wave scattering length a [Box 24.1]. Thus, one can write an effective interaction potential of the form $V(\mathbf{r} - \mathbf{r}') = g\delta(\mathbf{r} - \mathbf{r}')$ with $g = 4\pi\hbar^2 a/m$ and m the particle mass. This effective interaction potential is the basic ingredient for constructing the mean-field theory for condensed gases.

The starting point for the description of interacting atomic gases in an external potential, $V_{\text{ext}}(\mathbf{r})$, is the many-body Hamiltonian for an ensemble of N bosons written in terms of the Bose field $\hat{\Psi}(\mathbf{r}, t)$ (Pitaevskii and Stringari, 2003), i.e.,

$$\hat{H} = \int dV \, \hat{\Psi}^\dagger \left[-\frac{\hbar^2}{2m} \nabla^2 + V_{\text{ext}} + \frac{g}{2} \hat{\Psi}^\dagger \hat{\Psi} \right] \hat{\Psi}. \tag{24.20}$$

The basic and simplest description of a weakly interacting Bose gas relies on a mean-field approach. This formalism was set out by Bogoliubov (1947) and is based on the identification of the Bose field with $\hat{\Psi}(\mathbf{r}, t) = \Phi(\mathbf{r}, t) + \hat{\delta}(\mathbf{r}, t)$. The mean field $\Phi(\mathbf{r}, t) \equiv \langle \hat{\Psi}(\mathbf{r}, t) \rangle$ is a complex function that fixes the condensate number density $n_0(\mathbf{r}, t) = |\Phi(\mathbf{r}, t)|^2$ and has a well-defined phase. The field $\hat{\delta}(\mathbf{r}, t)$ represents the fluctuations of the system and its mean value vanishes. The mean-field approximation assumes that the condensate is characterised by the mean field and that the fluctuations give only small corrections.

The equation of motion for the condensate wavefunction is derived by averaging the Heisenberg equation of motion for the Bose field, i.e., $i\hbar \partial \hat{\Psi}/\partial t = [\hat{\Psi}, \hat{H}]$. This leads to the *time-dependent Gross–Pitaevskii equation* (GPE) (Pitaevskii, 1961),

$$i\hbar \frac{\partial}{\partial t} \Phi(\mathbf{r}, t) = \left[-\frac{\hbar^2}{2m} \nabla^2 + V_{\text{ext}} + \frac{g}{2} |\Phi(\mathbf{r}, t)|^2 \right] \Phi(\mathbf{r}, t). \tag{24.21}$$

This equation is valid for dilute gases and describes extremely well the macroscopic dynamics of the condensate wavefunction at low temperature when the ground state occupation is large.

> **Box 24.1**   **Elastic scattering in quantum mechanics**
>
> Scattering theory is a general framework within which scattering problems in many different contexts are solved. In quantum mechanics, an incoming particle is described by a wavefunction, ψ, associated with a probability density $|\psi|^2$ (Konishi and Paffuti, 2009). A free particle traveling along z has a wavefunction $\psi = e^{ikz}$ with energy $E = \hbar^2 k^2/2m$ and momentum $\mathbf{p} = \hbar k \hat{\mathbf{z}}$. The scattering process is typically studied under stationary conditions and the wavefunction describing the system is an expanding spherical wave superposed on the incident plane wave, $\psi = e^{ikz} + f(\theta, \phi)\frac{e^{ikr}}{r}$, where $f(\theta, \phi)$ is the scattering amplitude, so that the scattering differential cross-section is $d\sigma/d\Omega = |f(\theta, \phi)|^2$. A general method of analysis of scattering processes is *partial wave expansion*. Let us assume a central scattering potential for which the Schrödinger equation is written as
>
> $$\left[-\frac{\hbar^2}{2m}\nabla^2 + V(r)\right]\psi = E\psi \quad \text{with } E = \frac{\hbar^2 k^2}{2m} > 0,$$
>
> whose solution can be expanded in partial waves, i.e., $\psi = \sum_{l=0}^{\infty} A_l P_l(\cos\theta) R_l(r)$. Here, only $Y_{l0} \propto P_l(\cos\theta)$ appears, because of symmetry for rotations around the z axis. Similarly to the light-scattering case [Chapter 5], in quantum mechanics the radial wavefunctions are combinations of (scalar) spherical Bessel functions. In particular, the asymptotic behaviour of the radial wavefunctions is $R_l(r) \sim (2/r)\sin(kr - l\pi/2 + \delta_l)$, where δ_l is the *phase shift* for the l-wave. Thus, the scattering amplitude is written in terms of partial waves and phase shift, and the total cross-section σ_T is $\sigma_T = \frac{4\pi}{k^2}\sum_l (2l+1)\sin^2\delta_l = \sum_l \sigma_l$, where σ_l is the cross-section in the l-wave.
>
> A special case is a scattering potential with a finite range, d, beyond which the interaction can be neglected. Considering the low-energy limit, $kd \ll 1$, we have that $\delta_l \sim k^{2l+1}$ and the *s*-wave ($l = 0$) dominates. In this case, $\delta_0 \sim -ka$, where a is the *s-wave scattering length* and the cross-section is simply $\sigma_T \sim 4\pi a^2$. The scattering length depends on the interaction potential and can be positive or negative. The *s*-wave scattering length in the *atomic collisions* plays a crucial role in BEC in ultra-cold gases. Its sign and value are responsible for the re-thermalisation of atoms during the evaporative cooling stage and determine the different critical behaviour for different atomic species (Cohen-Tannoudji and Guéry-Odelin, 2011).

The ground state of the system is found by separating the space- and time-dependent parts of the wavefunction, i.e., $\Phi(\mathbf{r}, t) = \phi(\mathbf{r})\exp(-i\mu t/\hbar)$, where μ is the chemical potential and ϕ is normalised to the total number of condensed atoms so that $\int \phi^2 dV = N_0$. Thus, the time-dependent GPE simplifies to a nonlinear time-independent Schrödinger equation:

$$\left[-\frac{\hbar^2}{2m}\nabla^2 + V_{\text{ext}} + \frac{g}{2}|\phi(\mathbf{r})|^2\right]\phi(\mathbf{r}) = \mu\phi(\mathbf{r}). \tag{24.22}$$

The nonlinearity arises from the interaction term proportional to g. The case of large repulsive interactions, known as the *Thomas–Fermi limit*, is of particular interest because it is possible to find analytic solutions for the ground state and it applies very well to most dilute Bose gas experiments. In this limit, the interaction term is much larger than the

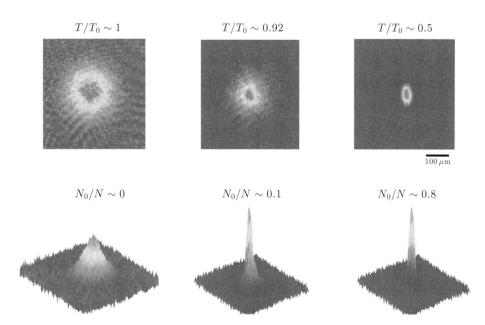

Figure 24.4 Bose–Einstein condensation: formation of a Bose–Einstein condensate in a cloud of rubidium atoms. The onset of BEC is detected by imaging the time-of-flight expansion of atomic clouds: the thermal atoms (left) feature an isotropic expansion, while the condensate (right) has an anisotropic profile corresponding to the anisotropic confinement in the magnetic trap. In the three-dimensional images (bottom), the sudden increase in the optical depth (corresponding to an increase of atomic density) as the temperature is reduced can be observed. Adapted from Maragò, Ph.D. thesis, University of Oxford (2001).

kinetic term and the ground state density distribution is simply

$$n(\mathbf{r}) = |\phi(\mathbf{r})|^2 = \frac{[\mu - V_{\text{ext}}]}{g}, \tag{24.23}$$

when $\mu > 0$ and $n(\mathbf{r}) = 0$ elsewhere. For a harmonic trapping potential this is an inverted parabolic profile and it is the striking signature of condensation in experiments. Indeed, the onset of BEC in ultra-cold atomic gases is quite dramatic and is observed through the change in the density profile of the confined system (Maragò, 2001), as shown in Fig. 24.4.

To investigate the dynamics of condensates containing a large number of atoms, the GPE can be written in terms of the density $n = |\Phi|^2$ and velocity field $\mathbf{v} = (\Phi^* \nabla \Phi - \nabla \Phi^* \Phi)/2imn$ of the quantum fluid. The velocity field is linked to the phase $S(\mathbf{r}, t)$ of the complex wavefunction, $\Phi = \sqrt{n} e^{iS}$, because from its definition $\mathbf{v} = \hbar \nabla S/m$. Thus, the GPE is completely equivalent to two coupled equations that in the Thomas–Fermi limit are written as (Pitaevskii and Stringari, 2003)

$$\begin{cases} \dfrac{\partial n}{\partial t} + \nabla \cdot (n\mathbf{v}) = 0 \\[6pt] \dfrac{\partial \mathbf{v}}{\partial t} + \nabla \left(V_{\text{ext}} + gn + \dfrac{mv^2}{2} \right) = 0. \end{cases} \tag{24.24}$$

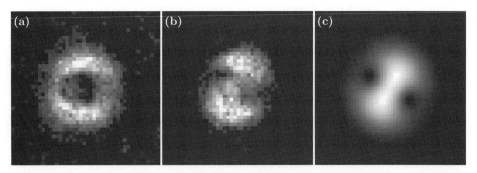

Figure 24.5 Transfer of orbital angular momentum to a Bose–Einstein condensate: creation of vortices by the transfer of orbital angular momentum from a Laguerre–Gauss beam to the matter wave. (a) Absorption image of an atomic cloud with a doubly charged ($+2\hbar$) vortex; the core is larger than that for a single charged state. (b) Absorption image of a cloud resulting from the interference between a doubly charged state and a non-rotating state. (c) Calculated interference pattern between a doubly charged rotating state and a non-rotating state. Reprinted figure from Andersen et al., *Phys. Rev. Lett.* **97**, 170406. Copyright (2006) by the American Physical Society.

These are the *hydrodynamic equations* of superfluids, which are a consequence of phase coherence over the whole system. The first is the continuity equation, whereas the second is a force equation that describes the irrotational dynamics of the superfluid velocity field.

The link between BEC and superfluidity is strikingly observed in the response of a Bose–Einstein condensate to rotations (Leggett, 2001). In fact, its velocity field is irrotational, $\nabla \times \mathbf{v} = 0$, since it is expressed as the gradient of a phase. Because the condensate's wavefunction must be single-valued, the phase change over a closed path must be modulo 2π. This yields the Onsager–Feynman quantisation condition (Pitaevskii and Stringari, 2003),

$$\oint d\mathbf{r} \cdot \mathbf{v} = 2\pi q \frac{\hbar}{m} \quad \text{with } q = 0, \pm 1, \pm 2, \ldots. \tag{24.25}$$

The $q = 0$ solution corresponds to a non-rotating condensate. However, this condition can also be satisfied by a *vortex* solution, i.e., a velocity flow for which $v \sim 1/r$, with r the distance from the *vortex core*, where the superfluid density vanishes. Vortices are common solutions also for classical fluids (tornadoes and whirlpools are well-known examples); the crucial difference is that superfluid vorticity is quantised. Superfluid phenomena in dilute gases have been the subject of much theoretical and experimental work related to the occurrence of a critical velocity (Raman et al., 1999; Onofrio et al., 2000), the generation of quantised vortices (Madison et al., 2000; Matthews et al., 1999; Abo-Shaeer et al., 2001; Hodby et al., 2001), the observation of the so-called scissors mode (Guéry-Odelin and Stringari, 1999; Maragò et al., 2000), Josephson's tunnelling (Burger et al., 2001; Cataliotti et al., 2001; Albiez et al., 2005) and persistent flow (Ryu et al., 2007). Creation of phase singularities such as vortices in a BEC can be also realised by phase imprinting (Andersen et al., 2006) using Laguerre–Gauss beams, as shown in Fig. 24.5.

24.6 Evaporative cooling and Bose–Einstein condensation in dipole traps

Techniques for approaching a regime of high atomic density in optical traps are desirable for reaching the goal of all-optical BEC. Adams et al. (1995) used an alternative configuration to the single strongly focused beam dipole trap consisting of two beams intersecting at right angles. Again, the beams are very far detuned from a resonance of the atomic species (in this case sodium), because a Nd:YAG laser was used with a maximum power in each beam of 4 W focused to a radius of 15 μm at the point of intersection. The maximum trap depth was ≈ 900 μK. This crossed-beam dipole trap was loaded from a MOT and evaporative cooling was implemented by reducing the power in the dipole trap beams (i.e., decreasing the depth of the potential well). The temperature of the atoms was reduced from 140 to 4 μK, corresponding to an increase in phase-space density by a factor of 28 at the expense of decreasing the number of atoms by a factor of 10.

All-optical BEC was ultimately achieved in ^{87}Rb by Barrett et al. (2001) using a CO_2 laser in a similar crossed-beam dipole-trap configuration. The dipole trap was loaded with 2×10^6 atoms at 75 μK from a MOT, a temperature which decreased to 22 μK due to loss of the more energetic atoms. Forced evaporation was implemented by lowering the trap potential depth, ultimately achieving a condensate containing 3.5×10^4 atoms. Unlike the case in a magnetic trap, the three internal states $F = -1$, $m_F = (-1, 0, 1)$ were all condensed. The possibility of optically trapping condensates in all magnetic sub-levels has started a new research field for the investigation of *spinor Bose gases* and their quantum dynamics (Stamper-Kurn and Ueda, 2013). The all-optical method finally provided a route to BEC for caesium, which, unlike other alkali metal species, had not previously been condensed in a magnetic trap because of the unfavourable scattering cross-section of its magnetically trappable state, leading to loss of atoms by inelastic two-body scattering. By using an optical dipole trap to confine and evaporatively cool caesium atoms, Weber et al. (2003) could exploit Feshbach resonances in the magnetic field dependence of the scattering length to tune the self-interaction to be favourable for condensation. In the optical dipole trap, the caesium atoms could be trapped in the lowest internal state, where inelastic two-body loss is suppressed. Again, the crossed-beam configuration was used, with two 100 W CO_2 laser beams producing a trap ≈ 10 μK deep, but with a relatively large volume of 1 mm^3. Evaporative cooling in this trap reached a temperature of 1 μK. The density of caesium atoms was then increased by applying an additional laser beam with a narrow waist (30 μm) to adiabatically deform the optical potential and tightly confine the atoms. Forced evaporation took place by decreasing the intensity of this beam until a condensate of 1.6×10^4 atoms was produced.

Recently BEC of strontium atoms has been realised by pure laser cooling of the atomic samples held in dipole traps (Stellmer et al., 2013). This has been made possible by exploiting a very narrow linewidth transition that enabled Doppler cooling below 1 μK. Atoms were then rendered transparent to the cooling light by inducing a large light shift with a far-detuned infrared laser. This enabled the increase in phase space density needed for quantum degeneracy.

24.7 Holographic optical traps for cold atoms

Holographic beam shaping and steering techniques have become a powerful addition to the range of applications of optical tweezers for microscopic particles, as we have seen in Chapter 11. Unsurprisingly, efforts to apply them to dipole trapping of cold atoms have also been productive. Bergamini et al. (2004) included a spatial light modulator (SLM) in the atom dipole trap experiment of Schlosser et al. (2001) and were capable of creating arrays of dipole traps with control over the relative intensities in order to produce traps of equal depth. With this apparatus, small-volume dipole traps for single atoms in a variety of geometries were produced.

Boyer et al. (2006) demonstrated that holographic traps could even be used for dynamic manipulation of an atomic Bose–Einstein condensate. In contrast to the single-atom trap experiments, a ferroelectric liquid crystal SLM capable of a higher refresh rate (500 Hz) was used. A significant technical issue in using the ferroelectric SLM [Subsection 11.5.1] is charge migration in the liquid crystal, which, when it is operated continuously, reduces the retardation effect and hence diffraction efficiency of the displayed phase pattern. This is prevented by regularly flipping the phase of each pixel, although, because switching the state of the pixels takes ≈ 0.25 ms, this causes the diffracted output to flicker. This problem was addressed by only flipping the phase of a small fraction of pixels each time the hologram is updated, resulting in a dynamic potential that varies smoothly enough to transport an atomic BEC over tens of micrometres with no significant heating.

An example of the capability of holographic optical tweezers to manipulate single atoms is shown in Fig. 24.6, where holographic patterns obtained with a spatial light modulator are used to create multiple traps where single cold atoms are loaded (Nogrette et al., 2014).

24.8 Optical lattices

As noted in Section 24.1, the observation of sub-Doppler temperatures in an optical molasses revealed the importance of optical interference and spatially varying polarisation in the laser beams as part of the cooling mechanism. The presence of a periodic potential for ultra-cold atoms can also lead to their localisation in a regular array called an *optical lattice*. The existence of such confinement was first noted by Westbrook et al. (1990), who used an optical heterodyne technique to measure the spectrum of resonance fluorescence from a sodium optical molasses. This showed a narrow peak with a width well below the natural linewidth, characteristic of *Dicke narrowing* – a reduction in the fluorescence linewidth as a result of the strong confinement of the radiating atom to a distance less than the optical wavelength (Dicke, 1953). A theoretical treatment by Castin and Dalibard (1991) predicted the quantisation of atomic motion in the optical molasses, which was later observed by Jessen et al. (1992) through the existence of well-resolved sidebands in the fluorescence spectrum and by Verkerk et al. (1992) using probe absorption spectroscopy

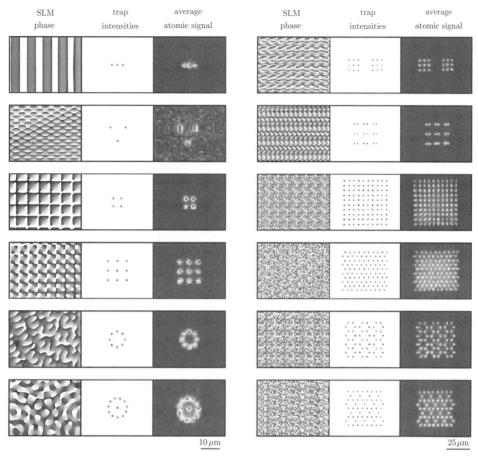

Figure 24.6 Arrays of holographically trapped single atoms: a gallery of single atoms in holographic microtrap arrays with different geometries. Each panel shows the calculated phase pattern used to create the array (left), an image of the resulting trap arrays (middle) and the average of about 1000 fluorescence images of single atoms loaded into the traps (right). Reprinted figure from Nogrette et al., *Phys. Rev. X* **4**, 021034. Copyright (2014) by the American Physical Society.

in one-dimensional optical molasses of rubidium and caesium, respectively. Indeed, in its simplest configuration, an optical lattice is generated using a standing wave. The confining potential follows the intensity periodicity dictated by the interference of light with wavelength λ, so that the resulting one-dimensional optical potential is written as

$$U_{\text{lat}} = U_0 \cos^2(\pi z/\ell), \qquad (24.26)$$

where $\ell = \lambda/2$ is the lattice spacing and U_0 is the lattice depth, which is often measured in terms of recoil energy, $E_R = \hbar^2 k^2/2m = \pi^2 \hbar^2/2m\ell^2$.

Localisation in a three-dimensional cubic crystal was demonstrated by Hemmerich et al. (1993) using six beams arranged as three counter-propagating pairs, similar to the configuration used for the MOT or optical molasses. The important difference, however, was that the time phases of the six beams were stabilised, resulting in a stable periodic optical potential. The transmission of a weak probe beam showed features characteristic of the strong localisation of atoms, namely a narrow peak arising from elastic (Rayleigh) scattering, and Stokes- and anti-Stokes peaks due to Raman transitions between well-resolved vibrational energy levels in a potential well. Verkerk et al. (1994) generalised the one-dimensional lin⊥lin configuration shown in Fig. 24.3b to three dimensions by splitting each beam into two in the plane perpendicular to the respective polarisation directions and making equal angles with the z-axis. Probe transmission spectra showed Rayleigh and Raman features similar to those in the experiments of Hemmerich et al. (1993) indicating localisation of the atoms in sub-wavelength potential wells. The advantage of this configuration, however, is that the beams do not need to be phase-stabilised, as the structure of the interference pattern is insensitive to phase fluctuations, which only produce a spatial shift of the pattern. Indeed, as shown by Petsas et al. (1994), the one-dimensional lin⊥lin scheme can be generalised to higher dimensions using $N+1$ beams for an N-dimensional lattice, with a variety of crystal structures possible depending on the polarisation and orientation of the laser beams. Depending on the detuning of the light, the atoms accumulate in the high (red detuning) or low (blue detuning) intensity region of the light field.

Optical lattices have been the subject of intense research because they allow many different ways to trap and manipulate cold atoms. They permit several geometries and configurations to be implemented that allow very accurate control over interaction between atoms in different lattice sites. The optical potential depth can be easily changed by changing the light intensity, whereas time-dependent optical lattices have been used to study quantum chaos by realising model systems such as the δ-kicked rotor or ratchet (Jones et al., 2004; Gommers et al., 2006).

Loading Bose–Einstein condensates into optical lattices – effectively perfect periodic arrays of optical dipole traps – has opened the way to investigate condensed matter physics and nonlinear physics with a huge control over experimental parameters (Bloch, 2005; Morsch and Oberthaler, 2006). Phenomena such as Bloch oscillations and Landau–Zener tunneling (Anderson and Kasevich, 1998; Morsch et al., 2001; Jona-Lasinio et al., 2003), instabilities and breakdown of superfluidity (Burger et al., 2001; Cristiani et al., 2004; Fallani et al., 2004) and Josephson physics (Cataliotti et al., 2001, 2003) have been investigated. A particularly important aspect is that when the number of atoms per lattice site is small, a regime where the system is well described by the Bose–Hubbard Hamiltonian is reached. This describes bosons with repulsive interactions in a periodic potential in the lowest energy level. First introduced by Fisher et al. (1989) in the context of liquid helium confined in porous media, it has been proposed by Jaksch et al. (1998) as a model for ultra-cold bosonic atoms confined in an optical lattice. This is related to the interplay between the tunnelling current between lattice sites, J, and the depth of the potential for a single site, U, as

$$\hat{H} = -\frac{J}{2}\sum_{i,j}\hat{a}_i^\dagger\hat{a}_j + \frac{U}{2}\sum_i\hat{n}_i(\hat{n}_i-1), \qquad (24.27)$$

where the sum is over nearest-neighbour pairs (i,j), $\hat{a}_i^\dagger$, $\hat{a}_i$ are the destruction and creation operators at lattice site i, and $\hat{n}_i = \hat{a}_i^\dagger\hat{a}_i$ is the number operator for the i-th well. The first term describes the tunnelling between neighbouring sites, whereas the second term describes the repulsive interaction within each site.

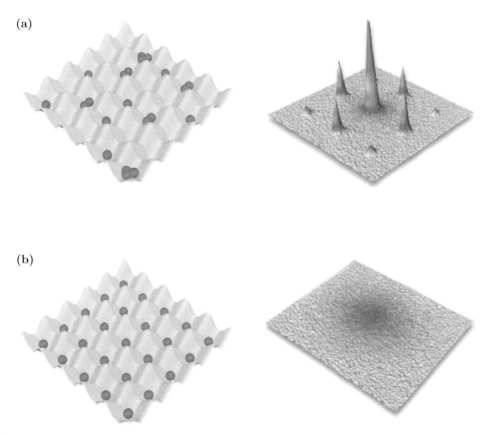

Figure 24.7 Detection of the superfluid-to-Mott transition. A Bose–Einstein condensate in a shallow two-dimensional optical lattice is described by a macroscopic wavefunction with a global phase. (a) When the condensate is released from the optical lattice, an interference pattern appears because of the phase coherence between different sites. (b) When the lattice depth is increased, the system supports a Mott insulator state: each lattice site is filled with a fixed number of atoms and phase coherence is lost, thus destroying the interference pattern when the atoms are released from the lattice. Reprinted by permission from Macmillan Publishers Ltd: Bloch, *Nature Phys.* **1**, 23–30, copyright 2005.

The Bose–Hubbard model predicts a quantum phase transition from superfluid to Mott insulator that is controlled by the ratio U/J, and it was first observed by Greiner et al. (2002). When the lattice depth is small, i.e., $U/J \ll 1$, tunnelling is dominant over site confinement and the system exhibits global phase coherence and superfluid behaviour, as shown in Fig. 24.7a. When the lattice depth is increased (e.g., by increasing the intensity of the light creating the optical lattice), i.e., $U/J \gg 1$, the phase coherence is broken and the system wavefunction is the product of the wavefunctions describing the individual site single-particle Fock states, as shown in Fig. 24.7b. In this strongly correlated Mott insulator state, tunnelling is suppressed and there is a fixed number of particles per site, i.e., *number squeezing* (Gerbier et al., 2006; Bakr et al., 2010).

Ultra-cold atoms in optical lattices have opened up exciting perspectives in research areas that include nonlinear matter waves, strongly correlated many-particle systems and quantum simulations and computation, where optical lattices are used to prepare engineered quantum states (Bloch et al., 2012).

24.9 Further reading

Many review articles have appeared in the last decade covering more specific aspects of cold atom physics. Of particular interest for the application of dipole traps, i.e., optical tweezers, to cold atoms are the reviews by Grimm et al. (2000) and Meschede and Rauschenbeutel (2006). Further experiments aimed at the on-demand delivery of single atoms using optical dipole forces can be found in, e.g., Kuhr et al. (2001) or Hill and McClelland (2003). The use of SLMs is permitting new physics to be probed using BECs as improved hologram calculation algorithms become avaliable to create arbitrarily shaped potentials for BEC (Gaunt and Hadzibabic, 2012; Gaunt et al., 2013). In all the experiments described, the dipole-trapping beam is detuned to the red side of an atomic resonance and, thus, forms an attractive potential. Blue detuning the beam create a repulsive potential by which atoms can be trapped away from regions of high intensity using multiple beams (Davidson et al., 1995) or Laguerre–Gaussian beams (Kuga et al., 1997). An advantage of this approach is that the atoms are trapped in a region where the photon scattering rate (and hence recoil heating and decoherence rate) is reduced. In another similarity to microscopic particles, atom traps based on evanescent fields have been demonstrated by Ovchinnikov et al. (1997). These also employ blue-detuned beams to create the evanescent field that provides a repulsive potential to keep the atoms away from the surface. BEC in this two-dimensional configuration has also been achieved by Rychtarik et al. (2004).

References

Abo-Shaeer, J. R., Raman, C., Vogels, J. M., and Ketterle, W. 2001. Observation of vortex lattices in Bose–Einstein condensates. *Science*, **292**, 476–9.

Adams, C. S., Lee, H. J., Davidson, N., Kasevich, M., and Chu, S. 1995. Evaporative cooling in a crossed dipole trap. *Phys. Rev. Lett.*, **74**, 3577–80.

Albiez, M., Gati, R., Fölling, J., et al. 2005. Direct observation of tunneling and nonlinear self-trapping in a single bosonic Josephson junction. *Phys. Rev. Lett.*, **95**, 010402.

Andersen, M. F., Ryu, C., Cladé, P., et al. 2006. Quantized rotation of atoms from photons with orbital angular momentum. *Phys. Rev. Lett.*, **97**, 170406.

Anderson, B. P., and Kasevich, M. A. 1998. Macroscopic quantum interference from atomic tunnel arrays. *Science*, **282**, 1686–9.

Anderson, M. H., Ensher, J. R., Matthews, M. R., Wieman, C. E., and Cornell, E. A. 1995. Observation of Bose–Einstein condensation in a dilute atomic vapor. *Science*, **269**, 198–201.

Andreev, S. V., Balykin, V. I., Letokhov, V. S., and Minogin, V. G. 1981. Radiative slowing and reduction of the energy spread of a beam of sodium atoms to 1.5 K in an oppositely directed laser beam. *JETP Lett.*, **34**, 463–7.

Aspect, A., Arimondo, E., Kaiser, R., Vansteenkiste, N., and Cohen-Tannoudji, C. 1988. Laser cooling below the one-photon recoil energy by velocity-selective coherent population trapping. *Phys. Rev. Lett.*, **61**, 826–9.

Bakr, W. S., Peng, A., Tai, M. E., et al. 2010. Probing the superfluid-to-Mott insulator transition at the single-atom level. *Science*, **329**, 547–50.

Balykin, V. I., Letokhov, V. S., and Sidorov, A. I. 1985. Radiative collimation of an atomic beam by two-dimensional cooling by laser beam. *JETP Lett.*, **40**, 1026–9.

Bardou, F., Bouchaud, J.-P., Aspect, A., and Cohen-Tannoudji, C. 2002. *Lévy statistics and laser cooling: How rare events bring atoms to rest*. Cambridge, UK: Cambridge University Press.

Barrett, M. D., Sauer, J. A., and Chapman, M. S. 2001. All-optical formation of an atomic Bose–Einstein condensate. *Phys. Rev. Lett.*, **87**, 010404.

Bergamini, S., Darquié, B., Jones, M., et al. 2004. Holographic generation of microtrap arrays for single atoms by use of a programmable phase modulator. *J. Opt. Soc. Am. B*, **21**, 1889–94.

Bloch, I. 2005. Ultracold quantum gases in optical lattices. *Nature Phys.*, **1**, 23–30.

Bloch, I., Dalibard, J., and Nascimbène, S. 2012. Quantum simulations with ultracold quantum gases. *Nature Phys.*, **8**, 267–76.

Bogoliubov, N. 1947. On the theory of superfluidity. *J. Phys. (U. S. S. R.)*, **11**, 23–32.

Bohr, Å., and Mottelson, B. R. 1998. *Nuclear structure. Vol. II: Nuclear deformations*. Singapore: World Scientific Publishing.

Bose, S. N. 1924. Plancks Gesetz und Lichtquantenhypothese. *Z. Phys.*, **26**, 178–81.

Boyer, V., Godun, R. M., Smirne, G., et al. 2006. Dynamic manipulation of Bose–Einstein condensates with a spatial light modulator. *Phys. Rev. A*, **73**, 031402.

Burger, S., Cataliotti, F. S., Fort, C., et al. 2001. Superfluid and dissipative dynamics of a Bose–Einstein condensate in a periodic optical potential. *Phys. Rev. Lett.*, **86**, 4447–50.

Castin, Y., and Dalibard, J. 1991. Quantization of atomic motion in optical molasses. *Europhys. Lett.*, **14**, 761–7.

Cataliotti, F. S., Burger, S., Fort, C., et al. 2001. Josephson junction arrays with Bose–Einstein condensates. *Science*, **293**, 843–6.

Cataliotti, F. S., Fallani, L., Ferlaino, F., et al. 2003. Superfluid current disruption in a chain of weakly coupled Bose–Einstein condensates. *New J. Phys.*, **5**, 71.

Chu, S., Hollberg, L., Bjorkholm, J. E., Cable, A., and Ashkin, A. 1985. Three-dimensional viscous confinement and cooling of atoms by resonance radiation pressure. *Phys. Rev. Lett.*, **55**, 48–51.

Chu, S., Bjorkholm, J. E., Ashkin, A., and Cable, A. 1986. Experimental observation of optically trapped atoms. *Phys. Rev. Lett.*, **57**, 314–7.

Cohen-Tannoudji, C., and Guéry-Odelin, D. 2011. *Advances in atomic physics*. Singapore: World Scientific Publishing.

Cristiani, M., Morsch, O., Malossi, N., et al. 2004. Instabilities of a Bose-Einstein condensate in a periodic potential: An experimental investigation. *Opt. Express*, **12**, 4–10.

Dalibard, J., and Cohen-Tannoudji, C. 1989. Laser cooling below the Doppler limit by polarization gradients: Simple theoretical models. *J. Opt. Soc. Am. B*, **6**, 2023–45.

Davidson, N., Lee, H. J., Adams, C. S., Kasevich, M., and Chu, S. 1995. Long atomic coherence times in an optical dipole trap. *Phys. Rev. Lett.*, **74**, 1311–14.

Davis, K. B., Mewes, M.-O., Andrews, M. R., et al. 1995. Bose–Einstein condensation in a gas of sodium atoms. *Phys. Rev. Lett.*, **75**, 3969–73.

Deng, H., Haug, H., and Yamamoto, Y. 2010. Exciton-polariton Bose–Einstein condensation. *Rev. Mod. Phys.*, **82**, 1489–537.

Dicke, R. H. 1953. The effect of collisions upon the Doppler width of spectral lines. *Phys. Rev.*, **89**, 472–3.

Einstein, A. 1925. Quantentheorie des einatomigen idealen Gases: Zweite abhandlung. *Sitzungsber. Preuss. Akad. Wiss.*, **18-25**, 3–14.

Fallani, L., De Sarlo, L., Lye, J. E., et al. 2004. Observation of dynamical instability for a Bose–Einstein condensate in a moving 1D optical lattice. *Phys. Rev. Lett.*, **93**, 140406.

Fisher, M. P. A., Weichman, P. B., Grinstein, G., and Fisher, D. S. 1989. Boson localization and the superfluid–insulator transition. *Phys. Rev. B*, **40**, 546–70.

Foot, C. J. 2005. *Atomic physics*. Oxford, UK: Oxford University Press.

Frese, D., Ueberholz, B., Kuhr, S., et al. 2000. Single atoms in an optical dipole trap: Towards a deterministic source of cold atoms. *Phys. Rev. Lett.*, **85**, 3777–80.

Gaunt, A. L., and Hadzibabic, Z. 2012. Robust digital holography for ultracold atom trapping. *Sci. Rep.*, **2**, 721.

Gaunt, A. L., Schmidutz, T. F., Gotlibovych, I., Smith, R. P., and Hadzibabic, Z. 2013. Bose–Einstein condensation of atoms in a uniform potential. *Phys. Rev. Lett.*, **110**, 200406.

Gerbier, F., Fölling, S., Widera, A., Mandel, O., and Bloch, I. 2006. Probing number squeezing of ultracold atoms across the superfluid–Mott insulator transition. *Phys. Rev. Lett.*, **96**, 090401.

Gommers, R., Denisov, S., and Renzoni, F. 2006. Quasiperiodically driven ratchets for cold atoms. *Phys. Rev. Lett.*, **96**, 240604.

Greiner, M., Mandel, O., Esslinger, T., Hänsch, T. W., and Bloch, I. 2002. Quantum phase transition from a superfluid to a Mott insulator in a gas of ultracold atoms. *Nature*, **415**, 39–44.

Grimm, R., Weidemüller, M., and Ovchinnikov, Y. B. 2000. Optical dipole traps for neutral atoms. *Adv. Atom. Mol. Opt. Phys.*, 95–170.

Guéry-Odelin, D., and Stringari, S. 1999. Scissors mode and superfluidity of a trapped Bose–Einstein condensed gas. *Phys. Rev. Lett.*, **83**, 4452–5.

Gustavson, T. L., Chikkatur, A. P., Leanhardt, A. E., et al. 2001. Transport of Bose–Einstein condensates with optical tweezers. *Phys. Rev. Lett.*, **88**, 020401.

Hänsch, T. W., and Schawlow, A. L. 1975. Cooling of gases by laser radiation. *Opt. Commun.*, **13**, 68–9.

Hemmerich, A., Zimmermann, C., and Hänsch, T. W. 1993. Sub-kHz Rayleigh resonance in a cubic atomic crystal. *Europhys. Lett.*, **22**, 89–94.

Hess, H. F. 1986. Evaporative cooling of magnetically trapped and compressed spin-polarized hydrogen. *Phys. Rev. B*, **34**, 3476–9.

Hill, S. B., and McClelland, J. J. 2003. Atoms on demand: Fast, deterministic production of single Cr atoms. *Appl. Phys. Lett.*, **82**, 3128–30.

Hodby, E., Hechenblaikner, G., Hopkins, S. A., Maragò, O. M., and Foot, C. J. 2001. Vortex nucleation in Bose–Einstein condensates in an oblate, purely magnetic potential. *Phys. Rev. Lett.*, **88**, 010405.

Hopkins, S. A., and Durrant, A. V. 1997. Parameters for polarization gradients in three-dimensional electromagnetic standing waves. *Phys. Rev. A*, **56**, 4012–22.

Jaksch, D., Bruder, C., Cirac, J. I., Gardiner, C. W., and Zoller, P. 1998. Cold bosonic atoms in optical lattices. *Phys. Rev. Lett.*, **81**, 3108–11.

Jessen, P. S., Gerz, C., Lett, P. D., et al. 1992. Observation of quantized motion of Rb atoms in an optical field. *Phys. Rev. Lett.*, **69**, 49–52.

Jona-Lasinio, M., Morsch, O., Cristiani, M., et al. 2003. Asymmetric Landau–Zener tunneling in a periodic potential. *Phys. Rev. Lett.*, **91**, 230406.

Jones, P. H., Stocklin, M. M., Hur, G., and Monteiro, T. S. 2004. Atoms in double-δ-kicked periodic potentials: Chaos with long-range correlations. *Phys. Rev. Lett.*, **93**, 223002.

Kasevich, M., and Chu, S. 1992. Laser cooling below a photon recoil with three-level atoms. *Phys. Rev. Lett.*, **69**, 1741–4.

Ketterle, W., and Van Druten, N. J. 1996. Evaporative cooling of trapped atoms. *Adv. Atom. Mol. Opt. Phys.*, **37**, 181–236.

Konishi, K., and Paffuti, G. 2009. *Quantum mechanics: A new introduction*. Oxford, UK: Oxford University Press.

Kuga, T., Torii, Y., Shiokawa, N., et al. 1997. Novel optical trap of atoms with a doughnut beam. *Phys. Rev. Lett.*, **78**, 4713–6.

Kuhr, S., Alt, W., Schrader, D., et al. 2001. Deterministic delivery of a single atom. *Science*, **293**, 278–80.

Leanhardt, A. E., Pasquini, T. A., Saba, M., et al. 2003. Cooling Bose–Einstein condensates below 500 picokelvin. *Science*, **301**, 1513–15.

Leggett, A. J. 2001. Superfluidity. *Rev. Mod. Phys.*, **71**, S318–S323.

Letokhov, V. S., Minogin, V. G., and Pavlik, B. D. 1976. Cooling and trapping of atoms and molecules by a resonant laser field. *Opt. Commun.*, **19**, 72–5.

Lett, P. D., Watts, R. N., Westbrook, C. I., et al. 1988. Observation of atoms laser cooled below the Doppler limit. *Phys. Rev. Lett.*, **61**, 169–72.

Lett, P. D., Phillips, W. D., Rolston, S. L., et al. 1989. Optical molasses. *J. Opt. Soc. Am. B*, **6**, 2084–2107.

London, F. 1938. The λ-phenomenon of liquid helium and the Bose–Einstein degeneracy. *Nature*, **141**, 643–4.

Madison, K. W., Chevy, F., Wohlleben, W., and Dalibard, J. 2000. Vortex formation in a stirred Bose–Einstein condensate. *Phys. Rev. Lett.*, **84**, 806–9.

Maragò, O. 2001 (Trinity Term). *The scissors mode and superfluidity of a Bose–Einstein condensed gas*. Ph.D. thesis, University of Oxford, Oxford, UK.

Maragò, O. M., Hopkins, S. A., Arlt, J., et al. 2000. Observation of the scissors mode and evidence for superfluidity of a trapped Bose–Einstein condensed gas. *Phys. Rev. Lett.*, **84**, 2056–9.

Masuhara, N., Doyle, J. M., Sandberg, J. C., et al. 1988. Evaporative cooling of spin-polarized atomic hydrogen. *Phys. Rev. Lett.*, **61**, 935–88.

Matthews, M. R., Anderson, B. P., Haljan, P. C., et al. 1999. Vortices in a Bose–Einstein condensate. *Phys. Rev. Lett.*, **83**, 2498–501.

Meschede, D., and Rauschenbeutel, A. 2006. Manipulating single atoms. *Adv. Atom. Mol. Opt. Phys.*, **53**, 75–104.

Meystre, P. 2001. *Atom optics*. Heidelberg, Germany: Springer-Verlag.

Migdall, A. L., Prodan, J. V., Phillips, W. D., Bergeman, T. H., and Metcalf, H. J. 1985. First observation of magnetically trapped neutral atoms. *Phys. Rev. Lett.*, **54**, 2596–9.

Morsch, O., and Oberthaler, M. 2006. Dynamics of Bose–Einstein condensates in optical lattices. *Rev. Mod. Phys.*, **78**, 179–215.

Morsch, O., Müller, J. H., Cristiani, M., Ciampini, D., and Arimondo, E. 2001. Bloch oscillations and mean-field effects of Bose–Einstein condensates in 1D optical lattices. *Phys. Rev. Lett.*, **87**, 140402.

Nogrette, F., Labuhn, H., Ravets, S., et al. 2014. Single-atom trapping in holographic 2D arrays of microtraps with arbitrary geometries. *Phys. Rev. X*, **4**, 021034.

Onofrio, R., Raman, C., Vogels, J. M., et al. 2000. Observation of superfluid flow in a Bose–Einstein condensed gas. *Phys. Rev. Lett.*, **85**, 2228–31.

Ovchinnikov, Yu. B., Manek, I., and Grimm, R. 1997. Surface trap for Cs atoms based on evanescent-wave cooling. *Phys. Rev. Lett.*, **79**, 2225–8.

Petrich, W., Anderson, M. H., Ensher, J. R., and Cornell, E. A. 1995. Stable, tightly confining magnetic trap for evaporative cooling of neutral atoms. *Phys. Rev. Lett.*, **74**, 3352–5.

Petsas, K. I., Coates, A. B., and Grynberg, G. 1994. Crystallography of optical lattices. *Phys. Rev. A*, **50**, 5173–89.

Pitaevskii, L., and Stringari, S. 2003. *Bose–Einstein condensation*. Oxford, UK: Oxford University Press.

Pitaevskii, L. P. 1961. Vortex lines in an imperfect Bose gas. *Sov. Phys. JETP*, **13**, 451–4.

Pritchard, D. E. 1983. Cooling neutral atoms in a magnetic trap for precision spectroscopy. *Phys. Rev. Lett.*, **51**, 1336–9.

Prodan, J. V., Phillips, W. D., and Metcalf, H. 1982. Laser production of a very slow monoenergetic atomic beam. *Phy. Rev. Lett.*, **49**, 1149–53.

Raab, E. L., Prentiss, M., Cable, A., Chu, S., and Pritchard, D. E. 1987. Trapping of neutral sodium atoms with radiation pressure. *Phys. Rev. Lett.*, **59**, 2631–4.

Raman, C., Köhl, M., Onofrio, R., et al. 1999. Evidence for a critical velocity in a Bose–Einstein condensed gas. *Phys. Rev. Lett.*, **83**, 2502–5.

Rychtarik, D., Engeser, B., Nägerl, H.-C., and Grimm, R. 2004. Two-dimensional Bose–Einstein condensate in an optical surface trap. *Phys. Rev. Lett.*, **92**, 173003.

Ryu, C., Andersen, M. F., Cladé, P., et al. 2007. Observation of persistent flow of a Bose–Einstein condensate in a toroidal trap. *Phys. Rev. Lett.*, **99**, 260401.

Schlosser, N., Reymond, G., Protsenko, I., and Grangier, P. 2001. Sub-Poissonian loading of single atoms in a microscopic dipole trap. *Nature*, **411**, 1024–7.

Stamper-Kurn, D. M., and Ueda, M. 2013. Spinor Bose gases: Symmetries, magnetism, and quantum dynamics. *Rev. Mod. Phys.*, **85**, 1191–244.

Stellmer, S., Pasquiou, B., Grimm, R., and Schreck, F. 2013. Laser cooling to quantum degeneracy. *Phys. Rev. Lett.*, **110**, 263003.

Tilley, D. R., and Tilley, J. 1990. *Superfluidity and superconductivity*. 2nd ed. Boca Raton, FL: CRC Press.

Verkerk, P., Lounis, B., Salomon, C., et al. 1992. Dynamics and spatial order of cold cesium atoms in a periodic optical potential. *Phys. Rev. Lett.*, **68**, 3861–84.

Verkerk, P., Meacher, D. R., Coates, A. B., et al. 1994. Designing optical lattices: An investigation with cesium atoms. *Europhys. Lett.*, **26**, 171–6.

Weber, T., Herbig, J., Mark, M., Nägerl, H.-C., and Grimm, R. 2003. Bose–Einstein condensation of cesium. *Science*, **299**, 232–5.

Westbrook, C. I., Watts, R. N., Tanner, C. E., et al. 1990. Localization of atoms in a three-dimensional standing wave. *Phys. Rev. Lett.*, **65**, 33–6.

Wineland, D., and Dehmelt, H. 1975. Proposed $10^{14}\Delta\nu < \nu$ laser fluorescence spectroscopy on Tl$^+$ mono-ion oscillator III. *Bull. Am. Phys. Soc.*, **20**, 637.

Wing, W. H. 1984. On neutral particle trapping in quasistatic electromagnetic fields. *Prog. Quant. Electron*, **8**, 181–99.

25 Towards the quantum regime at the mesoscale

Recently, much effort has been devoted to developing techniques that bridge the gap between laser cooling of atomic species and optical trapping of colloidal materials in order to study quantum phenomena at mesoscopic length scales. The aim is to explore and exploit quantum effects, such as entanglement, quantum superposition of motional states and long quantum coherence, in systems larger than atomic species. Experimentally, the first step towards this goal is the trapping and laser cooling of nanoparticles in vacuum, extending the methodologies used for neutral atoms, which we discussed in Chapter 24. This trend fits naturally within the miniaturisation of optomechanics, i.e., the study and control of mechanical motion induced by optical forces. In particular, in optomechanical systems, thermal fluctuations of a mechanical oscillator can be reduced by interaction with an optical field, and thus the system can be effectively cooled by a controlled back-action of the light, as shown in Fig. 25.1. In this chapter, we will explore these novel experimental schemes and how they open up new prospects for the cooling of colloidal particles to their quantum motional ground states in vacuum.

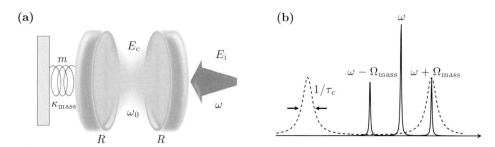

Figure 25.1 Cavity optomechanics. (a) An example of an optomechanical system is a cavity formed by a fixed mirror and a moving mirror and pumped by an incident field: a small displacement of the moving mirror changes the electromagnetic field present in the cavity and the field feeds back through its radiation force on the mirror position. (b) The mechanical oscillator modulates the pumping light (ω), producing two sidebands (the *Stokes sideband* $\omega - \Omega_{\mathrm{mass}}$ and the *anti-Stokes sideband* $\omega + \Omega_{\mathrm{mass}}$). As shown in the picture, when the input field is detuned on the red side of one of the cavity resonances (dashed line), the anti-Stokes sideband gets enhanced and, because anti-Stokes photons have more energy than the pumping field's ones, energy is extracted from the mechanical oscillator into the cavity field (*optical cooling*). For blue detuning, the Stokes sideband is enhanced and, thus, energy is transferred from the field to the mechanical oscillator (*optical amplification*, not shown).

25.1 Cavity optomechanics: The classical picture

The first system where optical and mechanical degrees of freedom were coupled by radiation pressure was investigated by Braginskii and Manukin (1977) in the context of experiments that aimed at the interferometric detection of gravitational waves. They considered an optical cavity of length L with one fixed mirror and one moving mirror of mass m attached to a mechanical oscillator characterised by a stiffness κ_{mass}, as shown in Fig. 25.1a. The quality of the cavity is defined by its *finesse* (Born and Wolf, 1999), i.e.,

$$\mathfrak{F} = \pi \frac{\sqrt{R}}{1-R}, \tag{25.1}$$

where R is the reflectivity of the mirrors. Transmission of light through the cavity occurs at resonance; i.e., when $2L = q\lambda_q/n_{\text{m}}$,[1] where q is an integer, λ_q/n_{m} is the *intracavity wavelength*, n_{m} is the refractive index of the intracavity medium and λ_q is the wavelength of the cavity mode with frequency[2]

$$\omega_q = q \frac{\pi c}{n_{\text{m}} L}. \tag{25.2}$$

For high reflectivity, i.e., for large finesse, the intracavity field is highly enhanced at resonance whereas it is suppressed out of resonance, so that the cavity transmission is peaked at ω_q with a width given by (Born and Wolf, 1999)

$$\Delta\omega_{\text{out}} = \frac{\pi c}{\mathfrak{F} n_{\text{m}} L}, \tag{25.3}$$

which is the same for all the cavity modes, as it does not depend on q and is related to the cavity quality factor

$$Q_{\text{c}} = \frac{\omega}{\Delta\omega_{\text{out}}}. \tag{25.4}$$

Therefore, the light intensity (or photon number) decay time inside the cavity is

$$\tau_{\text{c}} = \frac{1}{\Delta\omega_{\text{out}} + \Delta\omega_{\text{int}}}, \tag{25.5}$$

where $\Delta\omega_{\text{int}}$ is related to the cavity internal losses. Now, suppose there is an incident field with (complex) amplitude E_{i} going into the cavity with a coupling constant K.[3] For a frequency close to the fundamental resonance at ω_0, the dynamic equation for the cavity field $E_{\text{c}}(t)$ can be derived from the Helmholtz equation (Novotny and Hecht, 2012), obtaining[4]

$$\frac{dE_{\text{c}}(t)}{dt} = \left[i(\omega - \omega_0) - \frac{1}{2\tau_{\text{c}}}\right] E_{\text{c}}(t) + K E_{\text{i}}. \tag{25.6}$$

---

[1] At resonance, the phase shift acquired by the intracavity field in a round trip is $2\pi q$ and, therefore, the fields add up in phase.
[2] Note that the cavity modes can be changed by altering either the cavity length (L) or its refractive index (n_{m}).
[3] K takes into account, e.g., the losses due to partial transmission into the cavity of the incoming field.
[4] The physical field amplitude is $\mathcal{E}_{\text{c}}(t) = \text{Re}\{E_{\text{c}}(t)e^{i\omega t}\}$.

A small displacement of the moving mirror, $x(t)$, from its equilibrium position at $x = 0$ changes the cavity length by $\Delta L(t) = x(t)$ and hence shifts the resonance frequency by

$$\Delta\omega_0(t) = -\omega_0 \frac{x(t)}{L}. \tag{25.7}$$

Therefore, the intracavity field is coupled to the mirror displacement as

$$\frac{dE_c(t)}{dt} = \left[i(\omega - \omega_0) + i\omega_0 \frac{x(t)}{L} - \frac{1}{2\tau_c}\right] E_c(t) + KE_i. \tag{25.8}$$

The mechanical motion of the moving mirror driven by the radiation pressure from the cavity field can be described by the Langevin equation for a damped noisy oscillator [Section 7.2.2],

$$m\frac{d^2 x(t)}{dt^2} + m\Gamma_{\text{mass}} \frac{dx(t)}{dt} + m\Omega_{\text{mass}}^2 x(t) = \chi(t) + F_{\text{rad}}(t), \tag{25.9}$$

where Γ_{mass} is the mechanical damping coefficient, $\Omega_{\text{mass}} = \sqrt{\kappa_{\text{mass}}/m}$ is the oscillator's characteristic frequency, $\chi(t) = \sqrt{2m\Gamma_{\text{mass}} k_B T} W(t)$ is thermal noise, $W(t)$ is white noise [Box 7.3], $F_{\text{rad}}(t) = 2\varepsilon_m AR|E_c(t)|^2$ is the radiation force on the mirror and A is the illumination area of the mirror.

Therefore, the dynamics of the optomechanical system is fully described by the two coupled equations (25.7) and (25.9). Their solutions for $E_c(t)$ and $x(t)$ depend on the parameters of the system, e.g., the cavity resonance, the incident light intensity and frequency and the cavity and oscillator loss terms. The *dynamical back action*, i.e., the fact that a displacement of the mirror feeds back on itself through radiation pressure, is clearly visible in the fact that $x(t)$ depends on $|E_c(t)|^2$, which in turn is a function of $x(t)$. The radiation force acting on the mirror yields a change in the damping term and in the oscillation frequency of the mirror so that Eq. (25.9) can be re-written as (Kippenberg and Vahala, 2007)

$$\frac{d^2 x(t)}{dt^2} + (\Gamma_{\text{mass}} + \Delta\Gamma_{\text{mass}}) \frac{dx(t)}{dt} + (\Omega_{\text{mass}} + \Delta\Omega_{\text{mass}})^2 x(t) = \frac{\chi(t)}{m}, \tag{25.10}$$

with

$$\Delta\Gamma_{\text{mass}} = P_i \mathfrak{F}^2 \frac{8n_m^2 \omega_0}{mc^2 \Omega_{\text{mass}}} \frac{\tau_c/\tau_{\text{out}}}{4\delta^2 \tau_c^2 + 1} \left[\frac{1}{4(\delta + \Omega_{\text{mass}})^2 \tau_c^2 + 1} - \frac{1}{4(\delta - \Omega_{\text{mass}})^2 \tau_c^2 + 1} \right],$$

$$\Delta\Omega_{\text{mass}} = P_i \mathfrak{F}^2 \frac{8n_m^2 \omega_0}{mc^2 \Omega_{\text{mass}}} \frac{\tau_c^2/\tau_{\text{out}}}{4\delta^2 \tau_c^2 + 1} \left[\frac{\delta + \Omega_{\text{mass}}}{4(\delta + \Omega_{\text{mass}})^2 \tau_c^2 + 1} + \frac{\delta - \Omega_{\text{mass}}}{4(\delta - \Omega_{\text{mass}})^2 \tau_c^2 + 1} \right],$$

where P_i is the input power of the incident electromagnetic field, $\tau_{\text{out}} = \Delta\omega_{\text{out}}^{-1}$ and $\delta = \omega - \omega_0$. The first terms in the square brackets in the expressions for $\Delta\Gamma_{\text{mass}}$ and $\Delta\Omega_{\text{mass}}$ are associated with anti-Stokes scattering, whereas the second terms are associated with Stokes scattering.

The vibration of the mirror modulates the driving field in two sidebands, as shown by the solid line in Fig. 25.1b. In the case of $\tau_c^{-1} = \Delta\omega_{\text{out}} + \Delta\omega_{\text{int}} \gg \Omega_{\text{mass}}$, we are in the *weak retardation regime* and the optomechanical damping rate is proportional to $-8\delta\tau_c^2/(4\delta^2\tau_c^2 + 1)^2$. In analogy to atomic systems, the damping rate can be tuned to be positive (*red detuning*, $\delta < 0$), yielding effective cooling of the system, or negative (*blue detuning*,

$\delta > 0$), yielding amplification of the mechanical oscillations. The optimum coupling occurs for $\delta = -(2\tau_c)^{-1}$ (cooling) and $\delta = +(2\tau_c)^{-1}$ (amplification). Much stronger optomechanical coupling is achieved in the case of $\tau_c^{-1} \ll \Omega_{\text{mass}}$, i.e., in the *resolved sideband regime*. Also in this case, the damping rate can be positive for red detuning of the pumping light (cooling) or negative for blue detuning of the pumping light (amplification). Optimum coupling occurs for $\delta = \mp\Omega_{\text{mass}}$, i.e., when the incident pumping light is tuned to one of the sidebands; for example, when $\delta = -\Omega_0$, the incident radiation causes the system to create more anti-Stokes than Stokes photons, forcing a net energy flow from the mechanical oscillator to the optical field.

In the case of red detuning, enhanced damping yields an effective cooling of the system, and an effective temperature can be associated with the optomechanical Langevin equation [Eq. (25.10)], which can be written for $\Delta\Omega_{\text{mass}} \ll \Omega_{\text{mass}}$ as

$$T_{\text{eff}} \simeq T \frac{\Gamma_{\text{mass}}}{\Gamma_{\text{mass}} + \Delta\Gamma_{\text{mass}}}, \qquad (25.11)$$

where T is the equilibrium absolute temperature in the absence of radiation force and the relation holds for $\tau_c \Delta\Gamma_{\text{mass}} \ll 1$ and for $T_{\text{eff}} > 2T\Gamma_{\text{mass}}/\Omega_{\text{mass}}$.

25.2 Cavity optomechanics: The quantum picture

The classical treatment of optomechanics does not impose any lower limit on the temperature achievable in the cooling process. However, the quantum nature of light and the discrete energy spectrum of the oscillator with states separated by $\hbar\Omega_{\text{mass}}$ and zero point fluctuation amplitude[5] $x_0 = \sqrt{\hbar/(2m\Omega_{\text{mass}})}$ need to be taken into account to evaluate the ultimate cooling temperature.

The quantum theory of optomechanical cooling (Wilson-Rae et al., 2007; Marquardt et al., 2007) starts by treating the optically driven cavity coupled to the mechanical oscillator as an open quantum system. The total physical Hamiltonian is written in terms of the *quantum ladder operators* of the optical, $(\hat{a}, \hat{a}^\dagger)$, and mechanical, $(\hat{b}, \hat{b}^\dagger)$, quantum oscillators, respectively (Cohen-Tannoudji et al., 1992), i.e.,

$$\hat{H} = -\hbar\delta\,\hat{a}^\dagger\hat{a} + \hbar\,\Omega_{\text{mass}}\,\hat{b}^\dagger\hat{b} + \hbar\,\eta\,\Omega_{\text{mass}}\,\hat{a}^\dagger\hat{a}(\hat{b}^\dagger + \hat{b}) + \hbar\frac{\Omega_{\text{R,i}}}{2}(\hat{a}^\dagger + \hat{a}), \qquad (25.12)$$

where $\Omega_{\text{R,i}} = 2\sqrt{P_i \Delta\omega_{\text{out}}/(\hbar\omega)}$ is the driving amplitude of the input radiation and $\eta = \omega_0 x_0/\Omega_{\text{mass}} L$ is a dimensionless parameter characterising the optomechanical coupling via radiation pressure. The coupling is generally very small, as small as $\eta \approx 10^{-4}$ for typical

---

[5] A quantum harmonic oscillator has a zero point energy of $E_0 = \hbar\Omega_{\text{mass}}/2$. The zero point fluctuation amplitude can easily be calculated by equating this zero point energy to the average energy of a classical oscillator. In fact, for a classical oscillator the average energy is double the average potential energy (from the virial theorem); hence $\langle E \rangle = m\Omega_{\text{mass}}^2 \langle x^2 \rangle$. Thus, by posing $\langle E \rangle = E_0 = \hbar\Omega_{\text{mass}}/2$, the value of $x_0 = \sqrt{\langle x^2 \rangle} = \sqrt{\hbar/2m\Omega_{\text{mass}}}$ is found. Alternatively, one can use the expression of the position operator in terms of ladder operators, $\hat{x} = \sqrt{\frac{\hbar}{2m\Omega_{\text{mass}}}}(\hat{b} + \hat{b}^\dagger)$, and evaluate the positional fluctuations over the ground state, $x_0^2 = \langle \hat{x}^2 \rangle = \hbar/2m\Omega_{\text{mass}}$.

experimental conditions. This implies that the interaction between the mechanical oscillator and the electromagnetic fields can be treated pertubatively and, by the use of the generalised master equation (Gardiner and Zoller, 2004) for the density matrix of the system and the adiabatic elimination of the fast rotating terms ($\propto e^{\pm i\omega t}$), it is possible to identify the light-scattering contributions that yield the cooling and heating processes (Marquardt et al., 2007; Wilson-Rae et al., 2007). The cooling process is based on scattering Stokes and anti-Stokes photons with rates that are weighted by the sidebands' amplitudes,

$$A^{\mp} = \eta^2 \frac{4\Omega_{R,i}^2}{4\tau_c^2 \delta^2 + 1} \frac{\Omega_{mass}^2 \tau_c^3}{4\tau_c^2(\delta \pm \Omega_{mass})^2 + 1}. \tag{25.13}$$

The balance between these two scattering processes yields the final mean thermal occupation number of the quantum-mechanical oscillator that is related to the ultimate temperature limit. Thus, including also the contribution from the thermal reservoir given by Eq. (25.11), the final occupation that can be achieved is (Kippenberg and Vahala, 2007)

$$\bar{n}_f = \frac{A^+}{A^- - A^+} + \frac{\Gamma_{mass}}{\Gamma_{mass} + \Delta\Gamma_{mass}} \bar{n}_i, \tag{25.14}$$

where $\bar{n}_i$ is the initial occupancy at thermal equilibrium. The effective temperature limit is thus obtained as

$$T_{eff} = \bar{n}_f \frac{\hbar \Omega_{mass}}{k_B}. \tag{25.15}$$

To achieve ground state cooling of the mechanical oscillator, we need $\bar{n}_f < 1$ and hence $T_{eff} < \hbar\Omega_{mass}/k_B$, which, e.g., for a 100 kHz oscillator leads to a temperature below 5 µK. This regime is achieved in the resolved sideband cooling, for which $\Omega_{mass}\tau_c \gg 1$ and $\bar{n}_f \approx (4\Omega_{mass}\tau_c)^{-2}$ (Schliesser et al., 2008).

25.3 Laser cooling of levitated particles

Among the different schemes for optomechanical cooling, protocols for reaching the quantum regime at the mesoscale have been envisaged for ground-state laser cooling of optically levitated nanoparticles (Barker and Shneider, 2010; Chang et al., 2010; Romero-Isart et al., 2010). In these schemes a particle is held in a high-finesse cavity in vacuum either by the cavity standing wave field (Chang et al., 2010) or by a separate optical tweezers (Romero-Isart et al., 2010, 2011). The mechanical oscillator is created by a confining optical trap with spring constant κ_{trap}, oscillator frequency $\omega_{trap} = \sqrt{\kappa_{trap}/m}$ and zero point fluctuation amplitude $x_0 = \sqrt{\hbar/(2m\omega_{trap})}$. A radiation field excites a cavity mode that couples to the trapped particle's centre-of-mass motion. The presence of the dielectric particle in the cavity changes the cavity mode, yielding the position-dependent optomechanical coupling responsible for cooling. In particular, the dielectric particle alters the cavity resonance through the change in refractive index occurring within the small particle volume, V, so that the cavity mode, $E_c(\mathbf{r})$, where $\mathbf{r}$ is position, is frequency-shifted by an amount

$$\Delta\omega_0 \approx -\frac{\omega_0}{2} \frac{\int_V \Delta\mathcal{P}(\mathbf{r})E_c(\mathbf{r})dV}{\int_V \varepsilon_0|E_c(\mathbf{r})|^2 dV}, \qquad (25.16)$$

where $\Delta\mathcal{P}(\mathbf{r})$ is the modification of the polarisation produced by the particles in the cavity. For a sub-wavelength particle, we can use the dipole approximation for the polarisability, so that $\Delta P(\mathbf{r}) \approx \alpha_d E_c(\mathbf{r}_d)\delta(\mathbf{r}_d - \mathbf{r})$, where α_d is the dipole polarisability [Eq. (3.15)] and $\mathbf{r}_d$ is the centre-of-mass position of the particle. The zeroth-order contribution to the particle position yields a constant shift of the cavity resonance,

$$\Delta\omega_0 \approx -\omega_0 \frac{\alpha_d}{8\varepsilon_0 V_c}, \qquad (25.17)$$

where V_c is the cavity volume. The first-order contribution to the particle position gives the optomechanical coupling characterised by the adimensional parameter

$$\eta \approx \frac{x_0}{\omega_{\text{trap}}} \frac{\omega_0^2 \alpha_d}{4c\varepsilon_0 V_c}. \qquad (25.18)$$

Experimentally, optical trapping of particles can be achieved in vacuum as in the case of atoms, i.e., in a chamber with controllable vacuum pressure. In this case, as for aerosols [Chapter 19], the damping of the residual gas is so low that the ballistic regime and harmonic oscillations can easily be observed in the trap. Cavity cooling of a sub-micrometric particle was demonstrated by Kiesel et al. (2013). A high-finesse optical cavity ($\mathfrak{F} = 76\,000$) was used for both trapping and manipulation, demonstrating optomechanical control and cooling of the centre-of-mass motion of the particle. The final temperature was limited to about 64 K mainly because of the damping due to the residual gas pressure. To improve on this result, it has been necessary to use an active feedback cooling, which requires direct measurement of the particle state (position and velocity) in real time.

25.4 Feedback cooling schemes

Li et al. (2010) performed ultra-high-resolution measurements of a trapped particle's instantaneous velocity held in vacuum in an optical trap, directly showing that for different pressures the velocity distribution of the trapped bead is well fitted by a Maxwell–Boltzmann distribution and yielding an effective temperature measurement of the particle centre-of-mass motion in the confining optical potential. This has been the starting point for laser cooling of the centre-of-mass motion of trapped particles (Li et al., 2011; Gieseler et al., 2012). The scheme used by Li et al. (2011), shown in Fig. 25.2a, uses a counter-propagating optical tweezers at 1064 nm to create a harmonic potential where to trap a silica microparticle (1.5 μm radius) and pairs of counter-propagating beams at 532 nm with independently controllable power to cool down its centre-of-mass motion. External feedback adjusts the cooling beam power depending on the measured particle

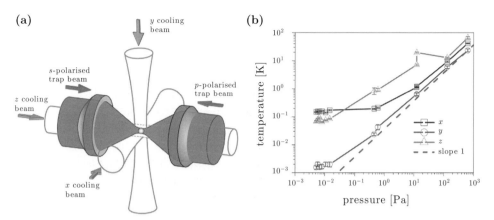

Figure 25.2 Laser cooling of a microparticle: (a) Experimental configuration for feedback cooling of a levitated microparticle. A counter-propagating optical tweezers (solid beams) creates a trapping potential where the particle is held. Three additional pairs of counter-propagating beams (outlined beams) provide the active feedback cooling. (b) Equilibrium temperature of a (1.5 μm radius) silica bead laser-cooled at different background pressures, achieving temperatures down to the millikelvin range. Reprinted by permission from Macmillan Publishers Ltd: Li et al., *Nature Phys.* **7**, 527–30, copyright 2011.

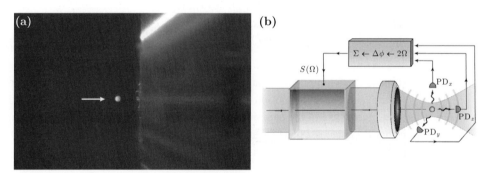

Figure 25.3 Laser cooling of a nanoparticle: (a) Image of light scattered by a laser-cooled silica nanoparticle confined in optical tweezers in ultrahigh vacuum. (b) This light is measured interferometrically with three detectors, labeled PD_x, PD_y and PD_z. Each detector signal is frequency-doubled and phase-shifted. The sum of these signals is used to modulate the intensity of the trapping beam. A feedback cooling scheme, where the same beam is used for trapping and cooling, is then used to reach temperatures as low as 50 mK for the particle centre-of-mass motion. Reprinted figure from Gieseler et al., *Phys. Rev. Lett.* **109**, 103603. Copyright (2012) by the American Physical Society.

velocity so that the excess radiation pressure of one beam counteracts the motion of the bead. The result is effective cooling that reaches temperatures in the millikelvin range, as shown in Fig. 25.2b, with occupation number of the centre-of-mass motion that drops from $\bar{n}_f \approx 6.8 \times 10^8$ at room temperature to about $\bar{n}_f \approx 3400$ at 1.5 mK.

Gieseler et al. (2012) demonstrated laser cooling of a silica nanoparticle (70 nm radius), shown in Fig. 25.3a. In this case, a single-beam optical tweezers is operated in vacuum at 1064 nm. Light scattered by the particle is monitored with photodiodes to drive a feedback loop that controls the trapping light intensity, i.e., the trap stiffness is increased when the particle moves away from its equilibrium position and reduced otherwise, as shown in Fig. 25.3b. With this configuration, an effective temperature as low as about 50 mK was reached, as measured by observing residual thermal fluctuations.

25.5 Below the Doppler limit

Differently from the laser cooling of atomic species, which, as we have seen in Chapter 24, takes advantage of near-resonant excitation of sharp transitions, the particle's intrinsic resonances and internal structure play no role in the feedback cooling schemes presented in the previous sections. Drawing inspiration from atom cooling, Barker (2010) theoretically proposed Doppler cooling of a dielectric microparticles by red detuning counter-propagating beams on a sharp Mie resonance (*whispering gallery mode*) with a Lorentzian profile and predicted a Doppler temperature limit in the tens of microkelvin for a 10 μm-radius silica particle. Later, Ridolfo et al. (2011) also theoretically demonstrated that, just as the multi-level structure of atoms enables sub-Doppler cooling mechanisms [Chapter 24], complex coupled or hybridised nanostructures [Fig. 25.4] enable the interaction with light to be modified by *quantum interference*. In the rest of this section, we give an account of near-resonant Doppler cooling of particles exhibiting *quantum interference*, and thus we will treat the case of a *Fano lineshape* resonance, which includes as a special case the Lorentzian lineshape treated by Barker (2010).

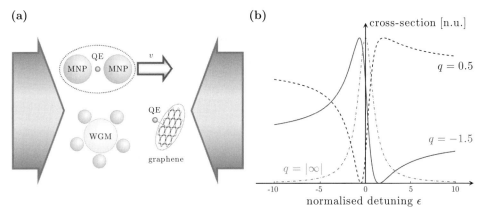

Figure 25.4 Laser cooling of hybrid nanostructures. (a) Sketch illustrating the one-dimensional laser cooling configuration for a number of levitated nanosystems. Two counter-propagating laser beams interact with hybrid metal–quantum emitter nanoparticles, silica microparticles sustaining Mie resonances hybridised with plasmonic nanoparticles, or graphene flakes hybridised with a quantum emitter. The near-resonant excitation close to a Fano resonance is tuned by changing the incident light detuning (Ridolfo et al., 2011). (b) Normalised radiation pressure cross-sections for hybrid nanostructures with Fano lineshapes. The lineshape and Fano factor, q, change with the coupling within hybrid particles. The plots for a Lorentzian, $|q| = \infty$ (dash-dotted line), and two Fano lineshapes, $q = 0.5$ (dashed line) and $q = -1.5$ (solid line), are shown. MNP = metal nanoparticle, QE = quantum emitter, WGM = whispering gallery mode. ϵ is the detuning normalised to the line width, i.e., $\epsilon = 2\delta/\Gamma$.

Quantum interference, i.e. the interference between two transition amplitudes in coupled quantum systems, is ubiquitous in physics and is often evidence of the quantum nature of a system (Miroshnichenko et al., 2010). Historically, quantum interference was independently pioneered by Majorana (1931a,b) and Fano (1935, 1961) in the context of autoionisation (Arimondo et al., 2010). The Fano lineshape [Fig. 25.4b] has been shown to occur in a large variety of systems (Miroshnichenko et al., 2010) as a consequence of coupling between a system with a discrete energy spectrum and a continuum or quasi-continuum of states. Of particular interest are hybrid nanostructures composed of a quantum emitter and a plasmonic nanoparticle, which exhibit resonances with Fano lineshapes in both extinction and scattering cross-sections (Zhang et al., 2006; Ridolfo et al., 2010; Koppens et al., 2011; Manjavacas et al., 2011) as the discrete levels of the quantum emitter are coupled to the quasi-continuum of the plasmonic nanoparticle.[6]

The starting point of our analysis of near-resonant laser cooling is the expression for the radiation force acting on a particle at rest in the propagation direction of the incident light beam, which is given by Eq. (5.124), i.e.,

$$F_{rad}^{\parallel} = \mathbf{F}_{rad}^{\parallel} \cdot \hat{\mathbf{k}}_i = \frac{I_i}{c}\sigma_{rad},$$

where $\sigma_{rad} = \sigma_{ext} - g_i\sigma_{scat}$ is the radiation pressure cross-section and we assume the surrounding medium to be vacuum ($n_m = 1$). The factor g_i [Eq. (5.69)] is the asymmetry parameter, which depends on the shape and optical properties of the particle; e.g., $g_i \approx 0$ for a Rayleigh scatterer. For resonant particles, such as quantum emitters or microparticles with Mie resonances, σ_{rad} exhibits a Lorentzian lineshape that determines the behaviour of near-resonance optical forces. Here, we instead consider a particle with a radiation pressure cross-section exhibiting a Fano lineshape resonance [Fig. 25.4b], i.e.,

$$\sigma_{rad}(q,\omega) = \sigma_a \frac{\left[q\frac{\Gamma}{2} + \omega - \omega_0\right]^2}{[\omega - \omega_0]^2 + \left[\frac{\Gamma}{2}\right]^2} + \sigma_b, \qquad (25.19)$$

where Γ is the full width at half maximum of the resonance, ω_0 is the Fano resonance frequency, σ_a and σ_b are the amplitudes of the resonant and background cross-sections respectively, and q is the *Fano factor*, which characterises the lineshape profile, so that for $|q| \to \infty$ it becomes Lorentzian (Miroshnichenko et al., 2010).[7]

As we did in Chapter 24 for the case of an atom, we now consider a particle with mass m moving in one dimension at velocity v under the influence of two counter-propagating laser beams with wavevector k, i.e., in a one-dimensional optical molasses, and detuned from the Fano (or Lorentzian) resonance by an amount $\delta = \omega - \omega_0$. This yields a Doppler shift $\mp kv$ from the stationary particle resonance peak and thus a net radiation force acting on the particle of $F_{rad} = (\sigma_{rad}^+ - \sigma_{rad}^-)I_i/c$, where '+' and '−' denote the cross-sections for

---

[6] Quantum interference between discrete energy levels (Lounis and Cohen-Tannoudji, 1992) also plays a special role in laser cooling of atoms, where the occurrence of coherent states uncoupled from the light is responsible for sub-recoil and electromagnetic induced transparency-assisted cooling (Morigi, 2003; Dunn et al., 2007).
[7] Taking this limit requires a normalisation of the line profile [Eq. (25.19)] by q^{-2}.

the wavevectors parallel and antiparallel to the velocity, i.e.,

$$\sigma_{\text{rad}}^{\pm}(q, k_0) = \sigma_a \frac{\left[q\frac{\Gamma}{2} + \delta \mp kv\right]^2}{[\delta \mp kv]^2 + \left[\frac{\Gamma}{2}\right]^2} + \sigma_b. \tag{25.20}$$

To examine the general features of the total radiation force and the Doppler cooling process, it is useful to define the reduced (dimensionless) quantities $\tilde{\sigma} = (\sigma_{\text{rad}} - \sigma_b)/\sigma_a$, $\epsilon = 2\delta/\Gamma$ and $\tilde{v} = v/v_c$, where $v_c = \Gamma/2k$ is the velocity capture range of the cooling process defining the range in which the force is linear with velocity. Thus, the reduced radiation cross-section is $\tilde{\sigma} = (q + \epsilon)^2/(\epsilon^2 + 1)$, which for small velocities ($\tilde{v} \ll 1$) yields a linearised (reduced) radiation force, $\tilde{F}_{\text{rad}} \approx -\tilde{\alpha}(q, \epsilon)\tilde{v}$, with

$$\tilde{\alpha} = \frac{4(q+\epsilon)}{\epsilon^2 + 1}\left[1 - \frac{\epsilon(q+\epsilon)}{(\epsilon^2 + 1)}\right] \tag{25.21}$$

being the damping coefficient of the process. Cooling of the centre-of-mass motion of the particle occurs when $\tilde{\alpha} > 0$, whereas heating occurs for $\tilde{\alpha} < 0$. In comparison with conventional Doppler cooling on a Lorentzian lineshape, which is only dependent on the detuning of the light from resonance, we now have an additional dependence of the damping coefficient on the Fano factor, q. This additional dependence gives rise to different behaviours in the light–particle mechanical coupling that correspond to either cooling or heating. In particular, the quantum interference giving rise to the Fano profile is responsible for unconventional Doppler cooling for a certain range of parameters (δ, q) that can occur for blue detuning (conversely to the Lorentzian case). Thus, whereas for a Lorentzian lineshape there is a maximum for $\epsilon = -1$, i.e., $\delta = -\Gamma/2$, for a Fano lineshape the detuning corresponding to the maximum damping is dependent on q and hence on the engineering of the hybrid system.

We now focus on the cooling process ($\tilde{\alpha} > 0$) and calculate the equilibrium temperature as a balance between the cooling process that damps the average energy of the particle in a characteristic time $\tau_{\text{cool}} = m/2\alpha$ and the heating by the random momentum diffusion due to photon scattering with rate $R(\epsilon, q) = F_{\text{rad}}/\hbar k$. The balance between these two competing processes yields the equilibrium temperature for laser cooling in one dimension, which can be written as $k_B T = (1 + \zeta)\hbar^2 k^2 R/\alpha$ [Chapter 24]. Here, ζ is a geometrical factor that takes into account the angular averaged redistribution of scattered photons; for a Rayleigh particle (isotropic scattering) $\zeta = 1/3$. For a particle with a Lorentzian resonance of width Γ, this yields the standard Doppler limit temperature of $k_B T_D = (1 + \zeta)\hbar\Gamma/4 = \hbar\Gamma/3$. Thus, following the same analysis done in Chapter 24 and using the expression of the total force and damping for the Fano–Doppler cooling process, we find that

$$\frac{T_{\text{Fano}}}{T_D^{(1d)}} = \frac{2}{\tilde{\alpha}}\left(\tilde{\sigma} + \frac{\sigma_b}{\sigma_a}\right). \tag{25.22}$$

Because of the quantum interference phenomena responsible for the Fano lineshape, as $\epsilon \to -q$ sub-Doppler temperatures can be achieved (Ridolfo et al., 2011).

As an example, Ridolfo et al. (2011) discuss the Fano cooling of a hybrid nanostructure composed of a quantum dot coupled to a silver nanoparticle dimer, as shown in Fig. 25.4a. In this system, the position and shape of the Fano resonance can be varied by changing the inter-particle distance and the energy difference between the quantum dot and the surface plasmon mode resonances, so that different Fano factors q can be engineered. The hybrid nanostructure considered is made up of two silver nanospheres with radius $r = 8$ nm separated by a gap of 8 nm and a quantum dot with dipole moment $p/e_- = 0.3$ nm in their centre (*hot spot*). The strong-field enhancement in the dimer gap enables strong Fano effects even for quantum dots with small dipole moments. The quantum dot is considered spherical with radius $r_{QD} = 2$ nm and an excitonic resonance shifted by 60 meV from the plasmon resonance peak. Because of its symmetry, a spherical quantum dot has three energy-degenerate bright excitons with optical dipoles parallel to the three directions x, y and z. Hence, the quantum dot can be described in the simplest way as an isotropic effective medium with a single resonance at the energy $\hbar\omega_0$ of the lowest energy exciton (Savasta et al., 2010). The radiative cross-sections σ_{rad} are calculated computationally using a T-matrix [Chapter 6]. The hybrid nanostructure is considered as embedded in a dielectric shell (e.g., a polymer) with refractive index $n_{shell} = \sqrt{\varepsilon_b} = 1.7$ in order to bind the quantum dot and metal nanoparticle and fix their distance in a core–shell structure. For a laser power of 10 mW focused on a spot size of 100 μm, cooling times below 5 s are obtained. However, the heating due to the non-radiative background, σ_b, limits the temperature to about half of the Doppler limit; i.e., $T \approx 450$ mK, corresponding to $v_{rms} \approx 13$ mm/s. When only radiative processes are considered, temperatures one order of magnitude lower are achieved in a few seconds, 45 mK corresponding to $v_{rms} \approx 1.3$ mm/s. Non-radiative processes can be significantly reduced by employing smaller metal particles or nanoshells (Fofang et al., 2011) and therefore in principle lower temperature could be reached.

Most of the ingredients needed for the application of this Fano–Doppler laser cooling scheme are within experimental reach with state-of-the-art technology. Nanostructures with Fano lineshapes have already been produced using wet chemistry approaches (Fofang et al., 2011). Optical trapping of nanoparticles and nanoaggregates has similarly been demonstrated [Chapter 23]. A possible experimental configuration would then involve different stages in a fashion similar developed for the creation of ultra-cold atomic samples [Chapter 24]. A spray of hybrid nanostructures dispersed in a liquid environment could be used in vacuum. The capture of a few particles in a far-detuned (with infrared radiation) optical trap or ion trap can spatially limit and increase in time the interaction with light, and thus near-resonant radiation can be added to cool and detect the velocity distribution of nanoparticles.

References

Arimondo, E., Clark, C. W., and Martin, W. C. 2010. Ettore Majorana and the birth of autoionization. *Rev. Mod. Phys.*, **82**, 1947–58.
Barker, P. F. 2010. Doppler cooling of a microsphere. *Phys. Rev. Lett.*, **105**, 073002.

Barker, P. F., and Shneider, M. N. 2010. Cavity cooling of an optically trapped nanoparticle. *Phys. Rev. A*, **81**, 023826.

Born, M., and Wolf, E. 1999. *Principles of optics: Electromagnetic theory of propagation, interference and diffraction of light*. Cambridge, UK: Cambridge University Press.

Braginskii, V. B., and Manukin, A. B. 1977. *Measurement of weak forces in physics experiments*. Chicago: University of Chicago Press.

Chang, D. E., Regal, C. A., Papp, S. B., et al. 2010. Cavity opto-mechanics using an optically levitated nanosphere. *Proc. Natl. Acad. Sci. U.S.A.*, **107**, 1005–10.

Cohen-Tannoudji, C., Dupont-Roc, J., and Grynberg, G. 1992. *Atom–photon interactions: Basic processes and applications*. New York: Wiley.

Dunn, J. W., Thomsen, J. W., Greene, C. H., and Cruz, F. C. 2007. Coherent quantum engineering of free-space laser cooling. *Phys. Rev. A*, **76**, 011401.

Fano, U. 1935. Sullo spettro di assorbimento dei gas nobili presso il limite dello spettro darco. *Nuovo Cimento*, **12**, 154–61.

Fano, U. 1961. Effects of configuration interaction on intensities and phase shifts. *Phys. Rev.*, **124**, 1866–78.

Fofang, N. T., Grady, N. K., Fan, Z., Govorov, A. O., and Halas, N. J. 2011. Plexciton dynamics: Exciton–plasmon coupling in a J-aggregate-Au nanoshell complex provides a mechanism for nonlinearity. *Nano Lett.*, **11**, 1556–60.

Gardiner, C., and Zoller, P. 2004. *Quantum noise: A handbook of Markovian and non-Markovian quantum stochastic methods with applications to quantum optics*. Heidelberg, Germany: Springer Verlag.

Gieseler, J., Deutsch, B., Quidant, R., and Novotny, L. 2012. Subkelvin parametric feedback cooling of a laser-trapped nanoparticle. *Phys. Rev. Lett.*, **109**, 103603.

Kiesel, N., Blaser, F., Delić, U., et al. 2013. Cavity cooling of an optically levitated submicron particle. *Proc. Natl. Acad. Sci. U.S.A.*, **110**, 14 180–85.

Kippenberg, T. J., and Vahala, K. J. 2007. Cavity opto-mechanics. *Opt. Express*, **15**, 17 172–205.

Koppens, F. H. L., Chang, D. E., and Garcia de Abajo, F. J. 2011. Graphene plasmonics: A platform for strong light–matter interactions. *Nano Lett.*, **11**, 3370–77.

Li, T., Kheifets, S., Medellin, D., and Raizen, M. G. 2010. Measurement of the instantaneous velocity of a Brownian particle. *Science*, **328**, 1673–5.

Li, T., Kheifets, S., and Raizen, M. G. 2011. Millikelvin cooling of an optically trapped microsphere in vacuum. *Nature Phys.*, **7**, 527–30.

Lounis, B., and Cohen-Tannoudji, C. 1992. Coherent population trapping and Fano profiles. *Journal Physique II*, **2**, 579–92.

Majorana, E. 1931a. I presunti termini anomali dell'elio. *Nuovo Cimento*, **8**, 78–83.

Majorana, E. 1931b. Teoria dei tripletti P' incompleti. *Nuovo Cimento*, **8**, 107–13.

Manjavacas, A., García de Abajo, F. J., and Nordlander, P. 2011. Quantum plexcitonics: Strongly interacting plasmons and excitons. *Nano Lett.*, **11**, 2318–23.

Marquardt, F., Chen, J. P., Clerk, A. A., and Girvin, S. M. 2007. Quantum theory of cavity-assisted sideband cooling of mechanical motion. *Phys. Rev. Lett.*, **99**, 093902.

Miroshnichenko, A. E., Flach, S., and Kivshar, Y. S. 2010. Fano resonances in nanoscale structures. *Rev. Mod. Phys.*, **82**, 2257–98.

Morigi, G. 2003. Cooling atomic motion with quantum interference. *Phys. Rev. A*, **67**, 033402.

Novotny, L., and Hecht, B. 2012. *Principles of nano-optics*. Cambridge, UK: Cambridge University Press.

Ridolfo, A., Di Stefano, O., Fina, N., Saija, R., and Savasta, S. 2010. Quantum plasmonics with quantum dot–metal nanoparticle molecules: Influence of the Fano effect on photon statistics. *Phys. Rev. Lett.*, **105**, 263601.

Ridolfo, A., Saija, R., Savasta, S., et al. 2011. Fano–Doppler laser cooling of hybrid nanostructures. *ACS Nano*, **5**, 7354–61.

Romero-Isart, O., Juan, M. L., Quidant, R., and Cirac, J. I. 2010. Toward quantum superposition of living organisms. *New J. Phys.*, **12**, 033015.

Romero-Isart, O., Pflanzer, A. C., Juan, M. L., et al. 2011. Optically levitating dielectrics in the quantum regime: Theory and protocols. *Phys. Rev. A*, **83**, 013803.

Savasta, S., Saija, R., Ridolfo, A., et al. 2010. Nanopolaritons: Vacuum Rabi splitting with a single quantum dot in the center of a dimer nanoantenna. *ACS Nano*, **4**, 6369–76.

Schliesser, A., Rivière, R., Anetsberger, G., Arcizet, O., and Kippenberg, T. J. 2008. Resolved-sideband cooling of a micromechanical oscillator. *Nature Phys.*, **4**, 415–19.

Wilson-Rae, I., Nooshi, N., Zwerger, W., and Kippenberg, T. J. 2007. Theory of ground state cooling of a mechanical oscillator using dynamical back-action. *Phys. Rev. Lett.*, **99**, 093901.

Zhang, W., Govorov, A. O., and Bryant, G. W. 2006. Semiconductor–metal nanoparticle molecules: Hybrid excitons and the nonlinear Fano effect. *Phys. Rev. Lett.*, **97**, 146804.

Index

4f configuration, 251–3, 337

Abbe's sine condition, **93–4**
aberration correction
 – with DOE, 332
 objective –, 230–1
aberrations, **33**
 – due to objective, 95
 – when focusing near interfaces, 100
 astigmatism, 33
 chromatic –, 33
 coma, 33
 comatic –, 33
 field distortion, 33
 focus deterioration due to –, 103, 241
 Petzval field curvature, 33
 spherical –, 33
 trap based on spherical –, 444–5
Abraham momentum, *see* Abraham–Minkowski dilemma
Abraham–Minkowski dilemma, **25**, 26, 110
absorption spectroscopy, *see* spectroscopic optical tweezers
ac Stark shift, 503
ACF, *see* autocorrelation function
achromatic objective, *see* objective
acoustic tweezers, 9–10
acousto-optic deflector, 252, 340–1, 350, 363
actin, 371, 379–82, 387, 389–92
actin cortex, 355
adaptive-additive algorithm, *see* holographic optical tweezers
addition formulas, *see* multipole, addition theorem
addition theorem, *see* multipole, addition theorem
adenosine triphosphate, 380, 390
aerosol, 439, 441–6
AFM, *see* atomic force microscopy
aggregates of spheres, *see* transition matrix, aggregates of spheres
angle of incidence, 20
angular Mathieu functions, 90
angular momentum, **111–15**, 158–60
 – operator, *see* ladder operators
 conservation of –, 107, 109
 orbital –, 6, 37, 114–15, 149–51, 332–4, 416, 512
 spin –, 6, 37, 114–15, 143–6, 378, 416
 sum of –, *see* Clebsch–Gordan coefficients
angular spectrum representation, **79–83**, 93, 97, 124
 – of a focused field, 95–6
 – of a focused field near an interface, 101
anomalous diffusion, 455, 458
AOD, *see* acousto-optic deflector
aplanatic lens, 31, 93
apochromatic objective, *see* objective
apodisation, 95
Asakura–Oosawa theory, 428
associated Laguerre polynomials, 87
associated Legendre equation, 117
associated Legendre functions, 117
astigmatism, *see* aberrations
asymmetry parameter, **126**, **142**, 532
 – for a Mie particle, 136
 transverse –, 126, 142
atom cooling
 1D Doppler cooling, 498
 1D Raman cooling, 498
 1D VSCPT cooling, 498
 2D Doppler cooling, 498
 3D Doppler cooling, 498
 Bose–Einstein condensation, 498
 Doppler cooling, 499–504, 513
 evaporative cooling, 507, 513
 magnetic trapping, 507
 spin-polarised BEC, 498
 sub-Doppler cooling, 498
atom optics, 505
atom trapping, 504–6
 strong-field seekers, 505
 weak-field seekers, 505
atomic force microscopy, 8–9, 296, 297
atomic momentum diffusion, 501
atoms, **498–518**
ATP, *see* adenosine triphosphate
autocorrelation function, 203–4, 300–7
 – analysis, *see* calibration, autocorrelation function
 intensity – of a speckle, 457, 458
 velocity –, 205, 206
average
 block –, 285
 ensemble –, 193, 277
 time –, 193, 277

Avogadro number, 191
axicon, 76, 90, 319

bacterial pili, 388–9
bacterial spores, 405
bacterium, 3, 193, 240, 358, 388–9, 392–3, 395, 405, 417, 476
Bauer's expansion, 123, 162
beam, 76, **83–92**
  Airy –, 104
  Bessel –, 90, 91, 443–6
  cylindrical vector –, **92**, 93, 104
    azimuthally polarised –, 85, 92, 93, 96–9, 103
    hybrid polarised –, 93
    radially polarised –, 85, 92, 93, 96, 98, 99, 103
  discrete –, 89, 91, 340
  doughnut –, 36
  Gaussian –, **83–5**
    – quality, 244
    Gouy phase shift, 85
    phase correction, 84
    Rayleigh range, 84
    wavefront radius, 84
    width, 84
  Helmholtz–Gauss –, 89
  Hermite–Gaussian –, **85–7**, 334
  Ince–Gaussian –, 104
  Laguerre–Gaussian –, **87–9**, 115, 149–51, 319, 333, 444, 445, 505
  Mathieu –, 90, 91
  non-diffracting –, **89–92**, 334
  transverse electric (TE) –, 85
  transverse electromagnetic (TEM) –, 85
  transverse magnetic (TM) –, 85
  Weber –, 90, 91
beam parameter product, 244
BEC, *see* Bose–Einstein condensate
Beer's law, 239
Beth's experiment, 2
birefringent particle, 377, 378, 392, 416
blazed diffraction grating, *see* diffraction grating
Bohr magneton, 505
Bose gas, 508
Bose–Einstein condensate, 498, 508–13
Bose–Hubbard model, 517
boson, 508
bound charges, 46
bound electrons, 68, 69, 470
boundary conditions, **49**
  – for a Mie sphere, 133
  – for a cluster, 174
  – for a metal sphere sustaining longitudinal fields, 170
  – for finite-difference time-domain algorithm, 182
  – for particles with inclusions, 176
bovine serum albumin, 404, 478, 479

Brewster's angle, 23, **23**
Brown, Robert, 189
Brownian diffusion, *see* Brownian motion
Brownian dynamics simulations, **199–205**
  – for non-spherical particles, 212–16
  – in a diffusion gradient, 209
  – of optically trapped particle, 202–5
  – of white noise, 200–2
Brownian motion, 6, **188**
  – in a diffusion gradient, 207–11, 452–5
  – in a random potential, 351–3, 455–9
  – in a viscoelastic medium, 211–12
  – of optically trapped particle, 202–5
  – simulation, *see* Brownian dynamics simulations
  free diffusion, 195–6
  inertial regime, 205–7
  mathematical models of –, 191
    Fokker–Planck equation, *see* Fokker–Planck equation
    Langevin equation, *see* Langevin equation
  the physical picture of –, 189–90
Brownian ratchets, 459
BSA, *see* bovine serum albumin

calibration, **273–93**
  Allan variance, 293
  autocorrelation function, 255, **280**, 281, 282
  crosstalk analysis, **280–3**
  drag force method, **291–3**, 473
  equipartition method, **276–8**, 378, 473, 502
  mean squared displacement function, 255, **278–9**
  potential analysis, **274–5**
  power spectrum analysis, 255, **283–90**, 473
capture velocity, 501
Carnot, Sadi, 466
CARS, *see* coherent anti-Stokes Raman spectroscopy
cavity enhanced Raman scattering, 434, 439
CCD, *see* charge-coupled device
CCF, *see* cross-correlation function
CDM, *see* coupled dipole method
cell (biological), 39, 240, 312, 349–50, 354–5, 358, **385–93**, 395, 398, 401, 404, 405, 409, 410, 412–14, 434, 438
cell membrane, 355, 389–91, 405, 438
cellular adhesion, 386–9
CERS, *see* cavity enhanced Raman scattering
CFL condition, *see* Courant–Friedrichs–Lewy (CFL) condition
charge-coupled device, 261
Chinese hamster ovary cell, 240, 410
Chu, Steven, 3
circular unit vectors, 115
cladding, 357
Clausius–Mossotti relation, **47–50**, 52, 54, 58
Clebsch–Gordan coefficients, 157, 163, **168–9**, 503

cluster of spheres, *see* transition matrix, cluster of spheres
CMOS, *see* complementary metal oxide semiconductor
Cohen-Tannoudji, Claude, 3
coherent anti-Stokes Raman scattering, 399
coherent anti-Stokes Raman spectroscopy, 401, *see also* spectroscopic optical tweezers
colloid, **422–30**, 449
coma, *see* aberrations
comet, 1, 2
complementary metal oxide semiconductor, 261, 365
complete unified device architecture, 335
contour length, *see* worm-like chain model
corkscrew cooling, 503, *see also* sub-Doppler cooling
counter-propagating optical tweezers, 29–31, 73, 312, 315, 342, 353–6, 372, 412, 417, 444, 445, 472, 499–504, 530, 531
coupled dipole method, *see* discrete dipole approximation
Courant–Friedrichs–Lewy (CFL) condition, 182
coverslip, 227
critical angle, 23, **23**, 104, 357, 358
cross product of a scalar and a vector, 46
cross-section, **54–6**
  – in quantum mechanics, 510
  absorption –, 55, **56**, 69, 126
  differential scattering –, 125
  extinction –, 54, 55, **56**, 69, 126
    – and optical theorem, 56–8
    – for a Mie particle, 136
    – in optical force, 60, 63
  radiation pressure –, 142, 499, 531
  scattering –, 54, **56**, 125
    – for a Mie particle, 136
crosstalk analysis, *see* calibration, crosstalk analysis
crystal, colloidal, 455
CUDA, *see* complete unified device architecture
curl of the curl, 46
cytoskeleton, 386–7, 389–92

DAQ hardware, *see* data acquisition hardware
dark spectrum, 290–1
data acquisition hardware, **273**
  – analogue input range, 273
  – bandwidth, 273
  – input channels, 273
  – resolution, 273
  – sample rate, 273
  single-ended vs. differential inputs, 273
DDA, *see* discrete dipole approximation
de Broglie wavelength, 486, 509
density
  melamin resin –, 237
  poly(methyl methacrylate) –, 237
  polystyrene –, 237
  silica –, 237
depletion forces, *see* depletion interactions
depletion interactions, 426–30
Derjaguin–Landau–Verwey–Overbeek theory, 425–6
detuning, 499–503, 505, 506, 531, 533
  blue –, 501, 516, 518, 524, 527
  red –, 501, 516, 524, 526, 531
diaphragm, 227, 228, 234
  condenser –, 232
  field stop –, 234
DIC, *see* differential interference contrast
dichroic mirror, 249
Dicke narrowing, 514
diffraction grating, 76, 323, 325, 326, 333, 335–7
diffractive optical element, 6, **319–23**, *see also* holographic optical tweezers
diffusion, 6
diffusion coefficient, 193, 195, 197, 274, 449
diffusion constant, *see* diffusion coefficient
diffusion equation, 191
diffusion gradient, *see* Brownian motion in a diffusion gradient
diffusion tensor, **212–13**
digital camera, **261–2**
  – interlaced vs. progressive scan, 261
  – sensor, 261
    charge-coupled device, *see* charge-coupled device
    complementary metal oxide semiconductor, *see* complementary metal oxide semiconductor
  area of interest, 262
  binning, 262
  deinterlacing, 262
  field of view, 262
  gain, 262
  gamma, 262
  subsampling, 262
digital video microscopy, **256–61**, 449
  comparison of – and interferometry, 268
  correction of artefacts when particles are close, 426
  dilation filter, 256
  erosion filter, 256
  feature point detection, 257
  holographic –, 261
  thresholding, 256
dipole
  – approximation, *see* optical trapping regime, Rayleigh –
  – moment, 43, 67, 399
    induced –, 45
  – nanoantenna, 476
  – traps for cold atoms, 506–7
  induced –, 42, 50–2
  oscillating –, 42, 50–2
    optical force on –, 58–65, 97–8, 106, 504–6

dipole (*cont.*)
  size parameter, *see* size parameter
  static –, 43–5
    torque on –, 43
dipole trap, 504–6, 513–14
discrete dipole approximation, **179–80**
dispersion, 33, 246
  – relation, 77
divergence of vector cross product, 46
DLVO theory, *see* Derjaguin–Landau–
    Verwey–Overbeek theory
DNA, **371–9**, 382, 403
  – Young's modulus, 373
  – force–extension curve, 373
  – stretching, 372–4
  – thermal fluctuations, 374–6
  – torsion, 376–9
  – torsional modulus, 378
DOE, *see* diffractive optical element
Doppler cooling limit, 502
  below the –, 531–4
Doppler effect, 499, 501
dot product of a scalar and a vector, 46
drift, 193, 207, 209
  – current, 197, 198
  – force, 309
  – velocity, 197, 309, 457
  spurious –, *see* spurious drift
droplet, 433–4, 438, 439, 442–4
  liquid crystal –, 417
Drude–Sommerfeld model, 67, 69
DVM, *see* digital video microscopy
dyadic
  – product, 108, 423
  unit –, 108

EBCM, *see* extended boundary condition method
elastic scattering, *see* Rayleigh scattering
electromagnetic wave, 20, 77–8
  plane –, *see* plane wave
electrostatic interactions, 425–6
energy density, electromagnetic, 55
entanglement, 524
entropic elasticity, *see* worm-like chain model
entropy production, 463
equipartition theorem, 197
ergodic system, 193, 277
evanescent focus, *see* focal field
evanescent tweezers, 356–8
evaporative cooling, 508, 513
extended boundary condition method, 155
extended optical potentials, **350–3**, 514–18
  periodic –, 350–1
  quasi-periodic –, 350–1
  random –, 351–3, 455–9

extracellular ligands, 386
extracellular matrix, 386

Fano lineshape, *see* Fano resonance
Fano resonance, 531–4
Faxén formula, 208, 289
FDTD, *see* finite-difference time-domain algorithm
feedback cooling, 529–31
fibroblast, 386–7
field distortion, *see* aberrations
field programmable gate array, 365
filter, **248–9**
  cut-off wavelength, 249
  dichroic mirror, *see* dichroic mirror
  neutral density –, 248–9
  wavelength-selective –, 249
    bandpass –, 249
    edge –, 249
    longpass –, 249
    notch –, 249
    shortpass –, 249
finite difference simulations, 199
finite-difference time-domain algorithm, **180–3**, 472
FJC, *see* freely jointed chain model
flexural persistence length, *see* worm-like chain model
fluctuation–dissipation theorem, 199, 462
  – for a non-spherical particle, 213
  – in a diffusion gradient, 208
  – in a viscoelastic medium, 211
fluorescence tweezers, *see* spectroscopic optical tweezers
fluorite objective, *see* objective
focal field, 244
  – amplitudes (for T-matrix), 160–2
  – of a Gaussian beam (approximated), 61
  electromagnetic calculation, **92–7**
  – near interfaces, 100–2
  – near interfaces (aberrations), 102
  – near interfaces (evanescent waves), 102–4
focusing
  – in the ray optics regime, 31, 32
  – near an interface, 100–2
  filling factor, 34–5
  numerical aperture, 34–5
Fokker–Planck equation, **197**
force
  – measurement, *see* force measurement
  binding –, *see* optical binding
  optical –, *see* optical force
force measurement, *see also* photonic force microscope
  direct –, 312–16
  drift method, 309–11, *see also* spurious force

equilibrium distribution method, 308, *see also* Maxwell–Boltzmann distribution
  non-conservative effects, 311
FPGA, *see* field programmable gate array
fractal plasmonics, 477
fractional optical vortices, 342
free diffusion, 201, 449, *see also* free diffusion equation
  – of a non-spherical particle, 213–15
  inertial regime, 205
free diffusion equation, 191, 195–6, 201
  – of a non-spherical particle, 213–15
free electrons, 67, 69, 470
freely jointed chain model, 374
Fresnel diffraction integral, **321**
Fresnel lens, 323, 325, 326, 337
Fresnel plane, 336, 338
Fresnel's coefficients
  – for electric fields, **101**
  – for intensity, **21–3**
friction coefficient, 193, 194, 197, 274
  – for a non-spherical particle, 213
  – for a spherical particle, *see* Stokes' law
fundamental theorem of vector calculus, *see* Helmholtz decomposition

galvo-mirror, 252, 340–1, 363
gauge
  – transformation, 112
  Coulomb –, 111–13
  Lorentz –, 112
  transverse –, *see* Coulomb gauge
Gaunt integrals, 163
generalised Lorenz–Mie theories, 155
generalised phase contrast, 341–2
geometrical optics, 19, *see also* optical trapping regime, ray optics –
Gerchberg–Saxton algorithm, *see* holographic optical tweezers
giant unilamellar vesicle, 433, 435
GLMT, *see* generalised Lorenz–Mie theories
Gouy phase shift, 84, 85, 264, 266
GPC, *see* generalised phase contrast
GPE, *see* Gross–Pitaevskii equation
GPU, *see* graphical processing unit computing
GPU computing, 335
gradient of vector dot product, 46
graphene, 398, 400, 402, 488–9, 492, 531
graphical processing unit computing, *see* GPU computing
Gross–Pitaevskii equation, 509
GUV, *see* giant unilamellar vesicle

half-wave plate, 249–50
haptic tweezers, 362–5
Hatano and Sasa equality, 467

heat engine (microscopic), 465–7
Helmholtz decomposition, 111
Helmholtz equation, 77
  homgeneous –, 77
  scalar –, 92, 116
    solution of –, 123
  vector –, 92, 123
    solution of –, 126
Helmholtz free energy, 428, 430
Hermite polynomials, 85
hologram, *see* holographic optical tweezers, hologram generation
holographic optical tweezers, 6, **319–43**
  – performance
    algorithm comparison, 328
    average intensity, 327
    efficiency, 327
    standard deviation, 327
    uniformity, 327
  applications, 349–51, 397, 411, 415–17, 434, 435, 443, 455, 514
  basic working principle, 320–3
  diffraction grating, *see* diffraction grating
  Fresnel lens, *see* Fresnel lens
  hologram generation, 324–35
    adaptive–additive algorithm, 331–2
    Bessel beams, 334
    continuous optical potentials, 334–5
    direct search algorithms, 332
    Gerchberg–Saxton algorithm, 329, 331
    Hermite–Gaussian beams, 334
    Laguerre–Gaussian beams, 332–4
    Monte Carlo algorithms, 332
    non-diffracting beams, 334
    random mask encoding, 327–8
    random superposition of gratings and lenses, 329
    simulated annealing, 332
    single trap, 325
    superposition of gratings and lenses, 328–9
    weighted Gerchberg–Saxton algorithm, 330
  set-up, 335–8
  spatial light modulator, *see* spatial light modulator
holographic video microscopy, *see* digital video microscopy
Hooke's law, 4, 297, 311, *see also* trap stiffness
HOT, *see* holographic optical tweezers
hot spot, 361, 471, 476, 534
HWP, *see* half-wave plate
hydrodynamic equations of superfluids, 512
hydrodynamic forces, *see* hydrodynamic interactions
hydrodynamic interactions, 207–11, 355, 375, 423–5, 452–5
  – to the power spectral density, 288–90
hydrodynamic memory effect, 288
hyper-polarisability tensor, 399, *see also* polarisability

hyper-Raman scattering, 399
hyper-Rayleigh scattering, 399

illumination scheme, 232–6
   bright-field microscopy, 232–4
   dark-field illumination, 235–6
   differential interference contrast, 235, 236
   Köhler illumination, 233–4
   Kretschmann geometry, *see* Kretschmann geometry
   phase contrast, 234–5
immersion oil, 227
   – properties, 229
Ince polynomials, 104
inelastic scattering, *see* Raman scattering
information-to-energy conversion, 465
integrated device, 354, 358, 409–17, 492
integrin, 386–7
intensity law of geometrical optics, 94, 95
intensity, electromagnetic wave, 77
interferometry, **262–73**
   comparison of digital video microscopy and –, 268
   crosstalk, *see* calibration, crosstalk analysis
   position sensing detector, *see* position sensing detector
   quadrant photodetector, *see* quadrant photodetector
intrinsic elasticity, *see* worm-like chain model

Jabłoński diagram, 396
Jarzynski equality, 464–5

Köhler, August, 234
Kepler, Johannes, 1, 2
Kramers rates, 449–51
Kramers–Kronig relations, 212
Kretschmann configuration, *see* Kretschmann geometry
Kretschmann geometry, 359, 470, 474, 475

lab on a chip, 409–17
ladder operators, 120
   quantum –, 527
Landé gyromagnetic factor, 505
Landauer bound, 468
Landauer principle, 467
Langevin equation, 191, **194–5**
   – for optically trapped particle, 202
   – in a diffusion gradient, 207–11
   – in a potential, 194
   – in a rotational force field, 301
   – in a viscoelastic medium, 211
   hydrodynamic interaction between two particles, 423
   inertial regime, 205–7
   overdamped –, 195
laser, **244–6**

– beam quality, 244
– classification scheme, **245**
– frequency stability, 246
– noise, 246
– pointing stability, 245
– power stability, 246
– safety, **244**
laser cooling
– of atoms, **498–504**
– of levitated particles, **528–34**
Doppler –, 498–502, 531–4
feedback –, 529–31
sub-Doppler –, 502–4, 531–4
LCoS, *see* liquid crystal on silicon
leapfrog algorithm, 190
lens, **246–8**
– anti-reflection coating, 247
– material, 246
– shape, 246–8
– surface quality, 247
light shift, *see* ac Stark shift
light spectrum, 290–1
like-charge attraction, 426
linear polarisability tensor, *see* polarisability
linear Raman effect, *see* Raman scattering
liposome, 435
liquid crystal, 339, 410, 417
liquid crystal on silicon, 339
localised surface plasmons, 470, 471
Lorentz force, 59, 65, 107
Lorentz–Lorenz relation, 179, *see also* Clausius–Mossotti relation
Loschmidt's paradox, 463
low-Reynolds-number regime, 6, 193, **193**, 195, 197, 215, 288, 392
LSP, *see* localised surface plasmons

magnetic induction (formula), 77
magnetic trap, 507–8
magnetic tweezers, 9
magnetisation, 48
magneto-optical trap, 505–6
maximum permissible exposure, 245
Maxwell fluid, *see* viscoelastic medium
Maxwell stress tensor, 106, **107–11**
Maxwell's demon, 465
Maxwell's equations, **47**
   boundary conditions, 49
   differential form of –, 47
   integral form of –, 47
Maxwell–Boltzmann distribution, 197–8, **198**, 209, 274, 452–5, 529
mean squared displacement, 193, 203
– analysis, *see* calibration, mean squared displacement function
– in a speckle optical field, 458

– of particle with inertia, 206
– of trapped particle, 204, 205
melamin resin, 237
metal sphere, *see* transition matrix, metal sphere
MF, *see* melamin resin
micromachine, 409, 415–17
microrobot, *see* micromachine
Mie coefficients, **134**
    Wiscombe formula, 134
Mie resonances, **136–7**, 151, 531
Mie scattering, 7, 116, **132–7**
    boundary conditions, 133
    extension of – to particles sustaining longitudinal fields, 170–2
    extension of – to radially symmetric particles, 172–3
Mie theory, *see* Mie scattering
Mie, Gustav, 132
Minkowski momentum, *see* Abraham–Minkowski dilemma
mirror, **248**
    dichroic –, *see* dichroic mirror
molecular dynamics simulations, 190
molecule, 3–4, **371–82**
momentum of a photon, 2–3, 24–5, *see also* Abraham–Minkowski dilemma
    – for direct force measurement, 312–14
    angular –, 131
    laser cooling, 499–504
momentum relaxation time, 205
monochromatic plane wave, *see* plane wave
MOT, *see* magneto-optical trap
motor protein, 371, 379–82
Mott insulator, 517
Mott transition, 517
MPE, *see* maximum permissible exposure
MSD, *see* mean squared displacement
multimode fibre, 456
multipole
    addition theorem, *see* multipole, translation theorem
    rotation theorem, 165–8
    translation theorem, 162–5
multipole expansion, **126–31**
    – for fields regular at the origin, 129
    – for fields satisfying the radiation condition at infinity, 129–30
    – of a plane wave, 130–1
myosin, 371, 379–82, 390

NA, *see* objective, numerical aperture
nanofibre, 490, 492
nanotube, 396–8, 401–2, 417, 488–9, 492
nanowire, 216–17, 316, 349, 358, 397–8, 404, 471–3, 480, 485–6, 488, 490–2
Navier–Stokes equations in the low-Reynolds-number regime, 288
NESS, *see* non-equilibrium steady state
neuron, 385, 389–92
    – directed growth, 385, 389–92
Nichols and Hull's experiments, 2
niosome, 435
nocodazole, 390
Nomarski prism, 236
Nomarski, Georges, 236
non-equilibrium steady state, 467
normal dispersion, 33
numerical aperture, *see* objective, numerical aperture
Nyquist frequency, 284

OAM, *see* orbital angular momentum
objective, **226–32**
    – abbreviations, 228
    – aberration correction, *see* aberration correction, objective
    – immersion medium, 227
    – magnification, 230
    – parfocal distance, 231
    – screw threads, 232
    – transmission coefficient, 95
    – transmission efficiency, 230
    – tube length, 232
    – working distance, 227
    achromatic –, 226, 230
    apochromatic –, 226, 231
    filling factor, 34–7
    fluorite –, 226, 230–1
    numerical aperture, **35**, 76, 227
    phase contrast –, 228, 235
    total internal reflection –, 100, 102, 103, 227, 356
OD, *see* optical density
optical binding, 42, **70–3**, 354–6, 358–9, 444, 445, 472, 474, 479, 489
optical bottle, 441, 445
optical chromatography, 410, 446
optical conveyor, 417
optical density, 249
optical elevator, 342
optical fibre trap, 353–6, 412–14, 416
optical force
    – (electromagnetic theory), **109, 137–9**
    – from plane wave, 139–43
    – in optical tweezers, 146–9
    transfer of orbital angular momentum, 149–51
    – calculated with T-matrix, 156–8
    – calculated with dipole–dipole approximation, 180
    – calculated with finite-difference time domain, 180–1
    – exerted by a focused beam on a sphere (ray optics), 32
    – exerted by a ray at an interface, 26

optical force (*cont.*)
- exerted by a ray on a mirror, 24
- exerted by a ray on a sphere, 27
- exerted on a dipole, 60
- on a dipole, 58–65
- on a dipole near focus, 97–8
conservative –, 29, 61, 63
gradient –, 2, 26–9
gradient – (ray optics), 27
gradient – exerted on a dipole, 61–3
non-conservative –, 29, 63, 64, 311
regime comparison, 106
scattering –, 2, 26–9
scattering – (ray optics), 27
scattering – exerted on a dipole, 63–4
spin-curl – exerted on a dipole, 64, 65
optical force lithography, 490–2
optical lattices, *see* extended optical potentials
optical lift effect, 39
optical molasses, 499–504, 514–18, 532
optical potentials, *see* extended optical potentials
optical pumping, 502
optical sorting, 410–12, 417, *see also* optical chromatography
optical stretcher, 354–5, 412, 413
optical theorem, **56–8**, 58, 126
optical torque, 6
- (electromagnetic theory), **109**, **137–9**
- by a ray on a convex particle, 38
- calculated with T-matrix, 158–60
- calculated with dipole–dipole approximation, 180
- calculated with finite-difference time domain, 180–1
optical torque wrench, 371, 378–9
optical trapping regime, **6–8**
- comparison, 106
  intermediate –, 7, 106
  ray optics –, 7, **19–41**, 106
  Rayleigh –, 7, **42–74**, 106
optical tweezers, invention of, 3
optoelectronic tweezers, 9
optofluidic device, 354, 358, 409–17, 492
optomechanics, 524–34
- with levitated particles, 528–34
  cavity
    optical amplification, 524
    optical cooling, 524
  cavity –, **524–8**
  resolved sideband regime of –, 527
  weak retardation regime of –, 526
orbital angular momentum, *see* angular momentum
organelle, 395, 405
Oseen tensor, 423
overdamped regime, *see* low-Reynolds-number regime

parametric resonance, 443
paraxial approximation, **83**
paraxial beam, 83–92
  focusing of –, 81
    electromagnetic theory, 92–7
    ray optics theory, 31, 32
particle sustaining longitudinal fields, *see* transition matrix, metal sphere
partition function, 198
PBS, *see* polarising beam splitter
PCA, *see* principal component analysis transform
Perrin, Jean, 188
persistence length, *see* worm-like chain model
Petzval field curvature, *see* aberrations
PFM, *see* photonic force microscopy
phase singularity, 87
phasor, 50, **51**, 55, 77, 110, 114
Phillips, William D., 3
photocurrent, 271
photodamage, 230, 239–40, 346, 349, 397
photodetector, **271–3**
- bandwidth, 272
- dark current, 272
- junction capacitance, 272
- material, 272
- mode of operation (photoconductive or photovoltaic), 272
- noise equivalent power, 272
- quantum efficiency, 272
- responsivity, 271
photoluminescence tweezers, *see* spectroscopic optical tweezers
photon, 24
photonic force microscopy, 3, **296–300**
photonic jet, 405
photonic torque microscopy, **300–7**
  measurement of non-conservative radiation forces, 311
photopolymerisation, 490
plane of incidence, 21
plane wave, 20, 77, 79
  evanescent –, 79
  expansion in multipoles of a –, *see* multipole expansion
  magnetic induction of a –, 77
  propagating –, 79
plasma frequency, 67, 68, 170
plasmon, 67
plasmonic aperture, 477–9
plasmonic optical tweezers, **359–61**, 412, 414, 474–7
plasmonic particle, 8, **67–70**, 170–2, 470, **471–4**, **485–6**, 492, 531, 532, 534
  heating, 403, 473, 474, 480, 485–6, 492
plasmonic resonance, 397, 402–3, 470–4
plasmonics, **470–80**
PMMA, *see* poly(methyl methacrylate)

point matching, 183
polarisability, **45**, 136, 399
   – of a plasmonic nanoparticle, 69
   – of a small dielectric sphere, *see* Clausius–Mossotti relation
      radiative correction to the –, *see* radiative correction
      atomic –, 65–7
      nonlinear –, 45
polarisation, 21, 45, 48
   p- –, 21
   s- –, 21
   circular –, 78, 115
   elliptical –, 78
   linear –, 77
polarisation charges, 46, 48
polarisation control, **249–50**
   half-wave plate, *see* half-wave plate
   polarising beam splitter, *see* polarising beam splitter
   quarter-wave plate, *see* quarter-wave plate
polarising beam splitter, 250
poly(methyl methacrylate), 237
polymerosome, 435
polystyrene, 237
position sensing detector, 263, **266**, 271
   – for direct force measurement, 312–16
   crosstalk, *see* calibration, crosstalk analysis
   transverse forward position detection, 265
power spectral density, *see* calibration, power spectrum analysis
power spectrum, *see* calibration, power spectrum analysis
Poynting vector, 20, **55**, 60, 63
Poynting's theorem, 55, **55**, **110**
Poynting, John Henri, 2
principal component analysis transform, 281, 307
prism, 19, 235, 236, 250, 359, 413, 474
probability density distribution, 191
   equilibrium –, *see* Maxwell–Boltzmann distribution
product rule for the gradient, 46
PS, *see* polystyrene
PSD, *see* position sensing detector *or* power spectral density
pupil function, 95

QPD, *see* quadrant photodetector
quadrant photodetector, 263, **266**, 271
   crosstalk, *see* calibration, crosstalk analysis
   longitudinal forward position detection, 267
   longitudinal forward scattering, 267
   transverse backward position detection, 270
   transverse backward scattering, 270
   transverse forward position detection, 265
   transverse forward scattering, 265
quadrupole trap, 507
quantised vortices, 512

quantum coherence, 524
quantum dot, 184, 487, 531, 533–4
quantum efficiency, 272
quantum interference, 531
quantum superposition of motional states, 524
quantum well, 487
quantum wire, 487
quarter-wave plate, 250
quasicrystal, colloidal, 455–6
QWP, *see* quarter-wave plate

Rabi frequency, 499, 503
radially symmetric particle, *see* transition matrix, radially symmetric sphere
radiation condition at infinity, 121
radiative correction, 52–4
radiative reaction, 54
Raman cooling, 498, 504
Raman scattering, 397, 398
Raman tweezers, *see* spectroscopic optical tweezers
random mask encoding, *see* holographic optical tweezers
random number, 192, 193, 200, 202, 214, 284, 285
random walk, 188, 191, **192–4**
   – in momentum space, 502
   biased –, 193
ray, 19, 20
   – reflection, 19, 21
   – transmission, 19, 21
Rayleigh scattering, 397, 401
Rayleight spectroscopy, *see* spectroscopic optical tweezers
RBC, *see* red blood cell
recoil limit, 504
red blood cell, 354–5, 358, 395, 401
reflection coefficient, 22, 23
refractive index, 20, 25
   air –, 8
   immersion oil –, 229
   lens material –, 246
   melamin resin –, 237
   poly(methyl methacrylate) –, 237
   polystyrene –, 237
   silica –, 237
   water –, 8
regime, optical trapping, *see* optical trapping regime
residual colour, 230
Reynolds number, 6, 193, **193**, 195, 197, 215, 392
RMS standard, 225
rotor of vector cross product, 46

SAM, *see* spin angular momentum
sample preparation, **236–9**
saturation intensity of the atomic transition, 499
scalar potential, 111, 112

scanning tunnelling microscopy, 8, 297
scattering amplitude, 57, 82, 116, 135, 510
    normalised –, 125, 126
scattering in quantum mechanics, 510
scattering problem, 106, 110, 116, **123–6**, 154
scratch-dig, 247
SDE, *see* stochastic differential equation
SDS, *see* sodium dodecyl sulphate
self-diffusion time, 449
self-field interaction, 54
self-induced back action, 345, 361–2, 477–9
semi-apochromat objective, *see* objective
SERS, *see* surface enhanced Raman spectroscopy
set-up construction
    digital video microscope –, **256–61**, *see also* digital video microscopy
        digital camera, *see* digital camera
    holographic optical tweezers, **335–8**, *see also* holographic optical tweezers
        spatial light modulator, *see* spatial light modulator
    inverted microscope –, **222–6**
        illumination scheme, *see* illumination scheme
        objective, *see* objective
    optical components
        filter, *see* filter
        lens, *see* lens
        mirror, *see* mirror
        polarisation control, *see* polarisation control
    photonic force microscope –, **262–71**, *see also* photonic force microscopy *and* interferometry
        data acquisition hardware, *see* data acquisition hardware
        photodetector, *see* photodetector
        sample preparation, *see* sample preparation
    spectroscopic optical tweezers, **346–7**, *see also* spectroscopic optical tweezers
    stability, 222
        noise tests, 290
    standard optical tweezers –, **239–44**
        laser, *see* laser
        operation, 250–1
        steerable –, 251–3
SIBA, *see* self-induced back action
silica, 237
size parameter, 8, 98, 134, 135, 142
skin depth, 68
SLM, *see* spatial light modulator
Snell's law, **21**, 100, 101
sodium dodecyl sulphate, 239
SOT, *see* spectroscopic optical tweezers
spatial light modulator, 338–40
    – damage threshold, 340
    – phase resolution, 339
    – pixel filling factor, 339
    – pixel number, 339
    – pixel size, 339
    – polarisation sensitivity, 339
    – wavelength range, 339
    comparison with acousto-optic deflector and galvo-mirror, 340–1
speckle optical tweezers, 351–3, 411–12, 446, 455–9
spectroscopic optical tweezers, 346–50, 395–405, 409, 412, 433, 435
    absorption spectroscopy, 396–8
    basic set-up designs, 346–7
    coherent anti-Stokes Raman spectroscopy, 402
    fluorescence tweezers, **347–9**, 397, 409, 434, 438
    photoluminescence tweezers, 348, **349**, 396–8
    Raman tweezers, 348, **349–50**, 395, **398–402**, 409, 435
    Rayleigh spectroscopy, 402–4
    surface-enhanced Raman spectroscopy, 402–4
sphere with inclusions, *see* transition matrix, sphere with inclusions
spherical Bessel equation, 121
spherical Bessel functions, 121, 122
spherical Bessel functions of the first kind, *see* spherical Bessel functions
spherical Bessel functions of the second kind, *see* spherical Neumann functions
spherical Hankel functions, 121, 122
spherical harmonics, 117–20
    – rotation, 165
    – translation, 162
    addition theorem, 119
spherical Neumann functions, 121, 122
spin angular momentum, *see* angular momentum
spinor Bose gas, 513
SPP, *see* surface plasmon polariton
spurious drift, 207, 209–11, 309–11, 452–5
spurious force, 309–11
stationary phase method, 82
statistical physics, 448–59
Stirling cycle (microscopic), 467
STM, *see* scanning tunnelling microscopy
stochastic activation, 459
stochastic differential equation, 195
    numerical solution of –, 199, *see also* finite difference simulations
stochastic integral, 209
    anti-Itô –, 209
    isothermal –, 209
    Itô –, 209
    Stratonovich –, 209
stochastic resonance, 451–3
stochastic resonant damping, 459
Stokes flow, 423
Stokes' law, 194, 198
sub-Doppler cooling, 502–4, 514, 531–4
subdiffusion, *see* anomalous diffusion
superdiffusion, *see* anomalous diffusion

superfluidity, 512
surface enhanced Raman spectroscopy, *see* spectroscopic optical tweezers
surface plasmon, 170, 470, **474–7**
surface plasmon polariton, *see* surface plasmon
surface plasmon polaritons, 470
Sysyphus cooling, *see* sub-Doppler cooling
Szilard engine, 468

T-matrix, *see* transition matrix
Talbot effect, 104
tapered fibre, 357
telecentric imaging system, 93
thalassaemia, 395
thermal energy, **6**, 189, 372, 390, 465, 505
thermodynamics, 462–8
    first law of –, 462
    second law of –, 462
    violations of the second law of –, 462–4
Thomas–Fermi limit, 510
time-averaged orbiting potential trap, 507
TIRM, *see* total internal reflection microscopy
TOP trap, *see* time-averaged orbiting potential trap
torque
    optical –, *see* optical torque
total internal reflection, 23, 100, 102, 103, 356, 414
total internal reflection microscopy, 296, **298**
tractor beam, 417
transition matrix, 116, 131–2, **155–79**
    addition theorem, *see* multipole, addition theorem
    aggregates of spheres, 174–6
    cluster of spheres, 174–9
    convergence, 178–9
    metal sphere, 170–2
    particle sustaining longitudinal fields, *see* transition matrix, metal sphere
    radially symmetric sphere, 172–3
    rotation theorem, *see* multipole, rotation theorem
    sphere with inclusions, 176–8
transmission coefficient, 22, 23
trap stiffness, 4, 32
    – calibration, *see* calibration
    autocorrelation function as a function of –, 204
    experimentally measured –, 255
    mean squared displacement as a function of –, 204
    numerically calculated –, 106
    relevance of non-conservative optical forces on –, 311
trapping efficiency, **27**, 34, 36
Triton-X, 239

vaterite particle, 392, 416
vector identities, 46
vector potential, 111, 112
    electric field in terms of –, 114
vector spherical harmonics, 126–8
    – rotation, 165
    – translation, 163
velocity-selective coherent population trapping, 504
vesicle, 433, 435–8, 486
vesicle fusion, 438
vesicle membrane, 433, 435–8
virus, 3, 395, 405
viscoelastic medium, 211–12
VSCPT, *see* velocity-selective coherent population trapping

wave equation, 77
whispering gallery mode, *see* Mie resonance
white noise, **195**
    simulation of –, 200–2
Wigner rotation matrices, 165, **166**
windmill effect, 6, 39, 139, 416
WLC model, *see* worm-like chain model
Wollaston prism, 236
worm-like chain model, 372–4
    contour length, 372
    entropic elasticity, 372
    flexural persistence length, 372
    intrinsic elasticity, 372

yeast cell, 401
Yee cell, 181, 182
Yee lattice, 181
Yukawa model, 449

Zernike polynomials, 332
Zernike, Frits, 234